UNIVERSITY OF GLAMORGAN
LEARNING RESOURCES CENTRE

Pontypridd, Mid Glamorgan, CF37 1DL
Telephone: Pontypridd (01443) 482626

Books are to be returned on or before the last date below

PRODUCTION PROCESSES

The Productivity Handbook

FIFTH EDITION

ROGER W. BOLZ

President, Automation for Industry, Inc.

Registered Mechanical Engineer, Ohio

Certified Manufacturing Engineer

INDUSTRIAL PRESS INC.

200 Madison Avenue

New York, New York 10157

Library of Congress Cataloging in Publication Data

Bolz, Roger William.
    Production processes.

    Originally published: 5th ed. Winston-Salem, N.C.:
Conquest Publications, 1977.
    Includes bibliographical references and index.
    1. Production engineering.   2. Manufacturing pro-
cesses.   I. Title.
TS176.B64   1981   658.5   81–6494
ISBN 0–8311–1088–0   AACR2

PRODUCTION PROCESSES: The Productivity Handbook,
Fifth Edition

Foreword to Fifth Edition

During the past twenty-seven years this handbook has seen five editions and revisions. Even though most manufacturing processes have retained their fundamental character, a large number have been further developed and refined to a singular degree. As a result, it has been necessary to revise and update the entire handbook.

As a part of the total updating of the contents, many new chapters have been added to cover the manufacturing area more completely. Along with new and updated process data, new topical areas of singular importance have been introduced. Such areas include designing for assembly, measuring and gaging, design for numerically controlled machining, and in addition the subject of designing for recycling has been introduced.

Because of the expanded reach of the current text, for the first time it will be noted that many new chapters have been authored by experts in many fields where rapid development has been taking place. The author and editor is deeply indebted to these expert contributors whose efforts have been extremely valuable in making this text the most extensive and authoritative available.

Without question, it is still a truism that objective careful design for production still can provide products of maximum value and acceptable functional utility at lowest possible cost. Thorough study and application of the data contained herein will provide the most cost effective results in the development of not only new products but in the redevelopment of currently produced products.

Roger W. Bolz, P.E.

Fifth Edition
Winston-Salem, N.C.

This book is dedicated to Ruth whose inspiration
and continuing help made possible its preparation
and production.

Preface to First Editions

THIS BOOK presents a broad study of the many manufacturing processes utilized in the production of any and all types of products and components. Production design of parts, to be both economical and practical — commonly referred to as producibility — must be based upon a thorough working knowledge of all the various processing methods employed in manufacture. As a general rule, a part designed for or suited to one method of manufacture need not and most often does not suit another. Each production method has a well-established level of precision which can be maintained in continuous mass production without exceeding normal basic costs. Again, each method has its economical minimum production level below which quantity cost per piece may be excessive. Cognizance of the multitude of processing factors which influence the practical design of manufactured parts will go far toward achieving the advantage of maximum economy, accuracy and automation in everyday production.

Much of the material included in this book has been available from widely scattered sources and also has been in use to a certain extent knowingly and in many instances unknowingly. For example, whenever a manufacturer requests that certain changes be permitted by the purchaser of parts or components, almost invariably the reason is to adapt the design to more economical or more automatic production. True, these changes generally appear to be small and trivial in many cases as to warrant little or no attention, yet profit or loss and the ability to meet a practical selling price for the final product may depend entirely upon such changes. Maximum value from the parts dollar can only be realized when design is set up specifically to accommodate the most practical method of automatic processing.

Much material has been published on direct comparisons of one method versus another, disparaging one to emphasize the advantages of the other. Based on the firm belief that each method of processing fills a specific niche in the panel of production methods, all processes treated in this book have been approached in a uniform manner, scanning the field for their major advantages and applications, presenting each on the same general basis so as to permit a broad, fair analysis of all from the same "dollars and sense" vantage point. Only by such an overall treatment can a true picture and, most important, the desired results be obtained.

Other published material on methods of processing provides general descriptions of equipment utilized, information on how the equipment is operated, what speeds and feeds are available, what devices are standard, and similar general data. In this book, only sufficient information is presented regarding the actual process and equipment with which it is carried out to provide a basic understanding of the techniques employed and an appreciation of the general field covered. Major stress and discussion are directed toward the design of individual component parts to emphasize those features which insure simple and economical processing.

Design details, though of major importance, must be accompanied by adequate consideration of the material to be used and also the tolerances to be specified. Selection between one material and another may mean easy processing or excessively expensive production; selection of the right materials for the method required is therefore extremely important. Practical shop production tolerances have seldom been discussed and rarely appear in print. Admittedly, many widely varying values may be practical under an equally wide range of circumstances. Here, however, the major goal has been to present the most practical ranges which can be readily held under average conditions, not those only possible with highly specialized or expensive tooling conditions. While this may mean somewhat wider tolerances in certain cases, such specifications assure the most economical conditions. Extremes have been indicated wherever practicable to make possible their consideration in special cases.

To simplify the study of modern producibility, the presentation has been arranged in basic sections covering the fundamental categories of manufacturing processes. No attempt is made to cover the more limited and highly sophisticated process refinements that are available. Section one titled, "Producibility, Automation, and Design Principles," has been included to emphasize and outline some of the basic rules which should be observed in the attempt to attain the goal of better, more economical parts and components. This fundamental attack on the producibility problem can achieve tremendous results. When combined with the opportunity to automate operations, even more spectacular dollar results can be attained. The basic philosophy is one of manufacturing systems engineering — engineering the product simultaneously with the production equipment system.

If this book begins in a small way to bring about a better understanding and a fuller realization of the economical importance and sound engineering approach of design based upon a knowledge of manufacturing processes, the effort will have been well worthwhile. To his many friends through industry and schools of engineering the author is indebted for the encouragement in undertaking this preparation.

Roger W. Bolz, P.E.

Cleveland, Ohio

ACKNOWLEDGMENTS

IN THIS BOOK the variety of manufacturing processes treated represents a broad cross section of American manufacturers. The information on each process presented is that generally applicable throughout industry, rather than that of only a small segment or one company, in order to be of greatest overall value to engineers engaged in all walks of product design. To be truly representative of the general design practices found most successful and economical throughout industry requires the collaboration of many organizations. The author wishes to express his deepest appreciation for the generous cooperation and assistance freely given by management, engineers, specialists, and organizations noted in the text, too numerous to mention, without whose efforts this work would not have been possible. To the following contributors of new and specially updated chapters the author would like to give special tribute:

Omer W. Blodgett
Design Consultant
Lincoln Electric Company
Cleveland, OH

Joseph Harrington, Jr.
Consultant
Wenham, Mass.

Gordon A. McAlpine
Vice President
KLT Industries
Ann Arbor, MI

Norman N. Brown
Vice President
Wheelock, Lovejoy & Company, Inc.
Boston, MA

Charles E. Chastain
President
Plastics Technical Concepts Inc.
Chicago, IL

Henry S. Royak
Product Mgr., Roll Forming
The Yoder Company
Cleveland, OH

A. F. Scheider
Assistant Chief Engineer, Brush Div.
Osborn Manufacturing Company
Cleveland, OH

Clifford E. Evanson
President
TAB Engineers
Northbrook, IL

Special thanks are due also to *Metal Progress*, American Society for Metals, *Machine Design*, and *Automation* for permission to include special material that helps in materially rounding out the coverage of generally utilized manufacturing processes.

CONTENTS

Chapter   Page

Contents

Contents

SECTION
PRODUCTION

1

PROCESSES

PRODUCIBILITY
Automation and Design Principles

1

Processes, Costs and Producibility

PHENOMENAL growth and development of mass-production methods over the past half century has, without doubt, made the greatest contribution to the American standard of living. Few, if any, other factors have had so great an impact upon this present age of automation.

All too often, people fail to recognize the full importance and value of our many mass-production methods and the great proportion of our present-day necessities which they make possible. The highest conceivable overall standard of living eventually will be attained through the final development and perfection of processing methods which make available unlimited necessities and luxuries at the lowest practicable costs in manpower and materials.

Mass Production. Low-cost, mass-produced and distributed products have been the very heart and making of the America of today; a better tomorrow rests upon further development and exploitation of the untapped potentialities of the means of high-quality volume production. The manufacturing engineer of today and the aspiring manufacturing engineer of tomorrow justifiably must take cognizance of, and promote more economic production design.

Accompanying the developments and strides in the automotive industry from its infancy were parallel developments in the highest speed mass-production processing methods. This drive to achieve maximum output at lowest cost was consummated in the practical application of the principles of interchangeable manufacture and large-quantity methods of production. Almost without exception where mass output of products has been achieved successfully on a grand scale, the principles of "designing for production" have also been employed, although perhaps unknowingly in many cases.

At this point, however, it might be well to elaborate a bit upon precisely what is meant by the qualifying term "mass production." Although it is generally understood that by this term is meant output in thousands, perhaps millions as in the automotive industry, such indeed need not be the case. The principles of mass production may be applied to practically any production quantity of units regardless of type. The important difference, therefore, is the use of the principles of interchangeable manufacture rather than those of so-called "custom" or hand-fitted and selected assemblies. Mass-production in the case of one product might conceivably be five or ten units while with another the number might run into millions.

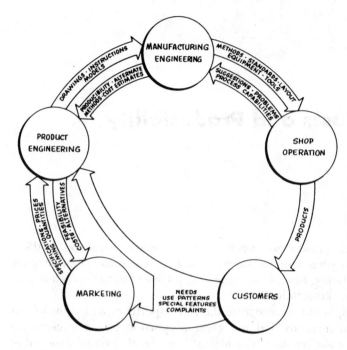

Fig. 1.1 — Representation, left, of product cycle shows that Manufacturing Engineering is an integrated part of the total business. Manufacturing Engineers must play an active role in determining product design with the emphasis on producibility. Chart, below, indicates steps leading to production of a new product. This is an example of the project procedure followed in a mass production business for the introduction into manufacturing of a product representing major differences from the existing products.

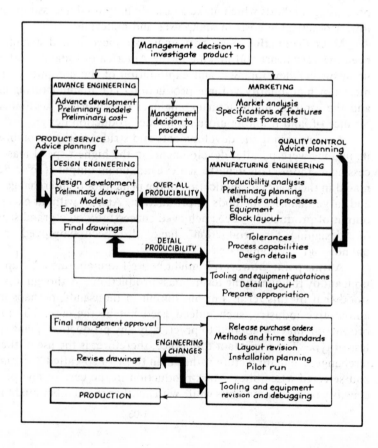

Production Economics. Quantity, thus, as well as general design of any product, is an important key to how that product may be produced most economically for the job at hand. Quantitywise and otherwise, each of the various methods of manufacture now available fills a certain void in the production picture and as developments proceed, the yet unfilled voids disappear one by one. Eventually a production method or combination of methods will be available to fit perfectly any specific set of design requirements that may occur. Until such a time, however, adequate thought given those methods now available will provide efficient and highly gratifying results.

To be eminently successful, any manufacturer must be original and visionary, but he must also produce final product designs which are practical. Difficult or expensive manufacturing characterictics in any product today must be classed as equivalent to functional mediocrity. To be eminently practical the designer must seek the creation of products capable of manufacture by processes incurring lowest possible costs. Competent analyses indicate that few, if any, designs are created that cannot be reduced to an assembly of parts producible by commonly available methods and techniques.

Design Influence. Practically speaking, as part designs are created, developed and detailed, the designer is actually selecting the process or processes by which the parts will be manufactured. Conversely, too, he may be limiting the useable methods to those least desirable. Drawing details and specifications actually place wide limits or narrow restrictions on practical processing. The design of the product often determines whether the shop can be set up to manufacture with relatively simple operations or will be strapped with costly, unpredictable or even unnecessary steps in production.

While numerous production innovations are generally adopted to attain acceptable manufacturing costs, in the overall analysis the greatest achievements in the direction of economy are most easily attained through proper design. Regardless of type of parts, quantities or tolerances involved, maximum quality and minimum costs can be obtained only if consideration is given the possible production methods available for their manufacture.

It may be argued that manufacturing methods, planning and cost analysis are not the responsibility of the product designer and, considered as specific on-the-job duties, this may be and most often is the case. However, the fact remains that after-the-fact planning and cost analysis seldom can effect economies in production in any degree comparable with those possible during basic design. It must be remembered that detail part drawings establish the number and complexity of fabricating steps required. Each feature and limiting specification beyond that necessary in the basic design increases total cost. By judicious forethought and co-operation with manufacturing engineers, the designer can easily arrive at the most reasonable specifications, *Fig.* 1.1.

Processes. In many instances, any one or several of a group of processes may be satisfactory for producing a part without affecting the functional characteristics desired. For instance, the stainless steel elbow shown in *Fig.* 1.2 was designed for machining from bar stock. Specification change to permit cast rather than wrought material permitted a savings of $2.18 per piece.[1]

The part illustrated in *Fig.* 1.3 was originally designed to be machined from

---

[1] *All References at end of chapters.*

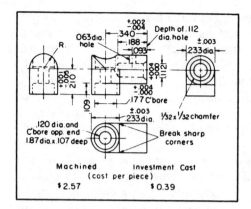
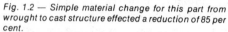

Machined   Investment Cast
(cost per piece)

$2.57 $0.39

Fig. 1.2 — Simple material change for this part from wrought to cast structure effected a reduction of 85 per cent.

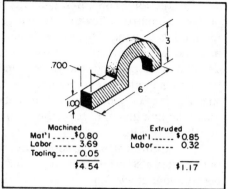

Machined		Extruded	
Mat'l _____ $0.80		Mat'l _____ $0.85	
Labor _____ 3.69		Labor _____ 0.32	
Tooling _____ 0.05			
$4.54		$1.17	

Fig. 1.3 — Change from plate to extruded material for this part produced a 74 per cent saving in quantities of 2000 pieces.

plate. However, a specification change to permit use of an extruded shape brought about a reduction in cost of $3.37 per piece.

With the foregoing parts, design changes required are insignificant but instances are equally numerous wherein considerable difference in design is necessary. One such is shown in *Fig.* 1.4. Here the functional end use is identical, but utterly different basic processing is employed to obtain the major decrease in cost with identical or improved characteristics.

Effects of Quantity. The influence of quantity on selection possibilities is an important consideration in design for economy. The case shown in *Fig.* 1.5 gives some indication of how quantity would influence the method of machining a part. Because of tooling and setup costs as well as overhead, quantity is important in determining the most suitable method. Use of high-output machines such as multiple-spindle screw machines also involves the factor of efficiency. It often requires hours of production for such a machine to "settle down" and for an operator to have all the tools adjusted properly and become accustomed to particular machining characteristics of the job.

Another factor is apparent from the data in *Fig.* 1.5. This is the important effect of materials selection. Machinability of the particular material utilized is important in its effect on both quantity and method.

Except for intermediate areas, quantity requirements are generally good indicators of suitable production methods. In these intermediate areas, all factors must be weighed carefully to assure clear-cut indication of where "breakeven" points occur. For instance, where stamping operations are concerned the range of

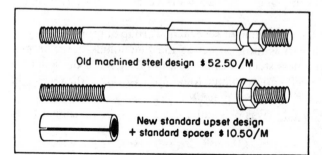

Old machined steel design $52.50/M

New standard upset design
+ standard spacer $10.50/M

Fig. 1.4 — Stud required production of 200,000 pieces per year. Automatic upset and roll threaded stud with a standard roll formed spacer offered worthwhile savings compared with conventional machined stud.

possibilities runs from machine-cut, through short-run to full production stamping on high-speed automatic presses. The charts shown in *Fig.* 1.6 indicate an area from 100 to 10,000 pieces in which this occurs. As quantity increases from the category of machine-cut to fully automatic production, tool and setup costs increase rapidly but labor costs decrease. Tool costs are easily amortized over a large number of parts and labor costs per part become negligible.

Design Details. Specification of minor design details on drawings is also critical regarding possible processes and suitability of parts for an indicated process. Specification of certain details may render a part difficult to produce by the most economical method or require manufacture by a method less than ideal.

Small variations in design requirements often may be overlooked. For instance, the choice of a parting line for the part in *Fig.* 1.7 — as dictated by whether or not allowance for draft could be permitted in the center hole — results in considerable cost difference owing to the method of coring.

Another small but recurrent case concerns machining. Where radii are designed to be tangent to adjacent surfaces, cutters used will most often fail to match perfectly and scrap or rework incurred is expensive, especially with critical stressed parts. This condition is illustrated at (*a*) in *Fig.* 1.8 and a preferred solution is shown at (*b*), with the tangent or matching condition eliminated. Machining as shown at *Fig.* 1.8 (*a*) costs three to four times that for one as detailed at (*b*).

Dimensions as specified often necessitate special tools which incur added costs. The case shown in *Fig.* 1.9 shows two types of accurate, mating parts with dimensions designated as *good* and *poor*. While little if any difference in manufacturing time would be evidenced in either instance, the special tools required must

THREADED BUSHING

Aluminum	17ST	Bar Steel	B-1113	Bar Brass	SAE 72
Materials ....235 lb @ 0.41/lb = $96.35		656 lb @ 0.076/lb = $49.86		720 lb @ 0.3385/lb = $244.72	
Turnings ....170 lb @ 0.05/lb = $ 8.50		....		520 lb @ 0.175/lb = 91.00	
Net material cost per M pcs ....= $87.85		= $49.86		= $153.72	

Method A		Turret Lathe			
Quantity ....Under 1000 pcs		Under 250 pcs		Under 1000 pcs	
Time per M ....25 hr		61 hr		22 hr	
Setup time .... 4 hr		4 hr		4 hr	
Tooling Cost ....$25.00 approx.		$25.00 approx.		$25.00 approx.	

Method B		Single-spindle automatic*			
Quantity ....1000 to 2500 pcs		250 to 5000 pcs		1000 to 10,000 pcs	
Time per M ....21 hr		55 hr		11 hr	
Setup time .... 4 hr		5 hr		4 hr	
Tooling Cost ....$75.00 approx.		$45.00 approx.		$45.00 approx.	

Method C		Six-spindle automatic†			
Quantity ....2500 pcs and up		5000 pcs and over		10,000 pcs and over	
Time per M ....4.2 hr		7 hr		2.7 hr	
Setup time ....16 hr		16 hr		16 hr	
Tooling Cost ....$150.00 approx.		$125.00 approx.		$75.00 approx.	

* One operator can handle three to four machines. †One operator can handle two machines.

Fig. 1.5 — *Threaded bushing with detailed costs for production by three methods and from three different materials.*

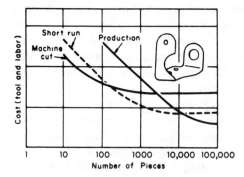

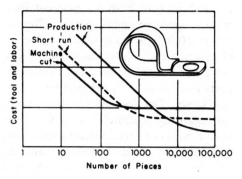

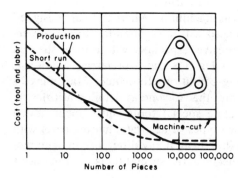

Fig. 1.6 — Charts showing quantity-cost relationship for three parts designed for stamping produced by machine cutting, short-run and high-production methods.

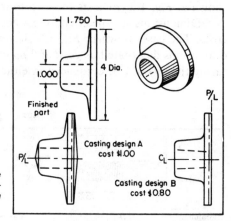

Fig. 1.7 — Additional cost for magnesium casting A is due to the need for a core to produce the hole. Since the finished hole requires machining, the design as at B with draft in the hole offers a saving.

be considered. With the hole at (*a*), Fig. 1.9, a special reamer, costing about $25.00, was necessary to hold the 0.877-inch diameter. For the slotting operation at (*b*), to produce the 0.261-inch width, a special milling cutter costing about $50.00 was needed.

Tolerances. Among the effects of design specifications on costs, those of tolerances are perhaps most significant. Tolerances in design influence the producibility of the end product in many ways, from necessitating additional steps in processing to rendering a part completely impractical to produce economically. In range, tolerances may cover dimensional variation, surface roughness range, allowable variation in such properties as heat treatment, endurance time of coatings against corrosion, etc.

Dimensional Limits. A simple example of how tolerances can increase cost is given in *Fig.* 1.10. Tolerances on dimensions specified at values smaller than functional requirements actually necessitate, increase costs through additional machine time, checking and gaging time, rejections, etc. In *Fig.* 1.10, *Design A* can be made from bar stock by sawing to length but tolerances for *Design B* are too close for sawing. To hold the tolerances on *Design B*, turning is required and for *Design C* grinding is necessary at an increase of 200 per cent over *A*.

In the same manner holes are affected by specifications. The cost data in *Fig.* 1.11 were found from an actual study made to determine the relative effects of tightening hole tolerances. Some 120 holes were used in determining the cost factors and these data do not include the additional costs of extra tooling, gages, etc., necessary.

Again, drilling to close tolerances is more costly in hard materials, materials with inclusions or hard spots, and in thin materials under 1/8-inch in thickness. Holes over 1-inch in diameter or four times diameter in depth are ordinarily difficult to hold to close tolerances and costs increase rapidly if extreme precision is demanded. *Fig.* 1.12 shows the comparative costs of holes with varying degrees of refinement. The solid bar indicates the most desirable range of values.

Some idea of costs relative to turning can be gained from the curves of *Fig.* 1.13. Long parts of relatively small diameter subject to deflection in turning are generally difficult to produce. Recesses requiring long or unsupported tools are also difficult to machine to close tolerances. Tapers, contours, large radii, and related areas too wide for single form tools require special machines or tools to achieve fine limits.

Straddle and face milling costs are illustrated in *Fig.* 1.14. Straddle milling with heavy cuts or hard materials may cause cutter deflection on entry and exit of cuts. Close tolerances may require additional cuts to remove taper. With face milling some heat warpage can result with the necessity of additional slow light "skim" cuts to achieve accuracy desired. With slotting and forming cuts, the basic form tolerance is ground into the tool. Close tolerances, however, can result in short cutter life between grinds with greatly increased maintenance and tool costs. *Fig.* 1.15 gives a comparative cost evaluation. Excessively close tolerances on slotting can necessitate a minimum of three passes — one rough and two finish — to hold size. Profiling and pocket milling costs are shown in *Fig.* 1.16. Here, tolerances are greatly limited by tendency of the cutter to spring from the work. Numerous passes

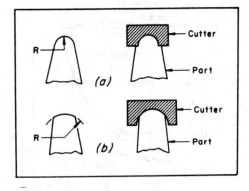

Fig: 1.8 — To produce a design with a tangent radius, a concave cutter usually creates a nick on one or both sides of the piece as shown at (a.) The non-tangent design shown at (b) offers considerable production economy.

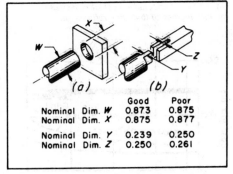

	Good	Poor
Nominal Dim. *W*	0.873	0.875
Nominal Dim. *X*	0.875	0.877
Nominal Dim. *Y*	0.239	0.250
Nominal Dim. *Z*	0.250	0.261

Fig. 1.9 — Good and poor design of fitted parts such as these is indicated by the need for special cutters which incur added costs.

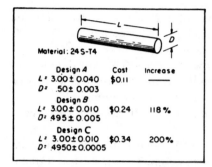

Fig. 1.10 — Tolerances have a profound effect on costs. Design A permits use of stock as drawn, B requires turning and C necessitates finish grinding.

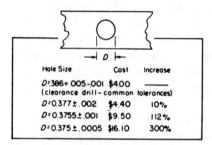

Fig. 1.11 — Cost relationship of holes with varying tolerances. Cost data shown was for 120 consecutive holes.

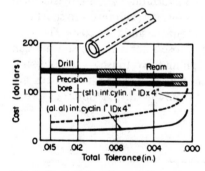

Fig. 1.12 — Comparative cost of holes with varying degrees of tolerance refinement.

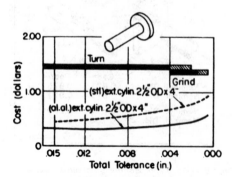

Fig. 1.13 — Cost relationship of turned parts with varying tolerances.

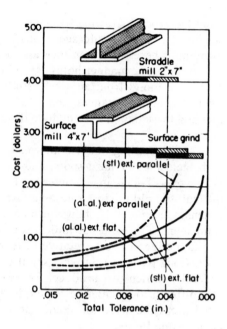

Fig. 1.14 — Straddle and face milling costs plotted against tolerance requirements.

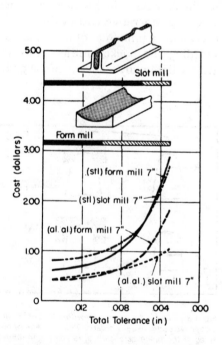

Fig. 1.15 — Tolerance effect on cost of slotting and form milling cuts.

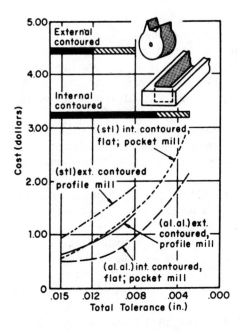

Fig. 1.16 — Plot of tolerance versus cost for profile and pocket-milled parts.

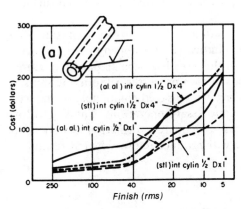

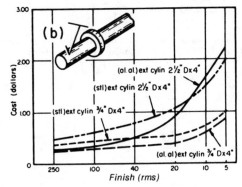

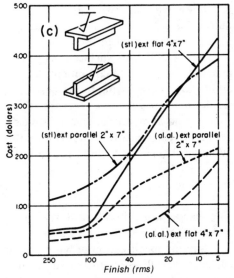

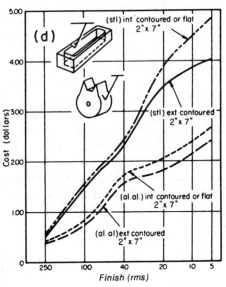

Fig. 1.17 (a) — Comparative costs for internal holes with varying degrees of surface finish refinement.

(b) — Plot of surface refinement versus cost for some turned and ground surfaces.

(c) — Cost values for surface roughness values on shapes similar to those shown in Fig. 1.14.

(d) — Surface finish versus cost of production for profiled and pocket-milled parts.

and light cuts are necessary for close limits and on enclosed surfaces smooth blending of surfaces is difficult when the direction of cut changes.

Surface Roughness Limits. Finish demands impose severe cost problems. As surfaces are refined, tool costs rise. Cutting tools used to produce a 250 rms surface roughness will last twice as long in production, before sharpening, as those required for a 100 rms finish. Those for 100 rms roughness will last three times as long as tools needed to hold a 40 rms finish.

Below a 40 rms surface it is generally necessary to move work from general machines to precision equipment, incurring greater costs. In addition, finer finishes add to total costs owing to increased inspection time, special handling and susceptibility to damage.

A general picture of comparative costs for internal cylindrical holes is given in *Fig.* 1.17 (*a*). To hold costs to a minimum it is suggested that finish on internal cylindrical surfaces be held to the following limits:

Drill .... To 250 rms minimum
Ream .... To 100 rms minimum
Bore .... To 40 rms minimum

The sharp upward break at 40 rms in *Fig.* 1.17 (*a*) reflects the addition of grinding to attain finishes below this value.

Similar specification of widest possible limits on finish of external cylindrical surfaces, as well as on milled and profiled surfaces, is desirable. At (*b*) are plotted some cost figures on examples of turned and ground surfaces. *Fig.* 1.17 (*c*) shows cost values for surface finish on shapes such as shown in *Fig.* 1.14. Here again, lowest cost is obtained by use of roughest suitable surface condition. Examples plotted at (*d*), similar to those of *Fig.* 1.16, cover end-milled and profiled surfaces as obtained without special attention.

Conclusion. Although the complete evolution of design costs is a considerably more involved problem than may have been indicated, it is hoped that the tremendous importance of serious consideration of this area has been emphasized. Effort and capital expenditure in streamlining production processes, simplifying steps in manufacture and speeding materials handling can be saved completely or reduced greatly by the elimination of all possible unnecessary operations. The key to low costs is in the evolution of final design detail. Adequate information as well as incentive and programming by management will produce unprecedented results.

REFERENCES

[1] Throughout the Handbook, all dollar costs quoted should be treated as relative costs since product manufacturing examples have been gathered over a period that has seen a doubling of "real" cost in production. The fact that many products have not increased in cost to the consumer, i.e., appliances, etc., attests to the viability of good "design for production".

2

AUTOMATION AND DESIGN PRINCIPLES

THE DESCRIPTIVE term "poor design," generally applied to mechanisms which fail to function properly or adequately, can just as well be used to label units found to be impractical or expensive to manufacture. In most instances, components which prove costly to produce are taken for granted. Similarly, many that have been manufactured in one specific manner for years are unquestioned. If service life is reasonable and there are few shop problems, no demand develops for re-evaluation of design.

On what basis, then, can a design be judged as poor or good? Too, how can manufacturing management be assured that poor designs are not released — that every component is adequately producible under the conditions prevailing at release? At least fair to good assurance can be had if adequate attention is given this problem at the management level. Even moderate schooling of manufacturing engineers in this all-important phase of producibility can produce real dollars-and-cents return.

Critical Need for Producibility. Under normal manufacturing conditions, producibility in design is a vital factor necessary to meet competition. Where automated production is concerned it is a "must." Also, it is important for conservation of materials, maximum output and maximum production speed as well as minimum cost. Organization for maximum producibility has now taken on a greater importance than ever before.

Automation in Production. Under an effective automation program, management has need to recognize that product design assumes a critically different role. It is possible to get by without a well developed product design policy with conventional production methods. Where automation technology is to be applied it is imperative that products be "designed for production." Unless product design is tailored to be compatible with the production scheme, there may be no significant rewards from efforts in designing new products or in improving production facilities[1].

Co-ordination of product design efforts with an experienced manufacturing engineering team is now a recognized necessity. Freedom from the restrictions of the usually limited number of "known" methods can effect amazing advances in quality, efficiency, and profitability.

Authorities in the field agree that success of any manufacturer in maintaining or establishing a sound financial position in the competitive market place is based only partly on the design and development of new products. Of equal importance, success depends also on the ability and efficiency of the manufacturing organization in producing the product to meet quality, volume, and cost standards that enhance the marketability of the product.

Basically, the product design engineer is called upon to create a desired product, using his ability to comprehend its end use and the conditions under which it must operate. His job is to plan, develop, assemble, and test one or more models or units meeting the desired product attributes. Process or manufacturing engineers will determine the necessary kinds and design of production facilities to manufacture the desired product advantageously, *Fig.* 1.1.

Probably the most potent factor in the rapid evolution of the manufacturing engineering function is the rise of automation in productive systems. Conventional approach to manufacturing had always permitted considerable flexibility and considerable correction of poorly planned operations without incurring extremely excessive costs. Machines could be rearranged, production steps added, process cycles readjusted, or personnel added or removed. Today's more complex production systems and automated lines have proved, conclusively, this practice is no longer either satisfactory or economical in most areas of industry.

Bringing the basic elements of automation technology together in the correct manner and balance requires engineers who thoroughly understand the principles involved and who have a wide knowledge of materials and the methods suitable for processing them. This group must analyze the design or character of the product, know what is expected of it, and sense whether the product is practical to manufacture with a known method or whether one can be evolved to do the job.

Manufacturing Engineering. Today, as never before, it is necessary to "tool up the team" to carry on profitable manufacturing operations. Manufacturing in the era of automation is no longer an art; it is fast becoming an engineering technology of broad, sweeping proportions. Truly, it can be said that the planning and devising of manufacturing operations have left the ranks of the shop man, the apprentice, the machinist, and millwright. To carry on competitive manufacture today requires an engineered approach, with full recognition that fundamental manufacturing problems are basic in all industries and eminently satisfactory solutions already exist for many problems encountered; it takes broad knowledge and engineering acumen to achieve practical success.

Fig. 2.1 — Group of typical office machine motor parts redesigned for lowest cost production almost entirely from stampings. With a minimum number of parts, simplified components and standardized end plates this open design eliminated need for a cooling fan.

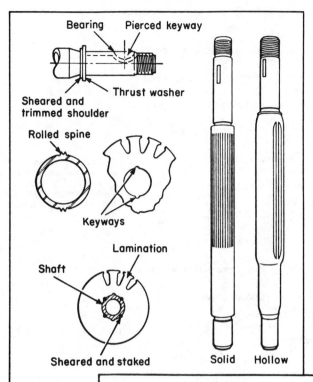

Bearing   Pierced keyway

Thrust washer

Sheared and
trimmed shoulder

Rolled spine

Keyways

Lamination

Shaft

Sheared and staked

Solid   Hollow

Fig. 2.2 — Change of this automobile generator shaft from conventional solid steel to hollow
welded steel tubing produced
substantial material savings. The
production sequence was made
automatic including transfer from
operation to operation. Even
though additional operations are
required, high-speed processes
ideally suited to continuous output
actually created savings greater
than expected. Use of unique
pierced keyway rather than a milled
keyseat created an effective retainer for the key and decreased
production cost.

SEQUENCE OF OPERATIONS

Solid Shaft	Hollow Shaft
Operation No.	Operation No.
1. Turn all diameters and cut off (an automatic screw machine)	1. Roll tube from hot rolled strip stock, weld, and cut to length
	2. Nose ends of tube
2. Centerless grind OD of shaft, rough, and finish	3. Centerless grind OD of tubes, rough, and finish
	4. Wash tube and coat with trisodium phosphate
	5. Extrude drive end of shaft, shear thrust shoulder, and extrude extreme end for later thread rolling. Press with indexing dial
	6. Machine drive end of shaft on special machine
	7. Extrude commutator end of shaft, and shear oil slinger on second press
	8. Machine commutator end of shaft
6. Mill keyway	9. Pierce keyway with press
4. Wash	10. Wash
5. Centerless grind both bearing diameters of shaft	11. Centerless grind both bearing diameters of shaft
7. Roll threads with thread roller	12. Roll threads with thread roller
3. Knurl OD of shaft	13. Roll splines on OD with special machine
8. Inspect	14. Inspect
9. Ready to have laminations assembled	15. Ready to have laminations assembled

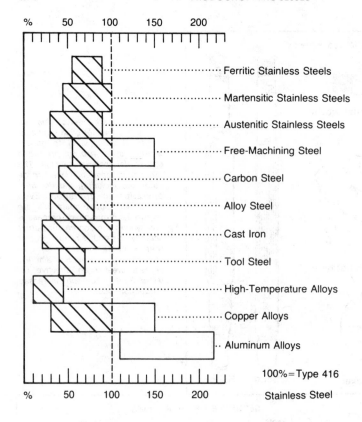

Fig. 2.3 — Chart showing the general comparative machinability rates of various metals. American Iron and Steel Institute data.

In manufacturing, the product produced is always a compromise between acceptable quality and acceptable cost. The seldom-reached goal is always one of maximum attainable quality for minimum possible cost. When the manufacturing engineering function is adequately recognized and properly implemented to fulfill its key part in the manufacturing cycle, amazing results are obtainable in terms of product quality and manufacturing economy.

In every instance, the well engineered production system amortizes the investment in engineering time and creates added profits. The problem of manufacturing economics arises regardless of whether a new product is to be manufactured at a specific price, an existing process improved, or manufacturing costs reduced to meet competition. What are the right steps to be taken? Should more automatic operation be considered? How much automation will be economical and how sophisticated should the systems and processes be to produce the product? What are the right steps processwise? How do you get there with the right answers or processes in the shortest period of time and with an adequate return for the dollars invested to get there?

Manufacturing engineering has evolved as the basic function to supply the answers to these types of questions. And, as industry becomes more diverse and products more complex, fulfilling this function calls for bringing into play the know-how from all facets of industrial enterprise. Tomorrow's effective solutions will result from crossbreeding of ideas from many areas.

Fundamental to manufacturing engineering is the fact that it encompasses thorough "before-the-fact" planning and detailed execution of the manufacturing process stages, not costly "after-the-fact" correction of poorly planned or unplanned operations. The word "engineering" implies just that — a broad, carefully engineered study and development of the entire processing operation, including study of product design suitability for maximum economy in processing.

Operational Study. What is the fundamental approach to producibility in manufacturing? Can a basic study reveal significant factors for practical guidance? The answer of course is yes.

Those who have worked seriously in the cost effectiveness/producibility field of endeavor have learned that low manufacturing costs do not automatically result as manufacturing experience and methods improvement begin to exert influence. Methods improvement simply may never take place, especially since products once tooled are difficult to retool and acceptance of change itself is always a major hurdle.

Foley[2] has pinpointed the responsibility for costs establishment at the principal design stage:

"Engineers and designers sometimes assume that someone else is causing product costs. But others only *count* these costs. The details of how a part is designed *generate its costs.* There are no other primary causes of costs. Product manufacturing costs begin with each and every drawing and detail specified. The basic responsibility for cost generation rests with product development and cannot truly be placed elsewhere. A design which misses its cost target is simply a bad design.

"The above guideposts may be summarized as follows:
a)   The lowest-cost part is the nonexistent part.
b)   The best way to achieve true reliability is by simplicity.
c)   Quality means strict technical adequacy — not frills.
d)   Simplify first — a simple prototype will be a simple product.
e)   The development of key parts includes their production tooling.

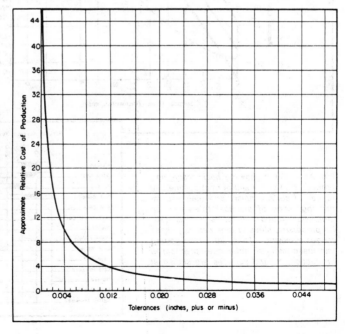

Fig. 2.4 — Chart of general cost relationship of various degrees of accuracy. Plotted from a variety of data, the trend is only indicative of the cost of refinement and does not take into consideration the possibilities offered by judicious designing for production.

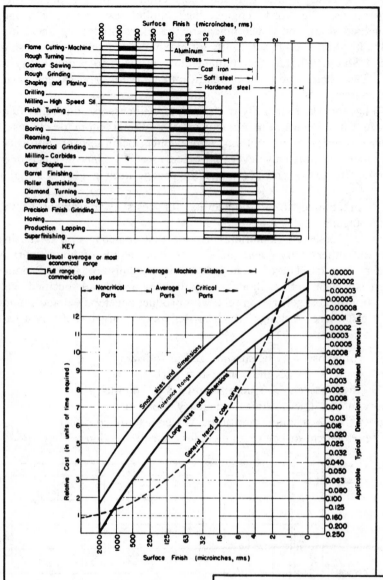

Fig. 2.5 — Chart showing basic machining processes, their full and normal commercial ranges of surface finish, and their accompanying range of practical tolerances on dimension. Added refinements in terms of extra processing steps or time can vary these ranges somewhat. Use of automatic gaging can make minimum tolerances practical in production. Natural surfaces are shown at right.

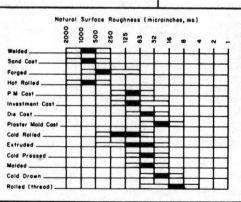

f)   Product costs originate in design and have no other true source.

"An engineering design is not finished — it does not represent a 'procurable reality' — until it has been evaluated in its tooled, production-made form."

While at this stage of evolution no sure-fire engineering approach is practical, it is possible to present some general rules for evaluation. Once the problem is reduced to finite proportions, work can be channeled into those areas which promise the most effective solution. What is being sought is adaptability of the product to permit integration of specific processing steps into an economical manufacturing process. Just which group or sequence of operations is most suitable depends upon the product and the process characteristics and other conditions of processing necessary.

Product and Process. Significantly important in any study of design producibility are the product and the processing steps. Ingenious developments in both these areas have revealed the effectiveness of good design. Thus, the two main considerations that must be brought into the picture are: 1. Design of the product. 2. Design for the process.

Both are critical. Without the first consideration, the product design may necessitate processing steps which are actually useless, thereby rendering the system more expensive or possibly impractical. Without the second, production is often difficult or inadvisable and any attempt at automation is hazardous.

General Rules for Design

It is no idle talk that, regardless of the principles developed, the skill exercised in creating products capable of inexpensive and easy manufacture often spells the difference between success and failure. Many otherwise excellent designs fail to succeed owing to manufacturing problems. With automation this factor can be greatly magnified in its overall importance. And, in many instances, the important adaptation leading to success may appear relatively of insignificant proportions.

Design of the Product. The character of the product is always important. Few products exist that cannot be improved or simplified. Manufactured products such as hardware and mechanical items are especially in this category. Final design should never be frozen until the production requirements have been evaluated.

In general a number of specific recommendations or rules for design can be made. No one rule can be said to predominate, each case must be studied carefully for best ultimate results.

Rule 1. Insure maximum simplicity in overall design. Develop product for utmost simplicity in physical and functional characteristics.

Production problems and costs have a direct relation to complexity of design and any efforts expended in reducing a product to its "lowest common denominator" will result in real savings. This applies as well to individual items, components, and assemblies or subassemblies. Actually, simplicity is a great deal more difficult to achieve than complexity, but production costs bear a direct relation to complexity of the product and efforts to simplify will be well rewarded costwise.

Simplification for Assembly and Maintenance. Perhaps the question regarding the parts of a new design should be, "Is it absolutely required?" The most obvious way to cut costs is to eliminate or combine parts to obviate the expense of attaching or assembling. However, the opposite possibility of breaking down an overly complex part into two simpler parts, cheaper to produce and assemble, should not be overlooked either, *Fig.* 2.1. The highly important requirement of fast satisfactory assembly, especially on the production line, must also be kept in mind. Machines or mechanisms which are difficult to assemble not only incur greatly added production costs but also, in certain cases, may result in considerably increased maintenance difficulty and too, as a result, receive poor customer acceptance.

Design Refinement Cost Reductions. The great importance of simplifying design and assembly has been shown extremely well in recent years. One eminent authority emphasizes this fact with this comment, "refinements of design and analysis are the key to sharp reductions in costs — not revolutionary improvements in production techniques."

Rule 2. Analyze carefully all materials to be used. Select materials for suitability as well as lowest cost and availability. Seek economy of use.

Selection of materials to be used for any product can play a key role in attaining lowest costs. Special materials may create a production problem regarding availability. As production increases, the material form and use assume a commanding role in the unit product cost, *Fig.* 2.2. Overall possibilities to reduce material costs deserve careful study.

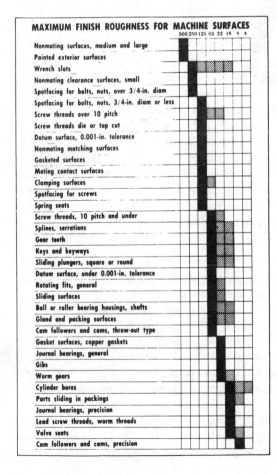

Fig. 2.6 — Some typical surface roughness values specified for precision machine part surfaces.

OLD DESIGN
Costing 0.057
Machining $0.076
$0.133

NEW DESIGN
Stamping $0.034

Fig. 2.7 — A change to high-speed stamping for producing this bracket reduced cost 75 per cent compared with the original cast style.

It is always necessary and extremely important that the designer carefully consider the materials to be used. Should a design be most adaptable for production by machining, the primary consideration must be a free machining material. For instance, though aluminum alloys and aluminum-alloy forgings cost somewhat more than steel for parts of similar design, time required for machining aluminum-alloy parts is often sufficiently lower (about one-third that required for similar steel parts) to offset the original additional material cost, *Fig.* 2.3. Where low-machinability material must be used, then thought should be given to producing the part by means of another process where machinability so necessary for turning operations is not a factor, if high costs are to be avoided on quantity runs. Perhaps the material has a high formability rating and the part can be cold headed, pressed from metal powders, stamped, or rolled; thus eliminating many or all machining or finishing operations.

Effect of Materials Requirements. Where design requirements allow little or no leeway as to material selection, the design will be influenced directly by the production methods by which the material can be readily worked, as in the case of jet-engine turbine buckets. Because of the high-temperature resistant materials used, these buckets often cannot be machined or forged and they must be produced by a means which requires little or no machining or working. Again, where powder-metal pressing is considered, the vast difference in the characteristics of the sintered parts must be kept in mind.

Machinability Ratings. With most machining operations — turning, broaching, gear shaper generating, etc. — the well-known machinability ratings of standard materials determine their suitability to a large extent. With alloy steels, however, structure and hardness have a direct effect upon the machinability of the material. Differing structures and hardnesses satisfy different types of machining operations. In some cases compromises must be made. In others, where a number of operations are to be performed, it is often found economical to use several treatments to condition the material for maximum ease of machining.

Table I—Preferred Tolerances and Allowances

0.0001	0.0010	0.0100
	0.0012	0.0120
0.00015	0.0015	0.0150
0.0002	0.0020	0.0200
0.00025	0.0025	0.0250
0.0003	0.0030	0.0300
0.0004	0.0040	
0.0005	0.0050	
0.0006	0.0060	
0.0008	0.0080	

RATIONAL TOLERANCES AND FITS

Tolerance is the amount of variation permitted in size or location.

Limits are the extreme permissible dimensions resulting from applying a tolerance.

Allowance is the intentional difference in the dimensioning of mating parts, that is, the minimum clearance space (or maximum interference) which is intended between mating parts.

Nominal Size is a designation given to the subdivision of the unit of length having no specified limits of accuracy but indicating a close approximation to a standard size.

Basic Size is the exact theoretical size from which all variations are made.

Table II—Recommended Standard Tolerances and Fits

Class 1—Loose Fit

Basic Hole System (left), Basic Shaft System (right)

Nom. Diam.	Standard Quality Hole	Standard Quality Shaft	Special Quality Hole	Special Quality Shaft	Allow.	Standard Quality Shaft	Standard Quality Hole	Special Quality Shaft	Special Quality Hole
1/32	.0312+.0050	.0292−.0050	.0312+.0020	.0292−.0020	.002	.0312−.0050	.0314+.0050	.0312−.0020	.0314+.0020
.0400	.040+.005	.038−.005	.040+.002	.038−.002	.002	.040−.005	.042+.005	.040−.002	.042+.002
.0500	.050+.005	.048−.005	.050+.002	.048−.002	.002	.050−.005	.052+.005	.050−.002	.052+.002
1/16	.0625+.0050	.0605−.0050	.0625+.0020	.0605−.0020	.002	.0625−.0050	.0645+.0050	.0625−.0020	.0645+.0020
.0800	.080+.005	.078−.005	.080+.002	.078−.002	.002	.080−.005	.082+.005	.080−.002	.082+.002
3/32	.0937+.0050	.0917−.0050	.0937+.0020	.0917−.0020	.002	.0937−.0050	.0957+.0050	.0937−.0020	.0957+.0020
.1000	.100+.005	.098−.005	.100+.002	.098−.002	.002	.100−.005	.102−.005	.100−.002	.102+.002
1/8	.125+.005	.123−.005	.125+.002	.123−.002	.002	.125−.005	.127+.005	.125−.002	.127+.002
5/32	.1562+.0050	.1542−.0050	.1562+.0020	.1542−.0020	.002	.1562−.0050	.1582+.0050	.1562−.0020	.1582+.0020
3/16	.1875+.0050	.1855−.0050	.1875+.0020	.1855−.0020	.002	.1875−.0050	.1895+.0050	.1875−.0020	.1895+.0020
1/4	.250+.005	.248−.005	.250+.002	.248−.002	.002	.250−.005	.252+.005	.250−.002	.252+.002
5/16	.3125+.0050	.3105−.0050	.3125+.0020	.3105−.0020	.002	.3125−.0050	.3145+.0050	.3125−.0020	.3145+.0020
3/8	.375+.005	.373−.005	.375+.002	.373−.002	.002	.375−.005	.377+.005	.375−.002	.377+.002
1/2	.500+.005	.498−.005	.500+.002	.498−.002	.002	.500−.005	.502+.005	.500−.002	.502−.002
9/16	.5625+.0050	.5605−.0050	.5625+.0020	.5605−.0020	.002	.5625−.0050	.5645+.0050	.5625−.0020	.5645+.0020
5/8	.625+.006	.6225−.0060	.6250+.0025	.6225−.0025	.0025	.625−.006	.6275+.0060	.6250−.0025	.6275+.0025
3/4	.750+.006	.7475−.0060	.7500+.0025	.7475−.0025	.0025	.750−.006	.7525+.0060	.7500−.0025	.7525+.0025
7/8	.875+.006	.8725−.0060	.8750+.0025	.8725−.0025	.0025	.875−.006	.8775+.0060	.8750−.0025	.8775+.0025
1	1.000+.008	.997−.008	1.000+.003	.997−.003	.003	1.000−.008	1.003+.008	1.000−.003	1.003+.003
1¼	1.250+.008	1.247−.008	1.250+.003	1.247−.003	.003	1.250−.008	1.253+.008	1.250−.003	1.253+.003
1½	1.500+.010	1.496−.010	1.500+.004	1.496−.004	.004	1.500−.010	1.504+.010	1.500−.004	1.504+.004
2	2.000+.010	1.996−.010	2.000+.004	1.996−.004	.004	2.000−.010	2.004+.010	2.000−.004	2.004+.004
2½	2.500+.010	2.496−.010	2.500+.004	2.496−.004	.004	2.500−.010	2.504+.010	2.500−.004	2.504+.004
3	3.000+.012	2.995−.012	3.000+.005	2.995−.005	.005	3.000−.012	3.005+.012	3.000−.005	3.005+.005
3½	3.500+.012	3.495−.012	3.500+.005	3.495−.005	.005	3.500−.012	3.505+.012	3.500−.005	3.505+.005
4	4.000+.012	3.995−.012	4.000+.005	3.995−.005	.005	4.000−.012	4.005+.012	4.000−.005	4.005+.005
5	5.000+.015	4.994−.015	5.000+.006	4.994−.006	.006	5.000−.015	5.006+.015	5.000−.006	5.006+.006
6	6.000+.015	5.994−.015	6.000+.006	5.994−.006	.006	6.000−.015	6.006+.015	6.000−.006	6.006+.006
8	8.000+.020	7.992−.020	8.000+.008	7.992−.008	.008	8.000−.020	8.008+.020	8.000−.008	8.008+.008
10	10.000+.020	9.992−.020	10.000+.008	9.992−.008	.008	10.000−.020	10.008+.020	10.000−.008	10.008+.008

Class 2—Free Fit

Basic Hole System (left), Basic Shaft System (right)

Nom. Diam.	Standard Quality Hole	Standard Quality Shaft	Special Quality Hole	Special Quality Shaft	Allow.	Standard Quality Shaft	Standard Quality Hole	Special Quality Shaft	Special Quality Hole
.016	.0160+.0012	.0152−.0012	.0160+.0008	.0152−.0008	.0008	.0160−.0012	.0168+.0012	.0160−.0008	.0168+.0008
.020	.0200+.0012	.0192−.0012	.0200+.0008	.0192−.0008	.0008	.0200−.0012	.0208+.0012	.0200−.0008	.0208+.0008
.025	.0250+.0012	.0242−.0012	.0250+.0008	.0242−.0008	.0008	.0250−.0012	.0258+.0012	.0250−.0008	.0258+.0008
1/32	.0312+.0012	.0304−.0012	.0312+.0008	.0304−.0008	.0008	.0312−.0012	.0320+.0012	.0312−.0008	.0320+.0008
.0400	.0400+.0012	.0392−.0012	.0400+.0008	.0392−.0008	.0008	.0400−.0012	.0408+.0012	.0400−.0008	.0408+.0008
.0500	.0500+.0012	.0492−.0012	.0500+.0008	.0492−.0008	.0008	.0500−.0012	.0508+.0012	.0500−.0008	.0508+.0008
1/16	.0625+.0012	.0617−.0012	.0625+.0008	.0617−.0008	.0008	.0625−.0012	.0633+.0012	.0625−.0008	.0633+.0008
.0800	.0800+.0012	.0792−.0012	.0800+.0008	.0792−.0008	.0008	.0800−.0012	.0808+.0012	.0800−.0008	.0808+.0008
3/32	.0937+.0012	.0929−.0012	.0937+.0008	.0929−.0008	.0008	.0937−.0012	.0945+.0012	.0937−.0008	.0945+.0008
.1000	.1000+.0012	.0992−.0012	.1000+.0008	.0992−.0008	.0008	.1000−.0012	.1008+.0012	.1000−.0008	.1008+.0008
1/8	.1250+.0012	.1242−.0012	.1250+.0008	.1242−.0008	.0008	.1250−.0012	.1258+.0012	.1250−.0008	.1258+.0008
5/32	.1562+.0012	.1554−.0012	.1562+.0008	.1554−.0008	.0008	.1562−.0012	.1570+.0012	.1562−.0008	.1570+.0008
3/16	.1875+.0012	.1867−.0012	.1875+.0008	.1867−.0008	.0008	.1875−.0012	.1883+.0012	.1875−.0008	.1883+.0008
1/4	.2500+.0012	.2492−.0012	.2500+.0008	.2492−.0008	.0008	.2500−.0012	.2508+.0012	.2500−.0008	.2508+.0008
5/16	.3125+.0012	.3117−.0012	.3125+.0008	.3117−.0008	.0008	.3125−.0012	.3133+.0012	.3125−.0008	.3133+.0008
3/8	.3750+.0012	.3742−.0012	.3750+.0008	.3742−.0008	.0008	.3750−.0012	.3758+.0012	.3750−.0008	.3758+.0008
1/2	.5000+.0012	.4992−.0012	.5000+.0008	.4992−.0008	.0008	.5000−.0012	.5004+.0012	.5000−.0008	.5008+.0008
9/16	.5625+.0012	.5617−.0012	.5625+.0008	.5617−.0008	.0008	.5625−.0012	.5633+.0012	.5625−.0008	.5633+.0008
5/8	.6250+.0015	.6240−.0015	.625+.001	.624−.001	.001	.6250−.0015	.6260+.0015	.625−.001	.626+.001
3/4	.7500+.0015	.7490−.0015	.750+.001	.749−.001	.001	.7500−.0015	.7510+.0015	.750−.001	.751+.001
7/8	.8750+.0015	.8740−.0015	.875+.001	.874−.001	.001	.8750−.0015	.8760+.0015	.875−.001	.876+.001
1	1.000+.002	.9988−.0020	1.0000+.0012	.9988−.0012	.0012	1.000−.002	1.0012+.0020	1.0000−.0012	1.0012+.0012
1¼	1.250+.002	1.2488−.0020	1.2500+.0012	1.2488−.0012	.0012	1.250−.002	1.2512+.0020	1.2500−.0012	1.2512+.0012
1½	1.5000+.0025	1.4985−.0025	1.5000+.0015	1.4985−.0015	.0015	1.5000−.0025	1.5015+.0025	1.5000−.0015	1.5015+.0015
2	2.0000+.0025	1.9985−.0025	2.000+.0015	1.9985−.0015	.0015	2.0000−.0025	2.0015+.0025	2.000−.0015	2.0015+.0015
2½	2.5000+.0025	2.4985−.0025	2.500+.0015	2.4985−.0015	.0015	2.5000−.0025	2.5015+.0025	2.500−.0015	2.5015+.0015
3	3.000+.003	2.998−.003	3.000+.002	2.998−.002	.002	3.000−.003	3.002+.003	3.000−.002	3.002+.002
3½	3.500+.003	3.498−.003	3.500+.002	3.498−.002	.002	3.500−.003	3.502+.003	3.500−.002	3.502+.002
4	4.000+.003	3.998−.003	4.000+.002	3.998−.002	.002	4.000−.003	4.002+.003	4.000−.002	4.002+.002
5	5.000+.004	4.9975−.0040	5.0000+.0025	4.9975−.0025	.0025	5.000−.004	5.0025+.0040	5.0000−.0025	5.0025+.0025
6	6.000+.004	5.9975−.0040	6.0000+.0025	5.9975−.0025	.0025	6.000−.004	6.0025+.004	6.0000−.0025	6.0025+.0025
8	8.000+.005	7.997−.005	8.000+.003	7.997−.003	.003	8.000−.005	8.003+.005	8.000−.003	8.003+.003
10	10.000+.005	9.997−.005	10.000+.003	9.997−.003	.003	10.000−.005	10.003+.005	10.000−.003	10.003+.003

Where heat treatments or hardening are found to be a necessity, careful consideration as to the material selected and its adaptability to processing prior to heat treating is imperative. With proper selection of heat-treatment procedure, a material which can be readily machined or formed can almost always be adopted. In fact, where close accuracy must be maintained and variations from heat treating distortion obviated, materials can be resorted to which offer good machinability following heat treatment. Just as simple weldments can be used to supplant more costly forgings where quantity is not sufficiently great, so low-alloy steels with suitable heat treatment can often supplant more costly and difficult-to-machine alloy steels.

Class 3—Running Fit

	Basic Hole System					Basic Shaft System			
	Standard Quality		Special Quality			Standard Quality		Special Quality	
Nom. Diam.	Hole	Shaft	Hole	Shaft	Allow.	Shaft	Hole	Shaft	Hole
.01	.0100+.0008	.0097—.0005	.0100+.0005	.0097—.0003	.0003	.0100—.0005	.0103+.0008	.0100—.0003	.0103+.0005
.0125	.0125+.0008	.0122—.0005	.0125+.0005	.0122—.0003	.0003	.0125—.0005	.0128+.0008	.0125—.0003	.0128+.0005
.016	.0160+.0008	.0157—.0005	.0160+.0005	.0157—.0003	.0003	.0160—.0005	.0163+.0008	.0160—.0003	.0163+.0005
.020	.0200+.0008	.0197—.0005	.0200+.0005	.0197—.0003	.0003	.0200—.0005	.0203+.0008	.0200—.0003	.0203+.0005
.025	.0250+.0008	.0247—.0005	.0250+.0005	.0247—.0003	.0003	.0250—.0005	.0253+.0008	.0250—.0003	.0253+.0005
1/32	.0312+.0008	.0309—.0005	.0312+.0005	.0309—.0003	.0003	.0312—.0005	.0315+.0008	.0312—.0003	.0315+.0005
.0400	.0400+.0008	.0397—.0005	.0400+.0005	.0397—.0003	.0003	.0400—.0005	.0403+.0008	.0400—.0003	.0403+.0005
.0500	.0500+.0008	.0497—.0005	.0500+.0005	.0497—.0003	.0003	.0500—.0005	.0503+.0008	.0500—.0003	.0503+.0005
1/16	.0625+.0008	.0622—.0005	.0625+.0005	.0622—.0003	.0003	.0625—.0005	.0628+.0008	.0625—.0003	.0625+.0005
.0800	.0800+.0008	.0797—.0005	.0800+.0005	.0797—.0003	.0003	.0800—.0005	.0803+.0008	.0800—.0003	.0803+.0005
3/32	.0937+.0008	.0934—.0005	.0937+.0005	.0934—.0003	.0003	.0937—.0005	.0940+.0008	.0937—.0003	.0940+.0005
.1000	.1000+.0008	.0997—.0005	.1000—.0005	.0997—.0003	.0003	.1000—.0005	.1003+.0008	.1000—.0003	.1003+.0005
⅛	.1250+.0008	.1247—.0005	.1250+.0005	.1247—.0003	.0003	.1250—.0005	.1253+.0008	.1250—.0003	.1253+.0005
5/32	.1562+.0008	.1559—.0005	.1562+.0005	.1559—.0003	.0003	.1562—.0005	.1565+.0008	.1562—.0003	.1565+.0005
3/16	.1875+.0008	.1872—.0005	.1875+.0005	.1872—.0003	.0003	.1875—.0005	.1878+.0008	.1875—.0003	.1878+.0005
¼	.2500+.0008	,2497—.0005	.2500+.0005	.2497—.0003	.0003	.2500—.0005	.2503+.0008	.2500—.0003	.2503+.0005
5/16	.3125+.0008	.3122—.0005	.3125+.0005	.3122—.0003	.0003	.3125—.0005	.3128+.0008	.3125—.0003	.3128+.0005
⅜	.3750+.0008	.3747—.0005	.3750+.0005	.3747—.0003	.0003	.3750—.0005	.3753+.0008	.3750—.0003	.3753+.0005
½	.5000+.0008	.4997—.0005	.5000+.0005	.4997—.0003	.0003	.5000—.0005	.5003+.0008	.5000—.0003	.5003+.0005
9/16	.5625+.0008	.5622—.0005	.5625+.0005	.5622—.0003	.0003	.5625—.0005	.5628+.0008	.5625—.0003	.5628+.0005
⅝	.625 +.001	.6246—.0006	.6250+.0006	.6246—.0004	.0004	.6250—.0006	.6254+.0010	.6250—.0004	.6254+.0006
¾	.750 +.001	.7496—.0006	.7500+.0006	.7496—.0004	.0004	.7500—.0006	.7504+.0010	.7500—.0004	.7504+.0006
⅞	.875 +.001	.8746—.0006	.8750+.0006	.8746—.0004	.0004	.8750—.0006	.8754+.0010	.8750—.0004	.8754+.0006
1	1.0000+.0012	.9995—.0008	1.0000+.0008	.9995—.0005	.0005	1.0000—.0008	1.0005+.0012	1.0000—.0005	1.0005+.0008
1¼	1.2500+.0012	1.2495—.0008	1.2500+.0008	1.2495—.0005	.0005	1.2500—.0008	1.2505+.0012	1.2500—.0005	1.2505+.0008
1½	1.500 +.0015	1.4994—.0010	1.500 +.001	1.4994—.0006	.0006	1.500 —.001	1.5006+.0015	1.5000—.0006	1.5006+.0010
2	2.0000+.0015	1.9994—.0010	2.000 +.001	1.9994—.0006	.0006	2.000 —.001	2.0006+.0015	2.0000—.0006	2.0006+.0010
2½	2.5000+.0015	2.4994—.0010	2.500 +.001	2.4994—.0006	.0006	2.500 —.001	2.5006+.0015	2.5000—.0006	2.5006+.0010
3	3.000 +.002	2.9992—.0012	3.0000+.0012	2.9992—.0008	.0008	3.0000—.0012	3.0008+.0020	3.0000—.0008	3.0008+.0012
3½	3.500 +.002	3.4992—.0012	3.5000+.0012	3.4992—.0008	.0008	3.5000—.0012	3.5008+.0020	3.5000—.0008	3.5008+.0012
4	4.000 +.002	3.9992—.0012	4.0000+.0012	3.9992—.0008	.0008	4.0000—.0012	4.0008+.0020	4.0000—.0008	4.0008+.0012
5	5.0000+.0025	4.9990—.0015	5.0000+.0015	4.999 —.001	.001	5.0000—.0015	5.0010+.0025	5.000 —.001	5.0010+.0015
6	6.0000+.0025	5.9990—.0015	6.0000+.0015	5.999 —.001	.001	6.0000—.0015	6.0010+.0025	6.000 —.001	6.0010+.0015
8	8.000 +.003	7.9988—.0020	8.000 +.002	7.9988—.0012	.0012	8.000 —.002	8.0012+.0030	8.0000—.0012	8.0012+.0020
10	10.000 +.003	9.9988—.0020	10.000 +.002	9.9988—.0012	.0012	10.000 —.002	10.0012+.0030	10.0000—.0012	10.0012+.0020

Class 4—Slide Fit

Nom. Diam.	Max. Clear.	Nom. Diam.	Max. Clear.	Nom. Diam.	Max. Clear.	Nom. Diam.	Max. Clear.
		1/16	.0003	⅛	.0003	2	.0006
.01	.0003	.0800	.0003	½	.0003	2½	.0006
.0125	.0003	1/32	.0003	9/16	.0003	3	.0008
.016	.0003	1000	.0003	⅛	.0004	3½	.0008
.020	.0003	⅛	.0003	¼	.0004	4	.0008
.025	.0003	5/32	.0003	⅞	.0004	5	.001
1/32	.0003	1/16	.0003	1	.0005	6	.001
.0400	.0003	¼	.0003	1¼	.0005	8	.0012
.0500	.0003	5/16	.0003	1½	.0006	10	.0012

Classes 5 to 8—Interference Fits

		Required Interference							
Size		Class 5 Wringing Fit		Class 6 Drive Fit		Class 7 Force Fit		Class 8 Heavy Force Fit	
From	To	From	To	From	To	From	To	From	To
.000	.599	0	.0003	.0003	.0005	.0008	.0012	.0015	.002
.600	.999	0	.0004	.0004	.0006	.001	.0015	.002	.0025
1.000	1.499	0	.0005	.0005	.0008	.0012	.002	.0025	.003
1.500	2.799	0	.0006	.0006	.001	.0015	.0025	.003	.004
2.800	4.499	0	.0008	.0008	.0012	.002	.003	.004	.005
4.500	7.799	0	.001	.001	.0015	.0025	.004	.005	.006
7.800	13.599	0	.0012	.0012	.002	.003	.005	.006	.008
13.600	20.999	0	.0015	.0015	.0025	.004	.006	.008	.010

Procedure to Results: First step is to consider selection of a preferred basic size or diameter. In Table III are listed sizes for diameters up to 4 inches. Sizes beyond 4 inches follow the same steps of ¼-inch up to 6 inches and reduce to ½-inch steps as size increases beyond 6 inches.

Preferred values of tolerances and allowances are shown in Table I and a simplified group of fits readily useable by any industry is given in Table II. Clearance fits are complete in two quality ranges—standard for maximum economy, and special for maximum precision where added cost is not too great a factor. No tolerances are given for critical fits. These should be dimensioned from data to suit conditions.

TABLE III—Preferred Basic Sizes

	0.0100	5/16	0.3125	1⅞	1.8750
	0.0125	⅜	0.3750	2	2.0000
1/64	0.01562	7/16	0.4375	2⅛	2.1250
	0.0200	½	0.5000	2¼	2.2500
	0.0250	9/16	0.5625	2⅜	2.3750
1/32	0.03125	⅝	0.6250	2½	2.5000
	0.0400	11/16	0.6875	2⅝	2.6250
	0.0500	¾	0.7500	2¾	2.7500
1/16	0.0625	⅞	0.8750	2⅞	2.8750
	0.0800	1	1.0000	3	3.000
3/32	0.09375	1⅛	1.1250	3¼	3.2500
	0.1000	1¼	1.2500	3½	3.5000
⅛	0.1250	1⅜	1.3750	3¾	3.7500
5/32	0.15625	1½	1.5000	4	4.000
3/16	0.1875	1⅝	1.6250		
¼	0.2500	1¾	1.7500		

Rule 3. Use the widest possible tolerances and finishes on parts. Be sure that surface roughness and accuracies specified are reasonable and in keeping with the product and its functions.

Tolerances on surface finish and dimensions play an important part in the final achievement of practical production design. It is imperative that the principles of interchangeable manufacture be observed for successful low-cost mass production. All components of products should be surveyed carefully to assure not only successful processing but also rapid, easy assembly and maintenance.

Here, it is also important to keep in mind the influence of practical production design and that of standard available materials. Tolerances on dimension and finish characteristics with standard materials as well as that regularly produced with conventional production methods must be observed to attain suitability and lowest costs.

Importance of Practical Tolerance Requirements. Knowledge of the inherent variations present in each process which set the economical dimensional-tolerance limits and surface finish obtainable completes the picture from a practical standpoint. It can be said, almost without reserve, that the specification of close tolerances without serious consideration of all the factors attendant to maintaining them in production will invariably result in needlessly excessive or prohibitive costs from slowed output, extra operations, high tool upkeep, and high rejection or scrap losses. Any tolerance less than plus or minus 0.005-inch should be scrutinized closely to be sure such is absolutely essential and then the tolerance selected should be compared with the normal average usually produced with the processing method contemplated. If the process cannot be relied upon for the accuracy or surface finish required, the part must be redesigned for another more suitable method or, where increased costs can be absorbed, additional refined processing operations can be utilized to obtain the desired result.

Dimensional Tolerances and Cost. A general idea of how machining tolerances compare as to relative cost of producing and holding them is illustrated graphically in *Fig.* 2.4. Naturally, as operations are added to gain refinement of tolerances and surface finish, costs increase. This chart is intended only to portray the general cost relationship of various degrees of accuracy, but the importance of judicious selection and specification of tolerances can be readily seen. Actually, each process must be analyzed individually along with the particular part, inasmuch as by designing particularly to suit a specific process, the desired degree of accuracy oftentimes may be obtained without undue cost penalty.

Fig. 2.8 — Careful study of a product can reveal most suitable production methods. Design of this sector gear was revamped to eliminate the two stop screws shown, both special. Process change to die cut the gear teeth instead of mill permitted eliminating the two end teeth so extra stops were not needed. Use of stamping and elimination of the screws reduced operations from 16 to 4, gave additional materials savings.

Generally speaking, fine surface finish and close tolerances do not result with operations where metal removal is extremely heavy. Once the functional specifications for a particular part and the accompanying accuracy necessary to assure satisfactory performance have been determined, it is relatively easy to ascertain the production method or methods that will be required to produce the part. Once the production quantities have been determined, the possibilities are usually narrowed to but one or two alternates, depending upon design flexibility.

Production Methods and Surface Finish. Neither dimensional tolerances nor surface quality should be specified to limits of accuracy any closer than actual function or design necessitates, to assure the advantages of lowest possible cost and fastest possible production. The roughest acceptable finish should be contemplated and so specified. If at all possible, the inherent average finish quality left by the basic productive method required should be acceptable. In *Fig.* 2.5, the basic metal removal and finishing processes are charted with their full range of surface finishes encountered in general commercial practice. Shown also is the narrower range of surface finish normally expected in average production which are not the very finest, but which can be anticipated with well-kept machines without incurring undue costs. Finish readings are in microinches, root-mean-square, taken on a Profilometer and occur in the accepted preferred-number sequence.

Relation of Dimensional Tolerances and Surface Finish. To provide at a glance a direct idea as to the general relationship of actual dimensional tolerances to surface roughness, the chart in *Fig.* 2.5 also indicates a range of accuracy for each processing method. That there must be some relationship between surface roughness and dimensional accuracy is, of course, obvious. To expect a tolerance of 0.0001-inch to be held on a part turned or milled to an average roughness of 500 microinches certainly is rather foolish. Since this reading indicates a roughness or actual peak-to-valley depth of about 0.002 to 0.0035-inch, not considering the waviness, chatter or feed marks, etc., this is self-evident. Likewise, to specify a finish of 10 to 15 microinches for a surface which is merely intended to provide proper size for locating for subsequent operations is also needless. A 40 to 60 microinch finish would be satisfactory and cost is at least 50 to 60 per cent less.

Owing to the many idiosyncrasies and ramifications which necessarily characterize all production methods, however, the values shown in *Fig.* 2.5 should not be used directly in specifying part tolerances. Rather, they are intended to be indicative of the practical or possible ranges. Each part designed should be considered separately along with the process contemplated in order to take into account such factors as size, shape or complexity and to obtain those benefits possible through adaptation to suit the peculiarities inherent in a particular method.

Finish Pattern. Again, in relation to surface finish it must be remembered that each process produces its own characteristic type or types of surface finish pattern. Finish specifications, where important, should not be general or contain loose phraseology susceptible of gross misinterpretation. Use of approved American Standard specifications will be found well worthwhile. Functional requirements for a specific part should be carefully analyzed — it is possible to have too perfect a surface! Table ways, for instance can be flat and smooth to the point that lubrication fails to separate the moving parts. Hydraulic valves have been fitted so closely by lapping that, they have, sans benefit of a definite lubricant retaining scratch pattern, failed to function under normal system pressure after several

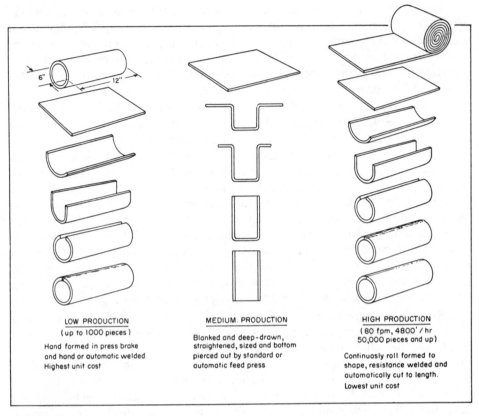

LOW PRODUCTION
(up to 1000 pieces)

Hand formed in press brake
and hand or automatic welded.
Highest unit cost

MEDIUM. PRODUCTION

Blanked and deep-drawn,
straightened, sized and bottom
pierced out by standard or
automatic feed press

HIGH PRODUCTION
(80 fpm, 4800'/hr
50,000 pieces and up)

Continuosly roll formed to
shape, resistance welded and
automatically cut to length.
Lowest unit cost

Fig. 2.9 — Simple example showing the influence of quantity on process selection for producing a tubular shape of large diameter and thin wall. Lowest unit cost is practical where substantial quantities are required on each production run. Welding need not be required and where high-speed press forming is used, tube joint can be interlocked.

shuttlings back and forth. In one case, only a special extreme-pressure lubricant saved the day.

Typical Surface Finishes. Characteristic range of surface roughness figures shown in *Fig.* 2.6 will help judge as to what constitutes a practical finish. Careful study of the basic characteristics of each production method will provide a clue to the scratch pattern to be expected, and the suitability of the part so produced for the intended use.

Rule 4. Standardize to the greatest extent possible. Use standard available parts and standardize basic parts for one product as well as across a maximum number of products.

Use of standard components in developing products needs no emphasis. Standard stock items cost less and require no development. Also, in seeking best production design, standardization of components across product lines offers tremendous savings. In manufactured products this is a fertile field for attaining lowest costs and maximum output. Study of this phase has revealed some impressive opportunities and, in many cases, showed improved functional characteristics as a result of closely controlled automatic production made possible by increased volume.

Design for the Process. It has been said that evolving the basic idea for a product constitutes about five per cent of the overall effort in bringing it onto the market; some 20 per cent of the effort goes toward general design of the product and about 50 per cent of the effort is necessary to bring about economic production of the product while the remainder goes toward removal of bugs from design and production, including servicing. The big job is in making the product economically — through design for more economic production, *Fig. 2.7.*

From the standpoint of automatic production, the president of one company makes this important observation:

"While the importance of tool design to successful automation is well understood, the importance of product design to good automation is not yet fully realized. The designers of parts must be thoroughly educated in automation. The designer who is familiar with automation and who designs his product with automated production in mind can make the work of the manufacturing staff easier and can save his company thousands and even millions of dollars. The simple precaution of designing a flange that turns out, instead of in, on an automobile body can make the difference between a manual and an automated operation on the production line."

While it is possible to produce almost any product conceived, it does not follow that production costs will be reasonable or acceptable. Design should be planned whenever possible to suit basic operations available through the common manufacturing methods which can be tooled and adapted with minimum effort. Design that centers only around the product and its function generally creates the need for entirely special production facilities. Because such facilities may not always be justified economically, the practical aspects of design modified for available automatic production facilities are readily apparent.

Rule 5. Determine the best production method or fabricating process. Tailor the product to specifically suit the production method or methods selected.

It is distinctly possible to tailor the details of a product to assure economic production and it is possible to have a basically similar part which is excessively expensive to produce. Small but extremely important details can create this difference, *Fig. 2.8.*

Every component and subassembly of a product should be carefully studied to insure close observation of inherent limitations imposed by particular processing procedures. For instance, it is important to keep in mind that every production method has a well-established level of precision which can be maintained in continuous production without exceeding normal costs. Actually, no amount of inspection will improve this level. For practical reasons it must be observed.

The practice of specifying tolerances according to empirical rules should be

Fig. 2.10 — Vacuum cleaner dusting brush body, originally produced from special deep drawing stock in 15 difficult operations, was redesigned for diecasting, right, with one press swedging or curling operation to achieve the spherical effect at considerably lower cost.

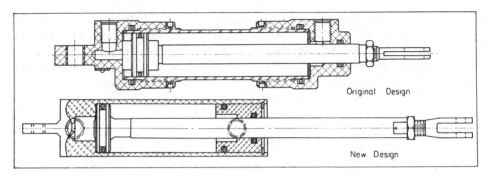

Fig. 2.11 — Conventional hydraulic actuating cylinder for tail wheel extension or retraction and the new, simplified design which costs approximately one-sixth as much. By designing each part with attention focused on economical production operations, the total number of operations and number of separate parts were reduced, the cylinder was 40 per cent lighter, yet provided the same operating capacity.

abandoned. Practical design for automatic production demands careful observance of the "natural" tolerances available with specific processes, *Fig.* 2.5.

Again, the process or processes selected may be directly influenced by the anticipated or required quantity output. Many processes have quantity limitations and parts can be tailored to permit production by the method offering optimum cost advantage for the quantity and output speed needed, *Fig.* 2.9.

Another important influence in process selection is that of the materials to be worked. The ideal material, acceptable costs and suitability for method of processing must be worked out. Material selection should be carefully made to insure maximum machinability, workability, castability, modability, formability or weldability for most economical processing conditions and suitable results.

To produce the original assembly of stampings for the dusting brush body shown at the left in *Fig.* 2.10, a series of fifteen difficult operations were required. Die maintenance was high and specially selected deep drawing material was necessary to stand the severe draws without cracking. The diecasting which superseded the stamping is shown to the right. Following casting, the part is finished by one simple punch-press swedging or curling operation to achieve the spherical effect. By selecting a malleable diecasting alloy, the operation can be readily performed and high cost incurred because of the original critical material requirements is overcome.

Rule 6. Minimize production steps. Plan for the smallest possible number or lowest cost combination of separate operations in manufacturing the product.

The primary aim here is to insure the elimination of needless processing steps. Even though processing is automatic, useless production steps require time and equipment. Few products are designed that cannot be improved in this regard. Small design details must be carefully weighed for their effect on processing. Needless operations of course create the need for unnecessary handling operations, hence added cost.

Actually, the greatest savings which can be obtained in designing for production usually result from reducing the number of separate processing operations required to complete a part. Next in importance is reduction in the number of parts used, and following that comes savings effected through the use of stock parts or interchangeable parts.

The hydraulic cylinder design shown in *Fig.* 2.11 is an excellent example of how this and the previous principles of designing for production can be applied. Actually, the number of operations required to produce the various parts of this item was reduced from a total of 45 to 21. The operations were simplified and greater use of automatic equipment made. Total number of parts was reduced from 14 to 10 and the resultant cylinder was 40 per cent lighter and had the same capacity and operating pressure, 1500 psi. Cost was reduced from over 40 to less than 7 dollars.

Where production can be automatic, it is desirable to recognize the exceptions to this rule where redesign changes the entire production sequence to attain other possible savings, *Fig.* 2.2.

Rule 7. Eliminate handling problems. Insure ease in locating, setting up, orienting, feeding, chuting, holding and transferring parts.

Where parts must be held or located in a jig for a processing operation, consideration must be given the means by which the operation will be accomplished. Lugs may be added, contours changed, or sections varied slightly to make handling a simple rather than a costly operation. Much as with tooling, jig and fixture problems resulting from some needlessly complicated designs become unsurmountable in an attempt to achieve mass production. Some degree of understanding of the processing functions involved in each method as well as an appreciation of the various aspects of fixturizing, holding or feeding the parts, can achieve vastly improved production, lowered jig or fixture costs, and, of course, reduced production cost per part.

Difficulty of handling certain parts in production is often overlooked. Pads or locating positions are almost invariably required in processing, regardless of quantity. Use of false lugs for locating and clamping often may be a necessity. Where conditions indicate, such lugs should be designed so as to be cast integral with the part and located so that, following completion of the machining operations, they can be readily cut off and destroyed, without affecting the part.

To simplify mechanized operation it is possible to tailor product design for most practical handling. It should be practical to transport parts, hopper feed, locate for assembly operations, etc., without undue difficulty. The product should be rigid and compact where possible to withstand mechanical handling. Fragile projections should be eliminated.

It is desirable to design parts subject to hopper feeding so that tangling is avoided. Wherever possible, overall design should be developed to permit ready handling of the product in simplified steps by means of complete and independent subassemblies amenable to automatic assembly. Uniform symmetrical pieces may pose an orientation problem requiring specific means for sure-fire orientation. Locating buttons or projections may be desirable for such parts.

While ingenious handling methods and mechanical production aids are continuously being developed, the designer who can eliminate the need for such equipment simplifies production problems and effectively reduces overall costs.

Conclusions

Achievements of the past few years have indicated conclusively that a thorough redesign analysis with the processing methods in view always results in low-cost assembly or product equal to and often better than the original with few, if

Fig. 2.12 — Wireform swivel loop provided a 64 per cent cost saving over the earlier, cast iron version. Made by Merrill Mfg. for the N. A. Taylor Co., of Gloversville, N. Y., to serve as the working end of an anchor buoy for the boating industry, the half-pound wireform is priced at 18 cents per unit, the 1 1/2-pound cast iron piece at 50 cents. Manufacture of galvanized steel provides the corrosion resistance necessary.

any, of the production handicaps, *Fig.* 2.12.

The possibilities of producibility are tremendous and are largely untapped. From a cost standpoint it becomes obvious that the advantages of good design for production should be sought seriously. Today, perhaps even greater stress must be placed on the conservation of our far from limitless engineering materials and upon achieving better results with smaller amounts. The practicability of the design for production approach has been validated. The accomplishments merely await the careful studied application of these principles.

REFERENCES

[1] Roger W. Bolz, *Understanding Automation*, Novelty, Ohio: Conquest Publications, 1966.
[2] Thomas P. Foley, "Development of a chain printer," *Computer Design*, March 1966.

ADDITIONAL READING

[1] Phillip F. Ostwald, "Product cost estimating," *Technical Papers*, pp. 105-116, May 24, 1973, Chicago.
[2] Phillip F. Ostwald, "Cost Estimating for Strategical Decisions in Manufacturing Group-Classified Designs," A.S.M.E. paper 74DE 7, 11 pp., April 2, 1974, Chicago.
[3] George F. Hawley, *Automating the Manufacturing Process*, New York: Reinhold Publishing Corp., 1959.
[4] Hugh D. Luke, *Automation for Productivity*, New York: Wiley-Interscience, 1972.
[5] Harley H. Bixler, *The Manufacturing Research Function*, New York: American Management Association, Inc., AMA Research Study 60, 1963.

3

DESIGN FOR ASSEMBLY

Most PRODUCT DESIGN activity always concerns some kind of assembly requirement which can involve component elements, materials, products, or other "assembled" goods. The combining of such basics into the desired end product or package presents a significant cost, time, and quality problem in attaining more efficient production and/or higher productivity. Concentration on the major efficiencies offered by a systematic approach to the opportunities of improved assembly should be a major concern in good design for productivity. In addition, the combination of the techniques of automation systems engineering with the necessary ingenuity in product design can effectively transfer marginal operations into highly profitable ones.

The Assembly Approach. Redesign for assembly can eliminate from 25 to 95 percent of the unnecessary and sometimes undesirable labor content of any product. Wherever labor content of a product comprises the largest portion of the cost, redesign to reduce assembly problems and/or automated assembly should be considered. New and unique assembly concepts which are conceived around optimized product design for automated assembly will result in improved products, conservation of valuable materials and highly attractive out-the-door

Fig. 3.1 — Actual historical cost experience in high-volume automated production of a complex sensing device. Courtesy Honeywell Government and Aeronautical Products Div.

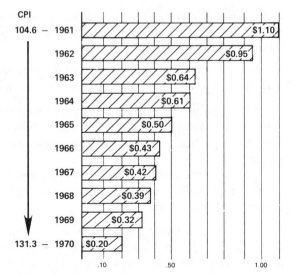

costs.

Innovative design for assembly can bring a new flexibility to the total manufacturing process. Design that accommodates automation manifests itself in the ability to perform a greater variety of manufacturing steps without delay or unprofitable handling of the product in process and, in addition, since automating operations *economically* requires a total rethinking of the entire process, it creates new and unexpected solutions to old and unrecognized problems.

As Peter Drucker[1] many years ago recognized the implications:

"...when we reach any degree of automation, the entire process becomes one. This means first that we can no longer consider a particular activity, phase or function to be an entity. It is only one aspect of a whole and incapable of being considered or managed as a discrete part. In particular the lines between *design, production* and *distribution* become blurred to the point of disappearance. *Automation* not only presupposes planned mass-distribution founded on systematic development of anticipated markets. It not only presupposes *rethinking* of design. It

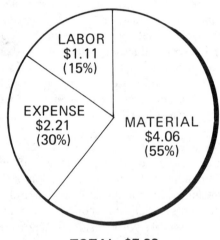

Fig. 3.2 — Analysis of typical cost elements involved in an automated assembly (500,000 units) highlight the importance taken on by the materials used.

presupposes that every major decision be simultaneously a *whole* decision and meaningful in terms of design, production and distribution (and also in terms of finance, purchasing, etc.)"

Successful application, thus, requires door-to-door study of the product manufacturing process in its entirety. It has been emphasized pointedly that:

"The best product design is one that anticipates automation... A design for automated assembly is most often a simpler design that can reduce (manufacturing) system complexity, thus directly lowering costs for both the system and the end product.

"Proof of the savings possible through assembly automation systems is shown in the accompanying actual case history (*Fig.* 3.1) representing experience in high-volume automation production."[2]

Therefore, whether a product is new or old, consider seriously the assembly problem as a vital part of product design or redesign. Rapid advance has been made in the use of unique methods such as adhesive bonding, untrasonic welding,

ASSEMBLING PROCESS OR OTHER WORK DONE AT WORK STATION	INSPECTION
• Embossing. • Stamping. • Printing. • Forming (washers, flat springs, etc.). • Press fitting. • Crimping. • Blanking and feeding material strips. • Coil winding. • Resistor trimming. • Metering. • Molding. • Curing. • Deburring. • Rolling. • Tapping. • Inserting components. • Soldering (flux and wire). • Welding. • Ultrasonic welding. • Lubricating (liquid or dry). • Adhesive application. • Electronic balancing.	• Probe for presence and orientation (air gaging, microswitches, transducers). • Weighing (to 0.1 ounce). • Check electrical circuits. • Check torque. • Check push and pull strength. • Check time delay. • Optically inspect (fiber optics). **FUNCTION TESTING**

Fig. 3.3 — Variety of production operations that are possible to perform during automatic assembly help reduce cost of separate preprocessing and handling. Courtesy Honeywell Government and Aeronautical Products Div.

heat sealing, and electron-beam or laser welding, etc., which enhance the opportunities for more economic assembly.

Design for Assembly. Good design for simplified assembly is difficult to separate from just plain good design for production with conventional methods. Where the latter leaves off and the former starts is often hard to discern.

To gain some idea as to just where to expend major effort in attaining lowest cost, and simplest assembly, it is useful to look at the cost of operations from the standpoint of automated assembly, whether it will be actually used or not. With automated production, unit cost elements show the largest proportion in materials and purchased parts as contrasted, *Fig.* 3.2, with that of manual operations — where labor and overhead take the lion's share of the unit cost breakout — so that total unit cost practically reaches little more than the cost of the materials.

It is well to note the cost of installing fasteners in any assembly — handling, orientation, feeding and driving. Experts indicate it requires 20 to 26 seconds to install an average screw with an ordinary hand tool, 5 to 8 seconds with a portable power tool, and but 2 to 4 seconds with automatic or semiautomatic equipment. Design for the method of fastening parts together thus can exert a powerful influence on the final production cost.

Taking actual costs from a typical highly automated assembly operation, here is how they break down: Originally, labor constituted 41% of the total cost, overhead 28% and materials 31%. Designed for automated assembly these changed to 19%, 21% and 60%, respectively. The change in emphasis can easily be noted and should be especially of interest in production analysis of any product for economy of production.

To achieve these dollar saving results calls for first the *right* methods, partial automation or automated assembly. Design for automation will provide an entirely new competitive posture.

Actually, the keys to automated assembly are to be found in these elements:

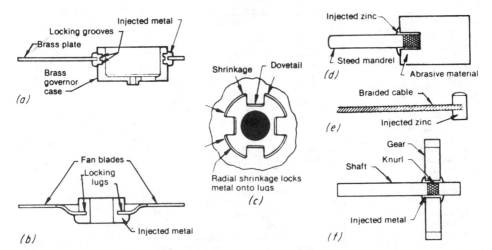

Fig. 3.4 — Injected metal assembly can reduce the costs normally encountered when using conventional approaches. Typical components shown here are provided with surfaces around which the injected metal can shrink. Governor case and plate for a telephone dial, a, has two locking grooves in the case; a cooling fan for a small power drill, b, has four locking lugs around which the center hub is cast. Mechanical lock is produced on damping-disc/shaft assembly, c, by providing dovetail shape in disc. Injected metal is forced into porous abrasive material, d, to lock the mandrel to the abrasive wheel. Similarly, as brake cable fastener is injected directly onto braided cable, e, and injected metal solidifies around irregular cable shape, thus locking itself to the cable. Knurled shaft, f, provides locking surface for small gear/shaft assembly. Below is shown an electric meter assembly. Courtesy Fisher Gauge Ltd., Peterborough, Ontario, Canada.

    1.   Specialization of labor, machines and management focused on the product design.

    2.   Integration of components so as to minimize the assembly steps or stages.

    3.   On-line or preliminary qualification of dimensions and variations of product components.

    4.   Simplification of both product design and assembly sequence.

    5.   Coordination of all assembly facilities required.

6. Reorganization of design and manufacturing activities to achieve automated assembly goals.

Design, which is of great importance, concerns primarily items 2, 3, 4, and 6. On some products there is no problem in design for automatic assembly. The product as designed can be assembled either way — manually or automatically. The steps are insignificant. On others the product is *frozen* as is — a *functionally acceptable* product not subject to any changes. On still others such as new or proprietary feature products, the design is critical if the final product assembly is to be economical and saleable. If real cost reductions are to be effected, three actions should be taken:

1. Design for the most suitable production processes with economical assembly as a goal.

2. Eliminate expensive operations not *really* needed to achieve function *and* simplify design details. Objective analysis is required here.

3. Redesign for automation, if necessary, even though this means a complete change in both processes *and* materials.

In general, some of the basic principles of product design are again important for automated assembly. These should be restated for careful study:

1. Design components to serve more than one function; make component parts and materials interchangeable across more than one product style.

2. Standardize quality and tolerances; eliminate highly precise fits wherever possible.

3. Design and plan different component parts so that production processing can be done *on* the assembly machine on demand, *Fig.* 3.3, by available methods.

4. Plan parts, materials and assemblies so they can be easily handled during assembly without damage.

5. Select materials or revise materials to suit each processing operation best. Evaluate cost of special materials and forms against cost of added operations needed with conventional types.

6. Use *building block* principles for product design and obtain each style by adding *standardized* parts.

To look at some of this interweaving of design and production characteristics, the only sound approach is to study each specific product component in detail.

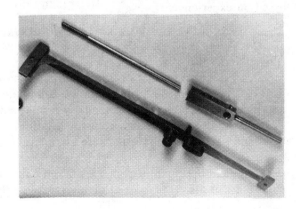

Fig. 3.5 — A die cast balance beam which replaced an assembled design not only eliminated expensive subassembly parts handling but simplified final assembly.

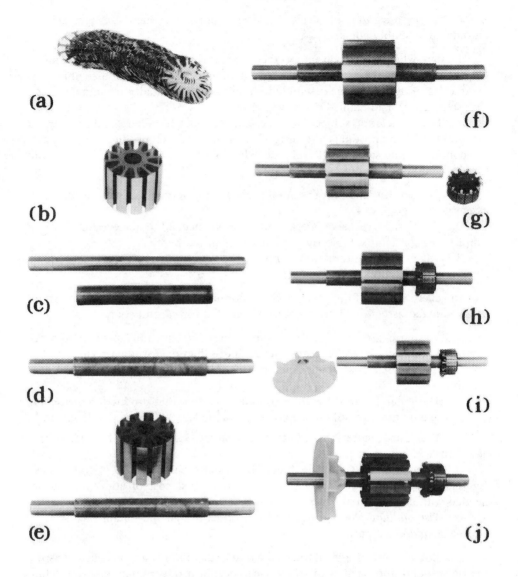

(a)

(b)

(c)

(d)

(e)

(f)

(g)

(h)

(i)

(j)

What are some of the possibilities? Some admonitions can be stated which are based on experience:

    1.   Avoid the slow processes; design for high-speed continuous processes.

    2.   Avoid machining if possible; simplify any necessary machining operations (this includes use of widest possible tolerances).

    3.   Look for unusual or *impossible* combination operations.

    4.   Redesign to simplify assembly. Eliminate or change parts hard to handle and place in assembly, either manually or automatically, *Fig.* 3.4.

    5.   Study the handling problem. Eliminate expensive and/or *useless* operations and handling (believe it, there are many, *Fig.* 3.5).

The final optimum in design is not only attainable by observing these general rules but it is also the result of highly creative effort. As Blumrich has written[3]:

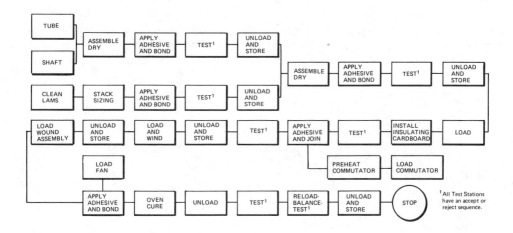

Fig. 3.6 — A radical redesign approach to reduce costly assembly can also produce a better product as with this electric motor.

(a) The unassembled laminations. The laminations are punched out and taken off the press with the same orientation — i.e., there is a burr on the sheared edge of the stamping which must be positioned correctly.

The laminations are stacked on a long bar and processed through an ultrasonic degreaser. From there the laminations go to a machine which builds the rotor stack. A three-foot long stack of clean laminations are placed in a vertical tube. A slicing device picks off the correct number of laminations and moves them out from under the stack. This lamination stack is then oriented so that all the laminations are in line. The stack is then transferred into a carousel unit for bonding. The rotor stack is indexed beneath the adhesive applicator head (bore coater) and the head advanced into the bore of the stack and floods the bore with adhesive. As the head retracts an automatic suckback feature built into the dispensing head removes all the excess adhesive from the bore. The lamination stack is then indexed into an induction coil where the head differential causes the adhesive to be sucked into the laminations and cured.

(b) The unitized rotor stack. Stack is indexed to a test station and off-loaded into a holding area.

(c) Tube and shaft unassembled. On a second, carousel type machine the steel shaft is manually loaded into a vertical collet. Onto this shaft the insulating tube is placed. (Tube is clamped — shaft positioned.) The tube and shaft are indexed to the position where the adhesive is dispensed into the clearance space between the tube and shaft (a counterbore is machined into the tube to provide a place for the adhesives). The shaft, during the adhesive dispensing cycle, is oscillated up and down. This effectively distributes the adhesive between the tube and shaft. The assembly is then indexed into an induction heater where the adhesive is cured.

(d) Assembled tube and shaft. The assembly is indexed to a test station and off-loaded into holding area.

(e) Unitized stack, tube and shaft assembly — unassembled. The two assemblies are slip fitted and installed in the vertical attitude in a third carousel type machine. Assembly is indexed into a preheat coil to warm the rotor stack. Assembly is then indexed to the station where adhesive is dispensed. The preheated rotor stack causes the viscosity of the adhesive to drop sufficiently so that it will run down the bore of the stack and effectively fill the clearance between the two parts. The assembly is then indexed into an induction coil to cure the adhesive.

(f) Assembled rotor/shaft and tube. Assembly is indexed to test station and off-loaded to a holding area.

(g) Commutator ring and rotor assembly, unassembled. Prior to assembly of the commutator the end fibers are automatically installed on the assembly which then moves to a position where adhesive is dispensed around the O.D. of the insulating tube. The unit is then moved to the station where installation of the commutator occurs. The commutator is fed into an oven from a vibratory feeder on a horizontal track. At the other end of the oven is a pick-retreat-index-advance-release-retract mechanism which picks the commutator out of the oven and places it on the assembly which has just had adhesive put on it. (Latent heat in the commutator ring cures the adhesive.)

(h) Assembled commutator. The unit is tested and off-loaded to a holding area. From here the rotor assembly is taken to machines where the toro is wound automatically.

(i) Assembly with fan unassembled. This is currently being automated. The assembly is placed in a horizontal attitude and adhesive dispensing heads place adhesive on the insulating tube as the assembly is turned by hand. The fan is manually installed and placed by hand in an oven where it is cured.

(j) Complete rotor assembly.

Courtesy Loctite Corp.

" . . . design in its essence is the opposite of analysis. I do not believe that there will ever be a substantial change in this situation because the design process is elusive, depends on, and expresses very personal qualities and at best is subject to rather general rules. . . . good design depends on very personal qualifications. This qualification consists of a combination of extensive engineering knowledge and creativity, plus — not to be overlooked — an open, unbiased mind."

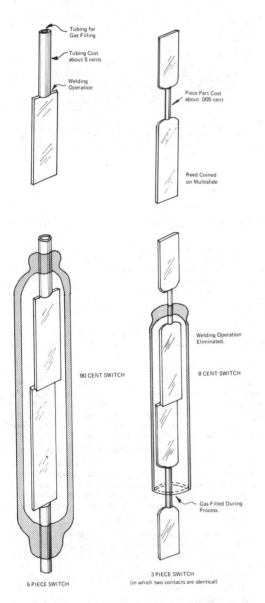

Fig. 3.7 — Dry reed switch redesign for assembly reflecting both parts and operational changes.

Outstanding advantages can be attained even though the design effort results only in justification of partial automation — the realization of an 80% reduction in unit cost in a typical instance. To be listed in the success column, merely recognize and hurdle the mental road blocks to this advanced design approach.

Design Serendipity. By approaching product design as an assembly problem

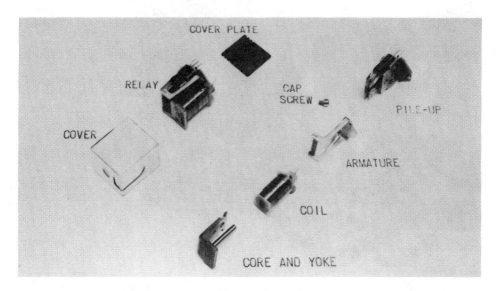

Fig. 3.8 — Major subassemblies of a typical MB miniature relay.

Fig. 3.9 — Riveted core and yoke assembly, top, with redesigned welded assembly shown below.

Fig. 3.10 — Eight-station automatic assembly machine which is used to produce the relay assembly.

some amazing improvements can be attained. Here the concept is based on recognition of the fact that, *as first designed,* significant manufacturing innovation is often limited insofar as cost effective results are concerned. However, once the total assembly concept is studied from the cost reduction standpoint the final result can change radically.

An excellent example is found in the redesign of small motors for automatic assembly. By eliminating the conventional precision press fits and substituting slip-fits with adhesive bonding for assembly, all previous manufacturing and assembly problems were eliminated and costs reduced more than 50 per cent, *Fig.* 3.6, while producing a superior product.

Actually, by means of a simple change in the method of manufacturing components parts and in the ultimate manufacturing assembly concept, even greater cost reductions are possible, *Fig.* 3.7.

The miniaturized relay, shown in *Figs.* 3.8, 3.9, 3.10 and 3.11, was reduced by 70 per cent in cost and a more efficient circuit was obtained, scrap was nearly all but eliminated and plating problems were ended by means of such redesign with automation in mind.[4]

Analysis of any end product as an overall assembly can result in what can be called *design serendipity.* To the engineer dealing in total concepts, the function of the end product may be better served by restructuring the final assembly so as to facilitate its manufacture. Combinations of assembled parts often prove to be producible by other, more inexpensive but *better* methods.

Even apparently complex products can often be redesigned to eliminate many of the costly manual operations. The roll support element, shown in *Fig.* 3.12, was revised to eliminate complex and time consuming manufacture and assembly operations and allow fast permanent assembly of only a few parts by welding. A significant 79 per cent reduction in cost could thus be obtained along with improvement in physical characteristics.

It has been observed that the successful industries of the future will have optimized their use of available resources by cost effective *improved* methods, management and organization. It is worth repeating that even today there are few products that cannot be improved measurably and/or reduced in cost significantly if only the ever-present problem of *not invented here* can be overcome.

The competitive race no longer will not be for those who merely get there first

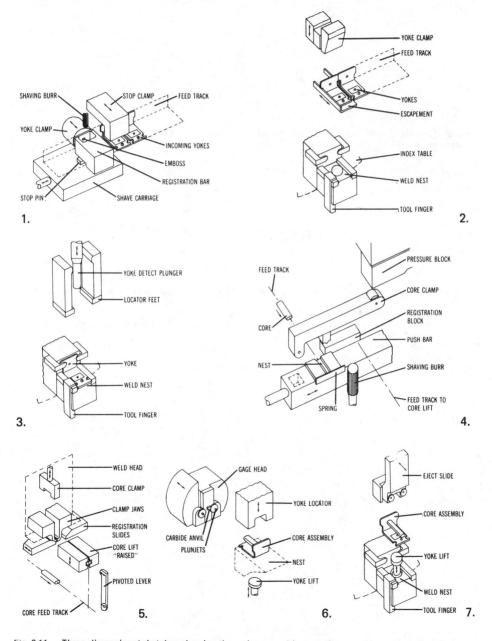

Fig. 3.11 — Three-dimensional sketches showing the major assembly operations used to produce the relay yoke: (1) Yoke clamped, (2) yoke feed sequence, (3) yoke detector, (4) core clamped in push bar, (5) core feed into weld jaws, (6) gaging core-to-yoke position, and (7) assembly eject.

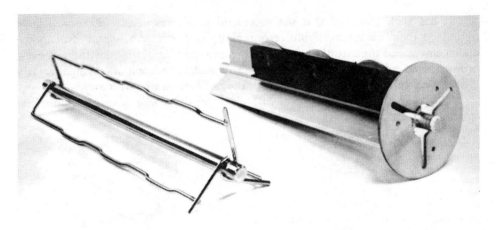

Fig. 3.12 — Steel paper roll support, left, costs 70 cents compared to the $3.34 it cost for the aluminum extrusion and stamping assembly, right. Courtesy Merrill Mfg. Corp.

or produce more but for those who also innovate and devise less costly and better methods of production, better use of materials and acceptable quality, long-service products.

REFERENCES

[1] Peter F. Drucker, "Management science and the manager," *Institute of Management Sciences,* Vol. No. 2, January 1955.

[2] *Honeywell Assembly Automation,* Hopkins, Minn: Honeywell Government and Aeronautical Products Div., April 1972.

[3] Joseph F. Blumrich, "Design," *Science.* Vol. 168, p. 1554, June 26, 1970

[4] Martin S. Kirwan, "Automatic assembly and welding of relay cores and yokes," *Western Electric Engineer,* Vol. XIX, No. 1, pp. 39-44, January 1975.

[5] J. G. Knapp, "Role of industrial design in cost reduction," paper presented at the ASME Design Engineering Conference, April 19-22, 1971, New York, N. Y.

[6] Ray J. Bronikowski, "Pareto's law for managers," *Machine Design,* Vol. 47, No. 18, p. 65, July 24, 1975.

4

Measuring and Gaging Parts

Gordon A. McAlpine

TODAY'S demand for complete reliability in everything from can openers to missiles suggests a quotation from *Arabian Nights* — "The priceless ingredient of any product is the honor and integrity of him who made it." And we could add that the lack of it is the surest way to lose a customer. A bad reputation will spread rapidly and in all directions like waves generated in a pail of water.

Early in this century the historian Henry Adams propounded his Law of Acceleration which he deduced from his exploration of the relationships between

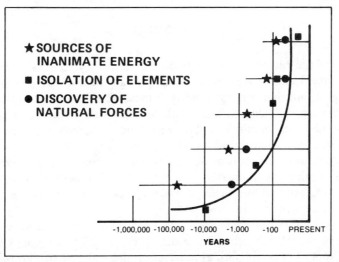

Fig. 4.1 — This composite chart is an extension of Adam's curve which plots the shrinking time intervals between successive discoveries. Reading from left to right the stars represent the development of the Grecian concept of natural forces instead of animistic causes, the discovery of the principles of gravity by Galileo and Newton, of electromagnetism by Faraday, of nuclear power by Fermi and others and the recent discovery of the "weak" force observed in the decay of elementary particles. The curve of squares shows the time sequence of man's discovery of the basic elements starting with man's early discovery of carbon, copper, iron, silver, and gold in prehistoric times. By Mendeleyev's time 100 years ago, over 60 elements had been identified and today all 92 elements of the periodic table plus some 10 trans-uranium elements are known. The curve represented by the dots shows the steady shrinkage of time intervals between discovery and harnassing of one form of energy after another — fire, water, wind, steam, internal combustion and only within the past 25 years, of nuclear energy. As the curves show, man's cumulative experience in the last 50 to 25 years is greater than that of all preceding history.

Gordon A. McAlpine is Vice President, Sales, KLT Industries, Ann Arbor, Mich. He was formerly Director of Marketing, ITT Industrial & Automation Products, Plymouth, Mich.

science and the history of society. As Gerard Piel, Publisher of Scientific American says, "Looking back over the tumult of modern history, he plotted the rising curves of the rate of scientific discovery, of coal output, of steampower, of the transition from mechanical to electrical power .... the curves followed the old familiar law of squares." Acceleration was the law of history, *Fig.* 4.1.

"The pattern of events shows history on a course of accelerating acceleration. The major developments in man's accumulative experience have occurred within the most recent times and these developments occur at shorter time intervals into the very present."[1]

As *Fig.* 4.1 shows, scientific and technological development in the past 50 years has been as great as that in all previous history. Another dramatic illustration of this law of acceleration is the rapidly decreasing time lag between a new discovery and its practical application:

Invention	Discovery	Application	Lag Time
Electric Motor	1821	1886	65 years
Vacuum Tube	1882	1915	33 years
X-ray Tube	1895	1913	18 years
Nuclear Reactor	1932	1942	10 years
Radar	1935	1940	5 years
Transistor	1948	1951	3 years
Solar Battery	1953	1955	2 years

Integrated Circuitry - Almost Simultaneously

Metrology

A trace of the historical development of metrology follows much the same curve as that developed by Adams. The origins of gaging are buried in the silence of the ancient past, but that some standards of measurement existed millenia ago we know from the biblical account which tells us that Noah was commanded to build an ark 300 cubits in length 50 cubits wide and 30 cubits high, and from the tremendous structures built by such civilizations as the ancient Egyptians. History is virtually mute as to the tools or devices that were used for measurement but we do know that Noah's unit of measure, the cubit, was the length of a man's arm from elbow to the tip of the middle finger and that the Egyptians used a cord knotted at regular intervals as their rule.

Such crude standards of measurement continued to be used for centuries. In fact, as late as the 15th century the width of the English monarch's thumb and length of his foot were established as the determining units we still know as the inch and foot.

Rough as these standards were, however, they were adequate for the requirements of the times. The first major impetus to greater precision in gaging came as a result of mass production requirements for guns with interchangeable parts in the late 18th and early 19th centuries in Europe and America. Eli Whitney (1830) was the pioneer in this effort in America. Innumerable gages were developed by manufacturers to make this interchangeability possible. However, whether the devices used agreed with standards of scientific measurement was relatively immaterial so long as all the parts for one and the same machined product were made in the same shop and all pieces were exact within the limits of

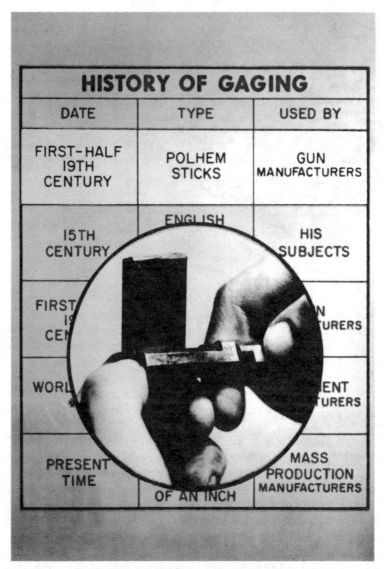

HISTORY OF GAGING		
DATE	TYPE	USED BY
FIRST-HALF 19TH CENTURY	POLHEM STICKS	GUN MANUFACTURERS
15TH CENTURY	ENGLISH	HIS SUBJECTS
FIRST		URERS
WORL		ENT URERS
PRESENT TIME	OF AN INCH	MASS PRODUCTION MANUFACTURERS

Fig. 4.2 — Polhem measuring sticks. These were combination measuring tools used by toolmakers during the period 1700-1850.

the prototypes of the producing shop. Surviving Polhem, or measuring sticks, *Fig. 4.2,* used as gages in the manufacture of guns in the first half of the 19th century show that absolute errors of 0.05 to 1.0 mm per inch were the rule and as much as 3mm were tolerable.

World War I demands and the widespread use of the universal sets of gage blocks developed by Johannson finally made interchangeability of parts produced by different plants practical.

By the second decade of the 20th century these sets were being made so precisely as to make possible comparative measurements accurate to one thousandth of a millimeter and, in effect, all shops using these sets were using the same yardstick. It may be of interest here to mention that Henry Ford hired Johannson and actually manufactured and sold these gage blocks, *Fig. 4.3.*

Fig. 4.3 — The first universal set of gage blocks sold by Johannson provided a means of measurement in intervals of five thousandths of a millimeter over the range of 0.0005 mm to 100 mm.

World War II with its tremendous demands for mass production of weapons and vehicles and the coordination of European and American manufacturers in producing interchangeable components of more precise tolerances than ever before in history resulted in a vast upsurge in the development of new gaging techniques and of high-precision production type machine tools. Among the most significant of these developments were numerical controls which were the outgrowth of the American Manufacturer's Cooperative program and the incorporation of electronic devices into machine tools as controls rather than as accessories for checking.

This trend successively gained momentum in the 50's and continues at an awesome pace through the present. During this period which saw the introduction and spread of extended warranties by the automotive industry, the quality control function gained enormously in stature and almost overnight in the terms of historical time automatic 100% gaging or inspection of precision parts became recognized as an important factor in assuring greater reliability, better utilization of manpower, manufacturing cost reduction and operation analysis.

This presents a fundamental requisite for development and application of predictability and reliability techniques. We realize how important this is when we look at statistics and find that today one out of every ten production workers is now engaged in some quality-control function. One facet of this is statistical analysis, or statistical quality control.

Statistical Analysis

We would like to present this aspect of quality control to establish the conviction that one hundred percent inspection is an essential part of product reliability.

Statistical analysis is absolutely essential in the development of machining techniques, and in proving the reliability and repeatability of machine tools, fixtures, and specialized operations.

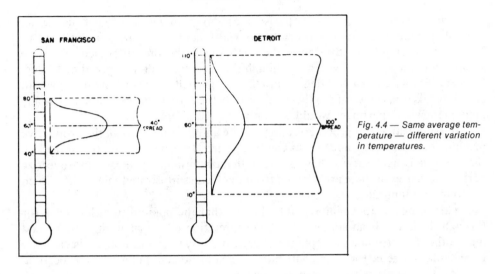

Fig. 4.4 — Same average temperature — different variation in temperatures.

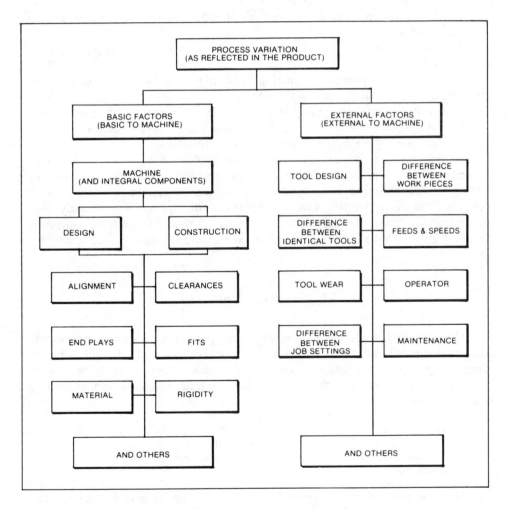

Fig. 4.5 — Basic causes for statistical spread of size values in manufactured components.

Many are familiar with comparisons as shown in *Fig.* 4.4 which use as an example the temperature in San Francisco, Calif., and the temperature in Detroit, Mich. The *averages* are the same; the *variations* make the difference. Note the similarity in these two curves, even though they represent a total spread of 40 deg. F. in San Francisco and 100 deg. F. in Detroit. The Chamber of Commerce would certainly take the center out of this and use only the "good" information, just as any manufacturing department would use only the "good" parts.

In establishing and predetermining the life and the reliability of the productive machine tool the very same curves can be applied. In order to get a full spread on the curve it is necessary to have some undersize, minimum, mean, and oversize parts. The causes of these are derived from the factors illustrated in *Fig.* 4.5, which no doubt are familiar.

This brings us to a major point which is that the possibility of human error through deviation from standards, misinterpretation of information, fatigue, and many other factors makes acceptance of parts directly from machine tools virtually impossible. These factors also contribute to the obsolescence of manual inspection.

Properly applied 100% inspection can:

1. Prevent the production of out-of-tolerance components by in-process application of size control.

2. Eliminate the need for expensive machining of parts to excessively close tolerances.

3. Make possible gaging, segregating and matchfitting of parts — the first and most important step to automatic assembly.

4. Permit simultaneous gaging of a multitude of dimensions of geometries.

5. Provide the mass of accurate data which must be available for computer control since most gage output is automatically recordable.

Among the various types of automatic gages, electronic and air-to-electronic capacitance type systems have gained wide acceptance within the last few years. This is because this type of equipment is exceptionally reliable and versatile. It assures 100% repeatability within specified limits, provides extremely high accuracy and utilizes standard interchangeable off-the-shelf modules. Currently these advantages have been further enhanced by utilizing solid state circuitry.

Motivating Factors

Management is continually faced with the responsibility of solving problems in its quest for profits satisfactory to company stockholders while strengthening the company's competitive position in the market. Automatic production is one way to help achieve these goals. Other forces which compel automatic manufacturing operations are:

1. Reduce general costs.	5. Overcome labor shortage.
2. Increase production.	6. Eliminate scrap loss.
3. Improve quality.	7. Reduce material costs.
4. Competition.	

In automated production, several elements make up the manufacturing cycle. Broadly, these elements are:

1. Raw materials handling and processing.
2. Production or component manufacturing.
3. Inspection and quality control.
4. Assembling or combining or mixing.
5. Performance testing and standards.
6. Packaging, warehousing and shipping.

Of the six elements which make up the manufacturing cycle, three of them involve measuring and gaging, inspection, assembly and test.

New gaging concepts are providing critical-tolerance assemblies from components manufactured under standard production methods and their accompanying economic tolerance ranges.

The problem of selective assembly is as old as manufacturing itself. It has become highly critical in recent years as the housewife demands a home refrigerator that will function for 15 years without any trouble or loss of efficiency from the motor/compressor unit. Her husband is equally demanding as he expects the engine, power steering, and air conditioning on his automobile to function flawlessly for the life of the car which, in this day and age, may be well over 100,000 miles. The consumer not only expects to get precision trouble-free products but he wants them at very attractive prices. The whole consumer movement is backing him up and no one knows this better than the automotive and appliance manufacturers, and other producers of consumer goods.

In reality, the consumer wants, and is actually getting, products manufactured to tolerances within 0.0001-inch at prices that ordinarily would not support manufacturing to 0.001-inch. In order to accomplish this, the major automobile and appliance manufacturers have evolved the concept of automatic gaging for selective assembly whereby parts manufactured to rather mundane tolerances fit and function as though they were slowly manufactured in a precision laboratory to "tenth" tolerances. To do this, the mass-production industries had to scrap the old idea of inspection that identified a part as "good" or "bad" and either sent it on for assembly or consigned it to the scrap bin.

The new concept of gaging for selective assembly actually adds considerable value to a component production lot by qualifying parts for very tight selective fits. This selective fitting is possible even though a whole production run may be manufactured to normal mass production standards at normal production costs — not the extremely high costs that would be involved in manufacturing to laboratory standards.

Two very important fundamentals are found in the concept of automatic gaging for selective assembly. The first involves the normal distribution curve of any production lot. Professional statisticians and quality control experts refer to this as the "bell-shaped distribution curve". It merely shows that in any sizeable production lot most of the components will cluster very close to the specified dimensions if the production process is basically capable of manufacturing to the stated tolerances. However, there will always be a few parts that are toward the extremes of the tolerance limits, as indicated in *Fig.* 4.6.

The second fundamental consideration is that in the selective fitting of components, such as a shaft in a sleeve, the relationship between the mating shaft and hole must be very close — possibly no more than a "tenth"; however, the actual size of the manufactured shafts or sleeves may have reasonably wide tolerances, such as a full thousandth of an inch range or even more.

Thus, if a production lot of shafts and sleeves were manufactured to normal 0.001-inch tolerances and both were within the curve of normal distribution, an

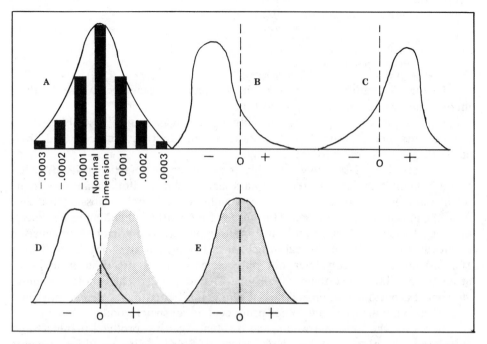

Fig. 4.6 — If a component, such as a shaft, is made to a specified diameter with a tolerance range of 0.001-inch, the normal distribution of acceptable diameters will vary above and below the nominal in a typical distribution shown in A. If the manufacturing process changes, the bulk of the workpieces may fall above or below the nominal dimension (shifted) as indicated by B and C. If selective fits of shafts and sleeves are to be made from workpiece lots that have different means in their "tenth" dimensions, as shown in D, most workpieces will not have mating parts. Modern concepts of automatic gaging for selective assembly combined with computer analysis allow the producer of an end product to manufacture a production lot of workpieces with dimensional distribution that matches that of mating parts being supplied by a vendor, as shown in E. When vendor and producer have identical dimensional distributions in their production lots, there are virtually no workpieces without an available mating part. This is why precision tolerance matching is possible from production lots machined to commercial standards.

automatic gage could segregate all components into ten different groups with a 0.0001-inch limit for selective assembly of the individual parts.

Of course, there is always the threat that the two component parts would not be within the normal distribution. The shaft producer may have his production lot mean toward the high side while the sleeve producer may have his production lot mean to the low side of the tolerance range. The net result would be a quantity of both parts which could not be matched; thus, the cost of selective fitting would be increased.

The answer, within the state of the art of current technologies — which have recently been developed— is combining automatic gaging with the minicomputer. Automatic gaging is an established technique for close-tolerance production parts. For quite a number of years, industry has used automatic gaging to determine acceptable parts and to classify rejects into categories; that is, oversize OD, undersize ID, thickness oversize, thickness undersize, or whatever the part configuration required.

With the advent of more accurate gaging techniques and solid state circuitry improvements, acceptable parts can be classified into very close tolerance groups — typically, 0.0001-inch.

This gaging procedure is being used very successfully today by the automotive industry. A typical application is the grouping of wrist pins into categories of

0.0001-inch diameter steps. As the bore in the piston is measured, the proper pin is automatically selected and inserted at rates up to 1800 per hour, *Fig.* 4.7.

Such technology is not limited to the automotive industry. Any product that requires tight tolerances at low cost is a candidate. This would include high-performance hydraulic pumps, air and refrigeration compressors, all kinds of servomechanisms and many valve components. The technology is actually being used to improve quality and reduce costs on many consumer products.

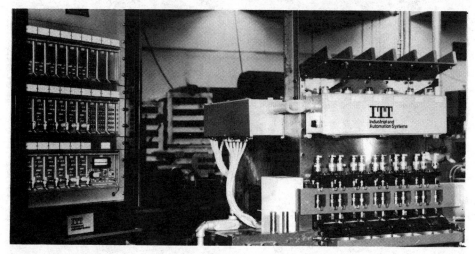

Fig. 4.7 — Wrist pins are automatically sized for selective fitting to automobile engine pistons. All gaging and selection of each item is electronic and fully automatic.

Automatic Gaging

The use of automatic gaging for selective assembly by contract shops in conjunction with an automobile manufacturer is very well shown by two contract suppliers who produce sleeves. As the sleeves are produced by the suppliers, they are automatically gaged to the nearest "tenth". The data is fed on a real-time basis to a mini-computer that "talks" to a computer in the automobile plant. The end result is a statistical curve of the production lot produced by each contract supplier. The computerized data is then used by the production staff of the automobile plant to properly control the production of the mating lot of shafts whose statistical curve matches that of the sleeves. With the two curves matching, the selective fitting process can be made with virtually no waste components since there are the same number of sleeves and shafts in each size increment throughout both production lots.

When the vendors ship the sleeves to the automotive plant, they are again automatically inspected and classified. At the same time, another portion of the automatic inspection system is measuring the shafts and marrying them to the appropriate sleeves with clearance fits of 0.0001-inch. Thus, parts that were originally manufactured to a much wider and more economical 0.001-inch tolerance are brought together with a very close and precise "tenth" allowance by the use of automatic gaging and selective fitting. The inspection and classification at the vendors' plants, along with reinspection and selective fitting at the

Fig. 4.8 — Automatic inspection system checks and matches power steering components at the rate of 1200 per hour. This photograph shows the sleeve gage only.

automobile plant, are performed at the rate of 1200 assemblies per hour. The concept greatly enhances the intrinsic value of the produced parts.

Such use of the latest gaging and computer technology insures the automobile buyer a reliable product, it reduces an extremely difficult manufacturing and warranty problem for the manufacturer, and substantially reduces the cost — and ultimately price — of the end product.

The use of automatic gaging for selective assembly is a major advance in manufacturing technology. Its potential has only been touched and it will affect all types of mass-production manufacturers and the thousands of suppliers who furnish component parts for them.

The concept of automatic gaging for selective fitting is not something confined to textbook theory. One very practical working example is the extremely fast precision matching of power steering shafts and sleeves to laboratory tolerances from batches of components manufactured to ordinary shop standards.

The function of the gaging system is the automatic inspection and classification of the sleeve and the shaft, and the final confirmation of the sleeve with respect to the shaft working clearance.

The process, shown in *Fig.* 4.8, starts with the entrance of the sleeve assemblies into a sleeve assembly gage. In this particular system example, shown in *Fig.* 4.9, the sleeves are furnished by an outside vendor and they are manually fed to the classifying gage in proper orientation. In the classifying gage unit the sleeves are inspected for a number of characteristics including out-of-round, taper, banana, hourglass, bellmouth and barrel to determine their true condition of cylindricality. The checks are made on four transverse and four axial planes, all equally spaced. Thus, there are 16 separate check points (refer to *Figs.* 4.10 & 4.11).

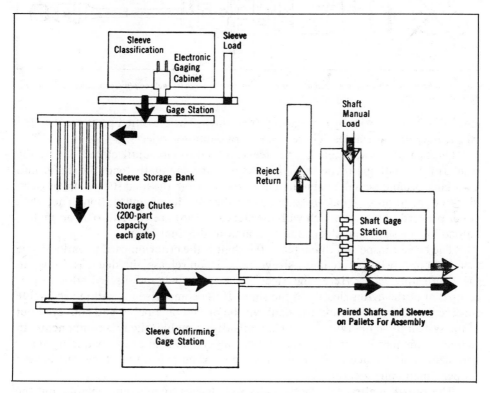

Fig. 4.9 — Both sleeves and shafts, machined to regular shop standards, are automatically brought together in selective fits that are controlled to "tenth" allowances.

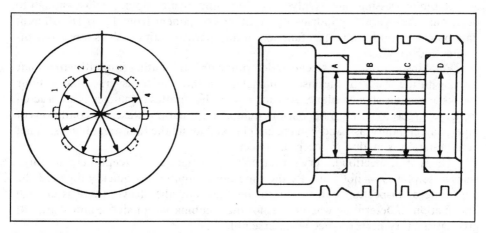

Fig. 4.10 — Sleeves are checked for all cylindricality and dimensional characteristics on four radial and four transverse planes.

The smallest inside diameter will determine the initial sleeve classification, which ranges from a large inside diameter (plus-4 class) to a small inside diameter (the minus-5 class). Thus, there are ten different classes — counting the nominal or

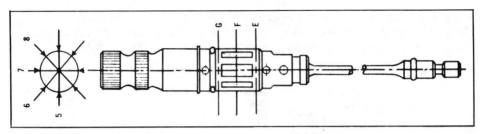

Fig. 4.11 — Shafts are checked for all conditions on four axial and three transverse planes for selective assembly with sleeves.

mean. Each class varies by a range of 0.0001-inch. The classified parts are then fed into ten acceptable chutes. There is an extra chute for rejects.

Once in the storage bank, the sleeves remain there until called out by the results of the shaft qualification gaging. An operator manually loads the shafts onto a walking-beam type indexing unit. They proceed from the load station, through an idle station, to an orientation station, and then to the gaging station. They then either proceed to a final match with the sleeve or they are carried to a "no-match, unload" if the shaft and sleeve fail to match in the final check.

The largest inspected diameter of the shaft is the criterion used to call a sleeve from the appropriate chute in the sleeve bank. Upon release, the sleeve is carried to the confirming gage. Here, the sleeve is reinspected and its cylindricality is compared to the cylindricality of the shaft. If the comparison is within acceptable specifications, the sleeve and the shaft will be placed on a pallet and conveyed out of the system to an operator who then manually assembles the two components. In summary, the purpose of the automatic gaging system is to properly qualify shafts and sleeves, and place them together so that, when assembled, they will have a proper functioning clearance.

The actual qualifying of both shafts and sleeves sounds quite simple, but the process is actually quite complex and requires the use of a computer to maintain acceptable processing rates. Only data processing by a computer is fast enough to determine the overall cylindricality of either component from 12 to 16 different data points. Cylindricality is a function of the characteristics such as taper, out-of-round, and so on.

Final inspection of the sleeve to determine its total mating suitability to a shaft starts with the check of the inside diameters on the four transverse planes. Four checks are made of each plane. At each plane, the smallest diameter is subtracted from the largest to obtain a net out-of-round determination. Next, the conditions of taper, barrel, hourglass, and banana are checked along the four axial planes. A net determination is made for accept or reject.

The shaft inspection is very similar to that for the sleeve except there are three transverse planes — not four. By the time the computer has quickly digested the multitude of inspection data for both the shaft and the sleeve, there is enough information to determine whether or not the combination of sleeve and shaft will give satisfactory performance when assembled.

Although the initial call-out of the sleeve from the storage bank was made on the smallest inside diameter, all diameters and other cylindrical characteristics of both the sleeve and shaft have to be met before final assembly is allowed.

While both components are incremented or classified to the nearest "tenth," the actual clearance factor is 0.0005-inch, plus or minus a "tenth."

When selective fitting is used to match close-tolerance parts for working clearances, it is quite possible to have large quantities of each part that would find no acceptable match with the other. In other words, it would be quite possible to end up with a large quantity of sleeves whose qualified diameters would be too great for the available supply of shafts, or the reverse could be true.

To avoid this type of a situation, the manufacturing of the sleeves is targeted for a range of only seven classes — the nominal or target dimension with a range of plus or minus 0.00035-inch. During the manufacturing process, the sleeves are automatically batched and counted for a particular shipment. The computerized printout shows the total number of sleeves in each classification. This information is then given to the grinding department that does the final operation on the shafts. The dimensional target for shafts is then set for that shown by the histogram for the sleeves. Thus, the two manufacturing processes are correlated to avoid the situation where the shafts might be shifted toward a "high" dimension while the sleeves are "low" or vice versa.

In addition to utilizing the gaging histogram of the sleeves to set the manufacturing target for the shafts, the gaging system has another feature to avoid large numbers of components that fail to find a match. That technique is known as the "multiple attempt" and it functions by allowing each shaft three attempts to find a match before a final reject is made. Most of the shafts that make an initial return on the reject index will find a match the second or third time.

Thus, not only are the components gaged and assembled to insure proper fits at a rate of 1200 per hour, but the number of final rejects is kept at a low, cost-acceptable figure.

The goal of any production system is maximum output and return on investment with minimum input. An automatic gaging, assembly and testing system is no exception. Systems could be designed and built so that no human hands would ever touch the workpieces or adjust a machine. Since the goal is reliable production at minimum costs, the really practical approach blends the best of automation with manual operations and interventions.

Continuing with the power-steering shaft and sleeve example described previously, automatic gaging and assembly has three prime goals:

1.   To reduce manufacturing costs by allowing production of components with a wider dimensional tolerance.

2.   To reduce warranty costs by bringing together components that have a very close dimensional match so that the resultant precision assembly will not malfunction once the product is in the customer's hands.

3.   To eliminate the prohibitive labor costs and inevitable errors that would occur if the fits between sleeves and shafts were manually determined.

All these objectives are met in systems now at work in both the manufacturing plants of major automobile manufacturers and the contract shops that supply component parts. So the whole concept of automatic gaging for selective assembly is not just something for the major mass-production manufacturers. The jobbing industry is being swept into use of the new concepts and it employs both the automatic gaging equipment and the minicomputers that tie the whole system together. Likewise, not everything is pure automation. In some areas it is far more practical to allow some degree of human intervention and judgment. The equipment becomes a powerful operating tool — not an end in itself.

Simply stated, the overall dimensional tolerance range of ground shafts and sleeves is 0.001-inch. This is a practical mass production grinding tolerance that can be met at normal production costs. There is, however, another very important consideration. The fit allowance between sleeve and shaft must be within a 0.0001-inch or "tenth" range if the power-steering unit is to function throughout the life of the automobile without leaking hydraulic fluid or failing.

The practical solution, then, is one of selecting sleeves and shafts from regular production lots so that the selective assembly has a precision fit. While it is somewhat impractical to mass produce low-cost items to "tenths," it is quite feasible to divide and classify each component of a production lot to its nearest "tenth" dimension. Such classifying work is totally impractical with manual gaging equipment. The equipment costs would be extremely low but the labor costs would be entirely prohibitive and there would be unacceptable errors in the work.

Automatic gaging and assembly units have been developed which range anywhere in cost from 20 to 300 times the cost of hand gaging equipment but they are well worth the investment in terms of speed (1200 assemblies per hour) and reliability of output (100 percent accuracy). A reduction in the number of one-hundred-dollar warranty replacements will justify the equipment on that basis alone — not to mention productivity increases.

The basic procedure whereby job shop component supplier and automobile manufacturer work from the same integrated data base is shown in *Fig.* 4.12. It all

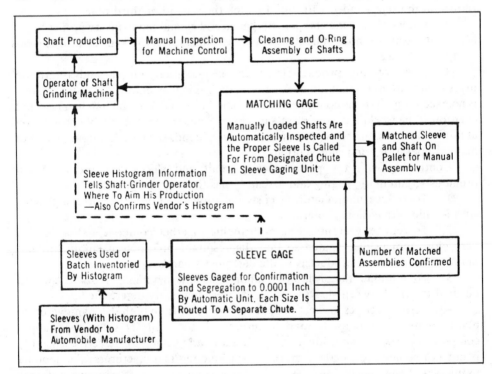

Fig. 4.12 — *Schematic of automatic gaging, testing and assembly system involving both the job shop and the automobile manufacturer.*

starts with the job shop suppliers who produce the sleeves. They set up their machine tools as they normally do and aim their production toward the middle of the 0.001-inch tolerance range. The bulk of the production will cluster at the midpoint with some workpieces drifting toward the high side and some toward the low side of the allowable dimensional tolerance span. In some instances, the bulk of the production may shift or "skew" toward either the high or low side, as shown in *Fig.* 4.13.

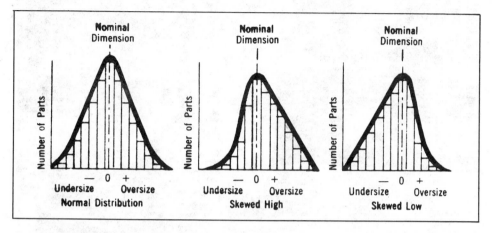

Fig. 4.13— A normal production histogram will look like the one on the extreme left. However, due to many conditions, one of the other two histograms may be representative of an actual production lot. To avoid having shafts or sleeves with no matching parts, production lots of shafts and sleeves with similar histograms are put into the gaging system to match parts.

With total production running into the millions and covering a period of years, the day-to-day output is gathered into specific production lots. A good working size is five thousand parts. As the sleeves come off the internal grinder, they are automatically gaged on four different radial and four separate transverse planes.

The cost for the automatic gaging equipment at the job shop is more modest since there is no component segregation at this point. The purpose of gaging at the job shop is to derive a histogram of the production lot which gives a graphic and statistical picture of just where within the tolerance range the sleeves fall. Should any sleeves fall completely outside the allowable tolerance limits, the automatic gaging unit will reject them.

It is important for the automobile manufacturer to know the character of the sleeve production lot so that he can machine shafts that will have the same tolerance distribution pattern. Unless the lots of shafts and sleeves that come together at the automatic gaging and assembly machine have the same dimensional distribution pattern, there will be many that simply will not have a fit because of too many large-diameter sleeves and small-diameter shafts or just the other way around.

So the capital outlay for the job shop is modest. An automatic gaging unit equipped with a minicomputer will weed out the completely out-of-tolerance parts at high speed and with 100 percent reliability while at the same time generating a histogram of the production lot in terms of its "tenth" dimension that will be used for assembly.

At the present time, it is the usual practice to ship the information histogram with the production lot. It is technically feasible, however, to transmit directly from the job shop's minicomputer to the auto manufacturer by utilizing a data terminal.

Meantime, back at the ranch (the automobile manufacturer) where a higher

Fig. 4.14 — Sleeves in the reconfirming station will be sent to the proper chute in the storage bank for call-out after the shaft is gaged; then the parts will be assembled.

level of sophistication prevails, the production of shafts has been taking place. The incoming shipment of sleeves are again gaged, *Fig.* 4.14, to confirm the vendor's histogram and here they are actually segregated into ten different classifications to the nearest "tenth" dimensional measurement. The histogram confirmation then tells the direction in which the shaft production should be headed. If the sleeves are centered about the middle of the tolerance range, the matching shaft production can also be so centered. The shaft grinding machine operator is responsible and he periodically checks his output to see that he is on target. This is largely a manual check and is normal in production of this type.

The shafts and sleeves are then brought together in the automatic gaging and assembly unit. The sleeves have already been classified and segregated and the assembly machine operator loads the individual shafts which are then automatically checked and verified by computer. The shaft inspection data is used to call out the proper sleeve from the storage bank. The matching sleeve is verified and it is then dropped with the shaft onto a pallet which goes to assembly with the full knowledge that a proper fit will be assured. The actual assembly of the component parts is a manual operation.

Both the shaft and sleeve grinding machines are manually set up and controlled. The operators utilize information fed back by the system about the production. If necessary, the machines can be adjusted to shift toward the high or low side of the tolerance. Parts are loaded into the automatic gaging and assembly machine by hand. The actual insertion of the shafts into the sleeves is a manual operation. But the determination of diameter, roundness, rainbow or banana, taper, and so on, and whether or not each shaft and sleeve meets overall dimensional requirements is the function and purpose of the automatic gaging units. Matching each sleeve with the proper shaft is also a function of automatic gaging.

With sustained operations, a number of shafts may accumulate that do not find matching sleeves. In any production lot of shafts and sleeves it may be possible to end up with better than a 99 percent match, but it is highly unlikely that a total 100 percent would be possible or even practical. If there is long-term accumulation of either large, or small, diameter shafts, the subsequent production can be shifted so that at no point in time is there ever an unduly large accumulation of shafts without matching sleeve fits. A few current shafts that cannot find a fit is an incredibly small price to pay for the speed, efficiency, economy and reliability of utilizing automation where it can work to best advantage.

Metrology Automation

In theory, the whole process from actual machine adjustment to physical assembly could be totally automated. As time goes on, it may become practical. In current practice, the application of automatic gaging and automatic inspection — where they work best — are bringing about economies, efficiency and the all-important product reliability within the framework of normal product production practices and costs.

An automatic gaging, assembly or testing machine or a combination unit that integrates two or three functions is tailored to a specific plant and functional requirement. No two situations are ever identical. Fortunately, it is possible to engineer a system for virtually any requirement by utilizing standard modular components for maximum economy.

The automatic unit that gages and selects power-steering sleeves and shafts — the example used throughout this chapter — utilizes eight standard modules. They include a CPU (central processing unit), basic power supply, digital input module, digital output module, Teletype, analog-to-digital converter, pressure transducer gage module and a pressure transducer signal converter. These are employed in the complete system as indicated in *Fig.* 4.15.

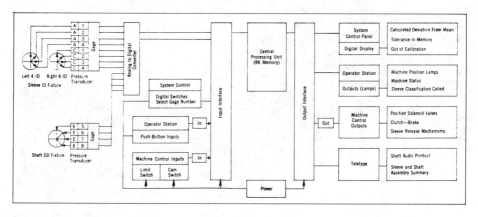

Fig. 4.15 — Automatic gaging systems, such as this one for matching steering-pump pistons and sleeves, are individually constructed for standard components.

The heart of the CPU is an 8K minicomputer with read-write memory, which is the average size and capacity and totally adequate for the job. If necessary, the unit

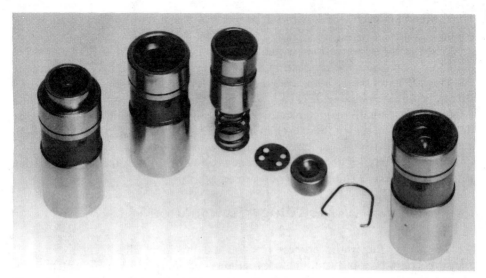

Fig. 4.16 — Valve lifter components gaged and assembled on automated equipment.

could be expanded to a 32K minicomputer. The CPU is the central controller and regulator of the entire system and contains all of the engineers' definitions, instructions, and controlling statistical and math data. It is the CPU that takes one segment of information, such as a dimensional measurement, and compares it against another or a standard to arrive at a decision to accept, reject, qualify, feed, store, control, and so on.

In no sense of the word does the CPU "think". It merely executes instructions based upon management's operational decisions. Thus, it has to be programmed with a clearly defined executive routine before it can properly function. The executive program is developed and made available by the equipment supplier. The advantages of the CPU are its incredible speed, unfailing accuracy of output,

Fig. 4.17 — Gaging, Assembly and Testing Valve Lifter Assemblies — Three basic sections comprise this system: 1) Plunger inspection and classification, 2) plunger body assembly and 3) leak down testing. Non Mar Hoppers feed and orient plungers and bodies to their respective sections. A plunger storage and classification release and overflow section monitors production flow. All sections are linked with power and gravity conveyors. Production rate is 3.000 parts per hour.

and its consistency of operation from the first minute of the work shift until the last.

It is the CPU with its minicomputer that can receive information from 16 different gaging points and determine within a thousandth of a second whether or not the shaft being gaged meets the requirements for roundness, taper, diameter, banana, rainbow, or any other condition that may be defined. The computer does not create the definitions or standards of acceptance or rejection — that is the function of management. But once the standards are determined and entered into the computer's executive program, the computer in the CPU can flawlessly accept, reject, quantify and qualify workpieces at extremely fast rates based on the gaging data input. The vast majority of the cycle time is not even taken up by the computer. Most of it is spent in the physical movement of the workpiece into the gaging station and out. Actually, the CPU in a normal system can handle 512 different input-output lines and this could be expanded to a total of 8,192 if necessary.

An essential part of the system, but not actually a part of the CPU, is the communications register or the interface. This unit takes data from various devices such as motors, pumps, transducers, valves, switches, photocells, solenoids, relays and so on and converts them into computer-compatible input signals.

A complete gaging, testing and assembly system does have one requirement — it needs a steady power supply with very little voltage fluctuation or line "noise". Therefore, the power module is used to supply the correct DC voltage and hold it to a plus or minus 0.1 percent variation even though the actual line voltage may vary as much as plus or minus 10 percent.

The input module converts 115-volt AC to a low-level DC that is utilized for electronic processing. A single input module can handle six separate 115-volt AC circuits. Each circuit has a signal light for trouble-shooting purposes.

The digital output module has just the opposite function. It converts low-level DC to 115-volt AC and can be used as a slave to drive a heavy relay. It, too, has six 115-volt AC circuits rated at 10 amperes peak and 1.0 ampere constant. It also has connections for machine and indicator devices.

The Teletype unit is merely the manager's or programmer's tool to interact and program the system. Unfortunately, no computer system can yet understand spoken commands beyond a special word or two nor can a computer compose thoughts in the English language. It can, however, print out information through such terminals as the Teletype which serves the dual function of receiving human instructions and printing out records needed for human decision making.

The analog-to-digital converter module is perhaps one of the most essential and least understood components of a complete gaging, testing, or assembly system. Unfortunately, the world around us is basically analog while people like to think in digital terms. Consider the temperature in a room. If it increases from 72° to 75° F., people like to say it went up three degrees — a digital term. Yet, the temperature did not move up in three even steps. It followed a curve of infinite increments from 72 to 75 degrees — an analog occurrence.

Manual gaging often requires very close interpretations. If a workpiece is very close to the specified dimension there is no problem, but problems do occur when the measurement is right on the line between accept and reject. It is entirely possible that two inspectors could take the same gage reading and arrive at two different conclusions or that the same inspector would arrive at one conclusion at one time and the other conclusion at another time with similar readings. The

Fig. 4.18 — Valve Lifter Assembly Plunger and Inspection Grading Machine — A Non Mar feeder feeds and orients plungers to a gaging station where diameters are classified into 20 grades. Each grading chute is capable of storing up to 350 plungers where they are selectively deposited onto a belt conveyor.

converter module can split an analog signal into neat digital packages with far greater speed, reliability and accuracy than any human could ever hope to attain. Thus, analog characteristics can be broken into logical digital packages that lend themselves to mathematical and statistical operating and management tools.

The pressure transducer gage module converts the input signal into a variable DC output and generates the RF source necessary for the capacitance gaging head. The output swings from minus 10 volts DC to plus 10 volts DC for full input range. The full output of the 20-volt swing can be set for either a fine 0.0005-inch displacement or a more coarse 0.010-inch displacement. It all depends on the gaging requirements.

The pressure transducer itself is where the action starts. In the shaft gaging unit, the pressure transducer consists of a fixed stator and a diaphragm that flexes under pressure. When the pressure in the air gage changes due to the dimensional characteristics of the workpiece, the diaphragm-to-stator air gap will also vary proportionately.

A system may have any number of different transducers depending upon the type of workpiece and functions to be achieved. Some are load cells, others detect pressure while others make direct contact and are displaced by the geometry of the workpiece. All of them in one way or another translate a dimension into an electronic signal for processing.

All modules are designed to standard dimensions for plug-in adaptability and all of them are keyed to prevent insertion in the wrong position. It is through the use of such modules that complete and automatic systems can be developed for individual, specific applications.

Gage, assemble and test system for valve lifter assemblies, *Fig.* 4.16, 4.17, and 4.18, make up this system:
1. Plunger inspection and classification,
2. Plunger body assembly and
3. Leak down testing.

Non-Mar hoppers feed and orient plungers and bodies to their respective stations. A plunger storage and classification release and overflow section monitors production flow. All sections are linked with power and gravity conveyors.

This system is perhaps the most precision high-volume assembly process in the automotive industry today. The valve lifter body is gaged and plunger sub-assembly is called for by a diametral clearance. This particular application happens to be in the 33 millionths category. This clearly illustrates the flexibility of gaging, and especially electronic gaging. Clearances required have a maximum of 66 millionths with categories of 33 millionths. Very high volume production is required. The assembly is then finish-assembled and tested on this overall system.

REFERENCES

[1]Gerard Piel, Publisher, "The Acceleration of History," *Scientific American*, 1963.

ADDITIONAL READING

[1]M. F. Spotts, "Eight easy ways to use statistics — Part 1," *Machine Design*, Feb. 20, 1975.
[2] , "Eight easy ways to use statistics — Part 2," *Machine Design*, March 6, 1975.
[3] , "Simple guide to TP dimensioning — Part 1, Loose bolts, datums at edges," *Machine Design*, November 13, 1975.
[4] , "Simple guide to TP dimensioning — Part 2, Loose bolts, datums at centerline," *Machine Design*, November 27, 1975.
[5] , "Simple guide to TP dimensioning — Part 3, Fixed studs, datums at edges," *Machine Design*, January 8, 1976.
[6] , "Simple guide to TP dimensioning — Part 4, Fixed studs, datums at centerline," *Machine Design*, January 22, 1976.

5

Designing For N/C Production

Joseph Harrington, Jr.

FOR half a century the management of manufacturing has differentiated sharply between the domain of the product designer and that of the manufacturing engineer. Each has been encouraged to "stick to his last" and let the other fellow run his own shop. Frequently design and production divisions have been located far apart. Sometimes designs are made before it is determined where or by whom they will be produced. Consequently, product designers go about the business of designing and dimensioning parts with all too little regard to what machine or machines will be used to make them.

All this is in a rapid state of change, partly brought about by the introduction of numerically controlled machine tools. Fortunately, the post-World-War-II trend of separation of the product design team and the production engineering team — organizationally and sometimes geographically — has been reversed. Manufacturing from product design to product delivery and service is a monolithic function, and is being treated as such. The division of technical skills is properly still practiced, but the management of the task is being re-integrated. No longer can we tolerate sub-optimization — the independent, autocratic and self-serving decisions at one stage of the process which can make work difficult in later stages. Wise management insists on optimization of the total manufacturing system.

This being the case, what should the product designer know about the manufacturing process and facilities, particularly when numerically controlled (N/C) machines are to be used? As production equipment has been improved, N/C as been introduced in order to add capability, flexibility or versatility. Many machines now utilize N/C control as will be seen in many chapters following. It behooves the designer to keep this opportunity in mind as his designs progress. There are a number of important considerations.

Know Your N/C Tools

Product designers are encouraged to keep the size of important parts within the working cube of the proposed machine tool. If a part is of such nature as to justify its production on an N/C tool, then it is self defeating to design a part just a

---

Dr. Joseph Harrington, Jr. is a Consulting Engineer, Wenham, Mass.

Fig. 5.1 — K&T Milwaukee-Matic traveling column machining center shown working on some unusual workpieces.

little too big for the N/C equipment in your shop. Frequently, a little ingenuity can keep part size within the capacity of the tools. The alternatives are, of course, using a non-N/C tool, or farming out the work at greater cost.

It is important here to distinguish between the outside envelope of the part and the volume within which machining is to be done, particularly on milling machines and N/C machining centers, *Fig.* 5.1. The maximum relative movements of the part on the table vs. the tip of the tool in the spindle prescribes the volume within which a cut can be made. It is immaterial for this determination whether the table moves or rotates relative to a fixed spindle, or the spindle on a column moves or rotates relative to a fixed work table. This "cube" of work space is usually specified for each N/C tool by the builder, but if not the manufacturing engineering department should have it tabulated for every tool they manage.

It is not necessary to confine machining to parts which will fit totally within this cube; it is merely necessary to confine required machining operations within

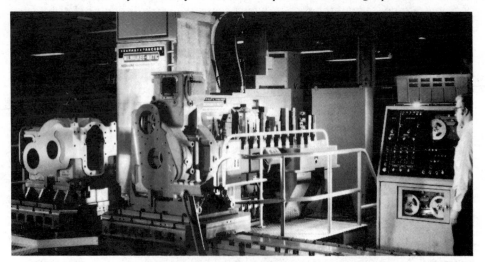

Fig. 5.2 — Some large sized cast housings showing fixture overhang on the K&T Milwaukee-Matic.

the volume of the cube. It is not unusual to see long parts hanging over at one or both ends of the table while work goes on opposite the spindle. If the N/C tool has a fixed table and a moving column, this type of localized work on large parts is easy to handle, *Fig.* 5.2. Improvised outboard support well beyond the end of the tool bed may be used if necessary to support the part in firm alignment. The manufacturing engineer will cope with this in his tooling design and process writing. Nevertheless, the product design engineer must be sure to keep the volume to be machined within the N/C tool's working cube.

Such outsize parts present more of a problem on N/C tools with moving tables. Any outboard support must follow all the movements of the table; even if only X-axis motion is involved, this still presents tooling and set-up problems. If no outboard support is used, then the part must be so balanced on the table that deflections due to part weight will not distort the position of points within the cube.

A second consideration in respect to outsize parts on a moving table is their weight. Accurate machining involves starting and then stopping the table and part at a desired position. Servo systems are designed for a maximum weight on the table, and excess weight may degrade the accuracy and repeatability of the system, particularly in contouring work. The excess weight may also severely overtax the table support bearings.

Careful consideration must therefore be given by the part designer to any design which calls for such capacity-stretching innovations. It certainly should call for a conference with one of the manufacturing engineers.

Many N/C tools have automatic tool changers, selecting one cutting tool from many stored in a magazine at a command incorporated in the control tape, and automatically placing it in the spindle and replacing the previous cutter in the magazine, *Fig.* 5.3. Other tools such as turret drills and many lathes have a turret of cutters set up at the beginning of a job. The appropriate cutting tool is indexed into operating position at a tape command.

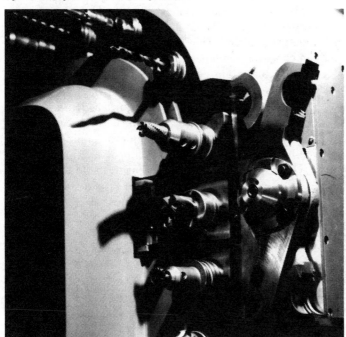

Fig. 5.3 — Tool changes showing a selection of the bank of tools available. Courtesy Kearney & Trecker Corp.

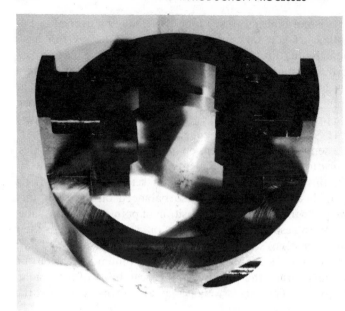

Fig. 5.4 — Pawl housing made in continuous production on a 40-tool capacity Numeri-Center. Face milling and pocket milling from the solid with drilling and reaming produce plus or minus 0.001-inch on slot width and hole locations, 0.001-inch on parallelism and perpendicularity of slot sides. Courtesy Giddings & Lewis, Inc.

Part designers are strongly encouraged to specify no more kinds of cutting tools than there are pockets in the magazine or sockets in the turret. To exceed that number means that the N/C tool operator has to stop the operation in the middle of each part cycle to remove a few cutters and substitute others to be used later in the cycle; at the end of the cycle for each part made he has to reverse this exchange before starting over. Such procedures are obviously error prone, and can be the source of human induced slow-downs in the production schedule. Changing cutting tools in the magazine or turret is a requirement to be avoided at all costs. To do so requires that the part designer know how big a tool magazine or turret head there is on the N/C tool, *Fig.* 5.4. This data should be tabulated and readily available in the design room from the manufacturing engineering department.

The product designer can contribute independently to the elimination of such problems by a tactic which is also just plain good design sense; cut down on the variety of sizes of threaded fasteners specified in the design of a part. Every tapped hole requires at least a tap drill and a tap; some will also require a clearance drill, or a counterbore, or a countersink, or a spot face in addition. Three different cutting tools is par for the course. Consider now a major machine part — a transmission housing, an engine block, or the main frame of anything from a steam engine to a computer. There may be a dozen different sizes of tapped holes called for, which means 24 to 36 different cutting tools just for threaded holes. This is more than enough to overtax the capacity of magazines of most small to medium sized machining centers.

But is all this variety necessary? The answer is clearly "NO". For example: A $1/4$-20 screw will hold an escutcheon plate as well as a #6-32 screw, and can be used also to attach brackets, bearing closure caps, and even cover plates. Furthermore the #7 (0.2010") tap drill will also make good vent holes, oil holes, cotter pin holes, and spring anchor stud holes, and is a good clearance hole for a 3/16" screw. Note that this is just an example, not an advocacy of $1/4$-20 screws. The moral is, pick as few different sizes of threaded fasteners as practicable.

The advent of metric fastener sizes will temporarily aggravate this problem if both metric and inch standard fasteners must be used on one part. Eventually the new metric fastener standards, with their smaller numbers of thread size standards, will help to reduce the variety of tapped holes now found in common practice. In the meantime the part designer will find many ways to reduce the number of tools, and hence tool changes required in N/C machining.

A dandy by-product of this practice can be the reduction of the number of wrenches a service man has to carry around with him. One astute designer produced a small stoker design using only 5/8''-11 machine bolts for assembly. And both nuts and bolt heads were the same size, so one pair of identical wrenches was all the field man needed to carry.

In the same way the designer can reduce the number of tools required if he will use fewer widths of keyways or slots in his designs. If there are a number of upstanding lugs on a casting which will be straddle milled, make them all the same width in so far as possible. If an end mill will be required to clean out the interior of a profiled opening, make all the inside corner radii the same, so one end mill will be needed, and there will be no need to contour some of the corners.

Aside from the tool magazine capacity, there are excellent economic reasons for reducing the number of cutter changes required. To change cutters manually can take 30 to 60 seconds, or more. Even to change cutters automatically will take from 5 to 9 seconds. Multiply this by the number of changes saved, and relate it to the parts cycle time, and the impact becomes obvious. If even 10 automatic changes at 6 seconds each in a part with a 10 minute cycle time, productivity can be upped by 10%!

A good mental rule for the part designer is to consider how he would make this piece if he had to do it in his own home workshop with a bare minimum of tools. Are all those variations justified?

Part Configuration Problems

In selecting part configurations, the designer must serve two masters with sometimes variant demands. First and foremost of course are the demands of the function of the part, but running a close second are the demands of the manufacturing department for improved producibility. Failure to give consideration to the latter demands has in the past put burdens on the shop to which the average production department has apparently become inured. The advent of N/C tools, however, has brought into sharp focus the importance of these bad habits. Three examples follow: The Weird Shape Disease, The Skewed Hole Syndrome, and the Inaccessible Spot Problem.

Weird Shapes. Even a cursory glance at the finished parts warehouse will reveal some really weird part shapes, presumably justified by the functions which they must perform. But many a manufacturing engineer suspects that the parts design boys are testing his forbearance. The problem of holding a piece of material on an N/C machine can be significantly simplified if the part designer will make the effort. This applies particularly to raw materials such as castings and forgings, and some weldments, with irregular external surfaces, *Fig.* 5.5. Parts made from slabs, sheets, bar stock and the like are usually easily clamped directly to the table of the machine tool, with simple modular clamps, or held in the standard chuck or collets of a lathe.

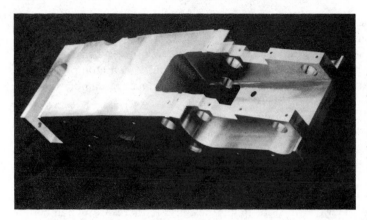

Fig. 5.5 — Three fixtures permit complete milling of this aluminum pick support with 20 operations. Tolerances are 0.0002-inch on hole size, plus or minus 0.002-inch on hole locations, 0.001-inch on squareness and alignment. Courtesy Giddings & Lewis, Inc.

But castings and forgings, and some semifinished parts are much more troublesome. Even ostensibly rectangular parts may have slightly angled surfaces resulting from draft on the patterns or in the die cavity, which makes it difficult to true them up. And some parts have such spidery, distorted shapes that they present no easy or obvious support face. In the past these parts were handled in holding cradles called fixtures whose interiors formed a socket for the parts, and whose exteriors had the requisite flat, square surface adapted for holding on the table or faceplate. The more weird the part shape, the smaller the lot size on the shop floor which would justify the cost of making the fixture.

Now, part shapes of this category are well justified in design, but there is no reason to make the machine tool set-up more difficult than absolutely necessary. Most castings and forgings get an initial operation which precedes the N/C machining center operations. This is a clean-up and square-up of the reference surface which is to sit on and be clamped to the N/C machine table. A few seconds on a big surface grinder is all that is necessary. If the product designer will add an extra lug or projection at some judiciously selected point, he can achieve a casting which will stand squarely on three points, and present itself substantially oriented with reference to its design coordinate axes. The cost of cutting off the added prop, if it is necessary to do so, as the last step of manufacture is offset by the reduction of tooling and set-up costs, and palletizing problems.

Another consideration to aid set-up is the provision of a surface or surfaces which, when the part is positioned flat on the table, can be pushed up against a stop on the table surface and thus rotate the part in the horizontal plane into proper orientation and position. This requires two things:

First, the product designer must provide one straight surface, or two bosses, whose face or faces determine a line parallel to the X or Z axis of the part. This slight extra bit of metal may be cut off later if its presence would interfere with the function of the part. Second, the manufacturing engineer provides a stop bar fixed to the machine table at one side, or one end and one side. These bars are placed parallel to the axes of the tool. When the part is flat on the table it has been oriented in two rotational directions and one linear dimension. Then, when it is pushed into contact with the stop bar or bars, it is oriented in the third rotational direction (about the vertical axis) and in the two other linear dimensions, and is thus completely and uniquely oriented. Spacer bars (parallels) may be used between the stops and the part to move it to a more central position, if the part is small.

Finally, whether the part is cut from a slab or sheet of rolled metal or from a casting or forging, it has to be held down on the table by clamps. Sometimes it is inevitable that a clamp intrudes into an area where machining must be done. For example, profiling around the outline of a flat plate. Clamps have to be removed and relocated to permit the cutter to have access to the part. But if the part designer has the opportunity on a casting, forging, or weldment he can provide a foot on which the clamp can be placed and be permanently out of the way of all machining cuts. Such clamping lugs can also be arranged directly over the point of contact with the table, so that tightening the clamp introduces a minimum of distortion in the part.

The Skewed Holes. Skewed holes are those whose axes make compound angles with the coordinate axes of the part. Consider the design of a piece of agricultural machinery; the rugged main frame of the vehicle is supported on sturdy wheels, carried on a big axle in cheap, reliable journal bearings. The bearings have to have an oil hole equipped with an oil cup and cap. The designer mentally approaches his brain child with an oil can; where would the operator find it easiest and most obvious to stick the oil can spout? Nine times out of ten the answer is a perfect response to an operator's psychology, but abysmal in so far as machine manufacture is concerned: a straight hole, from the hub boss down to the bearing, but on a compound angle! And tapped to receive the oil cup and cover.

N/C machining of surfaces cut, or of holes drilled, on a compound angle means part rotation about two axes, or spindle rotation about two axes, or a combination, plus the capacity to feed the spindle axially in the rotated position, *Fig.* 5.6. This is a capability of some large sophisticated machine tools, such as the Sundstrand Omnimill, but is lacking in many of the less versatile N/C machining centers and milling or boring machines. If a part designer has a predilection for skewed holes, he had better be designing for a shop with appropriate N/C tool capacity.

Some N/C tools have rotary tables, providing for rotation of the workpiece about one axis. Of these, some provide only 4 rotation steps of 90°, others 8 steps of

Fig. 5.6 — Steel ball nut produced with 56 operations in 1/2 hour in lot sizes of 50 pieces on the 40-tool Giddings & Lewis NumeriCenter-15V.

45° or 24 steps of 15°, while a few can be set to any desired angle. It is therefore essential that the parts designer know what angular rotational capabilities his shop's N/C tools have; if there is any doubt he should, if possible, stick to multiples of 15° in his designs.

The Inaccessible Spots. When machining on a compound angle is necessary (and there are such cases) then the designer should make sure that five axis N/C tools are available or anticipate that the compound angle work will have to be done on a second operation, guided by a fixture or jig made for the purpose and taking its position from other reference surfaces previously machined on the part. He may even have to provide such reference surfaces for that sole purpose.

There are some spots on a part which cannot be reached for machining. It is axiomatic that any part to be produced on a metal cutting machine tool must be held on the table or in the chuck or collet of the machine. N/C tools have the special merit that many parts may be held with simple modular clamps right on the milling machine table, without the need for expensive fixtures. Parts made from rolled metals — bars, sheets, slabs, rods, etc. — can frequently be clamped right to the table. Parts made from castings or forgings may have to have one surface cleaned up to insure that they can be clamped without introducing strains.

Furthermore, once clamped to the table of an N/C mill, or in the chuck of an N/C lathe, all the rest of the machining may well be completed in that one set-up. This is one of the features of N/C technology which make it so economically attractive, *Fig.* 5.7. Instead of a route sheet calling for operations on ten special-

Fig. 5.7 — Aluminum hydraulic valve body requiring 222 operations with tolerances of plus or minus 0.001-inch on bores, 0.003-inch on drilled hole center locations. Production time with 75 piece lots was reduced from 5.1 hours to 58 minutes each in a single setup. Courtesy Giddings & Lewis, Inc.

purpose conventional tools, the N/C machining center does all the work on one machine tool at one stop on the route sheet. This saves not only transportation costs and set-up costs, but more importantly it cuts down the lead time necessary to finish the machining of a part by two or three weeks.

Therefore it can be seen how irritating it is to the manufacturing engineer to find, in an otherwise "piece of cake" part design, a feature that requires machining in the middle of the side of the part which will be clamped to the tool table. It requires another entire set-up and operation to remove, invert, realign, and check out the part even if all that has to be done is to drill and tap one hole for an oil drain plug. To pursue this illustration, a little ingenuity and understanding of the shop floor problem can arrange to bring the oil drain hole out to one side, down near the bottom.

In lathe work it is not uncommon to have two chucking operations on a part, thus reaching every surface of the metal. But a little forethought can avoid designs which call for a third chucking operation. This kind of consideration becomes increasingly important as the versatility and capability of N/C tools increases.

Drafting and Design for the N/C Shop

The adoption of modern drafting room practices can be justified on its own merits without reference to how the parts are to be made, but the use of N/C manufacturing technology provides some additional incentives.

One consideration is the method of specifying the location of points on a part, as for example the centers of holes. It is good design practice to relate the positions of a hole or a group of holes to another feature to which they are functionally related. For example, the holes in a bolt circle for a bearing cap are all related to the center of the bearing. Tolerances tie these subsidiary locations to the main feature so that mating parts will assemble properly.

Nevertheless the N/C programmer converts all locations into absolute locations with reference to one datum, the origin of coordinates. Frequently this is not even within the bounds of the part, but is located well off to one side so that all coordinate measurements in that plane are positive numbers, *Fig.* 5.8. These coordinates become directly the entry for the X and Y dimensions of point to point commands on the N/C manuscript.

Modern drafting practice suits this situation well. Drawings are uncluttered by witness lines and dimension lines. Point locations are marked by a small cross, on which is placed the symbol for the machining to be done there (a drilled and tapped hole, e.g.). Beside the point is an alphanumeric label such as "D-17". And off in one corner of the drawing, or on another sheet, are the tabulated dimensions and specifications for that point:

D-16 | Center drill, Drill .245, Ream .249/.250 | X = 1.750 | Y = 6.125

If there are many features to be located, all like operations may be grouped, as:

C | ....
D | Center drill, Drill .245, Ream .249/.250
E | ....

and then the dimensions of the various features tabulated, as:

D-15 | .... | ....
D-16 | 1.750 | 6.125
D-17 | .... | ....

As will be discussed below, tolerances in location dimensions are covered by a

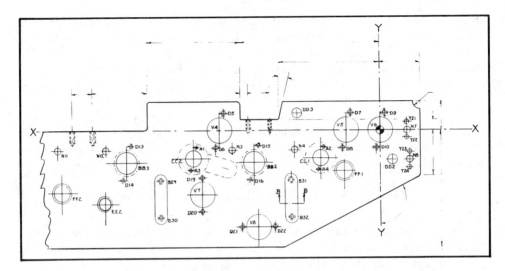

Fig. 5.8 — Portion of a drawing of a main frame plate, containing many holes. The alpha numeric symbols at each hole identifies it and also refers to the tables in which the hole dimensions and location are listed. These were listed on a separate sheet of the multi-sheet complete drawing. Notice the lack of clutter compared to how the same drawing would appear with witness lines and dimensions for each one of the 48 holes shown. Notice also the cleanly visible plate outline and the few witness and dimension lines necessary to define it.

The Table at left shows, for each type of hole (e.g., "V"), the number of holes of that type in the entire piece, and their dimension and tolerance. Tools — a drill, or a drill and tap, or a drill and bore or counterbore — will be needed for each type. The N/C programmer can easily see how many of each type must be programmed and how many tool changes to anticipate. The cup-shaped symbol means counterbore.

The Table at right shows, for each hole identified, by type and number, its X and Y coordinates measured from the origin at the center of hole V-6. Tolerances are included; when no tolerance is given it is understood to be ± .005. All dimensions are in inches.

HOLE	NO. OF HOLES	SIZE
A	7	4-40 UNC-2B
B	49	6-32 UNC-2B
C	15	8-32 UNC-2B
D	29	10-32 UNF-2b
E	2	⌀ .1869 +.0005/ -.0000
F	11	.250-20 UNC-2B
G	2	⌀ .166
H	31	⌀ .1245 +.0005/ -.0000
.		
.		
N	12	⌀ .185 +.005/ -.002
.		
T	24	⌀ .1245 +.0005/ -.0000
.		
V	8	⌀ .8750 +.0005/ -.0000
.		
.		
BB	3	⌀ .8125 +.015/ -.015 ⌴ .8750 +.0005/-.0000
CC	3	⌀ .5628 +.0005/-.0000 ⌴ ⌀ 1.062
DD	3	⌀ .265
EE	2	⌀ .500 ⌴ ⌀ .625
FF	2	⌀ .625 ⌴ ⌀ .750

HOLE NUMBER	DISTANCE FROM X-X	DISTANCE FROM Y-Y
A-1	- .578	- 6.500
A-2	- .578	- 2.000
A-3	- 1.302	- 6.694
A-4	- 1.302	- 2.194
B-29	- 1.700	- 7.766
B-30	- 3.000	- 7.766
B-31	- 1.700	- 3.134
B-32	- 3.000	- 3.134
D-5	.566	- 5.598
D-6	- .566	- 5.902
D-7	.566	- 1.098
D-8	- .566	- 1.402
D-9	.566	.152
D-10	- .566	- .152
D-13	- .548	- 8.848
D-14	- 1.680	- 9.152
D-15	- .548	- 4.348
D-16	- 1.680	- 4.652
D-17	- 2.246	-18.624
D-19	- 1.613	- 6.182
D-20	- 2.745	- 6.486
D-21	- 3.327	- 4.931
D-22	- 3.327	- 3.750
N-3	- .676	- 5.281
N-4	- .676	- 3.000
N-5		
N-7	.000	.938
N-8	- 1.020	1.045
N-11	- .676	-11.438
N-12	- .676	- 9.812

HOLE NUMBER	DISTANCE FROM X-X	DISTANCE FROM Y-Y
T-21	.250±.001	.935±.001
T-22	- .250±.001	.935±.001
T-23	- .676±.001	1.045±.001
T-24	- 1.364±.001	1.045±.001
V-4	.000±.001	- 5.750±.001
V-5	.000±.001	- 1.250±.001
V-6	BASIC	BASIC
V-7	- 2.179±.001	- 6.334±.001
V-8	- 3.327±.001	- 4.345±.001
BB-2	- 1.114±.001	- 4.500±.001
BB-3	- 1.114±.001	- 9.000±.001
CC-1	- .940±.003	- 2.097±.003
CC-2	- .940±.003	- 6.597±.003
DD-2	- 1.032	.450
DD-3	.550	- 2.929
EE-1	2.88 ±.06	-13.50 ±.06
EE-2	- 2.50 ±.06	9.75 ±.06
FF-1	- 1.38 ±.06	- 1.31 ±.06
FF-2	- 2.13 ±.06	-11.25 ±.06

single statement determined by the N/C machine capabilities, unless specifically noted. Not only does this technique unclutter the drawing and make it easier to read and to comprehend, but the tabulation can be typed instead of laboriously hand printed.

The N/C programmer will benefit directly from such practice. The operation list (the D's in the above example) tells at a glance how many tools will be required. The tabulated dimensions (X,Y) in the above example tell where all similar holes are and become direct entries into the manuscript. And if the designer has used an origin of coordinates which achieves all positive coordinates, then the programmer need not even put in a translation of coordinates when writing up the manuscript.

Specification of Tolerances

In N/C fabrication, tolerances are determined not by the designer, but by the N/C machine tool. Tolerances placed on a dimension on a drawing indicate the limits of variation within which that dimension of a part achieves the designer's purpose. The narrower the tolerance band, the more expensive it is to achieve the accuracy requested. Hence, in general, a designer should select as large a tolerance as he feels he can safely prescribe. No two sets of circumstances are exactly alike, and as a result we find a complete gamut of tolerance bands, from $\pm0.030''$ or more down to $\pm0.0001''$ or less. This is all well and good when a part will be manufactured on conventional machine tools and inspected with conventional gages. But if the part is to be made on an N/C machine, a new set of conditions obtain.

An N/C programmer writes into his manuscript the dimension he wants to achieve, but makes no mention of tolerances. The N/C machine will place the cutter at that dimension as closely as it possibly can. There is, of course, a variation in the location of the surface produced; it will lie within a tolerance band, but that band is a function of the make and model of the machine tool, its state of maintenance, tool loading, tool, workpiece and machine deflection under the cutting forces, and the accuracy built into the N/C servo system, *Fig.* 5.9. The tolerance achieved will have

Fig. 5.9 — Stainless steel pump housing highlights accuracy of N/C machining. Bores are held to 0.0005-inch and bore locations to plus or minus 0.0005-inch in lots of 25 on the NumeriCenter-15V.

no relation to what the designer put on the drawing. The N/C machine will do the best it can, every time; nothing will give it significantly better performance, and if properly maintained it should not do worse than its designed characteristics.

Hence the designer should learn what N/C machines may be used to make the parts he is designing, and specify tolerances to correspond. It is important that he specify tolerances, because the inspection department will need them, as will the subcontractor should these parts ever be built outside the home plant. For example, when dimensioning parts for a conventional machining center, one can ask for and expect ±0.002". However if the design calls for two bores to support shafts for high-precision gearing, and needs and calls for ±0.005" on the center to center distance and the bore diameters, one must expect that the work, or that part of it, will be assigned to a precision mill or jig boring machine. And every dimension that that tool produces will have that close tolerance. While close tolerances are not objectionable, they *can* be expensive.

It is conceivable that the designer, will incorporate two levels of tolerance in one part, one for the unimportant surface locations, and one for precise surface locations. In a gear reducer housing, for example, the shaft bearing bore holes may require a precision boring mill, while all the cover holddown bolts, escutcheon pin holes, oil holes, surface milling for the cover plate and the like may be done on a less accurate and less costly milling machine. For such parts the manufacturing engineer may elect to specify two operations on two different machine tools, using the time of his expensive jig mill for only the work which it alone can do, and using any of several less expensive but still effective tools for the routine work. Nevertheless, the two levels of tolerances should be suited to the two available classes of machine tools.

Very frequently, closely limited tolerances define surfaces on one part which will mate with surfaces on another part, also covered by toleranced dimensions. For example, the bore diameter in a casting which is to receive the outer race of a ball bearing. The difference in these two dimensions is called the allowance, and determines the type of fit between the parts. Tolerances and allowances have a well known mathematical relationship, and they are *not* independent.

When one of a pair of toleranced parts has its tolerances prescribed by circumstances outside the designer's control, as in the case of a purchased ball bearing, then the mating part's tolerances are thereby also predetermined, assuming the allowance is fixed. But when both of the pair are made in-house, the distribution of the tolerances is left to the designer. To carry on with our example, if the bore of a bearing is to be cut on an N/C mill, and the journal which is to run in it is to be made on an engine lathe, then the bore tolerance will be determined by the N/C mill's capability, the allowance will be the conventional amount for a running fit, and the tolerance on the shaft will have to be specified to complete the mathematical requirements of the design algorithm. Thus, the designer must anticipate the tooling in the dimensioning of his two parts.

Accuracy and Repeatability. In discussing the capability of an N/C machine, two terms are frequently used: Accuracy, and Repeatability. They are not the same, and the distinction should be clearly understood. To explain:

Let us assume that we have an N/C machine with a table moveable in the X direction. From the position $X = 0$ we will give the command to go to $X = 8.000$ inches, and give it a large number of times in succession, say 100 times. Because of the nature of the real world, the table will never come to rest in exactly the same spot twice. We will measure the actual position with great precision each time, and plot the frequency vs. position. The resulting curve will resemble *Fig.* 5.10. The peak of the curve represents the most probable position reached, and its departure

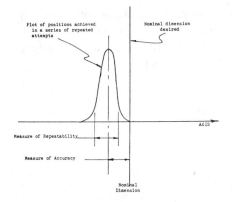

Fig. 5.10 — Plot showing repeatability characteristic of machines in real terms.

from the 8.000″ nominal dimension desired is a measure of the *accuracy* of the machine and control system; the smaller, the better. The spread of the curve is a measure of the *repeatability* of the system performance; the narrower, the better. It is common to find the repeatability to have a smaller measure than the accuracy, frequently in a 1:2 ratio.

Design Simplification. Patterns of holes (or other machining features) present an opportunity for the product designer to make N/C programming and manufacturing simpler, *Fig.* 5.11. A pattern, in N/C parlance, is an array of points. Once programmed, it may be used in several places without having to repeat the total programming operation again. The N/C programmer simply calls for the pattern by an identifying name and states where it is to be located each time it is repeated.

Fig. 5.11 — This line of 8 Sundstrand OM-2 Omnimill Machining Centers machine magnesium aircraft constant speed drive gearboxes and pump housings in sizes up to a 12-inch cube. Some 70 different part numbers are handled and the line processes 1200 housings per month in lots of 25 to 300 pieces. Courtesy Sundstrand Machine Tool Div., Sundstrand Corp.

The design engineer can contribute by reusing a few patterns rather than by introducing a completely new one each time the opportunity arises. The programmer can do a wide variety of things to a fixed pattern of which the designer can take advantage. Patterns may be of several basic designs:

1. Points in a straight line.
   a) Evenly spaced.
   b) Spaced at varying but specified distances.
2. Points in a circle.
   a) Evenly spaced.
   b) Spaced at varying but specified distances.
3. Points in a parallelogram lattice.
4. Points in a hexagonal lattice.
5. Groups of points in a random pattern.

Whole patterns may be:

1. Translated horizontally or vertically, or both, to new positions where they are repeated.
2. Rotated about one of the points, or some other reference point, and repeated in the new orientation.
3. Scaled up or down in size.
4. Two or three of the above, in any combinations.

Finally, patterns can be repeated with special exceptions to:

1. Omit certain specified points in the new position.
2. Retain only specified points in the new position.

For the designer, this suggests that he should favor:

1. Evenly spaced holes in either linear or arcuate patterns; if an uneven spacing is essential, achieve it if possible by omitting one or more holes in an evenly spaced pattern.
2. Once a pattern has been established, try to use it again. It can be put down anywhere, like a rubber stamp on a piece of paper.
3. In locating repeat patterns, dimension only the reference point.

Know the Capabilities of the
N/C Programming Language

Programming an N/C machine involves two steps by the manufacturing engineer, or programmer:

1. He describes the surface boundaries and important points of the part design, with dimensions.
2. He directs the motion of the cutting tools to produce these surfaces and dimensions.

Then the computer converts this input into a series of command symbols which, when transmitted to the machine tool, will cause it to make the part.

This involves two elements which in noncritical conversations may be regarded as one. They are:

1. The *language* of communication between the programmer and the computer. It consists of a fixed and rather precise vocabulary and a syntax or format in which the words must be arranged to be comprehensible to the computer. The analogy to the English language is obvious: a dictionary tells what various words mean, and a

grammar sets forth the rules by which words are formed into sentences which transmit information from one person to another. While humans are quite tolerant of aberrations in the English language, the computers are quite intolerant. The N/C programming languages must be carefully written, and conform absolutely to the right spelling and correct syntax.

2. The *processor* which receives input instructions, and converts them into machine tool control instructions. The input is called the "source code" and the output is called the "object code". (Code here means message.) The processor is a compendium of mathematical procedures — or "algorithms". Each feature of the part design, either surface or point location, is converted into a mathematical definition called the "canonical form". Tool motion instructions are converted into machine axis and spindle motion commands which will pass the tool through the raw material so that it will remove all the unwanted metal falling within the envelope of its cutting edges, thus leaving the desired part shape.

It must be obvious that the N/C language and processor should be able to cope with the part geometry to be described. This means that — for example — if the part has a spherical surface at some point, then:

1. The programming language must include a word such as SPHERE, with a format of symbols which will uniquely define a certain sphere. (There may be several such formats to describe spheres.)

2. The N/C processor must include a mathematical procedure which, with this input, can first translate the input parameters of any format into the standard canonical equation; next, figure out where the cutter should move to cut the metal and produce that certain sphere; and finally write out the moves in the N/C machine tool readable codes and output these instructions, frequently in the form of a punched paper tape. Furthermore, it should be obvious that the N/C machine tool to be used must be capable of obeying these instructions.

If all of the above three conditions cannot be met, then the spherical surface can not be cut on the machine available, to carry on with the above example. It, therefore, behooves the part designer to specify part surfaces which are included in the capabilities of the N/C language, the N/C processor, and the N/C machines at his disposal, if he intends the part to be made on N/C equipment. The Numerical Control Society conducted a study for the Electronics Command, Department of Defense, in 1974 tabulating the geometric capabilities of seven N/C languages and processors at that time. While a copy of this report treats this subject in great detail, the designer should consult his manufacturing engineer for current details of available capabilities; languages, processors, and computers change rapidly with the passage of time.

The Manufacturing Engineering Department should furnish information of this sort for the use and guidance of the Product Design Department. It is not unusual for a large N/C facility to be using two languages at any one time, particularly if N/C has been in use for a decade or more, or if the N/C Department is the result of the merger of two or more prior groups. It is also not unusual to find two or more computers available — possibly a small computer dedicated to N/C programming, and a much larger computer which the programmer may use if necessary, or access to a large computer network over telephone lines. In such multiple choice cases, consultation is recommended.

While generalities are always open to exceptions and argument, surface configurations can be ranked in ascending difficulty. It must be remembered that many N/C programs define a two-dimensional array of lines, curves, and points. Nevertheless, the parts are bounded by surfaces; the plane curves become right cylindrical surfaces as the contouring cutter moves along them. The ranking of programs can be taken in ascending order of difficulty:

   1. *Machining center work: 2$^1/_2$ axes with contouring:*
      a) Circular holes perpendicular to the base plane.
      b) Tapped holes perpendicular to the base plane.
      c) Plane surfaces parallel to the base plane.
      d) Plane surfaces perpendicular to the base plane.
      e) Segments of circular cylinders perpendicular to the base plane.
      f) Segments of 2nd order cylinders perpendicular to the base plane
          (ellipse, parabola, hyperbola).
      g) Segments of general 2nd order right cylinders.
      h) Segments of fitted curves.
      i) Canted planes.
   2. *Machining center work with 4 or more axes plus contouring:*
      a) The above list.
      b) Planes not parallel to the base plane.
      c) Cylindrical surfaces at skewed angles to the base (circular, elliptical,
          parabolic, hyperbolic).
      d) Cylinders based on fitted curves at skewed angles.
      e) Cones, spheres, and quadrics.
      f) Polyconics.
      g) Ruled surfaces.
      h) Sculptured surfaces mathematically definable.
      i) Sculptured surfaces which must be digitized.

   A very large fraction of the parts to be programmed can be completely described by the first six categories in the first list above; another large segment can be handled by the first four categories in the second list. That will handle most ordinary part designs. If it is really necessary to specify part surfaces such as are listed in the lower part of the second list, the part designer should consult with the manufacturing engineer to determine whether or not the parts can be made in house.

ADDITIONAL READING

[1] Dr. Joseph Harrington, Jr., *Computer Integrated Manufacturing*, 321 pages, 101 illustrations. Industrial Press, Inc., New York, 1973.
[2] James J. Childs, *Principles of Numerical Control*, 294 pages 167 illustrations. Industrial Press, Inc., New York, 2nd ed. 1969.
[3] James J. Childs, *Numerical Control Parts Programming*, 340 pages 340 illustrations., New York, Industrial Press, Inc.
[4] *Numerical Control Language Evaluation*, a Report by the Numerical Control Society for the U.S. Army Electronics Command. Reprints available from the NCS, 1201 Waukegan Road, Glenview, Illinois, 60025.
[5] Nils O. Olesten., *Numerical Control*, 646 pages, Wiley-Interscience, 1970.
[6] Many valuable articles have been published in the annual volume of the *Proceedings of the Numerical Control Society*, available if in print from the office of the Society, 1201 Waukegan Road, Glenview, Illinois, 60025.

6

Design for Recycling

C. E. Evanson and M. A. Rugo

WHAT FINALLY happens to the product that has been thoughtfully designed and economically produced? It is sold, used — then replaced and discarded. In effect, thrown away.

One of the most important attributes of American industrial life is that it is waste-producing. Products are manufactured toward a single end, and no thought is given to the final disposition — or recovery — of their materials. Often, according to E. R. Pariser of the Massachusetts Institute of Technology, materials are considered to be "waste" simply because they are not being utilized. "Waste," he suggests, "is matter out of place."

The figures are significant: each year in the United States 125 million tons of urban solid wastes are gathered and thrown away. These wastes are estimated to contain, in part, some 800,000 tons of aluminum, 10.6 million tons of ferrous metals, and 400,000 tons of nonferrous metals such as copper.

If recovered and recycled, they could yield 400,000 tons of aluminum, 6.9 million tons of ferrous metals and 100,000 tons of copper.

For decades we have accepted this creation of scrap and waste as the other side of our system of production and consumption of mineral resources... "the rich droppings of an affluent society."

Can we continue to pay for, haul and bury these mountainous piles of waste, while writing off the mineral riches they contain? Can we go on as we have always done, depending on our primary producers to find, refine, and produce high-quality metals for us?

The apparent answer is that we cannot. Shortages of materials which are vital to our technology have begun to appear. Other countries have sharply increased consumption of the same materials. These facts become of considerable importance when we realize that over 80% of our ten most important materials are imported.

In the short run we may insist that producers dig ever deeper, extract and refine from ever-leaner ores. We may pay the cost in greater power consumption and ecological damage. In the short run, we may outbid all other countries for the scarce materials needed to keep our system going as it has "always" gone.

But in the long run it must be accepted that our resources are finite. There are a finite number of metals in our world.

C. E. Evanson is President and M. A. Rugo is Administrative Manager of TAB Engineers, Inc., Northbrook, Ill.

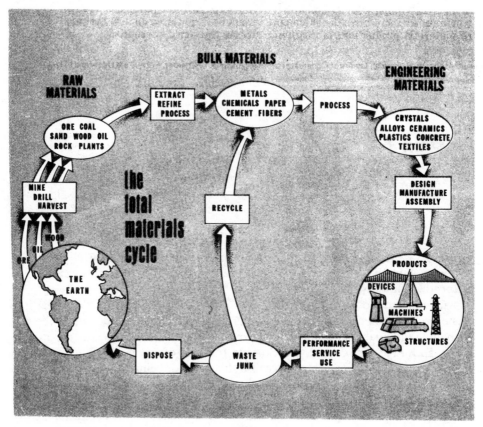

Fig. 6.1—The total materials cycle as shown in the COSMAT Report[1].

Where Are the Metals? In addition to what remains in the earth, an enormous quantity has been brought up, refined, used, and scrapped, into what the Danish engineer Niels Gram calls "the overground mine." The overground mine theoretically contains all the metal that man has extracted from the earth's crust — minus what has corroded away or been spread out so thinly that it cannot be reclaimed.

Much of our tin and zinc has been diluted and thinly spread as coatings, etc. Lead and copper are easier to reclaim, because they are usually not applied in thin surface layers, and they don't corrode easily. On the other hand, some quantities of lead and copper are lost in chemicals such as lead anti-knock compounds and copper-bearing fungicides and paints.

Junked cars (some 15 million or so on hand, with another 6 million being added in a normal year) contain valuable light and heavy steel, cast iron, copper, lead, aluminum, etc. Until recent development of giant wrecking machines that literally hammer a car to pieces, allowing economic recovery of its metals, much of this material was effectively lost.

Mineral resources still in the earth are thinner, leaner, and deeper than the supplies that we used to build our present industrial society. Harrison Brown of the California Institute of Technology asserts that, if some catastrophe erased our industrial civilization, it would be impossible to build it again, because the abundance of metals and easy-to-extract raw materials that formed the base of our modern technology is gone forever.

Recycling. So we find our metal resources in three places: still in the earth; being used in our industrial and consumer society; or on the scrap heap. Can we recycle the scrap heap back into useful life?

If we can, the environment should profit. Materials in use do not clutter or degrade the landscape. We may find our net energy use decreases: the Bureau of Mines estimates that many metals can be recycled at energy costs only 10-20% as high as was required to extract them in the first place.

An example is aluminum. "When you add up the cost per ton of installing the necessary mining, refining, smelting and fabricating equipment needed to establish primary production," says an Alcoa spokesman, "and compare that to the cost of setting up an aggressive recycling program — the comparison automatically justifies the recycling effort."

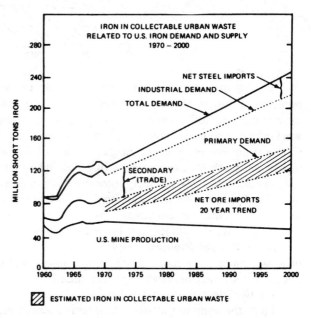

Fig. 6.2—Recycling iron from urban waste can increase the nation's metal supply and reduce environmental degradation. About half the iron in collectable urban waste can be recovered[2].

Another attractive notion is that cities and towns may be able to separate and reuse garbage and refuse. "Waste must be treated," says American Can chairman William F. May, "as the valuable resource it is — a resource containing metal, glass, and vast quantities of organic materials of high caloric value that can be converted to energy." American Can's own system, called "Americology" not only consumes raw garbage and recovers ferrous and nonferrous metals, glass, plastic, and fibers, but also generates energy from the combustible materials.

Writer Jane Jacobs suggests that in the highly developed economies of the future, large cities will become huge, rich, and diverse mines of raw materials. "These mines will differ from any now to be found," she says, "because they will become richer the more and longer they are exploited... in cities, the same materials will be retrieved over and over again. The largest, most prosperous cities will be the richest, most easily worked and most inexhaustible mines." Cities that take the lead in reclaiming their own wastes will develop related industries who manufacture the necessary gathering and processing equipment, and export it to other cities and towns.

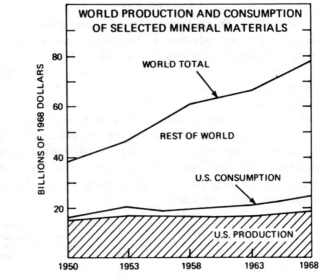

Fig. 6.3—Industrialization was responsible for expanded production and use of 31 major raw materials during the 1950-70 period. These materials range from asbestos, mercury, and salt to copper, coal, and zinc[2].

All this may suggest that we are well on the way to developing a thriving recycling industry. However, as yet, this is quite untrue. A few cities and towns have begun limited recycling operations; a few large and small companies are working on resource-recovery systems. (A particularly interesting application by Combustion Power, Inc., pioneers the use of the aluminum "magnet." Its dry separation method involves repelling the light metal from the waste stream: a special coil winding in a linear motor creates a magnetic field, which reacts with an opposing field in aluminum can stock to kick the metal out of the stream and into its own container or conveyor.) Most of our solid waste, however, continues inexorably into the dumps and landfills of America.

A Look at the Problem. Why this apparent contradiction: an enormous supply of waste resources, a continuing need for supplies of production materials, and no adequate recycling industry to bridge them?

We can identify several reasons:

First, the common attitude toward reclaimed materials is a serious obstacle. The industrial buyer, the technical user, and the general public all suspect somehow that recycled material is less pure, contains dross and dirt, suffers fatigue from previous use, or has even "lost its soul."

Many material specifications impede recycling. Whether they are national, international or company specifications, they are usually drawn up in cooperation with producers of the material in question.

Producers invariably means primary producers, who are not always aware of recycling, or who even try to defend "their" material against secondary metal by setting close tolerance limits for the industry that could not possibly be met by the reclamation companies. Certainly for many applications it's technically and economically right to use high grade primary metal, but with our present knowledge of materials and the effect of impurities in metals, it must be wrong to preclude a secondary material just because it is secondary.

Some governmental bodies have placed roadblocks in the path that leads to recycling and reuse. For example, in the United Kingdom the most recent revision of the British standard for zinc die casting alloys specified that "no zinc scrap or secondary zinc alloy shall be used." Why specify the origin of the zinc? It should now be possible to specify metals by their composition and properties and not by history. If an alloy maker could meet the purity specifications, using carefully segregated secondary zinc alloy, why shouldn't he be allowed to do so?

Resource development is regarded by many as a risky business, and money is a problem. Many state governments won't commit themselves to long-term contracts with marketers. This is a particular shame, because according to Mr. Abert of the National Center for Resource Recovery, a city could recoup its investment in pilot waste fuel projects in two years. In all these considerations, the same fact pops up again and again: it's just too expensive, in labor, energy, or both, to recover materials from wastes as they are today.

The Designer's Role

Well, what can be done about that problem for tomorrow? And who can do it?

One answer is to begin with the design process. The designer's role can hardly be over-emphasized. To quote Tom Schlabach of the Bell System:

"The impact of an industry on our resources and environment is determined largely by the materials and manufacturing processes that it selects."

The designer assumes a central role because his decisions, consciously or unconsciously, knowingly or unknowingly, influence the pattern of consumption, pollution, and recycling of the product.

The hard-working designer used to feel that he had done a pretty good job if he balanced function and cost in his product. Now we must ask, in addition, that he conserve mineral resources, and to consider also the near-and far-term effect on the environment.

What, specifically, can the designer do to enhance eventual recoverability and scrap value of the product on his drawing board? Here are some suggestions:

1. Parts made of different materials should be distinguishable and separable for scrapping. Or they should form a useful mixture when scrapped together.

2. If materials are surface treated, the surface layers must not be detrimental for recycling the metal.

3. Metal parts should not be so heavily coated with plastic that excessive burning is required to remove them. For example, there is a mechanical method for economically stripping the plastic insulation from copper wire when old cable is to be recycled. The mechanical method is obviously preferable to burning from an environmental viewpoint. But all too often mechanical stripping is impossible because of the way the cable was designed.

Richard Farmer of Indiana University concludes, "We could insist that cars be designed to be taken apart rather easily. So far, no one in designing a car gives any thought at all to how tough it is to separate all those things when the car finally gets junked. But it would be fairly easy to design a new car so that it could be taken apart easily — and if it were, our junkmen would do just that ." Many products fall into this category.

4. Some design problems are effectively solved by the use of composites — combinations of materials, often including a critical (or expensive) material, designed to exploit the best qualities of each. But here again, the solution can create a scrap problem. If the composite scrap cannot be recycled easily, the composite may be uneconomical in an overall sense. In addition, if the critical constituent is

not readily recoverable, the adulterated scrap may be worthless, or can even represent an absolute loss of the critical constituent for future use.

5. The designer can make an effort to specify the lowest grade of metal or other material that will meet his needs. Very many applications do not require the high degree of purity or quality that we often specify as a matter of course.

Many more ideas will occur to the designer who continues to think about the ultimate use — and reuse — of the materials his product will consume.

Management decisions on materials use can affect the course of a company or an industry. For example, a few years ago the Adolph Coors Co., the nation's fourth largest brewer, decided to package all its beer in aluminum cans. The decision was for aluminum because it is expensive enough to make recycling worthwhile. "We think the aluminum can is *the* environmental package for carbonated beverages," according to Bill Coors. (Makers of steel cans violently disagreed pointing out that an aluminum can that is not recovered and recycled will lie by the roadside forever, while steel cans at least will finally rust away.) Coors spokesmen claimed that in 1973 about two-thirds of their beer cans in some marketing areas were making it back to Coors' own recycling plant. Other beverage makers have converted some or all their production to aluminum. Primary aluminum producers have also set up a network of recycling centers. The operations of at least three industries — beverage, steel, and aluminum — have thus been affected.

The alert designer is accustomed to consulting with suppliers of the materials his new product will require. He discusses availability, cost, machinability and suitability of the suggested materials. He makes changes as the desirable alternatives become clearer.

Perhaps he should also consult with salvagers or recycling experts as the design evolves (assuming he can find them; they are not numerous as yet). Can the product as designed be salvaged? What are the problems? Would any component or combination be unrecoverable? Would recovery cost be too high? If more than one design is considered, the final decision may be in favor of the most "scrappable" version.

Today's designer may believe it unnecessary to weigh such considerations. (Yesterday's designer was completely oblivious to them.) Tomorrow's designer, whose products will be in use as we enter the 21st century, must incorporate them firmly into his design philosophy.

Failure to do so on a widespread scale will escalate the already serious problems of material scarcity and waste. If those problems are not solved by industry, legislative solutions will assuredly be attempted by governments.

In either case, it may well be, in the future, that materials specifications for massive applications (especially governmental work) will include "recyclability" and "scrappability" in their parameters. We may even reach the point where there will be a tax or penalty for failing to design "scrappability" into future projects.

Just as Oregon's law requires deposits on soft drink bottles and cans, a time may come when a deposit will be required on *all* products that contain recoverable materials. Your deposit might be refunded when you eventually turn in the item at the designated recycling center.

To briefly recap these ideas on design for recycling:

1. The supply of metals in the world is finite.

2. If we are to continue to use metals as materials, we must develop and use

efficient means of recycling.

3. The best place to plan for recycling is at the original design level.

4. In designing the product we should remember that its materials must eventually be reclaimed and reused, and we should utilize whatever techniques work best to assure that such reuse will be as easy and inexpensive as possible.

5. The engineer and the company who learn how to design for recycling and "scrappability" will have a real advantage in tomorrow's business environment.

For the next few years we still have a choice. We can design for recycling or we can keep on piling up waste. Eventually, that choice will disappear, because everything we use will ultimately have to be recycled. This must happen because we live on another finite resource — spaceship Earth.

REFERENCES

[1] "Materials and Man's Needs," National Academy of Sciences, 1974.
[2] "Can America Survive Materials Shortages?" *Mechanical Engineering*, pp. 29-33, January 1975.
[3] "Milwaukee To Reclaim Its Solid Waste," *Machine Design*, p. 18, February 20, 1975.

METAL REMOVAL
METHODS

7

Flame Cutting

Evolution and development of flame cutting — the severing of ferrous metals by rapid oxidation from a jet of pure oxygen directed at a portion heated to the fusion point — has been surprisingly rapid in the short period since its introduction. Wide acceptance and utilization of all-welded machines and machine parts have brought into sharp focus the mass production of component units by means of accurately controlled mechanical flame cutting machines. For the equipment designer this development eclipses the many other uses of gas cutting such as manual cutting, scarfing, gouging, descaling, lancing or piercing and cutoff.

Today, mechanical flame cutting machines have taken their place alongside other production machines in industry and have opened a vast field for quantity manufacture of many construction machines, cranes, ships, engine and other parts hitherto restricted to somewhat slower method, *Fig.* 7.1.

Fig. 7.1 — AIRCO flame-cutting machine with 20 computer-controlled torches for burning intricate shapes from steel plate within an 80 x 44 foot cutting bed. Photo, courtesy Joseph T. Ryerson & Son, Inc.

Fig. 7.2 — An Oxweld CM-21 cutting machine with a torch capable of cutting material up to 60 inches in thickness. Photo, courtesy Linde Div., Union Carbide Corp.

Dimensional Scope. Size of parts which can be produced is limited primarily, if not entirely, by the size of the table for the machine on hand. Likewise, metal thickness is limited only by the cutting heads available. With proper torches, materials ranging from light-gage sheet all the way to 60-inch billets, *Fig.* 7.2, can be readily cut or contoured. Quantity of parts which may be produced depends mainly upon the design of the part and the number of cutting heads on the machine. Simple parts such as annular rings and plain shapes may be cut from plate in considerable quantities, *Fig.* 7.3, if a multiple-head machine is available. By means of the stack method, even greater output can be achieved. With a combination digital-N/C machine cutting with plasma torches, high productivity cutting at speeds up to 250 ipm can be attained, *Fig.* 7.4. Plasma cutting is especially useful for high speeds in stainless up to 6 inches, high alloy steels, aluminum and carbon steel up to 2 inches in thickness.

Fig. 7.3 — Numerically controlled CM-80 shape cutting machine mass cutting 20-foot plates into identical components. Photo, courtesy Linde Div., Union Carbide Corp.

Fig. 7.4 — Single plasma torch on a CM-100 machine cuts $^5/_{16}$-inch low-carbon steel at a speed of 175 ipm. Photo, courtesy Linde Div., Union Carbide Corp.

Cutting Speed. Production speed is dependent also upon linear torch traverse speed, and torch travel necessary to completely skirt the perimeter of a part determines the time cycle. A part of complicated design which necessarily must be cut from a heavy billet naturally will limit the speed of production. Torch speed varies directly with the material thickness. Linear cutting speeds with ordinary machine gas cutting are indicated generally in TABLE 7-I. These speeds, though, are subject to modification by the existing operating conditions, material, machine, intricacy of contours, cutability of the material, etc., and may run from about 1 to 1.5 inches per minute with the heaviest sections to as high as 250 inches per minute — a speed which has been used successfully with N/C plasma torch shape-cutting machines.

TABLE 7-I — Machine Gas Cutting

Metal Thickness (inches, min to max)	Jet Size (drill no.)	Kerf Width (inches, approx.)	Cutting Speed* (inches per minute)
$\frac{1}{64}$ to $\frac{1}{4}$	72	$\frac{1}{32}$ to $\frac{1}{16}$	17 to 35
$\frac{1}{4}$ to $\frac{1}{2}$	65	$\frac{3}{64}$ to $\frac{5}{64}$	17 to 25
$\frac{1}{4}$ to 1	58	$\frac{1}{16}$ to $\frac{3}{32}$	17 to 20
$\frac{1}{4}$ to 2	54	$\frac{5}{64}$ to $\frac{7}{64}$	14 to 17
$\frac{1}{2}$ to 3	52	$\frac{3}{32}$ to $\frac{1}{8}$	11 to 15
2 to 5	50	$\frac{7}{64}$ to $\frac{9}{64}$	7 to 14
4 to 6	45	$\frac{1}{8}$ to $\frac{11}{64}$	6.5 to 9
6 to 10	40	$\frac{9}{64}$ to $\frac{3}{16}$	3.8 to 6.5
10 to 12	35	$\frac{5}{32}$ to $\frac{7}{32}$	3.5 to 3.8
12 to 14	28	$\frac{3}{16}$ to $\frac{1}{4}$	3.2 to 3.5
14 to 15	20	$\frac{1}{4}$ to $\frac{5}{16}$	3.1 to 3.2
14 to 16	$\frac{11}{64}$	$\frac{5}{16}$ to $\frac{3}{8}$	3.0 to 3.2

* Lowest speeds are for maximum quality cutting, intricate designs or low-cutability steels. Highest speeds are for lower quality cutting, straight-line work and high-cutability steels. Width of kerf will vary widely with pressure of oxygen, speed, orifice size, and height of tip above the plate. Speeds less than 1-inch per minute seldom used on machine cutting with even the thickest work and may run as high as 75 inches per minute in special cases.

With production in any quantity whatever, where ferrous materials to be cut are bulky or have heavy sections to be removed, the flame-cutting process offers many advantages. Relatively little power is required to operate the mechanisms of a mechanical flame-cutting unit and to remove or cut the metal. Cost is made up primarily of the cost of the gases used in cutting. Initial investment and overall operating cost of gas-cutting equipment, as compared with purely mechanical methods for cutting ferrous metals, are considerably lower, and speed of metal removal per unit volume is far greater.

Limiting Factors. Certain limiting factors or undesirable characteristics attendant to heating of the metal must be considered. The heat created during cutting tends to relieve or redistribute the locked-up stresses in steel plate or billets in addition to causing general expansion. This results in warpage or distortion in the finished part which is especially undesirable where highest accuracy is imperative. No specific rules can be set forth disposing of this matter and each case must be considered and worked out separately. A number of methods now in use for obviating or compensating for distortion are: (1) Adjustment of template size; (2) use of dual-torch operation, either 180 degrees apart on symmetrical shapes or operating in parallel on rectangular-type work; (3) clamping or wedging; (4)

cutting at elevated temperatures; (5) cutting under water; and use of plasma cutting.

Flame Cutting Equipment. Cutting machines available today might be divided into two general groups — portable and stationary. The portable machines, which are usually limited to operations on material up to 10 inches thick, can be further modified into straight-line cutting machines, cutoff machines and shape-cutting machines.

Portable Machines. Flexibility is probably the most outstanding advantage of the portable cutting machine. A number of economics result from the fact that the machine can be brought to the work rather than the work brought to the machine. Investment in equipment is lower than with other methods and specialized work can be handled as readily as production work. As a rule, portable machines, *Fig.* 7.5, are usually recommended for use only where commonly termed "structural

Fig. 7.5 — Portable shape cutting machine making hand guided cut. Photo, courtesy Linde Div., Union Carbide Corp.

accuracies" can be tolerated. That is, for steel structure fabrication such as required in shipbuilding, road machines, similar equipment where welding type edge preparation is of sufficient accuracy. A limited amount of contour cutting for plate-edge preparation can be done with this type machine. The smallest of the tracer type shape cutting machines will produce complex shapes within its size range, and is portable where crane or hoist facilities are available.

Stationary Machines. In general, stationary machines, i.e., those requiring movement of material to the machine, include: Straight-line, shape-cutting, production cutoff, flame planer, surfacing, and special-purpose machines. These machines can be used with multiple-torch arrangements, *Fig.* 7.6 and are employed widely for accurate plate edging.

Stationary shape cutting machines in a wide variety are available with hand-operated, mechanical, magnetic, or photoelectric tracing or N/C controls. Multiple torches can be used on these machines and the longitudinal working area can be

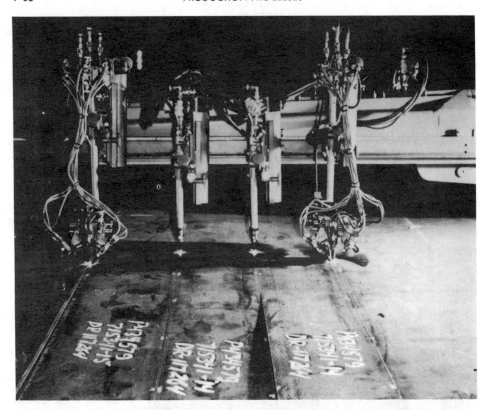

Fig. 7.6 — Rotating three-torches used on a machine for plate-edge preparation. Three cuts are made simultaneously in each traverse of the machine. Photo, courtesy Linde Div., Union Carbide Corp.

Fig. 7.7 — Group of frames being cut simultaneously from heavy plate in one operation. Photo, courtesy Linde Div., Union Carbide Corp.

increased to suit requirements by extensions to the permanent track and the cutting and tracing tables, *Fig.* 7.7.

The need for fast, automatic welding led to the development of the so-called flame planer type of machine for the rapid production of long, straight parts. The flame-cut edge produced is sufficiently accurate that further preparation is unnecessary. Three of these machines can operate on four sides of a plate simultaneously and triple-face cuts may be produced at one pass.

Special Machines. Many special-purpose machines have been developed and are used where production quantities warrant continuous operation on more or less standardized parts, *Fig.* 7.8.

Fig. 7.8 — A special production flange-cutting machine which cuts and bevels the inside diameter while also cutting the outside. Maximum diameter which can be cut is 36 inches, minimum diameter 8 inches and thicknesses up to 4 inches. Photo, courtesy Linde Div., Union Carbide Corp.

Design Considerations

Generally speaking, the design possibilities available by means of flame cutting are legion as has been exemplified by the multitude of all-welded machines produced in recent years. The inability to project a jet of flame and oxygen in other than a straight line imposes the primary limitation on design, since curvilinear contour cutting in more than one plane is thus impossible. Perhaps the closest approach to a solution to this problem has been the several multiface edge preparations such as double bevel to shoulder, and the formation of continuous J-grooves with the use of gouging nozzles. In gouging, the oxygen jet impinges upon the plate edge in such a way that the deflection of the jet results in a continuous curved surface. Applications of this phase of the process at present are limited to the preparation of the plate edges for welding.

Design Changes. Changes in the detail design of parts produced by flame cutting can be made rapidly and at little expense. Templates for magnetic tracer machines or drawings for electronic tracer machines can be replaced or revised easily. This design flexibility and low-cost operation offers many economies and opens a wide field of new possibilities. In many cases, even large parts can be cut within ¼-inch of finished size, eliminating a tremendous amount of difficult machining otherwise necessary, *Fig.* 7.9.

Fig. 7.9 — Marine fitting torch-cut from a solid forged-steel block 30 by 30 by 52 inches prior to finish machining. Center hole through section 36 inches thick is 14 inches in diameter. Photo, courtesy Bethlehem Steel Co.

Fig. 7.10 — Multiple-torch CM-56 machine producing complex components in heavy plate. Photo, courtesy Linde Div., Union Carbide Corp.

Design Intricacy. Introduction of electronically controlled tracing mechanisms has made possible extremely accurate reproduction of intricate designs, *Fig.* 7.10. These machines extend the possibilities of the process, which formerly were restricted somewhat by the inability of the follower wheel of magnetically controlled machines to track the most intricate details without the use of elaborate jigs and some additional machining of the cut piece. However, the sprocket wheels shown in *Fig.* 7.11 illustrate excellently the capabilities of electronic type machines with proper templates.

Fig. 7.11 — Sprockets being cut from a steel slab on an AIRCO machine. Photo, courtesy McNally Pittsburgh Mfg. Co.

Corner Limitations. Silhouette templates or pen or pencil drawings used with the photoelectric and N/C tracers can be used to achieve exacting results with square or unusual corners and sharp corners can be negotiated at speeds up to 30 ipm.

Range of Application. In designing to take advantage of low-cost hot or cold-rolled plate, forged slabs, rolled or drawn bar, etc., a great range of types of parts and the sizes of machines can be considered within the scope of economies afforded by flame cutting. Extremely small parts, including 1.125-inch thick bodies of 9-mm pistols as well as parts and 2.031-inch thick bodies for guns, have been flame cut from 3.5 per cent nickel gun steel bar stock in great quantities. In direct contrast is the cutting of two crankshafts from one forged slab 25 feet long, 10 feet wide and 25 to 28 inches in thickness with a 38 to 42-inch thick end flange. One flame-cut blank crankshaft cut from this slab after preheating to 400 F is shown in *Fig.* 7.12. Contours are clearly shown and cutting was carried out at approximately 2 to 2.5 inches per minute. Following cutting, the blank is annealed and then the bearings between throws are rough turned. After heating, the various throws are twisted to

Fig. 7.12 — General view of a gigantic crankshaft torch-cut from a forged slab. Throws are hot-twisted into desired angular positions after rough machining. Photo, courtesy Bethlehem Steel Co.

proper position. A final heat treatment is followed by complete finish machining all over.

Selection of Materials

Essentially all ferrous metals and alloys can be flame cut, although the low and medium-carbon types can be handled most easily and economically. Steels under 0.30 per cent carbon can be cut cold — without heat treatment before or after cutting. Steels in the higher alloy groups can also be cut with relative ease but require preheating before and annealing afterward.

Galvanized material, especially in gage thicknesses, is poorly adapted for flame cutting. Thin sheets buckle badly, the heat destroys the galvanizing, and the zinc fumes created are hazardous. Stack cutting is equally poor. Heavy galvanized stock can be cut but, if at all possible, its use should be avoided.

Heat Buckling Limitations. The problem of heat buckling is also present in plain steel sheet under 1/8-inch in thickness and cutting of single sheets is impractical with oxyacetylene means. These light gages, however, can be readily handled by means of the plasma torch or stack cutting, but of course the operation is only economical when a reasonable quantity of parts is required. Material ranging around 1/4-inch in thickness and upwards can be cut singly, with one or multiple torches, with excellent results.

Heat Effects. Flame cutting has no appreciable effect on the physical and chemical characteristics of low-carbon steels except in the case of forgings. These should be annealed and cut hot because of inherent internal strains. Steels over 0.30 per cent carbon tend toward some migration of carbon to the cut surface but preheating and annealing obviate this tendency. Preheating to a temperature somewhere between 200 to 800 F is also essential for maximum ease in the cutting of alloy steels and for heavy sections of large mass except where analysis makes such treatment inadvisable owing to metal affinity for the oxygen jet at elevated temperatures. With stock over 0.40 per cent carbon, total decarburization should not exceed 0.0006-inch per side to eliminate undesirable excess slag adherence.

Stainless Cutting. Ordinarily, the chromium and nickel content of stainless steels produces oxides which make ordinary cutting procedure slow and undesirable. However, as noted, the new plasma torch makes possible the cutting of stainless steels of as high as 50 per cent alloy content as readily as carbon steels by the ordinary torch. High quality cuts, equal to those regularly obtained with low and medium-carbon steels can be made at production rates in material up to 6 inches thick.

Preliminary Preparations. Material to be flame cut should have good surface conditions free from surface pitting, rounded corners, crowning, etc., to assure full utilization of the close tolerances available. Material should be pickled so as to be free from scale and all but moderate rusting avoided.

Specification of Tolerances

Final accuracy on a flame-cut part depends primarily upon a number of factors; (1) Accuracy and rigidity of the machine used; (2) accuracy and character of the torch jet; (3) thickness of the material; (4) cutability of the material; and (5) the clamping and distortion factors. Naturally, only the most accurate machines should be required to produce "precision" work. Thus, most portable straight-line machines are recommended only for "structural accuracies", that is, where welding type edge preparation is of sufficient accuracy. This usually works out to be within $1/8$-inch, plus or minus. Likewise with portable shape-cutting machines, similar and, in certain cases, somewhat better accuracy around plus or minus $1/16$-inch can be achieved.

Overall Accuracy. Stationary shape-cutting machines are far more accurate, and can be expected to follow a line or duplicate a pattern, guided by a magnetic follower, to within $1/64$-inch, plus or minus. N/C machines operate holding accuracy within plus or minus 0.002-inch of true path. Flame planers are designed to have a straight-line accuracy of plus or minus $1/64$-inch or less. As a rule, the oxygen jet will remain within about 0.006-inch of a desired line or contour. Consequently, a combined accuracy of plus or minus 0.021-inch in such cases or better can be expected, *Fig.* 7.13. To this must be added the usual work-distortion and work-movement factors which vary over wide limits from nothing up to a matter of inches, depending upon the cutting technique, material condition, plate size, and procedure control. These can be handled properly, however, and detrimental effects obviated to a great extent. One manufacturer has found that convex distortion on straight-line cuts will not exceed $1/16$-inch on a flat plate having a width of 3 feet or greater, irrespective of length. Where the cut length is 10 feet or less, or the thickness under $1/2$-inch, the distortion seldom exceeds $1/32$-inch.

Warpage. Materials $1/2$-inch and over in thickness are seldom warped or buckled unless unduly long or narrow. Cases on record show that with N/C machines, template or drawing contours can be reproduced accurately to within plus or minus 0.002-inch. Considering the jet inaccuracies, such machines can be expected to produce shapes in quantity with contours identical within plus or minus 0.008-inch, barring other factors. In flame planer work, the total inaccuracy can be consistently held within 0.030-inch on a side with $1/2$-inch plate up to 30 feet in length.

Close tolerances, of course, should only be required where absolutely necessary. Many applications do not require exacting dimensional tolerances and in such cases the usual procedure is to specify work on a weight basis of 5 per cent tolerance or greater.

Fig. 7.13 — Contoured hole, cut at 2 to 3 inches per minute in 18-inch armor plate, shows accuracy of cut and square corners of a 2-inch offset. Photo, courtesy Bethlehem Steel Co.

Fig. 7.14 — Bevel torch-cut 17$^1/_2$ inches long on a heavy die block shows fine quality surface finish obtainable even on extremely large parts. Photo, courtesy Bethlehem Steel Co.

Vertical Accuracy. In addition to variations of the oxygen jet from a line or contour are the variations in the vertical accuracy of the kerf produced by the jet. Ordinarily, cut surfaces true to within plus or minus 0.003-inch as to cross-sectional squareness for steel under 0.35 per cent carbon around 1-inch in thickness can be had. This becomes 0.010-inch for 2-inch material, 0.018-inch for 4-inch material, 0.031-inch for 6-inch material, 0.062-inch for 12-inch material, and 0.125-inch for 24-inch material. These tolerances should not be specified for ordinary work but in cases of necessity can be had and oftentimes bettered, *Fig. 7.13*, with proper handling. Vertical accuracy also limits the thickness of the stack possible in stack cutting to meet specified tolerances.

Kerf Allowances. Extreme accuracy in reproduction of shapes requires knowledge of kerf widths produced. Any template used must be made with the necessary allowances. For "precision" work the rule-of-thumb allowance of twice the oxygen orifice is not sufficiently accurate. Actual kerf widths are determined by the cutting conditions involved and vary appreciably with pressure, speed, tip size, and space from the tip to the plate. Consequently, exact kerf allowance must be obtained by actual test for each job and the conditions under which this test is made must be adhered to rigidly in order to avoid variations. An idea as to cutting tip size for various plate thickness, approximate kerf widths etc., is given in TABLE 7-I.

Surface Roughness. Surface smoothness depends primarily upon the rigidity of the torch holding mechanism and secondarily upon atmospheric conditions. Air movement at the cut may affect the smoothness somewhat. However, where the cut surface may constitute a bearing, gear tooth, or similar surface, finish machining — turning, broaching or milling — is usually necessary. In such cases allowance should be made on the template and with careful cutting it is necessary to remove only 0.020 to 0.030-inch of stock in finishing to final dimensions. The bevel cut shown in *Fig.* 7.14 illustrates the excellent surface quality which can be obtained even on extremely heavy cuts.

8

Contour Sawing

To ACHIEVE in many medium and low-production items such as those required for machine tools, scientific equipment, testing machines, transit cars, road machinery, and special machines the same qualities of low cost, design versatility and the all-important interchangeability as are obtained with high production processes, contour machining or sawing can often be employed. With low and medium-production lots the key factor which designates the most economical method or methods is the quantity involved. As indicated in *Fig.* 8.1, contour sawing will usually fall into the range below one thousand pieces.

Conventional Contour Sawing. With plain contour band sawing, it is possible to produce blank parts in quantities up to approximately five hundred pieces,

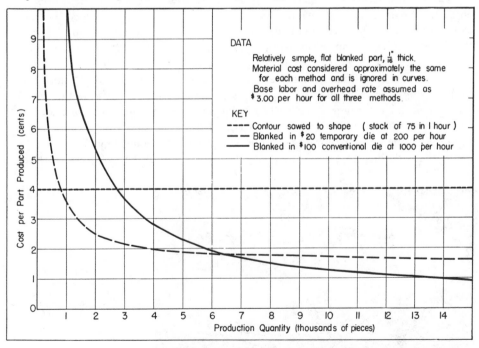

DATA

Relatively simple, flat blanked part, $\frac{1}{16}$" thick.
Material cost considered approximately the same for each method and is ignored in curves.
Base labor and overhead rate assumed as $3.00 per hour for all three methods.

KEY

---- Contour sawed to shape   (stock of 75 in 1 hour)
— — Blanked in $20 temporary die at 200 per hour
—— Blanked in $100 conventional die at 1000 per hour

Fig. 8.1—*Quantity-cost relationship of sawed blanks and those from temporary or permanent dies. Cost of temporary dies, with few exceptions, averages around 20 dollars while that for a permanent die to produce the same part would run about 100 dollars although the cost ratio may vary from 4-to-1 to 10-to-1 according to the complexity of the design. Ten thousand parts can often be run on temporary dies before die wear is excessive, and in many cases as many as 15,000 to 20,000 can be produced.*

Fig. 8.2—Contour sawing stacked sheets at conventional speeds, produced these fifty identical electric switch parts simultaneously. Photo, courtesy DoAll Co.

eliminating the usual blanking die costs for such parts, *Fig.* 8.2. Sheets are compressed and welded together to simplify the sawing process. Intricate shapes are cut with the same ease as simple ones, chief controlling factors being the length of cut and material used.

Three-Dimensional Shapes. To provide at low cost small quantities of complicated parts such as shown in *Fig.* 8.3, sawing direct from the forged block of steel can be readily accomplished. Much expensive processing and die cost otherwise incurred is thus eliminated. Again, sawing direct from forged or hot-rolled billets can be used to provide sample forgings, substitutes for forgings, etc. Contour sawing simplifies in many cases the design of large forgings. Simplified blank forgings can be rapidly cut to proper contours, making possible the use of such a blank for a number of basically similar parts. In certain cases, for instance, right and left-hand parts can be contour sawed from the same forged blank, not only eliminating an extra forging die but also much additional machining.

Without doubt, on such jobs, contour sawing is one of the fastest methods

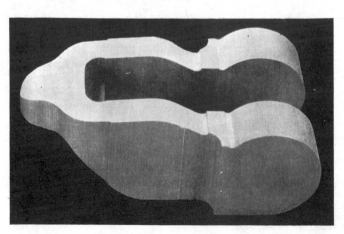

Fig. 8.3—Production of such parts as this locomotive pop-pet valve lever arm is simplified by contour sawing. Photo, courtesy DoAll Co.

known for removing bulk metal and leaving a relatively good surface. Subsequent finish machining time is thus reduced to a minimum, *Fig.* 8.4.

Conventional Sawing Machines. In addition to the wide variety of standard two, three and four-wheel band machines available for normal-velocity work, open-end band machines are available. These carry the toothed band helically wound on a drum or reel and require no blade welding or brazing, which is specially advantageous for internal cuts. Up to two and one-half minutes of sawing can be had before the end of the band is reached and rewinding to start position is necessary. The rewind is sufficiently rapid as to incur little delay. Extremely accurate and delicate contours are possible on such machines with, however, the disadvantage of the short duration of cut.

Photo, courtesy DoAll Co.

Fig. 8.4—Sawing reduced machining time for diesel engine connecting rods from 8 hours to less than 1 hour per connecting rod.

Fig. 8.5—Friction sawing extends the scope of contour work to many materials not possible to machine by other methods. Photo, courtesy General Motors Corp.

High-Velocity or Friction Sawing. During recent years, the contour sawing field has been broadened to a considerable extent by the perfection of high-velocity sawing, usually referred to as friction band sawing, *Fig.* 8.5. Basically a frictional melting or burning process, this superhigh-speed method of contour cutting has stepped up production of extremely hard and often otherwise unmachinable materials used in the manufacture of ships, planes, automotive vehicles, railway equipment, farm machinery, tanks, etc. The primary difference between conventional and friction band sawing lies principally in the surface speed of the band and feed pressure employed. While the conventional sharp saw tooth cuts chips and carries them out of the kerf oftentimes under a coolant, the dull, dry friction saw develops sufficient frictional heat at the tooth contact to bring a material instantaneously to melting temperature for easy removal. A fine jet of compressed air directed upon the blade often assists in the cutting of ferrous materials owing to oxidation much as with torch cutting, *Fig.* 8.6.

Machine Speeds. Ordinary contour sawing is normally accomplished at speeds from 50 to 3000 fpm depending upon the type of material and the thickness. High-velocity sawing, however, begins where conventional metal sawing leaves off and today machines are in use which deliver as high as 16,200 fpm. Blades with either raker or wave-set teeth are used with a temper slightly softer than that for wood saws, to avoid premature fatigue. For materials up to 1/8-inch in thickness, 18 teeth per inch are used; from 1/8 to 1/4-inch thicknesses, 14 teeth per inch; and from 1/4-inch on up, 10 teeth per inch are used. These blades are low in cost and in production may last up to 40 hours.

Thickness of Material. Most present-day friction sawing is performed on parts made from materials up to 1/2-inch in thickness. Light gages of ferrous or hard

Fig. 8.6—Typical cuttings resulting from friction sawing armor plate at 12,000 fpm with a 14-pitch blade. Photo, courtesy General Motors Corp.

materials (up to 1/8-inch) are cut using a band speed from 3000 to 5000 fpm. This speed increases to as high as 10,000 to 12,000 fpm with thicknesses up to 1/4-inch. Using speeds around 15,000 fpm, up to 1/2-inch sections can be easily cut. Blade speed, roughly speaking, will be determined by the melting point of the material to be cut — hardness actually helps a little. Some authorities place the limit on thickness for such materials at 1/2-inch. However, much heavier plate has been cut and in one instance at least, using a speed of 16,200 fpm, armor plate 2 inches thick has been easily handled. These heavier gages, and especially those over 1-inch,

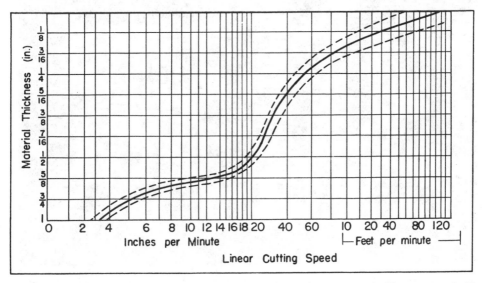

Fig. 8.7—Conservative estimate of linear cutting speeds in friction sawing carbon steels. Chart, courtesy DoAll Co.

often require a constant power feed to assure sufficient friction and in such cases torch cutting — if it can be used — is usually more economical. Nonferrous and nonmetallic materials, on the other hand, require no special techniques and can be cut, though finish may be poor, in sections up to capacity of the machine in use.

Cutting Speeds. Friction sawing cutting rates are considerably higher with materials up to 1-inch in thickness than rates obtained with various other methods. Speeds attained in production run from 20 inches per minute with 1/2-inch thick materials to as high as 120 feet per minute on 1/16-inch thicknesses, *Fig.* 8.7. Armor plate 3/4-inch thick has been cut at approximately 14 inches per minute and Hastelloy at 3 inches per minute, using a 1-inch wide, 10-pitch blade operating at 15,000 fpm. Using a 1/2-inch wide, 14-pitch blade operating at 12,000 fpm. 5/16-inch AISI 4140 chrome-molybdenum steel plate has been cut in aircraft production work at 20 inches per minute, 5/16-inch armor plate at 12 inches per minute, 1/8-inch stainless at 80, 1/4-inch Tantung G at 3, 1/2-inch Delloy tool steel at 1-inch, etc. Cutting speed, therefore, is not entirely dependent upon thickness of material, but to some extent upon the material hardness — the harder tempers having a

Fig. 8.8—Stroboscopic photo showing diamond band saw cutting circles in 1-inch thick glass. Photo, courtesy DoAll Co.

Fig. 8.9—Substitute for a forging sawed at conventional speeds from a solid bar of high-tensile alloy steel. Photo, courtesy DoAll Co.

higher coefficient of friction are heated more easily and cut more quickly.

In addition to speed, there is the advantage of negligible downdrag on the part being cut and ability to cut light-gage sheet metal on the scarf with wide angles in either direction. This particularly suits friction sawing for odd contour trimming operations such as in the aircraft industry where it is used almost solely for trimming light-gage formed work of aluminum alloys and steels.

Diamond Contour Sawing. To complete the contour sawing picture and make possible the cutting of those materials or designs which resist ready severing by means of either of the other methods, diamond sawing can be employed. By means of special diamond band saw blades, *Fig.* 8.8, hardened steel, tungsten carbide, stone, and various vitreous materials can be cut direct to a layout line with a precision comparable to that of conventional contour sawing. Machines developed specially for this purpose will handle work up to 10 inches in thickness and provide a throat depth of 16 inches. Blade speeds from 3000 to 8000 fpm are available. Normal kerf left by the blades is approximately $1/16$-inch in width.

Design Considerations

Although the contour sawing processes are limited primarily to the production of surfaces generated by a straight line, the latitude in design variations is extremely wide. Many parts required in only nominal quantities can be produced by this method to serve satisfactorily and economically in place of finished forgings. This is especially true in such instances as with industry work where design changes rapidly and also with experimental machines where time is a factor. The unit shown in *Fig.* 8.9 shows such a part and indicates the design possibilities.

Thickness and Radii Limitations. Except in the case of materials which cannot be sawed rapidly with a satisfactory finish at conventional speeds, ordinary contour sawing methods are indicated for parts over 1-inch in thickness. Generally speaking, however, the heavier the section to be cut, the wider the blade must be to maintain the maximum economical linear feed. Where intricate curves must be produced with thicknesses heavier than those indicated in TABLE 8-I, some sacrifice in output speed results. Open-end automatic-rewind saws operating at speeds up to 180 fpm can traverse the most exacting and delicate contours using the narrowest width blades. Using high-velocity friction cutting, a blade of 3/16-inch width will produce almost equally exacting results in most plastics.

Typical aircraft parts which have been friction sawed in production from AISI 4140 high-tensile steel are shown in *Fig.* 8.10. Smaller radii than those shown in TABLE 8-I were called for on these parts but nevertheless in every case output per hour was about six times that possible at conventional speeds.

In the manufacture of motor mounts and other parts to be fabricated from tubing, contour sawing has been found most economical and, in certain cases, shapes can be cut that would be impractical by any other means. Friction cutting of

TABLE 8-I — Saw Width Effect on Gage and Radius of Turn
(inches)

Saw Width	Maximum Material Gage*	Minimum Radius†
1/16	Under 1/16	1/16
3/32	Under 1/16	1/8
1/8	Under 1/16	7/32
3/16	1/16	3/8
1/4	1/8	5/8
5/16	3/16	7/8
3/8	1/4	1
1/2	3/8	3
5/8	1/2	5
3/4	3/4	7
1	1	9

* Maximum material gage indicates a safe limit for production considerations primarily in friction sawing hard steels, high-tensile steel alloys, armor plate, and other extremely hard materials to assure sufficient blade strength for maximum feed. Materials too thick to bring to melting temperature at available blade speeds can often be cut by using a "rocking" technique.

† Minimum radii indicated hold for any thickness of nonferrous or softer materials conventional or friction sawed. Radii can in some cases be reduced further by careful handling. Use of the largest radius possible naturally allows widest blade with maximum linear cutting speed.

tubing is somewhat restricted by present equipment as is sheet and plate material. For most practical purposes maximum size of tube is limited to about one inch of tooth contact where the blade momentarily breaks through the inner tube wall. Thus if tube diameter is large, the wall section must be thin and conversely if the diameter is small, wall section can be heavier.

Automatic Jobs Restricted. Shapes produced on automatic contour friction saws with hydraulic feed which are available for production quantities, are duplicated by means of a template. Forms having radii not less than 12 inches can be cut fast, clean and square with the entire cut completely automatic.

Selection of Materials

Most of the nonferrous materials such as brass, bronze, aluminum, magnesium, copper, zinc alloys, plastics, etc., can be handled with satisfactory feeds at conventional saw speeds. Certain types of highly accurate work such as is carried out in the making of dies also can be handled most easily at conventional speeds. However, for production-quality work, most steels and steel alloys can be handled at extremely high feeds and with relative ease by high-velocity friction

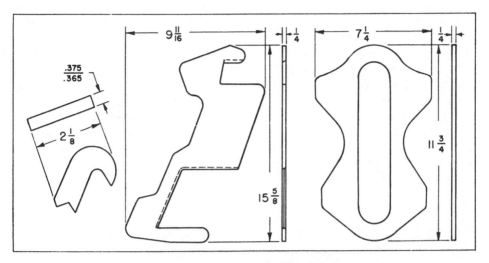

Fig. 8.10—Typical aircraft parts friction sawed from AISI 4140 high-tensile alloy steel. Output increased as much as 6 times over previous methods. Drawings, courtesy General Motors Corp.

sawing with a satisfactory finish. In fact, many metals that are difficult or impossible to saw by conventional methods — armor plate, stainless, hardened alloy and tool steels, etc — can also be cut with ease. Only those materials having extremely high melting points are exempted as are those with extremely low melting points which are prone to charring or excessive stickiness. Extremely brittle materials are as a rule not well suited — they usually shatter. Such materials as those, however, can be handled by means of diamond band cutting.

Aluminum in other than sheet form is not considered to be a suitable material for friction sawing. Finish is poor and welding or gumming sometimes causes difficulty with the heavier sections. Speeds around 3000 fpm produce optimum results in such cases. However, trimming formed shapes of aluminum and other sheet materials is done extensively in production by means of friction sawing and lack of downdrag as well as free cutting on such lightgage formed material makes this method ideal.

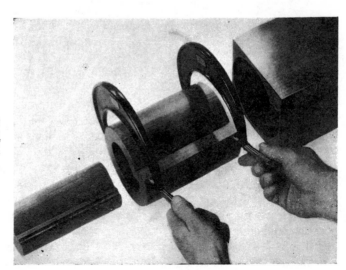

Fig. 8.11—Cam shape cut at conventional speeds shows a variation from end to end of only 0.002-inch. Photo, courtesy DoAll Co.

According to certain manufacturers, many plastics can be easily handled by first dulling the saw teeth by cutting up several old files or other hard material and then smoothing with a fine hard stone. Plastics can then be cut as fast as the material can be fed into the blade. Better finish is obtained with plastics than other materials. Certain of the newer plastics which cannot be cut with a carbide saw and those having glass, mica, and similar fillers can also be cut easily.

Plate glass can be cut by high-velocity sawing, but chipping and cracking often result. Heavy insulating glass block can be cut satisfactorily and pyrex glass will finish with practically no chipping. However, the amount of polishing necessary if a smooth edge is desirable is excessive and use of diamond band cutting is more desirable.

Fig. 8.12—Chrome-molyb-denum forging regularly run in production showing typical roughness and pattern left on friction-sawed surface. Photo, courtesy General Motors Corp.

Effect of Heat. Except in cases where heat penetration would be detrimental, with parts over one-inch in section thickness, torch cutting is the most economical method for reproducing contoured parts from ferrous materials. However, by high-velocity sawing of heat-sensitive steels or materials such as Hastelloy, subsequent machining is simplified since heat of friction sawing seldom penetrates to a depth greater than 0.002-inch. On certain alloys and high-carbon steels this hardening may run to as much as 0.005-inch in depth. It is good practice to anneal friction-sawed parts prior to profile milling or otherwise machining, in order to increase tool or cutter life.

Ferrous castings of any variety may be friction sawed efficiently. Section thicknesses are restricted as shown in TABLE 8-I.

Specification of Tolerances

Conventional contour sawing produces parts with comparatively clean, uniform surfaces, requiring but little finishing. Although feed is slow, results obtained on highly exacting contours are exceptional, especially on extremely intricate designs produced with open-end machines which obviate the welded blade joint. A layout line can be followed within 0.008 to 0.010-inch without much trouble. By the use of a magnifying arrangement, a scribed line can often be followed within 0.003-inch. Vertical accuracy with conventional sawing — important especially on stack cutting — is excellent. The 6-inch high cam shape shown in *Fig.* 8.11 cut by this method showed a variation from one end to the other of only 0.002-inch.

Tolerances maintained in friction sawing are about the same as those for general low-speed work. A good operator in production can hold the contour of a part to within 0.007 to 0.010-inch of the scribed line. In fact, on special jobs operators have been able to halve this figure, holding to around 0.010-inch overall variation, as against a normal 0.014 to 0.020-inch. Vertical runout is comparable to that of low-speed sawing.

Surface Roughness. In most cases of contour sawing work, the parts are finished finally by some other means such as milling, filing, grinding, or polishing. The surface finish in low-speed sawing may vary from pronounced shallow scratches to scarcely visible ones. In general, the surface left by the friction saw is comparable, *Fig.* 8.12, but in some cases is smoother, and, with plastics, almost polished.

A burr forms on the underside of the cut in friction sawing steels and some other metals, especially on those in the lower hardness ranges. The burr is not a problem and can be removed easily by filing. Burrs produced are thin and may project $1/32$ to $1/16$-inch below the cut. Also, with parts from softer materials which produce the larger burrs, cuts are often made over a bar support which raises the plate from the machine table, both facilitating manipulation and preventing rapid dispersion of the heat created in friction sawing.

Tolerances held in diamond sawing follow very closely those held in conventional sawing both in regard to dimension and surface finish.

9

Planing, Shaping and Slotting

BASIC among the important metal-removal methods are the closely related processes of shaping, planing and slotting. Primarily designed for producing flat or plane surfaces or related flat surfaces, the means by which metal is removed is fundamentally the same with each method, namely, with a reciprocating tool. However, distinct differences in the machine tools used with each method satisfy widely differing machining needs. Together, a broad field of design, not only in low but also high-production quantities, is directly concerned.

Production Quantities. Although shaping, planing and slotting work is in a great many cases confined to the production of one or several identical pieces, a tremendous number of high-production machine parts are also processed by each of these methods, *Fig.* 9.1. Typical of production parts produced on a shaper is that shown in *Fig.* 9.2. A shaft for a lathe, this part is produced in lots of 250 to 3000 pieces. The variety of parts which require the use of these machines ranges all the way from the very largest to extremely small ones.

With each of these methods straight, flat surfaces are produced by passing the tool over the work or by moving the work past the tool. Even though many such

Fig. 9.1 — Rough machining the base for a textile machine. Interrupted $^3/_8$-inch deep cut is made at 200 sfpm with Carboloy tools. Photo, courtesy Warner & Swasey Co.

surfaces can be produced by means of milling, broaching or grinding, each of these alternates require special tooling. Too, the simple, readily ground single-point tools for the shaper, planer and slotter can easily reach into small corners and out-of-the-way places not possible with other types of tooling. The capacity to remove larger amounts of metal in widely varying thicknesses makes them indispensable in

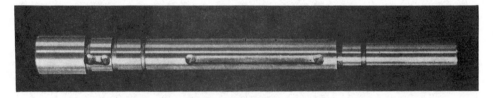

Fig. 9.2 — Production part for a lathe having a 1/2-inch long blind keyway cut on a hydraulic shaper. The keyways are cut in an average time of 30 seconds each. Photo, courtesy Rockford Machine Tool Co.

finishing castings and similar parts to size. Thus it is that these processes are widely used in the production of ships, locomotives, road machinery, engines, machine tools, and a tremendous variety of other machines and special equipment.

Shapers. The shaper is primarily intended for rough metal removal and finishing to precision flat or adjacent flat related surfaces on parts which can be readily handled with a fairly short stroke. Reciprocation of the ram on standard machines ranges from extremely short strokes, *Fig.* 9.3, of less than 1/2-inch to approximately 36 inches on the largest machine, *Fig.* 9.4. Parts requiring longer strokes must be relegated to the planer, which is somewhat slower in operation.

Production speed is not often an important item with shaping work but where quantities of duplicate parts are to be produced speed is dependent upon the amount of metal to be removed, setup time per piece, etc. Modern shapers are designed for speeds suitable for carbides and consequently are fast.

Although the shaper is best suited for producing precision flat surfaces, by proper use of the combined horizontal and vertical table feeds an unlimited variety

Fig. 9.3 — Hydraulic shaper cutting 1/2-inch long keyway on part shown in Fig. 9.2. Only 0.008-inch is allowed at blind shoulder at each end. Photo, courtesy Rockford Machine Tool Co.

of contours can be produced, *Fig.* 9.5. These, however, are not automatically produced but require manual control. Holes or portions of holes from 8 to approximately 18 inches in diameter can be finished automatically with a special attachment.

One of the many unusual production problems specially suited to the shaper is

Fig. 9.4 — A 36-inch stroke heavy-duty shaper finishing blind slots in a 10,000-pound steel machine base plate. Photo, courtesy Cincinnati Shaper Co.

Fig. 9.5 — Contouring a cam of irregular shape. Manufactured in small lots, the shape is finished by following a scribed outline with automatic cross feed and hand down feed. Photo, courtesy Cincinnati Shaper Co.

shown in *Fig.* 9.6. Such parts, usually required in small quantities, are difficult to finish by other means.

Planers. Suited primarily for producing flat surfaces on parts both too long and too cumbersome for the shaper, the planer covers a vast range of sizes. Almost any combination of width and height capacity is available. Too, stroke is practically unlimited with mechanical-drive models, *Fig.* 9.7. Hydraulic planers, of necessity, are designed with standard stroke lengths up to 20 feet. Table widths range from 2 feet to approximately 6 feet and capacity under the crossrail ranges from 2 feet to 6 feet. Larger special machines are employed, however, where heavy machine tool beds, *Fig.* 9.8, ship parts, and similar production is required.

Several styles are available, namely, openside planers, *Fig.* 9.8, and double-housing planers, *Fig.* 9.7. Openside planers are used primarily for machining parts which exceed the width limitations of the more rigid and stable double-housing planers. Whereas as many as four heads can be used on the double-housing, only three at the most are available for openside work. Some of the largest planers built operate as shapers and are known as pit planers. The work table is in a pit between two universal ways and the head carrying the tools travels over the work. One of the largest ever built is 42 feet wide by 18 feet high by 76 feet long. Great parts of

Fig. 9.6 — Special forged eccentric having a Kennedy keyway in a blind hole. Depth and width of keyway are held within 0.0001-inch. Photo, courtesy Cincinnati Shaper Co.

armor plate are machined on such planers because it is simpler to move the machine than the work.

Owing to the fact that most average jobs for the planer are heavy or cumbersome, notwithstanding usually being in rough-cast or rough-welded form, setup is difficult and often time consuming. Generally, work for the planer is conceded as being among the most difficult to set up for machining. This is of particular consequence when parts are in the small or medium-size range which necessitate setting up a number for planing at one time.

To improve output, "false" or setup tables are often used. Setup time is often as great or greater than the actual cutting time on many gang planing jobs and for this

Fig. 9.7 — Extra heavy-duty planer, 144 inches wide by 120 inches high by 30 feet long, finishing four heavy press beds at once. Photo, courtesy G. A. Gray Co.

Fig. 9.8 — Five large machine tool beds being finished on a 48-inch high by 16-foot long openside planer. Photo, courtesy G. A. Gray Co.

reason duplex-table planers have been developed. These machines have two tables to allow setup on one while the other is at work, eliminating idle planer time.

Shaper-Planers. Designed to fill the gap between the shaper and the planer, these machines provide shaper-like speed with planer accuracy. Made in three sizes (24 x 24, 32 x 24, 32 x 36) and five lengths ranging from 4 inches to 144 inches, these versatile machines are used for a tremendous variety of unusual work, *Fig.* 9.9.

Master forms mounted on the table can be reproduced by means of a duplicating attachment, *Fig.* 9.10. Complex shape production for unusual machine elements is virtually unlimited, *Fig.* 9.11, within the dimensional capacity of the machines. Where required, such equipment could undoubtedly be used with larger openside planers.

Slotters. Differing radically from shapers and planers, the slotter carries a vertical ram which, with its tool or tools, reciprocates up and down to remove metal. Offering as standard a rotating round table with longitudinal and transverse feed in lieu of a reciprocating one, a wide variety of work can be done, *Fig.* 9.12, especially with the new N/C type machines.

Standard strokes for slotters are 12, 20, 36, and 48 inches. Throat clearance on

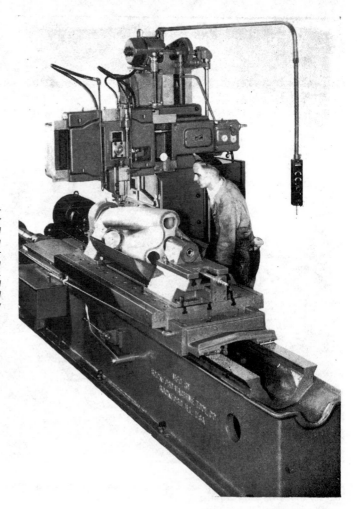

Fig. 9.9 — Largest shaper-planer with sine bar, head for generating lobed forms and special fixture, finishing diesel blower rotors. Generated lobe form is 19 inches in diameter by 27 inches long. Accurately finished and smooth, a rotor is completed in 4.6 hours. Photo, courtesy Rockford Machine Tool Co.

Fig. 9.10 — Master forms reproduced by means of a duplicating attachment on a shaper-planer. Photo, courtesy Rockford Machine Tool Co.

the 12-inch is 18 inches and maximum distance from the table to bottom of tool head is 30 inches. On the 20-inch this is 24 inches and 40 inches, on the 36-inch — 40 inches and 54 inches, and on the 48-inch — 46, 55 or 59 inches and 60, 69 or 73 inches, respectively.

Work for slotters is often of large size and can be classed as follows: (1) Work of irregular section where a clear pass over the face to be machined is difficult to set up on a planer, *Fig.* 9.13; (2) work requiring planing on internal sections such as splines, teeth or keyways, *Fig.* 9.14; and (3) work such as ratchet or gear rings which require primarily rotary feed, *Fig.* 9.15. Design possibilities are considerably increased by the fact that the ram may be tilted for travel at any angle from vertical to 10-degrees forward.

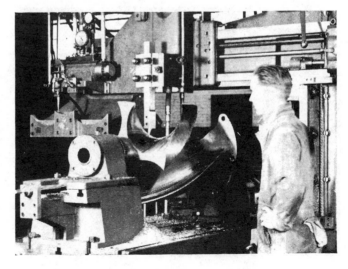

Fig. 9.11 — Shapes, which are extremely difficult to create by other means, can be produced rapidly and accurately on the shaper-planer. Photo, courtesy Rockford Machine Tool Co.

Fig. 9.12 — Morey Machinery, Inc., 5-axis manual and N/C shaper or slotter. The 15 hp drive permits cuts in hard materials at speeds up to 100 spm, strokes 3" to 24". N/C permits free form shaping without form cutters such components as turbine blades and elliptical gears.

Design Considerations

Design of parts which are to be finished on one or more major surfaces by means of shaping, planing, or slotting normally need not be restricted in many ways. The broad coverage as to size makes possible a tremendous size range. The primary observation, of course, is that each tool generates flat surfaces. Although special equipment is used with both shapers and planers, such is not readily available in all shops and this fact may deserve some consideration. Contours produced on the shaper or planer often require additional finishing for the desired

Fig. 9.13 — Steel casting set up for finishing on a 36-inch slotter. This part is a "natural" for the slotter, flat surfaces are 34 inches apart and 12$^{1}/_{4}$-inches high. Photo, courtesy Rockford Machine Tool Co.

smoothness or surface finish.

Design for Handling, Locating and Fastening. As can be seen in the accompnaying illustrations and as mentioned previously, the handling, locating and fastening of parts may often require more time than the actual surfacing. As few as four to as many as several dozen parts may be surfaced at once and the accompanying setup problems can be anticipated and simplified by the designer. Setup, locating and clamping for planing work, for instance, is probably the most fertile field of opportunity for increasing output. Even though the ingenious machinist, with a plentiful supply of all variety of clamps, etc., can manage to produce the part as desired, the results can be improved and cost decreased by forethought regarding these items. Locating pads, clamping brackets, bracing, stopping, measuring or clamping pieces etc., may be added with considerable cost justification. Where necessary, such pieces can be planed or cut off later. Holes for

Fig. 9.14 — Blind internal planing of ratchet teeth on a large, 0.55 carbon cast-steel gear for a ladle crane. Three types of hydraulic feed — longitudinal, transverse and rotary — were used in addition to the rotary motion for tooth spacing by the dividing head. Photo, courtesy Rockford Machine Tool Co.

Fig. 9.15 — Finishing four large external gears in one operation on a hydraulic slotter. Photo, courtesy Rockford Machine Tool Co.

lifting by chains, hooks, or clamps are often of value. The possibility of obtaining a stronger, more rigid design less costly to machine in two pieces rather than one, should also be investigated.

A typical planer setup is shown in *Fig.* 9.16 with the usual characteristic available equipment identified. Top and side of the casting are to be planed. Shaper parts, being smaller, seldom give any great holding problems. In the surfacing of long, thin parts on the planer, chisel points are usually used, *Fig.* 9.17. If cutting pressure is not too great, magnetic chucks can often be used. Taper gibs, using a wedge-shape support underneath, are normally finished on these chucks.

Tool Clearance. Inasmuch as table reversals on a planer are not often as precise or as instantaneous as on a shaper, adequate room for reversal of the tool should be contemplated. Where slides, pads or attachment surfaces are adjacent to an edge, wall or other interference, a relieved portion between should be allowed where no finishing is necessary. This relief can be several inches, or greater if possible, in width and of sufficient depth to clear all finished edges.

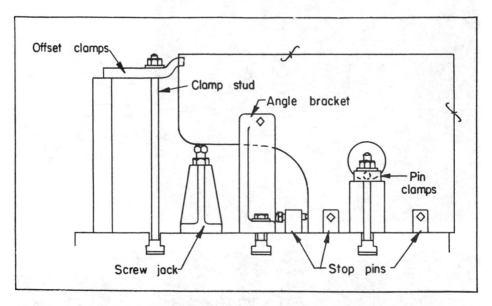

Fig. 9.16 — *End view of a typical workpiece set up for planing the top and right side simultaneously. Drawing, courtesy G. A. Gray Co.*

Rigid Parts for Economical Machining. To insure use of maximum possible finishing speeds, the parts should be designed for rigidity. Ability of the work to withstand the pressure of the cut often limits the feed and speed used. Again, the ability to hold the work to the table is a limiting factor. Strong, well-ribbed machine parts, bases, etc., allow use of carbide tools and surface planing speeds of 200 fpm, *Fig.* 9.1. Speeds as high as 315 fpm are often used, but the depth of cut and feed must be light — $1/16$-inch by $1/32$-inch, respectively. Generous ribbing and filleting should be used, the extra cost of the metal being more than offset by decreased machining time and improved accuracy of the part.

Simplifying Machining and Setup. As many surfaces as possible should be designed so as to fall in the same plane. In such cases all mounting pads can be completed at once without difficulty or special setup. Although apparently a minor item, pads are often located in such positions as to necessitate a considerable amount of special setup to allow the tools to clear. Pads in one plane, in addition, are simpler to check and gage.

Allowance for Finishing. Stock allowances normally left for shaping or planing operations vary with the part. Total depth of rough and finish cuts normally averages about 1/4-inch. Small parts or castings may require only about 1/8-inch of metal for machining to size whereas large castings such as engine beds, special machine frames, machine-tool housings and beds, etc., may require 3/4 to 1-inch for machining to assure a clean surface free from flaws or defects. Effect of stock allowed on speed of cutting can be seen in TABLE 9-I which shows average cutting speed relative to feed and depth of cut. General practice is to rough plane 1/4 to 3/8-inch average depth, semifinish 0.010 to 0.012-inch and finish with 0.001 to 0.003-inch depth of cut.

Selection of Materials

Practically any material that can be machined can be shaped, planed or slotted. The better the machinability rating, however, the more easily and

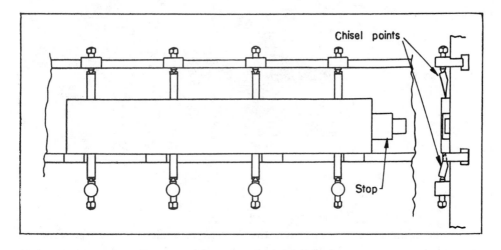

Fig. 9.17 — Long, thin plate set up for planing by means of chisel points. It is often necessary to turn the plate several times during finishing to obviate excessive warping. Drawing, courtesy G. A. Gray Co,

economically it can be worked, TABLE 9-I. The various materials require varying cutting speeds for fine finish, good tool life and maximum production. Use of carbide tools permits maximum cutting speed, better control of size and improved surface finish.

Free cutting steels and cast iron can be cut at up to 315 sfpm but bronzes and aluminum can be cut at maximum table speeds available regardless of feed or depth of cut. Copper can be finished to a precision flat surface with carbide tools with excellent results.

Specification of Tolerances

Close accuracy and uniformity are obtainable in shaping and planing work. With machines in good mechanical condition, tolerances possible depend to a great

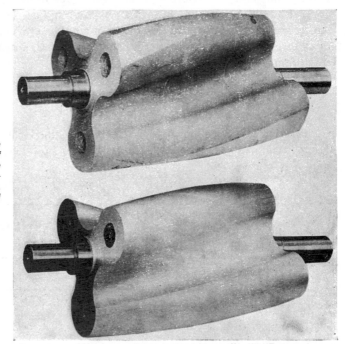

Fig. 9.18 — Diesel engine scavenging blower lobes of cast aluminum shown before (top) and after (bottom) finishing on the shaper-planer. Photo, courtesy Rockford Machine Tool Co.

extent upon the size of the piece and the operators. Precision flat surfaces ranging around plus or minus 0.005-inch with finish from 125 to 500 microinches are obtainable. On cast iron the finish is better and, in many cases, tolerances to plus or minus 0.001 to 0.002-inch with finish to 60 microinches can be had.

Shaper tolerances depend upon the design and its intricacy. Plus or minus 0.0001-inch to plus or minus 0.0005-inch, *Fig.* 9.3, can be held on small and medium dimensions. Those ranging up to the larger shaper parts require somewhat greater tolerances around plus or minus 0.001 to 0.002-inch, *Fig.* 9.4.

Similar accuracy can be held in shaper-planer and slotting work but more liberal tolerances naturally are more conducive to economy, especially on large parts, *Fig.* 9.7. The impeller for diesel engine scavenger blowers, *Fig.* 9.9, is shown before machining the contour on the shaper planer, *Fig.* 9.18 top, and after final finish *Fig.* 9.18, bottom. Surface finish quality is excellent and readily discernible as it is also on the rotor, *Fig.* 9.19.

Fig. 9.19 — *Templates mounted on the cross-rail of this shaper-planer guide the cutting tool while reciprocation generates the helix of this special rotor. Photo. courtesy Rockford Machine Tool Co.*

TABLE 9-1 — **Planing Speeds and Feeds for Various Materials†**

(surface feet per minute)

Tools	High-Speed Steel				Cast-Alloy				Carbides			
Depth of Cut	⅛	¼	½	1	⅛	¼	½	1	1/16	3/16	3/8	¾
Feed	1/32	1/16	3/32	⅛	1/32	1/16	3/32	⅛	1/32	3/32	1/16	1/16
Cast Iron (soft)	95	75	60	50	160	135	110	95	255	205	165	140
Cast Iron (medium)	70	55	45	35	125	105	90	75	205	165	135	110
Cast Iron (hard)	45	35	25	..	95	80	65	..	140	110	90	...
Steel (free cutting)	90	70	55	40	140	105	85	65	315	245	190	140
Steel (average)	70	55	40	30	105	80	60	45	270	205	160	120
Steel (low machinability) ...	40	30	25	..	65	50	40	..	195	145	115	...
Bronze	150	150	125	..	*	*	*	*	*	*	*	*
Aluminum	200	200	150	..	*	*	*	*	*	*	*	*

† Courtesy G. A. Gray Co. It is impossible to recommend definite speeds and feeds for all types of jobs. The size of the work often limits the speed and the feed which can be used. In other cases the clamping of the work on the table or in a fixture may be the limiting factor. The ability of the work to withstand the pressure of the cut often limits the amount of feed which can be used. The power of the drive motor may be the limiting factor when heavy cuts are taken with several tools at one time. The speeds and feeds listed represent approximate values. In actual practice the best speed and feed for the job must be determined by trial.

* Maximum table speed possible.

10

Automatic and Shape Turning

PLAIN turning operations, according to survey results, constitute one of the primary machining functions employed in manufacture. The useful and highly desirable attributes of the engine lathe — which generally have been supplanted by those of the turret lathe and automatic bar and chucking machine wherever possible because of production economies — today are available in a wide variety of automatic lathes. All the advantages of single-point tooling for maximum metal removal, finish accuracy, center turning, etc., are now at the designers fingertips, with production speeds on a par with the fastest processing equipment on the scene today.

Fig. 10.1 — Model LR-250 Seneca Falls mechanical cam operated production lathe set up for turning the ends of electric motor shafts with automatic magazine loading of the gravity type.

Machine Classifications. Turning equipment of this nature can be roughly subdivided into several classifications. Fully automatic types with automatic magazine or walking-beam feeds, *Fig.* 10.1, and semiautomatic hand-loaded types, *Fig.* 10.2, constitute the first category while full automatic N/C lathes, *Fig.* 10.3, and semi-automatic tracers or shape turners, *Fig.* 10.4, constitute the second. N/C lathes are available for chucking and center work, *Fig.* 10.5, with automatic bar

Fig. 10.2 — Model LR-250 Seneca Falls mechanical cam operated production lathe set up for turning, facing ends and chamfering ID of brake system cylinder housings.

feeds; and some models are virtually machining centers. Tracer lathes are available which use either a flat template for reproduction or an actual pilot model workpiece. Automatic lathes in the first group are available in a variety of standard or special models with mechanical, electrical or hydraulic actuation.

Automatic-Cycle Lathes. Employing primarily only simple, single-point turning operations, automatic-cycle lathes are ideally suited to the production of a wide variety of machine parts such as automotive and aircraft pistons, aircraft engine cylinders, transmission cluster gear blanks and shafts, camshafts, bearing races, spur and bevel gear blanks, etc. Unlike the turret lathe and automatic screw machine, these lathes are well adapted to the machining of forgings, castings and bars held between centers, or work supported on fixtures as well as ordinary chuck work. Plain bores on many parts, preferably for accuracy in chucking, are best finished in a preliminary operation on a turret lathe or by drilling and broaching.

Special bores or tapers, however, can often be more readily produced on automatic N/C lathes, *Fig.* 10.6.

Scope of Process. In general, parts which fall within the scope of automatic lathes would be:

1. Those which because of length or special requirements as to accuracy and finish can only be turned satisfactorily on centers, including work turned in preparation for grinding or other machining requiring the use of centers.
2. Those which because of having been forged, cast or welded cannot be cut from a bar or fed through a hollow spindle.
3. Forgings, castings, etc., which, on account of irregular shape, cannot be, readily held in chucking machines.
4. Those which require machining all over or at both ends.
5. Those parts which cannot be chucked and lend themselves to stacking for magazine feed.
6. Those parts required in small quantity not within the economical range of automatic bar and chucking machines.

Where parts are long and require machining at both ends, a center drive attachment can be employed which acts both as steady rest and driver. Large, long parts which require machining all over are usually driven by spur drives, *Fig.* 10.7.

Fig. 10.3 — Versatile N/C Turning Center which can swing a nominal 32 inch diameter and 72 inches between centers. A General Electric 7542 N/C control is used with operator's station on the carriage. Photo, courtesy American Tool Works.

Size Range of Machines. Automatic lathes range in size from small units designed to handle work 1/2 inch to 1-3/8 inches in diameter by 2-1/2 inches to 18 inches in length, to units capable of handling diameters up to 84 inches and lengths up to 10 feet or more. These machines are powered by motors from 5 hp on the small units up to 125 hp or more on the larger machines, with the full advantage of carbide tooling wherever practicable and maximum possible production speed. Automatic tool relief at the end of each tool stroke avoids scraping of the tools, increasing tool life and providing better finish.

Fig. 10.4 — Semiautomatic hydraulic tracer reproduces a variety of designs using an actual part as a pattern. Photo, courtesy American Tool Works.

Fig. 10.5 — Tape Turn II arranged for between centers operations on complex shaft designs. Tool path travel is compensated so as to hold ±0.0005 inch tolerance. Photo, courtesy LeBlond, Inc.

Machine Limitations. Although the lengths which can be turned or shaped are necessarily limited by the carriage travel of specific models of any of these machines, long lengths of turn can be handled by special machines or by multiple tooling. Truck steering knuckles, for instance, are produced to 0.9 minute or 66 per hour, *Fig.* 10.8. Large draw nuts, *Fig.* 10.9, are produced on a Tape Turn CX from a rough blank in 13.2 minutes vs. a previous time of 22.5 minutes. Setup time is only a fraction of that for a regular automatic chucking lathe.

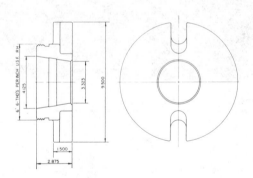

PART	MATERIAL	OPERATION N°	SET—UP		CUTTING TIME		N° OF OPERATION	
			PREV.	CX	PREV.	CX	PREV.	CX
FACE PLATE	CI	SIDE 1	330 MIN	5 MIN	10.4 MIN	6.4 MIN	2	1
		SIDE 2	261 MIN	5 MIN	12.1 MIN	9.1 MIN	2	1
PREVIOUS MACHINING: AUTOMATIC CHUCKING LATHE, THREAD MILL AND ENGINE LATHE								
NEW MACHINE: TAPE TURN CX								

Fig. 10.6 — Cast iron face plate with threads and special taper produced on Tape Turn CX in 15.5 minutes with two operations compared with conventional method requiring 22.5 minutes and 4 operations. Setup was reduced from 591 minutes to 10 minutes. Courtesy LeBlond, Inc.

Changeover of automatic lathes is relatively simple, few models requiring any new cams or equipment beyond the tooling for each new job. Setup time is short and flexibility is sufficient to encompass a broad field of design. In many cases design changes and improvements on parts being run involve little or no cost over setup.

Automatic tracers. Similar to the automatic-cycle lathes, automatic-tracer lathes employ single-point tooling but, ordinarily, only one tool. Primary

Fig. 10.7 — One of the largest of the automatic loading hydraulic lathes, this 150-hp axle lathe turns both ends simultaneously and drives the part by means of a spur center hydraulically loaded. Courtesy, Morey Machinery, Inc.

limitation to design is that all details must be possible to reach with a sharp-point tool. In general, parts which are considered suitable for obtaining the advantages of the automatic tracer are somewhat similar to those mentioned previously but usually contain a complicated form, taper, or other details which cannot be reproduced except with complicated banks of tools or special attachments. Characteristic parts are valve plugs, valve bodies, nozzles, orifices, stepped shafts, contoured rolls, etc.

Scope of Tracers. Employing low-cost templates for reproduction, automatic tracers are ideally adapted to production quantities ranging from only a few pieces to large-quantity output. Compressor crankshaft, *Fig.* 10.10, has five stepped diameters, two-tapers, radii, and shoulders turned in 24 seconds from the rough castings. Automatic tracer lathes range in sizes paralleling those of the automatic cycle lathes and offer a broad coverage of machine parts.

Automatic Shape Turning. Equipment such as this offers possibilities for design which cannot be obtained by any other means. These machines often find use in the efficient, economical production of parts for some of the newer, more complex mechanisms, *Figs.* 10.11 and 10.12.

N/C Lathes and Machining Centers. Today, the most versatile automatic lathe is the numerically controlled machine. Changeover is rendered simple and tooling arrangements offer everything from the single-point cross slide tool to entirely automatic tool changers wherein as many as 16 tools in a magazine are available. Thus, without transfer of the piece part, hole patterns can be drilled, keyways milled, steps milled, or boring, reaming and tapping perpendicular or parallel to the centerline done. These machines fill the production gap where small to moderate lots of parts of some complexity are required.[1]

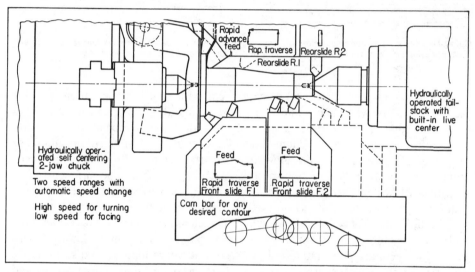

Fig. 10.8 — Layout of tooling for steering knuckles. Production speed of 0.9-minute floor-to-floor with 0.6-minute machining time is attained. Drawing, courtesy Morey Machinery, Inc.

Design Considerations

Parts which are "naturals" for automatic lathe production, because of one or more of the various factors outlined previously, can be designed to assure minimum production cost by observing a few basic rules. Naturally, production time is controlled to a great extent by the amount of turning and consequently the turned portions of a part should be kept to a bare minimum wherever possible to do so.

Handling. Second controlling factor is handling time. Owing to speed of turning operations, parts should always be considered from the possibility of hopper or chute feeds and automatic loading to attain maximum production. Such parts, of course, should be symmetrical to a degree, offer few problems as to orientation and be easy to chuck, grip with a collet, hold on an expanding mandrel, or center.

Chucking. Economical design dictates easy chucking or holding for all parts as well. Complex holding devices often slow down operations drastically. Because

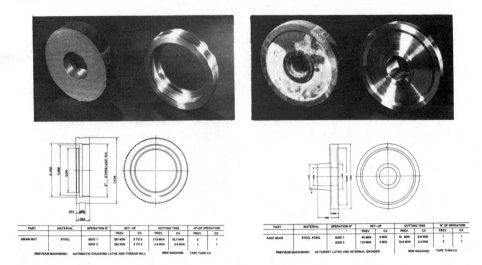

PART	MATERIAL	OPERATION Nº	SET-UP		CUTTING TIME		Nº OF OPERATION	
			PREV.	CX	PREV.	CX	PREV.	CX
DRAW NUT	STEEL	SIDE 1	297 MIN	3 TO 5	17.9 MIN	10.7 MIN	2	1
		SIDE 2	234 MIN	3 TO 5	4.8 MIN	2.8 MIN	1	1

PREVIOUS MACHINING: AUTOMATIC CHUCKING LATHE AND THREAD MILL   NEW MACHINE: TAPE TURN CX

PART	MATERIAL	OPERATION Nº	SET-UP		CUTTING TIME		Nº OF OPERATION	
			PREV.	CX	PREV.	CX	PREV.	CX
FACE GEAR	STEEL FORG.	SIDE 1	44 MIN	3 MIN	10 MIN	5.9 MIN	1	1
		SIDE 2	113 MIN	3 MIN	12.5 MIN	4.3 MIN	2	1

PREVIOUS MACHINING: #3 TURRET LATHE AND INTERNAL GRINDER   NEW MACHINE: TAPE TURN CX

Fig. 10.9 — Draw nut and face gear designs produced on a Tape Turn CX. Secondary operations requiring thread milling or grinding are easily handled with N/C machines. Illustrations, courtesy LeBlond, Inc.

an operator often handles several machines, chucking time is at a premium and this factor should always be kept in mind. Light or thin-wall parts often cannot be held by conventional means without distortion and unless some other means is readily adaptable, such as perhaps vacuum chucking, turning at optimum speeds and feeds may not be practicable. Again, long slender parts may not allow maximum feeds and speeds and where production speed is a factor, parts should be sufficiently stiff to withstand heavy cutting pressures. Experience has proved that the extra metal required for adequate stiffness is more than compensated for in production savings.

Complicated Forms. Wide forming tools such as are common with turret lathe and screw machine production are not often employed on automatic lathes. Forms suitable for a single-point lathe tool are used, but where long formed sections are required single-point generation is used. Generation beyond simple tapers and curves is generally not suitable on ordinary automatic lathes, but where complex generated forms are required the automatic tracer lathe or the N/C lathe must be considered for production of the parts.

Fig. 10.10 — Requiring no rough-turn operation, this automatic tracer completes a crankshaft in 24 seconds. Photo, courtesy Monarch Machine Tool Co.

Fig. 10.11 — Helical pump rotor and screws can be produced on the automatic shape turner. Courtesy, Myers Tool Co.

Fig. 10.12 — Lindsinger shell end milling equipment used to produce the screw shapes of Fig. 10.11. Courtesy, Myers Tool Co.

Corner Radii. Square shoulders and wide faces pose few problems owing to use of single-point tooling and automatic tool relief on return. Likewise, relatively sharp corners can be produced but to favor tool life and production, corner radii should generally be not less than 0.005-inch and preferably should be somewhat greater.

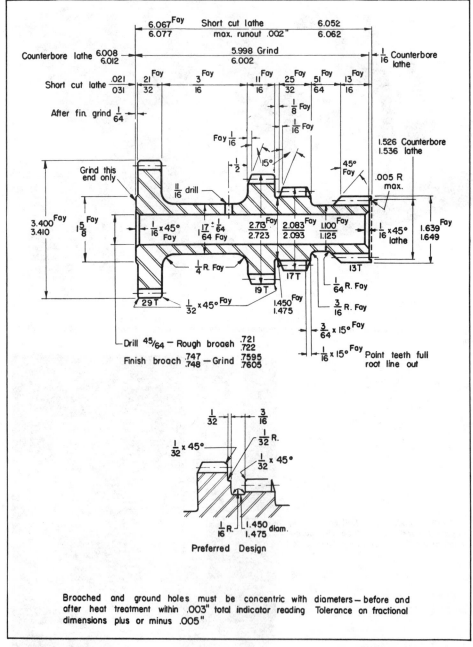

Fig. 10.13 — Transmission cluster gear as designed for automatic lathe production and a suggested improvement for longer tool life. Dimensions marked "Fay" indicate those to be finished on the Fay automatic. Drawing, courtesy Jones & Lamson Machine Co.

Undercuts. Necking or undercutting often presents a problem in automatic lathe production. As designed, the cluster gear, *Fig.* 10.13, requires that the undercutting tool face as well as neck. To prolong the usefulness of the shortlived necking tool, the alternate design shown, which relieves the tool of this extra duty, is preferred. Function is not altered and especially since the neck is only for purpose of clearance in gear cutting, the odd-shape design is best eliminated.

Bores. Quality of finish is important and at maximum production speed fine finish is not always possible. Small bores, for instance, usually cannot be produced without a mark when the tool is retracted. Tool relief overcomes this but cannot be used generally on bores under about 3 inches. Consequently, unless a tool mark is permissible, two operations must be employed, increasing costs and handling. Stepped or complicated cylindrical bores should be considered for one of the precision boring machines.

Finish. On external parts where maximum production is subordinate to quality of finish special attention to tooling can usually provide the desired results. The same situation is prevalent, however, in that fine finish usually requires two operations to assure elimination of tool marks. Occasionally, rough and finish turning operations can be done by proper tooling in one setup if the work permits the heavy cutting loads.

On sections which are to be finish ground after turning, only a rough turn is required and coarse feeds are desirable to reduce the abrasive action between the tools and the work. Ordinarily, about 0.015-inch is allowed for the grinding operation and a satisfactory rough turn can have a surface finish of about 400 microinches. Where finer finish is used, grinding allowances can be reduced on centered work to an amount just sufficient for removal of tool marks and, if

Fig. 10.14 — Diesel engine injector body showing quality of surface finish obtained in average automatic turning with a single operation and multiple tools.

Fig. 10.15 — Inner and outer ball bearing races showing the high quality surface finish attained in finish turning operations.

hardening is involved, distortions. Careful design can thus make substantial savings in grinding time. On those portions of parts having no precise fit for mating, a fairly good finish should also be specified, possibly from 75 to 150 microinches.

Selection of Materials

As with other turning operations, high machinability rating is an all-important factor to be taken into consideration when specifying materials. A fairly complete table of machinability ratings is contained in Chapter 68 and to insure maximum tool life and good finish, materials with the highest possible machinability rating should be favored.

Specification of Tolerances

Surface finish depends greatly upon the material turned, tooling, and speeds and feeds employed. As mentioned previously, on rough-turn operations surface finish may run to 400 or more microinches with coarse feeds down to possibly 75 to 150 with finer feeds, *Fig.* 10.14. For light finish turning operations on work such as bronze bushings, small cast-iron parts, or bearings races, *Fig.* 10.15, which require a fine finish with a polished appearance, surface quality around 30 microinches can be obtained. The railway axles in *Fig.* 10.7, for instance, must meet rigid finish requirements. The journal bearings are burnished after turning and no tool marks or chatters can be tolerated. Too, finish on the wheel seat must be suitable for press fitting and no burnishing by the tool can be permitted.

Attainable Tolerances. Extremely close dimensional tolerances can be attained readily when desirable. Truck, *Fig.* 10.8, or automotive steering knuckles, for example, are held to plus or minus 0.002-inch in continuous production using but one cut. Automotive and aircraft pistons are produced with groove widths held within plus or minus 0.0002-inch. Bores drilled and single-point finished can be held to plus or minus 0.0005-inch.

Suggested Tolerances. In general, on high-production runs where maximum output is desirable, a tolerance of plus or minus 0.0005-inch is desirable on both diameter and length of turn. Where no fit or succeeding operation requires otherwise, tolerances of plus or minus 1/64-inch are conducive to low cost and high output speed. Where design necessitates, tolerances can be held to plus or minus 0.001-inch and on occasion plus or minus 0.0002-inch to 0.0005-inch.

Concentricities. Concentricities normally can be held to close limits, especially when centers or precision bores are concerned. Ordinarily concentricity will be within a total dial reading of 0.002 to 0.003-inch, maximum.

Tracers and N/C Machines. Accuracy of work produced on automatic tracers parallels closely that of the automatic lathes and in many cases can be relied upon to produce complex contours with even smaller part-to-part variation. Compressor crankshafts, for purpose of illustration, on lots of 2000 were within a maximum error of 0.0005 inch on all diameters. Like accuracies are characteristic of the shape turners. Pump rotors are turned within plus or minus 0.001 inch on diameter and lead of the helix is such as to provide a medium fit with the bore of the pump body produced on the same type machine. N/C machines and machining centers easily hold such tolerances. Finish is in the low microinch (32 rms) area; and since repeat accuracy falls into the area that normally calls for grinding, savings can be significant, *Fig.* 10.9.

REFERENCES

¹For a complete study of the application of numerical control principles to these and other machining operations, see *Production and Numerical Control* by William C. Leone, The Ronald Press Co., New York.

11

Turret Lathe Machining

PRODUCTION machining equipment must be evaluated more than ever before in terms of "ability to repeat accurately and rapidly". Applying this criterion for establishing the production qualifications of a specific method, the turret lathe merits a high rating. A natural outgrowth of the old, well-known engine lathe, the turret lathe can handle either bar or chucking work. Although adapted to produce parts in quantities too limited to be economically produced on the automatic screw machine, the turret lathe is primarily suited for parts not within the capacity of the screw machine, *Fig.* 11.1. These might be unusually shaped or heavy forgings or castings requiring large holding fixtures such as shown in *Fig.* 11.2, long complicated shafts, parts where excessively heavy cuts are necessary, and complicated designs with a large number of finished surfaces.

Basic Operations. Basic turret lathe operations and their general sequence of occurrence are illustrated in *Fig.* 11.3. Combinations and specialized variations of these operations are used to provide rapid, economical production.

Machines and Capacities. Made as hand-operated, semiautomatic or N/C machines, turret lathes fall into two classes — horizontal and vertical. Horizontal machines can be either bar or chucking type with ram or saddle style turrets. Ram

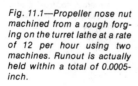

Fig. 11.1—Propeller nose nut machined from a rough forging on the turret lathe at a rate of 12 per hour using two machines. Runout is actually held within a total of 0.0005-inch.

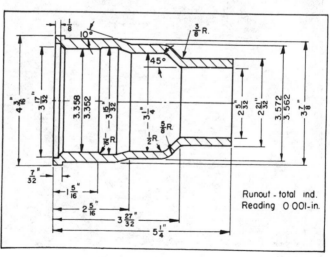

style machines have the shortest travel, are lightest in construction and suited primarily for handling bar work or light chucking jobs where ram overhang can be kept short, *Fig.* 11.4. Fastest in operation and designed to turn at high speeds, ram style machines can handle with maximum economy bar work from approximately $5/8$-inch diameter by 4 inches in length up to as high as $2^1/2$-inch diameters with turret cuts up to 14 inches and side-carriage cuts up to 24 inches. Chucking work up to about 15 inches in diameter can be handled (nominal swing may run to 20 inches over bed). Maximum travel at the square turret cross slide is about 10 inches.

The longer stroke, more rigid construction and side-hung carriage of the saddle style machine makes it more suitable for longer and heavier chucking or bar work which requires long turning and boring cuts, *Fig.* 11.5. Available with a cross sliding turret, the saddle style machine is used for facing deep holes in chuck work where tool overhang would be excessive with other arrangements. It is well adapted for producing internal tapers, contours and threads. Handling capacities may run as high as 12-inch diameters with either turret or side-carriage cuts up to 93 inches for bar work, and up to 32-inch diameters with a maximum swing of approximately $36^1/2$ inches over the bed for chucking work. Maximum cross travel of the square turret is about 16 inches and that of the hex turret, 12 inches.

The vertical turret lathe is similar in principle to the horizontal but is designed for considerably larger and heavier work. Not in any way adapted to bar work, machines of this type are designed for and will handle complicated chucking work up to 84 inches in length with a facing capacity of about 96 inches, *Fig.*11.6.

A number of special machines are in use and of these some are larger in capacity and others, such as the electric turret lathe, much smaller. One of the newest machines is the numerically controlled turret lathe, *Fig* 11.7. Two-axis machines provide the maximum in economy for medium to large-diameter chucking work, and 4-axis machines offer real economy on long bar parts and shaft jobs. Many jobs can be run with merely a change of chuck jaws or collet and a tape. Newer N/C models feature radically different turret design, *Fig.* 11.8.

Fig. 11.2—Turret lathe tooled up for production of a solid cast-steel valve wedge. Photo, courtesy Warner & Swasey Co.

Production Quantities. Large-quantity manufacture, usually requisite for maximum economy in most methods of production, is not always a necessity. Part adaptability — design for production with minimum tooling and set-up costs — plays the leading role in economical and efficient production.

Since tooling and set-up costs must necessarily be divided between the number of pieces to be produced, small and medium-quantity lots require the maximum in design consideration. These costs in low production are, roughly speaking, inversely proportional to the ease with which the part design can be reproduced or sized by standard tools using standard tool holders and work-holding devices. Small lots (3 to 15 pieces) seldom can justify any special tools or holding fixtures. Naturally, medium-lot work (15 to 100 pieces) allows a considerably wider range in setup to reduce production time per piece, but standard tools and holders are most economical wherever design will permit their use rather than the more costly special ones.

On continuous or high-production runs, tooling cost becomes an insignificant part of the total cost per piece. Consequently special tools, cutters, clamps, holding fixtures, etc., can be utilized to reduce machining, handling and indexing time. Where unusually shaped pieces are involved, entirely special units are often designed for the job. Bulky steel valve forgings are handled in special chucks, *Fig.* 11.9, which allow indexing the part to a number of different positions in order to finish the bore and flange faces, machine and thread the angular seat portions, insert bronze seat rings, and finish machine the seat inserts at the proper angle all at one chucking.

Certain parts, often possible to machine by other methods, can be produced more economically on the turret lathe owing to the secondary operations which are practicable. The stainless-steel control valve seat, *Fig.* 11.10, is such a part. Owing to the fact that the cross-hole had to be located accurately in relation to one face and be perfectly straight, a jig was necessary. Readily adapted to carry both the necessary jig and an auxiliary drilling motor, the conventional turret lathe provides for completion of the part at one handling.

Design Considerations

Whenever quantities of individual parts are limited but a great many pieces of a similar nature are necessary, designing for identical tooling can reduce costs effectively. Following this procedure, the cast-iron valve pistons, *Fig.* 11.11, make an ideal turret lathe set-up. Identical design details on all the pistons facilitate tooling and production.

Finished Surfaces. A good objective for the designer to keep in mind during the evolution of a machine part is design for a minimum number of finishing operations. Machining to size consumes the greatest amount of production time and consequently reduction in the amount of metal to be removed will help considerably to reduce costs. This is true especially with frail parts having thin sections which must, as a rule, be finished with extremely light cuts to avoid distortion. Likewise, it follows naturally that the simplest of shapes require the least number of operations and are the easiest to reproduce. Cup-shape sections with flat bottoms are often difficult to machine. Tooling and machining can be considerably improved if a through hole can be provided to offer a place to readily start the bottom facing cut.

BASIC INTERNAL OPERATIONS

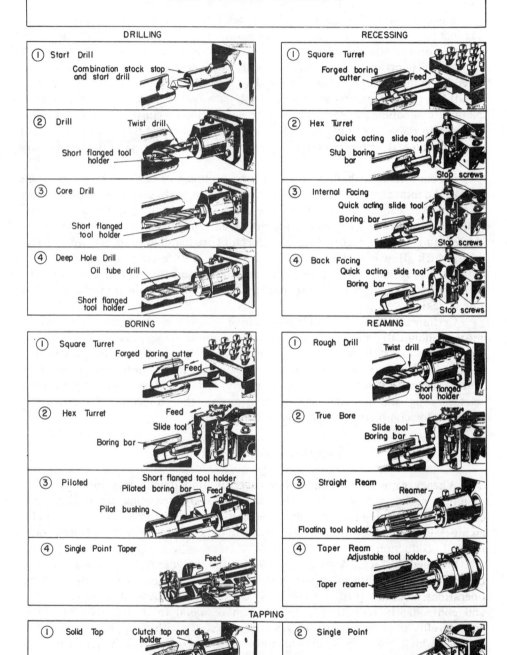

DRILLING

① Start Drill
Combination stock stop and start drill

② Drill — Twist drill
Short flanged tool holder

③ Core Drill
Short flanged tool holder

④ Deep Hole Drill
Oil tube drill
Short flanged tool holder

RECESSING

① Square Turret
Forged boring cutter — Feed

② Hex Turret
Quick acting slide tool
Stub boring bar
Stop screws

③ Internal Facing
Quick acting slide tool
Boring bar
Stop screws

④ Back Facing
Quick acting slide tool
Boring bar
Stop screws

BORING

① Square Turret
Forged boring cutter — Feed

② Hex Turret — Feed
Slide tool
Boring bar

③ Piloted
Short flanged tool holder
Piloted boring bar — Feed
Pilot bushing

④ Single Point Taper
Feed

REAMING

① Rough Drill — Twist drill
Short flanged tool holder

② True Bore
Slide tool
Boring bar

③ Straight Ream
Reamer
Floating tool holder

④ Taper Ream
Adjustable tool holder
Taper reamer

TAPPING

① Solid Tap
Clutch tap and die holder

② Single Point

Drawings, courtesy Warner & Swasey Co.

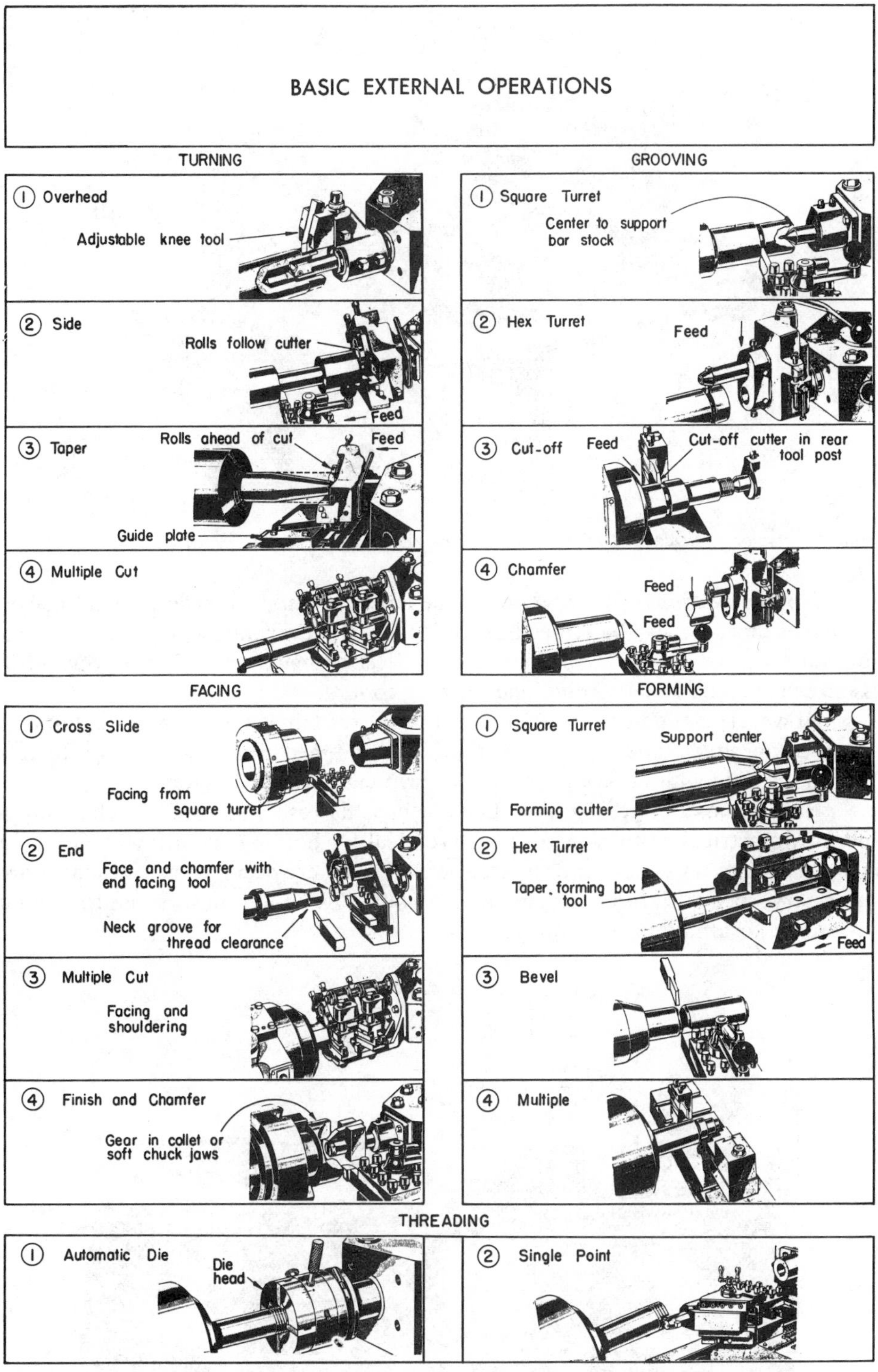

BASIC EXTERNAL OPERATIONS

TURNING

1. Overhead

   Adjustable knee tool

2. Side

   Rolls follow cutter

   Feed

3. Taper

   Rolls ahead of cut   Feed

   Guide plate

4. Multiple Cut

GROOVING

1. Square Turret

   Center to support bar stock

2. Hex Turret

   Feed

3. Cut-off   Feed   Cut-off cutter in rear tool post

4. Chamfer

   Feed

   Feed

FACING

1. Cross Slide

   Facing from square turret

2. End

   Face and chamfer with end facing tool

   Neck groove for thread clearance

3. Multiple Cut

   Facing and shouldering

4. Finish and Chamfer

   Gear in collet or soft chuck jaws

FORMING

1. Square Turret

   Support center

   Forming cutter

2. Hex Turret

   Taper forming box tool

   Feed

3. Bevel

4. Multiple

THREADING

1. Automatic Die

   Die head

2. Single Point

Fig. 11.3—Group of illustrations which show the basic internal and external turret-lathe operations performed on conventional machines. Many variations of these operations are used and many additional special-purpose tooling arrangements are used.

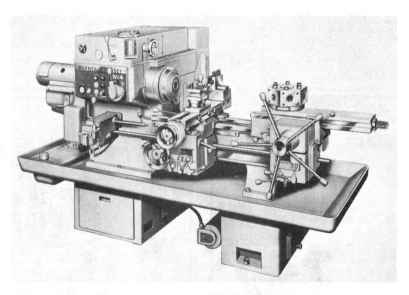

Fig. 11.4—Typical ram-type turret lathe. Photo, courtesy Warner & Swasey Co.

Formed Sections. Forming, one of the fastest methods of producing a finished diameter or shape, can be utilized to advantage wherever the length of forming cut does not exceed about 2-1/2 times the smallest diameter of the work, *Fig.* 11.3. Longer cuts require more than one forming tool.

However, if the numerically controlled turret lathe is available, single-point tools can produce tapers, contours, radii, or combined contours from two turrets simultaneously without need for special form cutters or templates.

Holding and Chucking. Another objective is ease and speed in chucking a part. Symmetrical castings or forgings are ideal for holding and allow the use of a scroll chuck or universal air or hydraulic chuck. Where peculiar or frail shapes are necessary, special fixtures or gripping methods, *Fig.* 11.12, must be used to avoid distortion, crushing or movement.

Fig. 11.5—Saddle type turret lathe is designed to machine longer and heavier parts than the ram type. Photo, courtesy Warner & Swasey Co.

Handling and Gripping. Designing for ease and speed in gripping work is of great importance in production, applications. Any handling or loading motion saved is directly multiplied by the number of pieces to be produced. Parts such as five-spoke wheels or pulleys are difficult to grip, therefore if at all possible either four or six spokes should be used. Likewise, hole circles with odd numbers of holes are often costly. Even numbers of holes can usually be producted with multiple drills in several indexings. The hand-hole cover, *Fig.* 11.13, as designed allows little leeway in chucking. Redesigning to provide for internal gripping solves the problem.

So-called chucking extensions are often required in order to provide a practical method of holding. The cast steel fork, *Fig.* 11.14, is an example showing the use of chucking extensions. These are removed after machining if necessary. Such chucking extensions or, in certain cases, flanges are used on thin-wall parts to eliminate distortion. In many cases, provision of an extension makes possible completion of the part before cutoff, thus speeding production *Fig.* 11.15. Parts which form a portion of a circle sometimes can be produced in multiples to increase output and simplify chucking. The type bar segment, *Fig.* 11.16, designed as a single-piece casting presented both chucking and machining problems. Cast in pairs, however, the blanks are easy to hold. Machining and accuracy were

Fig. 11.6—Giddings & Lewis VTL Toolchanger N/C lathe. The G & L toolchanger can handle tools up to 15 inches diameter and 20 inches in length.

Fig. 11.7—SC-28 2-axis numerically controlled turret lathe. The 4-axis machine permits contouring from both turrets simultaneously. Courtesy Warner & Swasey Co.

improved and production rate was doubled.

Special Operations. Production economy is possible through the utilization of another feature common to the turret lathe — assembly and finish machining of bulky work at one handling. Parts requiring a combination of dissimilar metals finished all over can be handled easily. A bronze seat ring on the cast iron disk shown in *Fig.* 11.17 is rolled into a machined locking groove under 20,000 pounds pressure and finally finished all over before removal from the machine. Assembly and rolling in of the bronze ring is performed as merely one of the normal turret operations. Broaching from the hex turret has been attempted and is fairly successful with free machining materials such as brass or magnesium. However, it is nominally a hand operation and consequently is not recommended for inclusion in high-production work.

Fig. 11.8—Cincinnati N/C Turning Center features crown turret with 8 tool positions. Universal models have additional front turret.

Fig. 11.9—Special indexing chucks provide for economical production of odd-shaped valve bodies which require accurate, related surfaces. Photo, courtesy Warner & Swasey Co.

Knurls. Knurls which are generally required can be produced readily but are for the most part limited to bar work. Open-end knurls are most economical and are produced with standard adjustable tools. Proportions of such knurled portions are limited to approximately $2^1/_4$ inches in length on bar $3^3/_4$ inches in diameter (maximum). Standard pitches available range from 10 to 34 teeth per inch in any one of the common straight, diamond or spiral varieties of knurled surface.

Finish. Quality of finish is important. If extreme accuracy or fine finish is necessary the production time may be increased. Fine feeds are required to produce a high-quality finish. If parts are to be finished by grinding, heavy cuts and coarse feeds can be used with good results. With tough materials, fine surface finish may necessitate several finish cuts and in the case of threading, a roughing and a finishing die head.

Drilling. Provided the stock is fairly homogeneous and without hard spots, drills are machine sharpened, and feeds are moderate, holes reasonably straight

Fig. 11.10—A swinging jig provides for completion of an accurate and straight cross-hole, eliminating a more costly separate second operation. Details of the part are shown below.

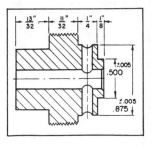

Fig. 11.11—Designed for identical tooling, these various valve pistons can be produced economically in small quantities.

and round can be produced in depths up to 5 times the diameter. Deeper holes may not be sufficiently accurate unless a subsequent reaming or single-point boring operation can be used. Stock allowance for reamed holes generally is 0.010-inch for a $1/4$-inch hole, 0.015-inch for a $1/2$-inch hole, graduating on up to about 0.025-inch for a $1 1/2$-inch hole. Insufficient stock for reaming may result in burnishing rather than cutting, which is undesirable.

Threads. Die-cut threads are usually preferred to those made with a single-point tool especially when speed and production is imperative unless an N/C machine is used. Consequently, it is good practice to avoid parts with threads too long for standard collapsing taps or self-opening dies, not within their standard range or in inaccessible positions. If extremely accurate threads must be produced, the taps or dies can be led on to the work by a lead-on attachment for threads from 4 to 32-pitch, thereby retaining the production advantages of automatic tools with

Fig. 11.12—Odd-shape casting requires the use of special chuck jaws to provide for proper holding. Photo, courtesy Warner & Swasey Co.

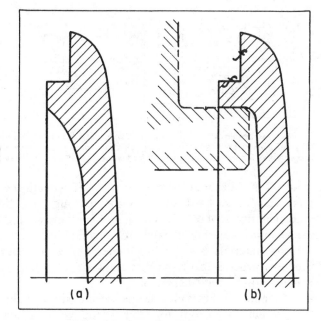

Fig. 11.13—Hand holes cover (a) redesigned as at (b) allows for simple, inexpensive chucking.

the precision of the single-point chasing attachments. All threaded parts designed for production with automatic taps or dies should provide a clearance or neck groove at the end of the thread of at least $2^1/2$ threads, and preferably 3, in width and 0.010 to 0.020-inch under the minor thread diameter. This allows the self-opening dies or collapsing tap chasers to snap free of the work under no load, obviating breakage or chipping of the chasers otherwise often encountered.

Selection of Materials

Parts made from castings, forgings, etc., should in all cases provide sufficient

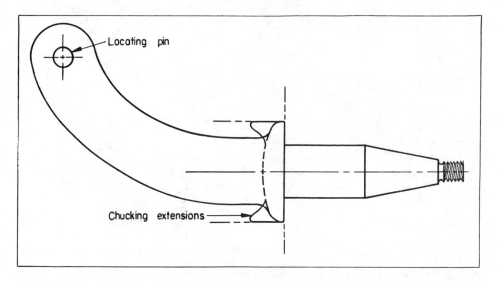

Fig. 11.14—Designed for minimum weight, this landing gear fork has chucking extensions which are removed after machining.

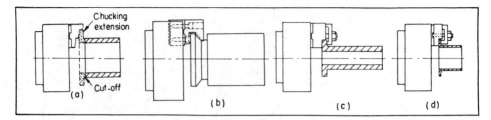

Fig. 11.15—Chucking extensions, (a), are often necessary to provide a means of holding a part; setup for gripping extremely heavy work, (b), using a chucking recess; chuck jaws with holding clamp (c), are used for long work; and soft jaws with clamps, (d) are used for frail work. Drawings, courtesy Warner & Swasey Co.

material in order that the tools can rough beneath the surface inclusions of sand, pits, scale, and hard spots to insure long tool life and good finish. High machinability is, of course, a primary factor and should be taken into consideration in specifying materials. A fairly complete table on relative machinability ratings of various materials is given in Chapter 68. Heat treatment and characteristics of previous processing of the materials are also important in their effect on the machinability. (See Chapter 68).

Commercial brass may be machined at speeds as high as 1000 fpm with carbide tooling but this drops to around 300 with ordinary high-speed steel tools. Aluminum and bronze likewise can be finished at speeds as high as 1000 fpm. However, aluminum drops to around 400 and bronze to 150 with high-speed steel tools. High machinability steels such as AISI 1108, etc., are preferred for production jobs owing to excellent finish at high cutting speeds. Hard, semisteel or

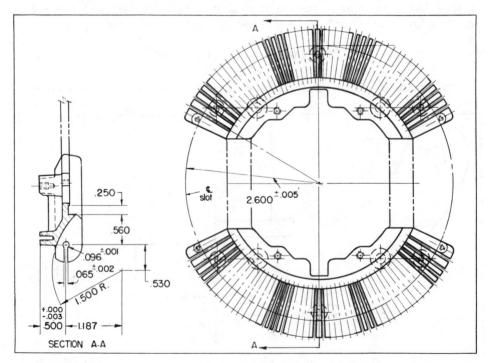

Fig. 11.16—Produced as a double casting, two type bar segments can be finished complete at one handling. Space between halves allows for entry of special form tooling for the circular groove. Drawing, courtesy Warner & Swasey Co.

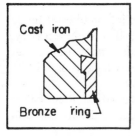

Cast iron

Bronze ring

Fig. 11.17—Locking groove is finished, bronze ring rolled-in, and entire part machined all over at one handling in the turret lathe. Groove detail is shown above.

medium-carbon steel castings may require reduction in cutting speeds to as low as 40 fpm with accompanying lower output and higher tool costs.

Specification of Tolerances

In general, on high-production runs where maximum output is both necessary and desirable from a cost standpoint, a tolerance of plus or minus 0.002-inch on diameters is considered a minimum and generally can be maintained with a rough and a finish cut, *Fig.* 11.16. It is possible to hold overall limits of plus or minus 0.001-inch on average turning and boring operations but this is generally not recommended for production runs because of the necessity for additional cuts over the same surfaces and for fine feeds. However, closer tolerances are often accepted, *Fig.* 11.18, but before demanding such, the absolute need should be definitely established.

Long Cylindrical Sections. On parts suitable for the use of a single-cutter roller turner *Fig.* 11.3, either piloted on the cut being made or on a previously turned surface, a tolerance of plus or minus 0.001-inch can be held. Occasionally it is possible to reduce this to a minimum of plus or minus 0.0006-inch. Multiple-cutter roller turners can be used for turning a number of surfaces simultaneously but the diameter tolerances required are usually wider, being on the order of plus or minus 0.003-inch or greater.

Concentricity. In normal set-up, concentricities of diameters turned can be held within a total dial reading of 0.006-inch where machine and tools are in good condition. Roller turners can often be employed to reduce this total runout to around 0.002-inch or somewhat less. Special handling on chucking jobs can also be expected to hold to a total runout of 0.002-inch. Where adequate care in tooling and setup can be exercised, runout can often be reduced to a total of about 0.0005 to 0.001-inch, *Fig.* 11.1. Under normal conditions threaded portions will be concentric with other diameters within a total of 0.004-inch or better.

Threads. Pitch and lead accuracy of threads can usually be held to that required by a class-3 fit as established by accepted standards. If more accurate threads are necessary, the previously mentioned lead-on attachments or N/C can be utilized to attain them.

Lengths and Depths. Standard stop equipment can be relied on to hold limits on lengths within plus or minus 0.001-inch. However, if closer limits are necessary, additional equipment in the form of dial indicators, dial stops or special micrometer stops are required. Wherever possible, depths and lengths should be specified as plus or minus 0.005-inch or greater for economy.

Drilling. Tolerances normally attainable in drilled holes will be identical to those outlined in Chapter 15 on Drilling and Boring. The same conditions exist and should be observed in practice.

Reaming. A tolerance of plus or minus 0.001-inch can be held on ordinary reaming operations. With additional accurate boring operations preceding, reaming tolerances can be held to plus or minus 0.0005-inch.

Finish Quality. Depending upon the type material being cut, the surface quality left in turret lathe operations will be about 60 microinches, rms, or less, *Fig.* 11.19. Surface patterns depend upon the type of tooling and cuts employed.

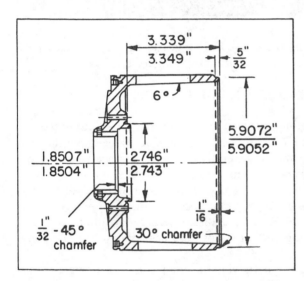

Fig. 11.18—Generator front bracket finished in 4.5 minutes has the bearing bore held within 0.0003-inch with a roughing, semifinishing and finishing cut.

Fig. 11.19—Group of finished parts produced on the Cincinnati N/C Turning Center shows finish quality attained.

12

Automatic Bar and Chucking Machining

ORIGINALLY designed for the rapid, automatic production of screws and similar threaded parts, the automatic screw machine has long since exceeded the confines of this narrow field and today plays a vital role in the mass output of precision parts in endless variety. Capable of producing turned and formed parts of both simple and intricate design, the basic machine utilizes a variety of tool slides, automatically sequenced, each of which performs a portion of the necessary machining on a rotating bar and cuts off finished pieces in rapid succession as the bar is continually fed out at each cutoff. By no means limited to the production of threaded parts, the wide range of possibilities afforded by this method of production both enhances its usefulness and makes imperative its consideration as a means of manufacturing machine parts in quantities.

Machines Available. Generally, automatic screw machines fall into several categories; single-spindle automatics, multiple-spindle automatics, and automatic chucking machines. Single-spindle machines are usually designed to produce parts

Fig. 12.1 — Standard No. 2 Brown & Sharpe Ultramatic single-spindle screw machine which handles up to 1-5/8-inch diameter stock, turns up to 3-1/2 inches at one turret movement and has a collet feed up to 4 inches.

in rapid succession from a length of bar stock fed through the machine spindle, whereas the multiple-spindle machine is available both as a bar machine and as a chucker. On the chuckers automatically operating chucks grip cast, forged or other single parts and carry them through the cycle, ejecting the parts on completion. A wide variety of models are available in each basic type of machine as well as special adaptations.

Single-Spindle Automatics. Several basic types of single-spindle bar machines are available. The Brown & Sharpe utilizes a cam-actuated tool turret, the axis of which is normal to the bed ways, having six radial tool positions which can be successively brought into working position. The standard machines have, in addition, two independent cam-actuated cross slides disposed at right angles to the spindle, *Fig.* 12.1. The machine shown in *Fig.* 12.2 utilizes a cylindrical drum turret, the axis of which is parallel to the bed ways, with five tool positions in the end of the turret which can be automatically brought into working position and cam-fed into the end of the rotating bar to perform the desired machining

Fig. 12.2 — Model 135 3-1/2-inch single-spindle automatic with electronic programming for setup and control of speeds, strokes and machine functions. Photo, courtesy New Britain Machine Co., Div. of Litton Ind.

operations, *Fig.* 12.3. Front and back cross slides are standard equipment.

Turret and cross-slide cams for actuating the tools on the Brown & Sharpe are specially designed for each part produced. Cams used on machines such as the Cleveland, on the other hand, are universal and are adjusted to suit each job.

Variations of each basic type are also used. Automatic turret forming machines, for instance, will perform all normally required operations except threading and are used to produce simple parts requiring straight and taper form turning, drilling, reaming, counterboring, recessing, knurling, etc. Another variation, often termed an automatic cutting-off machine, is similar to the turret former except that in place of a usual 6-tool turret, only a simple slide for one tool is used along with the cross slides. One of the simplest of these machines, which uses swing tools rather than slides, is shown in *Fig.* 12.4. In the automatic screw threading machines for producing large quantities of threaded parts, the 6-tool turret is replaced by a single horizontal spindle which carries a die mounted in line with the work spindle.

From these simple types with their somewhat limited range of operations, the single-spindle machines available increase in size and complexity to the largest which will handle solid bar stock in diameters up to 8 inches, tubing to 9-1/2 inches, and will turn lengths up to 9 inches with one movement. The smallest machine has a capacity of 5/16 inch maximum diameter and a maximum length of turn of 3/4 inch.

Special attachments make it possible to perform a variety of auxiliary operations while the standard run of operations is being completed, eliminating the cost of extra secondary operations. Thus, in addition to the usual forming, facing, drilling, reaming, threading, knurling, etc., operations such as slotting, milling, burring, turning, thread chasing, index drilling, and cross drilling can be performed during the regular machining cycle.

A number of special machines make possible a still broader outlook. One special extra heavy-duty single-spindle model, *Fig.* 12.5, will handle diameters up to $10\frac{1}{2}$ inches but is restricted to relatively simple operations. That shown in *Fig.* 12.6 is similar to the horizontal-turret machine, having a 4, 5 or 6-face turret, but is specially designed for carbide tooling and simple operations. Another, a multiple single-spindle machine, has four spindles arranged vertically and will quadruple output of parts up to $2\frac{5}{8}$ inches in diameter with lengths to about 6 inches.

Multiple-Spindle Automatics. A variety of multiple-spindle bar machines are available. With these machines, instead of having each of the turret tools index into position so as to work on a single rotating bar, the rotating spindles index in a carrier about the nonrotating turret so as to bring the bars successively into position to allow each tool to perform its particular portion of the overall machining

Fig. 12.3 — Model 135 tooled up for production on a part requiring turning, drilling, threading and roller burnishing. Photo, courtesy New Britain Machine Co., Div. of Litton Ind.

Fig. 12.4 — Basic Standard automatic, one of the simplest machines, performs two separate forming operations or a roughing followed by a finishing cut, and a cut-off operation. With special attachments drilling and boring can be done. Photo, courtesy Standard Machinery Co.

operation, *Fig.* 12.7. All the tools in their respective positions are at work on the various bars at the same time. Thus the time required to produce one piece is only the time necessary to complete the longest single operation plus the time required to return the toolslides, index the spindle carrier, and feed the cut-off stock into position for a new part.

Multiple-spindle design provides the opportunity to divide long cuts between two or more operations, thus reducing the time the tools are in the work during the machining cycle. Standard models, *Fig.* 12.8, come in four, six and eight-spindle arrangements; eight-spindle models allowing the maximum number of separate operations in the minimum overall time per piece produced. Typical parts produced on multiple-spindle bar machines are shown in *Fig.* 12.9, some of which, however, require one or more special attachments or tool set-ups.

A variety of special attachments can be used to speed production, perform unusual operations, or obtain finer finish, *Fig.* 12.10. Special tapping and threading units can be used. High-speed drilling and recessing attachments are regularly applied to the end toolslide. Form turning, taper turning, combined taper turning and taper boring attachments can be used on any machine. More specialized operations, *Fig.* 12.11, such as form generating, slotting, cross drilling, chamfering, milling, knurling, external and internal burnishing, pickoff devices for finishing

Fig. 12.5 — Probably one of the largest automatics in use today, this model handles up to 10¹/₂-inch stock. Photo, courtesy Timken Roller Bearing Co.

back ends after cutoff, etc., can often be employed to economic advantage.

Work handled in multiple-spindle bar machines may range as high as 7 inches in diameter and lengths may range to as high as 20 inches. Machines of various capacities are available in each of the multiple-spindle types. Nonstandard special machines have been built to handle larger diameters, but normally larger work is done on the turret lathe unless quantity justifies an entirely special machine.

Chucking Machines. Basically similar to the multiple-spindle automatic, chucking machines utilize automatic chucks on each spindle for holding castings, forgings, pressings and other parts which cannot be made from bar stock, *Fig.* 12.12. Virtually all the standard operations, as well as the special ones, normally afforded by the multiple-spindle bar automatic are available on automatic chuckers.

Fig. 12.6 — Heavy-duty single-spindle automatic specially designed for carbide tooling and simple operations. Photo, courtesy New Britain Machine Co.

Fig. 12.7 — Six-spindle Acme automatic producing a ball stud in B-1113 steel. Nine operations are performed and a piece is produced every 10.5 seconds.

Fig. 12.8 — 1-1/4 inch RA-6 Acme-Gridley multispindle bar automatic.

Some models employ rocking or swinging type forming tool arms in lieu of the regular slides for application on jobs which necessitate a wide swing to clear the parts. Largest size machine available in each machine category is shown in TABLE 12-I. The smallest capacity machine, in terms of size of parts handled, is the 8-spindle, 6-inch swing chucker.

A number of special models are available such as double-ended, six-spindle chuckers for machining castings, forgings or pressings which are so designed that both ends can be finished at once. Other simplified models are designed for second-

TABLE 12-I — Typical Automatic Screw Machines and Capacities
(inches)

Machine Type	Maximum Stock Diameter or Chucking Capacity	Maximum Length of Turn with Main Slide	Maximum Stock Feed in One Movement
Single-Spindle Bar			
Brown & Sharpe	2	5	6
Cone (4 vertical)	2⅝	6	12
Cleveland	8 (9½ tubing)	9	13
New Britain	8	7	18
Standard Machinery	1		4
Four-Spindle Bar			
Acme	4¾	8	10
Cone	7	7⅞	8
New Britain	2¼	6	6
Six-Spindle Bar			
Acme	3½	8	10
Cone	3½	10	12
New Britain	2¼	6	6
Eight-Spindle Bar			
Acme	2⅝	8	10
Cone	2⅝	10	8
Four-Spindle Chucking			
Acme	10 (max swing)	6	
New Britain	12 (11¾ max swing)	8	
Five-Spindle Chucking			
Cone	8¾ (max swing)	6	
Six-Spindle Chucking			
Acme	12 (max swing)	8	
New Britain	9 (10¼ max swing)	8	
Baird	7 (7¼ max swing)	6	
Eight-Spindle Chucking			
Acme	6 (max swing)	6	
New Britain	8 (max swing)	8	

operation work where no forming is necessary.

Production. Specifically intended to satisfy mass-production requirements automatic screw and chucking machines are seldom used where quantity may run but 1000 pieces or less. As a rule, hand screw machines are used for such small lots. For medium-lot production a standard automatic would be used without special attachments. Second operation work would be used if necessary. High production, running to 100,000 pieces or more, would justify a totally automatic setup with all available attachments necessary to complete the part on the machine.

The particular type automatic used would be determined not only by quantity requirements but also by the particular design of the part — size, shape, intricacy, etc. — and the particular operations necessary to produce it. Thus, in some cases the single-spindle machine may often be found most economical for quantities of even 100,000 pieces or more, whereas, in other cases, the multiple-spindle bar automatic or perhaps the chucker may be the more economical though quantity is no greater. Generally, the multiple-spindle machine is designed for handling extremely large production runs; but when tooled properly, it can also be economic for small lots.

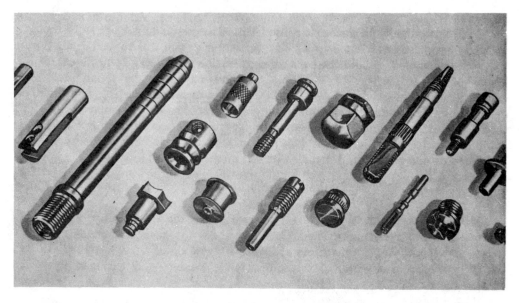

Fig. 12.9 — Group of typical machine parts produced on multiple-spindle bar automatics. Photo, courtesy National Acme, an Acme-Cleveland Co.

Design Considerations

Design of parts for manufacture by means of one of the various types of automatics is relatively unrestricted. Over and above the general limitations as to size and capacity of the machines available, recognition of a few factors relative to detail design of parts will assure reasonable economy and maximum production speed.

Shape. Wherever a design can be altered, without seriously affecting its intended function, to decrease the number of machining operations necessary or the time required to complete operations, total cost per piece can be reduced materially. As can be recognized readily, the simplest of shapes are reproduced most easily and, naturally, require the minimum in tooling.

Fig. 12.10 — Cast iron pulley being machined on an 8-inch RPA-8 chucker. Twenty-five operations are performed to complete the part in 49 seconds. Photo, courtesy National Acme, an Acme-Cleveland, Co.

Forming. Where forms are simple, one of the most efficient methods of production is by "form and cutoff", *Fig.* 12.13. Minimum tooling equipment is needed, setup is easy and a minimum of care is necessary in operation of the machine. To obtain the benefits of this method, however, the ratio of length of form to smallest formed diameter should not exceed 2.5 to 1, except in certain special cases where ratios as high as 3 to 1 have been used. These limiting ratios apply not only to total length and full diameter, but also to each separate length and diameter. For average cases 2.5 to 1 is considered most suitable except where diameter becomes small. Immediately below 1/8-inch a ratio of 2 to 1 is used and at about 3/32-inch, the length should not exceed the diameter.

Long forming cuts, where required, are usually produced by means of two or more forming tools, *Fig.* 12.14. Where deep holes or complicated formed bores along with an intricate or exacting formed exterior are necessary, the number of operations required to complete the part on an automatic may require the tooling capacity of an eight-spindle machine, *Fig.* 12.15. Long or heavy forming cuts which exceed the foregoing ratios require one or more back rests or roller supports to resist the thrust load of the tools, as shown in the first, second, third, and fifth positions in *Fig.* 12.16. It often may be advisable to be sure there is a place such rests or rollers can be used on a part to simplify the tooling problem.

Tapers. Where the outboard end of a part has a relatively short tapered or pointed design, such can often be produced with cutoff, pointing or box tools, where length of taper or forms exceed the practicable ratio mentioned previously, swing tools or form turning tools are usually used especially on single-spindle machines. Swing tools can often be employed to single-point turn slender tapered or formed parts on which a support cannot be used. Forms, in such cases, must be smooth continuous curves without abrupt rises. The swing tool is often used with a support in advance of the cutting blade. Taper turning on single-spindle machines is most often done with a swing tool. Tapers located behind shoulders or in other difficult positions also require a form tool or, where the length-to-diameter-ratio is

Fig. 12.11 — Steel worm form-generated with special attachment on a six-spindle automatic. Photo, courtesy National Acme, an Acme-Cleveland Co.

Fig. 12.12 — A group of typical automatic chucking machine jobs. Photo, courtesy National Acme, an Acme-Cleveland Co.

unsatisfactory for unsupported forming, a swing tool. It is well to keep in mind, however, that swing tools cut more slowly than form tools and are only desirable to use where a form tool will not perform properly.

Shoulders. Perfectly square shoulders can be produced on parts finished from one end with hollow mills, balance turning tools or box tools. Shallow grooves or shoulders produced with a form tool likewise can be made square. Where shoulders are wide or grooves deep, side rubbing on the tool would be objectional and finish usually poor unless side clearance is allowed. A minimum side taper of at least half a degree should be allowed, one degree being preferred for the shallower cuts to about 3 degrees for deep ones, *Fig.* 12.17. If perfectly square corners are a necessity, an extra squaring operation may be needed. In most cases, however, a minimum side angle will still maintain normal tolerances for the part and produce an acceptable commercial finish. Undercut shoulders are often produced with the swing tool or angular cutoff tool.

Corner and Edge Radii. Filleted corners and chamfered edges are preferred to sharp ones inasmuch as tool life is greatly improved. Wherever possible, corner fillets of at least 0.005-inch should be used. Generous fillet radii or smooth chamfers at intersections also may materially strengthen parts. Chamfers are preferred to rounded edges and are far less expensive to produce. Radii as small as 0.001 to 0.002-inch are held in some commercial operations but such sharp corners should not be used unless it is imperative, 0.005-inch or greater is preferable. If

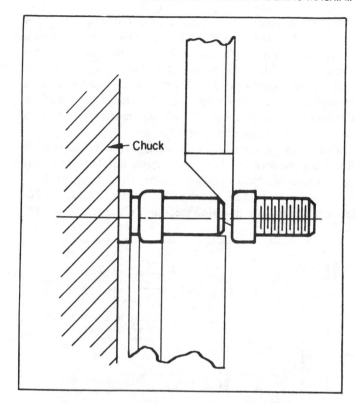

Fig. 12.13 — Simple part set up for single-spindle production via "form and cutoff". Finished piece is not cut off until stock is fed out for new piece.

cutoff burrs on solid or hollow parts will be detrimental to function or appearance, the drawing should call for their removal. In many cases, such burrs can be removed in the automatic without extra cost.

Holes. Holes, wherever possible, should be shown as drilled with a standard drillpoint at the bottom, *Fig.* 12.18*a*, for maximum economy. If flat bottomed holes, *Fig.* 12.18*b*, or square shoulders on stepped holes, *Fig.* 12.18*d*, are deemed necessary, a secondary or squaring operation usually is required. Tooling can be simplified by providing allowance for a drillpoint at the center of a blind hole, as at *c*, and a tapered step on through holes, as at *e*. Shallow, flat-bottom holes can be produced easily with simple tooling such as counterbores, etc. Hole shapes which can be produced with special step drills are shown in *Fig.* 12.18*f* to *k*.

Holes 3 to 4 times their diameter in depth can be produced in one operation, although holes 4 1/2 to 6 times their diameter in depth can be produced by utilizing a series of pullout spaces on the cam to clear chips, cool, and lubricate the drill. Holes of lesser depth are of course more economical. By utilizing more than one station, total hole depth ordinarily can be increased to 8 times the diameter and at the greatest ten. Where holes must be reamed for finish or accuracy, stock allowance for reaming should follow that indicated in Chapter 11 for turret lathe work.

External Threads. For maximum economy American National Standard thread always should be specified, the fine series of this standard being most satisfactory. A minimum of material is removed in cutting fine threads and consequently maximum threading speeds can be employed. Fine threads can be run up closer to a shoulder than the coarser ones, in some cases safely within 1 1/2 threads, *Fig.* 12.19. On ordinary threaded parts, *Fig.* 12.19, shoulder clearance *A* should be at least equal to 2½ threads in length and preferably 3 for maximum ease of set up. If the part must screw down tight, the neck should have an undercut, *Fig.* 12.19, at least 2½ threads in width. Maximum neck diameter at *A* should be about 0.010-inch under the minor diameter of the thread and it is recommended that a 37-degree bevel be used to reduce chaser breakage. To assure good lead-on and eliminate the first portion of incomplete thread, a 45-degree chamfer to 0.010 or 0.020-inch below minor diameter is recommended.

Use of proper chamfer or throat on threading dies materially increases die life and improves the thread finish considerably. The approximate number of threads affected by various angles of chamfer is shown in *Fig.* 12.20. The 30-degree, 1½-thread chamfer is recommended and used for free cutting steel or brass; the 20-degree, 2-thread chamfer is preferred for machine steel and alloy steels; the 15-

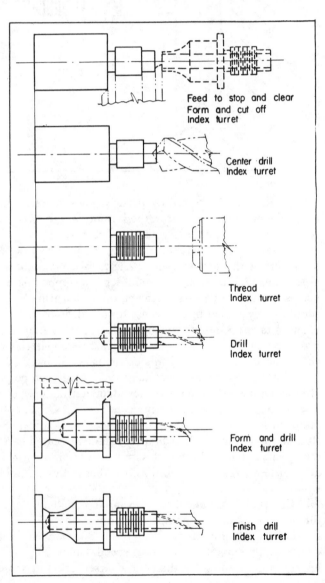

Feed to stop and clear
Form and cut off
Index turret

Center drill
Index turret

Fig. 12.14 — Single-spindle layout showing sequence of operations for a part using two forming tools. Drawing, courtesy Brown & Sharpe Mfg. Co.

Thread
Index turret

Drill
Index turret

Form and drill
Index turret

Finish drill
Index turret

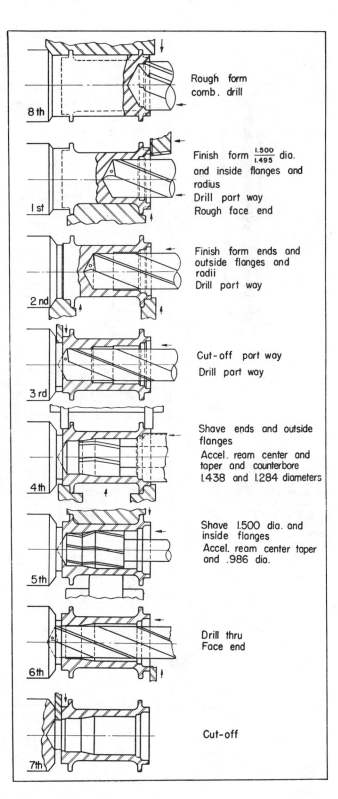

Fig. 12.15 — Eight-spindle layout showing sequence of operations and tooling for producing a steel hub. Drawing, courtesy National Acme, an Acme-Cleveland Co.

Fig. 12.16 — Layout for a steel ball stud, Fig. 12.7, showing positions in which supports are used to make possible long forming operations. Drawing, courtesy National Acme, an Acme-Cleveland Co.

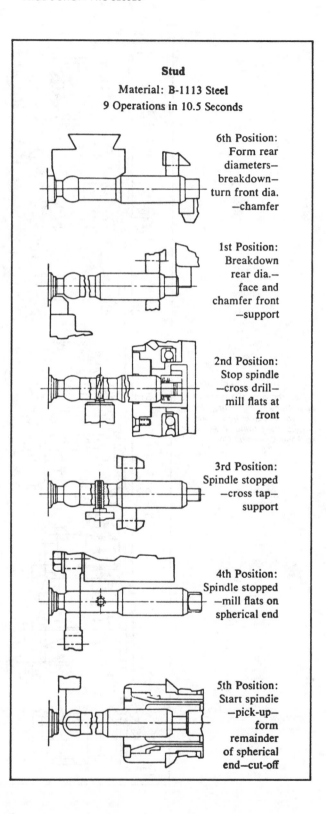

Stud
Material: B-1113 Steel
9 Operations in 10.5 Seconds

6th Position: Form rear diameters— breakdown— turn front dia. —chamfer

1st Position: Breakdown rear dia.— face and chamfer front —support

2nd Position: Stop spindle —cross drill— mill flats at front

3rd Position: Spindle stopped —cross tap— support

4th Position: Spindle stopped —mill flats on spherical end

5th Position: Start spindle —pick-up— form remainder of spherical end—cut-off degree, 3-thread chamfer is used extensively for tool steels and other tough steels. Where threads necessarily must be cut close to a shoulder, the 40-degree, 1-thread chamfer is used but is special, has low life, requires slower cutting speeds, and seldom produces as good a finish. As a rule such dies are employed following a roughing die.

Although National Standard threads are most economical to produce, other types of standard threads such as Acme can be obtained. Such threads should be completely dimensioned on the drawing, *Fig.* 12.21, whereas National Standard threads require only the class-fit specifications. Even specials such as worm threads, *Fig.* 12.11, can be generated. The tooling layout in *Fig.* 12.22 shows the details of the part as run on an Acme 1 1/4 inch RA-6.

Internal Threads. Specification of tapped holes also demands special design consideration. Blind holes, *Fig.* 12.23, require plenty of chip clearance. As shown at *A,* chip clearance should be equivalent to 5 threads in length (minimum) to allow the use of a standard tap. Otherwise, unless the hole is drilled through (which is preferable), a special secondary bottom tapping operation with somewhat inferior finish must be used to enable tapping within a 3-thread clearance. Where parts must thread and seat to a flat bottom, *Fig.* 12.24, a recess can be provided with a width of at least 3 threads. The minimum diameter of the recess should be 0.010-inch greater than the major diameter of the thread and tolerance limits for the recess diameter must be wide.

Thread Lengths. The required number of full threads should always be specified. A length of 5 full threads usually is allowed as a bare minimum for assuring sufficient strength. Leading threads on both male and female work should be chamfered to assure good entry and centering in machining and to eliminate burrs. Good practice is to countersink 0.020-inch over the major diameter of a tapped hole and chamfer 0.020-inch under the minor diameter of a male thread with an included angle of 90 degrees.

Thread Rolling. In many cases with parts, it is found desirable to complete the entire part in the automatic to obtain lowest cost per piece. When attachments are available, threads — taper or straight — which cannot be cut by a die can be advantageously rolled on single-spindle machines if the thread is short. Ordinarily thread rolling on these machines is considered for brass, aluminum, or similar soft materials only and length of thread in relation to its diameter must be considered under the same limitations as noted under forming tools. If rolling cannot be used, a rear-end threading unit or a special thread chasing attachment is available. Particularly accurate threads are also normally cut by this latter attachment. On multiple-spindle machines, rolling is often used for threads inaccessible to threading dies.

Knurls. Knurls for appearance, grip surfaces, press-fitted joints, anchor holds for inserts, etc., can be applied to any outside diameter surface, *Fig.* 12.25. Open-end knurls or those that can be produced by feeding in from the turret along the axis of the part are perhaps the most economical and easiest to produce. Those located in recesses or behind shoulders must be formed by a cross slide or swing tool and, consequently, total width of knurled surface is limited. The width of knurl should never exceed the diameter of the knurled section. Raised type knurls usually are specified for secure grip surfaces, press fits, and anchored parts while the depressed or female type is used only on parts requiring a small amount of roughness on narrow widths where the turned stock diameter must not be exceeded.

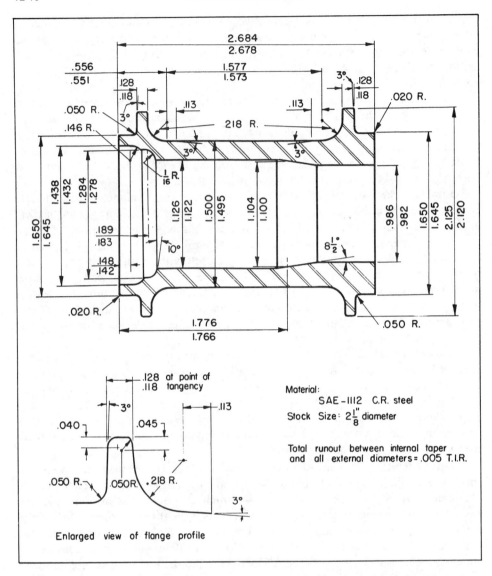

Fig. 12.17 — Detail drawing for steel hub shown in Fig. 12.15. Side taper of 3 degrees is used to assure good forming on the wide projecting radial flanges. Drawing, courtesy National Acme, an Acme-Cleveland Co.

Drawings of knurled parts should, for uniform quality, specify the type and pitch of the knurl desired as well as the diameter of the section either before and after knurling, preferably before. Types shown in *Fig.* 12.25 are produced in standard pitches ranging from 16 to 62 teeth per inch. Included angle of the knurl teeth is of value in determining the diameter before and after knurling. An included angle of 90 degrees works most satisfactorily on soft materials such as brass and hard copper; 80 degrees is recommended for iron; 70 degrees for wrought iron and machine steel; and for tough materials such as drill rod and tool steel an angle of 60 degrees is most efficient. Diameter increase due to knurling will be approximately 0.4 (minimum) to 0.7 (maximum) of the depth of knurl tooth calculated from the included angle.

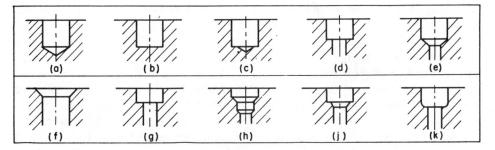

Fig. 12-18 — Holes as shown at (a), (c) and (e) are more economical to produce than those at (b) or (d).

Marking. In a manner not unlike the knurling procedure, names, trademarks, number identifications, etc., can be impressed upon external cylindrical surfaces of parts. Although a special marking tool is necessary, setup to utilize this feature usually results in a substantial saving over other methods.

Finish. If the generally accepted commercial finish left by forming tools, drills, etc., is not satisfactory for any reason, better surface finish can be obtained but at increased cost per piece. External surfaces can be skived (where finish is more important than exact sizing), or shaved after forming (where precision sizing and smooth finish are both important), and internal surfaces can be reamed. However, if still smoother finish is essential, burnishing can be resorted to as an added operation. Change in dimensions after burnishing is extremely slight, the process merely rolling over and smoothing out tool marks by cold working. Where burnishing is not feasible, grinding is usually specified to obtain the desired finish, although the regularly obtained commercial finish should be used wherever practicable for economical reasons.

Selection of Materials

Naturally, materials which afford easy machining at maximum speeds provide the lowest cost parts and least expensive tooling. Free machining brass is perhaps the most widely used material for general-purpose parts, especially those requiring

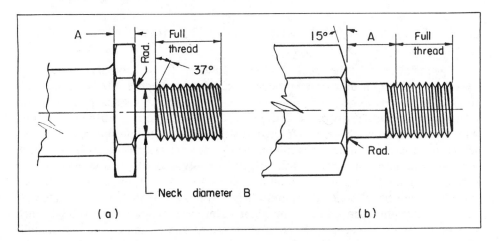

Fig. 12.19 — Neck clearance A of at least 2½ threads width is required for satisfactory threading operations.

an electroplated finish. Regardless of the high cost of brass stock, the extremely rapid rate of production coupled with the high return on brass scrap many times results in lower overall cost per piece. Free machining cold-rolled or cold-drawn screw stock (See Chapter 68) should be specified wherever and whenever design in steel permits. Highest machinability steels usually results in a minimum overall cost per part owing mainly to excellent finish and rapid cutting speeds. Where corrosion resistance is imperative and plating unsatisfactory, stainless steel of the free machining variety can be used but at some loss in output. Aluminum and magnesium alloys probably offer the maximum in machining speed where design is not too complicated. Chapter 68 shows the machinability ratings for some of the more common metals.

Hollow Parts and Special Shapes. Where hollow parts are relatively large, steel tubing stock sometimes can be utilized to advantage. Stock removal and boring operations are minimized, effecting considerable economy in time and material. In nonferrous materials even small sizes of tubing can effect worthwhile economics. Special extruded or drawn shapes often can be used with special collets to eliminate secondary operations.

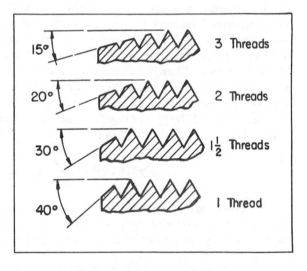

Fig. 12.20 — Shoulder clearance in threading is affected by the chamfer of the tap or die. Proper consideration of their application to various materials is imperative.

Specification of Tolerances

Specification of tolerances smaller than necessary inevitably reduces the maximum speed of production with resultant needless expense. Normal practice is to call for plus or minus 0.003-inch on important dimensions and plus or minus 0.010-inch on noncritical ones. Ordinarily, single-spindle machines are credited with the ability to work within closer limits than the larger multiple-spindle types where accuracy of spindle spacings and bearing fits are more critical. However, in average commercial practice, limits of plus or minus 0.002-inch commonly are held on decimal dimensions for plain diameters up to 1-inch, plus or minus 0.003-inch on diameters up to 2 inches and plus or minus 0.005-inch above 2 inches.

Where required on special dimensions, tolerances to plus or minus 0.001-inch can often be held, even on large chucking machine work. In no case should tolerances under plus or minus 0.0005-inch be demanded, even on single-spindle work; it is not economically feasible.

Forms. In general, forming tools are usually manufactured accurate to one-half thousandth. It is impractical therefore, to anticipate closer tolerances on the turned parts. Lengths from shoulders to other points finished with form tools can be held to plus or minus 0.001-inch to 0.005-inch, depending upon length.

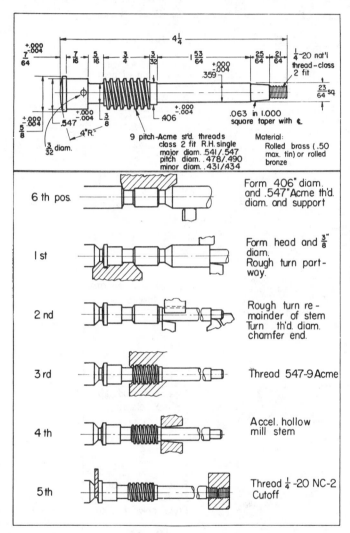

Fig. 12.21 — Acme-threaded globe valve stem and production sequence layout for a six-spindle automatic. Drawings, courtesy National Acme Co.

Diameters of noncritical portions usually are held to plus or minus 0.005-inch and lengths to plus or minus 0.010-inch. Naturally, closer tolerances can be held on some dimensions where it is essential, and good circular form tools used in a well-maintained single-spindle machine can be expected to hold plus or minus 0.002-inch on well-supported diameters. To hold such limits on large multiple-spindle parts, *Fig.* 12.17, or closer tolerances around plus or minus 0.001-inch on small multiple-spindle parts, *Fig.* 12.16, shaving tools can be resorted to in many cases. For extremely long, delicate parts 1/2-inch in diameter and under, requiring extremely close tolerances, a Swiss type automatic rather than the regular single or multiple-spindle types would probably be the most likely machine for the job.

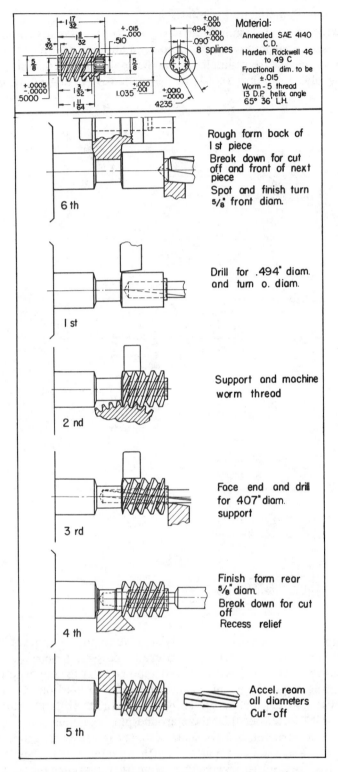

Fig. 12.22 — *Special worm and sequence layout of operations as arranged for production on a six-spindle automatic in commercial screw stock in 6 seconds with 7 operations. Drawings, courtesy National Acme Co.*

Holes. Drilled holes can normally be held to the same limits listed in Chapter 15. Larger holes and greater depths require greater limits than those specified. Smaller tolerances require an extra reaming operation and ordinarily an overall tolerance of 0.002-inch can be held. With high machinability material, reaming to plus or minus 0.0005-inch is commercially practicable. Close reaming tolerances on large holes and in poor machinability materials often requires both rough and finish reaming operations.

Stock Tolerances. Wherever possible, standard stock tolerances of plus 0.000 and minus up to 0.005-inch, depending upon the specific diameter should be used on the O. D. of parts. Plus tolerances over the nominal diameter necessitates using a larger diameter stock and machining down to size with considerably more expense and material waste.

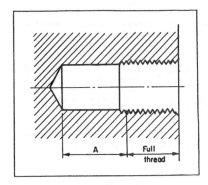

Fig. 12.23 — When tapping blind drilled holes, plenty of chip clearance A at the bottom is necessary to avoid tap breakage.

Screw machine operations   from NSMPA

Operation	Category	Special Fixtures	Commercial Tolerance (in.)	Commercial Finish (mu in. rms)	Remarks and Limitations
INTERNAL MACHINING OPERATIONS					
Drilling	Common	No	±0.003	63 to 125	Maximum practical depth, eight diameters. Over five diameters in brass and over four in steel requires pullout of drill. Same general limitations apply to multi-diameter drilling.
Reaming	Common	No	±0.001	32 to 63	Maximum practical depth, approximately eight diameters.
Cross Drilling	Special	Yes	±0.003	63 to 125	Requires special, separately driven attachment, usually warranted only by long runs.
Cross Reaming	Special	Yes	±0.001	32 to 63	
Broaching	Special	Yes	±0.006	32 to 63	Not a common operation. Requires special tools. Most adaptable on single-spindle machines.
Counterboring	Common	No	±0.004	63 to 125	Can be performed on all machines. Operation becomes special when depth is over four diameters.
Recessing	Common	No	±0.010	63 to 125	Standard on all machines.
Tapping	Common	Sometimes	Class 2	———	Both RH and LH tapping standard on all machines. Blind holes require 1/32 in. or more untapped length at bottom.
Boring	Common	No	±0.001	63 to 125	A standard operation. Hole depth over four diameters requires special attention.
Chamfering	Common	No	As required	63 to 125	Standard operation on all machines. No limitations, except back-end chamfering usually requires secondary, transfer operation.
Back-End Drilling	Common	Yes	±0.003	63 to 125	A common operation, but one requiring a special attachment. Limited to small holes on light parts.

Concentricity. Concentricity limits of 0.005-inch total indicator reading usually are maintained between holes and diameters where the holes are not excessively deep, *Fig.* 12.17. Where two sections of a part or two threads must be maintained to less than 0.002-inch, special consideration is necessary. With a single-spindle machine for instance, both thread blanks are formed at the same time, the first thread cut, and the rear one cut with a single-point swing tool or a rear-end threading attachment and pickoff finger to hold the required limits. Setup obviously is more complicated and production should justify the special attachments necessary to hold the limits imposed.

Where desirable and where cost can be justified, concentricity of turned diameters and reamed holes can be held to a total indicator reading of 0.001-inch, *Fig.* 12.22. However, such limits should be applied judiciously.

EXTERNAL MACHINING OPERATIONS

Operation	Category	Special Fixtures	Commercial Tolerance (in.)	Commercial Finish (mu in. rms)	Remarks and Limitation
Turning					
Rough	Common	No	±0.005	63 to 125	Standard on all machines. Only limitation
Finish	Common	No	±0.005 to 0.002	16 to 63	is machine capacity.
Forming (form turning)					
Rough	Common	No	±0.005	63 to 125	Requires special cutting tool. Length limi-
Finish	Common	No	±0.001 to 0.002	16 to 63	tation approximately four times smallest diameter.
Roller Shaving	Common	No	±0.001	16 to 63	Requires special cutting tool. Length limi- tation approximately four times smallest diameter.
Skiving	Special	Yes	±0.001 to 0.002	8 to 32	Requires special tool for long, slim work. Length limitation approximately six times smallest diameter.
Broaching	Special	Yes	As required	———	An uncommon operation requiring special attachments. For long runs only.
Threading					
Die Head	Common	No	Class 2 and 3	———	Only limitation is machine capacity.
Single Point	Special	Yes	Class 3	———	Either LH or RH threads can be cut.
Milling					
Straddle or Slab	Special	Yes	±0.005	63 to 125	Although considered special, these opera-
Cross	Special	Yes	±0.005	63 to 125	tions are commonly done by most shops. Only limitation is machine capacity.
Deburring	Common	Yes	As required	———	Limited only by machine capacity. Usually done in primary machining cycle; some- times secondary operation is required.
Slotting	Common	Yes	±0.005 to 0.010	63 to 125	Requires special attachment, but is a common operation.

METAL-FORMING OPERATIONS

Operation	Category	Special Fixtures	Commercial Tolerance (in.)	Commercial Finish (mu in. rms)	Remarks and Limitation
Punching, Staking	Special	Yes	As required	———	Requires special attachments. Most common on single-spindle machines.
Roll Tapping	Common	No	Class 2	———	A common operation to all types of equip- ment. Limited only by machine capacity. Results in a high-strength thread form.
Peening	Special	Yes	As required	———	Not a common operation. Not possible in all situations. Restricted to the more duc- tile materials.
Roll Threading	Common	No	Class 2-3	———	A standard operation, limited only by machine capacity.
Burnishing (OD and ID)	Special	Yes	±0.005	6 to 10	Limited only by machine capacity. Re- quires special attention in setup and origi- nal surface preparation. Not practical on all parts.
Knurling	Common	No	As required	———	A standard operation, limited only by by machine capacity.
Roll Marking	Common	Yes	As required	———	Special rolls required. A fairly common operation, limited only by machine ca- pacity.
Spinning (flaring)	Special	Yes	As required	———	Require special attachments. Not possible
Bending	Special	Yes	As required	———	in all situations. Generally warranted only
Twisting	Special	Yes	As required	———	for long runs.
Indenting, Upsetting	Special	Yes	As required	———	

Squareness. True squareness of turned faces normal to bores or turned surfaces can be held to a total indicator reading of 0.0005-inch on small dimensions, *Fig.* 12.22. More generous limits should be allowed as the face surface being toleranced increases in distance from the bore.

Angles. Taper angles in commercial practice are held to plus or minus one degree. Smaller limits can be maintained where essential however, especially on parts which can be finished with a shaving or taper reaming tool.

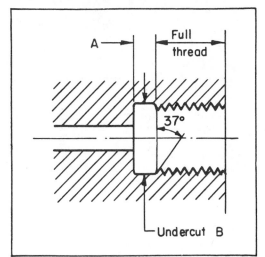

Fig. 12.24 — A recess used with tapped holes should be at least 3 threads wide to allow full threading.

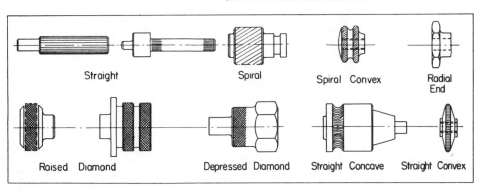

Fig. 12.25 — Various types of knurls utilized for appearance, grip surfaces, press-fitted joints, anchors, etc.

Threads. Fractional thread lengths unless otherwise specified are held to plus or minus one thread, *Figs.* 12.19*b* and 12.23. Unless the somewhat increased cost of producing and holding class-3 limits or even the greatly increased cost for class-4 limits are warranted, class-2 general practice fits are recommended for all commercial screw threads. Special extra long threads should have pitch diameter limits increased according to the American National Screw Thread Standards to account for additional lead errors and insure proper engagement, otherwise a special lead-screw attachment must be utilized.

Additional Reading

[1]"Engineering Design for Screw Machine Products", National Screw Machine Products Association, Cleveland, Ohio 44120.

13

Swiss Automatic Machining

S WISS principle of automatic machining was originally developed over fifty years ago for the clock and watch manufacturing trades in Switzerland. The American counterpart of the Swiss high-precision automatic screw machine has now become a vital and necessary part of the production picture. It is particulary well adapted to the turning of pinion blanks, studs, worm gears, indicator staffs, shafts, and other minute or slender parts used in clocks, meters, radio or electronic equipment, calculating machines, and a wide variety of industrial and laboratory instruments.

Method of Operation. The method of turning employed in these automatics which makes them ideally suited to the production of small and slender, finely finished parts is that of feeding a revolving piece of stock through a guide bushing past radially-fed tools, *Fig.* 13.1. Unlike the turret type single-spindle automatic screw machine, only single-point turning tools and narrow form tools are utilized. The tool slides are mounted in one working plane and disposed radially from the collet centerline in the upper half of a semicircle, *Fig.* 13.2. Ability to produce long, slender parts to a high degree of finish, accuracy and concentricity therefore

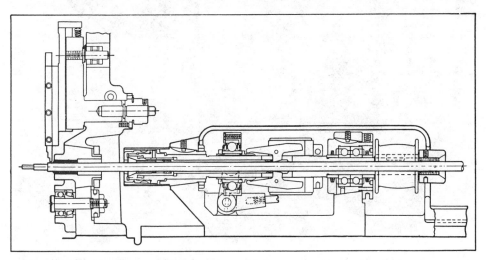

Fig. 13.1 — Stock in the Swiss automatic is rotated within and fed to the tools through a guide bushing collet. Drawing, courtesy George Gorton Machine Co.

may be attributed to the fact that the cutting faces of the tools are but a few thousandths of an inch from the stock support and at no time need be farther than $1/32$-inch away. Stock is fed through the guide bushing to the tools by means of a sliding headstock which grips and carries it by means of a collet feed mechanism.

Sliding headstock as well as the five radial tools are traversed by individual cam action. Independent cam control of each of these units eliminates the need for wide form tools since proper combination of headstock and tool movements can be had to generate pivot points, back shoulders, multiple diameters, tapers, etc. There is almost no limitation to the scope of the form generating, back shoulder and back recessing work which can be accomplished within the capacity of the machines. In fact, many parts incorporating pivot points, extensions, spherical ends, etc., can be completed and finished to final form in the cut off operation, *Fig.* 13.3, with one automatic cycle of the machine, thereby eliminating secondary machining operations and the attendant loss of close concentricity. Quality of finish and high degree of accuracy obtainable also obviate, in most cases, secondary operations such as grinding; burnishing or polishing of pivots or bearing surfaces usually being the only additional requirements.

Machines Available. Capacity of American machines now available ranges

Fig. 13.2 — Close-up of the tool slides with the five cam-actuated radial tools. Note the micrometer adjustments for accurate tool positioning and the readily adjusted ratio links of the slides. Photo, courtesy City Engineering Co.

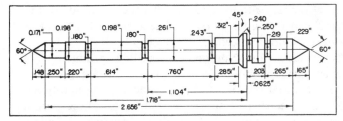

Fig. 13.3 — Intricate designs, of which this shaft is typical, can be produced in one cycle. Previously made in 14 separate setups, it was completed in 4 minutes, 36 seconds on the Swiss automatic at a tremendous saving in time. Drawing, courtesy Wickman Corp.

from 5/32-inch to 1/2-inch, maximum. Four machine sizes within this range are available for production and have 5/32, 3/8, 7/16 and 1/2-inch maximum chuck capacities, the 7/16-inch size is shown in *Fig.* 13.4. Principal advantages of the smallest machine are higher spindle speeds and proportionally smaller working parts which permit closer running fits and consequently production of parts to the greatest possible accuracy. Minute parts, *Fig.* 13.5, can be handled in this machine with ease and tool settings can be made with the exactness and delicacy necessary. Parts produced on the 5/32-inch machine, *Fig.* 13.6, are limited to a maximum length of 1 9/16 inches.

With almost equal capacity for intricacy and accuracy, the largest of the machines, having a 1/2-inch maximum chuck capacity, can handle the greatest range of sizes. Using the standard headstock-feed plate cam, parts up to 2 3/4 inches in length can be produced. By substituting a special bell cam, the maximum length

Fig. 13.4 — The 7/16-inch size machine. Opposing lower tools fitted on a single rocker for operation by one cam are visible at the left. Photo, courtesy City Engineering Co.

can be increased to 4 inches. However, with the addition of a special steady rest to provide adequate support and a special stock stop to assure accurate length, parts up to 9 inches can be produced by double feeding the stock, *Fig.* 13.7.

Special Machine Attachments. In addition to speeding production of intricate, slender parts through virtual elimination of secondary turning, polishing or grinding operations often required for finish and accuracy, many additional operations may be performed by means of special attachments. These permit drilling, chamfering, counterboring, tapping, threading, knurling, reaming, burring, and slotting, *Fig.* 13.8.

Production Speed. Suited primarily to the production of parts too small, slender or intricate for completion on a turret type automatic, the Swiss type automatic reduces the production time and improves the accuracy compared with the usual secondary operations and precision bench-lathe work, *Fig.* 13.3.

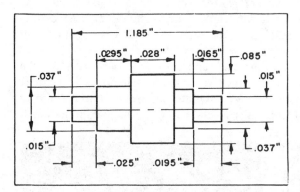

Fig. 13.5 — Minute parts such as this pinion blank are handled and finished to highly exacting tolerances on Swiss type machines. Drawing, courtesy George Gorton Machine Co.

Whereas, on most standard single-spindle turret type automatics it is common practice to turn out small screw machine parts of simple design in three seconds to possibly six minutes or so, cycle times on Swiss type machines average about 1 minute for most parts. Depending upon the size, design intricacy, accuracy and finish required, however, cycle times may vary from as little as 2 seconds to as much as 4 or 5 minutes per piece. Reported scrap losses as high as 50 per cent with other machining methods have been reduced to less than 10 per cent, enhancing production and reducing costs. Considerable increase in output can often be accomplished by designing parts so that as many as three or four complete pieces can be produced and cut off before the nonproductive time of retracting the headstock takes place.

Although normally most efficient and economical for large quantity production, standard cams, once developed and perfected, make it possible to simplify setup and reduce tooling costs to a point where quantities around 1000 pieces can be made at highly competitive prices.

Centerless Turning Machines. In principle not greatly unlike the Swiss automatic, the Taber Centerless Turning machine performs turning operations without the use of centers. Bar stock is continuously fed through the drive spindle and is supported by a carriage mounted rigid collet assembly. The machine turns shafts with contours by means of a template and direct tracer stylus or N/C in one pass. At any desired point the carriage can be stopped to permit facing, grooving or cutoff from a six- or eight-station tool turret.

The Centerless Turning Machine thus extends the capabilities of the Swiss method to larger diameter shafts — up to $2^1/4$ inches in diameter and up to 50 inches in length, *Fig.* 13.9.

Design Considerations

Inasmuch as both the headstock and tool movements are directly controlled by cams, cam grinding must be particularly accurate. Depth stops or limits are provided only on the rocker tools, *Fig.* 13.2. Accurate depth or diameter control is, therefore, dependent upon cam accuracy. Slight inaccuracies in cam base diameter can be corrected by micrometer adjustments at each tool slide. Generating cams and cylindrical cam segments, though, must be accurate to reproduce proper form. Consequently, once a set of standard cams is on hand, production costs can be kept to a bare minimum by designing all parts to the greatest possible extent around standardized dimensions and forms. One aircraft instruments plant has in this way bettered the unit cost of certain standard parts produced on the Swiss automatic over those made on small hand screw machines.

Radial tool arms, with the exception of the two bottom rocker tools, are provided with adjustment slots allowing the ratio of tool travel to cam travel to be

Fig. 13.6 — Smallest of the American Swiss machines is limited to maximum diameter of $5/32$-inch and length of $1^9/16$-inches. Photo, courtesy Wickman Corp.

varied from 1:1 up to 3:1, depending upon the particular type machine. Thus standard cam motions can be modified to suit a variety of jobs within a limited range.

Size Range. Standard collet capacity ranges from $1/64$-inch to $1/2$-inch diameter in increments of $1/64$-inch. Odd-size stock or special-shape cross sections require special collets. Cheaper cold-drawn stock in lieu of centerless ground stock

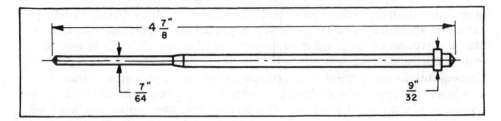

Fig. 13.7 — Part longer than normal feed of machine. Slender shafts up to 9 inches in length can be produced by double feeding the stock. Drawing, courtesy Wickman Corp.

requires a special self-adjusting type collet to accommodate the fluctuations in size and finish.

Diameters which may be produced satisfactorily run from 0.005-inch to 0.500-inch. Turn lengths may run from $1/32$-inch to as much as 9 inches, *Fig.* 13.10. As a demonstration of the turning possibilities, a pin 0.005-inch in diameter was machined to a length of one-inch successfully.

Forms. An idea as to some of the wide variety of forms which are possible to generate can be gained from the parts shown in *Fig.* 13.11. That shown in *Fig.* 13.12 illustrates counterboring, back recessing, centering, and drilling operations. In *Fig.* 13.13 a complete layout of operational sequence and rough plan of tool usage for a typical part is shown to help visualize the possibilities of designing to utilize simple standard tooling.

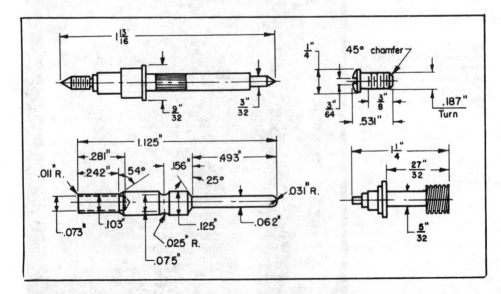

Fig. 13.8 — Typical parts which include some of the additional operations performed by special tooling.

Corners and Edges. Corner radii are necessarily sharp for close shoulder work and usually are on the order of 0.005 to 0.015-inch. Sharp corners or small radii can be held economically on soft materials such as brass, aluminum or "pinion stock", but where tough materials such as stainless steels are to be turned, tool grinding and setup costs are high unless carbide tools can be used. Wide allowable radii variations, of course, enable much longer production runs, *Fig.* 13.10.

Holes and Threads. Special triple-spindle drilling attachments provide for centering, drilling, counterboring, reaming, tapping, or threading. One spindle is stationary for centering and a second can be stationary or rotated left-hand

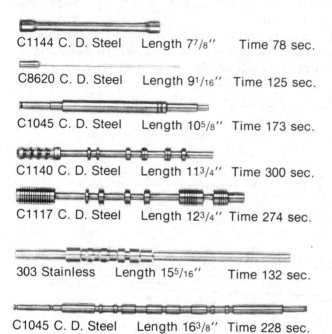

C1144 C. D. Steel   Length 7⁷/₈''   Time 78 sec.

C8620 C. D. Steel   Length 9¹/₁₆'' Time 125 sec.

C1045 C. D. Steel   Length 10⁵/₈'' Time 173 sec.

C1140 C. D. Steel   Length 11³/₄'' Time 300 sec.

C1117 C. D. Steel   Length 12³/₄'' Time 274 sec.

303 Stainless   Length 15⁵/₁₆''   Time 132 sec.

C1045 C. D. Steel   Length 16³/₈'' Time 228 sec.

Fig. 13.9 — Some typical parts produced on the Centerless Turning Machine. Centerless turning becomes economically attractive as length to diameter ratio becomes large and O.D. complexity increases. Photo, courtesy Teledyne Taber.

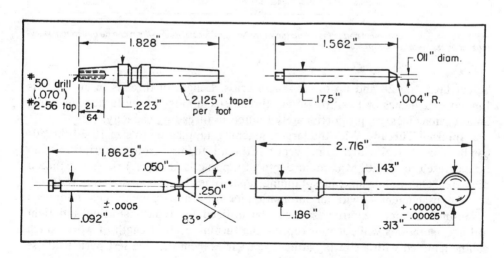

Fig. 13.10 — Owing to constant support of the stock at the tools, exceedingly small and delicate sections can be produced to fine accuracy.

(opposite the stock) for drilling or reaming. Third can be stationary for drilling or reaming or turned in a right-hand (with the stock) direction for slow threading or tapping operations. Maximum diameter of drill that can be used is 1/4-inch in steels and 3/8-inch in brass or other nonferrous materials. This three-spindle attachment is, of course, limited to the drilling of two diameters or to drilling one hole and tapping it. It is possible to thread one diameter with the end attachment and thread a second diameter with a thread-rolling tool from one of the radial tool slides. Deep holes up to 2 inches can be drilled by means of three or four pull-outs to clear chips. Owing to machine construction, however, as work length increases beyond 2 inches, depth of hole possible is reduced. For instance, a part 3 1/2 inches long allows a hole depth of only 1/2 to 5/8-inch.

External Threads. With acorn or button dies, threads from size 0-80 up to 10-32 can be made in steels and up to 1/4-28 in brass or other nonferrous materials. By means of a special attachment with a self-opening die head, threads in steel can

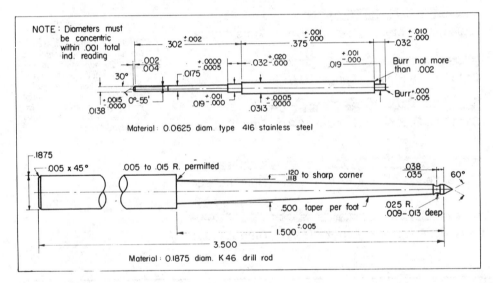

Fig. 13.11 — *Typical instrument parts illustrating the wide variety of special forms which can be generated. Drawings, courtesy George Gorton Machine Co.*

be cut up to 1/4-28 and up to 5/16-24 in brass. Length of threads in any case is limited to 2 inches or less. Maximum thread size applies to the largest machine. Smaller machines are proportionately limited in threading capacity.

Internal Threads. With the largest machine, tapping up to size 10-32 threads can be done in steel with a 75 per cent full thread. In nonferrous materials this can be increased to 1/4-20. Maximum depth of tapping is about 2.5 times the thread diameter. On smaller machines, tapping capacity is also proportionately reduced.

An attachment — with an automatic pick-off arm — which accommodates a 1.750-inch diameter slotting saw can be utilized for typical slotting and light milling operations without interrupting the turning cycle. Length of work which can be handled with a slotting attachment varies from the smallest parts to those 1 1/4 to 1 1/2 inches long, maximum, depending upon the machine.

Knurls. Plain or diamond knurling can be used much as in other screw machine work but, of course, only the very finest are practical. Single-roll contact knurling can be used but only where the stock is of sufficient size to withstand the pressure. Double-roll knurling is preferable in that standard roll widths can be used the desired length of knurl being obtained by advance of the bar. Size is limited to 5/32-inch diameter, minimum. Straddle knurling is used for diameters from 1/32 to 5/32-inch and because of the tool size more space is required for tool clearance. Length is obtained by advance of the stock. Knurls obstructed by shoulders, etc., are limited to diameters of 5/32-inch or greater inasmuch as end lead-on required by straddle knurling tools is not possible and single-roll contact knurling must be used. Other knurls must be completed in a secondary operation.

Selection of Materials

Both materials and surface speed of cutting have a direct bearing upon quality of finish and the accuracy obtained. For this reason, carbon drill rod (0.95 to 1.05c) is frequently used. Rods up to 0.125-inch in diameter are centerless ground and polished to plus or minus 0.0003-inch and those from 0.125 to 0.500-inch are ground to plus or minus 0.0005-inch. In standard lengths, small size drill rod (0.013 to 0.016-inch diameters) is costly but accuracy and finish obtained usually warrants its use.

Pinion Stock. Similar to drill rod both in tolerances and finish is the previously mentioned "pinion stock". This material is used widely in place of screw stock, which has about the same physicals.

Standard Stock. Nonferrous materials such as brass, bronze and aluminum can also be used. As previously noted these materials as well as various high machinability steels, Chapter 68, can be handled in the plain cold-drawn condition provided a self-adjusting collet is available and tolerances required on the finished parts are not too exacting.

Stainless Steels. Many types of precision instruments require stainless-steel parts and this material, usually centerless ground to the same limits as drill rod and pinion stock, can be handled successfully provided feeds are slower. For instance, feeds for regular turning and forming cuts may range between 0.0004 and 0.001-inch per revolution whereas regular high machinability steels allow feeds per revolution of 0.0007 to 0.003-inch. Feeds and speeds depend greatly upon the finish and accuracy desired and vary with the width or depth of cut.

Cemented carbide tools are imperative in working stainless steels to maintain sharp corners and extend production runs. Likewise collets and bushings with carbide inserts are particularly recommended to obviate seizure and galling on stainless and similar types of tough or "gummy" stock.

Proper Specification. For the most accurate parts, centerless-ground stock should be specified. Such stock should be within a maximum out-of-roundness tolerance of 0.0003-inch. Uniformity of size throughout the length of a bar should be plus or minus 0.0003-inch or less and uniformity from bar to bar not over plus or minus 0.0005-inch.

Specification of Tolerances

As noted previously, size accuracy and fine finish are dependent to a large

Fig. 13.12 — Right — Back-of-shoulder recessing is easily accomplished by reversal of the headstock feed during the machining cycle.

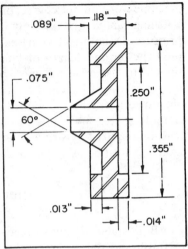

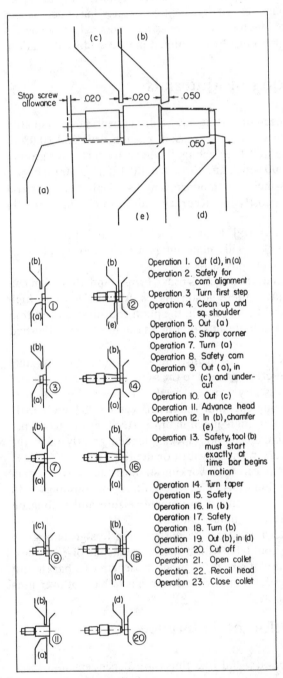

Operation 1. Out (d), in (a)
Operation 2. Safety for cam alignment
Operation 3. Turn first step
Operation 4. Clean up and sq. shoulder
Operation 5. Out (a)
Operation 6. Sharp corner
Operation 7. Turn (a)
Operation 8. Safety cam
Operation 9. Out (a), in (c) and under-cut
Operation 10. Out (c)
Operation 11. Advance head
Operation 12. In (b), chamfer (e)
Operation 13. Safety, tool (b) must start exactly at time bar begins motion
Operation 14. Turn taper
Operation 15. Safety
Operation 16. In (b)
Operation 17. Safety
Operation 18. Turn (b)
Operation 19. Out (b), in (d)
Operation 20. Cut off
Operation 21. Open collet
Operation 22. Recoil head
Operation 23. Close collet

Fig. 13.13 — Left — Operational sequence in machining a typical Swiss automatic part. Rough plan of individual tool usage, top, shows shape of generating tips and work accomplished by each of the five radial tools. Drawings, courtesy George Gorton Machine Co.

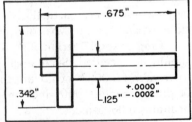

Fig. 13.14 — Excellent finish — 5 microinches on this piece — eliminates need for grinding on certain parts such as this 303 type stainless-steel pinion shaft blank.

extent upon materials, feeds and cams. The accuracy of parts produced is affected directly by the accuracy of the stock and its fit in the guide bushing. Centerless-ground material makes possible a close push fit for each bar. Cemented carbide guide bushings and collets with carbide inserts provide for closer adjustments to centerless ground stock without galling and thus assure maximum accuracy.

Diameters. Normally, using centerless-ground stock, parts produced on the larger machines can be held to plus or minus 0.0005-inch in large-scale production without noticeable tool trouble. However, with light cuts and slow feeds to favor accuracy, tolerances of plus or minus 0.0002 to 0.0003-inch can be maintained. Certain forming cuts can be held to even closer tolerances, *Fig.* 13.11. The smaller machines producing minute parts, can with favorable conditions hold limits of plus or minus 0.0002-inch or less in production.

Lengths. Turned lengths to shoulders or other locations can normally be held to plus or minus 0.0005-inch. Normal tolerances specified are plus or minus 0.002 or 0.003-inch with plus or minus 0.005 to 0.010-inch wherever design permits, *Fig.* 13.10, for maximum economy.

Surface Roughness. With proper feeds and speeds, surface finishes from around 50 microinches to as fine as 5 microinches can be had, *Fig.* 13.14. In the manufacture of instrument parts, surface accuracy of 12 to 16 microinches is reported as an average obtained in regular production.

Concentricity. The precision with which axes of related diameters or forms coincide is one of the primary features of the Swiss machine. Concentricity naturally is dependent greatly upon the accuracy or uniformity of the stock as well as that of the guide bushings. With centerless-ground material and carbide-insert collet guide bushings, total runout will not exceed 0.001-inch and as a rule will be considerably less. Plus or minus 0.005-inch is not uncommon in regular production. On the 5/32-inch size machine runout under these conditions will usually be plus or minus 0.0002-inch or less.

Economical Tolerances. Use of highly exacting tolerances, of course, should be limited to critical dimensions and wherever possible wider tolerances should be used to aid production. If concentricity and diameter tolerances can be somewhat liberal, costs can be reduced considerably by making possible the use of cold-drawn material.

14

Production Milling

OF ALL the production methods for metal removal with the exception of turning, milling is undoubtedly the most widely used. Well suited and readily adapted to the economical production of any quantity from a few parts to a multitude, the almost unlimited versatility of the milling process today commands attention and serious consideration of designers concerned with the manufacture of machines or mechanisms.

Scope of Process. Parts which fall within the scope of production milling may range from a coupling shaft for a tiny mechanism, to massive cast housings, *Fig.* 14.1. Finished surfaces likewise may range from simple, plain geometric shapes to extremely complex shapes and irregular forms. However, in the design of parts which are to be produced in very large quantities to close tolerances, where surfaces to be produced can be adapted for broaching, production speed will usually be higher and cost per piece lower. Otherwise, where repetitive production machining is concerned, milling provides an almost universal method of creating accurate parts for interchangeable manufacture.

Milling Methods. Milling of metal is performed in a multitude of ways on a wide variety of machines — standard and special — with an equally wide diversity

Fig. 14.1 — Massive cast housings being machined on a large horizontal spindle Milwaukee-Matic.

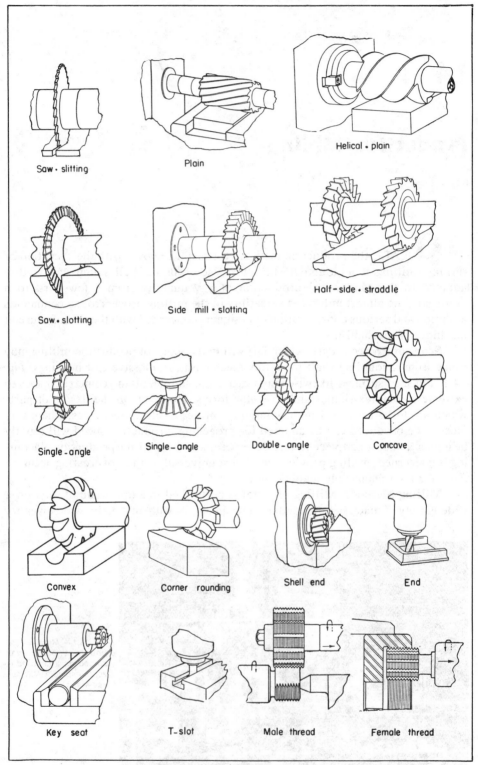

Saw · slitting

Plain

Helical · plain

Saw · slotting

Side mill · slotting

Half–side · straddle

Single - angle

Single – angle

Double – angle

Concave

Convex

Corner rounding

Shell end

End

Key seat

T–slot

Male thread

Female thread

Fig. 14.2—Basic milling operations and cutters, illustrating a few of the wide variety of surfaces and surface combinations which can be generated. Drawings, courtesy Illinois Tool Works.

of cutters, *Fig.* 14.2. Inserted or integral teeth on the cylindrical body of the milling cutter remove excess metal in small individual chips as a portion of a part is passed through the path of the cutter teeth. In the majority of cases, the surfaces generated in milling may be classified by the basic method used. Those generated in *peripheral* milling are the result of cutter rotation about an axis parallel to the finished surface while those generated in *face* milling result from cutter rotation about an axis perpendicular to the finished surface, *Fig.* 14.3. As shown, peripheral milling operations can be performed in what is known as "conventional" or "up" milling as well as "climb" or "down" milling while in face milling these two methods are usually combined.

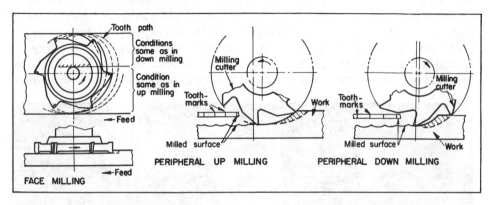

Fig. 14.3—Views showing the type of surfaces generated in both peripheral and face milling. Drawings, courtesy Cincinnati Milacron Co.

Up and Down Milling. Conventional milling methods are widely employed in manufacture owing to the fact that the natural separating forces created between the work and cutter tend to minimize rigidity, accuracy and safety demands. Climb milling, on the other hand, requires well-built machines without backlash in the table mechanisms to assure accurate results and safe operation. Much attention, however, has been given climb milling in recent years and with improved machines now available this method has gained wide acceptance. In some cases climb milling simplifies fixture design and makes practical the holding of slender and intricate parts. Power consumption is somewhat lower, permitting removal of more metal, and considerably increased cutting speeds and feeds are practical owing to thicker chips and lack of lifting forces on the workpiece.

Face Milling. Undoubtedly, the most superior surface finish is obtained with face milling, *Fig.* 14.4, especially where high-speed milling with carbides is concerned. In addition, when compared with other types of cutters, face mills permit the greatest possible feeds and speeds commensurate with maximum cutter life and, since these mills also cost the least to manufacture and maintain, they afford the maximum in production at lowest cost per part.

Form Milling. In the many cases, however, where surfaces to be machined prove to be other than the straight-line simple geometric forms within the scope of standard cutters, special form mills of almost unlimited variety can be utilized, *Fig.* 14.5. Form milling cutters of such complexity are either of the form-relieved or shape-profiled variety; the former must of necessity be made of high-speed steels with accurately generated form-relieved teeth while the latter are made with high-speed steel or carbide inserts which are formed to the desired shape in a contour grinder, *Fig.* 14.6. Sharpening problems with the shape-profiled cutters are not as great as with form-relieved mills but naturally somewhat more so than with plain standard types. Shape-profiled cutters, however, present the opportunity for economical production of relatively simple geometric forms especially with carbide-tipped blades which can operate at high speeds and feeds. Carbide blades are often used with a negative rake for cutting the tough ferrous metals. For cast iron and the nonferrous alloys the rake angle for carbides usually varies from 0 to 10 degrees positive.

Cutters and Design Requirements. Part design, as a rule, dictates to a great extent the general type of cutter which is most suitable. Production quality, costs, available production equipment, etc., control the final selection. Where processing costs are found to be excessive, redesign to utilize less expensive cutters or to improve metal removal efficiency is sometimes necessary. An idea as to the range of special cutter possibilities can be gained from the sketches shown in *Fig.* 14.7. Also apparent are some of the relative advantages and limitations of solid high-speed

Fig. 14.4—Face milling cast-iron automotive clutch housings on an index base fixture.

steel, inserted-blade and carbide-tipped cutters with respect to design of parts to be milled.

Milling Speeds. Much has been learned in the field of milling relative to optimum cutter speeds and feeds. Although apparently no upper peripheral speed limit seems to be found for milling such materials as aluminum and magnesium, tool wear increases with cutting speed and accuracy of the parts to be produced often will not permit speeds in excess of 3000 sfpm. Practical machine and spindle limitations also set practical limits since gain in speed without a corresponding gain in feed would be of little advantage. Numerous careful investigations seem to indicate that with low-carbon steels, stainless steels and wrought iron, peripheral speeds in excess of 1000 surface feet per minute and probably 500 for high-carbon steels and alloy steels even in heat-treated state would not be advantageous.

Fig. 14.5—Left — Surfaces of this welded gear and flange housing require a special form-relieved milling cutter. One cut approximately 0.090-inch deep is required to remove the bead and finish the surfaces. Slightly less time is required than in turning similar parts. Photos, courtesy Kearney & Trecker Corp.

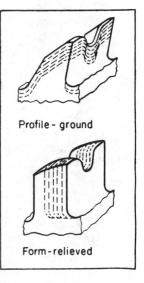

Profile - ground

Form - relieved

Fig. 14.6—Right — Difference in tooth structure between form-relieved and contour-ground or shape-profiled cutters. With profile-ground cutters, desired profile of tooth occurs at cutting edge only. Proper peripheral and side clearance and adequate chip space are easily maintained. With form-relieved cutters, however, desired profile must be carried back into tooth. It is impossible to use adequate clearance angles, especially side clearance. Drawings, courtesy Cincinnati Milacron Co.

TABLE 14-I—Feeds per Tooth and Surface Speeds for Milling*

MATERIAL		SPINDLE H.P. (RATED CAPACITY)											CARBIDE CUTTERS						HIGH SPEED STEEL CUTTERS					
		5	7½	10	15	20	25	30	40	50	75	100	FACE MILLS	SLAB MILLS	END MILLS	FULL & HALF SIDE MILLS	SAWS	FORM MILLS	FACE MILLS	SLAB MILLS	END MILLS	FULL & HALF SIDE MILLS	SAWS	FORM MILLS
MALLEABLE SOFT/HARD	CU. IN. PER MIN.	5.5	8.6	12	19	25	32	36	52	65	110	150												
	FEED PER TOOTH												.005 015	.005 015	.005 010	.005 010	.003 004	.005 010	.005 015	.005 015	.003 010	.006 012	.003 006	.005 010
	FEET PER MIN												.200 300	.200 300	.200 300	.200 300	.200 300	.175 275	60 100	60 90	60 100	60 100	60 100	60 80
CAST STEEL SOFT/HARD	CU. IN. PER MIN.	3.2	5.2	7.5	12	16	20	24	32	43	68	95												
	FEED PER TOOTH												.008 015	.005 015	.003 010	.005 010	.002 004	.005 010	.010 015	.005 015	.005 010	.005 010	.002 005	.008 012
	FEET PER MIN												150 350	150 350	150 350	150 350	150 300	150 300	40 60	40 60	40 60	40 60	40 60	40 60
100 150 BR. STEEL	CU. IN. PER MIN.	3.5	6	9	15	20	25	30	40	55	90	130												
	FEED PER TOOTH												.010 015	.008 015	.005 010	.008 017	.003 006	.004 010	.015 030	.008 015	.003 010	.010 020	.003 006	.008 010
	FEET PER MIN												450 800	450 600	450 600	450 800	350 600	350 600	80 130	80 130	80 140	80 130	70 100	70 100
150 250 BR. STEEL	CU. IN. PER MIN.	3.2	5.2	7.5	12	16	20	24	32	43	68	95												
	FEED PER TOOTH												.010 015	.008 015	.005 010	.007 012	.003 006	.004 010	.010 020	.008 015	.003 010	.010 015	.003 006	.006 010
	FEET PER MIN												300 450	300 450	300 450	300 450	300 450	300 450	50 70	50 70	60 80	50 70	50 70	50 70
250 350 BR. STEEL	CU. IN. PER MIN.	2.8	4.5	6.5	10	14	17	21	28	37	60	85												
	FEED PER TOOTH												.008 015	.007 012	.005 010	.005 012	.002 005	.003 008	.005 010	.005 010	.003 010	.005 010	.002 005	.005 010
	FEET PER MIN												180 300	150 300	150 300	160 300	150 300	150 300	35 60	35 50	40 60	35 50	35 50	35 50
350 450 BR. STEEL	CU. IN. PER MIN.	2.5	4	5.5	9	12	15	18	24	33	54	80												
	FEED PER TOOTH												.008 015	.007 012	.005 010	.005 012	.001 004	.003 008	.003 008	.005 008	.003 008	.005 008	.001 008	.003 008
	FEET PER MIN												125 180	100 150	100 150	125 180	100 150	100 150	20 35	20 35	20 40	20 35	20 35	20 35
CI HARD 225 350 BR	CU. IN. PER MIN.	3.5	6	9	15	20	25	30	40	55	90	130												
	FEED PER TOOTH												.005 010	.005 010	.003 008	.003 010	.002 003	.005 010	.005 012	.005 010	.003 008	.005 010	.002 004	.005 010
	FEET PER MIN												125 200	125 200	125 200	125 200	125 200	100 175	40 60	40 60	40 60	40 60	35 60	35 50
CI MED. 180 225 BR.	CU. IN. PER MIN.	5.5	8.6	12	19	25	32	36	50	65	110	150												
	FEED PER TOOTH												.008 015	.008 015	.005 010	.005 012	.003 004	.006 012	.010 020	.008 015	.003 010	.008 015	.003 005	.008 012
	FEET PER MIN												200 275	175 250	200 275	200 275	200 250	175 250	60 80	50 70	60 90	60 80	60 70	50 60
CI SOFT 150 180 BR.	CU. IN. PER MIN.	8.5	13	18	29	36	45	54	72	92	142	200												
	FEED PER TOOTH												.015 025	.010 020	.005 012	.008 015	.003 004	.008 015	.015 030	.010 025	.004 010	.010 020	.002 005	.010 015
	FEET PER MIN												275 400	250 350	275 400	250 350	250 350	250 350	80 120	70 110	80 120	80 120	70 110	60 80
BRONZE SOFT/HARD	CU. IN. PER MIN.	5.5	8.6	12	19	25	32	36	50	65	110	150												
	FEED PER TOOTH												.010 020	.010 020	.005 010	.008 012	.003 004	.008 015	.010 025	.008 020	.003 010	.008 015	.003 010	.008 015
	FEET PER MIN												300 100	300 800	300 1000	300 1000	300 1000	200 800	50 225	50 200	50 250	50 225	50 250	50 200
BRASS SOFT/HARD	CU. IN. PER MIN.	6.5	10	14	22	30	38	45	62	80	120	170												
	FEED PER TOOTH												.010 020	.010 020	.005 010	.008 012	.003 004	.008 015	.010 025	.008 020	.005 015	.008 015	.003 005	.008 015
	FEET PER MIN												500 1500	500 1500	500 1500	500 1500	500 1500	500 1500	150 300	100 300	150 350	150 350	150 350	100 300
ALUM. AL. SOFT/HARD	CU. IN. PER MIN.	15	25	35	55	80	100	120	170	220	330	450												
	FEED PER TOOTH												.010 040	.010 030	.003 015	.008 025	.003 006	.008 015	.010 040	.015 040	.005 020	.010 030	.004 008	.010 020
	FEET PER MIN												2000 UP	2000 UP	2000 UP	2000 UP	2000 UP	2000 UP	300 1200	300 1200	300 1200	300 1200	300 1000	300 1700

Maximum Cu. In. Per Min. for rated horsepower shown. Generally lower end of ranges used for inserted blade cutters, higher end of range for indexable insert cutters. Courtesy, Modco Tools, Div. Valeron Corp.

To provide at a glance a general impression of present-day overall practice in production milling, the data in TABLE 14-I have been compiled. Necessarily these data are not specific but rather indicative of what can be expected. Naturally, optimum efficiency of metal removal depends greatly upon the machine capacity, cutter design, part design, fixture design, etc.

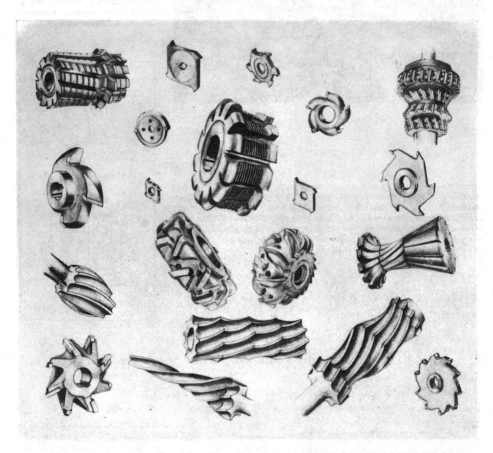

Fig. 14.7—A selection showing a few of the wide variety of special cutters used in production milling.

Machinability. Whereas, in general practice with high-speed steel mills and standard machines, not more than 25 cubic inches of cast iron or 12.5 cubic inches of steel could be removed per minute, the use of carbide-tipped mills make possible the removal of 150 cubic inches of cast iron, 75 cubic inches of steel and in aluminum or magnesium alloys as high as 500 cubic inches per minute. The latter work, however, requires a high-powered miller such as the present-day machines. However, it is interesting to note that results with the newer accurate, rigid and high-powered N/C millers show that "machinability" or efficiency of metal removal may be largely influenced, perhaps completely determined, by the cutting materials and machines utilized!

Designing for Efficient Metal Removal. The obvious advantages of high-speed milling, again, are subject to limitations. Naturally, the cutter design limitations mentioned previously come into play, and where part contours cannot be simplified or partially altered to permit using plain or shape-profiled carbide or similar mills, special mills are necessary even though these may be more expensive in first cost and reconditioning as well as lower in production speed. Also, if the part is thin or frail, it will not withstand the strains of a heavy cut and as a result high speeds and feeds cannot be used. Proper design of holding fixtures will, in many cases, permit the high-speed milling of small or light parts. However, additional ribbing or buttressing to "beef" up a part to resist heavy cuts often involves a relatively small increase in material costs which can be readily offset by drastic reduction in processing costs.

Fixture Design. Of primary importance along with massive, well-built fixtures is the necessity of designing such fixtures to require the very minimum in loading time. Few savings can accrue from a drastic speedup in machining time if handling time is out of proportion, *Fig.* 14.8. Consequently, parts which are complicated and

Fig. 14.8—K & T Milwaukee-Matic setup showing double fixturing which allows reloading during operation.

Fig. 14.9—Milling a forged fork on a 50-hp vertical heavy-duty miller. Chuck type table simplifies set-up, reduces handling time, improves accuracy, and rigidity. Photo, courtesy Kearney & Trecker Corp.

difficult to handle or require excessively intricate fixtures may prove worthy of consideration for redesign in view of the foregoing. Another point of interest is in regard to machine spindles — a horizontal spindle may be found preferable for heavy milling cuts. Naturally some operations require and can be accomplished best by a vertical spindle and cutter, but a smaller rate of metal removal and a less satisfactory surface finish are the usual penalties unless the machine has the required rigidity for good performance, *Fig.* 14.9.

Machines Available. An interesting fact which is undoubtedly well-known and recognized is the wide diversity of milling machines which can be called into play by the designer. Actually, with the exception of extremely specialized applications, practically any reasonable milling operation can be handled economically and rapidly when due consideration is paid the selection of the right machine for the job at hand. However, where quantities to be processed are only nominal and purchase of the right machine cannot be justified costwise, knowledge of the capabilites and limitations of the machines at hand will enable the engineering of usually reasonable and economically satisfactory designs.

Column and Knee Millers. Primarily intended for general-purpose work in low and medium production, column and knee millers usually fall into several groups

Fig. 14.10—Column and knee miller available in plain or universal styles. Photo, courtesy Cincinnati Milacron Co.

— plain or universal with horizontal spindles, *Fig.* 14.10, vertical with vertical spindles, *Fig.* 14.11, and bench type universals, *Fig.* 14.12. Plain types permit longitudinal, transverse and vertical table motion, while universal types permit swiveling of the table in addition. Vertical table motion is attained through actuating either the table or cutting head unit. Horizontal spindles have rotating motion only while vertical spindles need not be fixed but can have hand or power feed. Vertical machines are used on face and end-mill work, *Fig.* 14.13, and frequently have rotary tables for generating cylindrical surfaces. All three types are available in a wide range of light and heavy-duty capacities and varieties of construction.

Fig. 14.11—Column and knee miller of vertical style. Photo, courtesy Cincinnati Milacron Co.

Fig. 14.12—Standard bench type universal, equipped with a special fixture for milling gun barrel rifling cutters.

Bench type universals incorporate all the features of larger floor models, but of course are limited to somewhat smaller parts where power demand does not exceed one horsepower. For production work, models are available with automatic table feed making it possible to mass produce small, accurate parts with minimum cost in machine and tooling.

Hand Millers. These machines are usually of the column and knee type with one or two spindles. The dual-spindle machine can have the spindle spacing adjusted to suit. Table movements are hand controlled, providing high sensitivity which makes them ideal for delicate operations on small parts. Use is confined principally to the manufacture of small quantities of light work.

Fig. 14.13—Face and end mill work is often performed on vertical millers. Chuck tables used to hold this casting. Courtesy, Giddings & Lewis, Inc.

Fixed-Bed or Manufacturing Millers. Manufacturing type machines, as can be surmised, are designed primarily with an eye toward efficient production operations rather than overall general utility. They provide an extremely rigid construction and simple automatic operational characteristics for maximum economy in the mass production of small, medium and large-size parts. Light, medium or heavy-duty machines are available in either simplex or duplex models. The simplex miller, has but one column on which the vertically adjustable spindle housing is mounted or in some cases the bed is made adjustable while the duplex has two opposed columns with vertically adjustable spindles. As indicated, fixed beds provide controlled longitudinal travel only. Some plain automatics can be provided with a rise-and-fall movement to further add to the flexibility in production, *Fig.* 14.14.

Fig. 14.14—Plain automatic production miller for small to medium-size parts can be provided with a rise-and-fall cycling mechanism. Photo, courtesy Cincinnati Milacron Co.

Automatic Rise-and-Fall Millers. Because of the limitations of the automatic manufacturing millers another type of machine has become available and is known as the automatic rise-and-fall miller, *Fig.* 14.15. These machines perform an

Fig. 14.15—Automatic rise-and-fall production miller. Cutter can be made to rise and fall as required with standard equipment. Photo, courtesy Cincinnati Milacron Co.

additional classification of operations not possible on regular manufacturing machines — those operations which require the automatic raising and lowering of the spindle carrier to feed down to depth, rapidly traverse the cutter up or down to clear an obstruction, series of obstructions, or combinations of these requirements. Automatic vertical movement of the spindle from $1/4$-inch to 6 inches is available, all of which movement can be used for rapid advance of cutter to work, down feed at slow rates, varying combinations of both, or for rapid retraction of the cutter from the work. Cycles required can be quickly changed by means of preset cycle-selector camshafts. These machines permit the mass production of difficult parts such as those required for machine guns, rifles, and side arms as well as many automotive, aircraft, business machine and other parts.

Fig. 14.16—Rise-and-fall millers can be set to operate with completely automatic cycling. Photo, courtesy Cincinnati Milacron Co.

Automatic Profile Millers. By means of automatic cycle selection, rise and fall millers can be employed to produce profile shapes automatically, *Fig.* 14.16. The machines require only the setting of a cycle selector and a few dogs; then operation is automatic as desired.

Tracers and N/C Millers. Originally devised to produce difficult toolroom jobs, automatic tracer millers have been widely used for the rapid production of many difficult parts required in large or small quantities. Today, these machines are augmented or replaced, depending on the economics, by N/C machines of various types. Toolroom machines such as the Keller automatic which traces directly from a low-cost sheet metal template are widely used for producing complicated parts such as irregular machine cams, *Fig.* 14.17, or precision aircraft parts, *Fig.* 14.18, which require machining all over. Where specifications are stringent or repeat

Fig. 14.17—Large, irregular cam track accurately machined from a simple metal template by means of the Keller automatic. Photo, courtesy Pratt & Whitney Div.

identical pieces are a necessity, the N/C machine is ideal, *Fig.* 14.19. For mass-production jobs a number of regular heavy-duty production-type machines are available with automatic tool changers. Sensitive and accurate, these machines are designed for continuous operation in producing medium to large-size parts requiring accurate profile shapes, *Fig.* 14.20. A variety of either vertical and horizontal machines are available for the production milling of odd shape parts such as connecting rods, complicated gear housings, extremely large or small cam tracks of intricate shape, etc. Intricate machine parts can be profiled in stacks or groups where high accuracy for interchangeability is necessary but quantity does not warrant expensive tooling.

Planer Type Millers. The common makes of planer type millers are primarily intended for producing one or more long, straight surfaces on large, heavy machine parts, *Fig.* 14.21, or on a group of parts arranged in a series of fixtures on the table. Table travel on these machines resembles that of the ordinary planer. Designed for

Fig. 14.18—Keller automatic set up to finish aircraft propeller shaft reducer bevel gear housings to close limits, left.

Fig. 14.19—Feed box casting, with 83 operations using 22 tools, is machined on three sides in one operation saving almost 3 hours per unit in production. Photo, courtesy Cincinnati Milacron Co.

a wide variety of milling jobs, milling heads may be mounted on either of the two side columns or on the horizontal crossrail with common or individual drives. Those for gang milling usually employ a single horizontal spindle.

Rotary Millers. Usually of the vertical-spindle type, rotary millers, *Fig.* 14.22, are employed for contour and slab milling operations, channeling, milling tongues and jaws, trepanning, end milling, etc. These machines, however, are suited primarily to the production of large, heavy components in small or medium quantities. Where mass production is contemplated on parts with surfaces which can be face milled in a single pass, a single-purpose type of rotary miller is employed. Jigs or fixtures mounted on the table provide for virtually continuous cutting, parts being loaded, passed first beneath the roughing and then the finishing spindle, and automatically released for replacement with a rough part, *Fig.* 14.23.

An unusual departure from conventional rotary machines is the rotary-head machine illustrated in *Fig.* 14.24. Suited primarily for tool, die and small-quantity production, this miller adds a bit to the versatility of the rotary-table machine but is designed for much smaller parts. Intricate radial cam work can be produced readily from drawings without the use of templates.

Another rotary, the drum type machine, having as many as five horizontal spindles is designed for mass production of large parts such as motor blocks, gear cases, clutch housings, etc. On these machines, parts are carried in fixtures mounted on a drum which rotates continuously, carrying the parts between face mills which can be adjusted to suit, *Fig.* 14.25.

What might be termed another type of rotary machine is the vertical offset miller, which also uses a rotating table with fixtures. Cutters are carried on a short

Fig. 14.20—Large N/C production milling center at work on very large intricate castings requiring a multitude of operations. Photo, courtesy Kearney & Trecker Corp.

Fig. 14.21—Modern version of the adjustable-rail gantry machine. Eight axes of this 800-ton machine not only machine flat surfaces but the other surfaces required on large turbine parts. Photos, courtesy Ingersoll Milling Machine Co.

vertical spindle which rotates on an axis eccentric to the table axis. Suited primarily to production of small machine parts, the offset miller can perform facing, slotting, sawing, straddle milling and some simple form milling.

Fig. 14.22—Rotary milling machine for special contours has a 54-inch diameter table and separately controlled table and spindle housing adjustments.

Fig. 14.23—Two-spindle rotary miller provides continuous semi-automatic production. The oil pump casings shown are finished at a rate of 190 per hour. Photo, courtesy Ingersoll Milling Machine Co.

Eccentric Drum or Planetary Millers. Employing eccentric drums carrying the milling spindle, planetary milling machines, *Fig.* 14.26, derive their name from the automatic planetary action obtained, *Fig.* 14.27. Adapted to internal or external

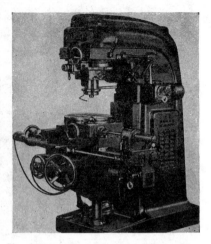

Fig. 14.24—Above—Vertical miller with powered rotary head in addition to usual functions. Photo, courtesy Kearney & Trecker Corp.

Fig. 14.25—Left—Three-spindle drum miller for rough and finish carbide milling cylinder-block end of a clutch housing and milling one cut on transmission end. Operation is continuous, 91 housings being finished per hour. Photo, courtesy Ingersoll Milling Machine Co.

Fig. 14.26—Eccentric-drum automatic thread and form miller. Photo, courtesy Hall Planetary Co.

thread milling where high production is contemplated, the planetary machine simplifies the manufacture of parts which are difficult to hold or swing. It also simplifies the production of concentric bores or diameters, either plain or threaded, *Fig.* 14.28. Both cutter rotation and feed are provided by the eccentric drums, the part milled being held stationary during the processing. For thread milling with multiple-thread mills, the drums are provided with a lead-screw to feed and finish an entire thread in one and a small fraction of a revolution. Range of work possible runs from about 5/16-inch to 20-inch external diameters and from about 5/8-inch to 20-inch internal diameters with thread leads up to 1.125-inches. Length of milled sufaces in most cases should not exceed 1 to 1 1/2 times the diameter. For special operations many single-purpose variations of the planetary principle in both horizontal and vertical models are employed. However, in most cases these machines require special cutters and fixtures and consequently are suited primarily to mass-production output.

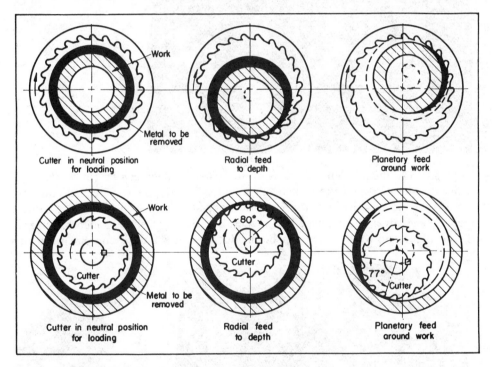

Fig. 14.27—Planetary action and cutter path of eccentric-drum miller on internal and external work. Drawings, courtesy Hall Planetary Co.

Offset Thread Millers. Somewhat less critical in their requirements and limitations regarding cutter design are the more common automatic offset thread millers, *Fig.* 14.29, in which both the part and the cutter are rotated. Unlike the eccentric-drum miller, the amount the cutter can be offset is fairly large. These machines handle a wide range of internal and external threads, utilize standard or special, single or multiple-thread mills, and are available with either hand or power collet chucking, air chucking, or with hollow spindles having pot chucks or fixtures. External thread length is only restricted by the particular machine size, a standard 4 1/2-inch machine handling diameters up to 4 1/2-inches by 36 inches in length. A 6-inch machine, on the other hand, handles diameters up to 6 inches and

Fig. 14.28—Aircraft propeller hub milled in planetary machine. Relation of tongues is held within 0.001-inch. Photo, courtesy Hall Planetary Co.

lengths to 120 inches. Thread forms up to $^{11}/_{32}$-inch deep can be cut on the $4^1/_2$-inch machine while on the 6-inch, up to $^{11}/_{16}$-inch depths can be had. Internal threads are more limited, maximum collet capacity on the 6-inch machine being 4 inches, swing over the carriage $9^1/_2$ inches, and over the bed, 16 inches. Internal thread depths up to $2^3/_4$-inch can be produced with standard equipment on this

Fig. 14.29—Steep lead, multiple-start screws being produced on an automatic offset thread miller. Photo, courtesy Pratt & Whitney Div.

miller. Other machines, standard and special, *Fig.* 14.30, allow precision external and internal milling and threading of diameters ranging up to cannon breech-lock threads, and milling of large screws, etc., *Fig.* 14.31. Primary limitation, however, is that parts must be chucked, colleted, held between centers, or swung in jigs or fixtures for the milling operation.

Fig. 14.30—Wanderer thread milling, spline milling and hobbing machine. Courtesy, Myers Tool Co.

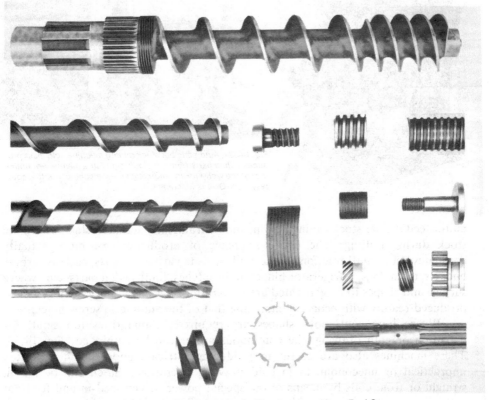

Fig. 14.31—Typical parts milled on the Wanderer thread miller. Courtesy, Myers Tool Co.

Special-Purpose Millers. The number and design of special-purpose millers is probably as unlimited as the design of machine parts themselves. However, with the exception of a few machines, special-purpose millers are designed to perform a single job made necessary by a particular machine part design or by failure to recognize the limitations of readily available standard millers. Where quantity is sufficiently large a special machine can, in many cases, offset its cost by virtue of increased output and automaticity. It is a fact, though, that many, many cases continue to recur which point up the value in considering machine capabilities during early design stages.

The counterpart of the automatic screw machine — the bar-stock automatic miller — for mass production of surfaces, forms, shapes, etc., adapted primarily to milling, utilizes bar stock of almost any cross section desired, mills it to a specified shape or shapes, and cuts it off. The part at this point is either completely finished or ready for final finishing.

Shown in *Fig.* 14.32, this automatic miller has a three-spindle head hydraulically actuated in a vertical direction. The cycle is continuous, including rapid advance down, feed forming cutters, jump feed, cutoff saw, reverse, and rapid return up. One lower spindle is adjustable. Rollers powered by a torque

Fig. 14.32—Automatic bar stock milling machine, left, mass produces milled parts from bar stock, right, in a manner not unlike that of the well-known automatic screw machine. Photos, courtesy Kent-Owens Machine Co.

motor feed the bar stock against a stop, and hydraulically actuated clamps hold the stock during milling. The entire sequence of producing one piece actually consumes but a small fraction of a second. A wide variety of parts, such as turbine buckets, latch dogs, refrigerator hinges, door latches, limit switch parts, etc., where surface and shapes to be generated are in parallel relation as in broaching, can be produced readily with economy matching that of the automatic screw machine.

Where long continuous shapes are required in quantities too small for extrusion or cold drawing, the continuous miller can be employed, *Fig.* 14.33. These machines also are useful in producing intricate undercut shapes often impractical or uneconomical in cold drawing operations. Stock can be fed in straight or from coils by means of the special power-driven feed-in and feed-out rolls.

Fig. 14.33—Continuous miller for production of shape profiled bars. Photo, courtesy U.S. Baird Inc.

Design Considerations

Manufacture of machine components to suit specific theoretical or calculated design requirements often, and justifiably, will require specially developed production equipment. This holds true with milling as well as with various other basic methods for processing parts. As a rule, however, such parts are usually special (as well as the machine into which they are assembled), quantities are small or design limitations so critical that changes are impossible or unreasonable.

Gang Milling. Quantity production, on the other hand, to be as economical as possible, dictates maximum simplicity in the design of parts, minimum number of processing operations, maximum ease of holding, locating and clamping, etc. With parts to be milled, this would indicate design for face or plain milling wherever they can be used.

Wherever possible in milling, part design should be such that the maximum number of surfaces can be milled at one pass. It is well to remember that once a part is in the fixture, it takes no more production time to mill three or four surfaces than it does for one, so long as the miller used can carry the cutters, *Fig.* 14.34. Also, relocating and reclamping a part is always a likely source of errors and consequently, accurately related surfaces are most easily produced at one setting.

Simplifying Fixturizing and Locating. Keeping in mind the problems incident to practical fixture design — rapid loading, accurate and easy locating, solid clamping without cutter interference, etc. — will assist materially in reducing many of the production problems commonly encountered. Elimination wherever possible of difficult, hard-to-reach, complex surfaces as well as all slow operations will help to achieve the economy of production so necessary to produce low-cost parts.

Optimum Metal Removal Rate. To insure advantages of optimum metal removal rate and surface finish, surfaces to be milled should be narrow and where possible designed contemplating the use of medium-diameter cutters. This

eliminates the necessity for superhigh spindle speeds which are usually required
with small-diameter cutters. Use of extremely large-diameter face mills and long
slab mills should be avoided. Again, flimsy or frail sections should be avoided
inasmuch as they usually necessitate slow speeds and extremely light cuts to insure
against breakage, distortion and warpage.

*Fig. 14.34—Gang milling a light cast-iron frame with carbide cutters. The 4-inch surface area is finished at 16
inches per minute, producing 128 pieces per hour. Photo, courtesy Midwest Tool & Mfg. Co.*

End-Milled Surfaces. As mentioned in the foregoing, horizontal-spindle
machines are preferable for heavy, high-speed work. Thus, it is well to avoid
designs which require end-mill work. Where end-milled surfaces are necessary,
however, cutters of 1-inch diameter or greater should be used if at all possible.
Small-diameter end mills are frail and require slow speeds as well as relatively
large tolerances on the parts milled. Radii produced by end milling — both those
created by the side of the mill as well as those produced by the edge or corner
radius — should be kept constant, *Fig.* 14.35, to obviate the need for two
operations. As a rule, corner or edge radii to be produced by end mills should be

limited to a maximum of 50 per cent of the cutter radius. Preferably, radii should be small to improve cutting action on the face or end of the cutter, 0.050-inch being an ideal minimum radius.

Narrow Slots. In specifying slot widths under one-inch it is usually preferable to use nominal fractions and assign a smaller dimension to the mating tongue. This avoids the necessity of special-width slotting cutters and the tongue part can be milled as desired with an adjustable straddle cut. Standard slotting cutter widths run in $1/16$-inch increments up to $1/2$-inch, in $1/8$-inch increments up to 1-inch and those over 1-inch are special but should be specified in $1/8$-inch increments.

Wide Slots. Where wide slots are contemplated — over one inch — design should provide for the use of halfside mills for maximum economy. Since these mills cut only on one side and periphery, relief or clearance at the middle of the slot is necessary, *Fig.* 14.36. Unless relief can be readily provided by casting, forging, etc., special-width or adjustable interlocking cutters must be provided, often at extra cost and slower production.

Fig. 14.35—*Proper standard radii for end-milled surfaces should always be used to assure the minimum in tooling and setup costs.*

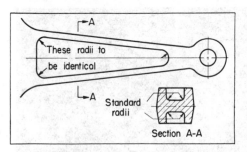

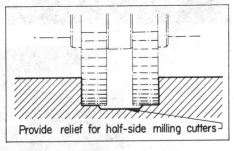

Fig. 14.36—*Wide slots should be specified as shown to simplify accurate milling.*

Slot Depths. Slots of greater depth than three times the cutter width should be avoided. If they have to be cut, they should be made on a machine with climb cutting provisions, good accuracy and rigidity. The power available should be high. It will then be possible to mill slots at comparatively high speeds and feeds. When milling slots 0.014-inch wide and 0.125-inch deep in 1330 steel, finish was rough and much breakage was encountered at a cutting speed of 100 sfpm. An increase in the cutting speeds of the 3-inch diameter, high-speed steel cutter to 235 sfpm with climb milling under a copious supply of cutting oil, permitted better tool life and higher accuracy at a feed rate of $12^{1}/4$ ipm. The aircraft part shown in *Fig.* 14.37 is typical of the deep slotting operations possible in aluminum alloys using high-speed milling machines.

In calculating maximum depth of slot for a specific standard peripheral cutter, allow $1/4$-inch clearance between the arbor collar and the work surface. End-milled slots using shell end mills can be much deeper of course, and are sometimes justified, even at some sacrifice in the cost of the milling operation, *Fig.* 14.38.

Projections or Lugs. Where lugs or ears are specified, advantage should be taken of the rapid combination surface and slot milling possible by means of cutter "gangs", *Fig.* 14.34. Bottom of slots as well as outer surfaces of such lug or ear projections should have the option of plunge or through milling, whichever is most economical, by allowing for either a radius or straight bottom edge, *Fig.* 14.39.

Also, allowance should be made for standard cutter corner radii which, naturally, should be specified to obviate special grinding or cutters. As a rule the corner radius for a 1/32-inch wide cutter is 0.005 to 0.015-inch; for a 1/16-inch wide cutter,

Fig. 14.37—Aircraft part deep slotted on a high-speed machine.

Fig. 14.38—Surface finish must be sacrificed where long spindle extensions are required, and such designs are well to avoid if possible. Photo, courtesy G.A. Gray Co.

0.010 to 0.030-inch; for 3/32 to 3/16-inch wide cutters, approximately 0.030-inch; etc.

Keyways. Regarding keyways, it is preferable to design for use of a slotting cutter rather than an end mill. Thus unless a Woodruff or a special key is necessary, keyway ends should be shown with a standard cutter radius runout, *Fig.* 14.40, although an actual runout radius need not be specified. As a rule, multiple keyways in a common line are preferred to one long plain keyway. Multiple plain keyways spaced radially around a shaft are not considered good practice for ordinary milling; angular tolerances are sometimes difficult to maintain unless a special spline miller is available. Standard splined sections should be specified in such cases and designed for production by hobbing according to the accepted standards.

Angular Faces. Angles between milled faces should be 90 degrees wherever possible. To guarantee easy cleanup as well as to avoid special form cutters where only one face requires machining, the angle should be greater than 90 degrees or the design should allow for a shoulder, *Fig.* 14.41. In such cases it is often impractical to attempt to match surfaces. Undercut shoulder portions should, of course, conform to standard 30-degree, 45-degree and 60-degree cutters available. Internal corners of such cuts should be shown with a 1/32-inch to 1/16-inch fillet rather than a sharp corner, both simplifying grinding and prolonging cutter life.

Corners and Edges. Where external edges are to be rounded, standard concave cutter radii should be specified. The cutter radius used, however, should be slightly greater than the tangent radius to avoid undesirable nicking of the sides of the part otherwise encountered. If a full 180-degree radius is desired, these also should conform to standard cutter dimensions. As a rule when plain external or corner radii are greater than obtainable with standard cutters, especially if production is to be high, economies of production speed as well as cutter and sharpening costs dictate the use of bevels in lieu of radii. This makes possible the use of ordinary face mills and often carbide face mills. Naturally, if the radius is along a compound

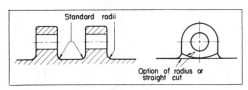

Fig. 14.39—To simplify production problems, lug design should have the option of flat or radius-milled corners.

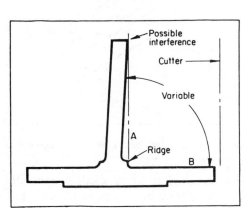

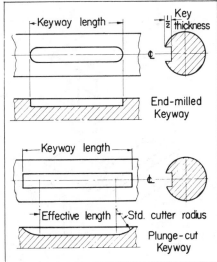

Fig. 14.40—Keyway which allows standard cutter runout is preferable to one end-milled.

Fig. 14.41—Where surfaces need not be matched a shoulder adjacent should be allowable.

surface a special mill can seldom be avoided.

Cutters and Radii of Cuts. As mentioned previously, avoidance of cuts requiring small-diameter cutters is advantageous, especially with high-speed milling where the cutter body acts somewhat as an added flywheel. Thus 6 and 8-inch diameter cutters are widely used and preferred. Obviously, the radii generated by such cutters are 3 and 4 inches, but in general by specifying radii 2 inches or greater, production can be greatly facilitated.

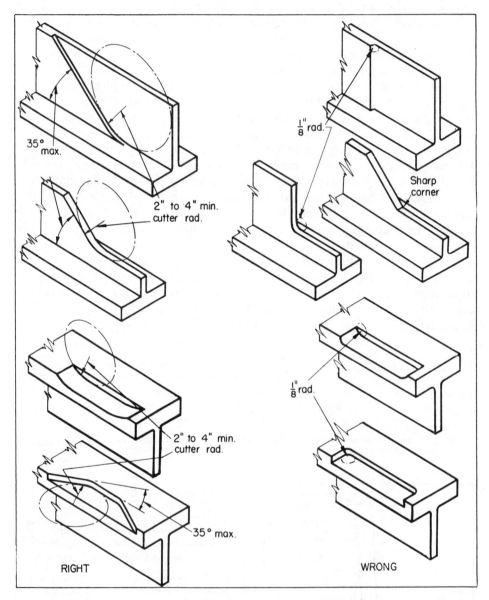

Fig. 14.42—Some of the right and wrong methods of designing long, complicated parts to be produced on a tracer or N/C miller.

Careful consideration of the machine and cutters to be used is especially valuable on intricate or complicated work such as long spars, etc., which normally necessitate an automatic tracer miller or N/C machine. The details shown in *Fig.* 14.42 indicate good practice in designing such parts. Since it is undesirable to stop the cutters until the entire part has been finished, design details should be such as to allow the cutter to plunge into and out of the work as required in following the pattern. As noted, it is impossible to obtain the exact cutter radius if more than 35 degrees of cutter circumference is required. In fact, it is best to avoid producing the exact cutter radius; it is much more satisfactory if a larger radius is allowable.

Locating Surfaces. With most small parts trouble is often encountered in providing accurate location in fixtures. Serious consideration should be given this problem, and where parts to be milled are large, say automotive engine blocks, large machine beds, etc., and all surfaces available are curved or slanted, it is advisable to consider the addition of cast or welded-on pads or projections to provide the locating surfaces. Similarly in certain cases such pads are valuable for holding or clamping.

Complex Surfaces. Specific limitations of design inherent in each of the methods and types of milling previously outlined are apparent to a certain extent from the discussion. Of primary interest in this respect is the versatility of the automatic-tracer type millers mentioned. An almost unlimited range of complex surfaces can be produced in quantity with these machines and at reasonable cost. Parts for guns and other machines necessarily designed for functional perfection can be readily produced by means of one of the methods of complex surface milling: (1) Form-relieved or shape-profiled cutters, the contours of which are reproduced in reverse on the workpiece; (2) standard cutters, the paths of which are varied relative to the workpiece to reproduce the desired contours, and; (3) a combination of these two methods in which the profiled cutter is made to follow an irregular path. Along with the automatic tracing or N/C machines, rotating or oscillating fixtures, synchronized rotation of fixture with table or spindle, gangs of plain, formed, or plain and formed cutters, etc., all add to the design possibilities.

Use of form milling, however, should be carefully considered in the light of the production difficulties with such cutters. It is often difficult to obtain desirable tooth-angle relationship in form-relieved cutters; they require painstaking care in design, are difficult to check dimensionally and, most difficult of all, require infinite care in maintaining cutter form and dimensions through a series of resharpenings. Probably the greatest limitation of the form or profile cutter is its single purpose. Such cutters can be obsoleted by a single necessary design change and consequently should be avoided wherever possible for maximum economy in overall production.

Thread Milling. Thread milling is exceptionally well adapted to production threading of blind holes or large and coarse thread forms as well as for any threading job where high accuracy, alignment and finish are necessary. Although most multiple-thread milling cutters are necessarily special, they make possible the removal of the imperfect start portion of both male and female threads of any type, a feature of design not economically available by means other than some form grinders. Too, by means of eccentric-drum millers cylindrical surfaces can be milled concentric with threaded portions, *Fig.* 14.43.

Special Machines. Touching upon the design possibilities and limitations of the automatic bar stock miller, the parts shown in *Fig.* 14.44 are a few that have been produced on these machines and present a fair idea of the range and versatility. As a rule, gangs or stacks of cutters rather than form cutters are preferred for this machine; staggering the individual cutters obviates chatter by maintaining two or more teeth in the cut at a time. The designs shown in *Fig.* 14.45 indicate a few of the shapes ordinarily possible to produce on the continuous form miller.

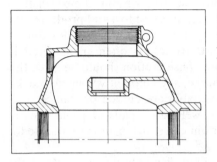

Fig. 14.43—Left— Differential carrier on which thread and two surfaces are milled simultaneously. Drawing, courtesy Hall Planetary Co.

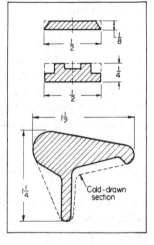

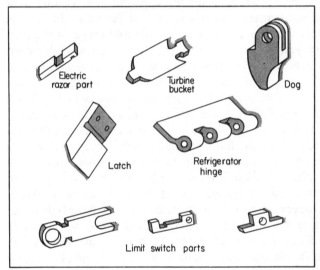

Electric razor part

Turbine bucket

Dog

Latch

Refrigerator hinge

Limit switch parts

Fig. 14.44—Left—Parts produced on the automatic bar-stock miller indicate the range of design possibilities. Tolerances range around plus and minus 0.001-inch.

Fig. 14.45—Above—A few of the many shapes possible to mill in continuous bar form. Drawings, courtesy Kent-Owens Mfg. Co.

Selection of Materials

Normal rates of metal removal at accepted speeds and feeds common to high-speed steel cutters have long determined to a considerable extent the suitability of a material for milling. Naturally, this eliminated a good many of the low-machinability or totally nonmachinable metals from serious consideration unless their physical characteristics were imperative. The advent of high-speed milling, however, has brought into the practical production range most of these materials, some of which can now be machined at high rates.

Low-Carbon Steels. The low-carbon steels ranging up to 10 points of carbon are highly desirable for many electrical applications and are considered most suitable for weldments, but necessary milling feed rates around one-inch made their use impractical in many cases. With proper technique these materials can now be milled successfully at high feed rates and speeds ranging up to 1000 surface feet per minute while that for ordinary steels ranges in the neighborhood of 500 surface

Fig. 14.46—Milling the surface of 0.40-carbon steel rack forging with negative rake, 12-inch diameter flywheel type carbide cutter at 24 ipm feed and 475 sfpm speed. The 50-hp head removes 50 cubic inches of metal per minute. Photo, courtesy G. A. Gray Co.

feet per minute, *Fig.* 14.46. In addition, surface finish, while not as good as with harder materials, is highly satisfactory. Also included in this low-carbon category is wrought iron, once considered an incorrigible as far as machining is concerned.

Alloy Steels. Milling techniques likewise affect the alloy steels. All alloy steels such as the chrome-manganese and chrome-moly series can be milled in the heat-treated state, ferrous materials brinelling as high as 400 being successfully milled. In one pass it is sometimes possible to mill AISI 4340 steels, in heat-treated state, to superior finish and size specifications with little or no heat effects since in high-speed carbide milling, often performed dry, chip removal is so fast that the heat of cutting is retained primarily in the chips, *Fig.* 14.47.

Work Hardening Steels. This speed of metal removal makes possible the economical milling of another group of metals — those with work-hardening characteristics. Thus the armor plates, stainless steels and similar materials can now be milled at feed rates up to 25 inches per minute and cutter speeds as high as 900 surface feet per minute. Even corrosion-resistant materials such as Ni Resist and type 304 stainless can be readily handled with carbide cutters.

The important factor in milling these work-hardening materials is that the cuts taken for the most part must be short so that the cutting tips are in the work the shortest possible time. In this manner the chips are removed before any great amount of work hardening can take place and minimum heat is transferred to the

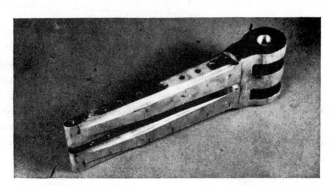

Fig. 14.47—Aircraft wing hinge forging of 4340 heat treated to 402 brinell, with a slot 10-inches deep by 0.505-inch wide, milled at 13 inches per minute with a carbide cutter to plus or minus 0.005-inch on width and 0.002-inch on parallelism. Finish ranges from 8 to 10 micro-inches.

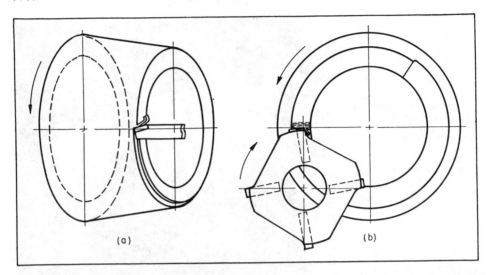

Fig. 14.48—Cast steel garbage disposal grinder ring turn milled successfully as at (b) whereas plain turning (a) work-hardened the material to the extent of severe tool breakage. Drawing, courtesy Kearney & Trecker Corp.

cutting tip. Designing to suit these conditions will allow use of such materials where design requirements so demand. The cast-steel ring shown in *Fig.* 14.48, having a composition comparable to high-speed steel, proved impractical to finish in a lathe as would normally be done inasmuch as the tools failed after finishing but one piece. Machining the same piece on a rotary-table vertical milling machine as indicated was accomplished without trouble or tool breakage. The detrimental work-hardening of the material is not carried as far as in a continuous lathe cut and tool tips have a chance to cool in air between cuts.

Lightweight Alloys. A general idea of where some of the various engineering materials fit into the milling picture may be gained from TABLE 14-I, which also gives the relative machining characteristics with high-speed cutters or carbides. From this table the ideal characteristics of the lightweight nonferrous alloys are quite apparent, hence these metals deserve serious consideration in design of all types of machine parts. At surface speeds of 12,000 to 15,000 feet per minute at the cutting tips, surface finish can be obtained ranging from 10 to 15 microinches even at feed rates as high as 300 inches per minute. In addition, under proper conditions tool life is extremely long.

Specification of Tolerances

Wherever cutters with carbide or cast-alloy tipped blades can be utilized along with higher speeds, milling tolerances which can be readily maintained in production reach a minimum as do variations from piece to piece. With ordinary milling methods using high-speed steel cutters, tolerances generally range around plus or minus 0.005-inch with surface finish from 50 to 250 micro-inches. To obtain the finest finish or dimensional accuracy, both roughing and finishing cuts are usually necessary. With carbide-tipped cutters, however, average surface finish obtainable in production ranges from 20 to 40 microinches and dimensional tolerances as close as plus or minus 0.0005-inch can be readily held in many instances.

Surface Finish. Generally, as hardness increases, the finish produced with carbide cutters improves and surfaces under 10 microinches have been produced, *Fig.* 14.47. The value of fine surface finish as well as virtual nonexistence of tool marks on such highly stressed parts is self evident and deserves serious consideration.

Flatness. The surface flatness obtainable at high production speeds is of much value. Leakproof joints are possible without the use of gaskets — flatness being of an order such that nongasketed hydraulic joints operate satisfactorily under 1200-psi pressures. Grinding operations ordinarily required on such surfaces can thus be eliminated, reducing cost considerably. An idea of surface finish possible can be gained from *Fig.* 14.49, the low-carbon weldment shown having been finished by a series of 4-inch passes to an accuracy such that the matching lines are visible but of such extremely minute character as to be virtually nonexistant. Naturally, such

Fig. 14.49—Low-carbon weldment showing a large surface carbide milled area. Joining marks cannot be detected by profilometer.

results can only be had on machines in first-rate condition and with proper operation.

Grooves. In producing long, accurate semicircular grooves, the shape-profiled inserted-blade cutter shown at left in *Fig.* 14.50 — in effect a slab mill — presents difficulty because cutter runout, revolution marks, etc., are hard to eliminate. Where accuracy and finish must be of a high order, say in the neighborhood of 100 microinches, the type cutter shown at the right will produce a groove which is straight within plus or minus 0.002-inch in a 10-foot length while diameter can be held within 0.001-inch.

Fig. 14.50—Cutter in right view will mill a long, semicircular groove to greater accuracy than the profile slab mill shown at the left. Photos, courtesy G. A. Gray Co.

Cams. Radial cams of various sizes can be produced on rotary-head or rotary-table machines with cam groove radii held to plus or minus 0.001-inch. Special tracer millers have regularly produced aircraft propeller cylindrical cams to within an accumulated error of less than 0.0005-inch.

Threads and Forms. Diametric accuracy of forms and threads produced on eccentric-drum thread millers is unusually good, breech ring lock threads of 3 to 4-inch diameter being readily produced to plus or minus 0.001-inch on dimensional limits and parallelism. Coarse and fine-thread forms up to 10 and 12-inch diameters in both straight and tapered types have been milled in production to within plus or minus 0.003-inch on the pitch diameter. Closer tolerances can often be held but naturally necessitate more frequent cutter grinding and much higher costs. For instance, aircraft propeller threads are produced to plus or minus 0.0005-inch on diameter and on occasion ordnance threads have been held to plus or minus 0.00025-inch. On offset millers breech-lock threads of 7-inch major diameter, 3/4-inch pitch, are produced to plus or minus 0.002-inch on major and minor diameters while lead error per tooth is held to plus or minus 0.0002-inch and accumulated lead error per foot is within 0.002-inch, plus or minus. Where design demands, male and female threads can be produced with matching points varying less than 0.031-inch peripherally after assembly.

Aluminum Forgings. An important factor often overlooked in designing is that many aluminum forgings shrink lengthwise after the skin is cut away; if the shrinkage is nonuniform and the tolerances too close, the part may have to be rejected. Thus with aluminum forgings and also extrusions for that matter, limits should be as liberal as possible, especially between rough and machined surfaces, the best results being obtained by machining all over.

15

Drilling and Boring

T HOUGH the specification of holes may seem commonplace and appear to deserve small attention, the increased accuracy of today's machine components demands serious consideration of hole requirements. The area covered in the field of precision drilling, reaming and boring is extremely wide both as to range of practical hole sizes and production tolerances commensurate with the design requirements. Too, hole spacings complicate the picture and should not be overlooked since accuracy of drive elements such as zero-backlash gear trains and other precision machine elements is rendered valueless where center-to-center distances cannot be held to the necessary limits.

Precision drilling, reaming and boring, performed with standard and special

Fig. 15.1 — Typical heavy-duty radial drill. Courtesy, Cincinnati Gilbert Machine Tool Co.

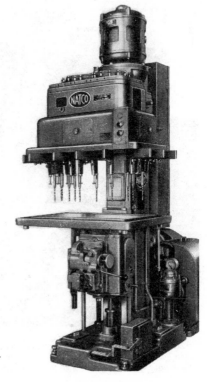

Fig. 15.2 — Power operated multiple-spindle driller with adjustable spindles. Courtesy, National Automatic Tool Co.

tools of almost unlimited variety on an equally diverse array of standard and special machines, ranks alongside milling in importance. Whether few or many parts are to be processed, equipment necessary for manufacture is readily available. Granting proper consideration of practical tolerances which can be held in continuous production, and also of the effects of quantity to be produced or of production speed, maximum economy and satisfaction can be assured.

Drilling Equipment. Machines available for performing drilling operations are many and varied. The common drill press is widely used for drilling plain holes in simple, small and medium-size parts while for larger size work the heavy-duty radial drill is usually employed, *Fig.* 15.1. Where mass production of holes is required automatic machines are normally the only economical solution. These may be in the form of multiple-spindle units for producing many holes at once, *Fig.* 15.2, or for small, intricate parts, an automatic-sequence type machine capable of being retooled to suit numerous design requirements, *Fig.* 15.3. Many additional operations such as tapping, light milling, reaming and screw insertion as well as assembly by spinning over or staking can be performed with these machines.

Complicated drilled parts or those requiring a large number of drilled holes accurately located are produced very often on "package unit" or transfer machines which allow positioning of each component for automatic drilling as required, *Fig.* 15.4. Similar machines are also used widely for large and intricate parts to produce all related holes at one loading. With these machines many additional operations such as reaming, counterboring, light milling, etc., can also be performed.

Where abnormally long holes are needed such as oil passage holes in camshafts, etc., special deep hole drilling machines are used. Such machines range in size up to those utilized for drilling gun barrels.

Fig. 15.3 — Automatic dial-feed multiple-spindle drilling machine. Part is a plate for a mechanical time fuse. Seven holes, from 0.5 mm to 2.8 mm are drilled to center distance tolerance of 0.0004-inch at 600 pieces per 50 minute hour. Photo, courtesy The Bodine Corp.

Fig. 15.4 — Special dial-indexing-type machine for manufacturing operations such as milling and drilling of holes and pads on automatic transmission cases. Photo, courtesy Cross Co.

Hole Sizes. With most mass-production automatic drilling equipment, drilled holes ranging from that left by the smallest drill available, 0.002-inch, up to about 1-inch in diameter can be produced. With the heavy-duty machines larger drills can be used. Ordinarily, standard drills up to about 3 1/2 inches in diameter are available. Holes of such size, as a rule, are only drilled to remove stock. Reaming or finish boring is used to obtain a satisfactory hole finish and size.

Boring. While reaming is widely used to size holes more accurately than is possible with drills, boring as a rule must be resorted to in order to obtain the maximum in precision sizing, location and surface finish accuracy. Machine parts designed for boring may range from small pump units to extremely large parts such as frames, bases, flywheels, etc., *Fig.* 15.5. Although most bored holes present few problems, their relation to the remainder of the design is highly important and it is in this phase that difficulty is often experienced. In fact, it is because of this that many boring machines available today, excepting possibly highly specialized ones, are extremely versatile and include provisions for performing many additional operations such as milling, facing, drilling, tapping, etc., *Fig.* 15.6. Too, provisions are made for rapid, accurate location of bores without special fixtures or measuring devices.

Boring Machines. Available boring equipment may be divided into two distinct categories, the horizontal and the vertical. With horizontal boring machines, the work is held stationary on a table having in and out as well as back and forth movements. A vertically adjustable head carries the boring spindle which can be fed into the work, *Fig.* 15.6. Large, intricate, or bulky parts such as shown in *Figs.* 15.7 and 15.8 are normally machined on the large universal horizontal

Fig. 15.5 — Boring an 80-ton arc welded flywheel on a vertical boring mill. Photo, courtesy Westinghouse Electric Corp.

Fig. 15.6 — Boring, face milling and drilling gas turbine components using two large five-axis numerically controlled horizontal boring, drilling and milling machines. Photo, courtesy G.A. Gray Co.

machines. These are usually of the planer type, table type with standard or revolving tables, floor type, or multiple-head crossrail type machines. Many adaptations of these models and new N/C machining centers are also available, extending the scope and capacity range of hole boring considerably for the designer, *Fig.* 15.9.

Fig. 15.7 — *Large housing being bored in production. Photo, courtesy Lucas Machine Div., Litton.*

Fig. 15.8—*The 9.287/9.281 bore is milled in one pass using the PMC-35 machine's contouring capability. Courtesy, Giddings & Lewis Machine Tool Co.*

For convenience, where length or height is less than the diameter, the usual practice is to machine in a vertical boring mill. Vertical boring mills, however, revolve the work on a circular table while the tools, much as in a vertical turret

Fig. 15.9 — Large electric motor frame being bored on a rotary table type boring machine. A 30-inch extension of the spindle is used to finish a bore diameter to 10.501 inches. Courtesy, Giddings & Lewis Machine Tool Co.

lathe, are fed into and away from the work, *Fig.* 15.5. Clamped to the rotary table of these mills, parts such as plates, turbine bucket wheel forgings, flywheels, and the like can be turned and faced as well as bored.

Smaller parts, to be produced in low, medium or large quantities, are handled more economically on single or double-end automatic machines, one version of which is shown in *Fig.* 15.10. These machines are easily adapted to design changes and provide increased production speed. Other higher precision machines accommodate a range of bores from around 10 inches on down to extremely small-bore parts such as valve bodies, pistons, etc., *Fig.* 15.11. Units may be of standard design with single-spindle single-head construction, multiple-head or, in some high-production cases, of special multiple-spindle design such as that shown in *Fig.* 15.11. On standard machines of this type, table travel ordinarily is about 18 inches maximum.

Special Boring Machines and Jig Borers. Entirely special machines for extremely large-production precision boring are designed primarily for maintaining exact hole spacings and alignment, *Fig.* 15.12. Inasmuch as many parts, both by design and quantity, fall outside the scope of economy considering totally special machines, the jig borer is normally resorted to for precision results, *Fig.* 15.13. Where accuracy is such as to be beyond the scope of drills and drill jigs, the part too fragile to withstand heavy cutting loads, or the material subject to poor finish by ordinary methods, the jig borer is especially well suited. Placed in a fixture, the part can be finish-bored in the sequence required, by means of the gaging features on these machines, without relocation of the part during the process. As can be seen in *Fig.* 15.13, interchangeable heads are provided where bores vary more than 1/8-inch to simplify and speed production.

Contour Boring Machines. In addition to the standard and special boring machines for small and medium-size parts, contour boring machines are also available, *Fig.* 15.14. Whereas most machines produce plain bores or stepped plain

Fig. 15.10 — Double-end machines, foreground, set up for rough boring and, in background, for finish boring a range of 8 sizes of motor shells in production. Photo, courtesy National Automatic Tool Co., Inc.

Fig. 15.11 — Versatile special 8-station dial index that produces 14 sizes of ball check bodies in two different designs. One valve body style, left, is made from solid carbon steel or stainless castings or forgings in 8 sizes from 1/4-inch to 2 inches. The other is made from cored carbon or stainless forgings or castings in 6 sizes from 1/2-inch to 2 inches. Production rate is from 36 to 138 pieces per hour. Photo, courtesy Snyder Corp.

bores, undercut portions are often produced by means of special automatic boring tools. The contour boring machine, however, will produce lands, recesses, flanges, steps, counterbores, and radii as well as straight precision boring, *Fig.* 15.15.

Single-Point Boring. Fine boring on most of the precision machines is based primarily upon the facility with which a single-point tool will generate a straight,

Fig. 15.12 — Precision cylinder boring machine which simultaneously finish bores eight cylinders in automobile blocks in a fully automatic sequence. Photo, courtesy Ex-Cell-O Corp.

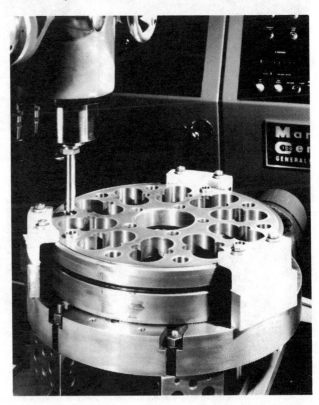

Fig. 15.13 — Precision boring a gear pump housing on an N/C jig borer with special tooling for holding hole size within 0.0005-inch and locations to within 0.0001-inch. Courtesy, Moore Special Tool Co.

round hole. Too, if the tool is tipped with diamond or cemented carbide, high speeds and fine feeds will produce bores to limits of accuracy and surface finish unobtainable by other means. Bores for parts such as bearings, valves, piston pins, connecting rods, etc., where size, roundness and alignment are paramount, preferably should be generated by single-point high-speed tools. Cutting speeds of about 450 surface feet per minute on cast iron and up to 1500 sfpm on aluminum and magnesium alloys are regularly used with feeds from 0.002-inch up.

Fig. 15.14 — Contour boring machine set up for boring automotive air conditioner — clutch pulleys to 0.0005-inch tolerance. Photo, courtesy Ex-Cell-O Corp.

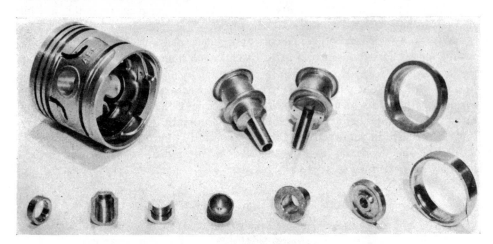

Fig. 15.15 — Assortment of typical machine parts bored on the precision contour borer shown in Fig. 15.14.

Jig guided drills and ordinary boring bars produce far less accurate results largely because of the running clearance necessary with the bushings that are used to guide them though, of course, they can produce more complicated bore designs and remove metal at a much faster rate. Thus on large work, *Figs.* 15.9 and 15.10, single-point tools are often not used when production speed and quantity justify more expensive arrangements. Multitooth cutters and form cutters are often employed in order to complete an entire part at one or two settings. However, practical hole tolerances, more expensive upkeep and the obsolescence factor must be contended with in such arrangements.

Production. Like the large universal boring machines, the jig borer is not limited to small-lot production merely as a jig eliminator. It can be employed on medium and even large-quantity work when, as noted previously, the required

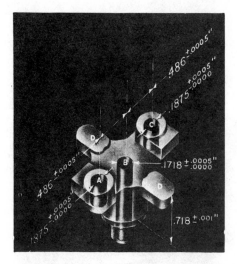

Fig. 15.16 — Aluminum stud jig bored on a production basis to limits shown required but 6 minutes per piece. Photo, courtesy Moore Special Tool Co.

accuracies cannot be had by means of a conventional jig. The aluminum stud, *Fig.* 15.16, is a good example of such a part. Required in a quantity of 5000 pieces, the holes were first drilled in jigs allowing about 0.012-inch for jig boring and then finish bored in a fixture to the limits shown. Total time per piece including milling the pads to height was 6 minutes, floor-to-floor.

With most of the universal or nonspecial drilling machines, however, production speed is limited to a great extent when numerous locating operations are necessary. Consequently, where maximum quantity production must be attained with machine parts involving numerous precisely located holes which may not be parallel and which are not within the scope of the average "package" type automatic machines, entirely special totally automatic machines are used, *Fig.* 15.17. Likewise, the same is true of boring, especially where a maximum number of additional operations can also be handled at one setting to assure the desired accuracy of location, *Fig.* 15.18.

Fig. 15.17 — Automatic six-station multioperation machine which drills, bores, reams, counterbores, and taps automotive front wheel spindles. Photo, courtesy Baker Bros. Inc.

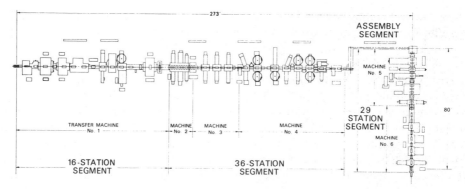

**SNYDER DIESEL ENGINE BLOCK
MACHINING SYSTEM**

Fig. 15.18 — Huge 353-ft.-long cylinder block machining system made up of 6 transfer machines and a machine for bearing cap assembly. The unique machine performs all finishing operations on block ends, head surfaces, etc., and also drills, reams, taps, rough bores, finish bores and hones. Circles show where blocks — three different designs — are indexed into new positions for machining. Photo, courtesy Snyder Corp.

Fig. 15.19 — Special multiple-spindle gun drilling transfer machine for connecting rods and caps. The system uses 52 single-flute carbide gun drills in six opposed multiple-spindle heads. Photo, courtesy Snyder Corp.

Design Considerations

Primary design consideration with respect to holes is largely whether such can be readily produced to satisfactory limits by drilling or whether the accuracy, size, or contours require a reaming or boring operation.

Drilling. Wherever possible, especially with holes 1-inch or under, drilling should be contemplated unless the method by which the part is processed originally permits production of the hole along with other features, as with die casting, molding, etc.

With drilled holes the major consideration is hole depth. As mentioned earlier, single-operation drilling with ordinary drills allows the production of hole depths up to about 5 times the drill diameter. Where a series of drilling spindles can be used successively, holes can often be drilled to depths of 8 to 10 times diameter. Such holes, however, often may not be sufficiently straight to be satisfactory. By using chip breaker attachments or internal oil hole drills with black oxide surface treatment, holes to as much as 20 to 25 diameters in depth have been produced. Conventionally, however, holes over 1/8-inch diameter to 50 diameters or more in depth are produced on deep hole drillers with deep-hole, crankshaft or carbide gun drills. Such drills run up to 2 inches in diameter and can produce holes 60 to 70 diameters in depth using high-pressure coolant flushing. Maximum depth for the larger sizes has run to 12 feet. Multiple spindle gun drilling concepts are used in special cases, *Fig.* 15.19.

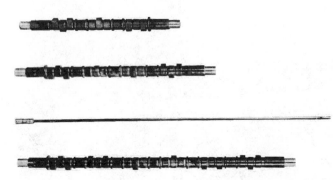

Fig. 15.20 — Drilled camshafts, longest of which is 43 inches, carburized prior to drilling to a hardness of approximately 260-270 Brinell. Crankshaft or gun drill used for boring oil holes through is shown in comparison. Photo, courtesy Carboloy Systems Dept., General Electric Co.

Deep holes which can be drilled without trouble from a small amount of runout normally encountered are typified by fluid ports, oil ducts, and similar holes, *Fig.* 15.20. Camshafts, such as those shown in *Fig.* 15.20, are drilled from each end to reduce the possibility of runout and insure central location of the hole at each end.

Drilled holes, wherever practicable, should be produced with standard-diameter drills. Drills are available in five series:
1. By decimal size in flat drills (from 0.002-inch to 0.080-inch in half-thousandth steps)
2. By numbers from 80 to 1 (0.0135 to 0.228 inch diameters)
3. By letters from A to Z (0.234 to 0.413-inch diameters)
4. By fractions from 1/64-inch to 3 1/2 inches (in steps of 1/64-inch from 1/8-inch to 1 3/4-inch diameters, steps of 1/32-inch from 1 25/32-inch to 2 1/4-inch diameters, and steps of 1/16-inch from 2 5/16-inch to 3 1/2-inch diameters)
5. By millimeters (available in steps of one millimeter from 1 to 76 millimeters, steps of one-half millimeter from 0.50 to 45 millimeters, steps of one-tenth millimeter from 0.10 to 10 millimeters, and half-tenths from 0.10 to 2.80 millimeters).

Preferably holes should be drilled through. Where a blind hole is necessary, the standard drillpoint should be allowed to eliminate need for extra bottoming drills or counterboring. The chart in *Fig.* 15.21 can be used to estimate roughly the length of drill point which must be accounted for.

Reaming. Where holes are to be reamed, common practice is to drill the hole about 1/64-inch undersize. The allowance generally depends upon hole size and the amounts usually allowed for removal range from 0.010-inch for 1/4-inch holes, 0.015-inch for 1/2-inch holes, graduating on up to 0.025-inch for 11/2-inch holes. For hand-reamed holes the allowance seldom exceeds 0.001 to 0.003-inch. Where through holes cannot be provided, allowance at the bottom of blind bores for the chamfered tip of the reamer will obviate additional operations with shouldering or bottoming reamers to completely finish the entire length of a hole.

Boring. Boring is usually an enlarging operation, sizing and finishing an existing hole that has been drilled, forged, punched or cored. Drilling is usually considered for producing holes up to about 3 inches in diameter, and where closer tolerances are necessary reaming is used along with the necessary jigs or fixtures to attain accuracy of spacings or positioning. Single-point boring, from about 1/4-inch diameter and up, can be used to assure much closer tolerances even with long bores without the use of jigs, only holding fixtures being necessary to locate and retain the part. Small parts often need no further finishing while large bores may require

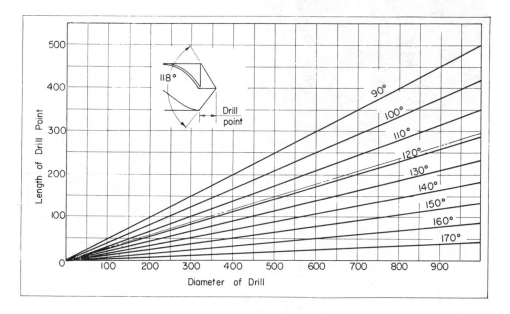

Fig. 15.21 — Chart for estimating roughly the length of drill point to expect.

Fig. 15.22 — Brass radar wave guide, 6000 of which were finished, poses a difficult boring problem because of extremely thin walls. Photo, courtesy Moore Special Tool Co.

burnishing, grinding or honing.

When specifying bores for any size machine part it is well to keep in mind the means by which the particular bore size, finish and accuracy can be produced. Of equal importance are spacing limits on multiple bores and also stepped bore complexity. Although, as indicated in the previous discussion, almost any design can be produced, whether the cost per part and normal production speed meet the practical requirements or not is another matter. The designer can do much to eliminate production problems by simplifying the design requirements as much as possible and recognizing the limitations imposed by too stringent tolerance limits.

Fragile Designs. Fragile, thin-wall designs should be avoided wherever possible to allow maximum cutting speeds and feeds and obviate inaccuracies from distortion. Where such designs are necessary, though, the jig borer provides a ready

Fig. 15.23 — Bronze hydraulic pump body with 3/16-inch diameter holes 6 times the diameter in length. Carbide boring bars, with one rough and one finish pass, hold tolerance on holes to 0.0005-inch, plus or minus.

Fig. 15.24 — Below — Rough and finish boring with a single tool. Piston pin bore is held to 0.0002-inch for size and alignment.

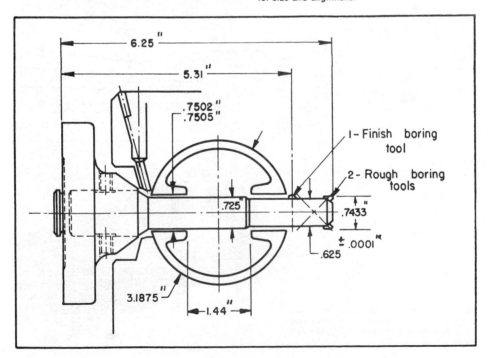

means for holding extreme accuracy on small parts by using single-point tools, *Fig.* 15.22. The brass wave guide shown in this illustration not only posed a clamping problem but required the internal semicircular surface to be machined to a radius of 1.419-inch, plus 0.001-inch and minus 0.000, offset 0.062-inch from the hole centerline.

Length Limits on Bored Holes. Ordinarily the practical limits of hole length to diameter for boring is about 4 to 1 with unpiloted or unguided bars. Lack of necessary stiffness in boring bars where longer length ratios are attempted accounts for the difficulty in maintaining size, roundness and finish. Use of solid cemented carbide boring bars, having some 2.8 times the stiffness (modulus of elasticity) of steel, permits precision boring of holes up to 8 times diameter in length or depth, *Fig.* 15.23.

Hole Characteristics. Good production design of small bored holes allows for a through hole which can be rough-bored on one stroke of the machine table and finish bored on the reverse stroke, the machine table centering for unloading. A faster method, when length of the hole permits, is to rough and finish with one tool. Generally, the roughing tool must finish cutting before the finish tool point begins its cut and consequently only relatively short holes can be handled. However, with carbide bars the practical length is doubled, *Fig.* 15.24.

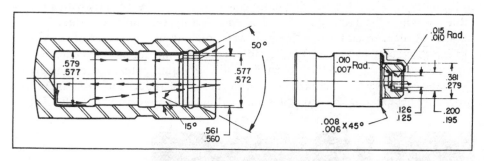

Fig. 15.25 — Above — Typical parts produced on the automatic contour borer showing drill-point at bottom of bore. Drawing, courtesy New Britain Machine Co.

Fig. 15.26 — Vertical-spindle borers feature rapid spindle feed, simplifying boring at different levels. Photo, courtesy Kearney & Trecker Corp.

Blind Holes. If at all possible, small blind holes should be avoided inasmuch as such holes necessitate automatic control of table travel both in and out and where two operations are necessary for accuracy, cross control of table travel is necessary. Where a flat bottom is required and must be machined, allowance should be made for a drillpoint at the center, *Fig.* 15.25, to provide a starting point for the cut. Undercuts which are not too deep or complicated can be readily produced.

Locating and Holding. The problem of locating and holding a part in a fixture should always be kept in mind. Where ready, positive location is not possible, lugs should be provided for this purpose. Locating lugs are machined along with the largest flat surface and, after serving to locate and hold the casting for finish boring, are removed.

Bore Spacing. Closely spaced bores on different levels require consideration since for maximum accuracy, long tool extensions from the boring spindle should be avoided. It is well to ascertain beforehand the clearance available on the machine to be used, *Fig.* 15.26.

Dimensioning. In dimensioning a multiplicity of bored holes and especially hole circles, *Fig.* 15.27, rectangular coordinates should always be used in preference to polar or angular coordinates. Although polar coordinates are simpler to lay out and calculate and often more convenient to use, especially with a rotary table machine, the rectangular ones are invariably the most accurate. This is because of inaccuracies which creep in when picking up the center of the rotary table with respect to the machine spindle, errors in picking up the center of the work piece with respect to the table and spindle, dividing system errors, and errors in reading. Calculation of coordinates, however, can be reduced to a relatively easy procedure by applying the principles outlined in the Woodworth "Circular Tables

Fig. 15.27 — Boring 24 holes on an 8.125-inch circle. Holes are located with respect to each other within a tolerance of 0.0002-inch. Photo, courtesy Kearney & Trecker Corp.

of Coordinate Factors and Angles"[1]

Selection of Materials

For the most part the relative suitability of materials for economical drilling, reaming or boring parallels that discussed in the previous chapters on turning, turret lathe, screw machine work, and milling. Of interest is the fact that many nonferrous bearing metals, aluminum, etc., do not finish to desired surface quality with drills or reamers but spall and pick up. Normally these materials require heavy cutting pressure for best results and consequently, where maximum surface quality and accuracy are desired, single-point boring may be necessary and preferable. In such cases the speeds and feeds common to flycut milling would apply for optimum results.

Extremely hard materials such as Ni Hard, ordinarily considered unmachinable, can be readily bored to sufficient accuracy to eliminate a grinding operation, *Fig.* 15.28, by using sintered carbide boring tools.

Fig. 15.28 — Ni Hard, heat-treated white iron, six-inch mine pump casing being bored with a Carboloy tool. The 20-inch bore is finished to plus or minus 0.001-inch. Photo, courtesy International Nickel Co.

Specification of Tolerances

Naturally, as with any production job, widest possible tolerances provide maximum economy in manufacture. Extremely close tolerances can be held in boring but usually incur additional operations, many times necessitating one or two reaming operations or a rough boring, semifinish and finish boring operation. Final precision is often dependent upon uniform stock removal in the finishing pass. Jig borer or fixture and gage tolerances should seldom be demanded, except in legitimate cases of precise gear trains and like design problems. Where only a few pieces are to be produced, the extra time involved is of small consequence but becomes a costly feature with large quantities.

Drilling. Twist drills on average equipment produce holes from 0.002 to 0.003-inch oversize on small sizes to as much as 0.010-inch on sizes 1-inch diameters and over. Under average conditions, drilled holes preferably not over about four times the diameter in depth can be held within the following tolerances:

Number 80 to number 71 .... plus 0.002 ... (minus 0.001)
Number 70 to number 52 .... plus 0.003 ... (minus 0.001)
Number 51 to number 31 .... plus 0.004 ... (minus 0.001)
1/8-inch to number 3 .... plus 0.005 ... (minus 0.001)
7/32-inch to size R .... plus 0.006 ... (minus 0.001)
11/32-inch to 1/2-inch .... plus 0.007 ... (minus 0.001)
33/64 to 23/32-inch .... plus 0.008 ... (minus 0.001)
47/64-inch to 63/64-inch .... plus 0.009 ... (minus 0.001)
1-inch to 2-inch .... plus 0.010 ... (minus 0.001)

These tolerances are sufficiently liberal for general production and, although a definite set of limits suitable for all cases is out of question, will provide a good conservative working base. Where more exacting specifications are desirable redrilling or reaming will usually be necessary. Tolerances on drilled holes are normally plus owing to the fact that drills invariably cut oversize from one to five thousandths depending upon the accuracy of sharpening. For this reason the above tolerances are often specified as all minus 0.000-inch. Straightness is dependent upon the hole depth and the homogeneity of the material.

Hole straightness in depths over four times the diameter may not be satisfactory. Likewise, finish may be rough. Where better finish and straightness is necessary, carbide gun drills can be used. Diameter tolerances within plus or minus 0.0003-inch can often be obtained. Hole finish with these drills is excellent — averaging 5 microinches on aluminum alloys and 7 to 8 on cast iron in production drilling.

Reaming. Improvement in finish and accuracy of drilled holes can be made by reaming. Reaming, however, does not improve the position or alignment of a drilled or turned hole to any great degree. Ordinarily, reamed holes under 1/2-inch can be held to 0.001-inch total tolerance on size, those from 1/2 to 1-inch diameters can be held to 0.0015-inch total and over 1-inch to about 0.002-inch total. Reaming in many cases does not produce a perfectly round hole and hole diameter may vary out-of-round about as much as the size tolerance. For instance, 2-inch reamed holes often are out-of-round as much as 0.002 to 0.003-inch.

Boring. Boring by various means, especially single-point, carbide and diamond, offers the maximum in accuracy, roundness, alignment, straightness, and finish. On large table type machines, bores as large as 24 inches can be readily held

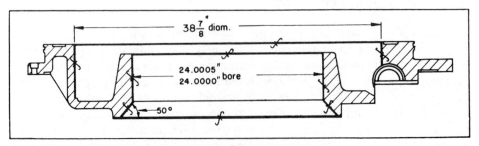

Fig. 15.29 — Cross section of a machine base showing large bores finished on horizontal table type machine. Drawing, courtesy Giddings & Lewis Machine Tool Co.

to plus 0.0005-inch and minus 0.0000, smaller bores to even closer limits, *Fig.* 15.29. Spaced bores as large as 10 inches are line bored to plus or minus 0.0005-inch, *Fig.* 15.30. However, plus or minus 0.001-inch on holes up to 6 inches and

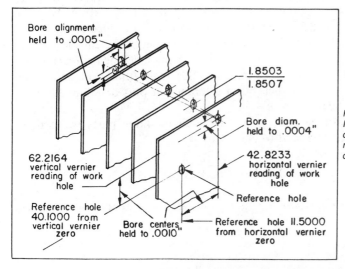

Bore alignment held to .0005"

1.8503
1.8507

Bore diam. held to .0004"

62.2164 vertical vernier reading of work hole

42.8233 horizontal vernier reading of work hole

Reference hole

Reference hole 40.1000 from vertical vernier zero

Bore centers held to .0010"

Reference hole 11.5000 from horizontal vernier zero

Fig. 15.30 — Sketch showing limits held in hole location and diameter on a horizontal boring machine. Drawing, courtesy Giddings & Lewis Machine Tool Co.

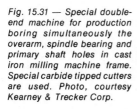

Fig. 15.31 — Special double-end machine for production boring simultaneously the overarm, spindle bearing and primary shaft holes in cast iron milling machine frame. Special carbide tipped cutters are used. Photo, courtesy Kearney & Trecker Corp.

greater limits with increasing diameter are much more readily produced. Bores in *Fig.* 15.31 are held to plus or minus 0.0002-inch in production.

Production Machines. Single-point boring of small holes on machines such as shown in *Figs.* 15.11 and 15.14 can be held to plus or minus 0.0001-inch to 0.0002-inch, *Fig.* 15.24, even on thin-wall parts. Piston pin bores are regularly produced to these limits while being held perpendicular to the skirt within plus or minus 0.0002-inch and in alignment within plus or minus 0.0001-inch. With single-point carbide or diamond tipped tools, these bores have been produced repetitively to 0.0001-inch maximum variation from piece to piece. Larger bores with double-end machines, *Fig.* 15.10, ranging up to 15 inches in diameter can be held to plus or minus 0.001-inch on diameter and concentricity in mass production with a rough and finish operation using special multiple-tool boring heads.

Jig Borers. Jig bored parts such as shown in *Figs.* 15.13, 15.16 and 15.22 are readily held to limits indicated generally in *Fig.* 15.16. Again, however, it should be stressed that only those dimensions absolutely requiring extremely close tolerances should be so specified to simplify production and reduce costs.

Hole Location and Depths. Hole location where necessary can be held to within plus or minus 0.0005-inch on any precision boring machine ordinarily and, in some cases with jig borers, to plus or minus 0.0001-inch. Blind hole depths can be held to plus or minus 0.0005-inch but a tolerance of plus or minus 0.001 to 0.005-inch is more practical.

Threads. Threads can be held to a class 3 fit on large parts while those produced on jig borers can be held to class 4 fit.

REFERENCES

[1] Included in the book *Precision Hole Location,* by J. Robert Moore, published by the Moore Special Tool Co., Bridgeport, Conn.

16

Hobbing

THE basis for the hobbing principle is said to have been discovered in 1856 by one Christian Schiele. By the year 1887 the first hobbing machines for cutting straight-tooth spur gears were built. In the intervening years the hobbing method has achieved a highly important and necessary place in the production process picture. Although originally conceived and still most widely used for the production of worms and gears, *Fig.* 16.1, the hobbing process today has reached a broad field of application. In addition to the many and varied types of gear tooth forms which can be generated, an almost unlimited variety of ratchet, sprocket and other accurately spaced toothed forms can be produced, *Fig.* 16.2. In addition, special hobs can be used to generate formed sections such as squares, fluted octagons, cam shapes, fluted squares, ribbed sections, and almost any type of splined shaft section, *Fig.* 16.3.

Fig. 16.1 — Setup showing the hobbing of helical countershaft drive gears. Photo, courtesy Barber-Colman Co.

16-01

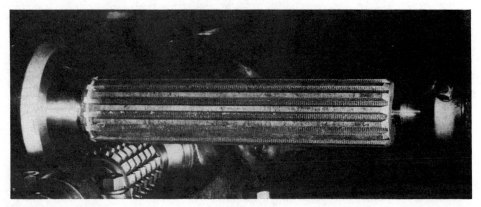

Fig. 16.2 — Setup showing the hobbing of a long spline on a rear axle shaft. Photo, courtesy Barber-Colman Co.

Generating Method. Unlike a milling cutter, a hob is designed so that all the cutting teeth lie in a helical path about the periphery of the tool, *Fig.* 16.4. Under ideal conditions of operation the teeth of the hob lie along a true helix of the proper lead, concentric with the axis of its rotation. By virtue of the construction and gearing of a hobbing machine, the rotation of the hob is maintained in the exact desired mathematical relation to the rotation of the workpiece. Thus, in the course of this uninterrupted working or cutting process, the teeth of the hob pass in positive progression through the workpiece with a thick network of cuts, each of which removes a small shaving to finally generate a smooth form, *Fig.* 16.5. The cutting action in hobbing is continuous and forms are generated in one passage of

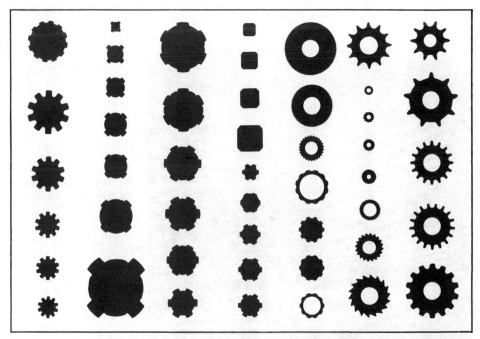

Fig. 16.3 — Typical cross sections of forms which can be produced by hobbing. Drawing, courtesy Barber-Colman Co.

Fig. 16.4 — A typical standard gear hob.

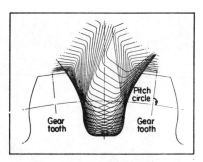

Fig. 16.5 — Layout showing the generating action of a hob in cutting the profiles of a gear tooth. Drawing, courtesy Barber-Colman Co.

Fig. 16.6 — Special elongated-tooth hob and cross section of form generated.

Fig. 16.7 — Model 10-20 Barber-Colman horizontal for hobbing multithread worms, worm gears, spur and helical gears up to 10 inches diameter and 20 inches length.

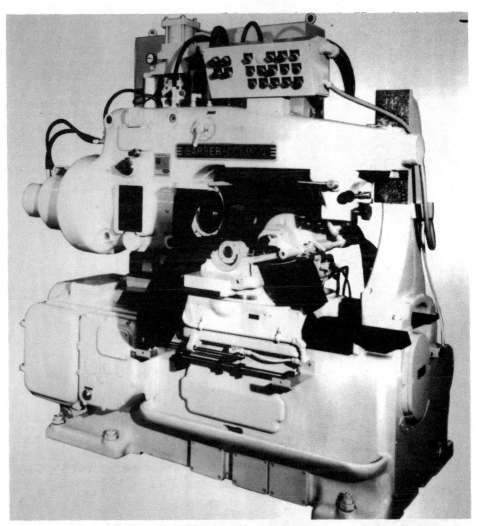

Fig. 16.8 — Barber-Colman Model 14-15 high-production hobber for gear and spline parts up to 14-inch diameters with 3.5 NDP (7.25 modules) single cut and automatic double cut cycles.

the hob through a blank. Since the hob advances uniformly as the blank rotates, indexing is automatic and continuous and a full form outline is generated at one revolution of a blank.

Scope of Process. In general, the hobbing process may be employed for producing any form that regularly repeats itself on the periphery of a circular part. It is not necessary that the form be symmetrical on an individual axis but each form must be a duplicate of any other or in alternate groups, *Fig.* 16.6. Certain exceptions, however, are present and occur with shapes which are deep in proportion to the width and where certain undercuts are required.

Machines Available. A wide variety of machines are available on which gears with almost any desirable tooth form can be hobbed in production for applications ranging from tiny instrument pinions to gigantic marine drive gears as much as 10 feet in diameter. One of the newer machines, *Fig.* 16.7, is capable of hobbing at a maximum metal removal rate of 3 cubic inches per minute on AGMA 9 class or 2 cubic inches on 10 class gear accuracy.

Fig. 16.9 — Fly-hobbing a worm gear on a Gould & Eberhardt machine. Four-thread carbide hob is used.

Typical of the larger high-production machines is that shown in *Fig.* 16.8. Work diameters up to 14 inches, maximum, can be handled with tooth forms 6-pitch and finer. Depending upon diameter, forms to as coarse as 3.5-pitch can be produced. Face widths up to 14 inches can be hobbed. All varieties of automotive spur, worm and helical gears, sprockets, splines, etc., are regularly produced on such machines.

Numerous other machine arrangements and sizes, with capacities ranging up to that required for 10-foot gears, are used although the principle of generation is basically similar, *Fig.* 16.9.

Limitations. Owing to the character of the hobbing process only a very small amount of undercutting can be produced on the generated forms. Actually where special tooth forms are required with radial, straight or undercut teeth, sharp at the root, a single-position hob must be utilized, *Fig.* 16.10. These cutters are much more critical as to setting for obtaining true tooth form. Regarding sharp roots or corners, regular hobbing inherently produces a filleted corner; it is impossible to cut sharp corners owing to the generating action, *Fig.* 16.5.

Close examination of a hobbed form reveals that it is comprised of a multitude of tiny scallops rather than a smooth, even surface or surfaces, *Fig.* 16.2. These tiny cuts made by each tooth as it generates a portion of the final form vary in depth with the number of teeth in the hob and the size of the part. Where the absolute maximum in smoothness and geometrical accuracy of surface or form are required the forms must be rough and finish hobbed or the part must be finished by shaving or by grinding, as is often done with both splines and gear teeth.

Design Considerations

Symmetrical forms which repeat regularly about the periphery of a central axis may range from a square up to many-toothed forms such as serrations, fine-pitch gears, ratchets or splines. Various types of spline designs which can be hobbed are shown in *Fig.* 16.10.

Splines. Ordinarily, in hobbing splines a fillet develops at the roots of the teeth, *Fig.* 16.10*a*. As with gear teeth, this is a basic characteristic of the generating action of a hob and is impossible to avoid. As a rule, the fillet radius developed is approximately one-fourth to one-fifth of the depth of the key. Where parallel keys straight to the root diameter are necessary, splines with four to six keys or a sufficiently wide space at the root can be designed to have clearance grooves at the root, *Fig.* 16.10*b*. Where bearing is necessary only on the outside diameter of splines with eight or more keys, a round-bottom form can be used, *Fig.* 16.10*c*. In *Fig.* 16.10*d* is shown the effect of a shouldered spline end and emphasizes the need for caution where the splines must run full to a shoulder. Where splines of the type shown in *Fig.* 16.10*b* are to be finish ground, a protuberance hob tooth is sometimes used to undercut the sides and protect the grinding wheel. These hob teeth are critical as to wear on the small protuberance and should be avoided.

Special elongated-tooth hobs, however, can sometimes be used to obtain parallel keys or parallel slots with root radii as small as 0.005 to 0.008-inch, *Fig.* 16.10*f*. For applications requiring only bearing on the sides of the keys, angular key splines are often used, *Fig.* 16.10*g*.

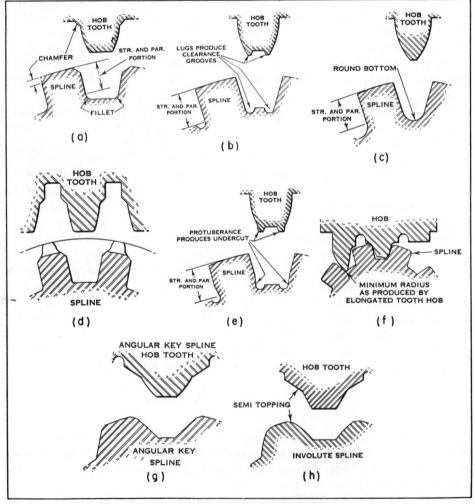

Fig. 16.10 — *Various types of splines normally produced by hobbing. Drawings, courtesy Barber-Colman Co.*

Fig. 16.11 — Taper splined shaft and pinion broached to fit. Photo and drawings, courtesy Barber-Colman Co.

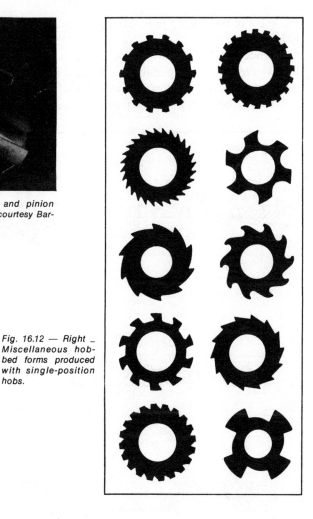

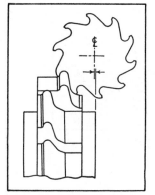

Fig. 16.12 — Right _ Miscellaneous hobbed forms produced with single-position hobs.

Fig. 16.13 — Setup showing undercut tooth form as produced by single-position hob.

Involute Splines. Because of the problem posed by the unavoidable generation of the fillet at spline roots, the involute spline has gained wide acceptance. The root fillet on involute keys can be entirely eliminated if desired, *Fig.* 16.10*h*. In designing involute splines or serrations, the American and SAE Standards should be followed.

Taper Splines. If mating splined parts require permanent, accurate location and maximum concentricity after assembly, the taper spline, *Fig.* 16.11, can often be used in lieu of the standard SAE tapered serration drive. Permitting positive metal-to-metal contact on the shaft root diameter as well as the customary fit on the sides of the keys, the taper spline is exceptionally well suited for reversing drives and applications where the assembled part requires minimum wobble. Torsional strength of taper splined connections for all practical purposes are equal to at least 85 per cent of the torsional strength of a shaft at full diameter. Taper splines may have either four or six parallel-sided keys, four-key splines ranging from 1/2 to 1-inch shaft diameters and six-key splines from 1 to 4 inches. Standard serrated connections have a somewhat broader application, being adaptable to diameters from 0.10-inch to 10-inch diameters. The length of spline can be varied to suit the thickness of the mating part. Mating parts are produced by broaching a ground or reamed taper hole.

Undercut Tooth Forms. Special parts with undercut, radial or straight tooth forms, *Fig.* 16.12, impossible with regular hobs, can be produced with single-position hobs. Tooth design must be such that the cutter can be positioned for hobbing without interference from the undercuts, *Fig.* 16.13.

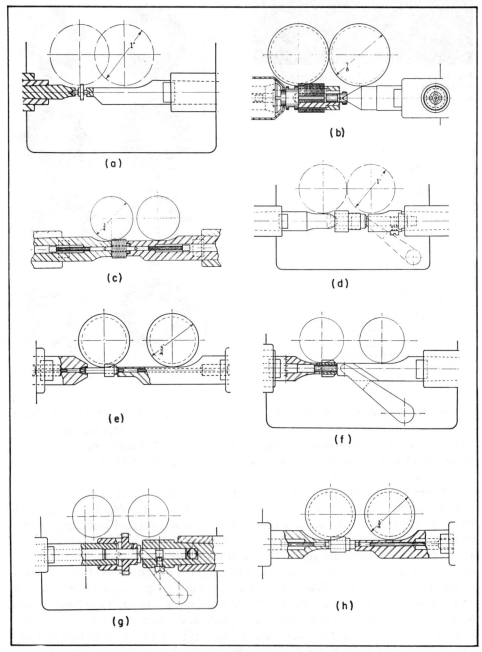

Fig. 16.14 — Group of machine setups showing various types of blanks arranged for hobbing. Drawings, courtesy Barber-Colman Co.

Fine-Pitch Gears. In detail designing of regular and fine-pitch gears, the established American Gear Manufacturers Association standards should be closely followed. As with the SAE and American Standards on splines, these gear standards represent the overall general practice of gear manufacturers and users and provide the best possible quality at reasonable production costs.

Blanks. Gear blanks as well as blanks with special tooth forms must be properly designed to assure satisfactory results. Accurate parts cannot be produced from blanks that are poorly proportioned or incorrectly dimensioned. Where the part requires a hole, the hole should be sufficiently large to provide adequate support but yet not oversize in relation to the remainder of the design. Face width of blanks should be sufficiently wide to provide proper rigidity in cutting. Thus, very short bore lengths as well as face widths should be avoided, unless the design permits machining the parts in stacks which can be gripped as one large blank, *Fig.* 16.14*b*. Blanks to be hobbed in stacks, however, must be flat and parallel as is practicable.

Where blanks with hubs are to be utilized, the hubs must be sufficiently large to permit proper clamping, *Fig.* 16.14*g*. Where no hole is required in the blank, such as with integral-shaft blanks, *Fig.* 16.14*a, e* and *h*, the shaft should be well proportioned in relation to the remainder of the blank to assure the increased accuracy usually obtainable with such designs. In *Fig.* 16.15 are illustrated some of the good and poor practices in designing blanks. Some of the problems encountered with unusual blanks can be visualized from the machine setups shown in *Fig.* 16.14.

Shoulders. Regardless of whether gears or special toothed forms are being designed for hobbing, the primary characteristic must be observed, namely that the hob must be traversed parallel to the axis of the rotating blank. From the layouts

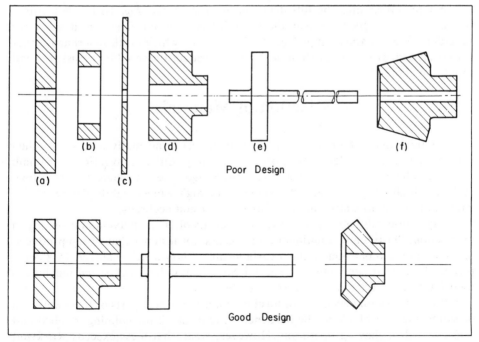

Fig. 16.15 — *Good and poor practices in designing blanks for hobbing. With blank (a) bore is too small in relation to the OD, bore (b) is oversize, face (c) is too narrow and OD too large, bore of (c) is too short also, hub walls (d) are too thin, integral shaft at (e) lacks stiffness, and bevel gear blank (f) is too long in proportion to the diameter.*

Fig. 16.16 — Splined drive shaft being hobbed. The A4140 steel, Rockwell C28 to 32, is rough and finished hobbed in 30 minutes. Photo, courtesy Barber-Colman Co.

shown in *Fig.* 16.14 it readily can be seen the problems posed by shoulders or projections adjacent to the portion of the blank to be hobbed. Wherever possible such shoulders should be below the root diameter of the form or sufficiently removed to allow full runout of the hob, *Fig.* 16.16. If shoulders interfere "plunge" hobbing may be required in the case of some gears and a subsequent shaving operation is then necessary to correct the form. With splines a special cutter may be needed.

Where splines must be full depth up to a shoulder, *Fig.* 16.17, as a rule a special cutter designed to clear the shoulder is required. The condition of the shoulder is readily apparent in *Fig.* 16.17, and cases which result in a sharp corner at the outer edge of the large shoulder diameter reduce the shoulder thrust area and should be avoided.

Selection of Materials

Applicability of hobbing to the various materials normally used for toothed and splined parts parallels that of ordinary milling. Either conventional or climb cutting can be utilized in hobbing. The entire range of ferrous materials from low-carbon to high alloy are regularly hobbed. The higher machinability materials, of course, offer the maximum in cutter life, output and accuracy.

The high-machinability nonferrous materials offer the maximum economy in production. Plastics and aluminum timing gears, for instance, have been produced at the rate of 200 per hour using a carbide-tipped hob which removed 15.8 cubic inches of material per minute. A special extra heavy-duty high-speed machine is used for such production with carbide cutters.

Where maximum accuracy of form is required on steel parts of Rockwell C36 or softer, the part blank can be hardened, leaving stock for hobbing, hobbed and finished with a shaving operation. However, where hardness exceeds Rockwell C36, the part should be hobbed allowing stock for shaving, shaved, hardened and lapped where the extreme in finish and accuracy are necessary.

Specification of Tolerances

With fine-pitch hobs, forms and comparable gear teeth from 30 to 268 pitch can be produced with profiles accurate within 0.0005-inch. Pitch diameters can usually be held within 0.0003 to 0.0005-inch and runout within 0.001-inch total indicator reading. A typical steel pinion with 9 teeth, measuring 0.250-inch long, 0.1578-inch outside diameter and 0.1039-inch face width, is held to a pitch diameter of 0.1378-inch, plus 0 and minus 0.0015-inch.

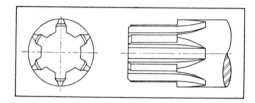

Fig. 16.17 — Spline with a backing shoulder requires a special hob to allow a full-depth cut up to shoulder. Drawing, courtesy Barber-Colman Co.

Large gears, as large as 10 inches in outside diameter, are held to pitch diameter tolerances of 0.001-inch total. Generally, accuracy may range up to 0.005-inch, 7¾-inch 60-tooth gears for instance being hobbed in stacks of seven to a tolerance of plus 0.003-inch and minus 0. Circumferential lead accuracy is usually within plus or minus 0.0002 to 0.0003-inch over a 2-inch face width. Pitch diameter runout can be held to 0.0015-inch.

TABLE 16-I — Production Tolerances Held on Straight Splines†
(inches)

Outside Diameter of Spline	Splines with 3 and 4 Keys		Splines with 6 Keys and More		Splines with 10 Keys and More—Round Bottom	
	Key Width	Root Diam	Key Width	Root Diam	Key Width	Root Diam
Single-Thread Ground Hobs						
Under 1½	0.001	0.002				
1½ and over	0.002	0.003				
Under 2½			0.001	0.001		
2½ and over			0.002	0.002		
2 or under					0.0010	*
Over 2					0.0015	*
Single-Thread Unground Hobs						
Under 1½	0.002	0.004	0.002	0.002		
1½ to 2½	0.003	0.005	0.002	0.004		
2½ and over	0.003	0.006	0.003	0.005		
2 or under					0.002	*
Over 2					0.003	*

† Courtesy Illinois Tool Co. Round-bottom teeth approximate a semicircle which extends below the diameter set by the minimum length of straight side on the keys, and splines marked * have no fit on the root diameter.

Where closer accuracies on size and surface finish are required, both a rough and finish hobbing operation are often necessary or a final shaving, shaping or grinding operation.

Ordinarily "open" gears (root diameters larger than adjacent shoulders, etc.) present no difficulties and can be hobbed in conventional manner. However, where adjacent shoulders prevent a gear from being hobbed with traverse of the cutter, plunge hobbing must be used whereby the hob is fed straight into the work without axial traverse. Plunge hobbing produces a concave tooth and root with the ends thicker than the center. Likewise, plunge-ground teeth are not completely generated. With plunge-hobbed or ground gear teeth, shaving must be resorted to for correcting involute form, tooth spacing, eccentricity and, where necessary, the helix angle.

Spline tolerances — straight or involute — should be specified to conform with the ANSI or SAE standards. Typical tolerances held on straight splines in production as shown in TABLE 16-1. Taper splines can be held to spacing accuracies comparable to those of straight or involute splines. The splines are parallel with the axis of the taper within plus or minus 0.0005-inch in the splined length. Tolerance on the taper is plus or minus 0.012-inch per foot.

Additional Reading

¹Paul D. Anderson, "Hobs and Hobbing", S.M.E. paper MF68-402, Detroit, MI.

17

Broaching

BROACHING has risen during the past half century from a relatively unknown production process to one of major importance in present-day mass-production manufacturing. The rise of broaching into industrial prominence is due primarily to its two salient characteristics — extremely high speed of production, which makes possible low cost per piece of extremely complex machine parts, and high repetitive accuracy.

Although broaching is recognized principally for its value as a production method suitable for manufacture at extremely high rates of output, it is being applied to considerable advantage in medium-volume work where production lots of 2500 to 5000 pieces can be run. Where multiple jobs can be tooled up to run in succession to make up such quantity lots, relatively smaller quantities of certain parts can thus be designed to obtain the benefits of broaching. Cases are on record where as many as 12 different setups are used in this manner.

Field of Application. Developed from the so-called "drifts" of a century ago, broaches were first used to cut keyways and were driven through the part by means of a hammer. Round holes followed later and by the turn of the century surface broaching was in limited use. Today, holes of any cross section from plain round or oval to inverted or helical spline can be readily produced to exacting tolerances. Over and above a wide variety of flat, related flat or flat and curved surfaces, compound shapes such as gear teeth, chain sprocket teeth, sinuous slots, turbine blades, etc., comprise a broad field of possibilities for economical production. The principal limitation of the process, however, must be observed in judging the

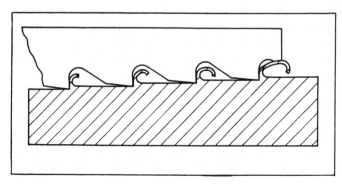

Fig. 17.1 — Cross section showing action of broaching tool teeth. Drawing, courtesy Cincinnati Milacron Inc.

advantages of broaching or in designing specific parts for such production. Owing to the necessity for each tooth gullet of a broach *Fig.* 17.1, to retain the chip created until the entire pass across the part is completed, the process is not particularly suited for removing large amounts of material. Stock removed per tooth and length of broach must be economically balanced to achieve optimum results.

The Broaching Process. Unlike turning, milling, shaping, planing or similar tools on which the most favorable operating conditions can be secured by simple adjustments in feed or speed, a broach has a predetermined fixed feed or cut which is represented by the rise in height of each tooth, *Fig.* 17.2. This, of course, is usually established by the broaching tool designer but also must be recognized in designing parts to be broached in order to assure practicable manufacture.

Because each tooth gullet of a broach must retain the chip created until the tooth has finished the cut, obvious limitations are present in broaching tool design, *Fig.* 17.2. Depth of cut or rise per tooth is thus broadly restricted by the practical size of gullet allowable between teeth. Size of tooth gullet, however, is directly related to tooth size or pitch, tooth depth usually ranging around one-third the pitch or slightly greater. As a rule, a minimum of two teeth should be cutting at all times and preferably three; a minimum of three teeth per inch being recommended for parts having a length of 1-inch or less.

Material Removal. In hole broaching, *Fig.* 17.3, i.e., with any type of round broach, the limitations on stock removal depend not only on size of tooth gullet but also upon the total effective length of tool possible. Broach length, of course, must be reasonable, within available machine capacities and of sufficient minimum cross section — at the root of the first roughing tooth or at the pull end — to withstand the operation. Standard machines available have strokes up to 66 inches and are capable of handling broaches up to 62 inches in length. Larger machines are in use, however, and some standard broaches available run up to as much as 74 inches in length. For special jobs such as rifling, even much longer broaching tools are used. Length, naturally, should always be held to a minimum for lowest overall costs, especially where push broaches are concerned. Push broaches with a length-to-diameter ratio greater than 25 act as a long column — ordinarily critical regarding broaching loads — and are best avoided. Reasonable broach proportions are also a must for best results in sharpening.

Tool strength in surface broaching is not a critical factor; machine capacity, however, is and again relates back to tooth pitch and rise inasmuch as volume of material removed is limited by both stroke and power. The total amount of material which can be removed for each available ton of ram pressure per foot of stroke is 0.06 cubic inches for cast iron, 0.04 cubic inches for medium-hard steel, and 0.03 cubic inches for tough alloy steels having Rockwell hardnesses over C28. In TABLE 17-I, the relationship between tooth pitch and chip area is correlated for the various commonly used pitches. Since these chip area figures represent the product of chip thickness and length of cut, the maximum stock removal for any size part can be approximated for any machine available.

Material removal in single-pass surface broaching thus is usually limited to a maximum of about 1/2-inch or less in thickness, depending to a large extent upon the capacity of the machine to be used. Standard surface broaching also allows for as much as 62 inches of broach while continuous machines run up to 140 inches with standard models and specials have been made carrying up to 240 inches of broach. Normally, surface broaches are made in sections from 6 to 15 inches in

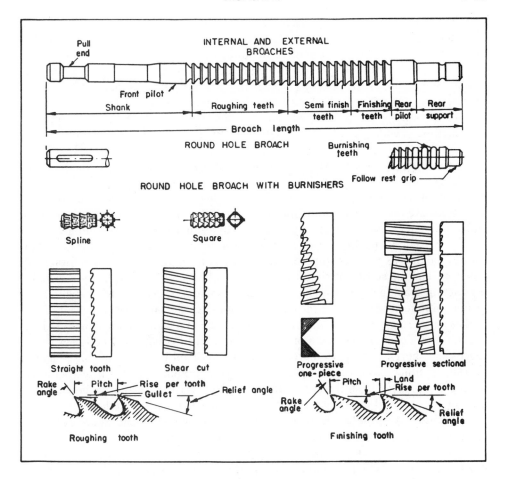

Pitch
Pitch is determined primarily by the length of cut and thickness of chip removed per tooth, giving consideration to type and condition of the material to be broached. Formula sometimes used to estimate approximate tooth spacing for round broaches is $P = 0.35L$ where P = tooth pitch in inches, and L = length of cut to be made. Actual pitch used must be modified to suit necessary chip gullet capacity. Table 17-1.

Gullet
Area of gullet space should be ten times the cross sectional area of the chip for steels and five times for noncontinuous chip materials such as cast iron. Table 17-1.

Rise per Tooth
In free machining steel the rise per tooth is usually 0.0015 to 0.003-inch for round broaching; 0.004 to 0.006-inch for spline; 0.001 to 0.006-inch for keyway and surface broaching. Practical minimum rise is about 0.00025-inch. Rise per tooth as high as 0.015-inch can be used on narrow trimming broaches where the cut is across the piece and the chip need not be contained in the tooth gullet. Slotting broaches should not exceed 0.00275-inch rise to avoid jamming of heavier chips against the sides of the slots. Preferable rise is 0.0025-inch. Values can often be doubled for cast iron, malleable iron and brass. For burnishing or sizing, rise per tooth is not over 0.001-inch and often 0.0001-inch or less.

Rake Angle
Rake angle, sometimes termed face angle, varies with the material and its condition. It is usually 6 to 8 degrees positive for cast iron; 8 to 12 degrees positive for hard steel; 15 to 20 degrees positive for soft steel; approximately 10 degrees positive for aluminum; and from 5 degrees positive to 5 degrees negative for brittle brasses.

Relief Angle
Normally the relief angle falls between $1/2$ and 3 degrees for internal broaches and up to $3^{1}/_2$ degrees for surface broaches. For steel, the relief angle may vary from $1/2$ to 2 degrees; for cast iron, from 2 to 5 degrees, and for brasses and bronzes, from $1/4$ to $1/2$-degree. Relief may vary from maximum at one end to minimum at the finishing end to minimize loss of size in resharpening.

Land
The land varies from zero to $1/_{16}$-inch in width and is usually graduated in width from the first finishing tooth to the last on internal broaches. Land is often calculated as 0.2 of the pitch. The land varies from zero to 0.005-inch for surface broaches and is usually backed off about $1/_4$-degree.

Shear Angle
Shear cutting teeth are used wherever possible to obviate chatter and improve finish, but should not be used on deep slotting. Angle of teeth relative to direction of cut is usually from 5 to 20 degrees. Minimum angle with surface broaching is usually 15 degrees.

Fig. 17.2 — Internal and surface broaching tools with typical tooth sections and definitions.

length primarily for economy, *Fig.* 17.4, but special sections such as for gear teeth are often made longer. Where amounts exceeding the removal capacity or stroke of the machine are necessary, repetitive or multiple-stroke cuts must be made, *Fig.* 17.5. Rise or cut per tooth can be made heavy if desirable by ingenious broach design. Staggering of the cutting portions of each tooth so as to achieve complete overlap over the entire surface or hole often can be employed to keep the total volume of material being removed at any instant well within machine horsepower. This method is widely used in surface broaching as the "progressive" broach, *Fig.* 17.6.

Broaching Speeds. Cutting speed or travel of a broach across the work is usually 15 to 25 times faster than in milling. Steel castings and forgings are normally broached at speeds of 20 to 33 feet per minute while cast iron, brass or aluminum are often broached at speeds up to 40 feet per minute. Although brass and aluminum can be satisfactorily cut at speeds as high as 80 feet per minute, most modern broaching machines have a much lower maximum ram speed, the fastest of which is about 43 feet per minute. Faster speeds normally are found impractical owing to lack of sufficient time for fixture loading and unloading. In the high-production manufacture of locks, however, broaching speeds up to as high as 59 feet per minute are used to cut the narrow sinuous key slots in brass, *Fig.* 17.7. Even at these high rates, actual tooth travel through the material is approximately half that normally encountered with most production milling cutters. The lower tooth speeds and short cutting period create a minimum of abrasive action, consequently heating of either the broach or the work piece is minimized. Work temperature during the process is sufficiently uniform so that errors created by heating are seldom encountered as is often the case with other high-speed methods.

Machines Available. Two broad classes of broaching machines are available for performing almost any variety of broaching operations; the vertical and the horizontal, either of which type of machine may have one or more rams. Although most modern broaching machines are hydraulically powered, a few are driven mechanically. Plain horizontal machines, *Fig.* 17.8, are normally of the pull type

TABLE 17-I — Relationship of Tooth Pitch and Chip Area per Tooth*

Pitch of Teeth (inches)	Chip Area (thickness × length, sq in.)	
	Steel	Cast Iron
0.2500	0.0014	0.0028
0.3125	0.0022	0.0044
0.3750	0.0031	0.0062
0.4375	0.0042	0.0084
0.5000	0.0055	0.0110
0.5625	0.0070	0.0140
0.6250	0.0086	0.0172
0.6875	0.0104	0.0208
0.7500	0.0124	0.0248
0.8125	0.0145	0.0290
0.8750	0.0168	0.0336
0.9375	0.0193	0.0386
1.0000	0.0220	0.0440
1.2500	0.0330	0.0660
1.5000	0.0495	0.0950
1.7500	0.0673	0.1346
2.0000	0.0880	0.1760

Courtesy Cincinnati Milacron Inc.

and are available in capacities ranging from 2¹/₂ to 50 tons or more with maximum strokes from 30 to 76 inches. Vertical machines, however, are available as push-down, pull-down, or pull-up types. Push-down models for simple push broaching operations are available in capacities ranging from 5 to 15 tons with strokes from 12 to 36 inches, maximum. Pull-down machines, *Fig.* 17.9, range from 5 to 50 tons capacity with maximum strokes ranging from 36 to 68 inches. Designed to cut on the down stroke but with one-piece or segmental broaches attached directly to the ram face, surface broaching verticals in single or multiple-ram models are available in sizes from 3 to 25 tons capacity with maximum strokes from 30 to 66 inches, *Fig.* 17.10. Pull-up verticals are available in capacities from 6 to 25 tons with strokes from 36 to 60 inches, maximum.

Fig. 17.3 — Close-up of a large special spline ring job with blank and finished part on top of machine.

Where extremely large quantities are in view, continuous broaching is a possibility. In these machines the broaches are usually stationary, a continuous chain conveyor with suitable fixtures carrying the workpieces past the broaches, *Fig.* 17.11. Continuous broaching machines range from the smallest, *Fig.* 17.12, with a capacity per minute of 3.35 cubic inches in mild steel or 5 cubic inches in cast iron to the largest standard machine, capable of removing 26.6 cubic inches of mild steel or 40 cubic inches of cast iron per minute. Maximum broach length possible with these machines ranges from 20 inches with the smallest to 240 inches with the largest, although special machines have been built to handle longer broaches.

For the machining of external surfaces on large parts such as engine blocks in high production, special large horizontal broaching machines are widely used, *Fig.* 17.13. Machines in this class range to as high as 100 tons capacity and long, complicated progressive broach sections are built up to suit the particular job, *Fig.* 17.14.

Various other special machines are also utilized. Among these are rotary broaching machines for finishing radial slots or surfaces, *Fig.* 17.15. These have been designed for both horizontal and vertical broaching motions.

Fig. 17.4 — Right - Built-up sectional broaches are used for finishing multiple curved surfaces as on this large sprocket.

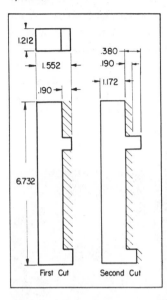

Fig. 17.5 — Winchester carbine receiver broached in two passes removing 0.190-inch of stock per pass

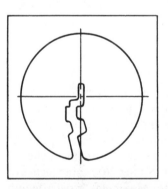

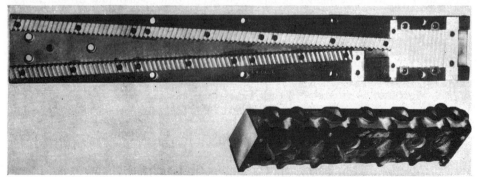

Fig. 17.6 — Below - Broach designed to take a progressive cut across the end of a cast iron cylinder head removing scale and taking a finishing cut in one pass. Photo, courtesy Colonial Broach Co.

Fig. 17.7 — Narrow, sinuous key slots of complex section, frequently less than 1/32-inch in width, are cut in solid brass in one pass. Photo, courtesy Ex-Cell-O Corp.

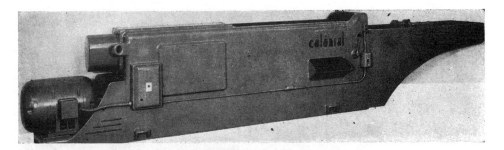

Fig. 17.8 — Typical universal horizontal broaching machine primarily designed for round holes, keyways, splines and some surface operations.

Fig. 17.9 — Steel yoke with a straight-splined hole produced on a dual vertical pull-down machine.

Design Considerations

The evident value of broaching for mass-produced parts often obscures the need for careful consideration in the design of such parts. Likewise, the proper consideration in designing families of parts can bring into the realm of economy the broaching of even comparatively small lots. The cumulative savings of broaching over other methods are often sufficiently large to make it imperative that the advantages to be gained be carefully studied. Grouping of all parts which are reasonably similar in machining requirements, possibly with a few minor design alterations, often will allow the use of one special broaching tool. Slight adjustments of the tools along with the required fixtures permits broaching of a wide variety of pieces in relatively small lots, some possibly as few as 150 pieces. On one such arrangement 24 different trip dogs and 13 different tongue strips are broached in this manner as inventories demand.

Broaching Forces. Idle time between broaching cuts is kept at a bare minimum by rapid return passage of the tool. Loading, unloading and clamping in suitable fixtures is usually simple, since the main force of the tool is in one direction only. Secondary forces created by this travel tend to separate the tool and work. Where the broach teeth have a shear angle a side force may be created. Actually, on most broaching fixtures, it is found necessary only to retain the work piece so that it cannot move; preloaded clamping, in the usual sense, is not always required and

Fig. 17.10 — *Typical single-ram vertical broaching machine with finished sprocket.*

Fig. 17.11 — *Continuous broaching machine showing fixture and connecting rod being finished. Photo, courtesy Footburt-Reynolds.*

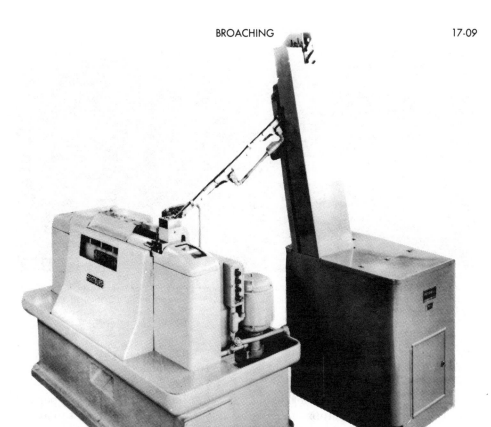

Fig. 17.12 — Smallest of the range of standard continuous machines. Fully tooled and automated, this Footburt machine produces 8000 pieces per hour.

some parts can be worked without any clamping arrangement on the positioning fixture.

Strength of parts. Because of the separating forces set up between the part and the broaching tool, the fixtures used must be sufficiently strong to resist deflection. It is advisable at all times to maintain uniform wall sections on parts to be broached and avoid frail or thin sections. Sections should be at least sufficient to

Fig. 17.13 — Special horizontal surface broaching machine for finishing engine cylinder blocks. Photo, courtesy Cincinnati Milacron Inc.

Fig. 17.14 — Close-up of the machine in Fig. 17.13. Two operations complete a block per minute; in the first the pan rail and half bore are finished and in the second the top, push-rod surface and distributor pad. Photo, courtesy Cincinnati Milacron Inc.

withstand the fixture retaining pressures without distortion. Where no locating surfaces are to be machined before the broaching operation, design of the part must be such that any piece can be easily located without severe misalignment resulting from casting, forging or forming variations. With long slender machine parts it often is desirable to "condition" the part by first broaching locating spots to assure clamping conditions conducive to straightness in the wide cuts which follow.

Fig. 17.15 — Colonial double-end rotary broaching machine.

Round Holes. This method is used for producing round holes of high accuracy and fine finish. Holes may be finish broached from drilled, bored, cored, punched, flame-cut or hot-pierced holes, *Fig.* 17.16. Starting hole size should be $1/32$-inch smaller than the finished hole for small diameters, increasing to as much as $1/16$-inch smaller on large diameters. To assure cleanup and good finish, the allowance on diameter should never be less than $1/64$-inch. Where cored holes are planned, draft angle and size variations must be taken into consideration to assure entry of the pilot and good cleanup. If extremely smooth finish is necessary, burnishing

Fig. 17.16 — As-cast hole, left, and flame-cut hole, right, can be broached but usually require a generating type broach. Photo, courtesy Colonial Broach Co.

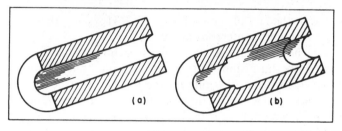

Fig. 17.17 — Redesign of long, plain holes (a) to be chambered or relieved as at (b) improves accuracy and finish.

Fig. 17.18 — Method of generating a square hole with a broach.

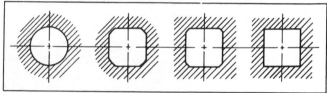

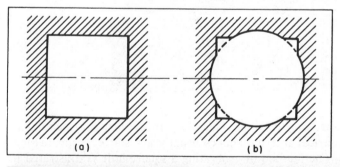

Fig. 17.19 — Redesign of square holes (a) to allow a slightly oversize starting hole as at (b) makes possible a shorter, more economical tool.

Fig. 17.20 — Push broach and die-cast frame. The 3³/₄-inch diameter on the broach acts as a pilot while the 7⁷/₈-inch diameter is cutting; the large diameter then pilots as the small end is cutting. Photo, courtesy Ex-Cell-O Corp.

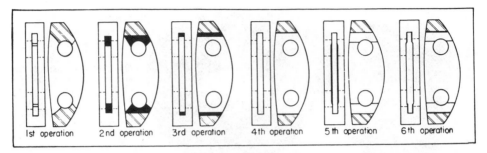

| 1st operation | 2nd operation | 3rd operation | 4th operation | 5th operation | 6th operation |

Fig. 17.21 — Sequence of operations in broaching an aircraft engine counterweight from a rough-milled slot. Drawings, courtesy Colonial Broach Co.

teeth or buttons can be used at the end of the tool, *Fig.* 17.2. However, these are recommended only for cast iron and nonferrous materials.

Blind or extremely long holes should be avoided. Broaching of blind holes requires a stripper broach having burnishing buttons to open the hole slightly in order to prevent the broach teeth from dragging on the finished surface as it is withdrawn. These tools are not satisfactory on high-production work. Long holes, of course, would require an extremely long broach and possibly a set of several. Such holes should be chambered as shown in *Fig.* 17.17 both to improve accuracy and finish as well as to reduce costs.

Fig. 17.22 — Slotted parts formed by broaching across part using an indexing fixture. Spacings are accurate to 0.0003-inch. Photo, courtesy Footburt-Reynolds

Regular and Irregular Shaped Holes. This phase of the process is similar to round broaching. Symmetrical internal shapes usually are started from drilled, punched or cored holes. The method of generating a square hole is illustrated in *Fig.* 17.18. However, where possible, an oversize starting hole should be used to make possible a shorter, more economical tool and much faster production, *Fig.* 17.19. Because a broach is a form generating tool, irregular holes of all varieties, *Fig.* 17.20, may be produced within the previously mentioned metal removal capacity.

Fig. 17.23 — A group of parts with straight and involute splined holes. Photo, courtesy Colonial Broach Co.

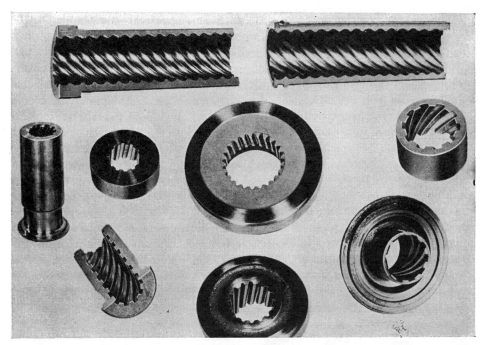

Fig. 17.24 — Gear blanks, bushings and sleeves with spiral splined through holes. Photo, courtesy Colonial Broach Co.

Round push broaching is employed to produce the same results as pull broaching. This method usually is restricted to parts requiring a relatively small amount of stock removal since the tool must be comparatively short in order to have adequate stiffness to withstand the force exerted by the press ram. Push broaches are used for sizing holes after drilling, reaming or pull broaching round or splined holes, and for sizing and smoothing die-cast and similarly produced holes, *Fig.* 17.20. Special sizing broaches can be used for removing scale and correcting distortion in the holes of hardened gears, correcting errors in spacing of splines, correcting distortion in splines, etc. Push broaches are widely used for forming or sizing "blind" holes through which the tool cannot be completely passed. To prevent the cutting teeth from dragging on the return stroke, burnishing buttons are sometimes used to open the hole slightly.

Burnishing. A cold-working operation, burnishing removes no metal but only compresses and smooths out fine surface irregularities. Contraction usually follows the expansion of a hole produced by burnishing. Amount of contraction varies with the material, size of the hole and construction of the part and can be determined

Fig. 17.25 — Group of complicated circuit-breaker arms and rectangular copper blank from which the arms are broached in stacks of eight. Contours are indicative of the possibilities of surface broaching. Photo, courtesy Colonial Broach Co.

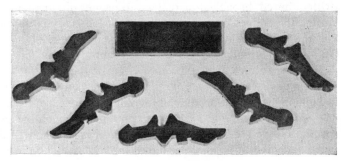

most easily by experiment. Only soft, ductile materials are well suited to the burnishing operation. Burnishers are usually short and consequently are pushed through the hole.

Keyways and Slots. Broaches can be used for producing single plain keyways, multiple plain splines by means of indexing fixtures, or for finishing round holes simultaneously with a plain keyway. A keyway type broach is adaptable for sawing or cutoff work and also, is useful in producing various slot designs especially where very narrow or sinuous key slots — frequently less than 1/32-inch wide — are required to exacting dimensions, *Fig.* 17.7. Effectiveness of coolant action with this type tool obviates heat distortion troubles otherwise present in such thin-bladed tools. Rise per tooth usually is restricted to a maximum of 0.00275-inch to avoid marking or scoring of the slot sides by the spreading action of chips formed with heavier cuts. This limits the depth of slot to about 7/16-inch on a 66-inch capacity standard machine. Slots 5/8 to 3/4-inch deep are possible on 120-inch machines. For internal slots, some means must be used to remove sufficient material to allow the broaches to enter for finishing. The aircraft engine counterweight shown in *Fig.* 17.21 is drilled and milled to allow the broaches to enter.

Fig. 17.26 — Sequence of operations in finishing connecting-rod forgings on a dualram surface broaching machine. Photo, courtesy Colonial Broach Co.

For slotted parts such as those shown in *Fig.* 17.22 either sharp corners or radii are equally satisfactory at the bottom of the slots. The broaching tool in either case can easily be sharpened. However, chamfered corners rather than rounded ones should be specified for the outer edges or corners. A chamfer at the outer edge will reduce the cost of the broaching insert required and make an almost impossible sharpening operation a simple one.

Splines, Serrations and Gears. Splines of any required shape can be broached and are usually generated rather than cut directly. Both straight and spiral splines are produced having straight, involute, angular-sided, radius-lobe, and inverted shapes, *Fig.* 17.23. Broaches are usually of the pull type owing to length of tool necessary for removing the large quantity of material. As a rule the work is first drilled, bored, broached or reamed to proper minor diameter and then broached, cutting the splines only. Where high concentricity is desired between the splines and the inside diameter, the hole can be prepared as for round broaching and a combination round-and-spline broach employed to finish the inside diameter as well as the splines. This type tool is particularly useful on high-production jobs requiring extreme accuracy. Present practice is to size the inside diameter with the broaching tool when producing involute splines.

Spiral splines, *Fig.* 17.24, are generated in much the same manner as straight splines. Where the helix angle of the spiral is less than 15 degrees and preferably under 8 degrees it is possible to pull the tool through the part with a ball-bearing puller, the broach revolving itself to generate the proper helix. If the angle is larger than 15 degrees, a lead screw is necessary to assure positive and accurate generation. It is not usually practical to broach splines with a helix angle greater than 60 degrees.

Involute spline or internal-gear broaches can be made to cut any number of teeth and are used for producing spur on helical gears with teeth as small as 48 diametral-pitch. In these gears the involute form is generated and checked from the base diameter. For extremely accurate parts or those heat treated after the first cut, a sizing or finishing operation should be used.

Serration broaches, similar to spline broaches, are usually made to SAE or ANSI standards although, of course, special designs can be made to suit. Sizes range from 1/8 to 3-inch nominal diameters with 36 to 48 serrations and may be tapered or straight. Serration drifts are used to produce taper serrations after broaching with a straight serration broach. A combination push broach and taper drift can be employed in some cases.

Fig. 17.27 — Large, flat surfaces (a) when reduced to a group of pads (b) are much more economical to finish.

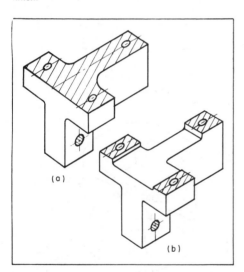

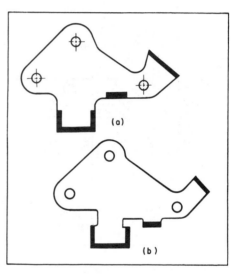

Fig. 17.28 — Relieved or undercut portions (b) simplify the broaching tools required for orginial design (a).

Surface Broaching. A fast, accurate method for producing flat, curved or irregular surfaces, this method of broaching is unexcelled for making a relatively smooth, even finish at low cost. One of the many combinations of tools available usually can be adapted to suit the job on hand, *Fig.* 17.25. Gear segments, racks, serrations of any variety, slots, teeth, curved guides, etc., can be machined at one pass with exceptionally accurate relationship between the various surfaces. Typical procedure for surface broaching a connecting-rod forging is shown in *Fig.* 17.26. Removal of the surface skin on such forgings relieves certain internal stresses which in turn cause the bore to open up slightly. A boring operation is necessary after broaching such a cap and rod to assure a perfectly circular bearing seat.

Hole broaching requires a definite cycle for starting the broaching tool through the hole and returning to starting position after a cut and even considering automatic broach handling turret type verticals, production output of the continuous surface broaching method is unexcelled. A wide variety of plain or irregular surfaces, so long as they are parallel on the part, can be machined by this method and particularly those parts which require a series of such related surfaces. Fixtures can be designed to index one or more times as the part passes through the stationary broaches to produce squares, hexagons, slots, bevel gears and similar forms on a single pass through the machine.

Consideration of large, flat surface-broached areas is of importance. Where possible such areas should be reduced to a series of bosses, *Fig.* 17.27, to simplify tooling and shorten the cutting cycle. Where sharp corners can be avoided, as mentioned in regard to slots, again the tooling and the production time can be improved. Use of a relief or undercut simplifies the design, *Fig.* 17.28. If blended contoured surfaces, *Fig.* 17.25, are required, a combination broach can be arranged to handle the job properly.

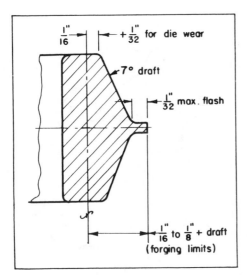

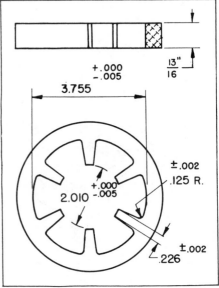

Fig. 17.29 — A typical forged section showing usual limits necessary to observe in order to avoid overloading of the broach on production runs.

Fig. 17.30 — Right — Inverted spline produced with a deep-spline broach. Heavy stock removal limits the use of such broaches to aluminum, brass and other easy cutting materials.

Forgings and Stampings. Where forgings are planned for broaching they should be held as closely to limits as possible in order to allow only a minimum of stock for finishing. A general idea as to the allowances for a typical forged section can be gained from the illustration in *Fig.* 17.29. On the other hand, castings require a greater stock allowance in order to allow the broach to rough beneath the inclusions, scale and hard spots. Cold-punched or pierced holes present much the same problem as castings. Such stampings that have been cold-worked or punched should be normalized before broaching. Parts having a nonuniform density such as that found in porous forgings should be avoided since bad tearing and drift of the broach often result.

Selection of Materials

Any material that can be machined can be broached. The higher the machinability rating, though, the more easily and economically it can be worked. For best results hardness should be held between Rockwell C25 and C35. Harder materials can be worked but sometimes involve problems of wear and lubrication. Softer ones are often too "mushy" or tend to tear, burr or adhere to the cutting teeth and overload the tool. In any case, hardness should be rather carefully controlled to prevent excessive variation because such variation often results in poor surface finish. Internal stresses due to heat treatment should be avoided in parts which must be accurate because the relief of such stresses in broaching results in warpage.

Various materials and hardnesses require different cutting speeds for fine finish, good tool life and maximum production. Nonferrous materials can be broached at much higher speeds than steels. Malleable and cast irons can be broached at speeds around 40 feet per minute and brass and aluminum at speeds as high as 80 feet per minute. However, available machine speeds are such that faster speeds, often accompanied by somewhat lowered tool life, are not justified. A maximum speed of 33 feet per minute usually is used for steels of normal hardness, dropping as low as 16 to 17 feet where a material hardness of Rockwell C40 to 43 must be utilized. Stainless steels can be broached but present a definite problem where fine finish is necessary. Stainless parts are successfully broached to close tolerances but surface finish as a rule is poor since the material tends to tear.

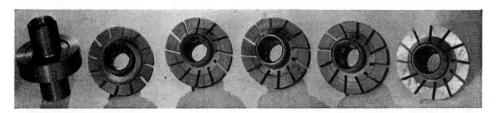

Fig. 17.31 — Hydraulic pump rotor slots roughed and finished in two passes to slot widths of 0.0780 to 0.0785-inch. Hole at bottom of each slot is 0.093 to 0.095-inch in diameter. Splined hub is broached first for locating in the slotting operation. Photo, courtesy Colonial Broach Co.

Specification of Tolerances

Broaching combines not only roughing and finishing operations, but as a rule removes stock to precision limits faster than any other known method producing similar results. In addition, sizing and burnishing teeth may be included on the broach to make possible even greater accuracy and the maximum in finish quality. Finish and dimensional accuracy are not affected by the accumulation of chips in the clamp and fixtures, a problem often encountered in machining operations.

Holes and Splines. Close accuracy and uniformity in surface broaching depends to a large extent upon heavy, rigid fixtures and a solid machine slide having a minimum of deflection. Internal broaching accuracy, however, is dependent more upon uniformity of material. Round or square holes can be held to within plus or minus 0.0005 to 0.001-inch in production. Plain splined holes can be held to plus or minus 0.001-inch to 0.002-inch on the major and minor diameters with width of splines within plus or minus 0.001-inch. Wider limits, however, are more economical to hold, *Fig.* 17.30. Inside diameter of serrations, pitch diameter and diameter measured between wires can be held to plus or minus 0.001-inch. The splines are normally not more than 0.006-inch per foot out of parallel with the axis of the shaft. Involute spline tolerances are readily held to the standards published in SAE and ANSI tables.

Surfaces. In surface work, straddle broached faces can be held to plus or minus 0.001-inch but, where design demands, plus or minus 0.0001-inch can be held on size and parallelism. Flatness can sometimes be held within a quarter-thousandth.

Slots and Lugs. Slots can be held to plus or minus 0.0002-inch, but plus or minus 0.001 to 0.002-inch are much more economical limits. Slot widths on automotive cylinder block bearing pads are regularly held to plus 0.0005-inch and minus zero. Narrow slots, *Fig.* 17.31, can also be held to a total variation of 0.0005-inch. Spacings of lugs, straddle-broached, can be held to 0.0003-inch spacings, *Fig.* 17.22. External gears and racks, broached one tooth at a time, can be held to 0.0005-inch on tooth spacing and form. On high-production runs where an entire segment is completed in one pass, even closer tolerances can be held. On large chain sprockets, *Fig.* 17.4, tooth form and spacing can be held to plus or minus 0.005-inch.

18

Gear Shaper Generating

$\mathbf{A}$LTHOUGH gear shaper machining is generally considered a special-purpose process primarily adapted to the generating of gears and gear teeth, its overall versatility and ability to produce a variety of machine parts, *Fig.* 18.1, places it among the important mass-production processes. The generation of eccentric, gap, metering, and special type gears by this method, as well as ratchets, clutches, cams, multiple cams, splines, and unusual shapes, presents not only a fascinating but also a highly practical study for designers interested in lowering production costs of such special machine parts.

Principles of Operation. Of great importance in connection with the design of parts to be produced on the gear shaper are the several basic methods of machining used and their principles of operation. With the Fellows method, *Fig.* 18.1, reciprocation of the cutting tool makes possible its application to the generating of external and internal surfaces by properly controlled movements of workpiece or

Fig. 18.1 — Close-up view of a tandem gear setup with cutters on Fellows No. 10-2 gear shaper. Cutting time is 1.25 minutes — two pieces per load.

Fig. 18.2 — Rectangular hole being finished by a special offset taper-shank cutter on a Fellows Gear Shaper.

tool, or both, and also permits operating the tool in comparatively narrow recesses. This combination of tool reciprocation with controlled generating motion make this method extremely universal in its application to the production of unusual machine parts.

Combining the basic principles of both the common engine lathe and the well known crank shaper, the gear shaper usually allows a selection between several methods of machining, simplifying production problems. Rotational movement as

Fig. 18.3 — Michigan "Shear-Speed" shaper being used to cut simultaneously all 47 teeth of an 8-inch diameter gear. Total operation takes only 52 seconds.

Fig. 18.4 — Below — Group of typical machine parts produced on the Michigan Shear-Speed shaper.

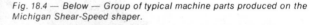

well as position variations can be imparted to the generating tool or to the work and, in some cases, a "conjugate" or mated surface can be used on the cutting tool to generate a desired shape. Often, combinations of movements and "conjugate" shapes can be employed to achieve results that would otherwise be extremely difficult or impossible to obtain, *Fig.* 18.2.

Generation, as utilized to some degree in other processes such as broaching, automatic shape turning and Swiss automatic machining, reaches a high degree in gear shaper machining. In general, it is the production of a surface — curved or straight — by means of the continuous motion of a point, a line, or a form. Generated profiles consisting of any combination of two basic surfaces — curved and plane — can be practically any shape desired.

With the Shear-Speed method the work is held firmly while the cutting head, with a complete set of individual cutting teeth to suit the particular form being generated, reciprocates rapidly while the individual tools feed in to produce the desired form, *Fig.* 18.3. By this method it is feasible to form any external shape that can be produced internally by broaching.

In order to produce almost any design of cam, ratchet, clutch, sprocket or similar form, as well as splines and gears, it is only necessary to substitute the necessary tool head in the machine. The primary limitation, however, is that with this method the work does not revolve and consequently parts must be of symmetrical form rather than compound, *Fig.* 18.4.

Machines Available. Eight distinct types of Fellows gear shapers are available in various configurations for handling a wide variety of work including internal and external spur and helical gears, tapered gears, crowned gears, face gears, racks and many noninvolute shapes, *Fig.* 18.5. These machines will cut gears ranging from small, fine pitch parts 2 inches in diameter 40 diametral pitch to parts 88 inches in diameter 2 diametral pitch. Width of shaped faces ranges from 3/4-inch to 16 inches. Many items of special equipment are available such as tilting work table, rack shaping attachment, quick-change tooling, automatic loading and automatic chip removal.

Fig. 18.5 — Front view of Fellows No. 10-2 gear shaper and a close-up of the 3-inch gear shaper with an automatic magazine feed for loading small pinions.

Fig. 18.6 — Shear-Speed gear shaper which will handle gears, splines, ratchets, cams, etc., up to 10 inches in diameter. Courtesy, Ex-Cell-O Corp.

Shear-Speed shapers, *Fig.* 18.6, are also available in a variety of capacities. Work with outside diameters ranging from 1-inch to 20 inches can be handled with shaped thicknesses up to $2^3/4$ inches to $6^1/2$ inches. With gear teeth, the coarsest and finest pitches recommended for diameters from 1 to 3 inches is 8 to 16, for diameters from 3 to 7 inches is 6 to 16, for diameters from 7 to 10 inches is 3 to 16, and for diameters from 6 to 20 inches is 1.9 to 12 DP.

Design Considerations

A fundamental requirement in the accurate machining of parts is that the work be located or supported from the same surfaces which are used as locating points when the part is assembled. Parts not readily adapted to such support, therefore, require highly specialized fixtures, increasing costs unduly where large volume production is not in order. Ordinarily gear blanks with a central hole are clamped on an arbor with support plates located both beneath and over the work, *Fig.* 18.7. Support plates are usually designed to be only slightly smaller than the contour of the part to lend support directly at the cut. Shank or shaft parts require other means such as clamp sleeves or centers. In producing large quantities of parts with odd contours, magazine fixtures and automatic arrangements are often used.

Shear-Speed Forms. Over and above the gear and spline forms possible to shape on these machines, any form which can be ground, as with broach teeth, into

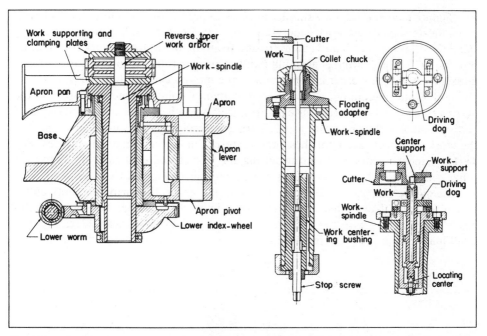

Fig. 18.7 — Section of gear shaper work-spindles showing typical methods for holding shank type and flat blanks. Work is supported at the cut, affording maximum rigidity and accuracy. Drawing, courtesy Fellows Corp.

the individual radial tools can be produced. Tools may vary in width as required, *Fig.* 18.8. Forms, however, cannot be undercut inasmuch as the blades are retracted to clear on each return stroke and are fed into final depth in small increments during the shaping cycle. Depth of forms can be judged from the suggested range of gear-tooth forms.

Fellows Generating. Because the workpiece can be rotated under control relative to the cutter rotation the Fellows method offers a wide range of both simple and compound forms.

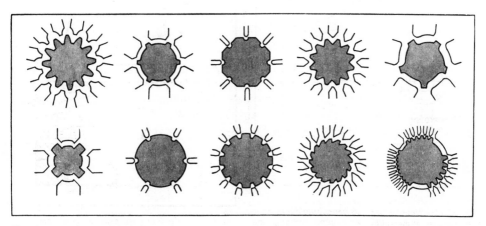

Fig. 18.8 — Typical designs which can be produced on a Shear-Speed shaper showing the various shaping tool teeth.

Simple Generating. Two simple motions imparted to a single-point tool with constant rotation of the work generate a simple shape, *Fig.* 18.9, contour depending upon the relation of tool movement (to and from the work) to the work rotation. By moving the reciprocating generating point in proper relation to the work center a constant-rise cam or other complicated cam contours can be obtained, *Fig.* 18.10.

For rapid machining, a surface rather than a point is generally used on the cutting tools. A simple irregular cam produced by such a cutter under the required radial control is shown in *Fig.* 18.11. Setting the reciprocating cutter at an angle to the work, however, accomplishes either external or internal conical surfaces, *Fig.* 18.12, shapes or gears. By locating cutting and work spindles at right angles to each other, throated or hourglass shapes can be produced, *Fig.* 18.13.

Conjugate Generating. To apply this molding-generating principle, both cutter and work must have pitch-line movement. This can be either two pitch circles rolling together as with the cutting of internal and external gears, *Fig.* 18.14, a combination of a pitch circle and a pitch-line movement as in cutting a rack, two

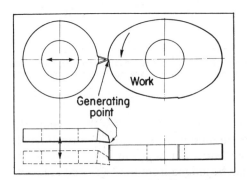

Fig. 18.9 — Layout depicting the generating of an elliptical surface by means of a single-point tool.

Fig. 18.10 — Top Right — Constant-rise cam generated by single point with controlled center distance.

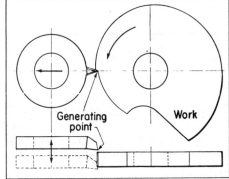

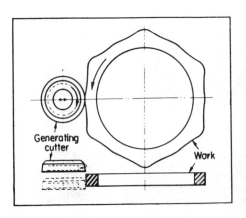

Fig. 18.11 — Using a disk type generating cutter, an irregular cam such as this can be produced at greatly reduced machining time over that needed with a single-point tool as shown in Fig. 18.10.

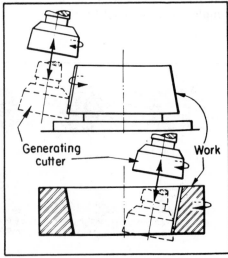

Fig. 18.12 — Conical surfaces or gears are produced by positioning the reciprocating cutter at the desired angle. Drawings, courtesy Fellows Corp.

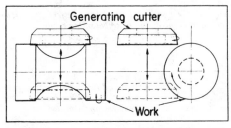

Fig. 18.13 — Throated or hourglass shapes are easily produced by locating the work and cutter spindles in offset positions at right angles. Drawings, courtesy Fellows Corp.

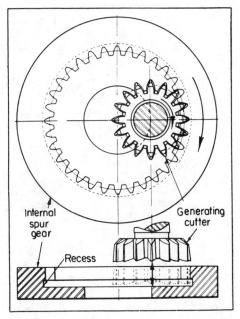

Fig. 18.14 — Molding-generating principle utilized to produce parts such as internal and external gears requires pitch-line movement of cutter and work.

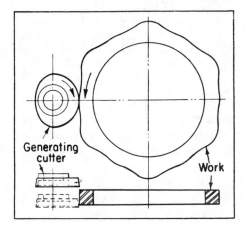

Fig. 18.15 — Left — Using molding-generating and a "conjugate" cutter, the cam shown in Fig. 18.11 can be generated without radial movement of either work or cutter.

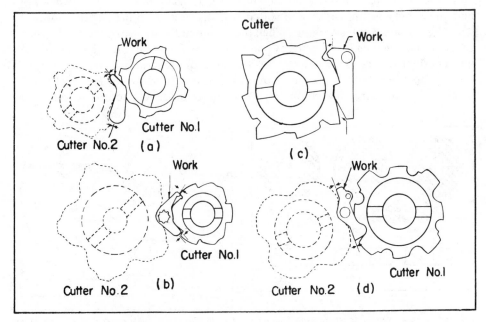

Fig. 18.16 — Irregularly shaped pawls can be shaped accurately and at higher production rates than that possible in milling by using two conjugate cutters.

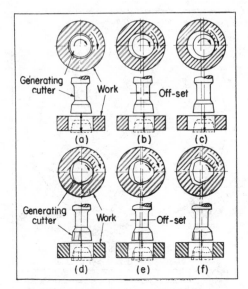

Fig. 18.17 — Left — Round or three-lobed hole generated in one revolution by offset of the cutter.

Fig. 18.18 — Square hole produced with square cutter by offset method has a corner radius equal to the offset. Drawings, courtesy Fellows Corp.

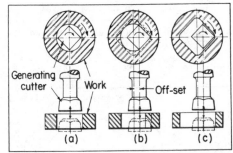

pitch circles as in cutting on-center or off-center face gears, or the pitch-line movement of irregular shapes. Using fixed work and cutter centers, the cam shown in *Fig.* 18.11 can be generated by a conjugate-shaped cutter, *Fig.* 18.15, with definite relationship between the rotation of both work and cutter. In this case the cutter makes six revolutions to one of the work.

Irregular shapes such as the special pawls shown in *Fig.* 18.16 are easily generated, two cutters as indicated being used in each case. Such parts can be produced at a rapid rate, accurately, and with a high degree of finish.

Warped surfaces similar to those of the teeth of a helical gear can be produced on a machine which not only has controlled relationship of rotation between work and cutter but also imparts a controlled twisting motion to the cutter spindle.

Offset-Conjugate Generating. When the cutter and work are of the same general shape, when one member — cutter or work — is external and the other internal, or when both members rotate at a one-to-one ratio, the generating is termed "offset". Some simple hole shapes produced by this method are shown in *Fig.* 18.17. Both work and cutter are rotated in the same direction at a one-to-one ratio, one

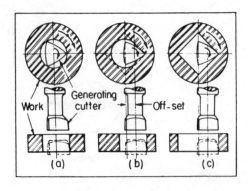

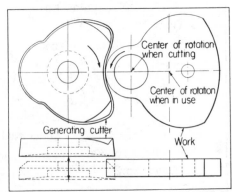

Fig. 18.19 — Square hole generated with triangular cutter by offset method with corner radius approaching that of the cutting edge of the tool.

Fig. 18.20 — To simplify cutter design, unusual and complicated shapes are located off-center. Drawings, courtesy Fellows Corp.

revolution completing a round as well as a three-lobed hole. The size of the hole is controlled by the amount of offset. Radius produced in a square hole, *Fig.* 18.18, is equal to the amount of offset and where such a large radius is objectionable it must be removed with a special tool or another method used. Sharpest corners possible are produced with a three-cornered cutter, *Fig.* 18.19, with the "rolling" circle used in generating passing through the corner. Multiple holes of the same or differing sizes of simple and complex shapes may also be produced simultaneously on equal or different radial and angular spacings.

Off-Center Conjugate Generating. In certain cases it is advantageous to locate work off-center in the fixture, especially to simplify cutter design. This is exemplified in *Fig.* 18.20. Rough formed by milling, this brake cam is finished in one revolution.

A unique application is shown in *Fig.* 18.21 where one set of teeth on a segment gear is concentric with the hole, and another set is on a cam surface. The concentric segment presents no problem but the cam section requires a special type fixture. To produce noncircular gears, three motions are usually required — a master segment gear and rack being necessary to impart the proper movement to the work.

Interrupted-Conjugate Generating. This method applies to unusual parts where the progression of cutting must be interrupted. The refrigerator pump piston, *Fig.* 18.22, is typical of a part successfully machined by an interrupted cutter. Various

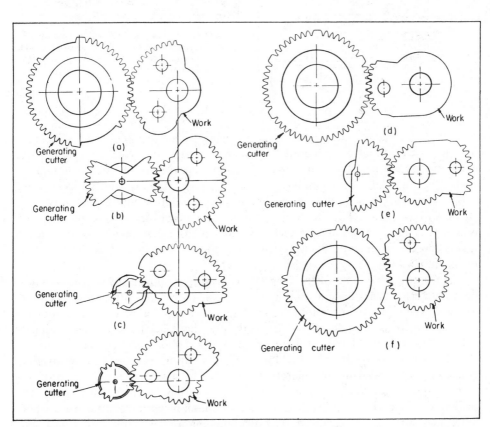

Fig. 18.21 — *Plan showing the sequence of operations used to generate a two-lobed segment gear. Two operations are performed on each segment — disking to blank size and cutting the gear teeth. Drawing, courtesy Fellows Corp.*

types of tooth forms can be generated on the same part, *Fig.* 18.23, by one cutter. A stack of such combination segment gear and ratchet parts for cash registers are produced in a single operation.

A change in position of cutter or work can be used to allow finishing of different or closely related portions of a complicated part. Twin cutters are used to provide for rough and finished cutting of accurate holes, *Fig.* 18.24. Table and fixture in this instance are geared in sequence with the cutter, the table rising to allow finishing of the hole by means of the upper portion of the cutter without disturbing the cycle.

Fig. 18.22 — Using an interrupted-conjugate cutter, a refrigerator compressor piston is finished in two cuts around.

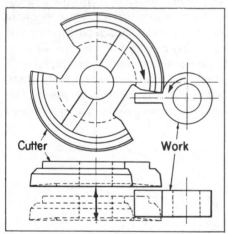

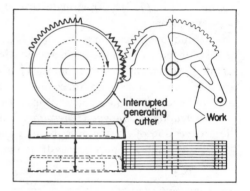

Fig. 18.23 — Both ratchet and involute gear teeth are generated on cash register segment gear by one inter-rupted-conjugate cutter shown above. Drawings, courtesy Fellows Corp.

Progressive cam cutting is achieved in a similiar manner, *Fig.* 18.25. Here six individual cam lobes are generated by a single cutter, the work being moved to bring the various cam lobes into position. Cam contours are identical and equally spaced but are not in consecutive order. The cutter is made with six generating projections spaced to produce each lobe in proper relationship to a keyway.

Gap-type cutter, *Fig.* 18.23, also can be designed to have roughing teeth (in one or two stages) and finishing teeth. By this means exceptionally accurate tooth forms can be produced especially when the softer nonferrous materials such as brass are used.

An unusual example of interrupted generating is shown in *Fig.* 18.26. These pump metering gears have a combination of concentric, eccentric and rack-tooth forms. Seven separate operations are necessary to complete the two parts, four for the one and three for the other. All teeth must have accurate mating conjugate action inasmuch as the pair of metering gears operate on fixed centers.

Describing Generating. This form of generating reverts to the original conception of producing or generating a surface by means of a point and differs from conjugate generating in that only one member has pitch-line movement. The principle is especially applicable to cutting of clutch faces and similar work where helicoidal surfaces are required. The airplane engine jaw-tooth clutch part, *Fig.* 18.27, illustrates this type of generation. Utilizing substantially a single rounded point of one tooth on the cutting tool as the work is rotated, the cutter advances

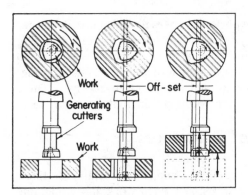

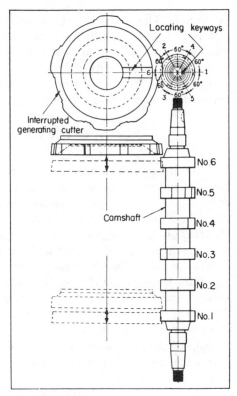

Fig. 18.24 — Twin cutters can be used to rough and finish machine highly accurate holes by elevation of the work table to position the part. Drawings, courtesy Fellows Corp.

Fig. 18.25 — Interrupted-conjugate method utilized to generate a series of six related cams. Work is elevated successively for the finishing of each of the cams.

into the work, generating the correct form and at full depth finishing the form. Such work is located at a slight angle to allow clearance for easy withdrawal of the cutter.

As can be seen readily from the general character of the unusual machine parts illustrated, few are produced with standard machine fixtures and tooling. However, a multitude of special attachments for the various machines is available and consultation with the manufacturer while unique or special parts are in the design stage will assure minimum costs and successful production in virtually every instance.

Cutter Clearance. As with any machining process, cutter clearance must be considered. However, the gear shaper requires less clearance as a rule than other methods and is admirably suited to parts which require close shoulders or projections. With external forms an undercut is necessary, *Fig.* 18.4, and with internal forms a recess, *Fig.* 18.14. With shallow forms such as fine-pitch gear teeth the width of recess or undercut can be as small as ¹/₁₆-inch. Width required increases with the depth of form to be generated.

Selection of Materials

Suitability or machinability of various materials with this process parallels somewhat that for broaching. With that Shear-Speed the number of strokes utilized can be varied as required to suit the material and the depth of form. On Fellows machines, cutting speeds normally available range from 36 to 2000 cutter strokes per minute. The slower speeds are used with long strokes and also for the harder materials. Considering high machinability stock only, Chapter 68, proper cutting fluid and a maximum feed stroke of one-inch, cast iron can be cut at a speed of 60 fpm, mild steel at 90 fpm, high-carbon steel at 65 fpm, tool steel at 70 fpm, chrome-nickel steel at 50 fpm, soft brass at 100 fpm, and aluminum at 200 fpm. These speeds are reduced about 10 per cent for each additional inch of cutter stroke.

Specification of Tolerances

To maintain accuracy on parts which cannot be supported solidly, the finest and consequently slowest feeds must be used. On generated work, more than one revolution is often required where accuracy and quality must be of the highest. Also, the accuracy of the machined blank is of primary importance, since the final accuracy of the part is governed to a large extent by the blank conformance as well as the method used in holding it. Naturally, demand for closer limits than those actually needed to suit the requirements increases considerably the cost of any part.

Bore Diameters and Face Runout. A study of manufacturing limits set and found to be practical by various industries reveals that they vary considerably, depending primarily upon the nature of the part to be produced and its application. Bore diameter, for instance, on parts with holes is controlled somewhat by the amount of runout allowed. On circular parts, the limit on hole variation is set at approximately one-half the total amount of runout allowable on the finished part:

Size Hole (in.)	Hole Tolerance (total, in.)		Max. Face Runout (total, in.)	
	General	Precision	General	Precision
0-1	0.001	0.0002	0.002	0.0005
1-4	0.002	0.0006	0.004	0.001
4-8	0.003	0.0010	0.006	0.002
8-12	0.004	0.0015	0.008	0.003
12-24	0.005	0.0025	0.010	0.005

Blanks produced for gear shaping can be readily held, where necessary, within a total variation of 0.0002-inch on bore diameters of 0.125 inch and over. Precision limits are usually applied where the face is used for important locating operations.

Outside Diameters and Runout. Outside diameters of blanks are normally held to different limits where parts are to be shaped on the outside and where the periphery is used for location rather than the hole. Typical limits specified are:

Outside Diameter (in.)	Diameter Tolerance (total, in.)		Outside Diameter Runout (total, in.)
	Locating	Nonlocating	
0-3	0.0005	0.001	0.001
3-6	0.0010	0.002	0.0015
6-8	0.0015	0.003	0.002
8-12	0.0020	0.004	0.003
12-20	0.0030	0.005	0.004

These tolerances are merely indicative of general practice; considerable variations are found in different industries. Tolerances on locational diameters in some fields are approximately those shown as nonlocating. Tolerances, of course, should be no smaller than considered practical for a particular design. Nonlocating tolerances are those used for outside diameters to be finished into gear teeth, clutch teeth, ratchets, or other forms and most often are applied minus. Outside diameter tolerance is also subject to consideration of the diametral pitch with gears, the tolerance normally ranging from 0.001-inch at 200 DP to about 0.0065-inch at 30 DP.

Cams and Radial Surfaces. Positioning of various cams in relation to each other, *Fig.* 18.25, can be held within 0.001-inch for alignment. Also, the radii of cams and other radial surfaces can be held to a maximum variation of 0.001-inch.

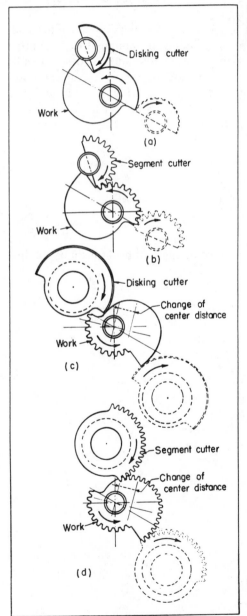

Fig. 18.26 — Left — Metering pump gears produced by interrupted cutters. Concentric, eccentric and rack teeth on these unusual gears are generated in seven operations. Drawings, courtesy Fellows Corp.

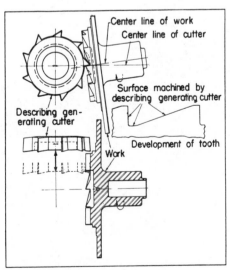

Fig. 18.27 — Describing generation utilizes a virtual single-point cutter to produce a true warped or helicoidal surface such as on the jaw-teeth of this airplane clutch part.

Instrument Gear and Similar Work. Limits held on ordinary fine-pitch gears in general production on standard or special gear shapers give a good indication of the accuracy which may be expected in the manufacture of a great variety of parts along the lines of those illustrated. Instrument gears for transmitting motion rather than power are regularly held to a total composite error of 0.006-inch for class 1, 0.004-inch for class 2, or 0.002-inch for class 3, depending mainly upon the machine requirements. Total tooth-to-tooth composite error for such "commercial" quality gears runs 0.002, 0.0015, and 0.001-inch, respectively. "Precision" quality gears of fine pitch for instrument work, however, are held to a total composite error of 0.001-inch for class 1, 0.0005-inch for class 2 and 0.00025-inch for class 3, tooth-to-tooth composite errors being 0.0004, 0.0003 and 0.0002-inch respectively.

Precision accuracy such as the latter requires the best possible machines and subsequent refining operations such as shaving, grinding, or lapping. Consequently, unless the machine requirements necessitate precision tolerance, widest possible commercial limits are preferable to speed production and assure lowest unit cost.

Power Gearing. In commercial production of gearing for automotive, ordnance, and general machine work, broader tolerances are specified for larger gears for power applications. These are as follows:

Gear Size (in.)	Pitch-Line Runout (total, in.)	Diameter Over Pins (in.)
3-6	0.002	0.002
6-13	0.003	0.003
13-20	0.004	0.004

Pitch-line runout is reduced about one third to one-half by a finishing shaving operation.

19

Abrasive Belt Grinding

ABRASIVE belt surfacing is a decidedly old processing method, developed originally in the woodworking industry where power sanding was and still is widely utilized. Use of these dry belts for production surfacing of metals, rubber, plastics, or glass, however, was out of the question owing to damage from generated heat. Advent of impervious, plastic-bonded cloth belts, with electrostatically-applied aluminum oxide or silicon carbide abrasive particles, ushered in the precision surface machining of these engineering materials in mass production by means of the wet-belt machine designed for precision finishing.

Scope of Process. Advancement of the wet-belt process was accelerated by the need for low-cost, fast finishing on a wide variety of machine parts, *Fig.* 19.1, where single high-quality flats, parallel or angularly disposed flats, round, and in some cases concave or convex surfaces were required, brought this method to the fore. Today, the wet-belt machine is finding its place alongside light milling machines, shapers, various types of surface grinders employing grinding wheels, and occasionally light lathes and cylindrical grinders.

Fig. 19.1 — Typical belt finished parts. Abrasive belt grinding provides low-cost, high-quality finishing for a wide variety of machine components.

Method of Grinding. Primarily, wet abrasive-belt grinding is a method of stock removal and surfacing performed by means of a tensioned abrasive belt operating over precision pulleys at a speed between 2500 and 6000 sfpm. Grinding or cutting takes place only where the belt passes over a sturdy support platen. Belt speeds as high as 10,000 sfpm can be attained on some machines when conditions so demand. Work is supported on a horizontal table, or on a conveyor belt, and contacts the abrasive belt with a motion against or normal to the belt travel, *Fig.* 19.2. Contoured parts, however, require a different technique since they must be fed directly into the belt to maintain the contour on the shaped platen. Accuracy of the table-to-platen relationship and also the accuracy of the platen surface of the machine itself largely dictate the accuracy of the finished pieces. Millions of uniformly graded and spaced abrasive grains on the belt move rapidly across the entire surface of the work at the same speed and, continuously flushed clean with coolant, impart a uniform surface finish without creating much pressure or heat.

Surface Quality. Work hardening or warpage due to generation of heat is seldom, if ever, a problem. Fusing or melting of soft and low-melting-point

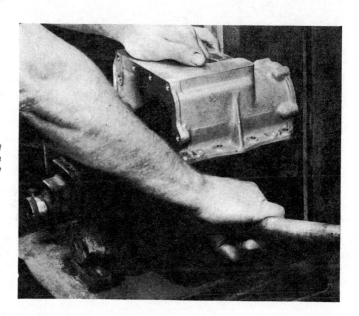

Fig. 19.2 — Horizontal feed movement of table assures a smooth, flat registry surface with an excellent finish.

materials such as plastics thus is not a problem. Stock removal on all materials is clean and no burrs are produced. Even though surfacing is fast, an excellent finish is obtained on most all materials and in many cases better than that in milling, planing or turning.

Production Speed. The unusually high rate of cut effected by this method makes possible an extremely high production speed — some operations requiring but 5 seconds. Since the cutting pressure of the belt at high speeds is low, there is no tendency to spring the work. Jigs and fixtures, therefore, can be simple and inexpensive to permit quick loading and unloading, *Fig.* 19.3. In fact, many components can be ground in a freehand manner to further simplify handling. However, where production runs into many thousands, multiple type fixtures are found expedient, *Fig.* 19.4. Such fixtures, although somewhat higher in cost than the simpler types, reduce overall cost per piece.

Material Allowance. While parts designed for shaping, planing or milling generally require at least 1/16 to 1/8-inch of metal for machining, only 0.015-inch allowance is necessary for belt finishing. This factor in conjunction with the high production speed and low cost of abrasive belts and machines can result in substantial economies through proper application.

Limitations. Quite naturally, as with other processing methods, designing to make the most of these advantages is subject to a number of limitations. These limitations sum up to: (1) Amount of stock removal, (2) size range of parts which can be handled most efficiently, (3) type of surface to be finished, and (4) accuracy.

Obviously, the lighter the cut necessary, the greater the advantage in time, belt life and fixture outlay. Although 0.015-inch stock allowance for hand surfacing and about 0.031-inch for automatic power-feed machines usually is considered

Fig. 19.3 — Simplicity of jig for belt finishing, right, compared with one for milling the same part, left. Piece at right was specifically redesigned for the wet belt.

Fig. 19.4 — Below — Mass-production fixture for finishing six stampings at once. One jig unit can be loaded while another is in the machine. Parts are ground on both sides to an overall flatness of 0.002-inch.

most efficient, as much as 0.125-inch can be handled on power-feed arrangements. However, about 0.062-inch for nonferrous materials and about 0.031-inch for ferrous materials is considered a practical maximum for production jobs.

Generally speaking, nonferrous materials permit the surfacing of larger areas than do ferrous materials, although the amount of stock to be removed is a deciding factor in determining the practical area limitation. The greater the stock removal the smaller the area which can be surfaced. To start, an area of 30 square inches generally is accepted as an economical limit before the restrictions of material and stock removal factors are considered. With the lightest possible stock removal, say 0.003 to 0.005-inch, on nonferrous parts, areas up to 52 square inches have been finished successfully. However, the usual top limit runs to about 40 to 45 square inches with a stock removal of 0.060-inch on aluminum and 0.030-inch on steel.

Fig. 19.5 — Meehanite cast iron grids loaded onto a swinging fixture board are transferred to a magnetic chuck for belt surfacing twenty-four at a time. Production was raised 50 per cent and labor reduced 80 per cent over previous methods.

Although many parts with large, solid flat areas have been ground, it is much more difficult to surface solid areas than dispersed or interrupted ones. A solid area of glass, cast iron or steel requires great quantities of water and a special corrugated support platen to allow the coolant to reach all the work surface in order to prevent firing.

Machines Available. Many simple vertical and horizontal type belt surfacers are available in various sizes. However, the vertical automatic power-table models and horizontal multiple-head flat finishers are best suited for high-production precision work. With these power models, production jigs and automatic fixtures make possible a wide variety of work, *Fig.* 19.5. Using the automatic feed table and a motorized fixture, even difficult turning jobs have been handled within extremely close dimensional limits, *Fig.* 19.6. The largest of these machines offers a working surface 10 inches wide by 17½ inches high. Use of a plain table in lieu of the automatic oscillating one increases this available working height to 24 inches. Thus, it is evident that only those parts that fall within this dimensional range and the preceding area limitations can be handled readily.

Fig. 19.6 — Grids, mounted ten at a time on a 40-inch power-driven arbor, are surfaced on the outside diameter of the three legs. Production was increased from 450 to 1100 per hour and burrs formerly encountered eliminated.

Design Considerations

From the foregoing discussion it can be seen that a wide variety of parts can be designed to utilize the low-cost production advantages of the wet-belt surfacer. Although many concave and convex type surfaces can be ground by using special formed platens, as a rule, surfaces generated by the belt operating against a flat platen are most practical and economical, i.e., flat surfaces, flat related surfaces, cylinders, etc.

Stock Allowance and Surfaces. In addition to the small material allowance of about 0.015 to 0.031-inch mentioned previously, still further material savings can be achieved by designing for the belt method. Whereas with milling, etc., a solid surface often is desirable for achieving a uniform cut, an interrupted surface is the ideal for belt finishing. Thus it is possible to core out large, flat areas and effect considerable savings in material, *Fig.* 19.7. Further, generation of accurate gasket and registry faces becomes a simple and highly economical operation.

Fig. 19.7 — Idler pulley carrier as originally designed for milling, left, and as redesigned for belt finishing, right. In addition to a substantial reduction in material required, production time for belt finishing was 6.26 hours per hundred pieces as compared to 32.5 hours per hundred for milling.

Faces and Bosses. Obviously in designing parts with mating surfaces and scattered bosses, such surfaces should be kept in a single plane wherever possible, *Fig.* 19.1. This procedure simplifies production whether the part is specifically intended for finishing on the belt grinder or not. Registry faces, as in *Fig.* 19.7, usually need be only about 1/4-inch wide except at bolt holes. Gasket or sealed faces should be kept as narrow as is practicable, *Fig.* 19.2, to assure adequate unit pressure and yet encompass the bolt holes.

End Radii. Parts which require the generation of end radii can be readily handled. A simple jig can be employed to generate true cylindrical radii up to about 180-degrees sweep.

Selection of Materials

Practically any of the metallic or nonmetallic materials used in the fabrication of product components can be finished on the wet or dry belt grinding machine. The various steels usually are finished with aluminum-oxide belts and the softer metals, plastics, hard rubber, glass, ceramics, fiber, etc., are ground with silicon-carbide abrasive. Water is used largely for cooling, but in some cases, as with magnesium, special coolants have been produced for safe operation.

Often such materials as aluminum alloys are finished dry. Critical materials such as magnesium and those that are highly heat sensitive are ideal for the wet method. As a general rule, with all materials, quality of finish is better when the wet-belt method is used.

Specification of Tolerances

As noted previously, many surfacing operations can be performed without jigs by merely holding the part against the belt and moving it back and forth. Naturally such operations involve the minimum in costs but, are recommended only where close tolerances, especially parallelism, are not at issue and where stock removal is light. For close tolerances and normal or heavy stock removal operations, special jigs or holding devices and controlled power feeds are necessary.

Flatness and Parallelism. For most operations normal tolerances on flatness and parallelism range from plus or minus 0.001-inch to plus or minus 0.0005-inch, the closest tolerances being economically maintained. In some cases, plus or minus 0.00025-inch tolerances have been held, although this of course means a somewhat higher cost per piece.

Radii. Parts on which end radii are generated can be held to plus or minus 0.002-inch in production using a simple fixture.

Finish. Surface finish is unusually good, little or no finishing marks mar the surface and many parts appear to be polished. With the proper grade of belt, belt speed and coolant, surface finishes from around 40 to 50 microinches to as fine as 3 microinches can be had. The finest finish with closest accuracy, however, may require two operations — rough grinding with a coarse abrasive, and finishing with a fine-grit belt. Needless to say, minimum cost and maximum production are most easily achieved by specifying the widest possible tolerances commensurate with the design contemplated.

20

Production Grinding

NEED for improved surface finish as well as facility in finishing parts too hard for ordinary cutting methods resulted in the development of synthetic abrasives about the year 1890. Natural abrasives, which for centuries served satisfactorily in simple sharpening and abrading operations, lacked uniformity and other characteristics required in mass-production grinding and as a result were soon supplanted. Early precision grinding thus had its beginning somewhere around the turn of the century and has since developed rapidly into one of the primary mass-production methods.

Although grinding had been thought of and used extensively as a means for

Fig. 20.1 — A typical cylindrical grinding operation on a Cincinnati Milacron Electronic Command plain grinder.

Fig. 20.2 — Bendix thread and form grinder set up to plunge grind from the solid a double-start, left-hand bronze worm with a crush-formed wheel. Complete operation requires but 35 seconds. Pitch diameter is held to 0.003-inch over wires.

finishing hard surfaces only, abrasive development and grinding machines have progressed until today even the softer materials can be ground efficiently. The much improved finish resulting from the grinding operation, offering surfaces with less tendency toward wear and dimensional change, has found wide application in machine manufacture. Developments and refinements to date require adequate recognition in obtaining maximum quality and economy in production of machine parts of all varieties, *Fig.* 20.1.

Grinding Process. Not unlike other machining methods, grinding is primarily a process of removing material in the form of chips. Abrasive grains of grinding wheels perform in much the same manner as single-point cutting tools, effecting by a multitude of contacts a virtual multiple-tool cutting action which produces extremely minute chips similar in character to those produced in turning. Since cutting edges of the grits are extremely thin, it is possible to remove much smaller chips and to refine surfaces to a much greater degree of accuracy with respect to both finish and dimension than with other methods of metal cutting presently available.

Wheel Dressing. Originally all wheels were dressed or prepared for grinding

Fig. 20.3 — Standard Cincinnati Milacron Electronic Command plain grinding machine. Axis of workpiece can be angled for grinding tapers.

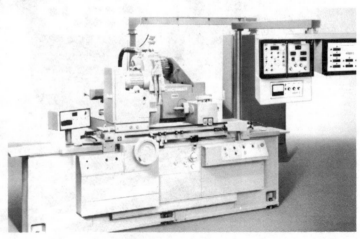

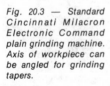

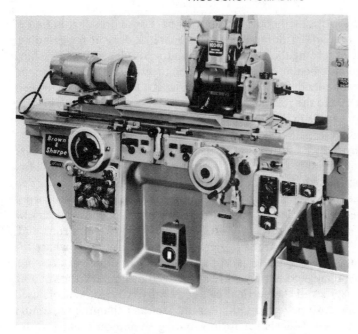

Fig. 20.4 — Standard Brown & Sharpe universal grinder. Unlike the plain grinder, these models allow both headstock and wheelhead to swivel.

by using the well-known diamond dressers. Production machines are now largely equipped with automatic dressing or truing devices for wheel preparation. A variety of mechanical form-generating diamond dressers are also available for producing other than plain faces. These, however, are somewhat limited mechanically in the forms which are practicable to produce, and until the perfection of crush dressing technique certain thread forms having angles with the vertical less than 25 degrees could not be produced on multiribbed wheels. In fact, it is usually necessary to employ "skip-rib" diamond dressing on multiform wheels, especially on the finer pitches although this increases the grinding time.

With the introduction of the crush dressing technique in production grinding, design possibilities have broadened considerably. Production rate can be 15 to 20 times faster. A vast variety of complicated forms and shapes can be ground or generated direct from the solid, rapidly and accurately, *Fig.* 20.2. Although machines have been especially adapted for crush-forming operation, some standard

Fig. 20.5 — Model 960 Ex-Cell-O automatic thread grinding machine for automotive steering gear ball tracks.

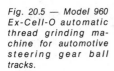

machines can be fitted for this method of wheel dressing. Crush dressing, in lieu of traversing a diamond across a grinding wheel to prepare or dress it for grinding, embodies the principle of forcing the abrasive wheel into a hardened steel crushing roll having a profile complementary to that desired on the wheel. Vitrified bonded grinding wheels are recommended with the crush process. Organic bonded wheels are not recommended due to the resiliency of the bond. Most plunge form abrasive machining is done with wheels in the range from 60 to 400 grit and falling within a hardness range from K to T.

Simplifying the procedure of dressing an abrasive wheel repeatedly to secure accurate duplication of complex forms, crush dressing thus enhances both quality and quantity in production. Although the somewhat faster, cooler cutting afforded by this method of wheel dressing is an advantage in production of complex forms, certain limitations exist which must be recognized. Extremely deep forms can be crushed but contribute somewhat to accelerated crusher wear on long runs and, too, it may be difficult to hold a square shoulder on the wheel especially where the shoulder must be normal to the wheel axis. An advantage over diamond dressing, however, is that worm, Acme, and buttress threads, as well as similar forms having flank angles within 5 degrees of the vertical can be readily crushed. Wheels used for close tolerance work must have individual grain size finer than the maximum root width or radius desired. Wheels for form grinding should be slightly wider than the width of the form to be ground.

Wheel Speeds. Proper grinding wheel speeds have been fairly well established and most machines available can be adjusted to attain the desired cutting efficiency. Major problem in the past has been with internal grinding where proper peripheral wheel speeds were difficult to obtain. Today, however, high-cycle, direct-driven wheel heads are used for extremely small wheels. Efficient peripheral wheel speeds, observing maximum manufacturers limits for specific wheels of course, are generally as follows:

Cylindrical grinding (organic)	5500 to	12,000 sfpm
(vitrified)	5500 to	6500 sfpm
Surface grinding	4000 to	5000 sfpm
Internal grinding	2000 to	6000 sfpm
(diamond wheels)	5000 to	6500 sfpm

Work Speeds. For obtaining the desired finish as well as maximum stock removal commensurate with good wheel life, proper work speeds are also necessary. Except for internal centerless grinding, at the line of cutting the work and the abrasive wheel pass in opposite directions. Work speed at this point should vary somewhat according to the material being ground, approximately as follows, the low range being used for roughing cuts and the high for finishing:

Soft steel	30 to 50 sfpm
Hard Steel	70 to 100 sfpm
Cast iron	200 to 400 sfpm
Aluminum	100 to 200 sfpm

Stock Removal. Most efficient stock removal is obtained with coarse-grit wheels and while surface finish is distinctly rougher, commercially acceptable dimensional accuracy can be maintained. Maximum economy in manufacture thus depends to a large extent upon judicious selection of surface finish. The following figures on stock removal per single pass indicate generally the comparative rates, though these will vary somewhat for different materials:

Rough Grinding .... 0.002 to 0.020-inch
Semifinish grinding .... 0.001 to 0.010-inch
Finish grinding .... 0.0005 to 0.005-inch
Plunge grinding .... 10 inches wide x 1-inch deep

Grinding Method. Particular design features of a part dictate to a large degree the type of grinding machine required. The material specified and its condition indicate, as a rule, the most desirable type abrasive wheel. Where processing costs are excessive, redesign, to utilize a less expensive, higher output grinding method may be found well worthwhile. For instance, wherever possible the production economy of surface, disk and centerless grinding should be taken advantage of by proper design consideration.

Types of Machines Available. Today, with but few exceptions, practically any reasonable grinding operation can be handled rapidly and economically, providing due consideration is given the design of the parts to be ground. Where projections are eliminated and surface area brought to a minimum, the surface grinder can be used to advantage in removing the smallest possible thickness of material to attain a true surface. Again, by proper design the speed and accuracy of the centerless grinder can help reduce costs significantly. Of greatest importance, where quantities to be produced do not justify purchase of a new machine, adequate knowledge of the machines at hand will make possible the design of functionally and economically satisfactory parts.

Cylindrical Grinders. The identifying term "cylindrical" is applied to that large group of grinding machines which employ centers for mounting the workpiece to be ground. As with the automatic lathe two centers are used, a dog or other attachment driving the workpiece. Where desirable, collets, chucks, or other holding devices can be substituted. For the most part grinding is confined to simple external cylindrical surfaces, tapers, faces, etc.

Most widely used of the external cylindrical machines is the so-called *plain* grinder which is a general-purpose production machine for the grinding of axles, shafts, spindles, etc., *Fig.* 20.3. Plain grinders are built with hydraulic, electric, mechanical, hand table traverse or a combination of all three, and feature various types of wheel and work drives such as electronic or electric. The work table, with power headstock and tailstock which can be positioned to suit, is mounted on the longitudinally traveling main table so as to allow angular adjustment relative to the

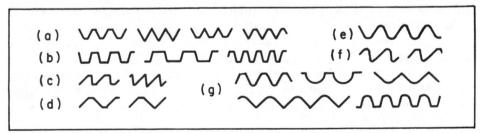

Fig. 20.6 — Variety of thread forms which can be ground: (a) Standard 60-degree, (b) Acme, (c) modified buttress straight-line, (d) special straight-line, (e) Whitworth, (f) modified buttress with radius, (g) special with radius. Drawing, courtesy. Ex-Cell-O Corp.

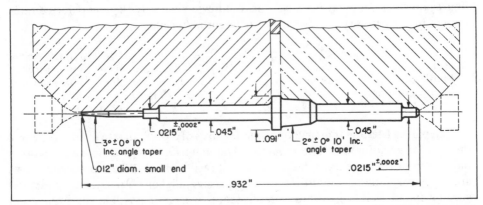

.0215" ±.0002" .045" .091" .045" .0215"±.0002"

3°±0° 10' Inc. angle taper

2°±0° 10' Inc. angle taper

.012" diam. small end

.932"

Fig. 20.7 — Instrument shaft center-ground on a small form grinder. Diamond dressing is used for delicate details. Drawing, courtesy Ex-Cell-O Corp.

Fig. 20.8 — Left — Standard Bryant automatic chucking grinder for parts up to 16 inches in diameter. Grinding stroke of holes is 9 inches and wheel speeds up to 100,000 rpm are possible.

table ways. Angular adjustment of the swivel table varies with different size machines. Cross feeds or wheel infeeds can be continuous or interrupted and are hand powered, independent automatic (with the table stationary) or full automatic (with the table traversing and a predetermined infeed at table reversal) or a combination of all three. The entire grinding cycle can be manual, semiautomatic or full automatic as desired. Plain grinders can handle parts from the smallest up to 36 inches in diameter by 192 inches in length, the lengths being proportioned to the swing capacity of each machine. For instance, a 6-inch machine is available in 18 to 60-inch lengths, a 10-inch machine from 18 to 100-inch lengths, a 14-inch machine from 20 to 168-inch lengths in 10-inch increments, while 20-inch and larger machines generally start at 48 inches and go up to 336 inches. Variations are suitable for almost any job.

Second most widely used machine is the so-called *universal* grinder which, though originally a tool-room machine, is now also a useful production unit, *Fig. 20.4*. In addition to the features offered by plain grinders, universal machines are

Fig. 20.9 — Abrasive Machine Tool Company internal finisher with compound workhead slides to enable grinding of two angularly related holes at one setting.

provided with a swiveling headstock and a swiveling wheelhead. Completely flexible in setup, these machines are ideally suited for production of small or large-quantity lots of parts requiring face as well as cylindrical and taper grinding. Complex parts are readily handled and where necessary internal grinding can also be performed. Universal machines are available to handle parts requiring swings up to 18 inches and center distances to 72 inches.

Another group of machines which falls under the external cylindrical category is the *thread* and *form* grinders, *Fig.* 20.5. Certain of these grinders are equipped for both single and multiform wheel operation and have both diamond and crush dressing attachments. Variety of thread forms which can be produced is shown in *Fig.* 20.6. One to eighty external threads per inch or two to eighty internal threads per inch can be ground on standard machines with diamond-dressed single wheels. Crush grinders generally available have capacities for work up to 24 inches in diameter by 96 inches long.

As a rule crushing can be used successfully for thread forms up to about 40 pitch which requires a 400-grit abrasive in order to maintain the necessary crest and root dimensions. However, 72-pitch threads have been ground with a 600-grit wheel, crush dressed. Diamond-dressed wheels can be single or multiribbed up to about 80-pitch and as a rule diamond dressing is recommended for special designs and forms requiring sharp angles, corners, extremely fine detail or minimum possible tolerances, *Fig.* 20.7. Machines are provided with manual or fully automatic operational cycles. External threads up to 12 inches in diameter can be ground in lengths up to 50 inches, measured from the headstock, on parts up to 115 inches in length. Internal threads from 1-inch to 9½ inches in diameter can be ground on standard internal thread grinders up to a maximum thread length of 5 inches. Maximum helix angle for internal threads is 10 degrees, right or left. For external threads helix angles up to 45 degrees left-hand or 30 degrees right-hand can be had.

Although most of the foregoing external-cylindrical machines can be fitted with internal grinding attachments, the grinders are designed primarily for external work. Another group of grinders, specifically designated as *internal* or *chucking* grinders, *Fig.* 20.8, are available for economical production finishing of holes in gears, bushings, bearings, etc., or more complex bores in precision machine parts, *Fig.* 20.9. Generally these grinders are equipped with a reciprocating wheel head and a rotary work head which is mounted on a movable cross slide with provision for angular adjustment. Face plates, chucks, collets or fixtures can be used for holding the work.

Adaptable for both manual or full automatic operation, certain machines feature in addition automatic rough and finish grinding, automatic dressing and sizing. Standard machines provide wheel speeds up to 150,000 rpm and will finish holes from 0.040-inch to about 20 inches in diameter. Straight, tapered, curved and irregular holes or a combination of these can be handled as well as face grinding where squareness is desirable. While standard machines have swing capacities ranging up to 30 inches, allowing a maximum hole of about 20 inches diameter 12 inches deep, *Fig.* 20.10, specials are in service providing swings up to 100 inches.

Centerless Grinders. As contrasted with regular cylindrical grinders, centerless machines develop cylindrical or formed parts without a centering or chucking operation and as a consequence have far greater output, *Fig.* 20.11. Actually, only one-half the amount of stock required in center grinding need be removed to produce a completely finished surface. Usually centerless machines utilize a large grinding wheel operating at 6000 to 8000 sfpm, a rubber-bonded abrasive regulating wheel rotating in a like direction at about 40 to 1200 sfpm, and an adjustable work rest or support. Contact on the slow-moving regulating wheel of the piece being ground maintains uniform rotational speed of the work in a direction opposite that of the grinding wheel.

Several classes of centerless grinding are normally employed depending upon the type of work to be processed. Straight, cylindrical surfaces having no

Fig. 20.10 — One of the largest of internal grinders, this Bryant Model 1130 will swing 30 inches and grind to a depth of 12 inches. Special versions extend this range to 36 inches swing.

Fig. 20.11 — Center-less grinders offer a range of production possibilities and out-standing economy. Cincinnati Cinco No. 15 for parts up to 3 inches is shown.

interfering shoulders or projections are ground by the *through-feed* method and machines available can handle diameters from about 0.020-inch up to about 12 inches and 20 feet or more in length at speeds up to 17 fpm. *Abrasive-belt centerless* grinders, also through-feed machines, are available to handle work from 3/32-inch up to about 9 inches in diameter and lengths from 3/4-inch up.

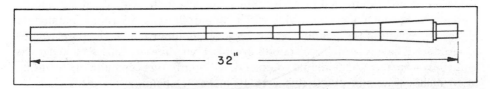

32"

Fig. 20.12 — Rifle bar-rels 32 inches long produced by in-feed or plunge grinding by centerless method on a Cincinnati Filmatic.

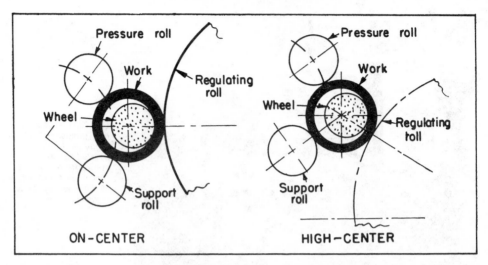

Fig. 20.13 — Principles of operation in centerless methods for internal grinding. Drawings, courtesy Heald Machine Co.

Straight threaded parts or shafts from 8 pitch to 80 pitch, 1/8-inch to 36 inches or longer and up to 5 inches OD can be rapidly ground on *centerless thread* grinders. Whether parts are long or short, through-feed grinding is continuous and production extremely high.

For formed or contoured pieces, parts having shoulders, heads, flanges, or designs requiring undercuts or tapers, *plunge* or *in-feed centerless* grinding, in which the work does not traverse relative to the wheels, is employed, *Fig.* 20.12. Many automatic controlled-cycle machines are available with diamond and/or crush dressing attachments for a wide variety of part designs. Maximum obtainable plunge movement of the grinding wheel, however, is about 1-inch and parts cannot be produced having radial variations greater than this amount. Part length is limited by wheel width; available equipment will normally accommodate forms up to 30 inches wide and diameters up to 10 inches. Depending upon the size and design, production can range up to 1000 pieces per hour or more. Occasionally belt centerless machines can be employed for plunge grinding tapers but length of parts is limited maximum belt width available, normally 6 inches.

In *end-feed centerless* grinding, used only for special taper work, the part is fed manually or mechanically between the wheels to a fixed stop, and ejected. Threaded parts with shoulders or heads can often be handled by this method.

Somewhat similar to external centerless grinding is that known as *internal centerless* grinding. Having the production advantages of ordinary centerless machines, internal machines allow the rough and finish grinding of inside surfaces of straight, tapered, interrupted, or blind holes with minimum chucking and handling. *On-center* method gives maximum support to the work and makes possible accurate grinding of thin wall sections without distortion while *high-center* method provides for accurate finishing of parts with slightly varying outside diameters or multiple work, *Fig.* 20.13. Inside diameters from 1/2-inch up can be ground in parts with outside diameters ranging up to a maximum 4 1/4 inches on the smallest machine, and 3 inches up with outside diameters to 24 inches on the largest. Maximum length of hole on the smallest machine is 1 5/8 inches and on the largest, 10 inches. Tapers up to 60 degrees included angle can be produced.

Fig. 20.14 — Brown & Sharpe No. 824 semi-automatic surface grinder.

Surface Grinders. There are three basic types of surface grinders — horizontal, vertical, and rotary. In the *horizontal* surface grinder, *Fig.* 20.14, the wheel axis is horizontal and grinding is done on either the face or the periphery of the wheel, while *vertical* grinders have a vertical wheel axis with the grinding taking place on the face of a cylinder or segmented-cylinder wheel. Each type is available with rotary or reciprocating table. The *rotary* surface grinder has a tiltable horizontal wheel axis and a rotary table for grinding flat, concave or convex surfaces. Chucking capacity, ranges up to 30 inches, maximum convex angle 17 degrees, maximum concave angle 15 degrees.

Rotary-table, vertical-spindle machines are adapted primarily to rapid production of flat surfaces on small and medium-size parts singly or continuously, *Fig.* 20.15. Vertical machines are capable of carrying segmented wheels up to 42-inch diameter and, with a 60-inch magnetic rotary-table chuck, will handle work

Fig. 20.15 — Variety of precision pump parts all surfaced on a vertical-spindle machine. Photo, courtesy Blanchard Machine Co.

Fig. 20.16 — Blanchard multiple-spindle automatic surface grinder for high production.

up to 96 inches in diameter. Vertical-spindle rotary machines are available also in multiple-spindle models for rough, semifinish and finish grinding at one pass with automatic sizing and wheel advance as well as part gaging, *Fig.* 20.16. Reciprocating grinders are primarily designed for producing long, straight surfaces of either flat, *Fig.* 20.17, or formed section.

Also important among the production surface grinders are the *disk* grinders. These are available in horizontal-spindle models with single or opposed spindles. Single vertical-spindle machines available are used mainly for surfacing flat areas without maintaining particular dimensions. Utilizing large, solid abrasive disks or cylinder wheels, horizontal single or opposed spindle machines have oscillating, reciprocating, rotary, or through feeds and are ideally adapted for the production of parts requiring parallel, flat faces such as piston rings, springs and bearing races, *Fig.* 20.18. Automatic work sizers are provided on many models and operation can often be fully automatic, *Fig.* 20.19. Work up to 14 inches in length can be ground in opposed-wheel grinders with wheels up to 30 inches in diameter.

Special Grinders. A special form of flat surface grinder, the *abrasive belt* machine, has been covered previously in Chapter 19. Naturally, many special

Fig. 20.17 — Reciprocating table horizontal-spindle machine for surface grinding large parts. Photo, courtesy American Engineering Co.

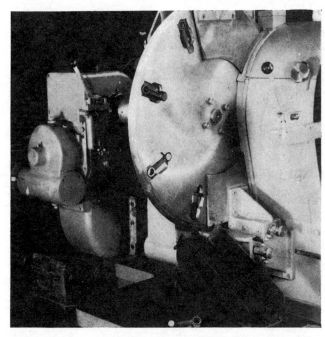

Fig. 20.18 — Opposed-spindle disk grinder with a rotary fixture for grinding small connecting rods. Photo, courtesy Hanchett Mfg. Co.

production adaptations of the various types of grinders discussed in the foregoing have been made but such applications require long runs in mass-production quantities to be practicable. Again there is a great variety of such machines as *roll* grinders which are capable of grinding all types of rolls (requiring swings up to 72 inches and 28 feet length between centers) with straight bodies or from 1/4-inch concave to 1/4-inch crown.

Also available are *crank-pin* grinders for finishing compressor and engine crankshafts. Machines carrying up to 9 wheels can grind, for instance, seven line bearings, fan pulley and gear fits, and OD of the flange in one operation. At the other extreme, using three wheels, center bearing, end-line bearing and flange OD can be ground. Other specials such as *multiple-wheel* cylindrical grinders allow the finishing of a number of diameters simultaneously. Parts up to about 10 inches in diameter, requiring wheels spaced or adjacent to grind up to 30 inches of length, can be produced with plain or formed contours.

Fig. 20.19 — Fully automatic disk grinder for ball bearing races with one wheel retracted to show path of travel. Photo, courtesy Hanchett Mfg. Co.

Single and multiple automatic *cam* grinders are available for finishing to size all varieties and combinations of cam shapes. Attachments for cam grinding can be used to convert many of the various standard cylindrical grinders for this purpose also.

A variety of *gear* grinders can be utilized in producing precision gears of all designs. Although small or medium quantities of certain gears can be produced on horizontal-spindle surface grinders with special indexing attachments, large-quantity production indicates a standard gear grinder for maximum economy and accuracy. Much the same is true of *splineshaft* grinding.

Design Considerations

Precision grinding permits the design and manufacture of machine parts to extremely close tolerances and quality of surface finish not readily attained with other machining methods. It oftentimes permits the designer to proportion a part in accordance with functional requirements and achieve size or weight savings made possible by the use of materials which cannot be readily machined by other methods. Too, the degree of accuracy now offered by modern grinding equipment has ushered in a new era of superprecision mass production and broadened the field of interchangeable manufacture, *Fig.* 20.20.

Fig. 20.20 — Automotive wheel spindles are abrasive machined from the rough casting in 42 seconds to finish tolerances of .0007" on the bearing diameters. High wheel speed and angle approach result in .330" depth of grind, and part finish of 35 RMS. Courtesy, Bendix Automation & Measurement Div.

Although some parts, as always, will necessitate specially developed production equipment, the great majority of parts, to afford maximum economy in production, should be designed to allow the use of standard equipment, *Fig.* 20.21. Though quantities to be produced may run from but few pieces to large quantities, the same advantages will accrue in the elimination of production problems.

Simplicity. Economy in production, as in other manufacturing methods, is to a large extent determined by part design. Maximum simplicity dictates a minimum number of separate grinding operations on each part, consideration of ease in holding, locating or clamping, a minimum thickness of stock to be removed, minimum variation in stock allowance within a production lot, minimum area of

Fig. 20.21 — Bryant two-spindle chucking grinder for high-production work requiring simultaneous finishing of hole and face to assure squareness relationship.

material to be ground, etc. With parts to be ground in large quantities, therefore, centerless and surface grinding merit consideration in achieving lowest production costs.

Fixtures and Loading. Careful attention to the multitude of problems incident to practical fixture design with parts not suited to center, centerless, magnetic chuck grinding, etc., will assist greatly in assuring rapid loading, accurate and easy locating, solid clamping without distortion, and no wheel interference, *Fig.* 20.22. Elimination of hard-to-reach, complex surfaces wherever possible as well as any slow loading or grinding operations will help insure the advantages of minimum production time and maximum quality.

Interrupted Surfaces. It is important in developing the design details of machine parts to give careful thought toward elimination of interrupted surfaces immediately adjacent to full continuous cylindrical surfaces which serve as bearings. The grinding action of the wheel engaged with both the full as well as the interrupted surface may cause a reflection of the interruptions into the continuous surface. Solution is to have the interrupted surface slightly smaller in diameter or have it separated from the adjacent surface by a clearance or undercut, *Fig.* 20.22.

Accessibility. Some care must also be given to the design of ground parts so that those surfaces which require finishing can be readily reached with a practical size of grinding wheel and, in some cases, a wheel of size suitable for the machine necessary to use. Projections on parts should never shroud over a surface to be ground in such a way as to require grinding with a small-diameter wheel approaching the work at an undesirable angle.

Shoulders. The preferable method of grinding a diameter and an adjoining square shoulder in one operation with the grinding head set at an angle of 30 or 45 degrees is indicated in *Fig.* 20.23. This method results in a concentric grinding pattern on the face which is desirable where that surface is to be used as a thrust

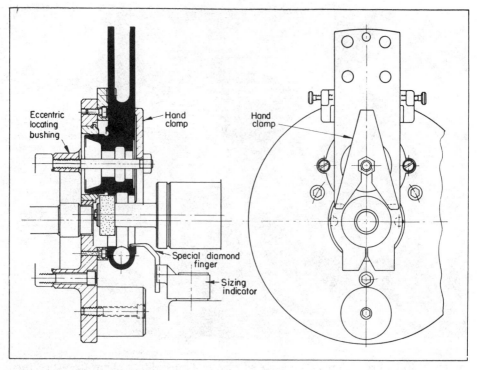

Eccentric locating bushing

Hand clamp

Hand clamp

Special diamond finger

Sizing indicator

Fig. 20.22 — Complex designs such as this crankshaft part often present difficult fixture and loading problems.

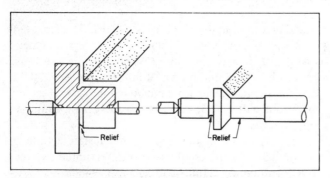

Relief

Relief

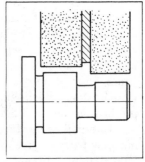

Fig. 20.23 — Where fillets are not a design "must", reliefs or undercuts improve production possibilities.

Fig. 20.24 — Relieved portions between adjoining diameters improve production and allow automatic grinding.

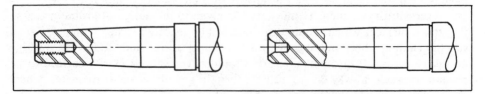

Fig. 20.25 — Largest possible standard center holes should be utilized for center-mounted parts.

bearing. If a universal machine is not available and a plain grinder is employed, the preferable method from an economy standpoint is to feed the wheel into the corner,

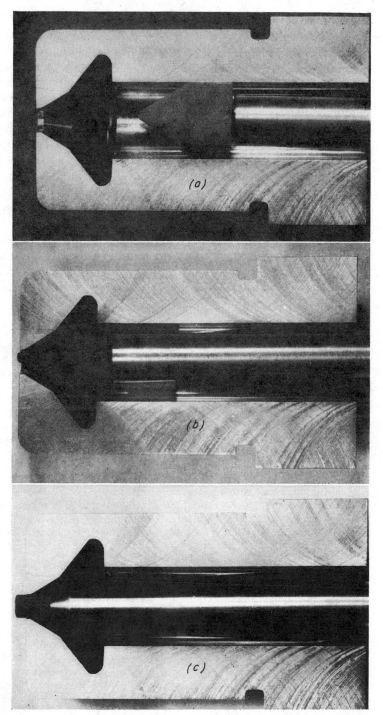

(a)

(b)

(c)

Fig. 20.26 — Views showing grinding of a fuel injection nozzle, (a) straight hole, (b) taper seat, and (c) 0.040-inch injection hole. Photos, courtesy Bryant Grinder Co.

thus finishing the shoulder, but this results in a criss-cross pattern which may be undesirable depending upon the application.

Grinding Reliefs. Precision cylindrical grinders permit establishing accurate radii for fillets as well as square or tapered shoulders adjacent to cylindrical surfaces. However, unless design dictates otherwise, it is preferable to substitute a relief for a fillet owing to the ever-present problem of maintaining the corner of a wheel trued to the accuracy limitations on radii. Wherever critical stress concentrations are not expected, a relief should be employed, *Fig.* 20.24. Production often can be increased to as much as double by this means.

To simplify production, reliefs should also be used in lieu of fillets where adjoining diameters are to be ground. Providing smoothly blended contours are not demanded for strength characteristics, the use of reliefs makes possible much greater output by elimination of extra wheel dressing and grinding operations. Often parts can be plunge-cut with several wheels or a wide wheel in but a fraction of the time, using automatic handling, *Fig.* 20.24. In all cases reliefs should be as wide as possible and preferably not less than 1/8-inch. In depth, the relieved portion should be recessed sufficiently so that the wheel does not contact on reaching minimum grinding size.

Generous reliefs should be used where ground threads end, especially adjacent to shoulders, to allow for wheel clearance. Where desirable, the leading and trailing portions of imperfect threads can be removed as in thread milling.

Fillets. Where fillets are necessary, consideration should be given to a number of important factors:

1. It is desirable to keep fillets as large as possible; a small corner radius on a wheel requires more frequent dressing and should seldom be less than 0.010-inch; on diamond-dressed formed wheels, fillet radii to be generated such as at the root of threads should never be less than about 0.017-inch inasmuch as smaller radii are difficult to retain on a fragile, sharp-pointed diamond.
2. If a radius must blend into adjoining surfaces to prevent formation of fatigue cracks, again the radius should be as large as possible; a large radius is much easier to blend.
3. Wherever possible the grinding of a fillet should be avoided and cost will usually be reduced, and in such cases the drawing should indicate how far the grinding must go to a forged, cast or machined fillet.
4. On a single piece, all fillets specified should be identical to make possible finishing of all without separate wheel truings.
5. Sharp internal corners should be avoided; if design does not require a fillet, an undercut or relief is the practical substitute.
6. In order to obviate the necessity for holding extremely accurate limits on fillet radii, attention should be given to the fit of mating parts so that ample clearance is guaranteed without close fillet tolerances.

Centers. Wherever center grinding is indicated, part design should

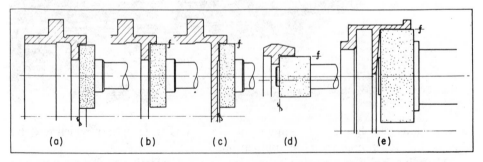

Fig. 20.27 — Provision of grinding relief on internal work should be as at (a) for face only, (b) diameter only, (c) face and diameter with bottom relief, (d) bottom hole in lieu of relief, and (e) taper clearance.

contemplate the use of center holes which can be retained throughout the processing of the part as this often offers an economy in manufacturing, *Fig.* 20.25. To insure accurate grinding, center holes must be in accurate alignment, perfectly round and, if necessary, lapped smooth. To minimize chatter, center holes should be as large as practicable. Again, to reduce handling difficulties and avoid the necessity for complicated counterbalanced fixtures, it is highly desirable that parts to be ground be naturally balanced about the center of grinding. Otherwise, complications from vibration may result, although fixture design often can obviate such problems where unbalanced design is essential, *Fig.* 20.22.

Holes. As a rule, in internal production grinding, holes under 0.040-inch are impractical and should be avoided, *Fig.* 20.26. Smaller holes are often ground in tool and die work but necessitate the use of a jig grinder with diamond-charged mandrels. Again, as in external work, provision of grinding relief for wheel clearance should be kept in mind, *Fig.* 20.27. Addition by the shop of an undercut which has been overlooked on a drawing may seriously weaken a part and may, in certain cases, amount to almost 50 per cent of the section. Another detail is relieving the center of a face, *Fig.* 20.27c, which again allows the wheel to finish off the surface and preserve the edge. As a substitute for a bottom relief, through holes or tapers can often be used to advantage in many designs, *Fig.* 20.27 *d* and *e*. Also, on deep holes it may be impractical to use a bottom relief because of design or conditions, *Fig.* 20.27b, in which case part design should take into consideration the fact that it is impractical to grind to full depth of blind holes, *Fig.* 20.28.

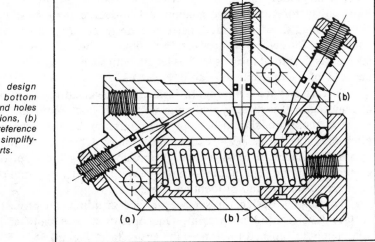

Fig. 20.28 — Part design should allow for bottom clearance, (a) on blind holes and external operations, (b) should be used in preference to internal ones for simplifying production of parts.

External Operations. Where external operations can be substituted for internal ones, production will be simplified and costs often reduced considerably, *Fig.* 20.28. Designing to keep internal finished holes to a simple cylindrical shape will prove most economical in production.

Dimensioning. In dimensioning a part, location of critical faces or forms must be properly shown if accurate grinding and correct angular relationships are to be expected. Only by locating critical points of a part from that one point which bears a specific relationship to the final assembly of parts, can complete interchangeability be assured, *Fig.* 20.29. Tolerances may be cumulative and result

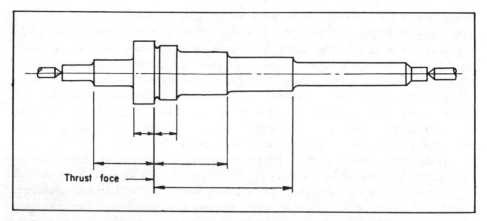

Fig. 20.29 — Key location point or points of an assembly should be used in dimensioning to insure accuracy.

in errors not anticipated. As a rule, critical locating points should be as close as possible to the surface or form that is to be ground.

Centerless Parts. In designing for centerless grinding, *Fig.* 20.30, certain limitations over and above those mentioned in the foregoing should be kept in mind. End shapes of cylindrical parts which must be ground along with the diameter require consideration. Although a full radius can be generated by the use of carefully diamond-trued wheels, this is usually expensive. To ease the problem a slight flat at the end should be acceptable. When crush-dressed wheels are used, a flat is unavoidable and a full spherical form is uneconomical in crusher life. The most satisfactory end design is about a 120-degree angle as shown in *Fig.* 20.31. Where long, slender formed parts are designed, *Fig.* 20.32, these should be considered for centerless production to obviate the difficulties in center grinding such parts. Depth of form as well as length with such designs should be suited to the capacity of the machine available, *Fig.* 20.33.

Surface or Face Grinding. In surface or face grinding, as in the other methods of grinding, cost is directly tied up to volume of stock removal. Surface area to be finished, therefore, should be kept to a minimum. Surfaces should be relieved to offer only the necessary area required for matching or fitting, *Fig.* 20.34. If opposing surfaces are to be ground, the possibility of grinding in an opposed-wheel grinder should be kept in mind. In this case it is usually essential that the opposing surfaces be similar and of identical area.

To adapt a design for surface or face grinding, no projections or steps in surface level should be present. One operation then can be used to machine completely to finished tolerances the entire face, *Fig.* 20.35.

For securing the finest finish and maximum flatness in surface grinding, parts should be reasonably thick with respect to their length and width or diameter. Thin sections may spring slightly during grinding, producing an uneven surface finish or undesirable variations in flatness. Where control of dimensions rather than surface finish is the prime objective, partly unfinished portions should be acceptable to make possible minimum stock removal and speed up production. Care should be exercised with hardened parts so as not to remove the hardened case in grinding to finish dimensions.

Stock Allowances. Stock allowed for surface grinding thus need only be a bare minimum. While cuts of 1/4-inch depth or greater are possible, for obtaining maximum economy in production, about 1/16-inch and preferably even less should be allowed. On cylindrical parts to be ground, external stock allowance of 0.005-inch is the best practical minimum amount. Additional allowance, however, must be made over this base figure to compensate for such unknown variables as normal material twist or warpage, part warpage in hardening, eccentricity in turning, turning tolerances, centering tolerances, etc. Once the full effects of any or all such variables are known, final minimum grinding allowance can be specified.

For holding the best possible accuracy in grinding, stock allowances must be

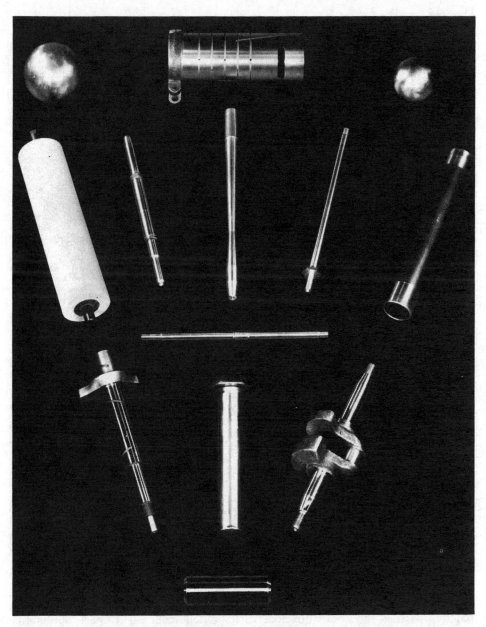

Fig. 20.30 — Variety of parts ground by the centerless method, both through-feed and in-feed, showing wide range of design. Materials include steel, cast iron and rubber. Courtesy, Cincinnati Milacron.

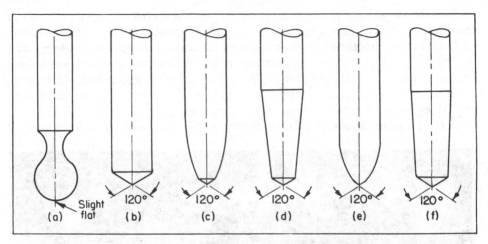

Fig. 20.31 — Alternative end forms for centerless-ground parts. Those shown in (b), (c), (d), (e), and (f) are most easily produced.

Fig. 20.32 — Long, slender formed parts are easily produced on centerless machines without difficulty from deflection during the grinding process. Courtesy, Cincinnati Milacron.

uniform from part to part and this factor should be kept closely in mind. In a single batch of parts, if the stock allowance varies widely from part to part, automatic-cycle machines may be grinding "air" on a large percentage of the parts at the beginning of the automatic-feed cycle inasmuch as the machine must be set to begin "feed" for the largest diameter blanks. Minimum variation from part to part can in this way effect considerable savings in production time. In centerless work, on average-size pieces, normal allowance for rough grinding can range up to 0.010-inch while 0.001 to 0.002-inch above the high limit should be allowed for finish grinding but here again, if variables are present, some additional allowance may have to be made.

Similarly, stock allowance for internal grinding depends to a large extent upon the bore and length to be finished. Typical total allowances are as follows:

Hole Diameter (in.)	Length (in.)	Allowance on Diameter (in.)
0 to $1/2$	1	0.004-0.008
$1/2$ to 1	4	0.008-0.012
1 to $1\frac{1}{2}$	8	0.012-0.018
$1\frac{1}{2}$ to 5	8	0.018-0.025
5 to 8	8	0.025-0.030

In any case, the smallest and shortest holes use the smallest allowance with the amount increasing to maximum with the largest and longest holes noted. As mentioned previously, unusual conditions or materials, improper centralization of the work or other conditions giving rise to runout, etc., will require some additional allowance.

Selection of Materials

There are practically no limitations to the kinds of materials which can be ground. Materials that can be machined by any method can be ground, as well as many that resist finishing by many other methods. The scope of grinding thus embraces all of the ferrous metals, including those hardened. All the nonferrous

Fig. 20.33 — Even unbalanced parts can be centerless ground. An outboard motor crankshaft is shown in position for final grinding. Tolerance held is 0.0003-inch with 0.002-inch stock removal and surface finish of 6 to 8 microinches is obtained. Finishing operation requires one-half minute and rough grinding operation, which removes 5/32-inch of stock, requires about three-quarters of a minute. Courtesy, Cincinnati Milacron.

Fig. 20.34 — Proper relieving for face grinding offers maximum benefit. Faces of this casting are held to within 0.003-inch on length and parallelism. Photo, courtesy American Engineering Co.

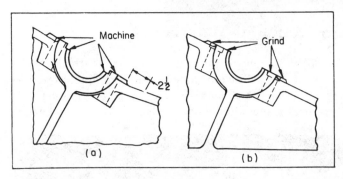

Fig. 20.35 — Stoker crankshaft bearing bracket (a) redesigned for surface grinding as shown at (b) reduced machining cost and speeded production. Drawing, courtesy American Engineering Co.

materials such as copper, brass, aluminum, magnesium, bronze, etc., also can be readily ground. Nonmetallics such as ceramics, carbon, glass, refractories, plastics, quartz, agate and sapphire have all been ground on a production basis.

Grinding Wheels. For cast iron, chilled iron, brass, bronze, aluminum, fiber and other rather low-tensile materials, silicon-carbide abrasive wheels are usually employed, while for soft low-carbon steels, annealed malleable iron, and certain high-tensile bronzes, tough regular aluminum-oxide abrasive wheels are usually used. On hardened or soft high-carbon steels, high-speed steels, and special alloy cutting tool materials, a highly refined white aluminum-oxide abrasive wheel is normally used.

Effect of Grinding. Even the most careful grinding often induces tensile stresses in the ground surfaces of through-hardened steel parts. Stress balance thus disturbed as well as the metallurgical effects of grinding temperature often result in structural change in the surface material and a decrease in impact and fatigue resistance. On the other hand, certain designs are strengthened by the effects of internal grinding. On critical parts which must withstand high stresses, it is thus often advisable to finish grind external portions prior to hardening. Where heat treatment cannot assure parts sufficiently clean and distortion-free, the lightest possible finish grind should be contemplated.

Specification of Tolerances

Dimensional control of parts being ground has now reached a point where arrangements are available which permit control of size from the work itself to a tolerance of less than 0.0001-inch. It is possible to hold diameters to within plus or minus 0.000050-inch to 0.000010-inch with automatic sizing devices. However, economic production is simplified if holes are sized to within plus or minus 0.0001-inch. The fastest method of sizing is done directly from the truing diamond and a hole 3/4-inch in diameter by 5/8-inch deep, for instance, can be finished in 15 seconds to plus or minus 0.0005-inch with a surface finish of 10 microinches. On a production basis, tolerances on diameter of 0.0001 to 0.0005-inch are entirely practical. Face runout or squareness in relation to bore or outside diameter on parts produced in chucking grinders can be held to within 0.00005-inch on special parts, *Fig*. 20.21. In production grinding engine crankshafts, for instance, 0.017 to 0.025-inch of stock is removed in finishing throws within 0.0005-inch. Actually, size of the throws is normally held to plus or minus 0.0003-inch without difficulty in regular production.

Surface Finish. Because of new developments in grinding machine design and grinding technique, parts such as gages, bearing races, etc., can be ground to a surface finish of 0.4 to 0.7-microinch. Where high output is desired in general production work, a surface finish of 4 to 5 microinches can be held with dimensional tolerances within 0.0001 to 0.0005-inch. Associated roundness and parallelism can be held within plus or minus 0.0001 to 0.0002-inch over lengths as great as two feet or more.

Centerless Work. Centerless grinding will produce parts within plus or minus 0.0001-inch on diameter and parallelism. Roundness is usually within less than 0.000050-inch and, where necessary, diameter and parallelism can also be held to this figure, but at increased cost. Stepped diameters can be held concentric within 0.0002 to 0.0005-inch. Surface finish can be on the order of 1.5 microinches although 5 to 6 or rougher is most practical for economy reasons. To assure

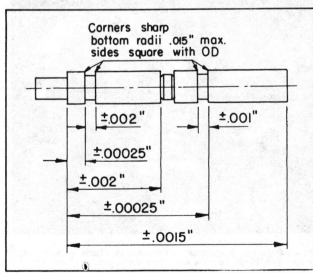

Fig. 20.36 — Hydraulic plunger of heat-treated stainless center ground with a wheel speed of 12,000 sfpm. Drawing, courtesy Cincinnati Milacron.

maximum possible accuracy, parts with keyways which would interfere with rotation on the work rest should be avoided.

Threads. Centerless-ground threads can be produced with lead error less than 0.0005-inch per inch. Pitch diameter of center-ground threads can be held from plus or minus 0.0002-inch to plus or minus 0.001-inch depending upon requirements and economy. Roundness can be held to within 0.0005-inch while concentricity of the thread form with the OD is well within plus or minus 0.003-inch on the largest sizes. Lead variation, when necessary, can be held to a total of 0.0002-inch per foot but in production plus or minus 0.0002-inch per inch or plus or minus 0.0005-inch in 5 inches on standard thread forms are generally considered more practical.

Grooves. Width of grooves can be held to close accuracy where necessary. Plus or minus 0.001-inch on width of grooves, plunge-ground in production, is readily held, *Fig.* 20.36.

Shoulders. Wheel spacings used in producing shouldered lengths can be held to a maximum variation of 0.0005-inch. Dimensions from shoulder to shoulder thus can be held to plus or minus 0.00025-inch, *Fig.* 20.36. In traverse grinding parts to a shoulder, machines will reverse within 0.001 to 0.004-inch.

Corners and Radii. External corners can be held sharp on parts which require such edges, *Fig.* 20.36. Otherwise, sharp corners should be avoided by providing a bevel in preliminary operations. Internal corner radii produced by the formed edge of a wheel are usually held within 0.005-inch in production. Such radii, however, should never be called for unless the extra expense of producing them is justifiable. Spherical sections, produced on oscillating grinders, can be held to a total variation of 0.0003-inch on diameter with centers located within plus or minus 0.001-inch.

Surfaces. Reciprocating-table surface grinders will produce surfaces with flatness varying but a few tenths over lengths of 20 feet or more. As a rule, ways and large areas, ordinarily given to hand scraping, can be finished to within 0.0003 to 0.0005-inch. Parts finished on rotary-table machines can be held to flatness of 0.0002-inch to 0.00005-inch, parallelism of 0.0004-inch to 0.00005-inch, and

Fig. 20.37 — Surface ground end plates, roll and blades of refrigeration compressors. Blades are held parallel within 0.00015-inch and square within 0.0001-inch. Photo, courtesy Blanchard Machine Co.

length or thickness to plus or minus 0.001-to 0.0002-inch, *Fig.* 20.37, depending upon the size and rigidity of the parts.

Flatness measured in light bands can be produced on disc and rotary-table, vertical-spindle surface grinders but metal removal must be low, wheel pressure light, and wheel face sharp. Best procedure is to grind to within 0.005-inch or less of final size and then finish grind. Material must be free from stresses and the section sturdy to resist distortion.

Practical Finish. Surface roughness is dependent to a certain extent upon the material ground, hardness of the material being a deciding factor. Materials such as hardened steel can be ground to a finish of 2 microinches or less in regular production while on softer materials, a finish of 10 microinches may be difficult to obtain. An approximate idea as to the best possible practical surface finish for various materials can be had from the following list:

Hardened Steel	2 microinches
Soft Steel	5 microinches
Cast Iron	5 microinches
Brass	12 microinches
Aluminum	15 microinches

As is evident, it is easier to produce a fine finish on the harder materials and in the case of iron castings the limit is primarily that which reveals the grain structure; beyond this the finish cannot be improved.

Tolerances as well as surface finish naturally should be no closer than actual function or design necessitates to assure maximum production at lowest cost. The roughest acceptable finish should always be specified and where surface finish and pattern are critical design characteristics, honing, lapping or Superfinishing should be considered. Finish specifications, where important, should not be general or contain loose phraseology capable of gross misinterpretation. Functional requirements for a specific part should be closely analyzed — it is possible to have *too* perfect a surface. Characteristic range of surface roughness figures for machine parts widely used, *Fig,* 2.6, will help to judge what constitutes a practical finish.

21

Mass Finishing

R ECENT developments and experience in the field of mass finishing have made practicable, if not highly desirable in many cases, the use of this process in the precision finishing of machine parts. Especially well adapted to mass-production work, *Fig.* 21.1, mass finishing requires considerably less time and results in surfaces far superior to those usually obtained with hand-finishing methods commonly used.

Fig. 21.1 — A few of the many precision machine parts which are mass finished in large quantities. Photo, courtesy Norton Co.

The Tumbling Process. In general, tumbling barrels have long been and still are widely used to remove burrs, fins, flash, scale, roughness, rust, oil, grease, sand and dirt on parts of such materials as steel, cast iron, aluminum, brass, plastics, etc. Some of the common media which have been used in tumbling are: sawdust, wood blocks, leather scraps, meal, steel balls, slugs, abrasive forms, emery, ashes, pumice, Vienna lime, white silica sand, gravel, sulphuric acid, soda cyanide, and potash. Sawdust and steel slugs are used to absorb oil and remove burrs; emery and white silica sand or gravel are used to remove flash from brass and steel stampings; sawdust is used alone to remove grease, rust and scale, to dry work after pickling

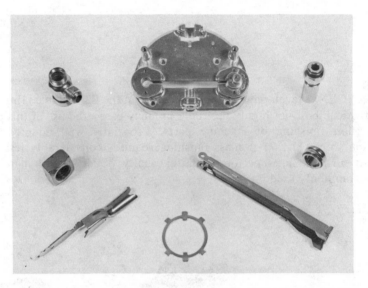

Fig. 21.2 — Group of typical machine components mass finished in large quantities with aluminum oxide chips. Photo, courtesy Norton Co.

and plating, and to clean as well as brighten brass and other nonferrous parts; fine sand, ashes or Vienna lime, and steel balls are used to clean and rough-polish pawls, studs and similar small parts. Sulphuric acid and water or water and egg of cyanide are often used to remove scale and roughness and to clean various parts; maple blocks and sawdust are used to remove flash and polish plastic parts. Most interesting perhaps is the fact that dry ice chips are used with great success for removing the flash from a variety of molded-rubber machine parts. The chips serve both to quick freeze the thin flash and also tumble it off smoothly.

Precision Tumbling. Though these and other similar tumbling techniques are widely employed to advantage in the processing of large quantities of parts, *Fig.* 21.2, the latest phase of importance to the machine designer is that in which specially prepared abrasives are employed to provide precisely controlled results not available heretofore.

Mass Finishing. This particular area of finishing, sometimes referred to as rotofinishing, is basically a mechanical grinding and honing process which depends upon the free-mass abrasive action of special abrasive compounds, quartzite or manufactured abrasive chips, stones, or pebbles under agitated impingement. Processing may be carried out either dry or flooded with water. Grinding, deburring, removal of metal fragments left from previous machining operations

(*Fig.* 21.3), producing of predetermined radii on edges, polishing (*Fig.* 21.4), honing and even coloring can be accomplished with closely controlled results in mass production of machine parts of all varieties.

Range of Parts. Practically any material — ferrous, nonferrous, plastics etc. — in soft or hardened condition can be finished with this process with proper

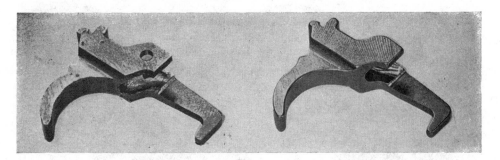

Fig. 21.3 — Removal of burrs and metal fragments left from machining this steel trigger cost only $0.0006 per piece, a saving of 95 per cent over previous low-production hand methods. Photo, courtesy Minnesota Mining & Mfg. Co.

technique. Parts may be rough castings or forgings (*Fig.* 21.5), stampings, spinnings, precision castings, or accurately finish-machined pieces such as cams, gears *(Fig.* 21.6), threaded units, gun mechanisms, housings, bearing rollers, cages and rolls, brackets, ratchets, etc., of practically any design and may range from instrument gears to spars 105 feet long.

Fig. 21.4 — Stainless-steel plate deburred, radiused and polished in two separate tumbling operations at approximately $0.0035 each, a saving of 86 per cent over previous methods. Photo, courtesy Minnesota Mining & Mfg. Co.

Range of Finishing. The wet method is recommended and best suited for grinding, deburring and finishing the majority of parts. Grinding utilizes the largest chips or stones for fast cutting in eliminating burrs, fragmented metal, sharp or rough corners and edges. Grinding also removes spinning, draw and die marks, tool marks on machined surfaces, etc. Honing action is obtained by the use of the smaller stones or chips and naturally results in a much finer surface finish approaching a polish. So-called "coloring" is an extremely fine finishing operation using the finest chips or, in some cases, nonabrasive compounds are especially well suited to the processing of small, irregularly shaped pieces.

Grinding Action. In a continuous finishing process, action of the abrasive stones to grind off burrs and sharp projections occurs because these projections rupture the protective water film and subject the burr or irregularity to the direct

Fig. 21.5 — Rough-milled casting mass finished with radii from 0.010 to 0.020-inch on all edges. Photo, courtesy International Business Machines Corp.

abrasive impact. Honing action on flat surfaces is limited primarily to the removal of raised portions or projections left by tools used in the previous processing, the water functioning as a lubricant to keep the stones from cutting deeply. This results in a controlled precision honing which rarely exceeds a few millionths of an inch on most surfaces while almost uniform radii are produced on all edges as exemplified by the gun trigger in *Fig.* 21.3. Different sections or edges of a part are usually affected by the chip action in the following order of heaviest cutting to the lightest: sharp corners, edges, convex surfaces, flat surfaces, concave surfaces, and last of all, deeply recessed or relatively inaccessible locations.

Advantages in Design. The value of mass finishing as a mass-production process over and above its well-known employment in burring and conditioning readily can be seen, *Fig.* 21.7. Not only can burring of sharp projections be accomplished at extremely low cost per piece, but tremendous improvement in surface quality can be produced simultaneously without significant reduction in finished dimensions or risk of additional stress-raising cracks or scratches. Mass finishing also provides an economical method upon which the designer can rely when designing intricate machine and instrument parts whose proper functioning depends upon the complete removal of flash, sharp edges and burrs — *Fig.* 21.8 — a condition which normally would place many parts beyond the reasonable cost bracket if hand methods were contemplated. In fact, the designer can safely specify uniformly rounded edges throughout a part without fear of complicated operations, *Fig.* 21.9.

The use of mass finishing, in addition, often makes it unnecessary to obtain the most exacting surface finish possible by means of the primary processing method; the honing action of the stones removes grinding wheel marks, machining tool marks, die scratches, etc., and where desired can improve surface finish to as fine as

3 microinches, depending upon conditions such as surface finish left by prior processing, physical design features of the part, edge radii desired, economical time cycle, etc.

Limitations. Naturally, some limitations exist which prohibit the use of mass finishing on certain types of parts. Edges which must be broken to a smooth radius, for example, would be impossible to finish when so located as to prevent direct action of the abrasive. Again, holes may be too small to allow free circulation of the media, or inner passages may be of such intricacy as to result in excessive cleaning problems. Here, liquid blasting with a fine abrasive may offer a good alternative.

Machines. The wide variety of barrels available for processing are of five general types: (1) tilting, (2) compartment horizontal, (3) tapered horizontal, (4) cylindrical horizontal, and (5) octagonal horizontal. Tilted barrels, with capacities from 1 to 14 cubic feet, are usually lined with seasoned hardwood and the tilting feature makes them highly economical of handling time. Compartment barrels are usually octagonal in shape and may have from one to five compartments, wood-lined. Capacities may run from around 3 to 20 cubic feet per compartment. Cast-iron tapered octagonal horizontal barrels, with capacities up to about 5 cubic feet, are primarily for water rolling and sand tumbling metal parts whose rough surfaces are to be cut down and smoothed. Steel cylindrical horizontal barrels, with capacities up to about 30 cubic feet, are primarily for removing fins and flanges

Fig. 21.6 — Mild-steel gear mass-finished at $0.018 per gear as compared to $0.06 previously, provided greater uniformity, increased production and tooth surfaces approaching the so-called run-in tooth condition. Photo, courtesy Minnesota Mining & Mfg. Co.

from large, heavy castings. Octagonal horizontal barrels, usually of wood with one or more screened faces, are for dry tumbling of plastics.

Large compartmented barrels and the tilting type are most widely used for precision finishing work, *Fig.* 21.10. Speed of rotation varies from about 10 to 50 rpm with 20 rpm being the most useful speed. The time required for processing parts varies considerably with the material, design and finish desired. Time for various metals, roughly, is as follows: for machined brass, 15 minutes to 2 hours; for machined aluminum, 15 minutes to 3 hours; for cast brass, 2 hours to 48 hours; and for deburring steel, from 1 hour to 8 hours. The harder the steel, the brighter the surfaces obtainable, but the longer the processing time needed. Production of simple bushings, which by former methods never exceeded 50 to 60, may be

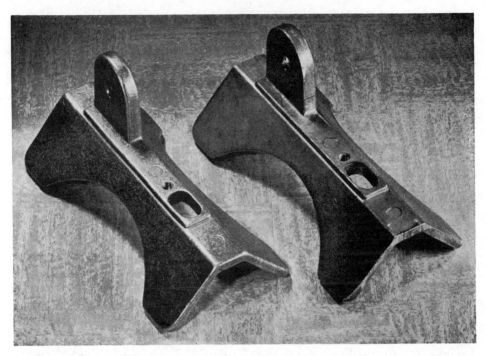

Fig. 21.7 — Die-cast cradle for Argus Spotting Scope tripod adapter before and after mass finishing which increased output from 120 per hour to 600. Photo, courtesy Argus Inc.

processed at from 12,000 to 15,000 per hour. With more complicated parts, such as the gun frame mentioned previously, finishing may run to several hours per piece. However, where high production rates must be met, hand methods would present an unsurmountable problem. The steel triggers shown in *Fig.* 21.3 were finished 4000 pieces per charge in 4 hours or about 1000 pieces per hour.

Vibratory machines, which come in a wide variety of styles, can be circular, compartmented or straight-line. Capacity is relatively unlimited and operation can be automated, *Fig.* 21.11.

Spindle-type machines utilize separate spindles to hold workpieces and provide precalculated plowing action between media and work.

Design Considerations

The design of any part is very important in the subsequent processing procedures, and the method of finishing required to produce an acceptable part may mean the difference between success or failure so far as economical production is concerned. A part best adapted to mass finishing should be simple and have no sharp corners or large flat surfaces. If design embraces principally sharp corners, edges and large flat surfaces, then the possibility of securing good results sometimes is problematical. Too, it must be remembered that intricate configurations, deep blind holes or slots, recesses, etc., will add to the problems of working out a technique.

With this process in mind the designer can specify smoothly rounded edges throughout a part and be certain these can be held within fairly narrow limits. Commonly, the practical range of radii is from 0.001 to 0.020-inch, but radii up to

Fig. 21.8 — Commutator holding ring mass-finished to assure proper functioning at minimum cost. Photo, courtesy International Business Machines Corp.

0.050-inch and greater have been produced but, of course, with much longer running times.

Where the design of intricate or highly accurate fitted parts is critical, specification of mass finishing guarantees removal of all loose particles, smoother surfaces which cannot be reached otherwise and greater uniformity, *Fig.* 21.12. Often it is possible to reduce parts cost considerably as well as to obtain the added feature of smoothly rounded edges, *Fig.* 21.13.

Selection of Materials

As previously noted, practically any material can be handled. The harder

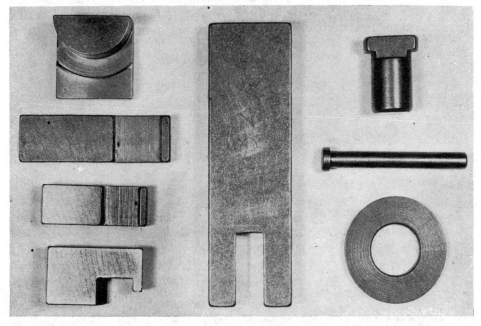

Fig. 21.9 — Large Link, key block, pins and blanks for production tools mass-finished to obtain uniform radii.

Fig. 21.10 — Large, octagonal compartmented barrels are used most widely for precision finishing. Barrel shown loading and unloading quantities of small parts. Photo, courtesy Roto-Finish Co.

metals may require a longer period for processing but present considerably less trouble with nicking. On materials such as die-cast zinc or aluminum, burrs are removed and radii produced quite rapidly. Brass and steel require a slightly longer period as a rule while materials such as stainless steel, being extremely tough, would necessitate the longest period.

In some cases substitute material can be employed with considerable reduction in costs. The bezel ring, *Fig.* 21.14, formerly made from a carefully hand-burred and polished stainless steel, was blanked in ordinary manner from cold-rolled steel

Fig. 21.11 — Spiratron mass finishing machine with compound recirculation, parts accumulator and feeder, and magnetic separation. Photo, courtesy Roto-Finish Co.

and mass finished to a highly satisfactory luster at a saving of several dollars per hundred pieces.

Specification of Tolerances

Normally, the grinding operation can be expected to produce a finish of approximately 25 microinches on unhardened parts, the honing operation will bring this down to about 10 microinches, and the wet coloring operation, to 2 or 3 microinches. Combinations can be used to round edges while at the same time honing the surface to a high quality.

Dimensional Change. Uniformity of dimensions and radii after mass finishing are in exact conformity with those before finishing; prior processing tolerances set the limits. Overall reduction in dimensions may run from a few millionths to a few tenths depending upon the material and design. However, each piece may be altered somewhat from the more exposed to less exposed portions as previously noted. Projections such as thin burrs are affected first and most; excessive variation in size of burrs from piece to piece may cause a similar variation after tumbling, although much reduced.

Finish tolerances also are affected by prior finish, actual cases showing microinch readings on parts being reduced from 15 before finishing to 3 after, from 60 to 15, from 500 to 80, etc.

Corner Radii. Tests on hardened steel samples show that corner spherical radii created increase rapidly to about 0.055-inch in the first 2 hours and then settle down to about 0.003-inch per hour of tumbling thereafter. Edge radii develop much slower, as can be readily noted in *Figs.* 21.3, 21.9 and 21.13, reaching about 0.015-inch in the first 2 hours and then settling down to a rate of 0.001-inch per hour. Dimensional change on flat, concave or convex surfaces reaches about 0.00045-inch during the first 5 hours, and thereafter usually proceeds at a rate of 0.0000375-inch per hour.

Naturally, part specifications must take into consideration this decrease in dimensions, since final dimensions of a certain percentage on close-tolerance parts may fall below the low limit. Likewise, tolerances on hardness of materials which work harden must take into account the several points in Rockwell hardness usually imparted by tumbling.

Fig. 21.12 — Bearing-cage stamping, mass finished to remove all particles, smooth the surfaces and round the edges. Charge of 2000 pieces was run 6 hours resulting in a piece cost of $0.0015 compared to $0.02 per piece by hand methods. Photo, courtesy Minnesota Mining & Mfg. Co.

Fig. 21.13 — Two tumbling operations, following broaching and hardening of this part, eliminated the need for more costly finish grinding operation. Photo, courtesy International Business Machines Corp.

Fig. 21.14 — Use of mass finishing on ordinary steel provided a high-luster finish equivalent to that obtained previously by hand methods on special stainless. Photo, courtesy International Business Machines Corp.

Production Tolerances. On aircraft engine parts such as gears, gear-and-spline shafts, bearing retainers, special bolts, housings, etc., surface finishes from 6 to 8 microinches and corner radii of 0.008 to 0.010-inch are consistently obtained.

Often the use of the mass finishing can produce finish results comparable to grinding at a much lower cost. The part shown in *Fig.* 21.13 was originally broached, hardened and finish ground. By application of two tumbling operations, one after broaching and one after hardening, it was possible to obtain the extremely satisfactory finish of 8 to 16 microinches along with smoothly finished edges and rounded corners.

22

Honing

PROPERLY classed as refined grinding operations, the method of processing commonly known as honing logically follows production grinding in the sequence of production methods classed in order of increasing accuracy and smoothness of surface finish. Although surface finish quality possible with finish grinding operations edges into the zone covered by honing and lapping, to demand the absolute minimum is usually not the economical answer in production. To obtain the finest surface finish and accuracy in grinding, greatly reduced speeds, extremely light metal removal, and special attention are required. Economy in production, for the most part, dictates the use of maximum speed in grinding and roughest satisfactory surface finish with subsequent operations to attain the desired refined surface quality and geometrical accuracy.

Fig. 22.1 — Group of precision machine parts with internal or external surfaces finished by Microhoning. Photo, courtesy Ex-Cell-O Corp.

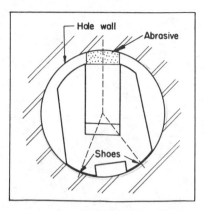

Fig. 22.2 — Large-area contact abrasive stone or stones are used in conjunction with unequally spaced guide shoes to generate true roundness in honing. Drawings, courtesy Sunnen Products Co.

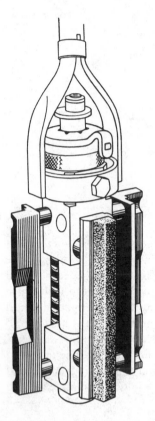

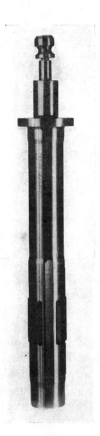

Fig. 22.3 — Right — Honing tool designed for hydraulic actuation in automatic machines.

Fig. 22.4 — Left — Portable honing tool for use on small lots or unusual parts not suitable for finishing on production machines. Photo, courtesy Ex-Cell-O Corp.

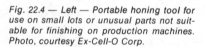

Purpose of Honing Process. Type and quality of surface finish produced by the basic metal removal methods is for the most part incidental. With the honing process, however, type and quality of surface finish is the primary purpose of the process. Where found desirable or necessary, material removal on the order of that possible with grinding can be had but of course at a sacrifice in finish quality. Most important is the fact that, beginning with honing, the major factor which places the refined grinding processes among the necessary production methods is their ability to correct the geometrical surface inaccuracies and inconsistencies normally left by metal removal operations.

Scope of Process. Need for maximum possible geometrical accuracy and quality of surface finish is most acute with designs requiring extremely precise interchangeable fits, bearing fits where run-in and wear are to be minimized, pressure fits where seals are impractical, etc. The major portion of these requirements falls into the realm of cylindrical or flat surfaces. Honing is now widely employed to produce such accurate internal and external cylindrical surfaces, *Fig.* 22.1, flat surfaces usually being finished to precision requirements by means of the lapping process.

Technique of Honing. Principal difference between honing and grinding is that, in honing, the abrasive stones have a large area of surface contact while in regular grinding a narrow-line contact is employed, *Fig.* 22.2. Like grinding, honing is usually performed by rotation of the abrasive stones but a long traversing motion is used providing a pattern of scratches with much greater lead. The finish developed in honing is a result of reciprocation of the rotating honing tool within a bore or reciprocation of the rotating work within the honing head, this reversal of motion developing a cross-hatch finishing effect which helps eliminate the greater portion of pattern defects left by previous production operations. Actually, modern honing techniques will provide practically any kind of controlled surface roughness, including the lay of the scratch pattern. Thus over and above the conventional cross-hatch pattern, multi-directional or random patterns and co-directional or straight-line honing patterns can be obtained.

Advantages and Limitations. Developed in the early 1920's because of the necessity for better finish and more accurate dimensional control of automotive cylinder bores, honing has now become a rapid, accurate, repetitive production

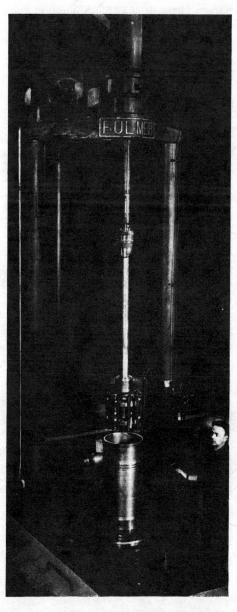

Fig. 22.5 — The XLO Micromatic 5VH-2 Multiple-Spindle Hydrohoner is a double-spindle machine equipped with automatic sizing controls. This machine may be set up to finish Microhone* two parts simultaneously, or one part progressively. Courtesy, Ex-Cell-O Corp.

Fig. 22.6 — Right — Chromium plated locomotive diesel engine cylinder liners being finish honed on a vertical machine. Only five minutes is required per liner. Photo, courtesy C. Allen Fulmer Co.

method. Capable of removing irregularities and high spots, correcting out-of-roundness, axial distortion or "rainbow", and taper, honing also generates fine surface finish much faster than does reaming or grinding. Honing does not, however, establish the position or alignment of a hole, bore or diameter. Little or no improvement insofar as alignment or location are concerned can be expected as the tool tends to follow the neutral axis of the bore or diameter and the preceding operations must establish this axis as accurately as is required.

Stock Removal. Although up to as much as 0.090-inch of stock can be removed in honing, stock removal should preferably be small as possible for maximum economy.

Hones. The honing tool generally floats or moves slightly with relation to its driver and has a body which carries several relatively long and narrow abrasive stones. These stones are mounted in metal holders carried in shoes which have free axial movement within certain limits and are arranged to be expanded or contracted radially by one of several methods. For mass-production work, full automatic hydraulically actuated hones are used with vertical machines, *Fig.* 22.3. For machines not equipped for hydraulic actuation of the tool, mechanically actuated 3-finger automatic hones are used. For limited production the so-called brake type hone is used, expansion and contraction of the abrasive sticks being

Fig. 22.7 — Barnesdril No. 5 horizontal hydraulic honing machine finishing a 4.750-inch gun bore 21 feet long to within 0.0007-inch for roundness and straightness with a stock removal of 0.015 to 0.020-inch per hour. Photo, courtesy Barnes Drill Co.

accomplished manually. Manual type hones are also available in portable models for work in small lots or bulky design not adapted for honing on production machines, *Fig.* 22.4. Portable hones can be used for bore sizes from 2 inches to $14^1/2$ inches.

Where extremely long parts are to be honed, horizontal machines normally are employed and special hones are required having fiber support guides to carry the weight of the tool and prevent marring of the finish on entrance to or retraction

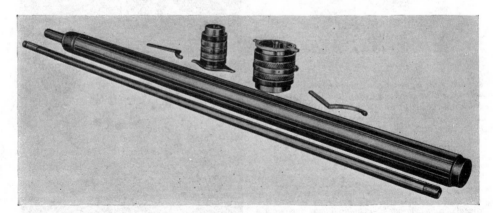

Fig. 22.8 — Piston rods $1^1/2$ and $4^1/2$ inches in diameter by 8 feet long honed within 0.0001-inch to 0.0002-inch for diameter and straightness, and external honing tools used. Photo, courtesy Barnes Drill Co.

Fig. 22.9 — Hydraulically operated special duplex automatic honer for mass production of engine blocks. Photo, courtesy Barnes Drill Co.

from the work. For external honing the work normally rotates between centers and reciprocates on a carriage while the hone is mounted in a fixed position near the center of the machine bed.

Honing Machines Available. A variety of standard and special honing machines are available in either the vertical or the horizontal models. Fully automatic vertical honers are available for large-quantity production of small and average-size work, *Fig.* 22.5. Verticals are used on work up to 6 or 7 feet in length and up to 30 inches in diameter, *Fig.* 22.6. One of the largest verticals built has a stroke of 8 feet and is capable of finishing diameters up to 30 inches by 90 inches long. Horizontal machines, in single and double-spindle models, are used mainly for finishing gun bores, cylinders, long tubes, etc., in diameters up to 50 inches and lengths up to 70 feet, *Fig.* 22.7. External work also can be honed, in diameters up to 6 inches or more and lengths from 2 inches to 12 feet, *Fig.* 22.8. Special mass-production machines also include a variety of multiple-spindle models in both vertical and angular setups, *Fig.* 22.9. In addition, vertical models can be used to hone flat parts to required surface finish and thickness, *Fig.* 22.10.

Honing Speeds. Honing speeds for cast iron range from 200 to 250 surface feet per minute in rotation and 50 to 75 feet per minute in reciprocation. Rotary speeds for steel range from 150 to 200 sfpm and reciprocation about 50 fpm. Bore size, character of material, stock to be removed, and finish desired sets the actual combination of speeds selected.

Machine Output. Production speeds are high. Crankshaft bearing holes in connecting rods, for instance , are finished four at a time at a rate of 300 per hour, while V-type, 8-cylinder motor blocks are finished at an average of 90 blocks or 720 cylinders per hour. Diesel engine cylinders 12.75 inches in diameter by 44 inches in length are completed in approximately 5 minutes floor-to-floor with a stock removal of 0.006 to 0.008-inch.

Fig. 22.10 — The XLO Micromatic 5HF-2 Microflat machine simultaneously and continuously generates two opposed surfaces parallel and flat, while Microhoning parts to required thickness. Courtesy, Ex-Cello-O Corp.

Design Considerations

Honing, in general, should not be considered for bores under 1/4-inch in diameter by 1/4-inch long. Maximum diameter of holes as well as rods feasible to hone have been outlined previously in relation to each type machine. In small-diameter parts, length of bore is limited to approximately 20 times the diameter. Bores over 5/8-inch, however, may be honed to any practicable length; holes 3/4-inch in diameter by 4-feet in length, for instance, in aircraft accessory drive shafts were honed to eliminate failures encountered from fractures which developed when the holes were reamed for finish. The primary limiting factor in the length of large-diameter parts is the machine tooling and handling equipment. For best results, the length of hole should be at least equal to the diameter. Work in which the length of hole is less than the diameter usually requires piloting of the hone and allowing the parts to float. Such parts are most often honed in multiple, *Fig. 22.11.*

Fig. 22.11 — Three-spindle Hydrohoner, adapted to hone ball-bearing inner races progressively. Parts are mounted 10 to a shuttle. Photo, courtesy Ex-Cell-O Corp.

Stock Allowance. For most economical honing, the roughing operation — drilling, boring, reaming, etc., — should remove the bulk of the stock leaving only sufficient material to clean up and correct inaccuracies. As a general rule, about twice as much stock should be left for honing as there is error present. For instance, slush pump liners from 60 to 64 Rockwell C showing 0.040-inch to 0.050-inch out-of-roundness from heat treating, have 0.080-inch of stock removed in honing to size. Thus, economical stock removal can range from 0.002-inch to 0.090-inch on diameters over 2 inches while below 2 inches the range is from 0.005-inch to 0.020-inch.

On parts processed for finish only or where the cost of accuracy on the preceding operation is low, the honing allowance can be from 0.001-inch to 0.004-inch to minimize honing time. However, if close tolerances on the preliminary operation are expensive, it will be cheaper to remove from 0.010 to 0.030-inch in honing. As an example, long tubes as received from the mill can have 0.040 to 0.075-inch of stock honed out faster than they can be bored close to size and finish honed. Again, on cast or forged cylinders where diameter tolerance is fairly liberal

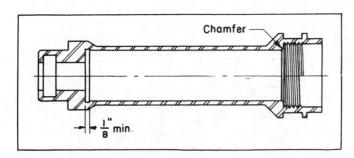

Chamfer

$\frac{1}{8}$" min.

Fig. 22.12 — Bore with flat bottom showing recess to simplify honing. Recessing permits holding closer accuracy. Drawing, courtesy Ex-Cell-O Corp.

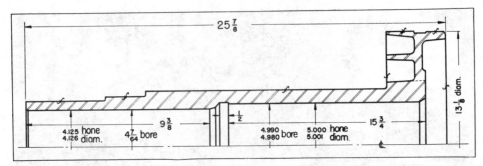

Fig. 22.13 — Motor armature spider with stepped, honed bores showing honing clearance at blind end. Drawing, courtesy C. Allen Fulmer Co.

it has been found advantageous to bore with a round-nose tool at a feed of about 0.050-inch so as to leave high ridges of about 0.010-inch because these can be honed down to a geometrically true bore surface with extreme rapidity.

For very fine finish and hard materials, rough and finish operations are often necessary if stock removal is great. Also, if a part is to be finish honed after hardening, costs can be considerably decreased by rough honing before hardening and allowing only sufficient stock to clean up distortions due to hardening. Precise bores, such as those for motor blocks, should be fine bored prior to honing, leaving 0.003 to 0.006-inch of stock to straighten and clean up the bore.

Interrupted Surfaces. Many complications such as bosses, struts, ports, slots at the end of a bore, oil grooves, keyways, shoulders, blind bores, etc., are unavoidable in design and serve to prevent the accomplishment of closest accuracy in honing but, in many instances, can be handled satisfactorily. The possible cost increases due to such complications, however, should be recognized. On

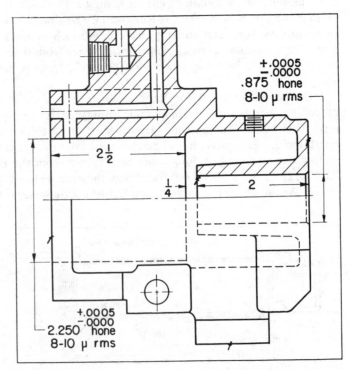

Fig. 22.14 — Switch control body showing two concentric bores. Adequate intermediate clearance for hone overtravel is necessary. Drawing, courtesy C. Allen Fulmer Co.

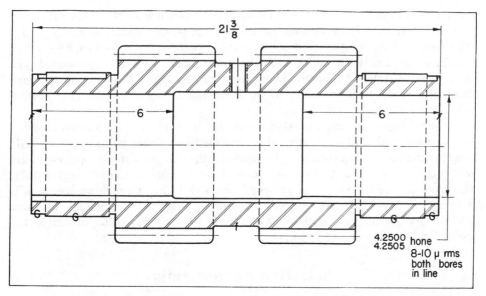

Fig. 22.15 — Forged AISI 4145 steel pinion sleeve with aligned honed bearing bores. Honing time with a stock removal of 0.005-inch is 1 minute per piece. Drawing, courtesy C. Allen Fulmer Co.

interrupted surfaces, alignment will be maintained and bore will be straight but some overcutting will be present at the edges of the reliefs. On relatively soft parts containing keyways, oil grooves, etc., honing usually is done prior to cutting the keyways or oil grooves whereas, on hardened parts, the job can be honed after these features have been added by using a special tool arranged so that the stone carriers straddle the relieved portions.

Blind or Special Holes. With blind-end bores, as in grinding, it is good practice to provide a slight undercut or recess at the bottom at least a few thousandths larger than the finished bore, *Fig.* 22.12, and at least 1/8 to 1/4-inch wide. A recess makes the honing operation much more practical and simplifies the holding of close tolerances. Stepped holes should follow similar practice, *Fig.* 22.13.

Where special spaced bores are required by design, adequate end clearance for the honing tool should also be allowed, *Fig.* 22.14, inasmuch as the achievement of straight, round holes without bellmouth is dependent upon a small amount of overtravel of the hone beyond each end of the hole. Relieved spaced bores, often desirable in many precision designs, are also practical and reduce the amount of metal necessary to remove, *Fig.* 22.15. Aligned bores up to 6 inches in length and

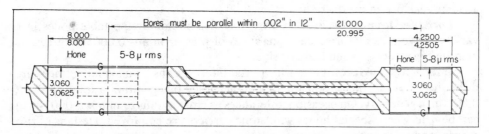

Fig. 22.16 — Large connecting rod showing honing tolerances held in production, and surface finish specified. Drawing, courtesy C. Allen Fulmer Co.

14½ inches in diameter with center distances to as much as 36 inches have been honed. Wherever aligned or concentric bores can be of identical size, *Fig.* 22.15, the honing can be completed in one operation whereas stepped bores, *Figs.* 22.13 and 22.14, require two handlings. Aligned bores should always be open ended. The open ends of holes should be chamfered wherever possible to eliminate sharp finished corners, facilitate entry of the hone and fitting of mating parts during assembly, *Figs.* 22.12 and 22.13.

Hole Alignment and Location. It should be kept closely in mind that the honing head is literally a floating tool. Finished bores thus will show very little improvement regarding location or alignment. Special preparatory operations are not necessary in order to hone — rough castings can be honed satisfactorily. However, if bore location or alignment *is* important, then a roughing operation is necessary to establish the desired location or alignment. Accuracy of these preliminary operations must be that desired in the final part with, of course, the exception of diameter and finish.

Selection of Materials

All ferrous materials — cast iron and steels, hardened or unhardened — as well as many nonferrous materials such as bronze, brass, and silver can be honed satisfactorily. Also, nonmetallics such as glass, some plastics, and certain ceramics can be processed. Quality of honed finish obtainable with the softer materials is somewhat inferior to that possible with the harder ones. Nonferrous materials, such as lead or babbitt, which tend to clog the pores of the abrasive stones are seldom suitable for honing.

Specification of Tolerances

Reproducible, uniform accuracy in most modern machine honing applications on diameters over 4 inches, *Fig.* 22.7, can be held within 0.0005 to 0.001-inch total variation for roundness and straightness or taper; on diameters from 4 inches down to 1-inch this can be 0.0003 to 0.0005-inch; and on bores below 1-inch it is usually within 0.0001-inch to 0.0003-inch. In production runs on Microhoning machines fitted with automatic Microsizing equipment, diameters under 2 inches can be held to less than 0.0003-inch total variation and over 2 inches to less than 0.0005-inch. In general, tolerances of plus or minus 0.0001-inch to 0.00025-inch are commonly attained and on special small-bore jobs where the maximum in accuracy is necessary, diameters can be held within a total of plus or minus 0.000025-inch but, at some sacrifice in production.

Portable Honing. On jobs unsuited to machine honing, portable hones in a variety of types can be used and tolerances of plus or minus 0.0005-inch can be held on taper, out-of-roundness or size.

Interrupted Surfaces. Parts with ports, interruptions or unequal surfaces are difficult to hold to the closest tolerances, accuracy being within a total of about 0.001-inch generally. On occasion, tolerances as close as plus or minus 0.0003-inch can be held where special hones with stone carriers are employed. Designs of this nature and blind holes without relief ordinarily should be avoided where tolerances must be held closer than 0.0005-inch.

Fig. 22.17 — Group of honed hydraulic cylinders showing typical scratch pattern. Photo, courtesy Barnes Drill Co.

External Honing. External cylindrical surfaces can be honed to equally accurate tolerances, even very long rods being finished to plus or minus 0.0001 to 0.0002-inch on roundness and taper, *Fig.* 22.8.

Surface Finish. As mentioned previously, the surface quality obtainable in honing can be held as required by the particular design application with only a few limitations regarding materials. In hard steel the surface finish attainable will range from 20 to as low as 1 microinch. In bearings with a hardness of 60 Rockwell C, removing from 0.006 to 0.014-inch of stock, finish may be held as low as 2 microinches in production. This accuracy requires a two or three-step honing operation, however, since the same abrasive that will remove stock will not produce the finest finish. In cast iron, soft steel or bronze, honed finish obtainable may run from 80 to about 3 microinches, 5 to 10 being average, *Fig.* 22.16. Fineness of finish with cast iron is limited mainly by the grain structure of the material. With aluminum, a finish of about 15 microinches is the best that can be produced.

Finish Pattern. Judicious selection of surface finish for honing will afford considerable economy in production and also will provide the most desirable surface functionally. For accurate pressure-lubricated bearings, fluid seals, etc., the most desirable surface finish is the smoothest possible, say from 1 to 5 microinches. Cylinder walls or other surfaces which are regularly wiped by another part will require quicker, faster wetting by a lubricant, and perhaps good heat dissipation in addition, which is afforded best by somewhat rougher surfaces, say 5 to 30 or more microinches, with the most desirable scratch pattern. A cylinder using leather or O-ring packings would indicate a finish from 10 to 15 microinches with a regular cross-hatch pattern, *Fig.* 22.17. Again, where a codirectional finish or surface with a minimum of pattern is indicated, straight-line honing can be employed to advantage.

23

Lapping

FOLLOWING the honing process in refined grinding operations classified in terms of increasing smoothness of finish and increasing accuracy is lapping. With lapping, parts can be produced to a range of accuracy approximating the very finest possible to obtain by means of honing and on occasion somewhat finer.

Hand Lapping. Hand lapping with abrasive-charged laps of soft metal, of course, is well-known but is extremely slow and requires considerable skill in obtaining consistent and accurate results. Consequently, hand lapping methods are

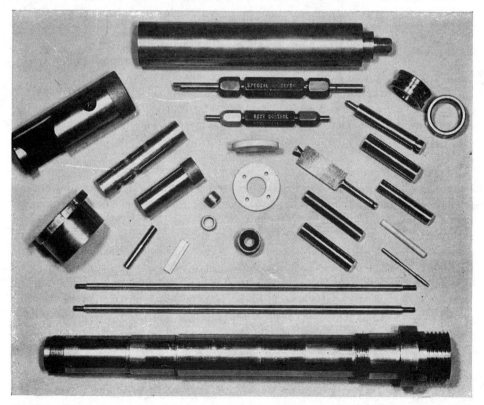

Fig. 23.1 — Variety of parts on which surfaces have been finished by centerless lapping with loose abrasives. Photo, courtesy Size Control Co.

Fig. 23.2 — External straight-line lapper finishing 3.350-inch recoil piston rods 57 inches long. Photo, courtesy Barnes Drill Co.

poorly suited for economical interchangeable manufacture.

Mechanical Lapping. Mechanical or machine lapping offers a degree of refinement over surface finish produced in ordinary honing operations and can be applied to bores, cylindrical surfaces and flat surfaces. It fulfills the requirements of mass production and eliminates the need for great skill on the part of the machine operator. Highly consistent and accurate results are obtainable by a variety of means. Lapping machines may employ bonded-abrasive laps, abrasive-charged laps, and in some cases abrasive-coated paper or cloth laps.

Metal Removal. With the random traverse of the lap over the work surface the tendency is toward reduction of surface irregularities left by previous operations. Being expressly a surface *refining* operation rather than a metal-removal operation, lapping is distinctly *not* intended for removing any sizeable amount of stock. Stock removal practicable in production never exceeds 0.0005-inch.

Scope of Process. Mechanical or machine lapping is primarily suited for refining surface finish on external or internal cylindrical and flat surfaces, *Fig.* 23.1. In addition, by means of semimechanical lapping a variety of surfaces such as external tapers, shouldered parts, spherical internal and external surfaces, etc., can be finished.

Straight-line honing equipment can often be used for lapping long parts. Lapping with solid abrasive stones is often desirable following the normal honing procedure to remove the cross-hatch pattern and produce an even finer lapping-stone finish with all scratch pattern codirectional with the operation of the mating part over the surface, *Fig.* 23.2. Solid abrasive sticks of somewhat wider face than regulation honing stones are used and a nonuniform indexing mechanism guarantees that the stones will not trace more than once in the same path. Since this lapping removes only an infinitesimal amount of stock, a honing operation must precede it, in most cases, to correct and perfect the geometry of the part.

Machines Available. For production lapping a variety of machines are available which use abrasive-charged laps, bonded-abrasive laps, and abrasive-coated cloth or papers. Where desirable, cast iron or other charged laps can be substituted for the honing tool on various standard honing machines to obtain a velvet lapped finish.

Vertical-spindle machines with parallel flat charged cast-iron laps will handle flat parts up to 4 inches thick and 9 inches in longest dimension or cylindrical work up to 4 inches in diameter by 9 inches long.

Fig. 23.3 — Dual-roll centerless lapping machine for both
rough and finish lapping
operations. Photo, courtesy
Size Control Co.

Centerless lapping machines are available which employ a charged lapping roll and a control roll both of which are driven, the charged roll normally being the faster. Where desirable, dual or triple-roll machines, *Fig.* 23.3, can be used to rough and finish lap parts with one handling. Either manual or automatic feeds can be used, depending upon production requirements, for lapping cylindrical, tapered or shouldered parts from 0.030-inch to 10 inches in diameter, *Fig.* 23.1. Contoured parts require special rolls.

Using the through-feed centerless principle and a solid bonded-abrasive wheel much the same as in centerless grinding, continuous mechanical lapping is obtained. On these machines both lapping and regulating wheels are 14 inches in diameter by 22 inches in face width, *Fig.* 23.4. Regular or grind-lapping is usually done with a 180-grain wheel and the finest lapping with a 500-grain. Where through feeding is used, only cylindrical parts can be handled, *Fig.* 23.5. Shouldered parts can be lapped by in-feed but at lower production rates. Short

Fig. 23.4 — Automatic centerless lapper designed to operate on the same principle as
the centerless grinder. Photo,
courtesy Cincinnati Milacron
Co.

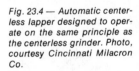

parts, 1/4-inch to 6-inch diameters by up to 15 inches in length, can be handled in these machines and long parts from 1/2 to 3 inches in diameter up to 15 feet in length.

One lapping machine utilizes charged laps and the feature of self-truing action in the machine. A continuously conditioned lap is assured by means of wear rings which wear the lap to the desired true curvature or flatness at a greater rate than the parts being lapped tend to wear it out-of-true. Parts requiring flat, concave or convex surfaces can be lapped in sizes which can be carried singly or in multiple within the wear rings which have internal diameters of 4 1/4 inches to 12 inches, *Fig. 23.6*. A variety of special machines are available for such operations.

Special machines are available for lapping main crankshaft bearings, crankpin diameters and thrust surfaces, etc., in one operation automatically using 320 to 500 grit aluminum oxide abrasive cloth laps. Some are completely automated.

Production Speeds. High output is normally possible with most mechanical lapping machines. Typical is the production by centerless lapping of such parts as wrist pins, rollers, pump plungers, etc., where rates ranging from 500 to 3600 parts per hour, depending upon size, are achieved.

Design Considerations

For lapped parts, restrictions on size and shape have been generally outlined in the preceding discussion and practical design must naturally observe such limitations. For most efficient lapping in production, a minimum amount of stock should be left for finishing. Stock allowance on diameter should never be more than 0.0005-inch, preferably not over 0.0002-inch for rough lapping or 0.0001-inch for finish lapping from a rough-lapped surface. The preliminary processing operation should either be grinding or honing to provide the geometrical and orientation accuracy desired. Lapping ordinarily provides little in the way of correction of out-of-roundness or alignment.

Shoulders. For maximum production, parts to be lapped should have no interfering shoulders. Less rapid to process, shouldered parts require in-feed handling but are limited in length to the maximum width of cylindrical lap

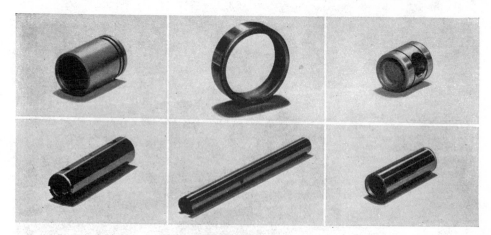

Fig. 23.5 — Group of centerless lapped machine parts ranging from 1/2-inch diameter by 6 1/4 inches long to 2.4375-inch diameter by 0.6875-inch long finished on the machine shown in Fig. 23.4. Photo, courtesy Cincinnati Milacron Co.

available — approximately 22 inches on the largest centerless lapper. Main bearings of cam shafts, for instance, are processed in this manner. Portions adjacent to the sections to be lapped should preferably be relieved slightly to aid the lapping operation and render unnecessary accurate endwise alignment in the processing.

Tapers. Vertical lappers allow the lapping of tapered pieces. Practical taper, however, is determined by the length of part, diameter and machine.

Fig. 23.6 — Lapmaster lapping machine which assures true surfaces by means of self truing action with wear rings. Accompanying parts shown are typical precision units produced on these machines. Photos, courtesy Crane Co.

Flats. Parts which require lapped flat surfaces can be of almost any design within the limitations on size of the various machines available. Parts can be lapped to achieve flatness or where necessary, lapped concave or convex. Valve disks, for instance, have been lapped concave 3 light bands (34.8 millionths concave).

Selection of Materials

Lapping with loose abrasives as a rule is somewhat unsatisfactory for finishing the relatively softer materials. The tendency is for the abrasive to charge or embed itself in the surface of the part being lapped. Also, there is a tendency to remove more of the softer portions of a surface than the harder with materials such as cast iron. This is also true with abrasive-coated paper lapping. With bonded-abrasive lapping, however, these tendencies are not present.

For best results with lapping, materials in the harder ranges should be specified, although any material from copper to quartz or hardened steel can be finished.

Specification of Tolerances

The primary purpose in lapping is to obtain surface refinement and, ordinarily, roughness is less than 2 microinches rms and at the best about 0.5 microinch.

Centerless Lapping. In lapping small parts by the centerless bonded-abrasive method, size accuracy can be held to a total variation of 0.00005-inch on diameter and 0.000025-inch on straightness. Preceding lapping, the parts should be ground to a good commercial finish, sized within about 0.00025-inch, and held as round as desired in the finished job. Lap-grind, a finish between lapping and grinding, is obtained in one operation with a maximum stock removal of 0.0005-inch and surface finish of 4 to 6 microinches. After the lap-grind operation, surface finish may be improved to 2 to 3 microinches with a maximum stock removal of 0.0001-inch using the 500-grit wheel. Finest finish of 2 microinches or better can be obtained in a third operation with practically no stock removal.

With centerless loose-abrasive lapping, size accuracy within plus or minus 0.000005-inch is said to be possible. Preliminary grinding operations must be accurate as for the bonded-abrasive method. Not more than 0.0002-inch of material can be removed and surface finish of 2 microinches or better can be produced with one to three operations.

Flats. Surfaces flat to within one light band (11.6 millionths) can be produced on vertical lappers. Finishes of one-microinch or less can be had and parallel flats can be held within 0.0001 to 0.00005-inch on parallelism.

Long Rods. Straight-line lapped piston rods and similar parts can be held to fairly close tolerances. For example, 8-foot rods are finished to within 0.0001 to 0.0002-inch for roundness and taper, *Fig.* 23.2.

Bores. Because internal honing permits extreme accuracies in the finishing of bores, hand lapping should seldom be required. Finish honing tolerances of plus or minus 0.000025-inch on diameter of small bores can be readily attained if necessary. Where lap finishes are wanted, use of charged or stone laps can often be substituted for the honing tool to produce the desired results.

Abrasive-Paper Lapping. Following a relatively smooth grind of less than 20 microinches, lapping with abrasive coated cloth or paper will produce surface improvement. However, little correction of surface irregularities or dimensional inaccuracies can be made. A bright polished finish can be effected on a variety of parts with contours which defy the use of previously mentioned methods. Internal contours of ball races, etc., are often lapped by this method. Smooth finishes of 4 to 7 microinches can be readily achieved on automatic machines.

24

Superfinishing

REPRESENTING the ultimate in present-day production methods, in terms of increasing refinement of surface finish, is that process commonly termed Superfinishing. As with the lapping process, Superfinishing makes a complete break with previous metal-removal methods, being used only for obtaining the finest surface finish and improved geometrical accuracy.

Basic Differences. Although honing and lapping methods discussed in the previous chapters apparently afford results very similar to those in Superfinishing,

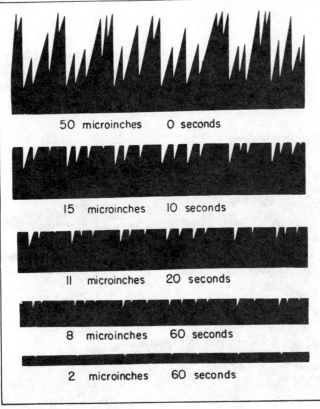

50 microinches   0 seconds

15   microinches   10 seconds

11   microinches   20 seconds

8   microinches   60 seconds

2   microinches   60 seconds

Fig. 24.1 — Diagram made from actual crankshaft operations showing ground profile to finished surface. Drawing, courtesy Chrysler Corp.

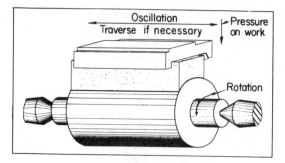

Fig. 24.2 — Schematic diagram showing arrangement and operation for finishing cylindrical pieces.

Fig. 24.3 — Schematic diagram showing principles employed in finishing flat surfaces.

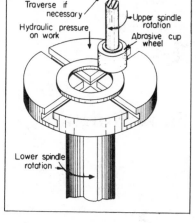

research has shown some basic differences are present. Because finishing is done at slow speeds and low pressures, all tendencies toward heating or burnishing are obviated. Owing to the fact that a large-area bonded abrasive of precisely controlled characteristics is employed, minute projections, waviness, etc., are removed and a clean-cut, undisturbed crystalline surface structure is produced.

Process Characteristics. In the process of Superfinishing, a large-area bonded-abrasive tool or stone is operated with multidirectional travel over the work surface. The stone is worn automatically to a true curvature or flat, as the case may be, and thus acts as a master shape to correct the geometry of the work surface. Actually, only irregularities, projections and inconsistencies over and above the desired true geometrical profile are removed, *Fig.* 24.1, a characteristic not found in other methods. In this respect the process resembles hand lapping to a metal master-shape, but is much faster and can never "charge" a surface as loose-abrasive lapping may do.

Fig. 24.4 — Superfinished shafts, the smaller of which operates through non-metallic packing and the larger through a cast-iron bore. Almost no wear has been found after seven years of operation. Courtesy Giddings & Lewis, Inc.

Stock Removal. Like lapping, there is no real stock removal in a sense, the processing merely refines the surface finish and increases the true bearing area. Actual dimensional change through removal of high spots seldom exceeds 0.0001-inch in most average production work.

Area of Abrasive Stones. Unique among processing methods, Superfinishing will remove the roughness and irregularities left by previous operations, produce any reasonably selected smoothness, then automatically cease cutting action. A bonded-abrasive stick, properly dressed, is used for cylindrical or cone shapes, *Fig.* 24.2, while a bonded cup wheel is used for spherical or flat work, *Fig.* 24.3. Large area is used; in cylindrical work, for instance, the stone is 60 to 75 per cent of the part diameter in width and often the full length of the surface to be refined.

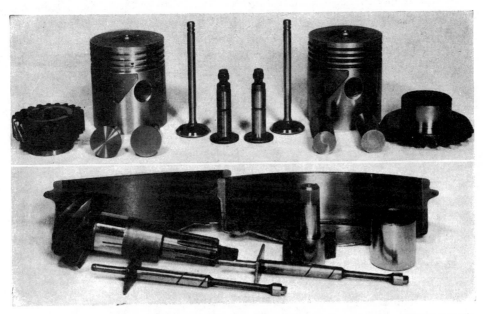

Fig. 24.5 — A group of representative parts which employ Superfinished surfaces. *Courtesy Giddings & Lewis, Inc.*

Superfinishing Speed. Abrasive speed seldom exceeds 80 fpm, with an average of 55 fpm, and the abrasive is held in contact with the work under a flexible pressure of 10 to 40 psi. Usual work speed is about 50 to 60 sfpm and operation is carried out under a flood of low-viscosity lubricant rather than a plain coolant.

Finishing Action. As the stone is applied to a rough surface, it contacts only the high points or ridges on waviness crests. Here the unit pressure becomes high owing to the small-area contact, and the stone easily penetrates a lubricant of any viscosity to begin abrading the peaks. A stone of the correct character breaks down somewhat, continually exposing new, sharp grits and is virtually self-dressing. As the peaks are reduced and the surface improves in smoothness, unit pressure decreases owing to increased contact. As this takes place, cutting and dressing decreases, the stone surface dulls or glazes progressively, and finally, as the perfection in shape of both stone face and work becomes great enough, pressure sufficient to force the grit points through the film of the lubricant is not available and further removal of material ceases.

Advantages of Superfinishing. First conceived in 1934 by D. A. Wallace of the Chrysler Corp., the process has been in a continuous state of development ever since. While there are several general sources of wear, through the results of comprehensive research it is becoming generally recognized that the welding together and breaking apart of minute high spots of rubbing surfaces is the major cause of metallic wear. Because of this fact, it becomes imperative that projecting defects be reduced as much as practicable. Hence the major purpose of Superfinishing.

Any metallic surface that rubs upon another, and certainly those parts that must operate with an absolute minimum of clearance or under heavy loadings, should be considered for Superfinishing. If such a part requires hardening, the need is more pronounced; because of heat produced in grinding, the surface hardness is often altered to some extent and removal of this fragmented layer is beneficial. One of the most important uses is on parts that operate within oil seals or packing glands. Maximum possible smoothness prolongs the life and effectiveness of such units immeasurably, *Fig. 24.4.*

Dimensional Change. While Superfinishing is primarily *not* a dimension creating process, it is obvious that in generating geometrical accuracy by removal of high spots, some small decrease in apparent dimensions will take place. This may amount to 0.0001 to 0.0002-inch on diameter as an average.

Fig. 24.6 — Group of flat, grooved and recessed parts which have been Super-finished. Courtesy Giddings & Lewis, Inc.

Range of Finishes. A range of surfaces from a mirror finish to about 30 microinches rms can be produced. Almost any reasonable scratch pattern is possible, ranging from none at all to intentional cross-hatch patterns.

Scope of Process. Representative of parts which are usually considered naturals for Superfinishing are crank pins, clutch plates, valve stems, piston pins, pistons, piston rods, tappet heads, tappet bodies, valve seats, brake drums, cylinder bores, bearing bores, etc., *Fig. 24.5.* Certain limitations, however, are present as to the shape of the parts which can be Superfinished readily and economically.

Surfaces to be refined must be those of symmetrical sections such as external cylinders, large internal cylinders, flats, spheres, or cones. Flat parts should have a reasonably round outside dimension. With the proper abrasive stone, fluted shapes, shafts with keyways, or other recesses can be handled, *Fig.* 24.6.

Machines Available. In the present state of development, a variety of both general-purpose and high-production machines is available for Superfinishing almost any reasonably symmetrical surface, *Figs.* 24.7 and 24.8. Cylindrical surfaces ranging from 1/8-inch in diameter up to as large as 14 feet can be processed. General-purpose Superfinishers, however, presently can handle parts up to about 41/2 inches in diameter by 36 inches in length. Range of dimensions for circular flat surfaces is from about 1/2-inch to the capacity of the largest machine (a boring mill, for instance) upon which a Superfinishing attachment can be mounted. Spherical sections may range from about 1/2-inch in diameter to about 4 inches on present equipment.

Production Speed. Time consumed in Superfinishing is extremely short and production consequently can be high. Small cylindrical surfaces up to about 1/2-

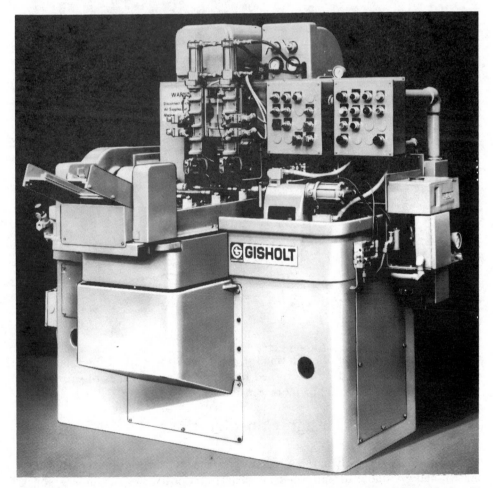

Fig. 24.7 — Model 93 Superfinisher for high production of small and medium-size cylindrical shaft parts. Of automatic modular design, these machines can be configured to do multiple diameters simultaneously providing average production of 200 pieces per hour or more. Courtesy, Giddings & Lewis, Inc.

inch in diameter and 1-inch in length are finished to 3 microinches rms in 30 to 45 seconds; those up to 2 inches by 2 inches, in about one minute. Larger surfaces, which require traverse of the stone, usually are completed at a rate of about 10 square inches per minute. Flat and spherical surfaces are finished in somewhat less time than cylindrical. Tappet heads, for instance, are usually corrected in geometry and finished to less than 2 microinches in a maximum of 30 seconds, *Fig.* 24.9.

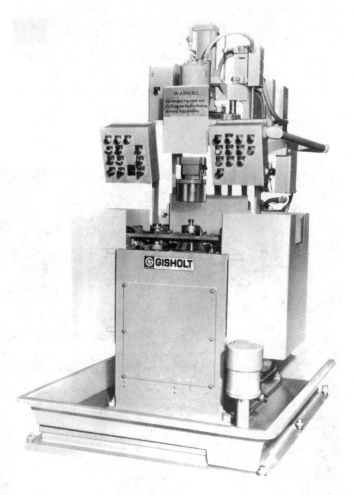

Fig. 24.8 — Model 91 Vertical Superfinisher for flat and cylindrical work. Surfaces are reduced to about 2 microinches on flat or inner or outer surfaces of cylinders automatically. Courtesy, Giddings & Lewis, Inc.

Cast-iron pump faces are often finished in less than 20 seconds. Clutch faces and brake drums are reduced from a turned finish of 200 microinches rms to 7 to 15 microinches at a rate of 100 or more per hour, *Fig.* 24.10.

Design Considerations

As outlined generally in the foregoing, surfaces considered for Superfinishing must usually be those of uniform symmetrical sections such as cylinders, cones, spheres, or flats.

Flat Surfaces. Where flat surfaces are contemplated, these should be designed

Fig. 24.9 — An aviation tappet head finished to approximately 2 microinches or less. Photos, courtesy
Giddings & Lewis, Inc.

Fig. 24.10 — Special brake drum Superfinisher
reduces a 200 microinch turned surface to 7 to 15 at a
rate of 120 or more drums per hour.

so as to eliminate the necessity for finishing projections and other features which
create a noncircular outside boundary. Portions which project beyond the normal
circular boundary should be relieved well below the main finished surface.

Accessibility. Complications such as shoulders, ports, flutes, slots, oil
grooves, keyways, blind bores, etc., are, of course, usually essential but these
features often serve to prevent ready access for Superfinishing. With the proper
width of abrasive stone, flutes, keyways, ports, etc., can be spanned and the main
surface finished. Blind-end bores, whether Superfinished on an internal or external
cylindrical surface, should provide an undercut or relief at least a few thousandths
larger than the finished surface, *Fig.* 24.11, and about 1/16 to ¹/₈-inch or more in
length.

Dimensional Change. For most economical Superfinishing the preliminary
metal-removal operations should remove the bulk of the material, allowing only
for the minute dimensional change which may take place. A reasonably accurate
rule which may be followed in ascertaining the average change which may be
expected is that 0.0001-inch will be removed on the diameter for each 10
microinches rms reduction in surface roughness. Of course, on flat surfaces this
figure would be halved.

Preceding Operations. Production economy in Superfinishing depends greatly
upon the degree of roughness left by preceding operations. It is seldom economical,
for instance, to attempt to obtain a surface of 3 microinches or less from a grind
rougher than 20 to 25 microinches; 10 to 20 microinches is the most desirable.
Degree of accuracy obtained in finishing also is dependent upon the preliminary

operations. Dimensional accuracy and uniformity should be held as close as possible in the preparatory operations.

Selection of Materials

Materials which can be Superfinished include practically all the ferrous and nonferrous metals and alloys and also many nonmetallics such as molded friction materials, glass, wood, fiber, etc.

Preparatory operations best suited for different materials will vary with the material used. Cast-iron parts such as brake drums, pistons or flywheels can be either turned or ground. Turned finish should not range over 150 to 200 microinches ordinarily. For soft-steel parts grinding is preferable as a rule. Hardened steel parts such as tappets, pins, gears, spindles, etc. definitely require a ground finish. Nonferrous materials can be turned or ground prior to Superfinishing but the primary requisite is, regardless of the preliminary processing operation, that surface roughness be uniform.

Specification of Tolerances

The primary purpose of the Superfinishing process is the creation or development of a geometrically true substantially *unworked* material surface free from fragmented, amorphous or smear particles. Not being specifically a dimension creating process, true Superfinishing does not completely remove gross out-of-roundness, taper conditions, deviations from true straightness, or misorientation. Although dimensional irregularities of some magnitude can be

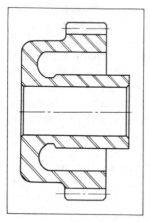

Fig. 24.11 — Engine reduction drive pinion presents a rather difficult grinding finishing problem. Relief is a necessity at the blind end of the bore.

Fig. 24.12 — Right — Range of finishes possible from cross directional scratch pattern to a scratchless mirror finish. Courtesy Giddings & Lewis, Inc.

Fig. 24.13 — An outboard motor crankshaft Superfinished to a mirror smoothness. Courtesy, Giddings & Lewis, Inc.

corrected, provided sufficient time and successive operations can be utilized, maximum efficiency in production requires that the desired uniformity and accuracy as to location, straightness, out-of-roundness, taper variations from piece to piece, etc., be provided by the preliminary operations.

Dimensional Tolerances. The only dimensional change or action incurred rests primarily with the removal of high spots, waviness, and other defects in reducing the curvature or flat to a reasonably true geometric surface. Thus the dimension resulting from a Superfinishing operation will depend upon that of the previous operation and the surface roughness it produces. Dimensional change to be expected has been covered in the foregoing. Actual tolerances on average production runs of parts normally can be held within the same dimensional limits produced by the previous operation. Where desirable, and providing preliminary operations can be properly controlled, dimensional accuracy can be held within closer limits than produced by the preceding process.

Scratch Patterns. As noted in the foregoing, any reasonable depth of scratch pattern can be produced, from virtually none to about 30 microinches rms, *Fig.* 24.12. Where a scratch pattern is deemed advisable, an intentional cross-hatch finish can be produced. Though no hard or fast rule is available as to the desirability or usage of scratch patterns, the favor most often lies with the smooth surface ranging from 1 to 4 microinches, *Fig.* 24.13. Where scratch patterns are desirable, however, the recommended limit seldom ranges beyond 4 to 12 microinches rms. Practical experience has established the fact that a cross-hatch pattern is equally as satisfactory as a random, multidirectional one and machines are no longer made for producing random patterns.

Surface Roughness. Normally, a finish ranging from 2 to 3 microinches is obtained in one operation of one minute or less duration. It is seldom economical to produce such a surface from a grind rougher than 20 to 25 microinches. Rougher surfaces can be used to start but the finish will range around 4 to 7 on the finished part. A turned surface on cast iron of about 200 microinches rms can be finished to 10 to 15 microinches in the same time. Where design demands, surface finish less than 1 microinch can be attained.

METAL FORMING
METHODS

25

Metal Spinning

$\mathbf{M}$ANUAL metal spinning, though firmly entrenched as a valuable small-volume production process, is in a large measure the work of a skilled craftsman rather than the mechanical output of a highly automatic production machine. Introduced in America sometime around 1840, the spinning of metal long remained an obscure art and was utilized almost exclusively in the fashioning of fine gold, silver and pewter ornamental hollow-ware and jewelry. The use of manual metal spinning in the forming of tougher materials, heavier gages and larger, more complicated parts had its beginning about 1920. The art of spinning in the ensuing years developed rapidly to a point where it now fills an important niche in the production picture. In certain instances the hand-controlled tools have been entirely supplanted by automatic mechanical and hydraulic means. Somewhat less versatile, however, but well adapted to a great many unusual forming jobs, automatic mechanical and power metal spinning rounds out the process possibilities to include large-volume applications. Flame spinning and hot flanging or dishing of heavy-wall parts, make possible the forming of unusually large, heavy parts.

Advantages. An almost unlimited variety of circular shapes can be produced by manual spinning on solid or collapsible forms of wood or steel and even "on air." Where nonstandard shapes subject to continuous change, parts for research projects or components for experimental machines are required, manufacture by manual metal spinning — wholly or in part — is often the most economical and many times the most satisfactory. This is especially true in cases where odd conical parts, exact parabolic contours, extremely large diameters or special types of material are involved.

Repetitive production of machine parts by manual spinning is a quick and, as a rule, an inexpensive operation for not only are the necessary forming tools of simple construction, but the usual hardwood forms are low in cost and can be readily reworked or otherwise changed to suit new developments in design. The advantages of manual spinning, therefore, lie in the speed and economy of manufacturing small-volume and special parts, *Fig.* 25.1, where tool and die costs for other forming methods are prohibitive.

Production Quantities. Cost comparisons show that production quantities are normally limited to 1000 pieces or less, especially for parts where design changes are not contemplated. If an order of 5000 to 10,000 pieces or more is to be filled,

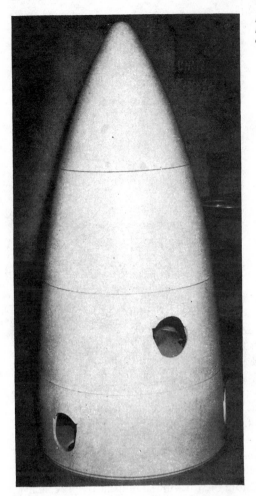

Fig. 25.1 — Left — Spun metal aircraft tail section 72 inches long and 40 inches in diameter. Photo, courtesy Benson Mfg. Co.

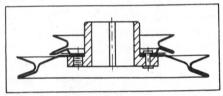

Fig. 25.2 — Mechanically spun machine pulley. Drawing, courtesy Spun Steel Corp.

frequently the cheapest and most satisfactory process would be automatic tracer controlled power spinning, mechanical spinning, stamping or drawing. In many cases, however, composite parts such as the machine pulley shown in *Fig.* 25.2, because of the quantity, size or design, lend themselves best to production by spinning. Large-volume manufacture of these and similar parts is usually by means of automatic mechanical spinning rather than the manual method. Thus, the process, manual mechanical or power, finds ready application in many phases of production and on a wide variety of parts.

Scope of Process. Conventional spinning has been used in fabricating odd-shape covers, venturi tubes, transmission housings, and crankcases both in steel and aluminum by cutting and welding together segments of spinnings or by welding spun sections to the stampings, castings, shapes, etc. Spun tanks, floats, shields, oval parts, dust covers, trays, and similar parts of aluminum or stainless steel are used widely in food machinery, bottle filling equipment, automatic processors and commercial laundry equipment, *Fig.* 25.3. Aluminum tubes may be flared, flanged, have the ends rounded or closed and sealed or tapered for a great variety of parts required in aircraft construction. Size of parts which can be handled on ordinary spinning lathes available today ranges from 1/4-inch to approximately 196 inches in diameter. The basic material is usually sheet of uniform wall thickness.

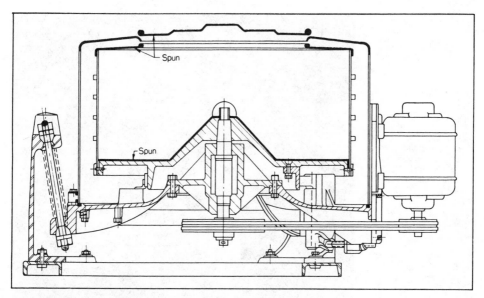

Fig. 25.3 — *Commercial centrifugal laundry dryers utilize many parts produced by manual spinning. Drawing, courtesy Milwaukee Metal Spinning Co.*

Fig. 25.4 — *Recent model Cincinnati Hydrospin spinning lathe especially designed for working parts up to 26 inch swing by 20 inches long.*

Conventional Machines. A variety of standard spinning lathes are available for the production of parts within the range from 1/4-inch to 72-inch diameters, *Fig. 25.4*. Special adaptations and pit lathes permit the production of parts ranging in diameter up to about 16 feet, *Fig. 25.5*. For standard machines which swing up to 36-inch diameters, standard oval chucks can be had for spinning oval parts having diameter differences up to 10 inches on the largest sizes.

Spinning Speeds. As is recognized readily, no hard and fast rules can apply to speeds for conventional spinning. Rotational speeds are normally governed by size or diameter, thickness, and temper or workability of the sheet of metal to be spun. The larger or heavier the blank, the slower the speed. Owing to the fact that spinning speed in surface feet per minute varies widely over a medium or larger size blank it is difficult to maintain an optimum speed on all portions. Consequently,

Fig. 25.5 — Spinning a parabolic antenna 102 inches in diameter from flattened expanded steel. Maple spinning block alone weighs two and one-half tons and is 104 inches in diameter. Spinning speed is 185 rpm. Photo, courtesy Milwaukee Metal Spinning Co.

Fig. 25.6 — Machine for hot spinning extremely large heavy-wall parts from plate up to 5½ inches in thickness. Photo, courtesy Birdsboro Steel Foundry & Machine Co.

maximum peripheral limiting speeds and the best lathe rpm are chosen depending upon temper, gage and size. Some speed data generally utilized are shown in Table 25-I.

TABLE 25-1 — **Approximate Peripheral Spinning Speeds** *

Blank Diameter (inches)	Speed (rpm)	Peripheral Speed (sfpm)
3 to 9	2000 to 2800	2000 to 6000
9 to 18	1500 to 1800	4000 to 8000
18 to 30	1200 to 1500	5500 to 10,000
30 to 50	800 to 1000	7000 to 12,000
50 and up	400 to 500	6000 to 10,000

Courtesy Milwaukee Metal Spinning Co. Speeds given apply generally to all metals commonly spun. Particular spinning speed in any one range depends largely upon the particular design, metal and gage utilized, speed decreasing as gage and size increases.

Small parts of course take the higher side of the speed range shown in Table 25-I. Large parts, especially those near the maximum size or thickness handled, must be worked at the lower speeds, *Fig.* 25.5. For instance, 1/32-inch thick steel disk normally spun at 600 rpm would, if increased to 1/16-inch in thickness, be spun at about 450 rpm, etc. Relatively large aluminum parts — 30 inches in diameter or over — require a peripheral speed under 3000 surface feet per minute, the rpm diminishing considerably as the size increases. Excessive speed will cause a disintegration of the metal in the early stages of breakdown operations.

Flame Spinning. An offshoot of the manual spinning process, flame spinning was developed to facilitate making pressure-tight tubing ends, for partially closing ends of tubes, or for reducing diameters at various points. Utilizing an oxyacetylene flame to heat the portion to be formed, this method permits very rapid forming operations on parts where size or wall thickness eliminates the use of cold spinning. Typical parts are converter tubes, bomb fuses, gas cylinders, shock absorber parts, and refrigeration units ranging from 1/2-inch to 4-inch diameters.

Heavy Spinning. Machines for the spinning of extremely large and heavy-wall parts are commonly termed dishing and flanging machines, *Fig.* 25.6. These machines will spin hot material up to 5 1/2-inches thick and cold material up to 1-inch. Cold material can only be flanged while hot material can be spun into flanged, dished, spherical and conical shapes, *Fig.* 25.7.

These mammoth spinning machines naturally are built to requirements only, the one shown in *Fig.* 25.6 is capable of handling a maximum blank of 230 inches. A typical large-size head is shown in *Fig.* 25.8.

Fig. 25.7 — Hot-spun conical head 99 1/8 inches in outside diameter by 56 inches in height of 2-1/16-inch thick steel. Photo, courtesy Worth Steel Co.

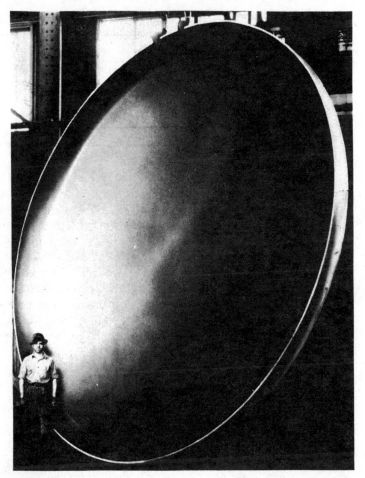

Fig. 25.8 — Flanged and dished head 228 inches in outside diameter with a 4¹/₈ inch straight flange and a 14-inch corner radius. Material is 1-1/16-inch steel. Photo, courtesy Worth Steel Co.

Fig. 25.9 — Floturn spun nose cone is indicative of the process capabilities. Courtesy Floturn Inc.

Shear Spinning. The most recent innovation in spinning is the Shear Spinning process. In conventional spinning the sheet metal blank is shaped over and around a mandrel or form but in shear spinning the *point-extrusion* process reduces the thickness of the starting blank or shape to produce the final form — conical, cylindrical, etc., *Fig.* 25.9.

Shear Spin Machines. Unlike conventional spinning lathes, machines for the shear process are large and rugged, providing spinning forces to as high as 250,000 pounds per tool slide, *Fig.* 25.10.

Fig. 25.10 — Closeup of vertical operation showing mandrel, rollers and finished part. Courtesy Cincinnati Milacron.

Fig. 25.11 — Cincinnati horizontal Hydrospin machine provides automatic cycling for efficient production of 1 or 1000 parts.

Machines of this type range from horizontals, *Fig.* 25.11, with swings up to 60 inches and length accommodation to 60 inches, to verticals, *Fig.* 25.12, with capacities up to swings of 75 inches and length accommodation to 100 inches. Machines are also designed for cylindrical or slightly conical parts up to 28 inches in diameter and 96 inches long (some machines will spin up to 200 inches long),

Fig. 25.13. Some are combination conventional and shear spinners. Machine feature multiple rollers with individual tracer control and automatic operation with optional loading and unloading equipment.

Spinning speeds range from as low as 17 rpm spindle speed to as high as 1500 rpm at the high end.

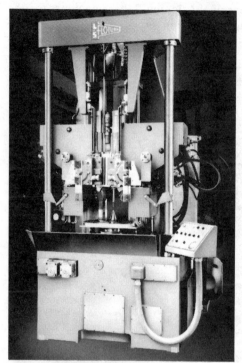

Fig. 25.12 — No. 12 Vertical Floturn with dual-syncro rollers handles work up to 15 inches long.

Fig. 25.13 — Model 3D50 Floturn machine can flow tubes up to 96 inches long.

Design Considerations

The most important points which must be considered in designing parts for conventional spinning include material thickness, bend radii, depth of spinning, diameter and steps in diameter, and workability of the material. General range of diameter limits has been covered in the foregoing. Conical shapes and variations

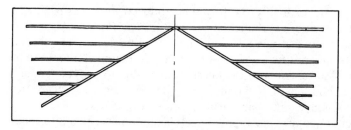

Fig. 25.14 — Conical shapes are simplest to reproduce by manual spinning.

Fig. 25.15 — Hemispherical shapes present no spinning problem until full depth parts are required to close tolerances.

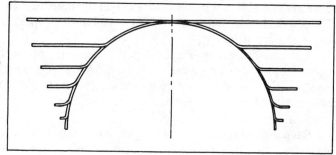

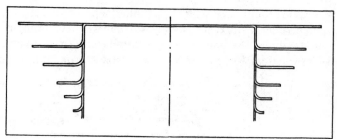

Fig. 25.16 — Most difficult to spin of the simple shapes are cylinders or cup-shape parts.

Fig. 25.17 — Breakdown sequence for a dished head spun from deep drawing steel sheet. Drawings, courtesy Milwaukee Metal Spinning Co.

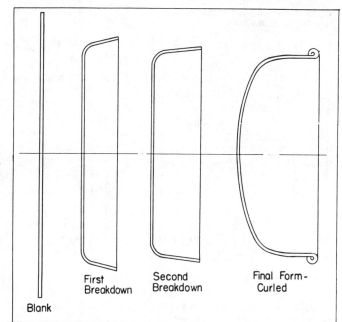

Blank

First
Breakdown

Second
Breakdown

Final Form-
Curled

illustrated in *Fig.* 25.14 are probably the forms reproduced most easily on the metal spinning lathe. Hemispherical shapes illustrated in *Fig.* 25.15 are somewhat more difficult to produce, especially in full-depth parts. Cylindrical or cup-shaped pieces, *Fig.* 25.16 are the most difficult of the simple shapes. Plain cylinders can be produced in small lots if sufficient wall thickness is allowable to provide enough metal to work. Diameter and length of cylinders must be reasonably balanced with the gage of the metal to assure minimum variation in wall thickness.

Oval Parts. As mentioned previously, oval parts can be spun with a difference between the major and minor diameters up to about 10 inches. The major diameter of simple elliptical shapes is limited to about 70 inches on shallow work and 54 inches on deep work.

Spinning of oval or elliptical parts in general production, however, is often impractical or too expensive. For economical and efficient adaptation to the process, it is readily apparent, therefore, that the general form of parts contemplated should be round, concentric, cylindrical, hemispherical, conical, or a part or segment of such shapes.

Deep Spinnings. With deep-spun parts requiring a great deal of working, metal strain or cold working is the greatest and, depending on the properties of the material, spinning to full depth requires a number of partial breakdowns with accompanying intermediate annealings. *Fig.* 25.17 shows such a breakdown sequence for a dished head spun from heavy steel. In a plain deep drawing steel, a head of this type can be spun economically from 10-gage stock in diameters ranging anywhere from 15 inches up to about 84 inches or so with comparative ease. On the other hand, it would be impractical to attempt the same shape from this gage

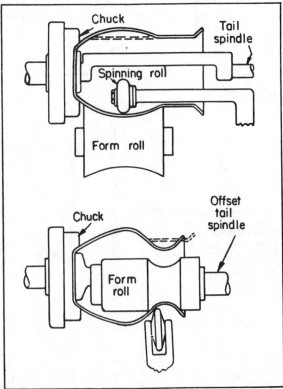

Fig. 25.18 — Off-center roll spinning for bulging and necking operations obviates need for collapsible forms.

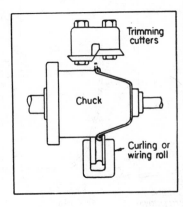

Fig. 25.19 — A combination trimming and wiring operation for finishing spun parts.

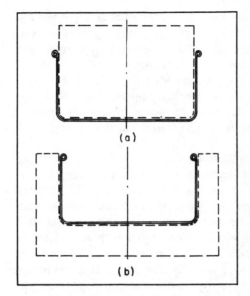

Fig. 25.20 — Outside bead shown at (a) should be used where possible to eliminate special form for inside bead (b).

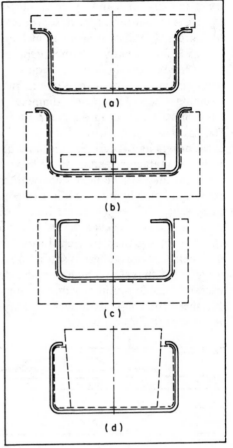

Fig. 25.21 — Simplest form of plain flange shown at (a), special squaring form (b), plain inside flange at (c) and form for exacting inside flange at (d). Drawings, courtesy Milwaukee Metal Spinning Co.

material, say in a 2 or 3-inch diameter. A lighter gage of the same stock — perhaps 16-gage — would be a practical maximum thickness.

Complex Forms. Intricate shapes or "closed" parts which cannot be spun manually on an ordinary open chuck (spinning block) often can be produced on a two-piece, on a sectional or on an off-center roll chuck, *Fig.* 25.18. Work such as that shown in *Fig.* 25.18 is first either spun or drawn to a plain cylindrical shape or cup and then mounted for the final inside spinning, bulging or necking operations. This procedure provides a practical method for obtaining unusual shapes of circular cross section combining expanded and reduced portions along the spinning axis.

Gage and Blank Areas. The area of blank stock for a spun part as a rule should be approximately equal to the surface area of the finished part. Normally the stock decreases somewhat in gage or thins out during the processing and naturally increases the stock area. This increase in area will usually provide sufficient metal for trimming, flanging, rolling or wire beading, *Fig.* 25.19. To compensate for this thinning out of the stock, the gage of the blank should be approximately 30 per cent greater than the thinnest section allowable on the finished part.

Beads and Flanges. Where parts require a strengthening bead or wired roll over the edge, the most economical form is the outside bead as shown in *Fig.* 25.20a. The inside bead, as in *Fig.* 25.20b, entails a greater cost since it involves the use of an extra form and a special setup. Similarly, the outside flange, shown in *Fig. 25.21a,* is the simplest type to reproduce. If the flange angle can vary from an exact 90 degrees by plus or minus 1/16-inch at the outer edge, a special or squaring form, *Fig.* 25.21*b,* to remove the metal spring-back is not necessary.

Where an inside flange is required on a part as shown in *Fig.* 25.21*c,* tooling requirements are about the same as that for the perfectly squared outside flange, assuming liberal depth and diameter tolerances. However, if such an internal flange must be exact in diameter and depth, a more complicated sectional form, *Fig.* 25.21*d* must be utilized.

Corners and Radii. Streamlined or smooth curves and large radii are an aid both to manufacture and improved appearance. Stepped sections can be much more easily produced if conical steps with large radii are specified, *Fig.* 25.22. With easily spun metals, inside radii formed over the spinning block can often be made fairly sharp by means of steel inserts or by using all-metal spinning chucks. It is more practical, however, to allow a minimum radius of 1/16-inch and wherever possible not less than 1/8-inch. Inside radii of 1/8-inch can be held with practically any metal. More generous radii, however, simplify the production problems, especially with metals which work-harden rapidly. In hot spinning magnesium, the minimum inside radius possible without cracking is usually limited to about twice the sheet thickness. However, specifying maximum allowable radii wherever possible simplifies production.

Fig. 25.22 — Smooth curves and tapered steps are an aid to manufacture.

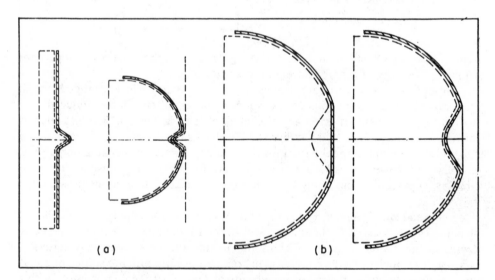

Fig. 25.23 — Deep reverse forms require several blocks (a) while simplified design (b) requires only one. Drawing, courtesy Milwaukee Metal Spinning Co.

In no case should severe reverse forming be required. Such forming often tends to thin out and stretch the metal excessively or requires too many extra operations. The form shown in *Fig.* 25.23*a,* for instance, necessitates two separate operations while a redesign of the same part, *Fig. 25.23b,* eliminating the deep form in favor of a smooth, shallow one permits spinning on a single block.

TABLE 25-II — Spinability Ratings in Various Groups of Metals*

Type of Material	Shallow Work	Deep Work	Group
Aluminum Alloys			
2S-0	100	100	100
3S-0	100	99	
17S-0	65	45	
24S-0	65	45	
52S-0	98	62	
53S-0	98	62	
Copper and Copper Alloys			
Admiralty brass	82	70	
Cartridge brass	99	91	
Commercial bronze (gilding metal)	100	100	
Copper (cold-rolled, annealed)	100	100	87
Copper (hot-rolled)	99	88	
Eyelet brass	99	91	
High brass (yellow spinning)	99	99	
Low brass	99	88	
Muntz metal	55	15	
Naval brass	75	25	
Nickel silver (up to 30 per cent)	98	87	
Phosphor bronze (5 per cent, soft)	85	45	
Red brass	98	89	
Silicon bronze (ASTM B97-41 type A)	96	85	
Silicon bronze (ASTM B97-41 type B)	97	87	
Steels			
Cold-rolled (deep drawing)	100	100	91
Enameling (extra deep drawing)	100	94	
Zinc-coated low-carbon (annealed)	100	impractical	
Galvanized sheet	91	impractical	
Hot-rolled (pickled drawing)	100	80	
Hot-rolled (copper bearing)	88	51	
Hot-rolled (dead soft)	88	51	
Lead coated (long ternes)	100	impractical	
Manganese silicon (high tensile)	45	15	
High-carbon (40 per cent and up)	25	10	
Stainless Steels			
AISI-302	98	50	
AISI-304	98	98	
AISI-316	98	50	
AISI-347	100	100	67
AISI-430	100	80	
Nickel and Nickel Alloys			
Inconel	95	88	
Monel (cold-rolled, deep drawing)	100	96	
Monel (standard, cold-rolled)	95	80	
Nickel	100	100	86
Miscellaneous Metals			
Lead	96	90	
Pewter	100	99	
Zinc	100	100	94

* Courtesy Milwaukee Metal Spinning Co. Compares the metals in each group, giving the rating 100 to the metal which lends itself most readily to forming by the cold metal spinning process. Ratings shown under the title "Group" compare the most satisfactory metal in each group with 2S-0 aluminum alloy which is rated at 100.

Flame-Spun Forms. Tubular parts for 1/2 to 4-inch diameters can ordinarily be handled in this manner. However, ratio of tube diameter to wall thickness should not exceed 50 to 1. Ends can be spun closed and pressure-tight with flat, oval, elliptical, or spherical shape. Where desired, ends can be reduced in diameter and walls thickened for tapping or other machining. For tanks and similar pressure chambers, both ends can be closed. When design is hemispherical, the original tube length is shortened approximately 0.2 of the tube diameter.

Shear Spinning. For design purposes it is useful to understand the basis for the shear spinning process. Shear spinning is predicated on the sine law — the relationship of the half angle, a, of the part, wall thickness, t, and starting blank thickness, T. The equation is $T \sin a = t$, *Fig. 25.24.*

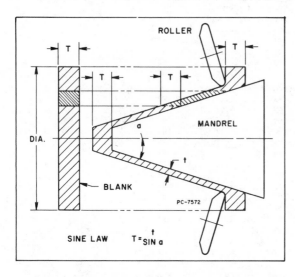

Fig. 25.24 — Conical shapes are usually spun according to the "Sine Law" t = T sin a. When the finished thickness t and side angle a are known, the blank thickness T can be precisely calculated by this law. Example: To spin a part with 0.125-inch thick wall at a 30° side angle, the beginning blank thickness T = 0.250 inches.

The simplest part for shear spinning is a cone with a uniform wall. Shear spinning can also form more complex shapes such as special conical shapes, *Fig. 25.25,* cones with tapering walls, curvilinear shapes such as nose cones with either uniform or tapering wall, and hemispherical and elliptical tank closures with either uniform or tapering walls.

Another variation of shear spinning is tube spinning. The sine-law relationship is not valid for this work. Instead, allowable wall reduction is based on the relationship between starting wall and finished wall thickness. Both forward and backward tube spinning can be done, *Fig. 25.26.* A wide variety of parts can be produced by this volumetric forming process to enhance spinning and offer new flexibility in design of special components.

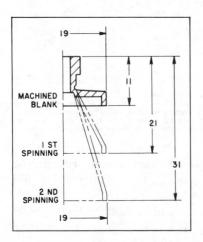

Fig. 25.25 — Starting with a high-alloy steel forging — machined, as shaded, to 3/4" thick at hub — the side wall was spun in two operations tapering from 0.225-inch at the hub up to 0.083-inch thick at the large end on a CINCINNATI 42" x 50" horizontal Hydrospin machine, with no heat treating between passes.

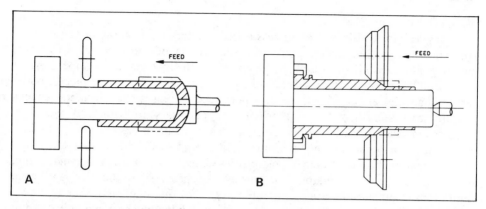

Fig. 25.26 — Cylindrical shapes or tubular parts are spun by starting with a shorter cylinder — machined, extruded, fabricated, drawn, cast, or forged. Sketch A illustrates the forward tube spinning technique where the metal is flowed ahead of the roller and in the same direction as the roller feed. This technique provides excellent control of accuracy on length, especially since the material is moved only once during the spinning pass. Sketch B shows backward tube spinning in which the metal is extruded beneath the roller and in the opposite direction of roller feed.

Selection of Materials

Practically any kind of sheet metal that can be formed or deep drawn can be spun. The most common of these are commercially pure aluminum, pewter, steel, copper, red brass, yellow brass, gliding metal, commercial bronze, tin, zinc, and nickel silver.

In addition to these common softer metals which are easy to spin, less ductile materials such as the chromium and chromium-nickel stainless steels, aluminum alloys, bronzes, nickel, Monel, Inconel, molybdenum, steel alloys, and high-carbon steels can be successfully handled by gradual breakdowns and intermediate annealings. Number of breakdowns, *Fig. 25.17*, is entirely dependent upon the rapidity with which a particular material work-hardens in processing. Chromium stainless, for instance, should range from 12 to 16 per cent chromium. Amounts of 17 per cent or greater are unsuccessful. The chromium-nickel stainless steels though higher in elongation, work-harden more rapidly and require frequent annealings.

Gage Limitations. In most cases, at least at present, spinning of the softer metals is limited to gages up to $1/4$-inch, low-carbon steels up to $3/16$-inch, cold-rolled steels up to $5/32$-inch, stainless steels up to $1/8$-inch and nickel alloys up to $5/64$-inch. Obviously, extremely large or deep parts cannot be spun from very light-gage materials. Deep parts require plenty of metal to work. Thus, for instance, the minimum gage for 24SO Alclad parts up to 2.5 inches deep is 0.032-inch, for parts 2.5 to 4 inches deep the minimum is 0.040-inch, and for parts 4 to 8 inches deep the minimum gage is 0.051-inch. Deeper parts requiring heavier gages should not be spun from this material. Gages as thin as 0.020-inch in 3SO aluminum alloy can be spun into parts up to 2.5 inches deep. For parts 2.5 to 4 inches deep the minimum gage is 0.025-inch, depths from 4 to 8 inches require a minimum of 0.032-inch, and 8 to 12 a minimum of 0.040-inch. Aluminum under 0.010-inch ordinarily cannot be handled successfully regardless of the alloy.

Perforated Metals. Any of the metals that can be spun can also be spun after perforating. Even expanded metals can be handled with good results when required for special designs, *Fig.* 25.5.

Magnesium. Unlike the other less ductile materials, magnesium can only be spun in small amounts and extremely shallow shapes while cold. However, spinning is successfully accomplished by working the material at a temperature of 450 to 600 F. Wooden forms or chucks are satisfactory for a few pieces, but if larger quantities are required, the cost per piece increases because metal chucks (usually cast aluminum or magnesium) are necessary to withstand the heat. Magnesium sheet 0.032-inch thick by 36 inches in diameter has been hot-spun into parabolic propeller spinner cowls 16 inches in depth by 24 inches in diameter at the base, indicating the wide range of possibilities with this material. As a general rule, if proper heat can be maintained during the processing, the size and gage limitations of magnesium are comparable to that of the other light sheet metals.

Molybdenum. As with magnesium alloys, molybdenum sheet can only be spun successfully while hot. The hot working temperature range is about 100 to 300 C depending upon gage. Ordinarily sheet up to 0.032-inch can be spun.

Beaded Forms. In wiring — beading a wire into a rolled edge — *Fig.* 25.19, use of a wire of the same analysis as that of the material from which the spinning is made is advisable. Otherwise the wire must be protected by some form of coating to prevent galvanic corrosion or precipitation action. Where steel wires are used in conjunction with other metals, especially stainless steel, present practice is to tin the wire and solder-seal the bead.

Spinability of Metals. To indicate the relative characteristics of various metals for conventional spinning, a good overall comparison, Table 25-II, is included. The rating 100 is given the material in each classification which is best suited and lends itself most readily to forming by cold metal spinning. The lower the rating, of course, the higher the production cost. Figures are based on actual experience and will vary somewhat with design, size, metal gage, handling, etc. In Table 25-II the ideal materials of each group are rated against aluminum in the same manner to show comparative adaptability. Here again variations will be found owing to the characteristics mentioned above.

Special Metals. A great many other less widely-used metals can be spun, but are only specified in unusual cases where special properties are necessary because of the special handling technique required and the high material cost. These include gold, silver, platinum, Kovar, Invar, ilium, and britannia or white metal.

Shear Spinning Metals. A wide variety of ferrous and nonferrous materials can be handled in shear spinning machines, many without heating between passes. These include aluminum alloys, steel and steel alloys, stainless alloys, etc. Ordinarily, material thickness may run from 0.20-inch to 0.80-inch in mild steel, 0.150-inch to 0.70-inch in stainless and 0.47-inch to 1.60-inch in aluminum.

Specification of Tolerances

During the spinning process the thickness of the metal decreases depending in most cases on the ability of the spinner to control the flow of metal. A good average for reduction in thickness at the edge of deep-spun cups is about 20 per cent. In other words, to retain a minimum wall thickness of 16-gage, the original sheet would be one of the next heavier gages depending on the amount of reduction which varies with different shapes, materials, sheet thicknesses, and metal tempers. Likewise variations in wall thickness — usually less than 25 per cent on good

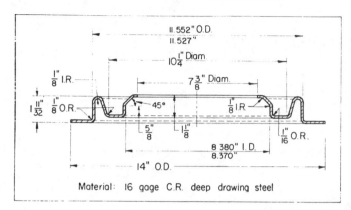

Fig. 25.27 — Spun from 16-gage steel, this dust shield is representative of close tolerances held on fitted parts. Drawing, courtesy Milwaukee Metal Spinning Co.

commercial quality spinnings — are dependent upon both the material and the spinner. Some materials, with proper handling, can be held to within 10 per cent of the desired gage.

Wall Thickness Variation. An excellent example of this effect of gage variation on tolerances is the spinning of a full 24-inch inside diameter hemispherical jacket, such as that shown in *Fig.* 25.15, from stainless steel with a minimum wall thickness requirement of 14-gage. Reduction in thickness at the edge of this shape proved to be such that 10-gage stock was necessary to assure the required minimum wall thickness. This heavy-gage stainless may be beyond the capacity of conventional spinning lathes and thus 11-gage plain deep drawing steel substituted for this part brings the job into the practical commercial spinning range.

Dimensional Variations. Notwithstanding these normal variations in material thickness, unusually close dimensional tolerances can be held on fitted work whenever it is found necessary. Outside diameter tolerances of plus or minus 1/64-inch are commonly held on small work, plus or minus 1/32-inch on medium-size work, and about plus or minus 1/16-inch on large sizes. Tolerances of plus or minus 1/64-inch are, where essential, generally possible for any size part without affecting the cost unduly. Less economical, but entirely within the scope of the process, tolerances of plus or minus 0.005-inch can be specified and held. *Fig.* 25.27 is a good example of a fairly large dust shield where such tolerances have been successfully held in production.

Fitted Surfaces. Critical fits should be so designed as to allow the fitted surfaces to be spun against the form. Wall thickness variations thus have no effect on fit and slight tool marks are immaterial. In any case, dimensions should be shown as applying to the part surfaces which contact the form contours. However, shear spun parts can be held to plus or minus 0.002-inch on ID and wall thickness although plus or minus 0.005-inch is more practical on final wall thicknesses. Surfaces formed on the mandrel are in the range of 5 to 20 microinches finish. Those formed by the tool range upwards from 20.

Length and Depth. Tolerance requirements naturally should be as liberal as is practicable on overall length or depth as well as on flange squareness. Accurate flange angles can be maintained but if a flange periphery can vary plus or minus 1/16-inch from true position, cost can be minimized. However, on length or depth of spun machine parts, a maximum variation of plus or minus 1/32-inch can be held and on occasion, where the design permits, smaller tolerances to perhaps plus or minus 1/64-inch can be maintained without excessive difficulty.

26

Brake Forming

RELEGATED primarily to the production of machine parts which are extremely long in relation to their cross section, press brake forming fills much of the production void between stamping and roll forming. The press brake was originally conceived to eliminate the limitations and expense of wide-bed presses and today offers a highly versatile, economical means for producing parts from sheet and plate which necessitate a long, narrow forming bed.

Scope of Method. Although the press brake is largely used and best suited for long bending or forming operations not adapted to regular presses, *Fig.* 26.1, a wide variety of blanking, punching, notching, coping, lockseaming, beading, wiring and similar operations is readily accomplished. Machine parts produced cover a tremendously varied field including such components as shells, frames, tubular sections either uniform or tapered in length, cabinets or housings, structural sections of almost any cross-section design, etc. Generally, the work must be such that it can be produced on a long, narrow press bed with a relatively short ram

Fig. 26.1 — Final stage in press brake forming a long tubular section. Photo, courtesy Cincinnati Inc.

Fig. 26.2 — Rib sections, 18 feet long, being formed from heavy plate on a 1500-ton hydraulic brake. Photo, courtesy Cincinnati Inc.

Fig. 26.3 — Press with a 42-inch special bed indicates the versatile nature of the press brake. Photo, courtesy Cincinnati Inc.

stroke, *Fig.* 26.2, and is normally confined to sheet metal or plate material although blanking of insulation board, pressed wood, rubber and similar materials can be done.

Machines Available. Many types of press brakes are now available, some being more accurately termed open-front presses. These range in bed width up to 42 inches, *Fig.* 26.3, with some special presses even greater. Press sizes range from small machines with a capacity around 20 to 25 tons for light bending — up to 4 feet in length with 14-gage sheet — to around 2000 tons for heavy work. Bed lengths respectively run from 6 to about 30 feet. A standard 900-ton press, *Fig.* 26.4, with V-die bending will handle 1-inch plate 12 feet long, 3/4-inch plate 20 feet long, or 1/2-inch plate 30 feet long. Special presses are available, however, which are capable of handling heavy plate over 50 feet in length. Where this extra length is necessary, presses are often used in tandem, *Fig.* 26.5, to achieve the desired results.

Limitations and Advantages. Ram stroke required in the forming of parts must be relatively small and generally should not be over 6 inches; the longest stroke available in present machines being approximately 8 inches. Standard brakes, however, are usually made with a 3 or 4-inch stroke. Simple adjustment provided on the ram and means for exactly recording such adjustment for duplication makes possible the forming of many parts not practical by any other method. This is especially true with work which must rise in front of or behind the dies during forming, *Figs.* 26.1 and 26.4. Again, work is not limited to the machine ram length, inasmuch as parts properly designed can be progressed through in stages when necessary by virtue of the open construction at either end of a press brake, *Fig.* 26.6. Where extremely wide work is to be produced, design of the part must be such as to allow the load to be reasonably well centered under the bed and ram.

Fig. 26.4 — Bending a heavy machine base with a 900-ton brake. Photo, courtesy Cincinnati Inc.

A press brake is considerably different from either a press or a roll former from the standpoint of tool cost. Most of the factors which are necessary to consider before tooling the latter machines need not be considered when designing for the press brake. A few standard dies will produce a great variety of parts and are relatively inexpensive, *Fig.* 26.7. By the use of "air bending", one set of acute-angle bending dies will form angles of any degree required by merely controlling the distance to which the punch enters the die opening.

Fig. 26.5 — Two 2000-ton press brakes being used in tandem to form tapered octagonal sections and similar parts up to 56 feet long from heavy steel sheet. Photo, courtesy Cincinnati, Inc.

For parts not suited to stamping, roll forming or similar methods, production on the press brake is highly economical even though in certain cases quantities may run to fairly high figures. Naturally where design and quantity are well suited to

Fig. 26.6 — Open-end construction of standard press brake makes practicable the forming of extra long parts. Photo, courtesy Cincinnati Inc.

either of these other methods, lowest costs and highest output speed can be realized.

Production Capacity. Standard brakes operate between about 5 and 50 strokes per minute. Handling of parts, however, sets the actual production capacity. Extremely large parts such as shown in *Figs.* 26.4 and 26.5 would probably reach only a few parts per hour with smaller parts, in some cases with automatic feed, production output may average around 500 per hour.

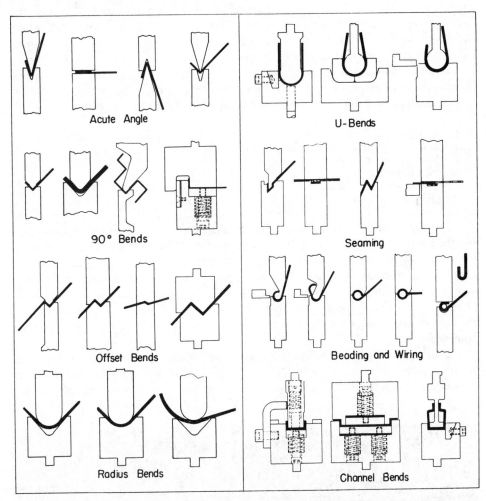

Fig. 26.7 — A group of representative press brake forming die sets which illustrate the multitude of forms which can be produced. Drawings, courtesy Cincinnati Inc.

Multiple operations also can be used on one machine to achieve rapid production, as many as ten separate functions being possible. Owing to the simplicity and low cost of the tooling on press brakes, trimming, punching, forming and similar operations are regularly performed in this manner in the production of cabinets, doors, tops, frames and panels on refrigerators and like equipment, *Fig.* 26.8. Too, by this method unusual freedom can be had in the way of design improvements since changes in tooling would incur relatively little cost. An outstanding illustration of this progressive production setup can be seen in *Fig.*

26.9, which shows the various steps in producing cabinets shown in *Fig.* 26.8. Not only economy in manufacture is achieved but also the merits of interchangeable mass-produced parts.

Design Considerations

Insofar as design of parts is concerned, only a few general rules can be applied. Press brakes and dies for performing ordinary as well as an unlimited number of special-purpose operations are extremely flexible and provide great freedom for the designer. The cut sections shown in *Fig.* 26.10 are but a small number of the overall range of possible design variations and indicate the general versatility.

General Rules. A number of normal limitations should be observed in design in order to simplify production and achieve low cost of parts. Several points which should be kept in mind and which have been indicated previously are:

1. Design should be such that work or operations to be performed will be fairly symmetrical under the centerline of the ram, meaning front to rear.
2. Material to be employed should be in sheet or plate form, preferably under 1-inch in thickness.
3. Design should be such that the stroke available on standard press brakes will give adequate clearance for easy and rapid handling.

Design for brake forming normally should avoid the necessity for making any but the simplest, shallow draws. Wherever possible, drawing should be avoided by using notched corners and welding, by adding welded ball corners, etc. Consideration should be given the distances between notches, holes, etc., so that simple tooling can be used as well as single-stroke production allowed. Since these conditions vary somewhat according to press size and design, doubtful designs should be discussed with the manufacturer or production department prior to completion.

Progressive Operations. Certain parts can be designed so as to lend themselves to progressive operations, *Fig.* 26.11, especially where quantity will warrant expenditure on a special die arrangement. For example, where two operations are

Fig. 26.8 — Cabinet parts produced on a press brake in relatively limited quantities. Photo, courtesy Cincinnati Inc.

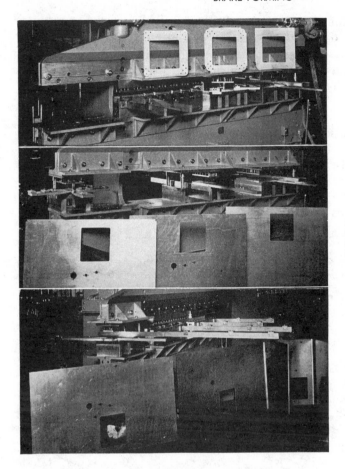

Fig. 26.9 — Progressive setup for economically producing cabinet parts shown in Fig. 26.8. Photos, courtesy Cincinnati Inc.

planned, these may be formed in double-deck dies in one handling. Here it is essential, however, that one operation be at the edge of the sheet so that it can be performed in the upper die, and the stroke required on both forming operations should be such that the press stroke will handle the two with reasonable room for removal, *Fig.* 26.12.

Flanges. Where length of work contemplated is well within the clear distance between the brake housings, there is relatively no limit to the size of flange that can be formed, but when parts require the total overall die surface or more, flange width is limited by the depth of the throat of the brake. For instance, the clear distance available on a 6-foot brake is about 4¹/₂ feet, that of a 10-foot machine is about 8¹/₂ feet, that of a 20-foot is about 16¹/₂ feet, etc.

The available throat clearance of presses up to 150-ton and 14-foot length is about 8 inches, from 200 to 250-ton in lengths from 8 to 14 feet is about 10 inches, and from 335 to 900 ton and 8 to 20-foot lengths is about 14 inches. Parts which can be formed at the extension as in *Fig.* 26.9, of course, are relatively unrestricted and can be completely closed if desired.

Bend Radii. Forming capacities of brakes are based on "air bends" where the material contacts only two lines on the female die and the extreme point of the male die. Such bends require the least pressure and normally are made on a V-die with an opening 8 times the metal thickness which results in an inside radius approximately equal to the metal thickness. Sharper radii necessitate the use of "bottoming" dies and a considerable increase in pressure, consequently radii less than gage thickness should be avoided if at all possible. Likewise, a female die opening of less than 8 times the metal gage requires increased forming pressure as the opening decreases. Thus, since the minimum flange that can be formed on any die is approximately one-half the die opening plus two times the metal thickness, flange dimensions should be based on the ideal die opening of 8 times gage or, when possible, made greater. Flanges smaller than these necessitate the use of confining forming dies similar to those used in stamping and naturally incur greater cost per piece. With plate over 1/2-inch in thickness and materials with tensile strengths greater than mild steel, die openings should be considerably greater than 8 times gage — thus radii much greater than the gage should be specified — to prevent the tendency toward fractures.

Blank Size. Numerous methods are employed to calculate the blank size of

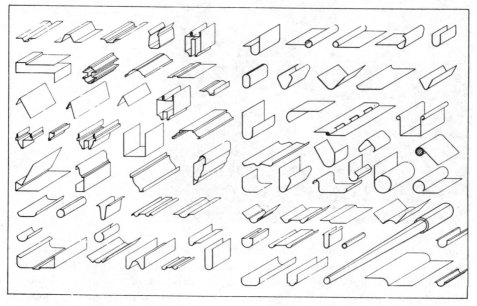

Fig. 26.10 — A representative group of cross sections showing the general type and range of brake operations. Drawings, courtesy Verson Allsteel Press Co.

metal, especially where 90-degree bends are used. Most are only empirical and depend upon uniform gage, hardness, direction of grain, and speed of forming. For accurate results, especially with large radii or unusual shapes, actual check is recommended. Developed width of simple sections and multiple-bend sections can be found from the formulas shown in *Fig.* 26.13. Note that simple bends having corner radii approximately equal to the gage are usually dimensioned from the inside and multiple bends having radii not equal to the gage are dimensioned from the outside. When the radii are over 3 times gage the blank width should be calculated on the neutral axis of the metal as little stretching occurs.

Holes and Hole Forms. As previously mentioned, a wide variety of punched, blanked, pierced, notched, formed, and combination designs can be produced on the press brake. Standard die sets as well as special ones are available in separate units which can be attached to the press, *Fig.* 26.9, in whatever relative relation desired to achieve an almost unlimited variety of designs. A few of these are shown in *Fig.* 26.14 to indicate the range available. Plain holes in nominal sizes from $1/32$ to $1 1/2$-inch can be specified in almost any arrangement. However, spacing between holes or hole to edge of the sheet should usually be $1/2$-inch or greater though in certain cases this can be reduced to about one thickness of the stock used.

Selection of Materials

In selecting materials for parts to be brake formed, strength and cost govern the selection to a great degree. Problems encountered in forming various materials are largely the same as those found with other forming processes. Mild steel is by far the most common material employed and is well suited to brake work.

Temper. Materials having a high degree of hardness present considerable difficulty in springback variations due to gage and temper variations. Hard cold-rolled sheet, high-tensile steels, heat-treated aluminum alloys and like materials usually offer many difficulties in brake forming which naturally involve increased production costs because of more expensive dies, continual adjustments, etc. For example, in forming the corrugated panel, *Fig.* 26.15, a great many adjustments were necessary on the die, especially when sheets were from different heats. Too, gage variations on thin materials, in this case 0.001-inch on 0.050-inch sheet, required a ram adjustment of 0.005-inch to hold pitch dimensions. Such

Fig. 26.11 — Gondola car stakes being formed progressively. Die at the right punches the holes, die at the left front makes the first forming operation while that in the rear finish forms. Photo, courtesy Cincinnati Inc.

difficulties, of course, influence production speed considerably.

Standard Dies. Press brake dies built for common mild steel seldom are satisfactory for other materials such as terne plate, zinc, bronze, copper, etc., without auxiliary operations or hand work. However, where quantities required do not justify special dies for a change in material, this procedure is employed as most satisfactory and economical.

Bending Considerations. Bending pressure varies directly as the tensile strength regardless of the type of material — straight-carbon steel, high-tensile alloy steel, stainless steel, etc. Where materials whose tensile strength is greater than that of mild steel are used, radii larger than the gage should be contemplated

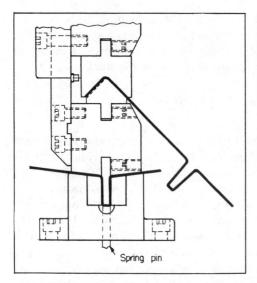

Fig. 26.12 — Cross section of a double-deck die for performing two operations.

Fig. 26.13 — Formulas and data for determining the blank width of single and multiple-bend sections where the bend radii are equal and not equal to the gage. Drawings, courtesy Cincinnati Inc.

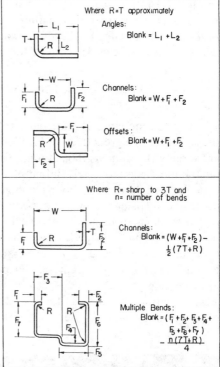

Where R = T approximately

Angles:
$$Blank = L_1 + L_2$$

Channels:
$$Blank = W + F_1 + F_2$$

Offsets:
$$Blank = W + F_1 + F_2$$

Where R = sharp to 3T and n = number of bends

Channels:
$$Blank = (W + F_1 + F_2) - \tfrac{1}{2}(7T + R)$$

Multiple Bends:
$$Blank = (F_1 + F_2 + F_3 + F_4 + F_5 + F_6 + F_7) - \frac{n(7T + R)}{4}$$

to ease the bending requirements, especially where parts are long. High-tensile steel plate, even in the thinner gages, usually must be formed over die openings ten to twelve times gage to prevent fracture. Often special flanging or boiler steel is employed to avoid breakage when heavy-gage material must be bent over a small die opening for proper radius.

Specification of Tolerances

As usual, consideration must be given to keeping tolerances no closer than the method of manufacture makes practicable. It will be readily recognized from the foregoing discussion that tolerances readily held in economical production on the press brake usually must be considerably wider than those normally specified in machining, stamping or roll forming.

Simple Sections. Unless relatively large quantities are involved and expensive special dies are warranted, tolerances on simple sections up to about 6 inches should be approximately plus or minus 0.031-inch, especially where length is considerable. Naturally where short parts are produced, say up to 6 feet in length, closer tolerances can be held but are dependent to a great extent upon uniformity of gage, temper, etc. Large sections or complicated designs, *Figs.* 26.4 and 26.5, for economical reasons should have wider tolerances, say plus or minus 0.062-inch.

Formed Sections. Parts completely formed in the dies, *Figs.* 26.2 and 26.9 can be held to limits normally expected in the usual run of similar stamping or drawing operations, say plus or minus 0.015-inch average. Dimensions of sections such as

Fig. 26.14 — Some of the various shapes and hole combinations which can be perforated with available standard dies. Drawing, courtesy S. B. Whistler & Sons, Inc.

Fig. 26.15 — Long, deep-corrugated aluminum panel brake-formed in large quantities. Photo, courtesy Cincinnati Inc.

the channels, bottom *Fig.* 26.7, will be as accurate as the dies but require allowance for springback.

Blanked Forms. Accuracies of blanked and pierced parts such as shown in *Fig.* 26.9 will depend upon the type of die used; wide tolerances allowing the use of low-cost "temporary" or "adjustable" dies in many cases. Normally, this would set the practical tolerances for blanked parts at, perhaps plus 0.10-inch minus zero with small parts and greater as size and gage increase.

Holes. Tolerances on pierced holes follow usual practice. Wherever possible hole diameters should be plus zero and minus 0.010-inch or greater. Hole spacings should be plus 0.010-inch minus zero or greater if possible, although in both cases extra care and cost will often permit somewhat closer limits.

Formed and Sheared Lengths. Length tolerances on parts with formed ends should always be generous. Although plus or minus 0.015-inch can be held in some cases where parts are short and of light gage, specification of at least plus or minus 0.031-inch on small parts, 0.062-inch on medium parts and 0.125-inch on large parts will afford maximum economy and ease in handling. Naturally, tolerances for parts sheared to length prior to forming will depend upon the care and type of shear used, though here again wide tolerances are conducive to maximum economy.

27

Roll Forming

Henry S. Royak

$\mathbf{C}$OLD Roll Forming parallels somewhat the stamping and drawing processes in its effects on the economical design of mass-produced products. Early fields for the use of cold roll-formed shapes have broadened during recent years to include automotive, aircraft, appliances, structurals, bicycles, and many other facets of our everyday lives. The first volume use of roll-formed parts began with the rapidly expanding automotive industry in the early twenties and has increased to the point where its outstanding versatility places it among the foremost mass production methods to be considered by the product designer.

Scope of Process. Aside from the decorative trim, finishing, architectural, and similar uses too numerous to mention, the uniformity, adaptability to assembly by modern welding methods, fine finish and low cost of roll-formed shapes have projected their use into the vast field of basic structural parts. Designers of

Henry S. Royak is Product Manager, Roll Forming, The Yoder Company, Cleveland, Ohio.

Fig. 27.1 — A group of representative cross sections showing the wide range of shapes which have been produced. Photos, courtesy Yoder Co.

Fig. 27.2 — Three sizes of standard roll forming mills with stock capacities of 0.063, 0.094 and 0.125-inch thickness and widths of 6¹/₂, 12 and 14 inches respectively.

automotive, farm, railroad, truck, bus, and aircraft equipment are rapidly taking advantage of the increased strength and decreased dead weight afforded by sections roll formed from alloy-steel sheet. The rigid requirements of design, weight, strength, finish, and cost are met by taking the flat steel sheet — hot or cold rolled — produced by the continuous mill in coils, slitting it into the necessary widths, and then cold roll forming these coil strips into the desired sections, *Fig.* 27.1.

Cold Rolling. Not unlike its larger steel mill counterpart, the roll forming mill utilizes a series of roll stations to progressively form a section to the desired shape. The mill takes the flat sheet or strip and in easy stages transforms it into a formed section by means of specially designed contoured rolls. Unlike the hot rolling mill, however, the metal gage or section thickness remains constant, only bending taking place.

Machines Used. Numerous machines are available for roll forming sheet materials and may be equipped with a few roll stations or as many as twenty or more depending upon the job, *Fig.* 27.2. Individual rolls on all stations are removable and only the number of pairs necessary for producing a particular section are used. As a strip passes through the machine each pair of rolls produces only a partial change in the cross section, consequently the number of rolls required depends upon the intricacy or general nature of the section, the type of material being formed and its gage. A simple angle or channel section can usually be formed in three or four pairs of rolls, the more difficult sections frequently require ten or more.

Two basic styles of cold roll forming machines are available. One type has overhung roll shafts, *Fig.* 27.3, both bearings of which are mounted within the drive

Fig. 27.3 — Ten-station cold roll forming machine having overhung roll shafts. Photo, courtesy Rafter Machine Co.

Fig. 27.4 — Rear view of a mill that will cold form shapes from plate up to 0.500-inch. Roll spacings on such mills preclude production of light sections.

housing. The second type, *Fig.* 27.2, has a double housing with inboard shafts. The overhung or outboard-shaft machine can, in many cases, be set up for production in the shortest possible time. However, on long production runs, setup time has little significance in the overall costs. The fundamental difference, therefore, is not in setup, tooling, or function but rather in rigidity, versatility and capacity and these three factors determine the usefulness and efficiency of the one type over the other.

Where extremely light or narrow width sections are required, the overhung shaft type machine is entirely satisfactory. On the other hand, if versatility is necessary or sections contemplated for future production would tax the maximum capacity, inboard shafts are desirable to maintain accuracy and economy.

Practical Machine Capacities. Sections can be roll formed cold from almost any metal that can be cold formed in widths from a fraction of an inch to 80 inches. Gage thickness of materials successfully handled to date ranges from 0.003 to 0.750-inch. A mill that will form structural and special shapes from sheet and plate

Fig. 27.5 — Detail view of cold roll forming head showing coated sheet being formed. Detail view of forming roll stand shows micrometer dial in upper right corner.

Fig. 27.6 — After forming tube, round electrodes weld open seam. Tube forming mills operate at speeds up to 300 feet a minute.

up to 1/2-inch in thickness is shown in *Fig.* 27.4. However, practical gage is 3/8-inch. The cost of machine and rolls after 3/8-inch is high.

Although the rolls on a machine can be adjusted to accommodate wide range of material thicknesses, *Fig.* 27.5, the indication is primarily one of versatility and adjustability. One set of rolls for any forming mill is, of course, not recommended for economical production over such a wide range of sizes inasmuch as a mill for roll forming 1/2-inch plate would be far too cumbersome and costly to operate for forming lightweight and smaller sections.

Production Speed. Early mills produced sections at a rate somewhere between 25 and 30 feet per minute. However, improvements and demand brought this speed up to around 75 feet per minute in the twenties. Today, most mills produce at speeds around 100 feet per minute although some have been operated at speeds as high as 300 feet per minute. Actual production speeds generally range from 50 to

Fig. 27.7 — Automatic cutoff and coiling unit for manufacturing bicycle wheel rims, and other circular sections.

200 feet per minute. Rolling at a rate of 100 feet per minute, mills equipped with automatic flying cutoff equipment have an average output in production of 80 feet per working minute or 4800 feet per hour and thus a single machine will produce more than 10,000,000 feet per year. These figures include downtime for roll changes, adjustments, etc.

Maximum economy in production, therefore, is dependent upon large quantity. Versatility is a necessity in most cases and one machine with five or more sets of tooling can accommodate short runs of 50,000 to 100,000 feet per year in many cases. Tooling upkeep is not great and some three to four million feet of material can be produced on the average set of rolls before regrinding is necessary. As a rule, several regrindings can be made before rolls must be replaced.

Where vast quantities are to be produced, single-purpose mills are used to advantage. Automatic piercing or blanking machines as well as automatic cutoff saws or flying shears can be used in the production setup. Use of automatic flying cutoffs has increased production almost 100 per cent over the older methods (running off the longest convenient length and then stopping the rolls while a swing or similar saw cuts off to length). Likewise the flying cutoff eliminates an extra press trimming operation.

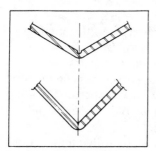

Fig. 27.8 — Creasing of the metal in forming is usually necessary to obtain sharp corners and is best avoided.

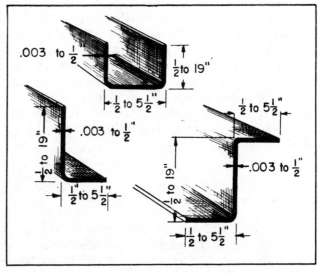

Fig. 27.9 — A number of simple sections produced on a mill with adjustable rolls.

Automatic Production. A good example of automatic section forming is the production of continuous piano hinge. Stock is fed from a coil to an automatic machine which pierces or notches the strip continuously and accurately. The notched or pierced strip is then roll formed complete and passed through a flying shear where it is cut to exact length. Completely automatic, the setup produces at a rate of 60 feet per minute.

Welding. In conjunction with the roll forming operations, welding can be used to great advantage on single-purpose machines, *Fig.* 27.6. The manufacture of electrically welded tubing bears this out. Edges of a steel strip are prepared for welding and passed through the forming rolls directly to the welding unit. The hot welding bead is trimmed off as the tube leaves the welder and passes through a cooling tank to a sizing and straightening mill and an automatic cutoff in one continuous operation. Inert-arc welding is used for stainless and high-alloy steels with production speeds from 10 to 90 inches per minute. Using 60-cycle or 180-cycle alternating current resistance welding on carbon steels, speeds up to 300 feet per minute have been attained.

In addition to welded tubing, roll formed nonwelded tubing can also be produced. Most nonwelded tubing is lock-formed, but substantial amounts are produced by butting two edges of the strip together. Aluminum T.V. antennae is a good example of butted, nonwelded tubing. It should be noted, however, that T.V. antennae tube is also welded.

Ring Forming. Machines can be equipped with a coiling attachment at the exit end for the rolling of ring sections in various diameters for ventilators, instrument bezels, bicycle wheels, light rims, guards, furnace rims, etc., *Fig.* 27.7, thus eliminating an extra operation. Painting devices can also be used.

Design Considerations

Although design variations in parts which lend themselves to this method of production are legion as exemplified by the sections shown in *Fig.* 27.1, there are a few limitations which designers should consider in planning new parts.

Corner Radii. Small corner radii on sections require sharp corners on the rolls and thus useful roll life is shortened considerably. For better forming characteristics and maximum roll life, the ratio between the radii of bends and the gage of the material should be kept as large as possible.

Many soft and annealed sheet materials can be successfully rolled into sections with minimum bend radii of approximately 75 per cent of the gage thickness. Good practice indicates that minimum radii should be at least equal to the gage and preferably two times gage thickness. Where tempered or tough materials are used radii must be greater. Recommended practice for various aluminum alloys, for instance, is as follows:

24ST	2.5 x gage	24ST81	4 x gage
24SRT	3 x gage	24ST86	5 x gage
R301T	3 x gage	75ST	5 x gage

Sharp corners can be made in soft materials by a grooving or beading operation, *Fig.* 27.8, and the groove must penetrate the metal from $1/3$ to $1/2$ the thickness. The main disadvantage of such bends, however, is the section weakness created. Forming of such corners is limited to gages over 0.031-inch.

Section Size and Gage Limitations. Sections of thin material should be kept as shallow as possible so that roll diameters can be as small as possible and less-expensive equipment necessary. The sections shown in *Fig.* 27.9 will give an excellent indication of the work that can be done in a machine equipped with adjustable rolls. This wide range of sizes, however, would not be considered for economical mass production on one machine only.

As previously mentioned, sections from a fraction of an inch to as much as 80 inches wide can be produced. Gages which can be handled range from 0.003-inch to 0.750-inch. Commercially, shapes up to 4 inches deep from sheet up to 40 inches wide can be most readily obtained; the heaviest gage for simple shapes generally being 9 and the heaviest for complex shapes being 11.

Lengths. Short pieces of material (lengths less than twice the center distance between spindles, usually 8 to 10 inches with a minimum on the smallest machines of about 4 1/2 inches) do not feed through a machine conveniently. Thus, parts adapted to continuous forming from coil stock strip with automatic cutoff (also blanking where necessary) are most economical. Length is unlimited. Only the available length of roll or flat sheet stock sets the practical length of section, but even then it is possible to weld successive coils together to produce virtually unlimited lengths.

Rolling Sequence. An all-round understanding of the factors which enter into the design of parts to be roll formed must of necessity include some details on the design of the rolls used. Good judgment is an important factor contributing to successful part and roll design. The process is not unlike that found in progressive die forming. If a shape has four or five breaks or bends across its width, it would be

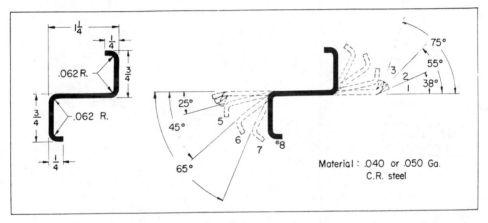

Fig. 27.10 — A Z-section layout and breakdown of the forming angles. The 65-degree flange and 75-degree leg bends are used primarily to allow clearance for rolling the leg corner radius.

logical to distribute the work of bending first of all singly and, at that, only in increments of 20 to 25 degrees at a time. After examining the sequence of working, *Fig.* 27.10, some of the steps often can be combined so that the number of operations or roll stations are at a bare minimum.

Bending Stages. By slowly unfolding or unforming a specified shape in the foregoing stages, the general design of each roll pass is determined, *Fig.* 27.11. The rule of restricting the work of bending to 20 to 25 degrees per roll pass is a conservative figure and will usually work successfully for a great many varying hardnesses and types of metals, except special aluminum alloys and stainless steel, to produce a section with minimum twist or distortion. Work of bending at the first pair of rolls may run to angles as great as 45 degrees but that at the remaining stations is usually less. As a section progresses through a machine, the amount of work done at successive stations is decreased to achieve accuracy.

Forming Angle. The largest possible satisfactory angle of bending at each stage must be modified somewhat by the forming angle — that compound angle

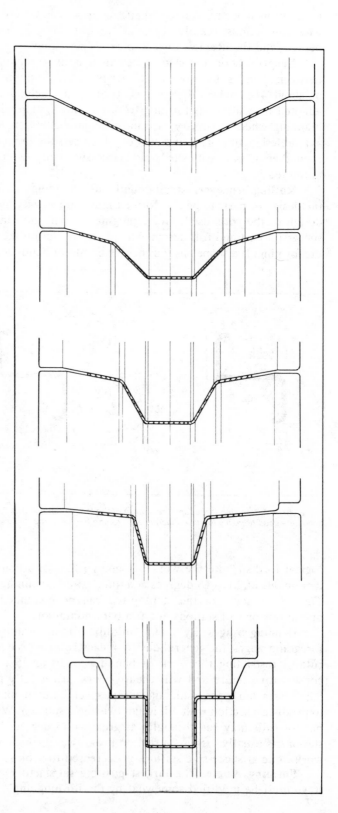

Fig. 27.11 — Sequence of roll stations showing flanged U-shape taking form progressing from the outside inward.

described by the edge of the sheet as it passes from one roll station to the next — which is limited by experience to an amount such that the stretch induced in the edge of the strip does not exceed the elastic limit of the material being rolled. Thus, the depth of a section may modify the maximum bending angle and actually determine the number of roll stations required. The total rise of the edge of a section through the mill or, in other words, the actual depth of the part which becomes the leg of a right triangle in conjunction with the horizontal spacing of the spindles, determines the approximate number of roll passes.

Twist. Bending or twisting during rolling which often results from slight inaccuracies or variations in the structure or hardness of the strip material is usually overcome by means of a straightener provided at the exit from the rolls, *Fig.* 27.12. Bow or twist resulting from internal stresses in the section which are not

Fig. 27.12 — A hat section emerging from the rolls, showing the straightening mechanism used to eliminate twist.

properly balanced about the neutral axis can sometimes be neutralized by preforming the strip in the first pair of rolls. This preform action sets up stresses opposed to those which produce the twist by creating a stretching effect on both sides of the neutral axis, *Fig.* 27.13. This method often can be used to eliminate undesirable springback in the harder metals.

Edge Conditioning. Where the sheared edges of strip being formed require conditioning or trimming, this can be done in the machine without additional operations. Where sharp sheared edges should be removed, the design should so indicate.

Embossing. In rolling, shapes can be embossed with designs, names or patterns. Where embossing must cover the entire width the operation must be performed in the flat at the first roll. Naturally in rolling, the sharpness of the pattern is lost to a varying degree. If the pattern must be sharp or of considerable depth, the operation must be done in the last roll to avoid the flattening action. The inherent problems and cost increase in such cases, however, is apparent and should be avoided if possible.

Selection of Materials

Practically any material that can be formed or bent by any other process can be roll formed, and frequently it is possible to roll form a section so intricate that it could not be produced economically by any other continuous process. The harder temper materials have a tendency to crack under severe high-speed forming,

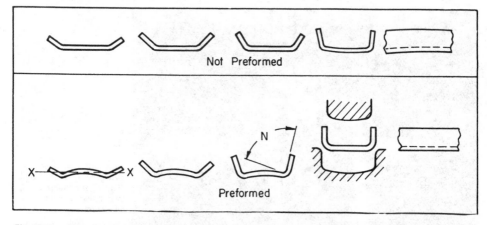

Fig. 27.13 — By preforming the strip at the first pair of rolls, stretching is produced on both sides of the neutral axis of a section to avoid twist, bow and springback. Angle is made less than 90 degrees, the exact amount depending upon the material characteristics.

consequently it is desirable as well as economical to use the more ductile or formable materials available. The extremely soft materials such as the nonaging aluminum alloys, copper, brass, and low-carbon steels can be handled successfully as well as magnesium-clad and other clad alloys. Stainless and high-alloy steels having a tensile strength of from 180,000 to 200,000 psi can be cold roll formed in high-speed production and, as previously noted, can be welded during the forming process. Perforated as well as notched or punched materials also can be rolled.

Material Combinations. In addition to the rolling of clad metals, several thicknesses of plain sheet or combinations of different metals can be formed simultaneously. Thin metal covers can be rolled over base metal in the same forming operation. Thin stainless or brass in gages as light as 0.004 to 0.006-inch can be rolled face-to-face with heavy-gage, low-cost steel to form almost any shape. Where necessary, fillers such as paper, fiber, felt, and rubber can be rolled-in during the forming sequence, *Fig. 27.14.*

Coated Stock. While most sections require no coating or plating, such materials can be handled. Electrogalvanized, prepainted or plastic coated stock is often used. Normally, rolling does not damage plating if proper lubrication and roll setup is used.

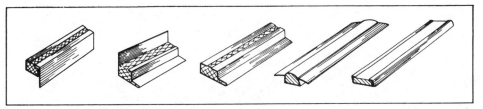

Fig. 27.14 — Fillers of almost any variety can be rolled-in during forming. Drawings, courtesy Tishken Products.

Specification of Tolerances

Uniformity or accuracy in the size of sections produced naturally is limited by the uniformity of the sheet material used. Cold rolled sheet material of course has the best surface finish and rolling characteristics as well as closest gage accuracy.

Surface. Runout in machine spindles, bearings and rolls has a considerable effect on the finished accuracy of rolled sections. Without question, the less the total of these runouts the better the forming characteristics. In forming the heavier metals from about 14-gage and up, two or three thousandths total runout will have little effect on the accuracy of the sections produced. However, on lighter gages, a few thousandths runout would amount to about 20 to 50 per cent of the metal thickness and cause waviness in the final part. This is particularly true of the softer metals, consequently maximum allowable total runout is limited generally as follows:

> For gages in the range of 0.005-inch, maximum runout allowable is 0.0008-inch
> For gages in the range of 0.015-inch, maximum runout allowable is 0.0013-inch
> For gages in the range of 0.031-inch, maximum runout allowable is 0.0018-inch
> For gages in the range of 0.040-inch, maximum runout allowable is 0.0024-inch
> For gages in the range of 0.062-inch, maximum runout allowable is 0.0026-inch

Lengths. Finished sections cut to length by automatic flying cutoffs can be held within normal commercial limits, obviating further trimming for interchangeability. On these cutoffs, the trips — either mechanical or electrical — are mounted on the runout table and are adjustable so that the length of cut can be set to repeat within plus or minus 0.015-inch. Generally, tolerances on cut lengths run about plus or minus 0.031-inch and, depending on the type of trip used and the length and cross section of the shape rolled, may vary from plus or minus 0.015-inch to plus or minus 0.062-inch.

One must remember as speed increases length tolerances will be more difficult to hold. This is especially true when running short lengths at high speeds.

Section Dimensions. Width of a simple channel section such as that shown in *Fig.* 27.9 can be held to plus or minus 0.005-inch or less in continuous production. In rolling, the sheet edges are confined by the rolls and ordinarily leg heights of sections can be held accurately. For instance, the leg heights of channel sections would not vary over plus or minus 0.005-inch. The same would be true of the leg heights of Z or angle shapes.

28

Section Forming

CONTOUR forming of metal sections, a processing method developed from a humble beginning on machines designed primarily for the forming of piping. Simple bending operations as performed on tubular and flat sections involve, essentially, little more than the application of a bending moment at the desired point. Contour forming as it is known today covers the forming of metal sections of all types — sections, shapes, etc. — normally received in straight lengths, into various contours.

Range of Application. Parts produced on bending machines start with the simplest forms of formed tubing and piping for an infinite variety of uses including machine frames, vehicle and equipment components, and lines for transmission of fluids. Production of complex sections with compound curves includes refrigerator cabinets, automobile rims, process equipment parts, jet-engine parts, railway-car parts, decorative trim moldings, etc.

Contour Forming Machines. Since the bending moment or forming action needed for the forming of sections can be applied in a great many ways, there is a wide variety of machines available to achieve these ends. Classified generally, these methods are as follows:

1. Draw	4. Compression
2. Roll	5. Stretch
3. Ram	6. Compression-stretch

Draw Machines. The so-called draw bending process for hollow sections is accomplished on a machine having a rotating die, to which the piece being bent is clamped, and a roller, against which it is drawn, *Fig.* 28.1. The neutral axis lies between the center of the section and the inner fibers, with resultant thinning of the outer portions and slight thickening of the inner walls at bends. In general, the strength of such a bend is lower than that of an unbent portion. Flattening effect on these machines is severe where sections are not fully supported, consequently tubular sections require an internal mandrel.

Design of mandrels for tubular sections is subject to infinite variation — plug, shaped, ball, and "snake" mandrels being the most common. These mechanical mandrels, however, are often difficult to set up and lubricate and in some cases require special means for withdrawal. To offset these difficulties and eliminate the 5 per cent or greater flattening effect at bends, a liquid or hydraulic mandrel is often used. Flattening is eliminated entirely and bending limits are governed largely by the "wrinkle point" of metals.

The draw machine is especially suited to contouring thin-wall tubing and similar sections to close radii in production and is frequently used for other sections. Power required for operation of draw machines is the lowest of all the types. For instance, pipe from 1/2-inch to 10 inches or more can be handled in such machines.

Fig. 28.1 — Draw bending machine fitted with a special profile attachment for the bending of sections. Photo, courtesy Pedrick Tool & Machine Co.

Roll Machines. The simplest roll machine comprises a pyramid of rolls, one of which is adjustable with respect to the other two, *Fig.* 28.2. A section to be bent is passed through the rolls which, acting somewhat like a small rolling mill, produce the desired curve or radius on the piece. Both vertical and horizontal models in a wide range of sizes are available. Rings or continuous coils may be produced and a variety of radii may be had with only a simple adjustment of the rolls.

Compared to the draw machine, loading and unloading operations are relatively more simple. In addition, much more accurate bends may be had, with little or no flattening or wrinkling, on a wide variety of materials as long as the radius of bend is not less than three times the diameter of the driving roll. Production rate is low on heavy sections.

Fig. 28.2 — Haeusler multiple-purpose roll bending machine equipped for long section forming. Courtesy, Myers Tool Co.

Fig. 28.3 — Grotnes Model No. 1300 Hydraulic Roll Former rolls and knurls rims up to 32 inches wide. Maximum rolling force is 100,000 pounds.

Fig. 28.4 — Three-roll Haeusler machine used for flanging outwards and inward, joggling and other work. Courtesy, Myers Tool Co.

An adaptation of the roll machine is often used as an adjunct to the roll forming process of producing sections and as such offers many economies where circular bends are desired, *Fig.* 28.3. Wider parts and end flanged parts are produced on horizontal roll machines, *Fig.* 28.4. Work up to 3 inches in thickness and 18 feet long can be handled on such machines.

Ram Machines. A ram bender consists of two supports for the section to be bent and a central die operated by a hydraulic cylinder for applying the load. Simulating the action of the common stamping press, the ram machine is used for bends on extremely heavy sections such as pipe up to 14-inch diameter, railroad rails, reinforcing bars, structural sections, etc. Accuracy is poor and production low, making such machines suitable more for structural and maintenance work. The principle of ram bending, however, is utilized in other forming machines.

Compression Machines. The compression process utilizes a stationary die onto which the part to be formed is held while a rotating arm wipes the piece into the die contour. Similarly, the tangent bender utilizes a rocking die plate to wipe the shape onto the master die, *Fig.* 28.5. Still another version utilizes a stationary arm and a moving die table, *Fig.* 28.6. Shoes, rolls or tangent dies mounted on the main ram of these machines are used to force the material into the die contour.

Compression forming results in a bend, the neutral axis of which is between the center and the outer fibers of the section. Thus the inner wall or, in some cases, flange is somewhat heavier and the outer wall substantially the same as the original which, with the action of cold working, results in a bend stronger than the unbent section. Owing to this minor stretching of the outer wall, chromium-plated sections can be formed to generous radii without damage to the plating. The same is true for nickel-plated and enameled parts.

Compression forming methods are highly satisfactory where bends are severe and heavy sections must be controlled, and also where metal must be formed to small radii or sharp offsets. Since mandrels are not generally employed on compression machines, thin-wall tube sections are seldom handled.

Fig. 28.5 — Double-wing tangent compression-stretch bending machine primarily designed for radius forming. Photo, courtesy Cyril Bath Co.

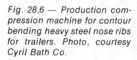
Fig. 28.6 — Production compression machine for contour bending heavy steel nose ribs for trailers. Photo, courtesy Cyril Bath Co.

Except on extremely ductile or workable materials, the ironing methods of compression forming often fail to reach the elastic limit of the material being formed. Consequently, compensation must be made in the dies to allow for springback. This problem of springback with materials having good elongation characteristics is easily overcome by resorting to the stretching method.

Stretch Machines. Although not as fast in production as compression forming, the stretch method overcomes the internal strains of prior cold working by stretching a section of metal to or slightly beyond its elastic limit during forming. Held in special gripping jaws, the piece is first stretched to approximately its elastic limit and, with constant tension, is then laid onto the die form, *Fig.* 28.7. Another type machine, without a rotating table and somewhat more limited in application, uses two stretching heads and a central die form, the stretch set being induced after forming.

Many combinations of curves are possible — constant radii, varying radii, single-plane or multiplane, simple or compound — in springy materials such as tempered aluminum alloys and stainless steels with part-to-part uniformity and negligible springback. Complicated reverse curves, of course, make for slower production, *Fig.* 28.8.

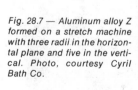
Fig. 28.7 — Aluminum alloy Z formed on a stretch machine with three radii in the horizontal plane and five in the vertical. Photo, courtesy Cyril Bath Co.

Fig. 28.8 — Severe reverse bends for this aluminum channel section necessitate reversal in table rotation during the stretch-forming process. Photos, courtesy Cyril Bath Co.

Compression-Stretch Machines. Where the elongation characteristics of a specified material are insufficient to withstand the severity of stretch forming, a combination of stretch and compression can be resorted to for maintaining uniformity and obviating excessive springback. The hat section shown in *Fig.* 28.9 required such a combination. Pressure forming on a machine such as that in *Fig.* 28.6 retained the section shape — the outer flange fibers stretching about 6⅞ inches over the original 36-inch length and the inner fibers compressing about 5⅛ inches. Long, difficult hollow-section forming, such as with rectangular tubes, also requires the combination method.

As noted previously, the so-called tangent bending process, which utilizes a rocking die to avoid the surface abrasion often present with confining rolls, also has been designed to include stretching, *Fig.* 28.5. A flanged sheet inserted between the dies is first stretched to crown the top and then carried around to form the sides, *Fig.* 28.10. Deep flanges on the parts are controlled to avoid buckling which normally would occur during simultaneous stretching of outer areas and compression of inner areas and flanges. Flanged sheets up to 42 inches wide with radii up to 6 inches can be handled in double-wing and up to 120 inches wide with radii up to 8 inches in single-wing machines.

Deeply curved heavy-gage parts such as automotive bumpers, roof trusses, etc., which require a controlled combination compression-stretch technique can

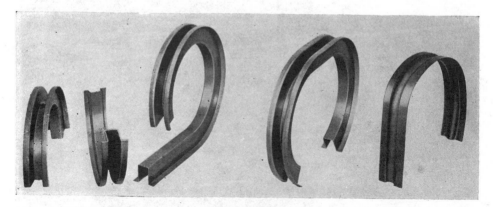

Fig. 28.9 — Heavy steel hat sections formed by the compression-stretch method. Photo, courtesy Cyril Bath Co.

also be handled on certain types of bending machines.

Production Quantities. Since all the bending techniques, excepting the roll method, necessitate the use of dies, it is usually more economical to hand form where quantities run less than 100 pieces. Most efficient quantities on universal machines run from 1000 to 100,000 pieces. On simple bends on 1-inch tube, a special compression bender handling fifteen at a time produces over 1000 pieces per hour, *Fig.* 28.11. Where tremendous quantities are required, automated machines can handle large parts at high speeds, *Fig.* 28.12.

Design Considerations

Inasmuch as all but the roll method require special dies, the subject of dies is highly important. Dies for most light compression and stretching work can be made of Kirksite, Formite, Masonite or similar material. Heavy forming, however, requires metal dies of adequate strength with the contour radii properly designed to eliminate springback discrepancies, unless stretching is used.

Springback. To account for normal springback, radii generally must be reduced by the following amounts:

Aluminum (soft) 3%
Copper (soft) 3%
Mild steel 5%
Stainless Steel (annealed) 5%
Aluminum Alloy (SO temper) 7%
Chromium-molybdenum steel (hard) 10%
Stainless steel (hard) 10%
Aluminum alloy (ST temper) 12%

Tube Radii. Standard minimum bend radii (to center of bend) for tubing are usually determined by the formula:

$$R = 0.775 \ (D^2 - d^2)/g$$

where R is the minimum bend radius, D is the outside diameter, d is the inside diameter, and g is the wall thickness, all in inches. Mandrel bends can be smaller but of course involve considerably greater production costs.

Offsets. In the design of contours, sections which require sharp or deep offsets naturally require consideration of the diameter of the rolls used. Radii must be generous to allow for easy traverse of reverse bends.

Holes. Where holes are to be punched prior to forming, the stretch method generally cannot be used owing to excessive distortion of the holes. In such cases design should be considered for the compression or roll process only.

Shape Bends. Preformed flanged sheet contoured with flanges out are usually stretch formed to shape. With about 0.036-inch sheet having 5/8-inch flanges, 1/2-inch inside bend radii can be produced on a tangent bender. If flange width is 7/8-inch, a 3/4-inch radius can be formed. With 0.060-inch sheet having a 1-inch flange, the minimum radius is also 3/4-inch. In each case flange width is reduced approximately 3/16-inch after stretching.

Fig. 28.10 — Refrigerator outer shell formed from flanged sheet on a double-wing tangent bender.

Fig. 28.11 — Production compression bender handling fifteen 1-inch tubes simultaneously produces a minimum of 1000 bends per hour. Photo, courtesy Pedrick Tool & Machine Co.

Flanged sheet preformed with flanges from 1/2-inch to 1 1/2 inches wide and inside radii of 3/16 to 1/4-inch can be tangent bent with outside contour radii from 1 inch to 6 inches respectively. Flanges are compressed or upset when bent inward and consequently the overall width remains the same.

For exceptionally wide flanges which must be welded, a design as shown at (a) in *Fig.* 28.13 is recommended. Welding, grinding and polishing are then confined to flat surfaces only. Where welding is not necessary, design as shown at (b) in *Fig.* 28.13 is recommended.

Cold Flanging. Where heavy gage material is to be dished and/or flanged, available machines will handle parts up to roughly 20 feet in diameter with mild steel up to 1 inch in thickness and stainless up to 1/2-inch. Flanges on tubular material can be cold rolled up to 1 1/2 inch on 10 and 12-gage materials; up to 2

Fig. 28.12 — Continuous automated line by Grotnes that produces auto rims at high speed.

inches on 3/16-inch materials and up to $2^1/2$ inches on 1/4-inch materials. Tube diameters up to 60 inches diameter by 60 inches in length can be flanged on standard machines while special adaptations will handle lengths as long as 40 feet. Flanging time required for such flanges on tubular parts is extremely short, a 2-inch flange on a 30-inch diameter by 1/4-inch wall tube taking from 1 to $1^1/2$ minutes only.

Selection of Materials

Some considerations regarding the bending of various materials have been covered in the foregoing discussion. In general, the various techniques of bending require a high-ductility metal for maximum economy. Where semi-ductile materials must be used, the compression or compression-stretch methods are usually necessary to squeeze the material into the new shape. Shapes of tempered aluminum alloy, hard-drawn shapes of K-Monel with a ductility of 2 per cent, and hard-drawn copper bar stock have been bent successfully by these methods.

With long shallow bends in soft materials, stretch forming produces the most accurate, constant results. Likewise high-alloy aluminum, stainless steel, etc., usually require stretching to achieve uniformity. Low-alloy carbon steels, aluminum, stainless steels, etc., would probably be relegated to the compression method for higher output.

Specification of Tolerances

As in all other work, the closer the tolerances the greater the cost in time, tooling, etc. As indicated previously, the stretch method is subject to the least effect from springback and logically can be relied upon to produce the most exacting results — dependent, of course, upon die accuracy where material worked has

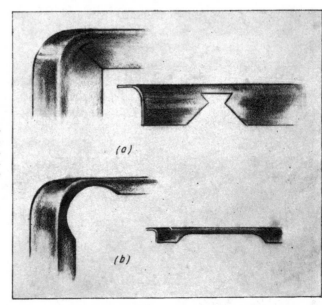

Fig. 28.13 — Where exceptionally wide flanges are required, design at (a) is recommended for welding, and design at (b) where welding is not necessary or where the finished notch must be open.

adequate elongation characteristics. Surfaces are smooth and free from wrinkles ordinarily produced in regular die forming on presses, *Fig.* 28.14. Although not as accurate in mass production from piece to piece owing to springback characteristics of materials, compression forming has a greater production rate.

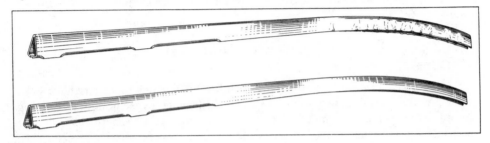

Fig. 28.14 — Stretch forming special aluminum-alloy sections results in a smooth, accurate part without the wrinkles ordinarily left by regular die forming. Brake-formed part, top, required 45.79 man-minutes per piece while stretch-formed part, below, required but 10.38 man-minutes.

Flattening of Tubular Parts. The compression machine produces a flattening effect of about 5 per cent on tubular cross sections unsupported by an internal mandrel, when the radius of bend is limited to a minimum of four times the diameter for 6-inch pipe, three diameters for 2 to 5-inch pipe, two diameters for 3/4 to 2-inch pipe, and one diameter below 3/4-inch. On tubing the minimum radius is about three diameters except for copper which may be appreciably smaller. Flattening effect with the draw process is about 7 per cent for the same limits but with a mandrel this can be reduced to around 3 per cent. As previously mentioned, a liquid mandrel offers no flattening effect within the above limits, wrinkling being the critical factor. No wrinkling and negligible flattening effects are experienced on the roll machine if the radius of bend is limited to a minimum of three times the diameter of the driving roll.

Square Ends. If the end of the bent tube must be maintained square without extra trimming, the length of straight end beyond the point of tangency with the bend must be at least equal to the diameter plus 0.01 times the radius of the bend.

29

Stamping

WIDELY recognized for its far reaching effects on product component design, the stamping process and its many ramifications provide in all probability one of the greatest sources of economical mass-produced machine and product parts. Practically every machine product, large or small, includes in its construction one or more stampings. Although stampings in the past have been responsible to a large degree for drastic reduction in costs of automobiles, refrigerators, washing machines, ironers, typewriters and other machines too numerous to mention, advances made in the process during recent years offer even greater possibilities.

Field of Application. Many machine parts previously designed to be made from forgings, castings, or bar stock have been redesigned for production as stampings or assemblies of stampings at great savings in time, labor and materials, *Fig.* 29.1. Conservative estimates show that at *least* three to five times as many finished stampings can be produced in the same amount of time as that required to produce similar finished parts which require preliminary casting or forging operations. Today stampings are made from a wide variety of metals — cold in thicknesses up to 3/4-inch and hot in thicknesses up to 3 1/2 inches. These may range

Fig. 29.1 — Assembled from 14 stampings, the 50-caliber machine gun feedway adapter here shown mounted on the gun frame cost one-third that of the original model made from machined castings

Fig. 29.2 — Right — A 1400-ton single-action press with automatic roll feed and levelers. Photo, courtesy USI Clearing Div., U.S. Industries, Inc.

Fig. 29.3 — Minster inclinable punch press for producing electric motor rotor and stator laminations at up to 170 strokes per minute. Photo, courtesy Reliance Electric Co.

in size from minute instrument parts all the way to machine frames 30 feet or more in length. However, the great majority of high-production stampings are made from sheet under $3/8$-inch in thickness.

Stamping Presses. A great variety of presses for performing stamping operations is available. These range from extremely small tonnage bench presses for tiny parts to very large single and transfer presses, *Fig.* 29.2. For many simple, single-operation piece-parts the inclinable type C-frame presses offer many possibilities for gravity feed and ejection, *Fig.* 29.3. These various machines may range from hand-fed types to those with completely automatic dial feeds, transfer arrangements, or multiple-station die sets devised for completion of the product, *Fig.* 29.4.

Fig. 29.4 — Fourteen-station die set and scrap strip showing the sequence of press forming operations for an electrical component. Skill in producing these dies offers highly accurate stamped parts at low cost. Photo, courtesy Oberg Mfg. Co., Inc.

Operating Speeds. From the relatively slow hand-fed and/or the extremely large mechanical or hydraulic presses, production speeds may range up to as high as 800 to 1000 strokes per minute (pieces per minute) with inverted presses. Inverted presses are ideally suited for multiple-station dies and today have up to 500-ton capacity, *Fig.* 29.5. By use of multiple dies on high-speed presses, parts requiring blanking, piercing and shallow drawing operations can be produced at rates as high as 10,000 pieces per minute. Equally as important as production speed, multiple-station dies permit a sufficient number of operations to virtually complete parts ready for assembly.

The Stamping Process. The process of stamping, as generally referred to in the foregoing, covers that portion of the press working and forming field which is considered as including the following operations:

1. Punching	2. Bending	5. Extruding
(a) Blanking	3. Forming	6. Swedging
(b) Piercing	(a) Beading or curling	7. Coining
(c) Perforating	(b) Embossing	8. Shaving
(d) Lancing	4. Shallow Drawing	9. Trimming

Special Operations. Coining and shaving are much more widely used today than ever before in producing fine finish and detail as well as high accuracy and sharp edges at lower cost than many of the machining operations otherwise necessary. However, in many cases where production is not sufficient it will be found desirable to resort to a light machine cut or grind for obtaining a desired fit. Comparative costs of stamping, sizing, etc., must be determined accurately by consultation with expert tool and die designers to assure a reasonable cost economy in unusual and borderline cases. Increased emphasis on low-cost mass production makes imperative close, friendly co-operation to achieve these ends.

Cost and Quality. Costs involved in the production of a great majority of machine part designs necessarily must remain secondary to quality and physical properties, and in some cases to rate of production. However, cost savings possible in redesigning for stamping with equal or improved conditions are highly desirable. These, as has been shown conclusively by many applications during recent years,

Fig. 29.5 — A 250-ton Henry & Wright double-crank machine in a large automotive plant producing door lock frames on a 17-stage die at a rate of 80 parts per minute. Photo, courtesy HPM Div., Koehring.

Fig. 29.6 — Distributor head insert made by mechanically assembling two stampings. The link shown is made from 0.156-inch soft-annealed Monel wire in six operations: (1) Coin form socket, extrude plug and coin center to thickness; (2) blank center bar square; (3) coin center bar to round form; (4) trim; (5) extrude metal around hole and countersink; and (6) pierce hole through. Cost of finished and assembled stampings was reduced 30 per cent over previous method of silver soldering two screw machine parts and a wire connector. Photo, courtesy Advance Stamping Co.

Fig. 29.6, can usually be realized by competent design. Per unit time, many times as many pieces can be produced by stamping as compared to various other manufacturing methods for obtaining similar end results. This is primarily due to the fact that stampings leave the machine essentially as a finished piece; separate surfacing, machining, drilling, etc., ordinarily being dispensed with.

Cost and Quantity. Generally speaking, the great advantage in plain and automatic stamping is the extremely low cost per piece of components produced in mass quantities. In some instances where design is relatively simple, a few thousand pieces can be made with the same advantage. Relatively simple designs involving punching, forming and drawing often can be produced most advantageously on "temporary" low-cost dies if quantities generally do not exceed 5000 pieces, *Fig.* 29.7. Even as few as 50 pieces can be produced economically by this method and small-quantity stampers offer this service for orders from 10 to a maximum of 50,000 pieces. Changes in design can be made for as little as one to two dollars tool cost in many instances. However, if tooling and die costs can be spread over many hundreds of thousands of parts, design can be as complex or intricate as conditions

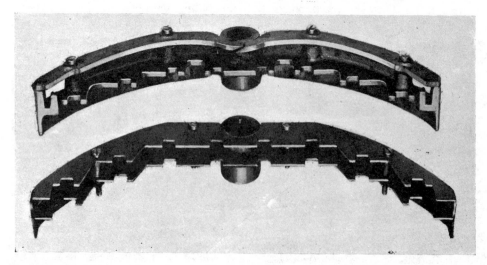

Fig. 29.7 — Parts of bottom blanked in temporary dies and assembled by means of welding provided a more uniform and economical design than those using machined castings, top. Photos, courtesy Dayton Rogers Mfg. Co.

Fig. 29.8 — By means of a combination of simple stamping operations, copper brazing and press sizing, all machining is eliminated on this damper hub pulley, right below. Weight is reduced by 40 per cent over that of the original casting, left below. Drawing, courtesy Mullins Mfg. Corp.

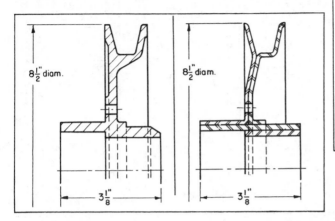

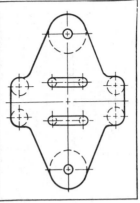

Fig. 29.9 — Above — True radii throughout on blanked parts provide maximum simplicity and lowest die costs.

require.

Limitations. A most essential factor to bear in mind in designing for stamping is that the basic material is in the form of sheets of substantially uniform thickness, and sections can be thinned or bosses raised only with considerable difficulty. Commercial tolerances on rolled sheet make it difficult to draw material lighter than 0.020-inch. Use of heavy plate may be necessary to overcome the lack of vibration damping qualities normally found in lightgage stampings.

Design Considerations

Although a part can be made as a single stamping or an assembly of stampings, if the number of operations required is excessive the overall cost may be such as to offer no advantage over other production methods. The particular operations and machine for producing the stampings must also be considered. Certain forms, shapes or sizes necessarily are limited to one or another type machine to assure economy in production.

Reasonable Design. Simplicity of design within reasonable limitations is essential to secure maximum value from the use of stampings regardless of method, *Fig.* 29.8. Designs should not follow that of complicated castings with many or tapering wall thicknesses and bosses. Costs mount rapidly if bosses, stiffeners, and similar features have to be added by welding, brazing, or riveting.

Blanking. Blanked shapes should always be designed with true radii throughout, *Fig.* 29.9, wherever possible. Rounded corners are, as a rule, preferable to square ones owing to the greater ease in die sinking. Blanks which can be produced by plain cutoff from a strip, as shown in *Fig.* 29.10, should *not* have full rounded corners. Such blanks can be designed to have square corners if desired. In addition to perfectly square corners on such parts, no scrap is produced from the sides or ends. Square corners can be produced easily with a sectional die wherever required, the type often used when the parts are unusually large or quantities extremely high (possibly a million or more pieces).

Blank Layouts. Blank developments for bent or formed parts should be

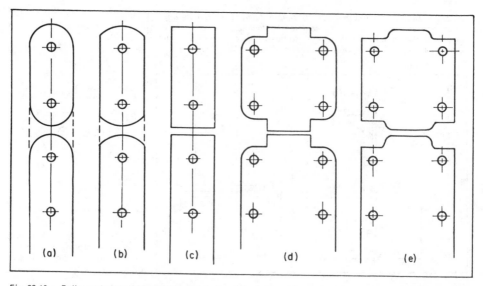

Fig. 29.10 — Full rounded ends blanked from strip are undesirable owing to feather edges and problems in blending at the edge (a), large-radius ends (minimum $5/8$ of the width) as at (b) are acceptable, but square ends as at (c) produce no scrap at all. Square corners along the blanked edge as at (d) are also to be avoided in favor of circular corners as at (e).

specified whenever the design is relatively simple and dimensions are not too exacting. However, where drawing as well as forming is required, a cut-and-try development is necessary to assure an acceptable job, *Fig.* 29.11. These can be most easily worked out in the plant where unusual metal flow or movements can be seen and taken into account before the production blanking die is made. Such designs are of course only practical where considerable quantity is anticipated. Otherwise the most economical procedure is to allow for the necessary variation in the final design.

Steel Rule Die Blanking. Often referred to as "skeleton" dies, steel rule dies are those widely used in blanking paper, rubber sheet, plastic foam and fiber forms. These dies are built-up by bending or fitting together, usually by screwing to a wood base form, steel rule-like cutting blades as required by the blank to be cut. Steel rule dies are extremely low in cost and easy to obtain. A wide variety of materials can be blanked, including thin-gage soft or half-hard sheet metals. These

Fig. 29.11 — Three press operations complete two halves of this rocker-shaft bracket. Proper blank development assures a reasonably straight finished edge without an additional trimming operation. Photo, courtesy Monarch Die Corp.

dies can be used in power presses fastened to the upper platen or merely set in place over the material for blanking as in a checker or dinker type cutting press, or between rollers. The blanks shown in *Fig.* 29.12 indicate the possibilities for design. Because cutting is often done over a piece of cardboard or fiber, the edges are pulled down slightly; normally this is an advantage inasmuch as in effect burrs are eliminated in the piece if final forming is required. The dies, however, are only suitable for blanking, scoring, creasing, or simple line embossing.

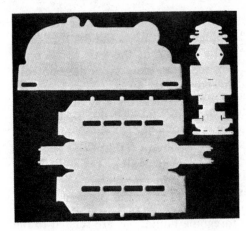

Fig. 29.12 — Three typical prepainted sheet metal blanks for mechanical toys produced with skeleton or steel rule dies.

Fine Blanking. A German development for producing extremely accurate finished parts in a single stroke, fine blanking is gaining in use in the United States largely because of cost savings in production. Costs can be reduced as much as 50 per cent on some designs. The process uses tools with virtually no clearance between mating parts of the punch and die (actually held to 0.0003-inch or less). Although similar to ordinary blanking, the key to success is in an impingement ring in the die that presses into the perimeter immediately around the outline of the part to be blanked. Thus, the punch extrudes the blank through the material under controlled pressure creating a fine finish and straight edges, *Fig.* 29.13.

Parts up to 5/8-inch in thickness and dimensions as large as 16 inches square or round are suitable. In addition to blanking, punching, coining, bending and countersinking can be produced in a single stroke. A minimum material thickness should be about 0.030-inch. Holes in steel and nonferrous cold forming materials can be as small as 1/2 stock thickness and 2/3, in high carbon steels. Radii on outside corners should be at least 10 per cent of stock thickness, inside corners at least 5 per cent.

Holes. In pierced or perforated parts, plain round holes are most economical

Fig. 29.13 — Typical single-stroke fine blanked automotive part produced on a 400-ton press at 350 per hour. Roughly 6 inches by 3 inches and 13/32-inch thick, the part is produced with tolerances on the round hole locations of ± 0.001-inch, shaped holes and to edges +0.0035-inch, -0.0015-inch. Sheared edges are perpendicular within 0.001-inch to 0.0005-inch, edge curvature has a radius under 0.030-inch. Photo, courtesy Monarch Die Corp.

Fig. 29.14 — Sheet perforated with staggered holes, 0.020-inch in diameter with 625 holes per square inch. Photo, courtesy E. W. Bliss, G & W.

to produce and should be used in preference to slots of irregular shape. Perforations or multiple tiny holes, *Fig.* 29.14, should not be smaller in diameter than the thickness of the material and preferably not under 0.030-inch. Smaller holes can be produced, however, and today punch arrangements are available which make possible the production in steel of holes one-half the gage in diameter; an 0.046-inch hole, for instance, being punched in a 0.093-inch sheet.

Perforations in high-nickel alloys usually are limited to a minimum diameter equal to the sheet thickness for gages down to 5/32-inch. Gages under 5/32-inch must, for economical production, have holes specified greater than the sheet thickness, say at least 1.5 to 2 times. It is likewise inadvisable to attempt to perforate chromium-nickel stainless steels where the hole diameters are less than 1.5 to 2 times the metal thickness. The same is normally true of piercing holes of other shape than round; in such cases the foregoing limitations are related to the least dimension.

Fig. 29.15 — Motor rotor end plate 3⁵/₈ inches in OD of 0.060-inch pure copper with 48 holes of 5/32-inch diameter is produced in progressive dies using "French" stripper. Holes are spaced 0.030-inch in from OD. Photo, courtesy Larson Tool & Stamping Co.

Fig. 29.16 — Three designs of lanced holes showing punch and die. Drawings, courtesy S. B. Whistler & Sons Inc.

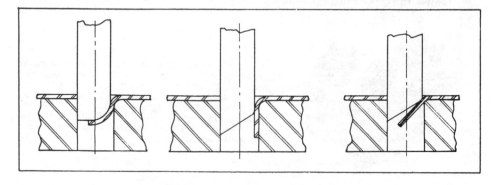

Hole Spacings. Spacing between holes or between the edge of a hole and the edge of the blank should be at least equal to the stock thickness and preferably greater to prevent distortion. However, where adequate tooling can be employed on quantity runs much closer spacings can be had with reasonable economy, *Fig. 29.15.* Holes should be located as far from bends as is possible to avoid the necessity for an extra operation or a special punch to assure a true hole. Ordinarily, holes located from bends, especially those close to bends will have to be punched after forming unless sufficient leeway is permitted. Accurately located holes should, if at all possible, be dimensioned from straight finished sides or edges in preference to formed or bent faces. Where holes must be located from bent faces, consideration should be given the fact that location will be most accurate from the side of the face which is formed against the die. The sheet variations will enter into the dimension to the outer face of such a detail and necessitate punching the hole after forming, causing extra tooling and fabrication expense.

Lancing. Internal tabs, louvers and similar formed designs ordinarily are produced by lancing. Where formed lanced sections or bent lancings are specified, generous radii are desirable but where the lanced portion is merely partially bent the radius can be fairly sharp, *Fig. 29.16.* Lanced internal tabs follow the same general rules as indicated under tabs.

Projections or Tabs. Width of any projection or slot should be at least 3/32-inch or greater for stock under 1/16-inch thick. In heavier-gage material the width should not be less than 1.5 times the stock thickness. Projections should be reasonably proportioned (as short as practicable) to assure reasonable die life and should be faired into the blank edge with as great a radius on each side as possible. Notwithstanding, bent tabs and formed portions of a blank should be designed so as to be at least 1.5 times the material thickness in length or height. Tabs shorter than this are exceedingly difficult to produce and bend.

Shallow Drawing. Shallow drawing operations on small parts performed progressively seldom exceed 3/16 to 1/4-inch in depth and ordinarily width and length or diameter should be at least two or three times the depth. Depth of draw on large parts is only a small fraction of the overall size of the piece. Depths equal to or greater than the diameter require deep drawing dies.

As a general rule, a round-shape die is the cheapest while tooling for a flanged box with sharp corners is excessively expensive. In designing box shapes with sharp corners, a welded construction will be found preferable and cheaper to tool than a stamped one. However, shallow boxes with generous radii all around — at least 1/4-inch to 1/2-inch — are practical so long as the tolerances on squareness of corners and parallelism of sides are not too stringent.

Trimming. On shallow boxes or covers of large size or unusual contour, trimming is sometimes difficult. As a rule, if such parts are designed as a flanged stamping they can be trimmed close to the sides, *Fig. 29.17,* most easily and with a satisfactory edge. To achieve streamlined design and good fit on some typewriter frames, the irregular contours were first broken down and then annealed before final forming. To maintain proper match of the special flanged sides as well as overall accuracy, these blanks were formed much deeper than necessary and the irregular flange was finished by a multiple trimming operation. In this setup the large amount of scrap encountered was justified by the total cost savings on the part itself.

Scrap Losses. Reduction of undue waste or scrap material can lower costs by

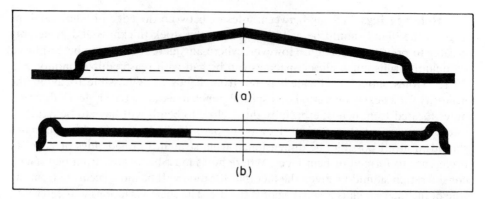

Fig. 29.17 — Trimming is simple for shallow formed parts with flanges (a). Flanges can be trimmed close (b) to give a satisfactory edge and avoid expensive dies.

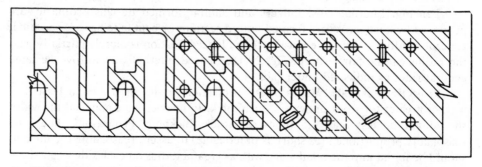

Fig. 29.18 — For economy, parts should "nest" effectively. Small parts can be made from scrap sections.

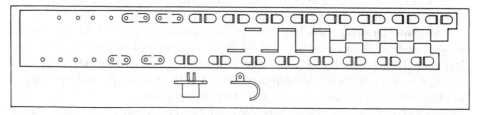

Fig. 29.19 — A spring brass clip design showing a high degree of material economy.

proper consideration during design. Slight changes to secure effective "nesting" of stampings in the strip to be used can minimize waste and often make possible the production of other parts from scrap centers, etc., *Fig.* 29.18. It is considered good economy if layouts utilize 75 per cent or more of the strip area. In many cases complete usage of the strip is possible by ingenious design planning, *Fig.* 29.19.

Forming. On parts being formed along with a sequence of stamping operations it is often difficult to hold exact right-angle bends. This is especially true with the harder materials and heavy gages. Unless slight deviations are allowable, more costly dies must be used. Bends should, if at all possible, be made across the grain or in other words at 90 degrees to the length of the strip to avoid cracking. Where this is not possible, a compromise between 90 degrees and 45 should be used to allow for optimum layout in the strip. Inside radii of bends should be equal to the material thickness or preferably greater, *Fig.* 29.20. Smaller radii tend to weaken the part to some extent and often require an additional press operation. On formed portions especially, generous radii are essential. Sharp corners will usually cause tearing. High-strength, low-alloy steels require large radii, in many instances as much as four times the material thickness for successful bending and forming. Outside radii are normally considered greater than the inside by the stock thickness.

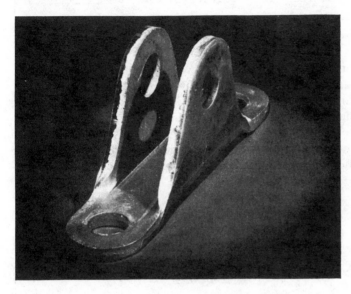

Fig. 29.20 — Sharp inside corners are sometimes a necessity but cause severe stretching of the adjacent material on heavy-gage parts. Photo, courtesy Dayton Rogers Mfg. Co.

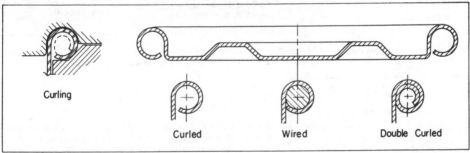

Fig. 29.21 — Bead formed by curling and various types of curls.

The severe stretching which takes place at the corner of bent-up portions is readily evident in *Fig.* 29.20. It is preferable with such parts to have the inside face of the bent sides flush with the sheared sides of the extensions. Thus, the bend is outside the base section and no corner stretching occurs. Where the flush design is desirable, an undercut at the corners will achieve the same end result and obviate stretching or tearing. The same is true of bent-up tabs or lanced tabs. A relief at either side of the bent-up or lanced tab is highly desirable. With tabs which must be lanced directly from the sheet, punched holes can be used at either end of the sheared section to relieve stretching and tearing.

Beading or Curling. Where straight or circular beads are desired around the periphery of a part such are produced by a curling operation, *Fig.* 29.21. Beads produced by curling are normally used for stiffening and providing a smooth, round edge. Curling operations are generally limited to gages under 0.031-inch and soft metals such as tin, steel sheet, brass or copper. Bead diameter must be in

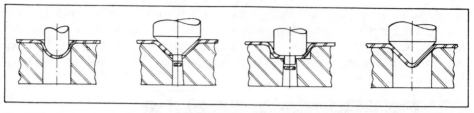

Fig. 29.22 — Series of typical embossings, plain and with punched centers. Drawings, courtesy S. B. Whistler & Sons Inc.

proportion to the size or diameter of the piece and normally ranges from 5 to 10 times the gage, but as a rule the diameter of beads seldom exceeds 3/16-inch.

Embossing. Ordinarily, embossing is used to obtain improved stiffness in thin-gage parts. Use of an angular embossing can be seen on the finished part shown in *Fig.* 29.11. Embossed portions can be specially shaped or plain, annular or button type. Specially shaped button type embossings are widely used for projections on stampings to be projection welded. Radii should be kept as large as possible and design should be simple for economy, *Fig.* 29.22.

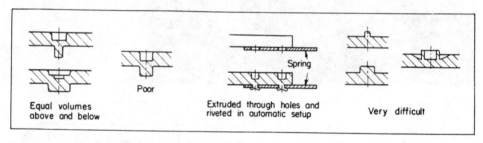

Fig. 29.23 — A variety of dowel embossings which can be used for location or fastening. Dowel diameter should always be slightly smaller than the section pressed down to eliminate any possibility of shearing out.

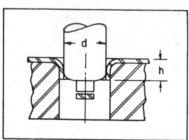

Fig. 29.24 — Extruding operation for drawing up a flanged hole or in some cases a hub. Height (h) ordinarily is about one-half the hole diameter (d). Drawing, courtesy S. B. Whistler & Sons Inc.

Dowel embossings, in reality cold-forged projections, *Fig.* 29.23, are often useful for assembly purposes and can usually be made of almost any reasonable length or projection. Embossings which can be incorporated directly in blanking, *Fig.* 29.23, are preferable unless the stations are available for longer or other proportions. Ordinarily for best results dowel height should not exceed 50 per cent of the gage although in actual cases this has been extended to as much as 80 per cent of the gage without trouble. Diameter should ordinarily be at least 2 times the height. Such dowels can be used to hold assembled parts as well as can lanced lugs. The former are normally riveted or staked while the latter can either be bent or staked.

Extruding. Design details requiring the drawing of stock out from a center hole, *Fig.* 29.24, are usually referred to as extruding. Extruded holes can be pulled to a height of about one-half the hole diameter produced. Larger pulls can sometimes be made but often result in splits especially where material is not in dead soft condition. Where additional operations can be utilized higher flanges can be obtained by ironing or thinning out the walls and also by first drawing the metal into a cup, punching a hole in the bottom and finally throwing out the remainder of the cup to finish. Where such holes are not to be threaded, folding back or countersinking can also be done, *Fig.* 29.25.

Swedging. Where the walls or hubs such as shown in *Fig.* 29.24 are too thin for any reason, swedging or upsetting to increase the thickness is often used. Such operations, however, should be avoided if possible. Small upsets can also be formed in various parts to provide guide bushings, etc., although such features require an extra powerful machine. Heads, keys and similar features can be swedged to shape.

Fig. 29.25 — Bracket made of 16-gage steel with five operations — piercing, forming, extruding, countersinking, and blanking.

Coining. As can be surmised the coining operation provides the means of obtaining smooth flat surfaces, fine sharp design details, closer accuracy, sharp corners, etc. Various forms can be produced in coining, *Fig.* 29.6. Gasketing grooves can be cold coined and the surface flattened sufficiently to eliminate further processing. Projection welding rings can also be produced in thick parts without embossing.

Gear Blanking. Many types of small spur gears for lightly loaded applications requiring primarily the transmission of motion can be produced by stamping. For accuracy, medium hard or annealed carbon-steel sheet is employed. Soft steels usually produce undesirable rough edges. Brass or aluminum sheet or strip also can be used. As a rule, thickness of gears which can be produced is limited to somewhere around 5/16-inch and at best 3/8-inch. Fine-pitch gear teeth should be restricted to light-gage blanks for good die life. Gear diameter should be at least 2 times the center hole diameter to avoid twisting out of shape during punching.

Gears such as those used in clock and watchmaking are ordinarily either pierced or blanked complete in pitches down to as fine as 0.011-inch pitchline width. Gear teeth above this size can be blanked providing the stock thickness does not exceed 1.5 times the tooth thickness at the pitch line. Commercially stamped gears up to a maximum of approximately 4 inches OD can be produced in diametral pitches from 6 to 72.

Shaving. Unless fine blanking is used, fine teeth and thin sections in gears will often exhibit considerable pull-down on the edges which is sometimes objectionable. Shaving can often be used to reduce the rounding effect and also obtain great accuracy. The minimum thicknesses for which shaving is practicable is 0.050-inch for 32 DP and 0.040-inch for 48 DP.

Built-Up Gears. Where limitations on gage preclude the stamping of gears with sufficiently wide faces for medium-power applications, built-up designs can be used. Such gears are often stamped from copper-plated low-carbon steel of 0.062-inch thickness, assembled to desired face width and furnace brazed, *Fig.* 29.26. Embossed dowels can be used for location in assembly and no jig is required. Following brazing, a shaving operation produces the desired accuracy and tooth face smoothness.

Fig. 29.26 — Group of gear tooth forms produced by stamping and brazing. Photo, courtesy National Stamping Co.

Selection of Materials

Although steel, aluminum, brass, and stainless steels are used most widely, practically any material can be stamped. However, economy and ease of production vary with the elongation characteristics of the material. Cold-rolled steel sheet and strip with a carbon content between 0.05 and 0.20 per cent is well adapted for general use. Steel with carbon content ranging over 0.20 per cent becomes increasingly difficult to form and shear. A maximum of 0.35 per cent carbon is usually recognized as the limit for reasonable cold-forming characteristics. Deep drawing steel — steel with less than 0.10 per cent carbon content — of course is the preferred forming stock, especially where the forming is to be severe. With the best deep drawing steels, manufacturers usually guarantee successful production only where the forming draw necessitates an elongation of the metal which is less than 33 per cent. Parts requiring more severe forming can often be produced but only with special handling at increased expense.

Cold-Rolled Sheet. Steels containing sulphur have undesirable cold-working characteristics. Hard temper low-carbon steel has approximately twice the strength of soft annealed steel but can be cold worked only in small amounts. Steel sheet or strip produced by cold rolling has the best surface finish and drawing characteristics as well as the closest gage accuracy. However, if design is such that satisfactory, economical completion of a part can be had only by utilizing an intermediate annealing process, the high surface quality afforded by using cold-rolled stock is impaired. Consequently there is little to choose between hot or cold-rolled stock on anneal jobs. Cold-rolled material is seldom obtainable in thicknesses over 3/16-inch and at best these heavier gages lack the fine surface finish of the lighter stock gages.

Stainless Steels. Stainless steel (18-8) is an excellent forming and stamping material. For best forming qualities it must be thoroughly annealed. It can then be worked severely — more so than ordinary steel — at one draw, but requires about double the press power and is somewhat harder on the dies. Ordinarily, stainless in gages over 1/4-inch is uneconomical costwise for stampings. Where properties of stainless steel are desired, however, clad steels can often be used. These are available in both single and double-clad types and have approximately the same cold working characteristics as the steel backing.

Nickel Alloys. Much like stainless steels, nickel alloys are more difficult to work and require greater press power than ordinary mild steels. Depending on the temper, anywhere from 13 to 30 per cent more power is required. Soft to skin-hard temper is preferable for nickel-alloy material to be perforated or punched. In punching, dies must be kept sharp as defective edges on holes will usually result in fatigue failures. Punching these alloys results in appreciable work hardening of the hole edges. Consequently, for perforated parts subsequently to be formed or bent severely, annealing is necessary.

Copper-Base Alloys. Excellent working characteristics of brass and copper alloys combined with good corrosion resistance make them highly satisfactory for stampings. Cartridge brass (deep drawing) is the most satisfactory but tobin bronze, phosphor bronze and other alloys of less ductility can also be used where design requires.

Aluminum Alloys. Most easily worked of the aluminum alloys are 1100, 3003 and 5052 (soft tempers). These are preferred where their strength is adequate. The less ductile heat-treatable and age-hardening alloys can be used if proper handling can be provided for working them. These are, as a rule, worked directly after heat treating before age hardening sets in. For severe forming operations these alloys must be worked in the annealed condition and the parts heat treated later.

Magnesium Alloys. Magnesium can be worked successfully while hot. As with spinning, the material must be stamped at a temperature of 450 to 600 F and can be worked severely in this temperature range.

Blanking. Relative ease with which the various materials can be sheared or blanked can be judged from the values in Table 29-I.

Specification of Tolerances

One of the most interesting and important developments in the greatly expanded use of stampings in recent years has been the achievement of close tolerances and machine-like finish on intricate parts. Coining, sizing and shaving operations made possible and practical accuracies heretofore out of reach. Spur

TABLE 29-I — **Shearing Strength of Materials***

Material	Tons (psi)
Aluminum	
Cast	6.0
Sheet	10.0
Brass	
Cast	18.0
Half-hard sheet	13.0
Copper	
Cast	15.0
Rolled	10.0–15.0
German Silver	
Half-hard sheet	16.0
Fiber	
Hard	9.0–13.0
Iron	
Wrought	25.0
Cast	10.0
Lead	
Ordinary	1.5
Leather	
Oak	3.5
Chrome	3.5
Rawhide	6.5
Paper	
Hollow die	1.5
Flat punch	3.2
Bristol board (flat punch)	2.4
Straw board (flat punch)	1.75
Steel	
Cold-rolled sheets 0.25 Carbon	25.0
Cold-rolled sheets 0.50 Carbon	30.0
Cold-drawn rod	29.0
Boiler plate and angle iron	30.0
Drill rod (not tempered)	40.0
Spring steel 1.00 Carbon	42.0
Tool steel 1.20 C (not tempered)	45.0
Tool steel 1.20 C (tempered)	95.0
Nickel steel 3 to 5% nickel	41.0
(Hot stock equal to 1/6 to ¼ the strength of cold stock.)	
Tin	
Sheet steel coated with tin	25.0
Zinc	
Rolled	7.0–10.0

* Courtesy Verson Allsteel Press Co. To find the pressure necessary to blank any material, multiply the length of the cut edge in inches by the thickness of the metal, times the tonnage given in the table. This gives the *tons* pressure with a flat punch.

gears normally are held to the tolerances shown in TABLE 29-II.

Commercial Stock Variations. Material variations sometimes present difficulties. Parts dimensioned with close edge tolerances usually require complete blanking since the width of slit-edge or edge-rolled stock varies considerably with the gage. Commercial variations in thickness of materials also have their effects on the depth of formed parts. A tolerance of plus or minus 0.010-inch on the depth of formed parts under 1/2-inch in depth and plus or minus 0.015-inch on round parts up to 1 1/2 inches deep can be held. Closer limits require an extra trimming operation. As a rule, plus or minus 0.010-inch can be held on the height *h, Fig.* 29.24, of extruded holes. Inside diameter of extruded holes can ordinarily be held to plus 0.0015-inch, minus zero and the outside of the hub, owing to material variations, can be held to within plus 0.0025-inch, minus zero.

Flatness. Hard-temper materials, of course, will retain flatness and parallel surfaces more readily than softer ones which pull down at the edges somewhat during blanking, punching, etc. Commercially flat parts — the condition found as parts leave the die — should be acceptable wherever possible to avoid extra

TABLE 29-II — **Tolerances on Stamped Gears** *

(For 32 and 48-pitch)

	Standard Tolerances (inches)		Close Tolerances (inches)	
	Plus	**Minus**	**Plus**	**Minus**
Tooth Thickness	0.000	0.002	0.000	0.001
Tooth Depth	0.005	0.000	0.005	0.000
Hole Diameter	0.002	0.000	0.0005	0.0000
Pitch-line Runout	0.005 Total Indicator Reading		0.002 up to 2-inch diameter, plus 0.001 each additional inch.	
Flatness	Within 0.005 for 1-inch diameter or less, plus 0.005 for each additional inch.		Within 0.003 for 1-inch diameter or less, plus 0.003 for each additional inch.	
Burrs	Approximately ⅛ the material gage, but not less than 0.003, minimum.		Shaved to ½ standard tolerances, but not less than 0.003, minimum.	

* Courtesy Shakeproof Inc. Tolerances on tooth thickness of Bakelite gears is plus 0.000 and minus 0.004-inch unless special tooling is used.

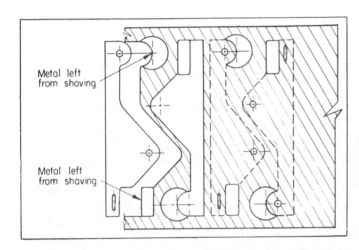

Fig. 29.27 — Removal of certain portions around a part prior to final blanking creates a shaving action to provide smooth, even and accurate edges, eliminating an extra operation which would otherwise be necessary.

Fig. 29.28 — Brass carburetor disks 2¹/₈-inches in diameter by ¹/₄-inch thick. Center hole, ¹/₂-inch in diameter, is shaved to plus or minus 0.0002-inch tolerance and the OD is held to plus or minus 0.001-inch. Cost was approximately 55 per cent of former machined parts. Photo, courtesy Farrington Mfg. Co.

operations. Surfaces of greater than several inches in area which must be flat within 0.002-inch or less will usually require a surface grinding operation. Generally, circular parts up to 1-inch can be expected to come within 0.003-inch, maximum out-of-flat. For each additional inch over 1-inch diameter, maximum out-of-flat is increased 0.002 to 0.003-inch.

Springback. Springback on formed parts is usually encountered with hard-temper materials and those that work-harden rapidly and necessitate wider tolerances or extra operations to overcome it. Tempers exceeding half-hard normally require an additional setting operation to hold accurate formed dimensions.

Burrs. Blanking and punching almost always give rise to burrs and on heavy-gage materials the sheared edge may present a rough surface. Where the amount of burr permissible on parts can be actually specified, expensive hand burring or tumbling operations can be eliminated. Burrs may run to 10 or 12 per cent of the material gage but in most cases a reasonable height of burr which can be held to in production is usually set up as 0.004-inch and in rare instances this can be reduced to 0.003-inch.

Shaving. Often smooth edges and close tolerances are necessary only on a part of a blank. Such cases can sometimes be covered by an integral shaving operation, *Fig.* 29.27, to hold the required accuracy on certain sections at little increase in cost while close tolerances overall would make the cost excessive. By removing a portion of metal adjacent to such specified surfaces, the edges opposite these portions are subjected to a shaving action during blanking, thus leaving them smooth and free from burrs and roughness.

Holes can be frequently held within 0.0005 and 0.001-inch and sometimes closer. Brass carburetor disks, for instance, are shaved to much closer tolerances, *Fig.* 29.28. Under ordinary circumstances, holes range from plus 0.000, minus 0.003-inch on 1/32-inch holes to plus 0.000, minus 0.010-inch on 1-inch holes.

Fine Blanking. In general, where fine blanking can be used, hole diameter and location tolerances of ± 0.0002-inch on light-gage parts can be held and ± 0.0005-inch with parts up to 1/2-inch thickness. Flatness of 0.001-inch per inch can be held on parts blanked from strip; about 0.002-inch per inch when blanked from coil stock. Edge finish in mild steel will range from 16 to 32 rms and in aluminum or brass alloys it will be even better.

General Dimensions. Production tolerances of plus or minus 0.005-inch can be held where essential, and with careful handling and inspection even plus or minus 0.001-inch can be obtained on circular shapes, hole locations, outside dimensions, width of small projections, etc. Production is simplified to a great extent if general tolerances can be opened to at least plus or minus 0.010-inch.

Owing to the fact that a blanking die determines the size of the blank produced and some taper relief is imperative, the die becomes larger after each grinding. Therefore, blanking tolerances should be specified as plus. Tolerances on pierced holes, for the same reason, should be given as minus. As the punch wears, the hole produced becomes smaller with each grind. With wide tolerances, say plus or minus 0.010-inch, the number of pieces produced before replacement of the punch or die becomes necessary is much greater than with small tolerances. Proper balance between cost and required accuracy must be determined to achieve the maximum benefit in quality and economy.

ADDITIONAL READING

[1] *How To Design Parts For Fine Blanking,* Designer's Guide MD-100, Mineola, NY: Monarch Die Corporation.
[2] *Fineblanking — A Periodic Technical Review,* No. 16, White Plains, NY: American Feintool Incorporated.
[3] J. V. Maleckar, "What the designer should know about short run stampings," paper presented at the ASME Design Engineering Conference & Show, Philadelphia, Pa, April 9-12, 1973, 73-DE-32, New York: ASME.
[4] E. W. Hand, "How to save money when specifying sheet steel," *Machine Design,* July 10, 1975.
[5] *Facts About Stampings,* Minneapolis, MN: Progressive Stamping, Inc.

30

Deep Drawing

$\mathbf{A}$LLIED closely with various punch press operations outlined previously under the general term "stamping" is the process of deep drawing. Unquestionably this production method parallels stamping in its importance as a vital source of low-cost machine parts. Often used in conjunction with one or more of the various stamping operations, the combination has become in many aspects synonymous. Effects of deep drawing requirements and limitations upon the design features of parts which fall within the scope of this method of forming metal consequently necessitate careful consideration by the designer to assure maximum economy in production.

The Process. Basically, drawing or deep drawing is a stretching process. Sheet is stretched over or drawn into a die to conform it to the required shape. Various means for accomplishing this drawing action are in use and also additional or companion operations. Generally these include:

1. Drawing
   (a) Deep drawing
   (b) Redrawing
   (c) Sizing
   (d) Ironing
   (e) Tapering
   (f) Upsetting
   (g) Flanging
   (h) Necking
   (i) Bulging
   (j) Trimming
2. Bulging
3. Rubber die forming
4. Stretch forming
5. Impact drawing

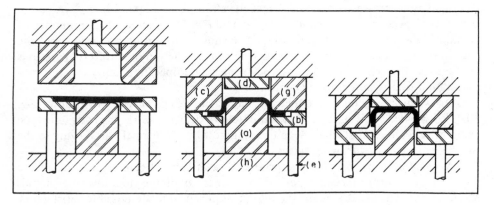

Fig. 30.1 — Simplified drawing sequence. Punch (a) has the shape desired on the inside of the shell or cup. Drawplate or blankholder is shown at (b), drawing die at (c) and knockout for ejecting the drawn part at (d). Drawings, courtesy Dayton Rogers Mfg. Co.

Deep Drawing. Most common and widely used, simple deep drawing consists of forcing a flat sheet into a cylindrical female die to conform it to the shape of the male die plunger, *Fig.* 30.1. A blankholder is usually used to retain the blank form while drawing. As this takes place, the portion of the blank clamped by the blank holder is drawn in and its circumference is reduced. Severe compressive stresses produced by this shrinking in the size of the blank tend to create wrinkles on light-gage sheet and so necessitate high holding pressures. Heavy-gage blanks, however, readily assimilate this shrinkage and consequently are easier to hold. In many cases heavy blanks require no blankholder at all.

Controlled stretching by means of controlled blankholding is highly advantageous and often necessary in drawing large-size irregular shapes, *Fig.* 30.2. In cases where stretching must be controlled to an extent greater than that normally available by means of the blankholder, draw beads or raised moldings are used to retard the metal movement. These are visible on the part shown in *Fig.* 30.2.

Fig. 30.2 — In the production of large, irregular shapes such as this turret automobile top, closely controlled stretching of the metal sheet is imperative. Photo, courtesy Ford Motor Co.

Drawing Presses. A tremendous variety of presses for performing deep drawing and associated operations is available. These range from small single-action machines that provide only a simple, single ram to multiple-action units with capacities in thousands of tons. Blankholding action on single-action presses is provided by springs, rubber bumpers, or pneumatic or hydraulic die cushions. Double-action presses provide a special slide for blankholding, *Fig.* 30.3. Mechanical presses for extremely large parts, *Fig.* 30.2, often have a number of blankholding slides to allow individual adjustment of holding pressure about the edge of the blank as required by the varying drawing action over different sections of the dies. Presses also range in type from single-operation units to multiple-station and totally automatic varieties.

Drawing presses may be either mechanical in design or fully hydraulic, *Fig.* 30.4. Each type is well suited to a specific field of drawing work, the hydraulic type being better suited for straight deep drawing of long cylindrical parts owing to the fine control over the drawing speed.

A specific drawn part may be produced, as a general rule, by various of the great number of different types of machines available for drawing metal. Modification will usually adapt the design required to make possible satisfactory manufacture by other available methods. However, careful consideration of all the factors involved will usually indicate one particular method or machine for maximum efficiency.

Large-quantity drawn parts of relatively simple design are most economically produced on the high-speed mechanical press. Plain shapes and sections with uniform and some with nonuniform curves or contours, which can be drawn in a single operation, are ideal for these presses. The more complicated parts — extremely deep parts, especially those with unusual shapes and difficult contours, *Fig.* 30.2 — are best suited for production on double or triple-action mechanical presses. Deep-drawn cylinders or parts with uniform cross sections, especially in stainless steels, are often produced most easily on hydraulic presses owing to the controllable, uniform drawing speeds.

Metals and Reductions. Drawing operations include the drawing and redrawing of all sorts of shapes and in addition sizing, ironing, and like forming work on certain parts. Machine parts are produced on the ordinary types of draw presses from practically all metals — cold formed in thicknesses up to about $3/4$-inch and hot up to about $3^1/2$-inches. In general, with a good deep drawing steel, a round blank can be drawn in one operation so that the resulting cup diameter is 40 to 50 per cent that of the original blank. Blanks to be subsequently redrawn are usually limited to 55 to 60 per cent or in other words a 40 to 45 per cent reduction in the first draw. Irregular shapes, such as the steel automobile top shown in *Fig.* 30.2, are limited to a draw which produces a maximum elongation of 40 per cent.

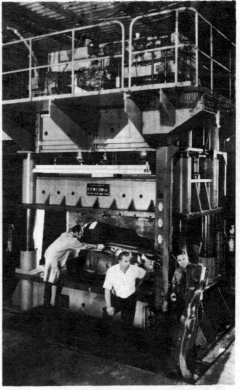

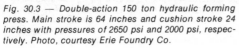

Fig. 30.3 — *Double-action 150 ton hydraulic forming press. Main stroke is 64 inches and cushion stroke 24 inches with pressures of 2650 psi and 2000 psi, respectively. Photo, courtesy Erie Foundry Co.*

Fig. 30.4 — *Self-contained 1250 ton multipurpose hydraulic press with 10 by 6 foot platens and 250 ton underslung cushion. An automotive prototype component is being checked. Photo, courtesy K. R. Wilson, Inc.*

Normally, draws requiring over 33 per cent elongation of the steel at any point require special handling or working.

On simple redraw work planned for single-action presses, reductions ordinarily considered as practical for production are as follows: For steel stock up to $1/16$-inch thickness, 20 per cent reduction in diameter over that left by the first draw; for stock between $1/16$ and $1/8$-inch, 15 per cent reduction; for stock between $1/8$ and $3/16$-inch, 12 per cent reduction; for stock between $3/16$ and $1/4$-inch, 10 per cent reduction; for stock between $1/4$ and $5/16$-inch, 8 per cent reduction. Large-production redraws which can be performed in more costly double-action dies allow an increase of approximately 50 per cent over these reductions. Quality deep drawing brass stock and like high-ductility materials drawn in single-action presses similarly show about a 50 per cent increase.

Drawing Depths. Ordinarily, cups or tubes can be drawn satisfactorily to a depth of 6 to 8 times their diameter in production. However, some special proprietary methods allow length to diameter ratios of 14 to 1 or more in aluminum or the newer deep drawing steels, *Fig.* 30.5. However, with ordinary parts, depth of draw should seldom be more than three-fourths its diameter in one draw, although in a double-action die it is possible to make the depth equal to the diameter in one draw. In hot drawing magnesium, cylindrical cups can be drawn to a depth 1.5 times their diameter. Physical properties of materials used, of course, may increase or restrict these depths. Transfer redraw presses will produce as many as 220 to 400 completed parts per minute. Because of the speed of transfer between redraws, no annealing is needed for draws at velocities over 200 feet per minute in

Fig. 30.5 — Special vertical hydraulic deep drawing press, 45 feet high with a capacity of 320 tons for use with the proprietary Metal Flo process. Horizontal version is also used to produce deep drawn shells as shown up to 14 diameters in length in one continuous multistage operation. Photos, courtesy Metal Flo Corp.

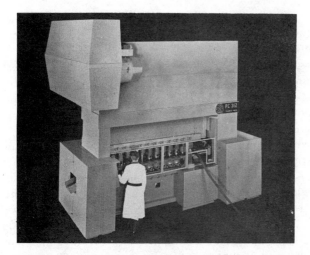

Fig. 30.6 — Typical sequence in the automatic drawing of a part on a multiple-station automated vertical press that blanks, draws, and forms while transferring the part from station to station until finished at production rates up to 550 pieces per minute. Photo, courtesy Waterbury Farrel, a Textron Co.

carbon steel. Typical diameters go up to 6 inches and depths to 9 inches with ratios as high as 12 to 1.

Multiple-plunger presses produce drawn parts with complex shapes and trimmed areas at rates up to 550 per minute. Blanking, cupping and finishing is done continuously so as to produce a finished part with every stroke, *Fig.* 30.6.

Tapered Shells and Walls. Deep tapered shells or those with tapered walls are usually difficult to draw and require numerous operations. The high output per hour and accuracy obtainable however, once dies are made, maintain low-cost production where large quantities are needed.

Forming Operations. In many cases production deep drawing of parts which would ordinarily be machined from solid stock can afford considerable savings in both time and materials, *Fig.* 30.7. On many parts such as that shown in *Fig.* 30.7, tapering, necking, sizing, profiling and upsetting can be used to advantage to

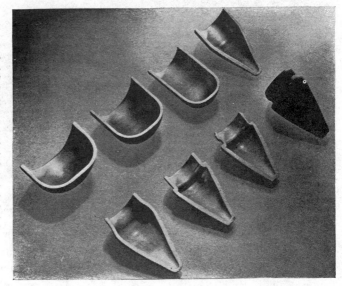

Fig. 30.7 — This dummy fuze for a 37-mm shell was converted from a machined aluminum casting to AISI C1010 steel drawn from a 3.25-inch diameter blank, 0.063-inch thick. Operations: Blank and first form; anneal; first redraw — correct thickness of material; second-redraw — thin walls; trim length; taper; first necking; second necking; third necking — size, profile, store metal in shoulder, and square corners; prepare for threading; thread. Photo, courtesy Nagel-Chase Mfg. Co.

eliminate machining operations. Although this part was threaded on the screw machine, roll threading can often be utilized to reduce costs still further when the wall thickness is sufficiently heavy to withstand the operation.

Expanding or Bulging. Where necking is undesirable or impractical on certain parts, expanding or bulging is often resorted to. This can be accomplished with metal dies, rubber dies or hydraulic dies. Bulging work is limited to a maximum increase in diameter of about 30 per cent in one operation. Bulges greater than this require additional steps and annealings.

Supplementary Drawing Processes. Logically falling under the deep drawing press operations are also those of rubber-die forming, stretch forming and impact drawing — all basically stretching processes. These operations are usually performed on special machines using special dies. Each, however, has assumed a major place in the mass production of parts which were unsuitable for the other methods.

Rubber-Die Forming. The rubber-die forming process, known as the "Guerin Process" utilizes a flat, thick pad of rubber and a single male die of any material with sufficient toughness to stand up under the working pressures involved. Plastics, low melting point alloys, impregnated fiber, and wood can be used. For greatest life as well as for intricate parts, metal dies are required. Dies are low in cost, easy to produce and require no fitting. The rubber pad on the upper press platen mates with the die regardless of form, shape or size. Blanking as well as drawing is possible. In some cases parts can be blanked and formed in one operation. Stock up to 0.040-inch can normally be blanked by this method.

Ideal where quantities of a variety of machine parts are needed constantly, a great many different parts can be formed or drawn at the same time, *Fig.* 30.8. At

Fig. 30.8 — Many different parts can be cut or drawn from one sheet by the rubber-die process. Sixty-six parts are blanked at once in the setup shown. Photo, courtesy McDonnell Douglas Corp.

each loading of the platen, a full-sized sheet of metal can be cut or pressed into parts. In normal operation over 15,000 parts can be produced in 8 hours by this method, *Fig.* 30.9. No lubricants are used and the parts emerge clean, eliminating a

Fig. 30.9 — *Typical variety of machine parts fabricated by the rubber-die process with low-cost tooling. Photo, courtesy McDonnell Douglas Corp.*

degreasing operation.

Rubber-die forming is successfully used for the drawing of aluminum and aluminum alloys, light-gage mild-steel sheet, stainless-steel sheet, copper and magnesium alloys. A minimum pressure of 1000 psi is required for working simple aluminum shapes. Greater pressures make for more perfect draws, especially with complicated parts. Hydraulic presses are considered best for this process and those in use range from 1000 to 5000 tons capacity.

Stretch Forming. Not unlike the rubber-die process is the drawing of metals on the stretch press. Flexible low-cost tooling for the production of quantities of parts which require continual design changes is the primary feature. Adapted to the economical fabrication of extreme double-curvature parts, this method is usually considered as intermediate between the drop hammer and the large, high-production draw press. Occasionally parts can be produced satisfactorily on no other machine.

Chief limitation is that only contours relatively symmetrical and with only slight reverse bends can be drawn, *Fig.* 30.10. In addition, a high percentage of scrap — as much as one-third the material used — is unavoidable. Each end must be clamped for stretching. Common practice dictates an average of 16 inches must be added to the length and 2 inches to the width of a part for trimming. Scrap loss can be reduced, however, by producing two or more parts simultaneously on the same die wherever possible.

Fig. 30.10 — Variety of stretch forming dies and typical aircraft parts produced on the stretch press. Photo and drawing, courtesy Chambersburg Engineering Co.

Drawing of sheet in the stretch press elongates the material and ordinarily reduces its thickness from 5 to 7 per cent. However, the increase in tensile strength resulting from stretching usually more than compensates for the reduction. Only designs which require about 8 per cent elongation (maximum 12 per cent) can be handled successfully. Nonuniform parts may be produced on the stretch press if the cross-sectional area remains constant, variation being in shape only.

Reverse contours, *Fig.* 30.10, can be formed if dies are designed to overcome wrinkling. Draw beads are used to hold the material from slipping at reverse curvatures. Proper use of lubricant also is effective in controlling reverse bending and wrinkling. Lubricant action such as that provided by vaseline or drawing compound is always required. However, rubber sheets can be used over the die with superior results. Marking or scratching and the problems of degreasing are entirely eliminated.

Impact Drawing. Unusual metal parts which are difficult to deep draw with regular draw presses or would require excessively costly dies can be produced by the impact method. Impact machines provide a wide range of blow intensities and difficult metal forming operations can be closely controlled with remarkable ease. This sensitive control and the fact that dies can be easily and rapidly produced from low-cost metals are the major advantages. Zinc is normally used for lower form dies and antimonial lead for the punches which are cast in the dies.

To handle the various types of work required, impact machines are available in sizes ranging from those with a work area of 12 by 15 inches to the largest with work areas of 120 by 120 inches and capable of handling dies weighing up to 53,000 pounds, *Fig.* 30.11.

Metals such as the aluminum and stainless alloys are particularly adaptable to this method of drawing and sheet materials can be redistributed into the most difficult forms with fairly uniform maintenance of gage. Even the most difficult

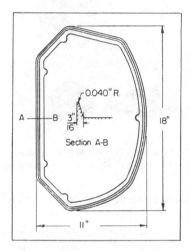

Fig. 30.11 — Typical Cecostamp impact machines and several drawn bus body parts. Plaster mold in foreground is used in casting dies from metals such as Kirksite. Photo and drawing, courtesy Chambersburg Engineering Co.

Section A-B

0.040" R.

A — B 3"/16

18"

11"

Fig. 30.12 — Narrow fillet section for a gas tank shell impact drawn. As many as 3500 pieces can be produced on such a notably weak die section.

draws, *Fig.* 30.12, can be made without die failures.

Hydroforming. Among the newer drawing processes is the Hydroform method. Somewhat akin to rubber die forming, this method utilizes a flexible die member, arranged as a seal across the bottom of a fluid-filled forming chamber. Only one male die is required, the flexible member serves as the upper blank-holder and the female die element. Uniform pressure acting all around the blank piece of metal "wraps" the metal around the punch and permits severe draws in one operation creating significant production savings and considerably less thinning of the metal gage during draws. With this process, production of very complex parts is possible, *Fig.* 30.13.

Production Hydroforming presses can be manual or automatic and some models match the output of conventional presses. High-speed automatics may range from 600 to 1800 strokes per minute. Automatic load and unload units offer high rates of production with the added feature of elimination of otherwise extra operations normally needed with conventional methods.

Explosive Forming. Occasionally a piece is required in the form of a large head or other deep drawn shape that is beyond the capacity of presses available to produce. To make such immense components the process of explosive forming has been perfected. By this method, high hardness materials, armor plate, titanium, etc., can be formed while producing superior physicals. Not only are relatively large parts produced requiring forming tanks as large as 30 feet diameter and 26

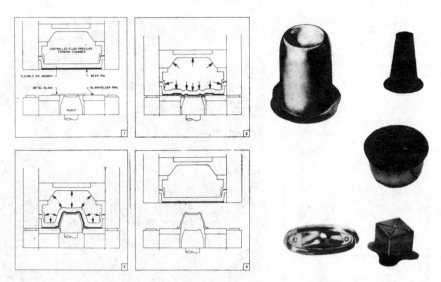

Fig. 30.13 — Four part sequence showing the Hydroform process. Pictures of some typical production drawn parts indicate the complexity of draw that can be made without marring the metal surface. Courtesy, Cincinnati Milacron Co.

feet deep — the process is performed under water — even smaller pieces are often produced, *Fig.* 30.14.

Design Considerations

Following rules somewhat similar to those for stampings, cost of drawn parts likewise can be kept to a minimum by maximum simplicity of design. Round cross sections are by far the simplest to produce while square, rectangular or irregular

Fig. 30.14 — Explosive drawn 12-inch diameter missile nose section of AM-350 alloy stainless 0.040-inch thick. Parts as large as 10 feet have been made by this process such as the hyperbolic antenna reflector units shown made in aluminum alloy 6061 0.190-inch thick within 0.005 T.I.R. of true contour. Photo, courtesy North American Rockwell.

cross sections involve greater difficulties in both die sinking and production. Shells having integral flanges generally involve higher manufacturing costs and tool maintenance than those without flanges. Shells without flanges create about one-half as much die wear and can be drawn at a much faster rate. To draw either flanged or plain shells near or at the maximum permissible reduction percentage, radii in the bottom and under the flange must be adequate.

Draw-Edge Radii. Radius of the drawing edge of a die — that radius under a flange — if too large tends to create wrinkles in the material while if too small will tend to tear the metal. This radius and that at the bottom of a shell should be governed largely by the gage of the material. Draw-edge or flange radii should be approximately 8 to 10 times the material gage. Sheet heavier than 16 gage (0.0625-inch) should have radii at least 6 times the gage size and wherever possible the radius should be increased somewhat for most satisfactory operation. Draw-edge radii for stainless steels may be reduced to 2.5 times the gage and for magnesium and most aluminum alloys it should be 4 to 8 times and preferably greater than 6 times. In Monel, nickel and Inconel, the radius should be from 5 to 12 times the gage. Large radii are especially desirable on parts drawn from heavy-gage plate. Where smaller radii are necessary, extra drawing or flattening operations are usually required to achieve satisfactory production. Deep drawings over one diameter in depth often can have these radii reduced on successive draws.

Corner Radii. Also important are corner radii or that radius on the nose of the punch used in drawing. Corner radius has considerable effect on the depth of draw possible to achieve in a single operation. The radius should, wherever design permits, be specified as four times gage, minimum, when shell depth exceeds one-third the diameter. Depending upon the corner radius and material ductility, shell depth may range from one-fourth to three-fourths shell diameter. Ordinarily, depths over 5/8 the diameter require more than one draw. With good drawing stock and generous punch-nose radius the operations are as follows:

1. Depth 5/8 to 1 1/8 times shell diameter — 2 reductions.
2. Depth 1 1/8 to 2 times shell diameter — 3 reductions.
3. Depth 2 to 3 times shell diameter — 4 reductions.

In no case should the punch-nose radius be less than the metal thickness; otherwise, wrinkles result, especially where stock is less than 5 per cent of the shell diameter, width or length. Internal or external corners can be made square in one or more extra setting or sizing operations, however, where metal thickness is 1/8-inch or greater.

Rectangular and Square Parts. Corner radii of rectangular or square shapes are of primary importance. Satisfactory production is entirely dependent upon these radii and the material used. Steel boxes over one-half their width in depth should be made from deep drawing material. Corner radii should be as large as possible and at a minimum five times the gage. Flange and bottom radii under 3/4-inch usually require more than one operation. In a good drawing material such as brass, boxes can be drawn to 6 times the corner radius in one operation. In other materials, depth in one draw will be somewhat less than 6 times depending primarily upon the particular elongation characteristics.

Parts designed for rubber-die forming also require generous corner, bottom and flange radii. Notches in the blank along outside bend contours are sometimes required, *Fig.* 30.9 to avoid wrinkling which results from lack of sufficient blank

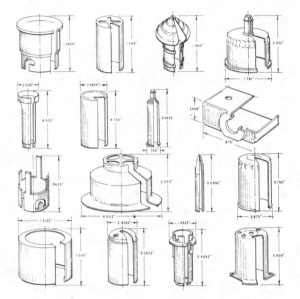

Fig. 30.15 — Group of fairly typical multiple-station drawn parts which show a variety of shapes and projections produced in the drawing sequence. Drawing, courtesy Waterbury Farrel, a Textron Co.

holding action exerted by the upper rubber-die pad.

Trimming. For relatively shallow shells where depth does not exceed one-third the diameter, it is possible to produce a finished edge within commercial tolerances without additional trimming operations. Deeper shells undergo sufficient stretching to produce an excessively uneven top edge. Such parts can be flange trimmed with one extra trimming die as follows: (1) The shell is partly drawn leaving a flange; (2) the flange is then trimmed accurately circular; and (3) finally the trimmed flange is finish drawn through the first die to produce the desired accurate shell length. Where a somewhat rounded inside edge can be tolerated, pinch trimming with two or three extra dies and operations can be used with medium and large lots. A variety of shapes and projections can be produced right in the drawing sequence to provide finished parts at high speed, *Fig.* 30.15.

Where large or continuous production is in view, trimming by means of the wedge or Brehm "Shimmy" die* can be utilized. Not only can ordinary plain edges

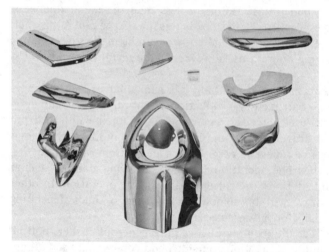

Fig. 30.16 — Group of plain and unusual parts trimmed with a shimmy die in one press stroke. Photo, courtesy Vulcan Tool Co.

*Manufactured by the Vulcan Tool Co.

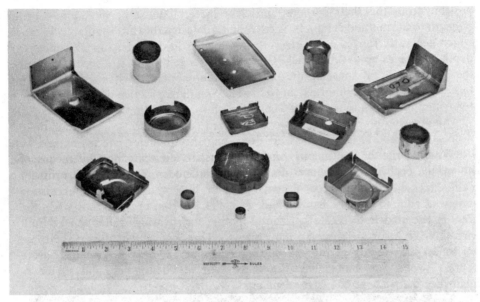

Fig. 30.17 — A variety of parts showing the trimming possibilities with Shimmy dies. Photo, courtesy Vulcan Tool Co.

be produced on drawn shells of almost any shape, but also a variety of contour designs can be had, *Fig.* 30.16. Action of these dies when the press is operated is to create horizontal motion for shearing. Dies can be built with three, four or six shearing movements. Three-movement dies make three horizontal movements of the die cutting edges at 120-degree spacings during a press stroke; four-movement dies make four at 90 degrees; and six-movement dies make six at 60-degree spacings.

Square, rectangular and most irregular shells are trimmed in four-movement dies. As a rule, the type of die is governed by the trimmed outline to be produced. Shells with one to three equally spaced notches or tabs can be trimmed in three-movement dies, one to four equally spaced in four-movement dies and one to six in six-movement dies. The projections, however, must be properly spaced, *Fig.* 30.17. Extremes of shape such as long narrow slots or projections often cannot be produced; the narrow slugs left are difficult to strip from the die openings. Variations in shape or number of projections or notches included on any one side can be made, *Fig.* 30.18. The main requirement is that tabs or slots be kept as far as

Fig. 30.18 — Shimmy die trimmed drawn part showing multiple tabs which can be produced. Photo, courtesy Vulcan Tool Co.

possible from the shell corners; sufficient material must be available for die strength or the method cannot be used. Standard design dies are adaptable for work which generally ranges in size from 2 to 7 inches in diameter, up to 3 inches in depth and gages up to 0.062-inch. Outside this range the dies are special. Because the hollow internal shearing die must be large enough to withstand the cutting forces, small parts are impractical to handle. Standard practice is to allow 1/2-inch for trimming to assure sufficient punch length for regrinding.

Depression or Pads. Depressions or raised pads should be no greater than is necessary. A good rule is to avoid exceeding the diameter width or narrowest dimension in depth. Deeper pads or depressions are often possible in deep drawing materials, *Fig.* 30.19, but generous radii all around are necessary. The absolute

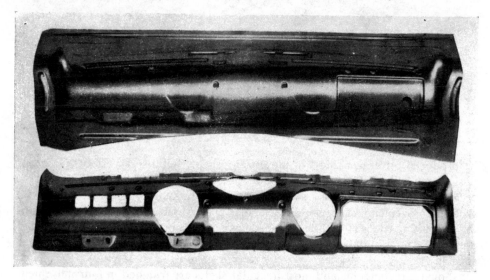

Fig. 30.19 — First draw and finished automotive instrument panel. Draw beads can be seen on first stage piece on sides and along bottom. Hole in section where glove compartment door is to be blanked out provides a spot for stretching to take place and obviate tearing. Numerous depressions and raised pads and radii used are readily visible. Photos, courtesy Ford Motor Co.

minimum radius which can be used is one equal to the gage. Preferably, the radius should be no smaller than 2 times gage; where 5 or 6 times or better can be used production is much easier. Where raised or depressed pads are plain circular types, *Fig.* 30.20, conditions for drawing depths on flanged cups apply, and also radii on die and punch, regarding single draws. Typical strengthening beads are also visible in *Fig.* 30.20 showing 90-degree sides and rounded ends. Die-edge radius for such beads should be at least two times gage for small bead widths, increasing to as much as four times gage with wide beads of 1-inch and over. The punch-nose radius in each case should be about double the die radius. Where two such strengthening beads cross, the joining radii should never be less than twice gage and preferably greater.

Unusual Shapes. Drawn parts having hemispherical or curved bottoms should be avoided in favor of those with substantially flat bottoms. Unless design benefits are gained by such deeply curved bottoms and elongations over 33 per cent, the higher production costs involved are hard to justify. Parts such as modern automobile fenders, doors, hoods, and tops fall into this category and usually entail much research and special handling technique for successful production. Such parts

Fig. 30.20 — Floor pressing for an automobile, drawn in a double-action press, showing raised and depressed pads and strengthening beads. Photo, courtesy Ford Motor Co.

are often tested out in quarter-size dies to determine the practicability of the design. In these test dies by drawing sheet which is about one-third the required gage of the full-sized part a very accurate check is made. Design changes can then be made before the final production dies are ordered.

For successful deep drawing of these irregular parts, the area of a blank at the holding plane must be less than the overall area of the part to assure the necessary stretch. Wherever even small amounts of compression are necessary wrinkling will usually occur. To avoid this, general practice is to provide take-up depressions to assure ample stretch, *Fig.* 30.21. Relief is often necessary for successful production of unusual shapes, *Fig.* 30.22. Shearing through sections to be blanked out on finishing operations relieves the strain of the draw sufficiently so that elongations much greater than 40 per cent can be handled. To obviate tearing of the entire sheet the shearing, however, must be timed properly to take place when the draw is from 60 to 70 per cent complete.

Blanked Details. Layout of holes, tabs, bends, perforations, etc, are governed by much the same procedure as was previously outlined in Chapter 29 on stamping.

Fig. 30.21 — Indentations or depressions provide the necessary stretch to take up excess metal and avoid wrinkling of the metal at what otherwise would be compression areas. Photo, courtesy Ford Motor Co.

Fig. 30.22 — Relief cuts at critical spots in sections to be blanked out reduces strain and makes ordinarily excessive elongations possible. Photo, courtesy Ford Motor Co.

Ironing. In plain cylindrical designs, as noted previously, ironing can be used to reduce gage evenly when necessary, *Fig.* 30.23. Ordinarily the blank must be one or two gages heavier than the final wall thickness but where smoothness and accuracy only are desired, about 0.004-inch reduction is satisfactory. The usual maximum reduction in wall thickness by ironing is limited to about 50 per cent although cases are on hand where this has been as high as 64 per cent.

Stretch-Formed Parts. Parts to be produced on the stretch press necessarily must be as simple as possible. Reverse curves parallel to the stretch are best avoided to simplify production, those normal to the direction of stretch being impractical. Inside radii of parts should not be less than 1-inch for successful stretching. Larger radii are more desirable. Edge condition of sheet or blanks to be stretched is often critical. Rough sheared edges allow only small stretches but polished or routed edges make possible maximum working of the metal without undue failure from stress concentrations.

Selection of Materials

Relative suitability of materials discussed in the chapter on stamping apply equally to drawing. Preferred steel for drawing, especially deep work, is of course high-quality automotive deep drawing sheet. Draws requiring up to 33 per cent elongation of the material are easily produced and depending upon the material, design and technique this can often be raised to about 40 per cent. Preferred hardness of this sheet is about 40 to 50 Rockwell B.

Drawability. For maximum drawability, 19-gage sheet when pressed by a 7/8-

Fig. 30.23 — Group of deep-drawn 6061 aluminum propeller domes. Drawn blank shown prior to machining at left indicates the variation in wall thickness obtained by ironing. Starting with a 23-inch diameter blank of 0.500-inch (plus 0.040-inch minus 0.000) gage, twelve successive press operations produce a dome thickness in the final piece of 0.520-inch, an ironed side wall of 0.385-inch and a heavy upper end of approximately 0.565-inch. Upsetting shapes the heavy end to a final thickness of about 0.672-inch and the form as shown. Cup on upper end is drawn down 31/32-inch with a punch having a nose radius of 1/8-inch. Bottom of cup, which is reduced to approximately 0.344-inch in thickness, is punched out with a diameter of 2³/₄-inches. Note drag down around hole which must be acceptable.

inch ball into a 1-inch female die having a 1/16-inch radius on the drawing edge, should draw to a depth of approximately 0.425-inch. This represents about a 40 per cent elongation. However, most production-run sheet now available will draw successfully only to about 0.400-inch. In normal production of irregular automotive parts a test depth of 0.415 to 0.420-inch is desirable.

Typical ductility (depth of cup) values at the point of fracture for 19-gage (0.0437-inch) soft sheet are as follows:

Deep-drawing steel 0.413-inch
Extra deep drawing steel 0.433-inch
Folding or tinned sheet steel 0.397-inch
Aluminum 0.402-inch
Deep drawing cartridge brass sheet 0.563-inch
Copper 0.472-inch
Zinc sheet 0.323-inch

Ease with which complex parts can be drawn will be in direct proportion to the ductility or drawability index of the material. Also, it should be noted that, in many cases, one alloy which offers excellent forging qualities may be poor for deep drawing. For instance, while 2017 aluminum alloy is a satisfactory forging alloy it will not stand repeated cold working and 6061 would be the practical substitute where heat-treated parts are desired. Automotive quality deep drawing aluminum sheet is available which offers an elongation value of either 25 or 30 per cent in 2 inches.

Specifying Materials. Cold-rolled sheet for drawing should be specified on a weight basis giving widths in inches, length in feet and gage in decimals of an inch.

Fig. 30.24 — Heavy-duty stainless part drawn with a peelable plastic coating which obviates die marks and shipping or handling scratches. Photo, courtesy Better Finishes & Coatings, Inc.

Type of edge and tolerances on width and thickness are important. Type of surface or surface finish required should be specified for the job — bright, extra-bright, nickel plated, blued, rustproofed, leaded, tinned, hot or electrogalvanized, or copper plated. Extra deep-drawn steel parts or parts requiring unmarred smooth surfaces are as a rule copper plated before drawing. The copper tends to lubricate the material during drawing and prevent scoring. It can be easily stripped after drawing if desired.

Magnesium. Magnesium can be handled hot easily by the rubber-die process using special heated platens on which the form dies are set. Heat is transferred to the die and then to the magnesium sheet. Special synthetic heat-resistant rubber is required in the upper die platen. Most synthetic rubber pads will withstand operating temperatures up to 450 F. Some simple deep drawn parts are being produced in special dies on mechanical presses. However, applications to date are rather limited. Reduction in one hot draw may run as high as 65 per cent.

Molybdenum. Sheet stock of molybdenum is available and in gages under 0.030-inch can be deep drawn successfully. Heavier gages must be heated and worked at temperatures ranging from 100 to 300 C depending upon the gage and shape of the part.

Hot Forming. Where steels over 3/4-inch thick are to be drawn the work may have to be done hot. Cold finished stock would have little advantage owing to surface scaling from the heat and consequently hot-rolled materials in thicknesses up to 3 1/2 inches maximum are specified for hot-formed parts.

Drawing Speeds. In single-action unrestrained drawing, brass sheet can be handled faster than any of the other metals. Speeds as high as 200 fpm can be used. This drops to about 100 fpm on double-action drawing and to about 70 fpm for ironing or burnishing. Steels can be handled at drawing speeds up to 55 fpm in single-action and 35 to 50 fpm in double-action dies. Ironing or burnishing is done at speeds up to 25 fpm. In steel dies, pick-up troubles are encountered at higher speeds. However, with carbide dies speeds can be considerably higher. Copper and zinc can be handled at drawing speeds up to 150 fpm in single-action and 85 fpm in double-action dies. For soft aluminum alloys drawing speeds are 175 and 100 respectively (the strong alloys about half). Stainless steels must be drawn somewhat slower than ordinary steels, say 20 to 25 fpm, to reduce work hardening effects. Magnesium is drawn at slow speeds of 7 to 17 fpm, the lower speeds for the thicker sheets.

Drawing Pressures. Pressure required for drawing various materials can be computed from the part size and material. Pressure in tons is roughly equal to the product of circumference of the draw in inches, gage in inches and tensile strength in pounds per square inch divided by 2000.

Die Marks. Many external parts require polishing to remove die marks often impossible to avoid in drawing operations. It is possible today, however, to draw parts without marks through the use of a plastic coating which makes practicable the use of prefinished or precoated sheet. Polished parts are thus obtainable without the difficulty of polishing complex contours by hand methods. Plastic coating available is called "Liquid Envelope" and withstands the abuse of almost any drawing operation, cost per square foot on the most difficult being only about 2.5 cents. Another advantage lies in its protection during shipment, *Fig.* 30.24.

Specification of Tolerances

Tolerances normally held in production of drawn parts are in many cases closely linked with those found under stamping. Widest practical tolerances, of course, promote economy in production.

Impact Drawing. The alloys used in impact work prevent any marring of the sheet and, under ordinary conditions, average parts can be produced regularly to tolerances of plus or minus 0.005-inch.

Hot Pressing. Hot pressed parts require generous tolerances. A typical part pressed from 3 1/2-inch stock having radii of 7 1/2, 17 and 26 inches, for instance, is held to minus zero and plus 3/8-inch. Overall dimensions of the part are 51 inches high by 59 inches wide by 20 3/4 inches deep.

Shells. Extremely close tolerances on shells demand tight dies which often result in excessive wear and difficult stripping where flanges are encountered. Tolerances of plus or minus 0.005-inch can be held on diameter of shells up to about 6 inches with sheet up to 1/16-inch in thickness, about plus or minus 0.010-inch on those from sheet up 1/8-inch, and plus or minus 0.015-inch on those from sheet around 1/4-inch. Irregular, rectangular, or square shells require tolerances about 20 per cent greater. Of course, closer limits can be obtained in any case by redrawing, ironing or sizing in extra operations. Weight tolerances under 0.1-ounce per ounce is held on the shell in *Fig.* 30.7 and dimensional tolerances are held within a total variation of 0.005-inch.

Ironing. Shells 4.250 inches in outside diameter drawn 4 inches deep from 0.050-inch stock are held in production to plus or minus 0.001-inch on the ID. Annealing and light ironing is required to hold this dimension. However, ironing adds considerably to the pressure of the punch on the bottom of the shell and generally reduces the possible diameter reduction. Since diameter reductions often must be reduced by one-half to achieve high accuracy and extra fine finish by ironing, production is drastically reduced and cost consequently is much higher.

Ordinarily, without ironing, wall thickness of shells drawn from regular steels may vary from 10 to 20 per cent from the bottom to the top, thickening especially in the top or flange. To reduce friction between the stock and the surfaces of a punch and die for high-production purposes, greater clearances are allowed and thickening at the top of shells may run over 30 per cent. Aluminum drawn freely without ironing usually runs approximately 25 per cent. Where walls must be uniform, ironing will produce exacting tolerances, *Fig.* 30.5, and wall thickness can be held uniform from end to end.

Lengths. Accuracy of drawn lengths is affected by the metal gage as well as depth of draw. Tolerances of plus or minus 0.010-inch on drawn shapes less than 1/2-inch in depth and plus or minus 0.015-inch on round shells up to 1 1/2 inches in depth is about minimum. Closer tolerances require a trimming operation.

Concentricity. A U-section oil seal retainer with an annular slot equal in width to the gage and about four times the width in depth is a good illustration of the achievement of high accuracy. Required total variation on the slot width was 0.003-inch. Total runout on the inside and outside flanges was to be no greater than 0.006-inch. To hold these tolerances the inside flange was drawn first, then the outside — in two operations to reduce tool pressures — and finally the sides were ironed to size. Cost was one-third less than that of the same part machined from a casting.

General Accuracy. Extremes of accuracy are found in such drawn parts as cylinder liners for internal combustion engines. Drawn and ironed to an accuracy and fine finish that requires no further processing besides heat treating, such parts obviously justify expensive dies and tooling. However, if plus or minus 0.001-inch is contemplated on average-size work, it should be remembered that plus or minus 0.002-inch or even 0.003-inch with larger tolerances on dimensions greater than 5 inches can greatly alleviate production problems. Tolerances around plus or minus 0.010-inch or greater, wherever possible, are much easier to handle and naturally make highly successful mass production at minimum cost.

31

Rotary Swaging & Hammering

BASICALLY a fast mechanical hammering or impacting method, rotary swaging affords many of the desirable characteristics inherent in the various hot and cold forming methods. Perhaps more accurately termed rotary reducing, the swaging process should not be confused with swedging which consists of forming or shaping with but a single squeeze or blow of a solid die. Rotary swaging is, broadly speaking, confined to the reducing of a symmetrically cross-sectioned metal blank to a desired shape or form. This factor places specific limitations upon the range of design possibilities, but savings in material, production economy and the fact that certain parts cannot be easily produced by other means makes it imperative that swaging be kept in mind by the designer in evaluating possible methods of production.

Basic Process. In swaging, the blank is fed by hand or mechanically into the die openings where thousands of hammer blows per minute conform the blank to the shape of the dies used. Blanks ordinarily may be of any symmetrical cross section, although the actual finished swaged form, being produced by a series of fast hammer blows accumulatively applied about a circle, must be substantially of circular cross section. Along the length of the part, however, an almost unlimited variety of diameters or forms can be produced.

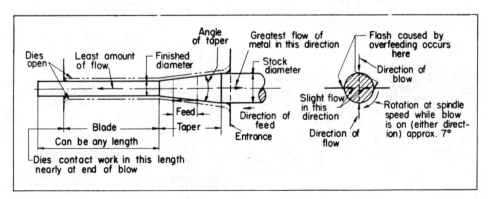

Fig. 31.1 — Characteristic metal flow conditions normally accompanying the cold-swaging operation. Drawing, courtesy Torrington Co.

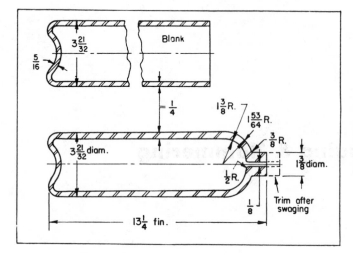

Fig. 31.2 — Heavy-wall, low-carbon steel bottle swaged hot in one operation. Swager equipped with a pneumatic feed holder and hand gripping device provides a production rate of 2 to 3 pieces per minute. Drawing, courtesy Fenn Mfg. Co.

Swaging Action. Under normal conditions each blow of the machine produces a flow of metal approximately as shown in *Fig.* 31.1. Where angle of taper is too great the metal feedback encountered requires the use of power feed devices. Flow, however, takes place in all directions but the amounts in each case depend upon such factors as the particular part design, metal being swaged, condition of the dies, etc.

Cold Working. Swaged parts may be either hollow or solid. In either case the working of the metal through swaging refines the grain structure and improves metal quality. Parts produced by cold swaging exhibit the highly desirable characteristics which normally accompany cold-worked metal. Extent of the effects of cold working depends largely upon the machine used. A light machine used on a heavy piece only has a slight surface effect while an adequate machine produces the maximum beneficial effect. Blanks cold worked on the surface only show concave end faces while those thoroughly worked show a definite bulge at the same point.

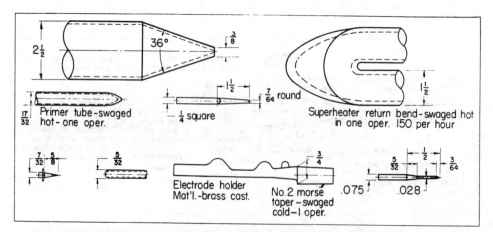

Fig. 31.3 — Group of cold and hot-swaged parts representative of pointing operations. Courtesy, Fenn Mfg. Co.

Hot Swaging. High-quality finish and increased strength are largely sacrificed when swaging hot, but hot working is necessary with highly work-hardening materials or extremely severe reductions, *Fig.* 31.2. Normally, therefore, hot swaging should be avoided whenever possible for maximum economy in production and where fine surface finish is desirable without machining.

Scope of Application. The field of design for swaged parts can be divided generally into three broad classifications: Pointing, forming and attaching. Examples of pointing are spouts, valve needles, bayonets, automatic pencil points, pen caps, and shell noses, *Fig.* 31.3. Forming includes a much wider field such as valve stems, roller chain bushings, bearing rollers, tubular parts, shell cases, axle housings, torque tubes, struts, axles, radio antennas, masts, etc., *Fig.* 31.4. Attachment applications involve the swaging on of fittings and other attachments to cables, wires, hose, etc., *Fig.* 31.5.

Swaging Machines. Consisting principally of a cage of rollers within which the swaging dies rotate, the swaging machine is relatively simple in design. Rotation of the die head causes the die shoes to strike the rollers, closing the dies against centrifugal force in rapid succession. Owing to slow rotation of the roller cage on contact with the dies, the actual number of blows per minute is somewhat less than the product of the rpm and the number of rolls. At 600 rpm, however, a 10-roll head produces about 3750 blows per minute.

Dies. Both two and four-die arrangements are employed in standard swagers. Because the four-die head practically doubles the working surface, speed of reduction is materially increased and heavier work can be handled. However, finish obtained with the four-die setup is somewhat inferior to that of the two-die arrangement. Too, inasmuch as die contact area is reduced to about half that with 2 dies, only large parts can be handled successfully. Generally, the four-die principle is not applicable where the reduced diameter is less than $3/16$-inch.

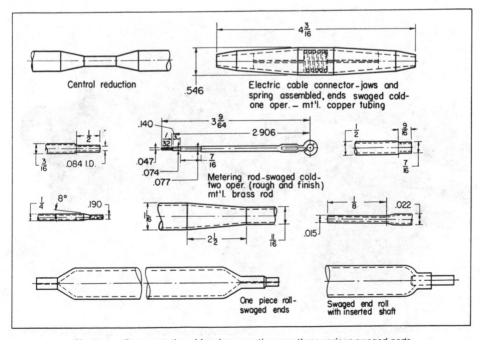

· *Fig. 31.4 — Representative of forming operations are these various swaged parts.*

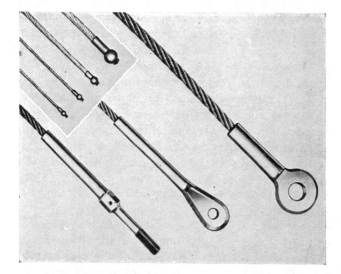

Fig. 31.5 — Group of machine fitting attachments produced by cold swaging. Photos, courtesy Standard Machinery Co.

Machine Capacities. Capacity of various machines is based upon the assumption that maximum diameter work specified is to be swaged the full length of the dies. Where length is less than die length, work diameter may be increased in inverse proportion limited only by the available die size. This also applies to hollow work and is based on tubing with wall thickness no greater than one-half the diameter of solid work capacity. Usually machine capacity on solid stock is roughly about one-half that on tubing.

Smallest machines available have a capacity on solid work of $1/16$-inch and smaller, on tubing of $1/4$-inch and under, while longest taper possible in one

Fig. 31.6 — A wide variety of standard swagers is available providing considerable leeway in design of parts. This heavy-duty hydraulic feed machine will handle up to 6-inch tubing. Courtesy, Fenn Mfg. Co.

operation is 7/16-inch. A wide variety of swagers is available, *Fig.* 31.6, carrying the tubing capacity range up to about 9 inches, solid round capacity up to about 6 inches and length of taper possible in one operation to about 24 inches. Models which carry up to three separate sets of dies located side-by-side are also available for multiple-operation work. Capacities are normally based upon mild steel, softer materials allowing a proportionate increase and harder ones a decrease.

Production Speed. In general swaging is a fast operation. Feed rates for reductions on either solid or tubular work range from 5 to 10 inches of swaging per second. Speed, of course, will vary from this figure according to design and other conditions but it is sufficiently close for all reductions up to a 15-degree included angle.

Fig. 31.7 — Fenn Model 8F Four-Die machine is equipped with a hydraulic feed mechanism for reducing forged axle housings from 6 inches to 5¹/2 inches in diameter in fifteen seconds.

Power Feeds. For large-quantity production swaging, and for work which cannot be handled by hand, various types of feeding devices — hand operated, semiautomatic or fully automatic — are available, *Fig.* 31.7. For sizing operations continuous roll feeds are usually used but for tapering and reducing, the hydraulic devices available generally provide the best results. For parts requiring a central reduction, *Fig.* 31.4, specially designed machines are required so as to allow entry and removal for undercuts, *Fig.* 31.8.

Hammering. An interesting adjunct to the swaging process is that known as hammering. Principle of operation is very similar to that of swaging and like swaging offers the advantage of cold working and economy of material. Hammering, however, is done by two dies operating in a vertical plane, the upper die reciprocating and striking the work held in the lower die, which is either mechanically or hydraulically fed to suit the work being produced, *Fig.* 31.9.

The hammering machine is well adapted for producing work impossible to swage. Parts can be formed from either solid stock or tubing, *Fig.* 31.10. Where round parts are produced a rotary attachment is normally used and where necessary, such as in producing balls, the rotary attachment includes a collet feed for continuous operation.

Design Considerations

A general idea as to the range of design application afforded by swaging can be gained from the foregoing illustrations. Normally, swaging an abrupt taper on a solid bar or swaging close to a shoulder is not easily accomplished. Generous fillet radii should be specified for all adjoining steps in diameter.

Taper Angles. For maximum production and economy the included angle on any reduced portion should not exceed 8 degrees. Steeper angles increase proportionally the feeding pressures required and for angles over 12 degrees, feeding fixtures usually are imperative. On tubing the included angle of reduction does not present quite as great a problem, but included angles should be held to 30 degrees or less for economy. In production, angles up to as much as 60 degrees are practical, *Fig.* 31.11. These large angles in conjunction with heavy reductions necessitate either hot working or more than one operation, *Fig.* 31.11, adding to production cost. Again, too great an angle on tubing often causes hard swaging and produces wrinkling or roughness in the finished part unless the work is performed in easy stages.

Pointing. On pointing operations the smallest radius of point that can be produced is about 0.010 to 0.015-inch. Good practice recommends a minimum radius of 0.015-inch be specified. Pointing on fairly heavy tubing normally requires

Fig. 31.8 — Hydroformer swager equipped with hydraulic wedge die openers for producing central reductions on parts or undercut designs. Courtesy, Fenn Mfg. Co.

Fig. 31.9 — A hammering machine equipped with a hydraulically operated lower die.

no mandrel for support but thin-wall tubing always does. Where a spherical pointed end is desirable and where such ends must be pressure tight, silver soldering or welding is recommended, *Fig.* 31.11. Tubular parts can be taper-pointed closed tight, *Fig.* 31.12.

Tubing. Normally, to avoid wrinkling or corrugating troubles on tubing, the OD must be less than 35 times the wall thickness. Pointing and reducing, of course, creates a thickening of the tube wall rather than merely a lengthening action as with solid stock, so for economy in production this condition should be acceptable. Of course, wall thickness can be maintained with a mandrel, provided such can be employed. If possible, in all cases where mandrels must be used, a taper should be allowed, however small, in order to facilitate removal of the parts from the mandrel in production. Representative of mandrel work are tube sizing, creating shaped or sized bores, pencil tips, torch tips, some types of thermostatic control bulbs, etc.

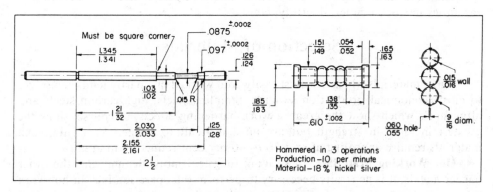

Fig. 31.10 — Several representative parts produced on the hammering machine with a rotary attachment.

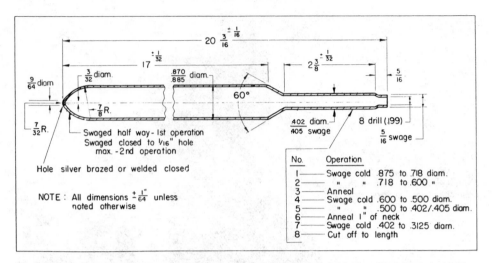

Fig. 31.11 — Bulb for temperature controller cold swaged from copper, stainless iron, Monel and steel tubing. Drawing, courtesy Taylor Instrument Co.

Lengths. Length of section which can be swaged, as well as diameter, depends upon the size machine available and normally falls within the general overall limitations covered previously. With straight cylindrical sections, of course, length is unlimited.

Diameters and Forms. Wire can be swaged to shape or size down to diameters as small as 0.005-inch. This, however, represents minimum possible diameter and is seldom, if ever, decreased. On all forms, wherever practicable, sharp corners should be replaced with generous radii to insure good die life and aid movement of metal.

Fittings and Sleeves. Ferrules or attachments can be swaged onto a cable, wire or other inner member tightly enough to withstand a tensile strain about equal to that of the inner member. Cross section of the ferrule or attachment, however, must be strong enough to provide the proper bond and gripping strength which is approximately proportional to the strength of the ferrule cross section used. Design of typical swaged-on cable fittings is shown in *Fig.* 31.13 and accompanying dimensional specifications are in TABLE 31-I. Where sleeves are to be swaged over a shaft, volume of metal needed to fill the recess completely, *Fig.* 31.14, must be computed and exact length for production should be ascertained by trial owing to variations in stock temper, ductility, etc.

Selection of Materials

Those materials which are most easily cold worked ordinarily lend themselves to most economical production swaging. Stainless steels, high-carbon steels and alloy steels which show the greatest work-hardening effects, of course, show the greatest increase in strength but are most difficult to swage. As a rule, such materials require annealing after a 30 to 40 per cent reduction in area.

Hot Working Materials. In swaging tungsten, molybdenum, and the other harder metals, it is generally necessary to perform the operation hot. In hot work there is no limiting ratio of tube OD to wall thickness or limit to the amount of

Fig. 31.12 — Group of tapered and pointed tubular parts. Courtesy, Fenn Mfg. Co.

reduction. However, as previously mentioned, hot working leaves a rather poor surface which requires machining if high accuracy is required. Steels such as chrome-moly which decarburize rapidly at hot working temperatures cause production difficulties especially if subsequent heat treatment is necessary.

Light and average-weight tubing in low and medium-carbon varieties, where OD to wall thickness ratio is under 35 to 1, can be reduced to 60 or 65 per cent of

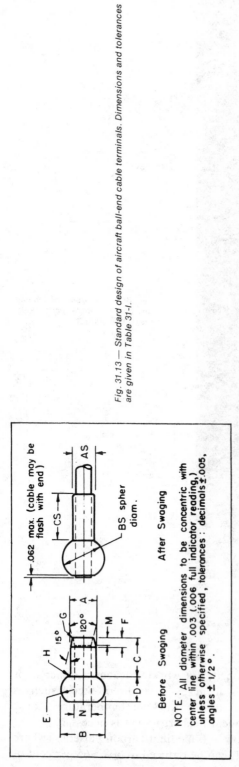

Fig. 31.13 — Standard design of aircraft ball-end cable terminals. Dimensions and tolerances are given in Table 31-I.

Before Swaging   After Swaging

.062 max. (cable may be flush with end)

BS spher diam.

NOTE: All diameter dimensions to be concentric with center line within .003 (.006 full indicator reading,) unless otherwise specified, tolerances: decimals ±.005, angles ± 1/2°.

TABLE 31-I — **Single-Shank Swaged Ball Terminals***
(inches, see *Fig.* 31.13)

Cable Diameter	Minimum Breaking Strength (lb)	A (Diam)	AS (Diam)	B (Diam)	BS (Diam)	C	CS +0.031 −0.000	D	E	F +0.006 −0.000	G +0.005 −0.000 (Rad.)	H +0.005 −0.000 (Rad.)	M +0.010 −0.000	N (Diam)
1/16	480	0.131 +0.000 −0.003	0.112 +0.000 −0.003	0.210 +0.000 −0.003	0.191 +0.000 −0.003	0.156	0.156	0.128 +0.003 −0.000	0.078	0.037	0.007	0.003	0.031	0.073 +0.003 −0.000
3/32	920	0.171 +0.000 −0.004	0.143 +0.000 −0.003	0.286 +0.000 −0.004	0.253 +0.000 −0.003	0.188	0.219	0.160 +0.005 −0.000	0.094	0.052	0.006	0.010	0.047	0.104 +0.003 −0.000
1/8	2000	0.222 +0.000 −0.004	0.190 +0.000 −0.003	0.364 +0.000 −0.004	0.315 +0.000 −0.003	0.250	0.281	0.187 +0.005 −0.000	0.109	0.060	0.010	0.010	0.063	0.139 +0.004 −0.000
5/32	2800	0.265 +0.000 −0.004	0.222 +0.000 −0.004	0.435 +0.000 −0.004	0.379 +0.000 −0.004	0.344	0.406	0.219 +0.005 −0.000	0.141	0.083	0.008	0.015	0.078	0.169 +0.004 −0.000
3/16	4200	0.295 +0.000 −0.005	0.255 +0.000 −0.005	0.500 +0.000 −0.005	0.442 +0.000 −0.005	0.438	0.500	0.278 +0.005 −0.000	0.172	0.079	0.007	0.025	0.094	0.201 +0.004 −0.000

Courtesy McDonnell Douglas Corp.

their original diameter without exceeding normal costs. This holds true for the entire range of sizes which can be handled in available machines and wherever possible these easily worked materials should be utilized for maximum economy.

Specification of Tolerances

Cold swaging normally results in a highly acceptable surface finish about equivalent to that of grinding. Properly polished dies produce a bright, smooth surface on parts and no further finishing is required. It must be remembered, though, blank stock must be of high quality for such results. Surface defects, if of any appreciable size, seldom are completely removed. This is especially true of tubing where pressures encountered are much lower than for solid-bar stock. The more thorough the working, the better the finish obtained. Surfaces are always improved considerably though not always perfected.

Diameters. Accuracy in many cases can be held to plus or minus 0.001-inch, but plus or minus 0.002-inch is much more easily attainable. Roller bearing rolls are swaged to size within plus or minus 0.002-inch in continuous lengths. Internal bores also have been mandrel swaged to within plus or minus 0.002-inch on diameters up to 1.500-inches. There is no guarantee in swaging tubing that the wall will thicken uniformly. Holes, therefore, are somewhat rough and of imperfect finish so that for accurate bores, swaging over a mandrel is imperative.

Wall Thickness. Wall thickness, even with a mandrel, will only be relatively uniform. Precise walls are not always possible inasmuch as there is no method of confining the mandrel to the center of the tube to obviate deflection. However, the *average* wall thickness will be as uniform as the difference in the OD and ID, which is usually within an acceptable tolerance range. Generally, the hole centerline will never reach an eccentricity greater than plus or minus 0.010-inch in relation to the swaged diameter centerline. Maximum angularity of the swaged diameter centerline with the true part centerline is about 1/2-degree. Where power feeding fixtures are used this angular condition is readily obviated.

Lengths. Owing to difference in metal movement in swaging heavy tubing, ends of parts usually show a bell mouth effect which may extend into the bore of the

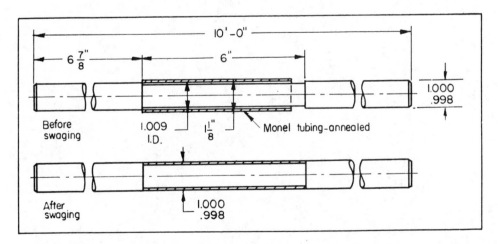

Fig. 31.14 — Monel sleeve cold-swaged onto pump shaft for corrosion protection.

tube as much as $1/4$ to $1/2$-inch on the largest parts. Because of this and the fact that ends usually come rough, a trimming operation often is necessary. This is especially true where length accuracy is desirable. A trimming cutoff operation is also desirable on solid parts where precise length and good end finish are necessary.

Hammered Parts. A good idea as to tolerances available with the hammering machine can be gained from the parts shown in *Fig.* 31.10 which are representative of parts normally produced on such equipment.

32

Wire Forming

I N THE continual search for the most economical and satisfactory method by which a specific part or parts can be manufactured, the capabilities of the wire forming machine and its many variations necessarily should not be overlooked. Wire and ribbon-metal forming machines rank high among our oldest mass-production machines, having appeared on the manufacturing scene about the same time as the first cold headers, somewhere around the middle of the nineteenth century. Known widely today by the general term "four-slide machines," those now available are the result of an evolutionary process which began with single-slide machines. Although many varieties of wire forming equipment are employed, ranging from relatively simple single-slide machines, *Fig.* 32.1, to the plain four-

Fig. 32.1 — Multeplex wire forming machine affords high production on simple single-plane parts. Photo, courtesy Economy Tool & Machine Co.

slide types, the combination four-slide machine embodies those features most universally adaptable and most useful in producing wire and ribbon parts, *Fig.* 32.2.

Fig. 32.2 — U.S. Multi-Slide No. 33 machine for ribbon or wire provides a feed length of 15^1/$_2$ inches and will handle ribbon up to 2^1/$_2$ inches wide by 3/32-inch thick or wire stock up to ^{1}/$_4$-inch in diameter. Photo, courtesy U.S. Baird Corp.

Scope of Process. A tremendous variety of machine parts falls within the scope of this versatile production method. Electrical components of all sorts, bushings, sleeves, rings, forms, clips, hooks, arms, spokes, and ribbon and wire spring forms of almost any design are characteristic of the type of parts well adapted to production by means of the wire forming machine, *Fig.* 32.3. Intricate parts, requiring many separate operations when produced on conventional presses, can be produced rapidly and with high accuracy, automatically and without handling. Setup flexibility of four-slide machines and ingenuity in tooling offers an almost unlimited field of design variations in forms and combinations of forms, *Fig.* 32.4.

Basic Machine. Essentially, the basic four-slide machine consists of a stock straightening arrangement, a feed mechanism for either wire or ribbon-metal, a cutoff unit, four horizontal slides designated (with the feed end of the machine at the left) as front, rear, right and left, and a vertical forming attachment with a stripping mechanism. Normally the stock is drawn from a horizontally positioned reel through the straightener rolls by the positive feed mechanism and is fed to the desired length as preset. The front tool, commonly referred to as the "binder," then advances to hold the part against the tool or stationary forming post on the head. The cutoff slide then functions, the horizontal slides following in the preset forming sequence, and lastly, where necessary, the vertical forming attachment and/or the stripper operate, *Fig.* 32.5. All the slides employed are usually cam actuated while the feed mechanism is usually crank or toggle actuated in conjunction with cam-operated feed and holding grip clamps.

Combination Machines. Adding to the capacity and field of application of four-slide machines is the feature of a horizontal press located between the feeding and holding grip mechanism on combination units. Thus, preforming operations can be done on wire parts prior to entry into the four-slide forming area. On ribbon-metal, operations such as piercing, blanking, forming, coining, embossing, stamping, etc., can be performed prior to final completion of a part by the four forming slides. In addition, the slides can be employed for producing flats, cutting or forming notches, necks, or slots, pointing ends, upsetting, or assembling. In *Fig.* 32.6 is shown the operational sequence layout of a representative part which

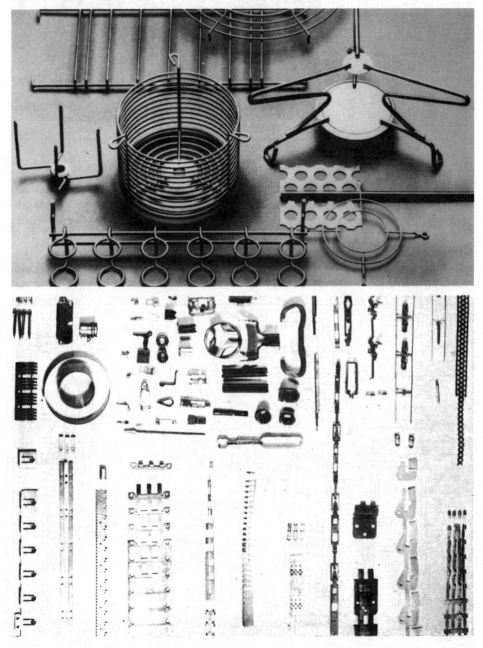

Fig. 32.3 — Machine parts and products formed from ribbon or wire stock on various types of four-slide machines. Top group courtesy Merrill Mfg. Corp. and bottom group courtesy U.S. Baird Corp.

Fig. 32.4 — Ingenious tooling coupled with setup flexibility of four-slide machines offers production economy not possible with conventional presses. Photo, courtesy Reliable Spring & Wire Forms Co.

indicates the complexity of design made possible by the combination press and four-slide machine, one version of which is shown in *Fig.* 32.2.

Complex Forms. Where extremely complex forms are involved, attachments are available whereby such designs can be readily produced. With a form raising attachment, parts necessitating a complete close-in are handled by raising the forming die entirely out of the plane of operation or to another position presenting a different cross section about which the final forming of the wire can be done. Even three die levels have been used for complex forming in certain cases.

Machines Available. A wide variety of plain forming machines, four-slide machines, attachments and adaptations are available for handling almost any

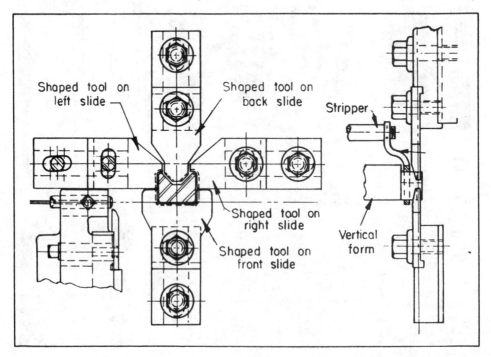

Fig. 32.5 — Simplified layout showing tooling of four-slide machine for forming a simple wire part. Courtesy U.S. Baird Corp.

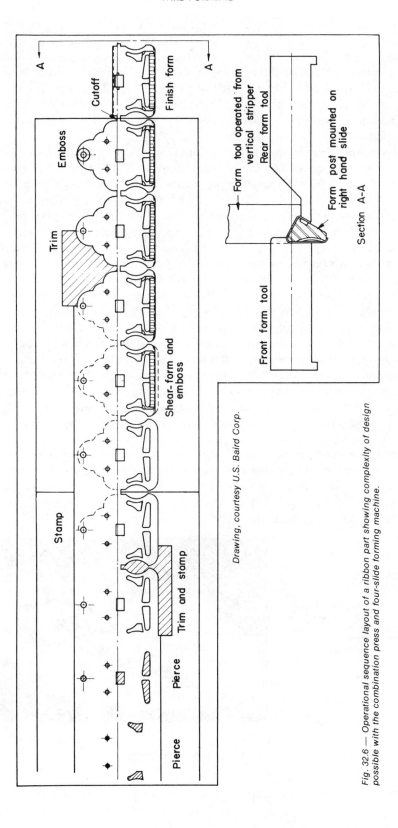

Drawing, courtesy U.S. Baird Corp.

Fig. 32.6 — Operational sequence layout of a ribbon part showing complexity of design possible with the combination press and four-slide forming machine.

design requirement in wire or ribbon forms. Plain formers and simple four-sliders will handle wire ranging from the smallest diameter available up to 1/2-inch diameter, maximum, in soft-steel wire. Wire can be of almost any available cross section, metal analysis or finish. Combination machines will handle, in addition, ribbon-metal stock up to 8 inches wide by 3/32-inch thick. Maximum length of stock which can be fed for producing a part ranges from about 2 inches with the smallest machine which handles 0.010-inch wire or 3/16-inch wide ribbon, to as much as 32 inches with the largest machines for 1/2-inch wire.

Horizontal Press Capacities. Machines with a horizontal press attachment have an available die space ranging from 3/4-inch to 1 1/4 inches with the smallest machine to more than 5 inches by 18 inches with the larger machines. Certain machines allow as many as three separate die heads to be mounted on the bed of the unit for handling the more complicated parts. Where design requires heavy duty coining or embossing operations, the larger machines available have lever or toggle press units capable of exerting 100 tons pressure or more for such operations.

Production Output. With the simplest plain forming machines, *Fig.* 32.1, production speeds range from 40 to 180 pieces per minute while with the four-slide machines production speeds may range all the way from about 500 pieces per minute with small parts to about 30 pieces per minute on large parts depending on part complexity. Ordinarily, machine production speeds are based on a maximum wire or ribbon feed speed of 100 feet per minute and consequently length of feed required for parts may have a direct effect upon output.

Special Designs. In addition to the various attachments previously mentioned, four-slide machines often are provided with unusual or entirely special equipment for extremely high-quantity production or unusual design requirements, *Fig.* 32.7. Hopper or magazine feeds can be used to handle separate parts previously cut-to-length for the purpose of performing preliminary operations such as end pointing, end milling, threading, etc. Occasionally such feeds are required where several forms are involved and arrangements can be made for the parts to be assembled in the machine as one piece is formed. Ring shapes often require welding and this feature has been included occasionally. Winding, milling, threading, secondary cutoffs, etc., have also been employed.

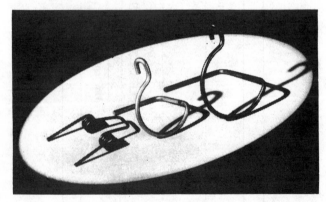

Fig. 32.7 — Rifle grenade safety pins and special coil-spring wire form produced on a four-slide machine with special production attachments. Photo, courtesy Reliable Spring & Wire Forms Co.

Design Considerations

The very simplest machines are limited to the production of forms having all bends in one plane. Bends should not be too severe and wire used preferably should be confined to the softer grades. Within the size and capacity range of the various types of four-slide machines few limitations are present on complexity of form. Actual limitations are found only in the total number of press and forming operations available and ingenuity in applying them in the most desirable sequence. Inasmuch as most wire or ribbon parts can be obtained readily and most economically from spring and wire forms producers who normally have available a wide variety of four-slide machines, consultation with these suppliers cannot be overemphasized in respect to obtaining satisfactory, low-cost parts. Too, most parts can be produced by using the adjustable cams regularly supplied with four-slide machines. Because of this feature, even smaller runs than would normally be expected often are economical.

Section Area and Complexity. Section area of wire or ribbon specified should be kept to a minimum to simplify the forming operation and minimize costs. In the case of spring-action parts, excessive thickness may operate only to render the part more likely to fail. For maximum strength and service life, forms should be as simple as design permits. Although extremely complex parts are within the scope of this method, the less the complexity the lower the cost and the greater the production speed.

Bend Radii. Radii of bends should never be less than metal thickness or wire diameter to avoid any possibility of cracking because of serious localized stresses introduced by sharp bends, especially with spring-action parts. Where metal to be used is in any other than the dead-soft annealed condition, bends closer than two times wire diameter or metal thickness should be avoided. If feasible, larger radii should be employed; a radius of four times thickness or diameter just approaches the point of being the optimum. Likewise, sudden changes in cross section and improperly placed holes or slots should be avoided not only to minimize stress concentration but also to obviate tool trouble and to minimize warpage encountered where heat treatment is necessary.

Closed Rings. It is impractical to produce with ordinary tooling a tightly closed ring of wire or sleeve of ribbon-metal. To obtain such parts a special setting-up operation is usually necessary to overcome the effects of the metal springback and in cases of average work the standard vertical forming slide can be used. By

TABLE 32-I — **Diameter Tolerances for Close Cold-Wound Springs** *

(inches)

Coil Diameter (mean)	Plus or Minus Variations In Diameter† D/d 3 to 6	D/d 7 to 10	D/d 10 to 12
1/8	0.0020	0.0030	0.0045
1/8 to 3/16	0.0030	0.0045	0.0060
3/16 to 1/4	0.0045	0.0060	0.0080
1/4 to 3/8	0.0060	0.0080	0.0100
3/8 to 1/2	0.0080	0.0100	0.0125
1/2 to 3/4	0.0100	0.0125	0.0150
3/4 to 1	0.0125	0.150	0.0225
1 to 1 1/2	0.0150	0.0225	0.0313
1 1/2 to 2	0.0225	0.0313	0.0470
2 to 3	0.0470	0.0625	0.0938
3 to 4	0.0625	0.0938	0.1250
4 to 5	0.0937	0.1250	0.1560

* Courtesy Reliable Spring & Wire Forms Co. Tolerances are based on standards set by the Spring Manufacturers' Association.

† D/d ratio is the spring index consisting of the mean spring diameter divided by the wire diameter.

using the setting operation on such parts, perfectly round parts with tight joints can be had. However, if metal is too light in gage, results may be unsatisfactory. From general experience it has been found that with wire rings, for instance, in order for setting to be effective the wire diameter should not be less than one-tenth the diameter of ring to be formed or, conversely, the ring should not exceed ten times wire size in diameter. Applied to bushings, this ratio would similarly assure the retention of a tight joint. Bushings produced by this method may be plain or pierced, lettered, oil grooved, indented, dovetail jointed, etc.

Upsets. Where upsets are desired, these should be designed for the smallest allowable upset of metal which will satisfy the design in view. Normally, only simple circular symmetrical upsets should be considered; complicated or special upsets cannot be handled.

Calculations and Tests. In designing wire or ribbon-metal parts, the standard beam theory can be utilized in calculating deflection and stresses

$$S = 6\ PL/bh^2$$

where P = applied load, L = moment arm, b = width, h = thickness, and S the bending stress in the extreme fiber which can be found in most available tables of physical properties. However, stress concentrations and cold working of the metal resulting from the forming operations complicate the analysis to such an extent that the only satisfactory plan is to confirm calculation results by experimentation with samples. This procedure is imperative where critical components with spring action are under consideration. Samples of parts are readily obtained from the producers of wire forms who usually maintain a sample department where even the most complicated form can be duplicated by hand. Except in the case of simple designs, tooling should never be begun until samples have been thoroughly tested and proved.

Selection of Materials

As mentioned previously, any type wire or ribbon-metal of any condition can be handled. However, the manufacture of many forms is rendered more difficult if hard wire or ribbon-metal is used. In fact, the hardness of the material has much to do with the layout and design of dies. For minimum die cost, soft or annealed wire should be used wherever requirements allow, especially where upsets, light extruding, embossing, or severe forming operations are contemplated. Certain complex designs, especially those with severe forming operations, may make hardening and tempering after forming imperative but it should be kept in mind

TABLE 32-II — **Variations in Spring Wound Length** *

(inches)

Number of Coils (per inch)	Length Variations per Inch of Spring D/d 3 to 6	D/d 7 to 10	D/d 10 to 12
5	±0.020	±0.030	±0.040
10	±0.030	±0.040	±0.050
15	±0.040	±0.050	±0.060
20	±0.050	±0.060	±0.070

* Courtesy Reliable Spring & Wire Forms Co. Based on unfinished, not ground springs. The D/d ratio is the spring index consisting of the mean spring diameter divided by the wire diameter.

that this often may be a more expensive method of manufacture. Capacity of most machines is based on soft-steel wire and where harder varieties are used capacity ratings must be reduced.

Materials, Tempers and Costs. Where no spring action is necessary, parts can be produced from the following materials listed generally in order of increasing costs:

1. Low-carbon steel (AISI 1005 to 1012)
2. Low-carbon steel (AISI 1013 to 1022 — may be heat treated)
3. Medium-carbon steel (AISI 1023 to 1040)
4. Soft to half-hard stainless steels
5. Nonferrous alloys in the soft to half-hard condition
6. Aluminum.

For various spring-action parts, the selection might be from one of the following groups of metals, also in order of increasing costs:

1. Hard-drawn medium-carbon steel (0.41 to 0.65 carbon)
2. Hard-drawn high-carbon steel (0.65 to 0.95 carbon)
3. Annealed medium-carbon steel (0.41 to 0.65 carbon)
4. Annealed high-carbon steel (0.65 to 0.90 carbon)
5. Pretempered or oil-tempered steel (usually 0.55 to 0.70 carbon)
6. Music wire (0.70 to 0.95 carbon)
7. Type 18-8 stainless steel in spring temper
8. Nonferrous alloys in spring temper

Protective and Decorative Coatings. For most economical forming into springs and wire forms, preplated wire or ribbon-metal may be used with commercial finishes such as cadmium, tin, copper, lead alloys, or zinc. These coatings, however, seldom exceed 0.0001-inch in thickness, are prone to scratching and marring in forming and of course are not plated on the sheared ends. Where heavier plating and full protection is desired, costs naturally are somewhat greater. Finished springs or forms may be electroplated with cadmium, zinc, chromium, copper, nickel, tin, or brass and generally these coatings offer corrosion resistance which can be rated from best to poorest in the order shown. As a rule, plated thickness ranges from 0.0002 to 0.0005-inch. Cadmium plate of 0.0003-inch provides a resistance equal to a 100-hour salt spray test and 0.0005-inch provides the maximum level of protection which is usually considered as a 200-hour test. Music wire and drill rod produce a much brighter plate than does basic, hard-drawn or oil-tempered steel wire.

For plain decorative finish and light protection parts may be coated with japans or other lacquers. For maximum economy color matching of the parts with other parts of a machine should not be required inasmuch as considerable expense is always involved in color matching. Many springs may be burnished bright and subsequently reheated to develop a decorative and protective finish of blue or straw colored oxide.

Specification of Tolerances

Most wire forming machines produce parts with an accuracy closely akin to that found with stamping. Machines can be set to feed lengths of stock within a few thousandths of being uniform. Formed rings and bushings, thus, often can be held to within an inside diameter tolerance of plus or minus 0.0005-inch where the diameter-to-gage is within the limits covered under the preceding data on design and a setting operation is contemplated. Where the extra expense incurred with holding such accuracies is not justifiable, wider tolerances should be employed, plus or minus 0.005-inch being an economical and readily obtainable tolerance.

Cutoff Length. Developed length can be held within plus or minus 0.005-inch in practically all instances with regular production practice. Therefore, for maximum ease in production and to avoid material variation problems, practical tolerances on rings or parts containing rings generally should be allowed to range from plus or minus 0.005-inch to about plus or minus 3/64-inch depending upon the ratio of radius to wire diameter or gage. Where coil spring sections are included on spring-action parts, *Fig.* 32.7, length tolerances on the finished part would range from plus or minus 1/32-inch to 3/32-inch depending upon size. Diameter tolerances on the coil section are covered in TABLE 32-I. Length tolerances over the wound coil section are shown in TABLE 32-II.

Formed Parts. Production parts which necessitate forming about a pin or form die naturally will have tolerances applied to those dimensions held by the form die. Some dimensions on parts will therefore have the standard wire or ribbon-metal tolerances to consider in determining the overall tolerances. Ordinarily, on average parts not too complicated in design, an acceptable tolerance on formed dimensions would be plus or minus 0.005-inch. Angular tolerances can be held to within a range of plus or minus 1/2-degree to plus or minus 10 degrees depending upon: (1) Ratio of radius to wire thickness at the bend, (2) ratio of length of material in the bend to material section area, and (3) quality of material.

TABLE 32-III — **Diameter Tolerances for Spring Wire** *

(inches)

Wire Diameter	Tolerance
Up to 0.016	+0.00025
0.017-0.062	+0.0005
0.063-0.080	+0.0010
0.080-0.104	±0.0010
0.105-0.144	±0.0015
0.145 and up	±0.0020

* Courtesy Reliable Spring & Wire Forms Co. Based on standards set by the Spring Manufacturers' Association.

Ribbon Parts. Parts designed for ribbon-metal manufacture ordinarily can be specified to carry overall length and width tolerances of plus or minus 0.005-inch. However, where holes or formed edges are included, tolerances on dimensions of these portions relative to the ribbon edges should not be required closer than plus or minus 0.010-inch to eliminate the necessity of special dies for blanking out the entire shape. Where full blanking is used, holes, forms, embossing and other features follow closely the tolerances obtainable with similar features in stamping.

Material Tolerances. Commercial wire and ribbon-metal tolerances should be acceptable on all parts to eliminate the need for special-size materials. Considering spring-action parts, commercial strip and wire tolerances also may enter the picture in still another way. Spring stiffness varies as the third power of the metal thickness so that within standard manufacturing tolerances there may be a variation in stiffness of three times the percentage variation in thickness or diameter. Similarly, safe spring load varies with the square of the stiffness, decreasing somewhat with the thinnest gages. Thus, for example, with 0.025-inch thick strip having a commercial tolerance of plus or minus 0.001-inch the stiffness will vary plus or minus 12 per cent and safe load will vary plus or minus 8 per cent. In the majority of cases these commercial variations are allowable but where greater accuracy is desirable, it often may be obtained by special selection of material or special processing, either of which incurs extra cost.

Tolerances on wire diameters generally used by wire formers are shown in TABLE 32-III. Ribbon or flat-wire stock which is available in widths up to 0.500-inch, normally is furnished with the following edge types:

No. 1 Edge — Filed edge, either round or square, of uniform width within 0.003 to 0.005-inch depending on size and gage. Used when edge finish is of primary importance and for all hardened and tempered flat wire.

No. 4 Edge — Round edge produced by cold working. Furnished when an approximately round edge is desired and finish is not important. Width tolerances range from 0.004 to 0.010-inch depending on width and gage.

No. 6 Edge — Square edge produced by edge rolling. Furnished where an approximately square edge is required. Tolerances same as for No. 4 edge wire.

Commercial thickness tolerances for ribbon or flat-wire stock are given in TABLE 32-IV. Finishes supplied are normally a No. 2 regular bright finish which is a semigloss finish but generally not suitable for plating unless polished and buffed, and a No. 3 bright finish (carbon 0.25 per cent, maximum) which has a high gloss particularly suited for electroplating.

TABLE 32-IV — Gage Tolerances for Flat Wire *

(Inches)

Thickness	Tolerance
0.225 to 0.0625 incl.	±0.0020
Under 0.0625 to 0.031 incl.	±0.0015
Under 0.031 to 0.010 incl.	±0.0010
Under 0.010	±0.0005

* Courtesy American Iron and Steel Institute.

Commercial cold-rolled strip is available in a wider range of edge types in widths from 1/2-inch to 6 inches and thicknesses from 0.005 to 0.2499-inch. Tolerances, however, are somewhat wider; on thickness the tolerances range from 0.0005 to 0.005-inch plus or minus and on width the tolerances depend upon the type edge, the most accurate being the No. 1 edge with plus or minus 0.005-inch and the lowest in cost being the No. 2 mill edge with plus or minus 1/32-inch up to 2-inch widths.

Heat Treated Parts. Where sections are light and forming not severe, pretempered material can be readily used providing maximum control of production tolerances within normal physical material variations. Otherwise, annealed materials must be employed and the parts subsequently heat treated to obtain the desired properties. Because of the final heat treatment, a certain amount of distortion is unavoidable and reasonable permissible tolerances with respect to dimensions must be allowed to avoid high costs due to rejects. In certain cases, if extreme accuracy of form is imperative, fixture heat treatment may be employed, but, of course, at somewhat increased costs.

METAL WORKING
METHODS

33

Die Forging

PRODUCTION by impact or pressure forging in closed impression dies merits serious consideration in the design and specification of machine parts which must withstand unduly heavy or unpredictable loads. Although a few other production methods such as centrifugal and permanent-mold casting produce in occasional cases comparable results, the uniform high quality and over-all dependability of properly designed die-forged parts are widely recognized. Such

Fig. 33.1 — Miscellaneous group of forged aluminum parts including an eccentric link and an aircraft engine mount. Photo, courtesy Aluminum Company of America.

parts manufactured from a great variety of metals embrace an unusually wide range of machine applications including cams, levers, gears, connecting rods, crankshafts, pistons, axle shafts, propeller hubs, and a multitude of other units which must resist high stress or be pressuretight, *Fig.* 33.1

Fig. 33.2 — Two pairs of production forging dies. Knowledge of plastic action of hot metals under impact affords uniform high quality in critical machine parts.

Designing for Forging. Accurate control of the forged metal in the two cavities of a set of dies, *Fig.* 33.2, is afforded mainly by good die design and a thorough understanding of the plastic action of various hot metals under impact or pressure. Knowledge of some of the basic principles of forging design, therefore, will enable the designer to utilize to the fullest extent those features which contribute to economy in production, affording maximum strength with minimum weight, high fatigue resistance, excellent surface finish and accuracy which eliminate or reduce appreciably machining operations that might otherwise be necessary.

Grain-Flow Orientation. To realize improved properties available through forging, consideration should be given to the direction in which previously rolled metal is worked in the dies. The forging flow lines should be oriented favorably in relation to the loads to be imposed. Resistance to impact is greatest along the grain rather than across. Tensile loads should be parallel to the grain and shearing loads perpendicular. Maximum refinement of grain structure and increase in density while plastic deformation takes place is readily apparent in the longitudinal cross section in *Fig.* 33.3.

Forging Equipment and Forging Sizes. In commercial forging practice the majority of small and medium-size nonferrous forgings are largely made in mechanical presses with capacities from 500 to 10,000 tons and usually are of materials such as copper, brass, copper alloys, etc. Die-pressed forgings may range in weight from less than an ounce to as high as thirty pounds, *Fig.* 33.4. Simple, relatively uniform parts without deep holes or difficult projections, *Fig.* 33.2, ranging in weight from an ounce or less to as high as fifteen pounds are usually forged from low-carbon steels, copper alloys, aluminum alloys, or magnesium alloys on the board drop hammer. Board drop hammers range in capacity from 200

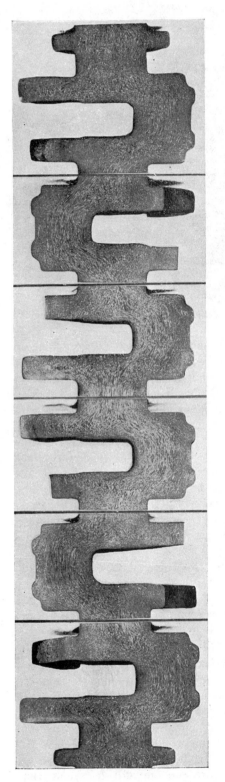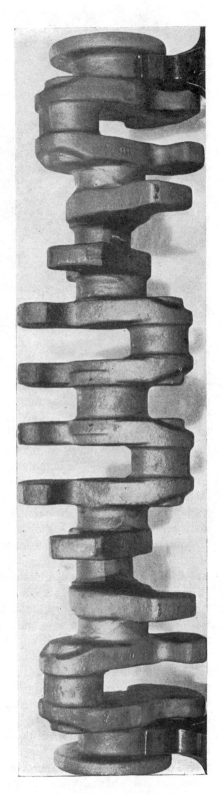

Fig. 33.3 — Engine crankshaft forging with six throws, seven bearings and twelve integrally forged counterweights. The etched longitudinal section shows the refined and compacted fiber-like flowline structure developed. Photos, courtesy Wyman-Gordon Co.

Fig. 33.4 — Indicative of the extremely intricate design possible in hot pressing non-ferrous metals, this forging in the rough weighs approximately twelve pounds. Photo, courtesy American Brass Co.

to 10,000 pounds impact and are used commercially to produce parts weighing up to as much as 100 pounds. However, the more intricate and nonuniform forgings ranging over fifteen pounds are often more economically produced on the steam drop hammer; those between five and fifteen pounds being about equally divided between the two machines. Particularly well adapted to producing parts with irregular shapes, thin or nonuniform sections in the less plastic metals, *Fig.* 33.5, the steam drop hammer is usually economical for parts weighing as much as 800 to 1000 pounds. Steam drop hammers range in capacity from 1000 to 50,000 pounds impact. Larger forgings such as propeller shafts for large ships, deep heavy cylindrical tubes and cups, heavy wide rings, etc., are produced on hydraulic forging presses, *Fig.* 33.6, capacities of which run to 18,000 tons. The largest press in existence today, an 18,000-ton hydraulic, extends the range for light-alloy forgings to about 550 pounds in weight and to sizes from 200 to 1000 square inches of projected area.

Effects of weight, design, material, quantity, forging method, tolerances, etc., on economy in production are considerable. After thorough study by the designer, these effects may well be discussed with the forgings producer to arrive at a satisfactory compromise between optimum design and inherent limitations of the process.

Production Quantities. Quantity runs are of course the most economical, die and set-up costs per piece being the lowest. Wherever quantities to be run at one time are small, say 200 to 400 pieces, die and set-up costs always assume a substantial portion of the total cost per piece. However, and most important of all, die life is affected directly by the number of times dies are refitted to the machine for producing a small lot. Thus the overall life of a die (in continuous production) may be reduced as much as 90 per cent by short-run production in small lots of 100 pieces or less. Consequently parts such as heavy shafts, large rings, odd shapes, or complicated crankshafts can be produced more economically as symmetrical

Fig. 33.5 — Magnesium
aircraft wheel brake forging,
one of the most intricate ever
produced. View of underside
is shown also. Photo, courtesy Wyman-Gordon Co.

rough-forged blanks with flat dies on steam hammers or hydraulic presses when quantities are extremely small or spread over a long period of time. Even though more machining is required for flat-die work, this is justified by the saving in die costs.

Trimming and Additional Operations. All drop and press forgings normally require a trimming operation to remove the excess metal or flash after leaving the finishing impression of the die. The sequence of steps in forging a typical part is illustrated in *Fig.* 33.7. Other finishing steps often required before completion include punching, forming, straightening, broaching, coining or sizing, planishing, and ironing. Trimming and punching operations are done either hot or cold. Small to medium size forgings ordinarily can be sheared or punched cold in low-alloy or carbon steels up to 0.35 per cent carbon. Large work and especially high-alloy parts are trimmed hot. The clearance for trimming usually runs 0.010-inch or a little greater depending upon the flash thickness.

The various other operations are primarily intended for further reducing machining work and in many cases can eliminate machining altogether. "Broaching" as termed above is used as a multiple hot punching operation to remove draft from a part of a forging, the tool having only two or three teeth. Forming is used extensively to produce a special shape on part of a forging not ordinarily possible in the dies. This usually consists of bending, twisting or drawing. Cold sizing or coining is used primarily to attain closer dimensional tolerances than are possible in the ordinary forging operation on relatively small areas. Hot sizing is used on large areas to achieve a considerably improved surface and attain close control of dimensions, alignment, and weight for larger parts. Ironing, performed hot or cold, is used to improve surface finish, remove draft

Fig. 33.6 — Ingot being forged into one of the largest generator shafts. Courtesy Bethlehem Steel Corp.

from certain portions of a forging or impress markings. Planishing, a combination of ironing and sizing, is used on rounded or spherical sections for improved accuracy and finish.

Design Considerations

Primary among the design factors to be considered is the draft or draft angle required for easy removal of a forged part from the die. Draft must be added to all surfaces perpendicular to the forging plane or parting plane of the dies. Since it must be added it should be shown on the drawing. A representative part exemplifying design for forging is shown in *Fig.* 33.8.

Specifying Draft. Draft angles on portions to be machined should be as small as possible, *Fig.* 33.9, but commensurate with increased die wear normally present with the smaller angles. Draft on the shallow portions, however, is often increased to match that on deeper portions, *Fig.* 33.8. The normal draft angle for external surfaces of parts designed for the board drop hammer is 7 degrees and for internal surfaces, 10 degrees. Normal external draft angle for the steam drop hammer is 5¹/₂ degrees and internal, 7 degrees. These can be varied according to the design and may run from 1 degree on shallow work to as much as 15 degrees on deep steel forging impressions.

Standard draft angles for forgings of nonferrous metals to be made in presses range from 2 to 3 degrees on all surfaces normal to the parting line, and for vertical extrusions as little as ¹/₄ to ¹/₂ degree can be used. Spherical, cylindrical, conical, and like surfaces often can be included so as to provide natural draft, *Fig.* 33.8. Ends of cylindrical or conical parts have draft provided at the flat ends either as a

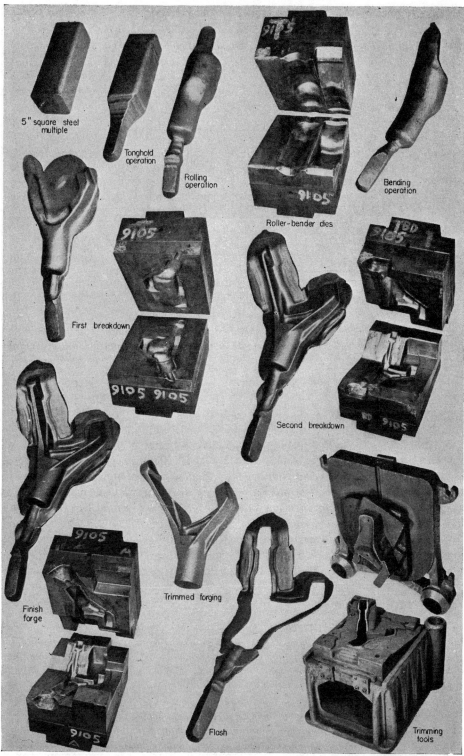

Fig. 33.7 — Complete tool equipment used in production of an airplane landing gear forging weighing 55 pounds. Four steam hammers and one trimming press were used. Elaborate locking of dies necessitated by offset arm can be seen. Photo, courtesy Wyman-Gordon Co.

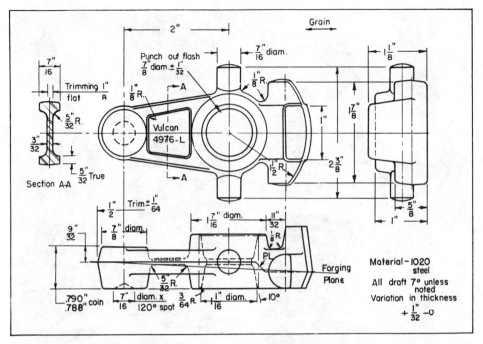

Fig. 33.8 — A representative machine part showing most of the general design details of impression-die forged parts. Drawing, courtesy J. H. Williams & Co.

flat cone or a spherical surface. Radial draft is usually assumed to be approximately twice the diameter of the projection but is seldom marked on drawings. Where draft must be otherwise it should be plainly marked.

Parting Lines. In laying out a part and establishing a parting line, *Fig.* 33.8, good practice is to have approximately equal volumes both above and below. Where a straight parting line is not possible, *Fig.* 33.10, an irregular one, *Fig.* 33.8, usually can be established though more often than not at a somewhat increased die cost. Particular attention should be paid to utilizing natural parting planes wherever possible and to insuring favorable grain flow structure inasmuch as structure is ruptured at the flash line.

Corners and Edges. Corner radii and fillets should always be specified as large as possible to assist the flow of hot metal and promote economical manufacture. Long sweeps and large radii promote sound forgings and prevent defects such as

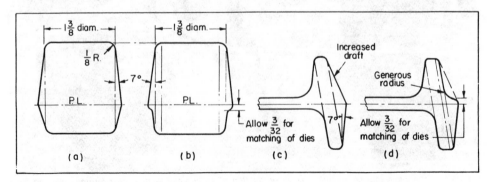

Fig. 33.9 — Proper draft on portions to be machined, (b) and (d), often result in greater economy than those at (a) and (c). Drawing, courtesy Aluminum Co. of America.

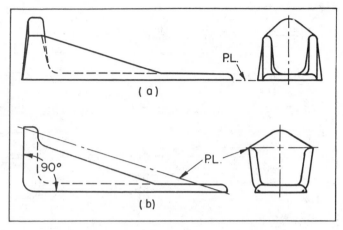

Fig. 33.10 — Designs can be altered slightly to provide various ways for parting. Considering die cost, (a) is most economical but (b) provides a lighter part without end draft. Drawing, courtesy Aluminum Co. of America.

coldshuts, laps, poor structure, etc. The minimum radius or fillet used runs from
3/64-inch on small parts to 1/8-inch on parts weighing 100 pounds, 3/32 to 1/4-inch
being preferred. All dimensions are normally made tangent to corner or fillet radii,
Fig. 33.8.

Sections and Ribs. Abrupt changes in section thickness should be avoided
inasmuch as the thin portions cool too rapidly and tend to set up severe stresses at
the junctions. Web or rib sections less than 3/32-inch in thickness should be avoided,
especially in alloy steels. Webs parallel to the parting line are difficult to produce
in medium or large areas and a thickness over 1/8-inch is preferable for economy on
all except certain small brass hot pressings, where 3/32-inch is practicable. In naval
brass the minimum is 1/8-inch, in aluminum-silicon-bronze the minimum web is 1/4-
inch as it is with copper and aluminum alloys, and with magnesium alloys the web
should not be under 3/8-inch. In addition to maximum rib widths it is also desirable
to keep rib heights as low as design will permit, attendant draft angles large as

Fig. 33.11 — Thin rib sections should be blended-in with generous fillets and should be no higher than necessary. Photo, courtesy Drop Forging Association.

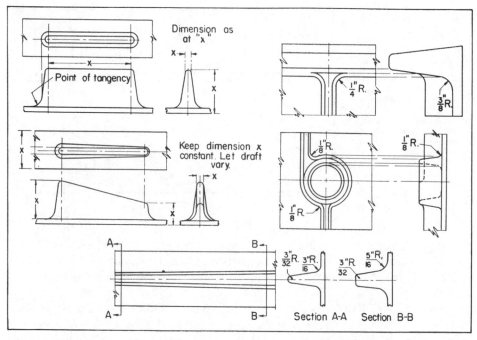

Fig. 33.12 — Rib designs should provide for constant edge radii in female die portions to simplify die sinking. Fillet radius produced by male die edge should also be kept constant but where absolutely required, can be varied more easily than the rib top radius. Drawing, courtesy Aluminum Co. of America.

practicable, and the rib tops a full radius, *Fig.* 33.11. Corner radii on rib sections should be constant with variations taken by draft, *Fig.* 33.12. Top radius for ribs up to 1-inch high should be 1/16-inch minimum; for ribs 1 to 1 1/2 inches, 3/32-inch; for ribs 1 1/2 to 2 inches, 1/8-inch; and for ribs 2 to 3 inches, 3/16-inch. Larger radii increase die life. Fillet radii between ribs and joining webs or other sections should be as follows: For ribs up to 5/16-inch in height, radius should equal height; from 5/16 to 1/2-inch, radius should equal 3/4 the height; and for higher ribs it should equal 1/2 the height, *Fig.* 33.13.

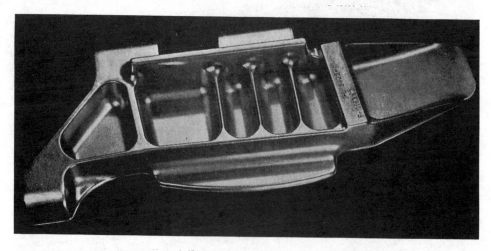

Fig. 33.13 — Aircraft part showing generous joining fillet radii on high ribs. Photo, courtesy Wyman-Gordon Co.

Fig. 33.14 — Group of master dies and sections produced by hot "typing". Photo, courtesy Ford Motor Co.

Recesses. Where pockets and recesses are necessary, it is generally advantageous to make all connecting sweeps and draft angles as generous as possible. Needless to say, such designs should be made no deeper than necessary to avoid undue die wear. Depth of a depression or forged hole should never exceed two-thirds of the diameter or narrowest portion.

Bosses or Stems. Projection of bosses (length of projection less than diameter) extending vertically into the die cavity should be held to a minimum wherever possible and in any case should never exceed two-thirds of the smallest diameter of the boss. Longer projections, usually termed stems, have been made, *Fig.* 33.5, but are extremely difficult to produce. However, those that lie parallel to the parting line may be considerably longer than bosses without incurring undue difficulty, *Fig.* 33.8. Where bosses or other flat pads are to be coined subsequently, boss or pad height above surrounding details must be at least $1/32$-inch and preferably greater.

Holes. To avoid deep small-diameter depressions, such holes should be drilled rather than forged. Where holes of $1/2$-inch diameter or over are to be drilled, accurately placed drill spots of 105 to 120 degrees (included angle) can be provided on any surface approximately parallel to the forging plane, *Fig.* 33.8. These not only aid in drilling but also help considerably to refine the grain structure. If tolerances on hole location are not too exacting, spots can be placed on both sides of the piece.

Locations. Locating points for machining operations should always be positioned away from the parting line. Die wear at the flash line results in a

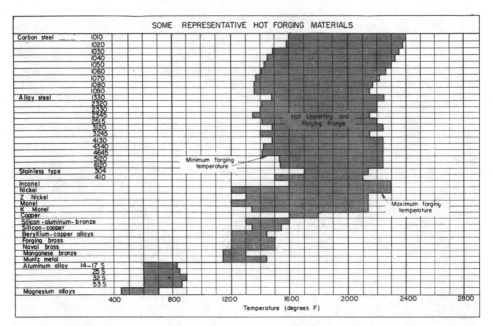

Fig. 33.15 — Forgeable metals possess an almost infinite capacity for hot working within a finite range limited by the upper critical temperature and the overheating point.

continuously changing width making it unsuitable for accurate locating.

Finishing Allowances. Material allowed for machining operations will vary somewhat with the size and shape of the part and with the type of machining operation. Sufficient metal should be provided so as to allow easy positioning but in any case machining economy dictates a minimum possible amount. Small and medium-size parts with simple finishing operations usually have an allowance of 1/32-inch to 1/16-inch for machining. Larger parts may have allowances of 1/4 to 1/2-inch on machined surfaces depending upon the design and intricacy. Allowance for

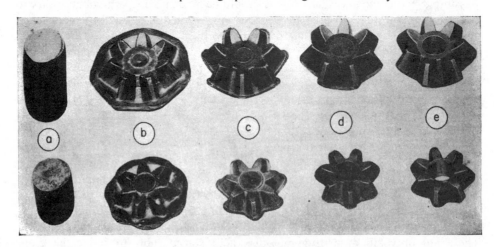

Fig. 33.16 — Cold coining of the teeth of these differential pinions obviated machining. Only the hole, chamfer and back face were machined. At (a) is shown the bar stock for two sizes weighing 2.12 and 0.92 pounds; rough forged pinions are shown at (b) before being trimmed; at (c) the pinions are shown after the second cold coining operation; at (d) the pinions have been trimmed; and at (e) the pinions are shown after machining. Stock saving is considerable over that required for cut pinions which was 4.77 pounds for the larger and 1.65 pounds for the smaller. Photo, courtesy National Machinery Co.

press forgings is usually $1/16$-inch. In most designs, even on small forgings, it is usually advisable to allow $1/16$-inch for finishing wherever possible to assure machining beneath the scale, pits and other irregularities. Accurate parts such as crankshafts should have an allowance of $1/8$-inch on the bearing surfaces. Forgings made with flat dies require a greater allowance and small parts up to about 8 inches in diameter should have an allowance of $1/8$-inch for machining or $1/4$-inch on diameter. This allowance for machining is increased to $1/4$-inch or $1/2$-inch on diameter for parts as large as 30 inches outside diameter. In most cases draft is considered as additional to the ordinary machining allowances.

Flash. Maximum flash thickness allowable should be specified in most cases, especially where excess flash will interfere with machining or locating. Thickness should never be less than $1/32$-inch nor greater than $1/4$-inch on the average run of impression-die forgings. The usual flash thickness* on nonferrous press forgings is as shown in TABLE 33-I.

TABLE 33-I — Commercial Tolerances for Hot Pressed Copper Alloy Forgings

FORGING TYPES:	COPPER	MUNTZ METAL	FORGING BRASS	NAVAL BRASS	LEADED NAVAL BRASS	MANGANESE BRONZE	HIGH SILICON BRONZE	ALUMINUM SILICON BRONZE	NICKEL SILVER 45-10
1. SOLID	0.010	0.008	0.008	0.008	0.008	0.008	0.012	0.010	0.008
2. SOLID with symmetrical cavity	0.010	0.008	0.008	0.008	0.008	0.008	0.012	0.010	0.008
3. SOLID with eccentric cavity	0.012	0.008	0.008	0.008	0.008	0.008	0.012	0.012	0.008
4. SOLID deep extrusion	0.012	0.010	0.010	0.010	0.010	0.010	0.014	0.012	0.010
5. HOLLOW deep extrusion	0.012	0.010	0.010	0.010	0.010	0.010	0.014	0.012	0.010
6. THIN SECTION, SHORT (Up to 6 in. incl.)	0.012	0.010	0.010	0.010	0.010	0.010	0.014	0.012	0.010
7. THIN SECTION, LONG (Over 6 in. to 14 in. incl.)	0.015	0.015	0.015	0.015	0.015	0.015	0.020	0.015	0.015
8. THIN SECTION, ROUND	0.012	0.010	0.010	0.010	0.010	0.010	0.014	0.012	0.010
Machining allowance (On one surface)	1/32	1/32	1/32	1/32	1/32	1/32	1/32	1/32	1/32
Flatness (Maximum deviation per inch)	0.005	0.005	0.005	0.005	0.005	0.005	0.005	0.005	0.005
Concentricity (Total indicator reading)	0.030	0.020	0.020	0.020	0.020	0.020	0.030	0.030	0.030
Nominal web thickness	5/32	1/8	1/8	1/8	1/8	1/8	3/16	5/32	1/8
Tolerance	1/64	1/64	1/64	1/64	1/64	1/64	1/64	1/64	1/64
Nominal filler & radius	3/32	1/16	1/16	1/16	1/16	1/16	1/8	3/32	1/16
Tolerance	1/64	1/64	1/64	1/64	1/64	1/64	1/64	1/64	1/64
Approximate flash thickness	1/16	3/64	3/64	3/64	3/64	3/64	5/64	1/16	3/64

*All tolerances are given in inches, plus and minus, and apply to forgings weighing up to 2 lb. For tolerances of forgings weighing more than 2 lb, consult producers.

Markings and Identification. Raised lettering, trade marks, part numbers, etc., can be furnished on any forging at little expense. These usually are made as an insert in the die and should be so placed on the part as to allow adequate area for such an insert and at the same time avoid subsequently machined surfaces. Marked surfaces must be on an area which lies approximately parallel to the forging plane, *Fig.* 33.8. Positive knockouts are often used on press forgings and require flat surfaces away from any complications. If the marks from knockouts will be objectionable the drawing should indicate their proper location.

Production Economy. Maximum economy in production naturally is achieved

* Suggested by Copper & Brass Research Association.

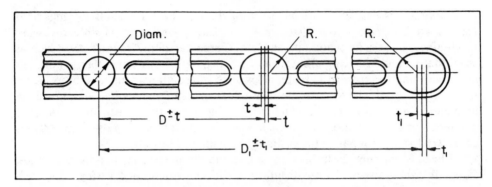

Fig. 33.17 — Shrinkage and die mismatch, whichever is the critical tolerance, should be allowed on bored bosses. Long parts should be dimensioned from one circular boss by proper shrinkage tolerances. Drawing, courtesy Drop Forging Association.

by reducing design to the simplest possible shapes and forms consistent with functional requirements. Die costs for average forgings vary roughly with the rectangular area which encloses the plan of the part. Straight or angular parting lines are far more economical than curved ones. Intricate or nonuniform shapes parallel to the parting line are often difficult to produce. Easiest to sink in a die and least expensive to maintain is the circular cross section. Elliptical cross sections are equally economical to reproduce and those parted along the major axis are much better than rectangular sections with minimum side draft regarding die life or die wear. Parts to be forged in great numbers often can be designed for multiple-cavity dies reducing cost per piece considerably. Also, use of multiple cavities simplifies forging by permitting balanced die forces.

Forging dies can be made by means of hubbing, the forcing of a master die form into a piece of die steel — cold or hot — to reproduce the reverse impression. Such dies are cheaper to produce, require little or no machine work and give greater life. Often termed "typing", this method is suited primarily for production of large quantities, the master "type" can be used for producing a great many dies. Smooth and flowing contours are highly important in such cases, *Fig.* 33.14.

Reduction in the number of machining operations necessary often can be achieved by utilizing the coining, forming, bending, twisting, broaching, drawing, ironing, planishing, and sizing operations available in the forge shop. Shackle ears, for instance, are produced in the conventional manner spread slightly and with draft on the inside of the ears. A restriking operation, after trimming, is used to flatten the inside ear faces parallel, increasing the outside draft to approximately 14 degrees and eliminating a machining operation.

Selection of Materials

Of the common forging materials, AISI 1020 usually is taken as standard and all other materials judged therefrom. As the carbon and alloy content of steels increases, forging becomes more difficult and additional working operations are necessary. Most of the nonferrous metals such as copper, forging brass, naval brass, bronze, and copper alloys are easily forged. Action of extruding and forging makes possible complicated pressings in one operation.

Materials specified are as a rule selected on a compromise basis. First, the required strength and physical properties must be met. Corrosion resistance, size, toughness, fatigue resistance, heat resistance, and section thicknesses must be balanced properly against forgeability and machinability to achieve maximum economy. TABLE 33-II outlines some of these factors with the comparison based on AISI 1020 steel. Materials with narrow forging ranges require careful, precise heating to avoid burning, *Fig.* 33.15. Where high strength is a factor, the forging range is important since it has been proved that thorough kneading of the metal in forging can be used to orient and form the metal structure flow favorably and unless the range is wide enough to allow adequate working, a great many reheats are necessary.

TABLE 33-II — Characteristics of Common Forging Metals

Metal	Die Life*	Pro- duc- tion*	Machin- ability	Use or Purpose	Remarks
Steel—AISI 1020	100	100	100	Small parts	Low physical properties; simple normalizing of forging required; for carburized parts
1030	98	98	104	Small parts	Better physical properties than 1020
1035	95	96	87	Medium-size parts	Moderate physical properties, heat treatment desirable
1040	95	96	81	Small and medium size parts	Heat treatment desirable
1045	90	93	78	Large parts	Avoid thin sections for heat treatment
4140	84	90	83	Aircraft parts	Caution required in quenching thin sections
5140	84	89	76	Gears and axles	Not recommended for water quenching
4340	78	84	67	Antifatigue forgings hardened gears, etc.	Ideal for parts subjected to severe conditions
Forging copper	110	125	125	Press forgings	
Forging brass	115	107	200	Press forgings	
Naval brass	110	103	160	Press forgings	
Monel	20–60	30–60	59	Corrosion and heat-resistance	

* Ratios shown as compared to standard 1020 forging steel.

Specification of Tolerances

Maximum permissible tolerances should be indicated wherever possible. Forging dies are machined to the low limits specified for a part and when the impression has worn to the high limits the die must be scrapped or resunk. It follows without question that wide tolerances extend the die life considerably, especially with the tougher alloy steels.

General Tolerances. The important tolerances to be considered by the designer are: (1) Thickness; (2) width and length as composed of shrinkage, die wear and mismatching; (3) trimmed size; and (4) draft angle. These are of two classes — regular or special. Special tolerances cover those so indicated on each dimension and where no tolerances are indicated the regular usually apply. Regular tolerances as set up may be either "commercial" for general forging practice or "close" where extra care and accuracy is desired at, of course, a somewhat increased cost. Commercial tolerances for hot pressed copper alloy forgings are shown in TABLE 33-I.

Thickness. Forging thickness tolerances, TABLE 33-III, apply to the overall thickness perpendicular to or across the parting line of the dies. Width and length

TABLE 33-III — Forging Thickness Tolerances*
(inches)

Net Weight (lb max.)	—Commercial— Minus	Plus		Close Minus	Plus
0.2	0.008	0.024		0.004	0.012
0.4	0.009	0.027		0.005	0.015
0.6	0.010	0.030		0.005	0.015
0.8	0.011	0.033		0.006	0.018
1	0.012	0.036		0.006	0.018
2	0.015	0.045		0.008	0.024
3	0.017	0.051		0.009	0.027
4	0.018	0.054		0.009	0.027
5	0.019	0.057		0.010	0.030
10	0.022	0.066		0.011	0.033
20	0.026	0.078		0.013	0.039
30	0.030	0.090		0.015	0.045
40	0.034	0.102		0.017	0.051
50	0.038	0.114		0.019	0.057
60	0.042	0.126		0.021	0.063
70	0.046	0.138		0.023	0.069
80	0.050	0.150		0.025	0.075
90	0.054	0.162		0.027	0.081
100	0.058	0.174		0.029	0.087

*Courtesy Drop Forging Association.

tolerances composed as in the foregoing apply to the portions that are formed in one die and that lie in a plane parallel to the parting line of the dies. Applying to these dimensions which do not cross the parting line, shrinkage and die wear

TABLE 33-IV — Shrinkage and Die Wear Tolerances*
(inches)

Shrinkage				**Die Wear**		
Length or Width	Commercial (plus or minus)	Close (plus or minus)		Net Weight (lb)	Commercial (plus or minus)	Close (plus or minus)
1	0.003	0.002		1	0.032	0.016
2	0.006	0.003		3	0.035	0.018
3	0.009	0.005		5	0.038	0.019
4	0.012	0.006		7	0.041	0.021
5	0.015	0.008		9	0.044	0.022
6	0.018	0.009		11	0.047	0.024
For each additional inch add:				For each additional 2 pounds add:		
	0.003	0.0015			0.003	0.0015

* Courtesy Drop Forging Association.

TABLE 33-V — Die Mismatching Tolerances*
(displacement of one die over the other along major parting plane)

Net Weight (max lb)	Commercial (inches)	Close (inches)
1	0.015	0.010
7	0.018	0.012
13	0.021	0.014
19	0.024	0.016
For each additional 6 pounds add:		
	0.003	0.002

* Courtesy Drop Forging Association.

tolerances, TABLE 33-IV, are used as a unit rather than separately and do not include draft. Mismatching or displacement of one die over the other, TABLE 33-V, is often a factor to consider in obtaining final total tolerances and may have detrimental effects if overlooked. Trimmed size of forgings is usually within the sum of the limiting sizes imposed by draft angle, shrinkage and die wear variations. Length of forgings may vary as much as plus or minus 0.003-inch per inch of length. Draft angle tolerances are given in TABLE 33-VI. Specified draft angles are normally held within plus or minus 1 degree.

TABLE 33-VI — **Draft Angle Tolerances***
(degrees)

Type	Nominal	Commercial	Close
Outside	7	0 to 10	0 to 8
Inside	7 or 10	0 to 13	0 to 8

* Courtesy Drop Forging Association.

Across Parting Line. If tolerances and surface finish closer than those ordinarily obtained are required between two flat surfaces parallel to the parting line, coining or cold sizing may be specified. A tolerance of plus or minus 0.005-inch can be held fairly easily on such dimensions and on some parts as little as plus or minus 0.001-inch is possible. Accuracy or improved finish can be obtained by cold coining and tolerances can be held to plus or minus 1/32-inch or less on diameter, *Fig.* 33.16. Planishing of spherical or rounded sections done at room temperature makes possible tolerances of plus or minus 0.002 to 0.005-inch. Similarly, hot restriking may be utilized on certain of the larger parts to obtain closer tolerances on weight as well as dimensions.

Punched Holes. As a rule, punched holes in the web or webs of parts can be held to plus or minus 1/32-inch on diameter. Tolerance on the location of punched holes should be no less than plus or minus 1/32-inch although plus or minus 1/16-inch is more economical to hold.

Bosses and Pads. Where bosses or pads are to be bored after forging, especially where the bore is large in proportion to the boss size, tolerances on shrinkage and mismatch should be considered. Long parts or distances between bosses must take into consideration the effects of length tolerances, the shrinkage portion of which becomes large and governs the design length. The preferable method is to make one boss circular and elongate the others by the proper forging length tolerance, *Fig.* 33.17.

Flatness. Forged flat surfaces, normally are flat within 0.005-inch per inch but individual variations depend largely upon characteristics of the particular material being forged.

ADDITIONAL READING

[1]*Aluminum Forging Design Manual*, 15, New York, The Aluminum Association, Inc., 1975.
[2]Paul M. Coffman, "Forging, stamping and extrusion of solid thermoplastics," talk given at the Annual ASTME Conference, May 2, 1968, New York: Shell Chemical Co., Plastics and Resins Division, SC:68-19.
[3]K. M. Kulkarni, "Finally, forged plastic parts," *Machine Design*, May 3, 1973.
[4]*Forging — P/M Forging Seminar*, Riverton, N.J., Hoeganaes Corporation, 1971.
[5]C. L. Downey, Jr. and H. A. Kuhn, "Application of a forming limit concept to the design of powder preforms for forging," prepared for presentation at the Winter Annual Meeting of ASME, New York, November 17-22, 1974. Paper No. 74-WA/Prod-11.
[6]*Forging — P/M Forging*, Hoeganaes Corporation, 1971.
[7]J. E. Jenson, Editor, *Forging Industry Handbook*, Cleveland, Ohio: Forging Industry Association, 1970.

34

Hot Upsetting

$\mathbf{A}$SSOCIATED and often used in conjunction with forging is that process known variously as hot upsetting, hot heading or machine forging. Hot upsetting of simple and large parts is widely performed on hydraulic presses, especially where extremely heavy sections are concerned. In addition, hot extrusion can be similarly performed and the use of extrusion is rapidly gaining favor. Where complex parts up to about 9 inches in diameter are concerned and production is high, the forging machine offers the greatest speed and flexibility. The original forging machines, like those used for cold heading, were developed primarily for the production of headed fasteners of various kinds. Today, however, fasteners are but one of a tremendous variety of machine parts which are manufactured rapidly and economically by means of the hot upsetting method.

Forging Action. In upsetting hot metal, the material — usually "cut-to-size" rounds rather than wire — is forged by dies moving along its axis to produce grain flow and refinement very similar to that in forging, *Fig.* 34.1. Except on the smaller sizes of material handled in automatic forging machines, straight lengths rather than coils are used and, in a great many cases, parts semifinished by preliminary processing methods are completed in the upsetting machine, *Fig.* 34.2. Some scaling often is encountered with heating unless induction heating is used. Hot upsetting offers a greatly increased field of design variations since hollow, bent and complicated formed parts can be produced easily from any material suitable for

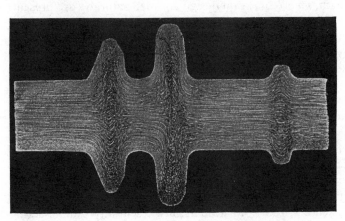

Fig. 34.1 — Hot upset cluster gear forging showing grain flow pattern produced. Photo, courtesy Wyman-Gordon Co.

34-01

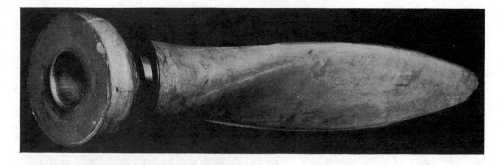

Fig. 34.2 — Above — Propeller blade of 25ST aluminum alloy showing heavy root completed on the upsetter. Photo, courtesy Aluminum Co. of America.

Fig. 34.3 — View looking into the split dies of an upset forging machine. The upsetting rams for the various dies can be seen in the background. Photo, courtesy Ford Motor Co.

forging. Much greater volumes of metal can be upset hot than cold and, in addition, any design that can be produced cold can also be made by the hot process but the reverse is not true.

Field for Hot Upsetting. Inasmuch as most metals have a somewhat limited capacity for cold working when subjected to pressure beyond the yield point and below the ultimate strength, severe limits on design and size naturally are present. Although fairly large wire sizes can be worked, standard machines for cold upsetting are limited in capacity to a maximum of two-inch diameter stock and only occasionally is this diameter limit exceeded. Mechanical restrictions, therefore, in addition to the fact that workability or plasticity of various metals increases as temperature increases, make it customary to turn to hot upsetting for mass producing many types of parts from stock one-inch diameter and over.

Hot Upsetting Equipment. In forging metals, either an impact hammer blow or a squeeze pressure is used. The largest upset parts, as with regular forgings, are produced on hydraulic forging presses and parts weighing hundreds of pounds can be upset. Smaller upsets are often produced on mechanical presses. However, where the diameter of the stock to be upset does not exceed 9 inches, the maximum capacity of modern forging machines, the maximum in production can be realized. Hot upsetting forging machines, like cold headers, are essentially double-acting horizontal presses, although as a rule not automatic in operation except on long-run production jobs. Where high production is desirable on a standardized part, straight bar stock for both large or small parts is fed through special furnaces to the machine. In some cases with small parts, stock in coil form with special continuous heating furnaces can be used. However, induction heating methods are rapidly coming into use and promise both increased production and better products, as precisely controlled heating easily is achieved and the steel remains nearly scale free.

Forging Machine Operation. After heating bar or billet stock for the forging machine, it is fed to a stop in the stationary half of a split grip die either automatically or by hand, and, when the cycle is actuated, the moving half of the grip die closes against the stock, holding it while the heading ram forces the exposed section of heated materials into the die cavities, *Fig.* 34.3. Although many upsets can be made in one automatic operation, most designs require movement or working of a large amount of metal and consequently the majority of parts are produced in a number of separate steps. These usually are set up in a composite die arrangement, *Fig.* 34.3, and the part moved progressively from the initial step to completion without reheating, *Fig.* 34.4.

Upsetting Restrictions. The limiting factors which largely determine the number of upsets necessary to shape the desired volume of material are:

1. Maximum length of unsupported bar that can be upset in one blow without buckling is three times the bar diameter.
2. Volumes or lengths greater than three diameters can be upset providing that the diameter of upset, i.e., die cavity, is limited to a maximum of 1.5 times the bar diameter.
3. Where the length of upset or volume is greater than 3 diameters the amount of unsupported stock extending beyond the die cavity cannot exceed one diameter.
4. Wall thickness of tubing cannot be increased externally more than 25 per cent at a blow, although internal upsets are almost unlimited since the arch effect prevents internal buckling.

Additional Operations Possible. In addition to plain upsetting, the transverse action of the gripper die and the longitudinal action of the punches can be utilized separately or simultaneously for swaging, bending, shearing, extruding, trimming, slotting, and punching as well as for internal displacement — an operation widely

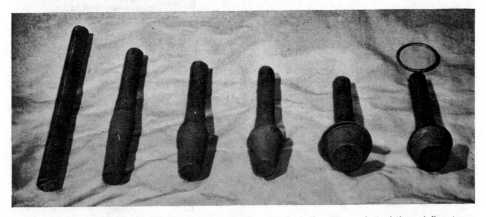

Fig. 34.4 — Pinion gear upset forging showing sequence from 1 1/2-inch diameter rough stock through five stages to the final piece ready for machining. Photo, courtesy Ford Motor Co.

Fig. 34.5 — Sequence used in machine upsetting steel marine engine cylinder barrels. Replacing conventional forging methods, these barrels were machine upset in four passes at one heat, reducing by almost 60 per cent the previous rough forging weight. Hole in end is for handling the $4^3/_4$-inch diameter by $17^1/_4$-inch long billet. Photo, courtesy Tube Turns, Inc.

used for producing hollow socket wrenches, cylinder barrels, and other deep-pierced jobs, *Fig.* 34.5.

Refinement of Metal Structure. Care in designing the dies and punches for each stage of upset results in optimum grain refinement and flow. Study of the four stages shown in *Figs.* 34.1 and 34.5 reveals this action. Internal working action provided by the punches makes it advantageous to utilize hollow designs rather than solid in many cases. Permissible heating temperature of the material, of course, dictates the high-temperature limit which can be used for maximum ease in forging and piercing. Ability of the piercing tools to withstand the tremendous pressures developed when working in the low-temperature range severely limits the practicability of low-temperature piercing in high production.

Extrusion Operations. Deep piercing of a representative part is shown in *Fig.* 34.6. The extrusion principle is not used in these operations on an upsetter because

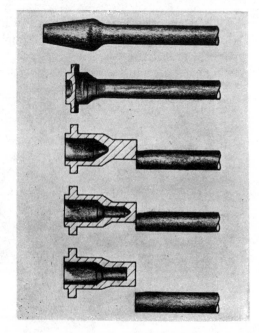

Fig. 34.6 — Progressive steps in deep piercing a representative upset part. Both outside diameter and bore can be stepped. The flange simplifies the piercing operation. Drawing, courtesy National Machinery Co.

wear on punches and dies makes it practical only for nonferrous materials such as aluminum, lead or brass. All straight-carbon and alloy steels, therefore, as well as all forgeable nonferrous metals are handled by piercing rather than extrusion. Hot extrusion of steels, however, is gaining in popularity as the technique of this phase of the forging process develops. Numerous extrusions are produced on large hydraulic forging presses and also cold and warm extrusion on the larger

TABLE 34-1 — **Requirements for Forging and Extruding***

Forged Front Spindle	Extruded Front Spindle
1—Shear billet (weight for 2 spindles—15 lb.)	1—Shear billet (5¼ lbs)
5—Ford slot type billet heat furnaces	1—Tumbler
5—4-inch Ajax upsetters	2—700-kva induction heating unit with two heaters
1—Grinder (snag when necessary)	2—No. 5 National Maxi-Press
4—Ford slot type billet heat furnaces	1—Trim press
4—2500-pound Erie steam hammers	2—Double-wheel grinders
4—55½ Toledo trim presses	
4—56 Toledo restrike presses	
2—spindles per forging	1—spindle per billet
Weight of forging—15 lb	Weight of billet—5¼ lb
Weight of machined spindle—3.70 lb	Weight of machined spindle—3.85 lb.
Personnel requirements (2 shifts)—46 men	Personnel required (2 shifts)—16 men
Floor space requirements—5000 sq. ft.	Floor space requirements—1000 sq. ft.

* Courtesy Ford Motor Co. Based on 9000 pieces per day.

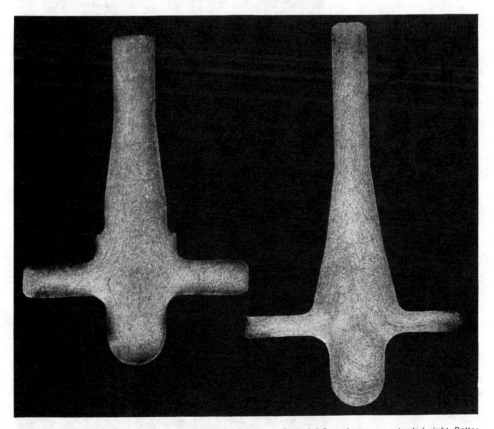

Fig. 34.7 — Front wheel spindle for a passenger car as hammer forged, left, and as press extruded, right. Better orientation of grain flow as well as reduced cost in time, men, machines, and materials as indicated in Table 34-I were achieved by designing for hot extrusion.

mechanical presses as well as on cold headers. One such hot extrusion is shown in *Fig.* 34.7 and the improvement obtained in material structure is quite evident. The tremendous reduction in production costs is shown in TABLE 34-I which accompanies *Fig.* 34.7.

Production Quantities. As with other mass-production methods, quantities of parts to be manufactured by hot upsetting or extrusion must be sufficiently great to offset the cost of the dies. Naturally, the die and die setup costs would reach an inordinate proportion of the total cost of each part on small-lot production, but where quantities are well into the thousands it becomes insignificant. In machine upsetting, however, where a standard design for which dies are available can be utilized, regardless of the overall length of the part which may vary considerably, quantities as small as 100 pieces can be economically produced.

Thus, for machine upsetting in a small-size range just overlapping that of the larger cold-heading machines, quantities normally duplicate those found economical for cold-heading practice, especially for similar designs. However, economical production rates for complicated or large parts may drop to a few pieces per hour. The cylinder barrels shown in *Fig.* 34.5 were produced at a rate ranging from 38 to 50 per hour on an 8-inch forging machine with only a nominal

Fig. 34.8 — So-called "electric gathering machine" gathers stock into cylindrical or spherical shapes which can be directly upset into finished or semi-finished parts. Courtesy, TRW Inc.

stock allowance for final finishing operations. Some 6000 to 8000 of these forgings are obtained from a set of dies before resinking of the cavities is necessary and several resinkings can be made before new dies are required. Die life, of course, depends greatly upon the forgeability rating of the material used. Ratings listed for forgeability give a good indication of comparative characteristics. As a general rule, piercing punches have a life of from 8000 to 12,000 pieces per redressing.

Electric Gathering. A highly interesting adjunct to the upsetting process is that known as "gathering". Developed specifically for speeding the production of extremely large-diameter heads such as those necessary for diesel engine valves, the so-called "electric gathering machine", *Fig.* 34.8, gathers stock into cylindrical or spherical shapes for subsequent upsetting. By this method the required volume of material is gathered into a satisfactory shape and passed directly to the upsetter where one blow finishes the part, *Fig.* 34.9. Thus, die design is simplified and furnace heating eliminated.

Electric Gathering Machine. Essentially, the gathering machine consists of a lower electrical contact of high strength and heat resistance, a crosshead with scissor type electrical contacts, means for adjusting the initial gap between upper and lower contacts, provision for automatically raising the cross-slide at a predetermined rate, and a source of power at low voltages and high current densities. The initial gap, the ratio of cross-slide travel to ram travel, the hydraulic pressure, and electrical power settings depend, of course, on the material, diameter and length to be gathered. By suitable selection of these factors, material may be gathered as desired at any point along a part, using round, square or hexagonal material from $3/16$ to $2^1/2$ inches in diameter or across flats, the main requirement being hot workability and proper electrical conductivity.

Large-Volume Gathers. Although with regular machine forging the limitations on length or volume of upset somewhat restrict the design of parts, in a few special cases, the gathering of as much as, say $2^1/2$ inches of $7/8$-inch stock has been accomplished with special sliding dies. However, with the electric gathering, up to 30 diameters easily can be made in production and indications are that this limit can be exceeded. Thus the cost of complicated dies as well as controlled furnace heating can be reduced to a point where even small quantities of unusual upsets are practical.

Electric Gathering of Tubing. Tubing also can be gathered in this way for upsetting. The critical factor here, however, is wall thickness rather than diameter

Fig. 34.9 — Large-volume upsets are electric gathered and finish upset in two simple operations. Photo, courtesy TRW Inc.

or *D/t* ratio. Stainless steels can be processed down to 0.065-inch wall but scrap is high and for maximum economy it is not advisable to use less than a 0.095-inch wall. On low-alloy tubing, 0.250-inch wall is about the practical minimum since this material has a greater tendency than stainless to develop folds during processing.

Electric Welder Upsetting. Not a great deal unlike the electric gathering machine, the standard butt welding machine is fast finding use as a production automatic upsetter of rod and tubing specialties. By this method blank shapes for cams, flanges, gears, and other parts to be made integral with a shaft can be easily and cheaply produced with all the assets of an upset forging.

Design Considerations

A part suitable for hot upsetting, much as with those for hot or cold forging, would require an excess of material, machining or forming if produced by other methods. Unlike cold forged designs, though, complexity and rearrangement of material can be great, taxing to the limit the hot workability of the metal used. Although uniform circular cross sections are most easily produced, practically any variety of cross section can be had — hexagonal, square, conical, oval, hollow, etc. It is practical to upset not only a flange on the extreme end of a part but also one or more at intermediate points along the part.

Designs which can be held within the three-diameter volume maximum outlined in the preceding limitations, of course, can be produced most easily; often with only a single blow. However, where maximum upset is desired to produce an unusual design, die cavities can be held to a diameter not more than 1.3 times the stock diameter for the first blow and for the second blow not over 1.5 times the diameter of the first upset. Extremely large flanged portions can be produced although as many as seven separate blows sometimes may be necessary. Naturally, the more cavities needed in a die, the greater is the cost and slower the production. Consultation with a reputable producer will assure a practical design as well as assist in obtaining simplification.

Corners and Edges. Inasmuch as split grip dies are used, tapers can be of any variety. Sharp corners should not be specified unless the anticipated finish machining operations can produce them. Generous radii or tapers at corners or stepped junctions will help to increase useful die and punch life considerably.

Hollow Parts. Hollow parts can have holes of practically any size pierced in the process. The pierced holes preferably should be of fairly symmetrical cross section. The pierced portion can be perfectly straight in bore or stepped, *Fig.* 34.6. The steps between bores can be square or tapered. Tapers produced directly by the piercing tool should not be over 60 degrees, included angle, inasmuch as this has been found desirable in practice. By the method shown in *Fig.* 34.6, flat sheared ends can be had and, if desired, pierced parts may have flat-bottom bores, the only restrictions being that the square-ended bottoming tool must have slightly rounded corners. Where the piercing tool is used to press the forging against the shear to square the bottom, the included angle on the piercer can be as much as 75 degrees but the length of stock being acted upon by the piercer cannot exceed one diameter.

Bores on pierced parts also may be carried through by a punch which operates simultaneously with the shear, *Fig.* 34.10. Scrap loss in the form of a slug, as can be

Fig. 34.10 — Flanged hub with stepped bore pierced through. The forging is sheared from the bar at the same time the small bottom slug is punched out by the last ram tool. Drawing, courtesy National Machinery Co.

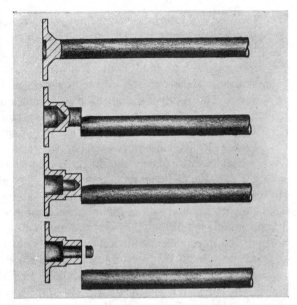

Fig. 34.11 — A few good examples of production upset parts help to visualize the wide field of application.

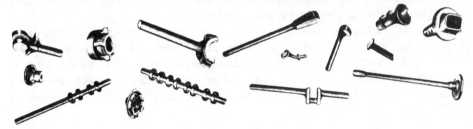

seen from *Fig.* 34.10, is insignificant. Depth of pierced holes is restricted somewhat, owing to the fact that at a single piercing the length of unsupported stock ahead of the piercer (between the upset position and the point where the stock is gripped) cannot exceed three times the diameter of the bar. Slender steel sleeve forgings having a pierced bore some 7.5 times the diameter are reported successfully produced at 200 per hour with a punch life of 8000 to 10,000 pieces per dressing.

Machining Allowance. A variety of unusual designs produced on a modern upsetter is illustrated in *Fig.* 34.11. From these an idea of the tremendous field of design applications can be gained. In each case a very minimum of machining is necessary to complete the part. Accuracy in detail obtainable necessitates the allowance of only about $1/16$ to $3/32$-inch for those sections requiring finish machining.

Selection of Materials

Of the common engineering materials available, any one suitable for forging is equally suitable for upsetting. Many metals unsuitable for cold forging are easily handled hot in production. Although hot forgings generally are more costly than cold, more intricate designs can be produced and, in certain cases such as with silicon-aluminum-bronze, stronger parts result having few or no dangerous internal mechanical stresses.

Nonferrous Materials. From the forging range of common metals, it can be seen that the temperature range through which many of the nonferrous alloys flow easily is relatively narrow. Danger of overheating with these metals is correspondingly high and control must be precise. Their high forgeability, corrosion and resistance and color, however, make their use in many instances imperative. The relatively high forging temperature of copper coupled with its tendency to form hard black oxide, which has serious erosive action on dies, makes it far better suited to cold heading wherever possible.

Steels. For maximum refinement of grain with steel parts the finishing temperature in hot working should be just above the critical range to assure only time for the austenitic grains to assume a fine and uniform normal structure. However, as this is seldom possible on fast production where die life depends upon working with the material in highly plastic condition, it is advisable to normalize all parts after forging to obtain maximum grain refinement. In addition, normalizing improves machinability and relieves internal cooling stresses.

Stainless Steels. Stainless steels present little difficulty in upsetting and handle very much the same as high-carbon steels. In sizes over $5/8$-inch with straight-chromium types and $7/16$-inch with austenitic chromium-nickel types these steels must be upset hot. Stainless steels are stronger at high temperature than ordinary steels and are therefore somewhat more difficult to work.

Electric Gathering. Although the electric gathering process has been used primarily for processing carbon and alloy steels, other materials such as brass, bronze, aluminum, magnesium, Monel, stainless, etc., also can be upset. Without doubt the process can be utilized to advantage on any material which can be hot worked satisfactorily and is a conductor of electricity.

Specification of Tolerances

As a rule, dimensional limits held in cold heading are closer than those held in hot upsetting. However, actual tolerances in most cases depend upon the finish machining operations involved. It goes without question that wide tolerances increase practical die life, in direct ratio, and costs in inverse ratio.

Standard Practice. Thickness tolerances listed for die forging, when pertaining to upset forgings, apply to those dimensions, in a direction parallel to the travel of the ram, of portions of a part that are enclosed and formed by the die cavities. Width and length tolerances, composed of shrinkage and die wear and mismatching tolerances, apply to the diameter, width or length in a direction perpendicular to the direction of travel of the upsetting ram.

In machine forging work, as a rule, some draft should be allowed on portions parallel to the ram axis to minimize die and punch wear. However, parts with no draft can be produced when it is found desirable to eliminate machining operations. Generally, from one to 4 degrees of draft are specified depending upon the material action and design. Standards adopted by the Drop Forging Association call for a nominal draft of 3 degrees for outside surfaces with a commercial maximum of 5 and a close limit of 4. On the inside of holes and depressions the nominal draft angle is 5 degrees while the commmercial maximum is 8 and closer limit 7. Fillet and corner radii tolerances are the same as those for drop forgings.

35

Roll Die Forging

PERHAPS least well-known among the various hot forging methods is the process of roll-die forging sometimes referred to by the trade as die rolling or crossrolling. An offshoot of the regular mill technique of hot rolling, die rolling offers the maximum in production speed, quality and low cost on machine parts which fall within its scope.

General Aspects. Unlike ordinary hot-rolling methods where multiple passes are employed to reduce a billet to the desired section, roll die forging is limited to one stand or pass. Because it is impossible to enter a previously die-rolled bar so that it matches the impressions in the rolls for a second reduction, the process is in reality a finishing roll pass. Much as in ordinary rolling, also, roll die forging is carried out with full length bars, eliminating much of the handling problem normally encountered in separate forging operations.

Inherently fast, roll die forging is well adapted to the production of preformed blanks for forging in closed impression dies or of parts which require only subsequent finish machining operations. Most widely used at present for preforming forging blanks, this method eliminates almost all except the final one or two drop forging operations, brings the blank to the final die impression much hotter, improves grain structure through better working of the metal, and increases forging output tremendously. The blank produced for forging a crankshaft and the

Fig. 35.1 — Complicated forgings like this crankshaft are produced with greater ease and better quality from die-rolled blanks, one of which is shown above. Photo, courtesy Republic Steel Co.

forged crankshaft are shown in *Fig.* 35.1 while the blank for a camshaft is shown in *Fig.* 35.2.

Machine Parts. Of primary interest to the machine designer, however, is the value of die rolling as a fast, efficient means for producing finished or semifinished machine parts, *Fig.* 35.3. Recognition of its place in the mass-production picture and design to utilize some of its many inherent advantages will assist greatly in achieving low-cost machines.

Die Rolling Equipment. A roll die forging stand or machine differs somewhat from an ordinary mill stand in that it must be extremely rigid with extra long windows to permit the use of rolls of varying diameters. It must have special fixtures to accommodate these various rolls and the drive arrangement must have variable speed; those in use today range from 20 to 60 rpm. The roll gearing must be such as to allow accurate registration of the impressions in the rolls. Rolls used at present are usually from 16 inches in diameter to a maximum of about 32 inches. Impressions to suit the design required are machined circumferentially about the

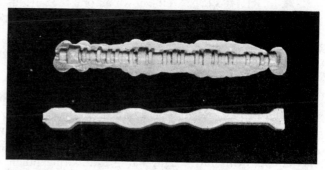

Fig. 35.2 — Intricate engine camshaft forged from a special die-rolled blank shown at bottom.

Fig. 35.3 — Light plow beam finished complete except for bending to shape, and set of rolls used for production die rolling these parts in quantity. Photos, courtesy Republic Steel Co.

rolls, side by side, as many as the part width allows, *Fig.* 35.4. Some current roll die forging machines are designed for single groove rolls and can efficiently produce runs as low as 100,000 pieces.

Limitations. Length of the part determines the number of identical impressions to be sunk in one circular impression. Short parts have multiple impressions while long parts often allow no more than one. Present equipment with 32-inch rolls allows for rolling of parts up to 96 inches in length although to date the longest part produced measures 75 inches in length and weighs about 330 pounds. Smallest part rolled measures about 1 1/2 inches in length and weighs about 1 3/4 pounds. Maximum cross-sectional area which can be handled is about 23 square inches. In calculating roll grooves the dimensions of a part must be figured to hot size, allowing about 0.015-inch per inch for shrinkage on cooling. In rolling the largest sections and rolling from a large reduction to a small reduction, a groove somewhat longer than the part is required owing to slippage in the rolls commonly referred to as contrusion. For sections running from a large area of small reduction to a small area of large reduction and straight sections of large reduction, the groove must be made shorter than required for the part owing mainly to forward slippage of the metal or extrusion.

Rolling Operation. In operation, a preheated billet is reduced by ordinary rolling methods in several passes to form a leader bar similar in shape and possibly

Fig. 35.4 — Set of roll dies, with five impressions, for producing a die rolled automobile axle blank. Photo, courtesy Republic Steel Co.

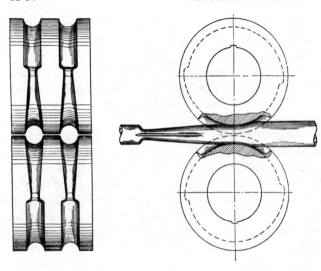

Fig. 35.5 — Leading rear axle blank emerging from roll dies. As one die groove wears, rolling is shifted to the next unworn groove. Drawing, courtesy Republic Steel Co.

about 16 to 20 per cent larger than the largest section of the part to be die rolled. From the last stand this leader bar passes into the die rolling stand and emerges therefrom as a string of parts, *Fig. 35.5*. Strings of rolled parts may run up to as much as 135 feet in length and, being accurately spaced along the bar, can be cold sheared to proper length rapidly.

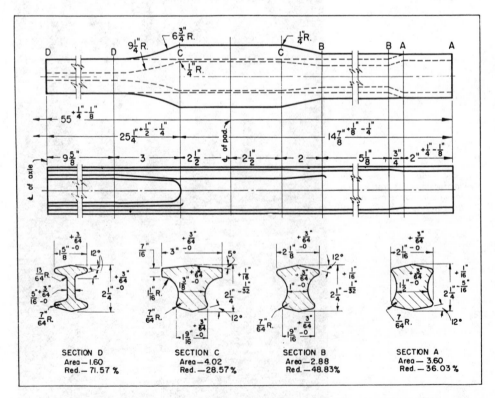

Fig. 35.6 — Die rolled front axle blank for a commercial ³/₄-ton truck. Drawing, courtesy Republic Steel Co.

Reductions. No upsetting of the metal takes place at any time in die rolling. As in plain rolling, working of the metal is entirely a matter of reduction. Reductions, however, can be extremely severe, *Fig.* 35.6, and can range up to as much as 75 per cent without complications. Largest reduction produced up to the present time has proved to be about 77 per cent.

Fig. 35.7 — Gun barrels in the as-rolled, sheared and machined conditions. Photo, courtesy Republic Steel Co.

Flash. Both flash and flashless parts can be rolled but reductions on flashless blanks are limited to a maximum of 48 per cent as against 77 for flash blanks. Too, flashed articles can be rolled to close tolerances owing to the fact that the metal is completely confined within and fills the roll grooves at all times, excess metal at points of reduction being squeezed out as flash. With flashless rolling, however, though the bar is within the grooves at all times, it never crowds the edges, no flash

Fig. 35.8 — Automotive front axle showing die-rolled blank at top and sheared blank at center. Finished axle shown below has been bent to shape, with opposite ends die forged to identical shape and machined. Photo, courtesy Republic Steel Co.

Fig. 35.9 — Forging roll or gap mill utilized for progressive forging of flashless machine parts. Automotive rear axle is shown being produced. Photo, courtesy Ajax Mfg. Co.

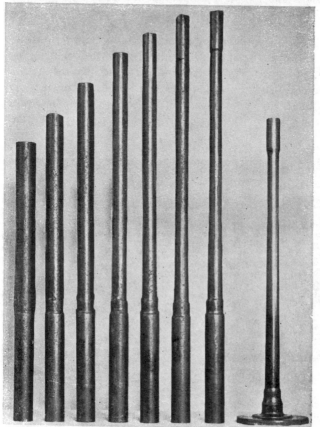

Fig. 35.10 — Progressive stages in gap mill forging a rear axle drive shaft. After one pass in each of seven roll grooves, it is straightened, sheared, upset, and trimmed. Photo, courtesy Ajax Mfg. Co.

Fig. 35.11 — Roll forged wheel blanks shown being produced automatically on a Bethlehem Steel Corp. Slick Mill.

is formed and the part is smooth. Necessarily, therefore, the metal does not always completely fill the die impressions and much greater tolerances are required for such parts and use is primarily for forging blanks.

Quantities. Being strictly a mass-production process, roll die forging quantities must be large in order to obtain the economical advantages available. Ordinarily a minimum quantity on light articles would range around 100 to 200 tons while for heavy or large parts this would increase to around 600 to 800 tons for an economical rolling. Gun barrels, for instance, can be rolled at a rate of about 20 to 22 tons per hour, *Fig.* 35.7, with about 40 to 42 barrels in a string. Automotive front axles, *Fig.* 35.8, being larger are rolled at a somewhat slower rate in number of pieces but by weight at a rate of approximately 40 tons per hour. About 250 tons of axles can be produced before redressing of the rolls is necessary. After one redressing rolls ordinarily must be resunk.

Gap Mill Forging. Of interest along with die rolling is the somewhat similar one known as gap mill forging. Performed on what is known as forging rolls, *Fig.* 35.9, flashless parts can be produced. Here, however, the roll dies are not continuous, the impression extends only over 50 to 75 per cent of the roll circumference and the remaining portion is cut away to form a gap for entry of the workpiece. Unlike die rolling the part is fed into the first impression and back, then rotated 90 degrees and moved to the second impression and back, etc., in a step-by-step squeezing action, *Fig.* 35.10. Because of the intermittent operation, speed of the gap mill is necessarily slow and output lower than with die rolling. As a rule, the minimum economical quantity of parts runs about 10,000 pieces. Smaller orders will not amortize mill costs.

Gap Mill Reduction and Shape. Because the work is rotated about a longitudinal axis during forging, the gap-mill process is limited to those parts having a symmetrical shape throughout. Any symmetrical cross section that can be turned 90 degrees in forging can be handled — such as a square part incorporating circular shoulders or necks. These, however, cannot be nearly as severe in reduction as with die rolling.

Slick Mill Forging. Another process somewhat akin to the foregoing processes is known as upset roll forging or Slick Mill forging, *Fig.* 35.11. Using again a rolling forging action, the Slick Mill produces economic and high quality circular forgings as shown in *Fig.* 35.12. These circular forged parts, especially important for wheels, turbine discs, compressor wheels, etc., may range from 16 to 52 inches in diameter with weights from 200 to 2250 pounds and minimum thickness of section 7/16-inch.

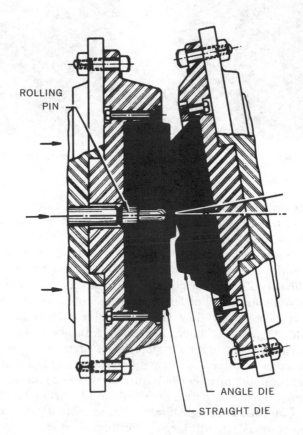

ROLLING PIN

Fig. 35.12 — Unusual arrangement of the Slick Mill with second roll die set at 10 2/3 degrees off horizontal to add roll forging to the upsetting action on the hot blank. Drawing, courtesy Bethlehem Steel Corp.

ANGLE DIE

STRAIGHT DIE

Comparative Advantages. Although definite advantages are gained with flashless forgings, slower output as compared with die rolling and the limited range of application makes die rolling more promising for parts other than simple forging blanks. Again, experience shows that die-rolled gun barrels permit accurate drilling with free floating gun drills, a characteristic not always present in gap-mill forgings.

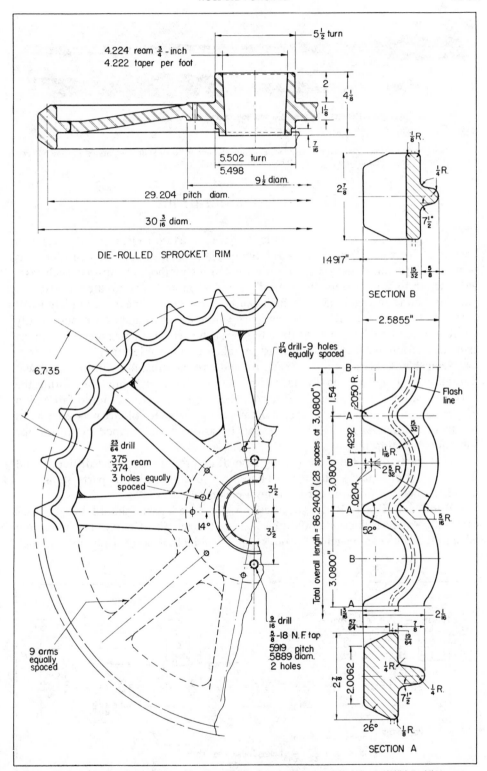

Fig. 35.13 — Track sprocket rim produced by die rolling indicates the unusual design possibilities of the process. Finished rim is subsequently bent into circular shape and welded to the hub section to form a completed sprocket as shown.

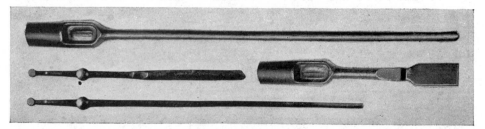

Fig. 35.14 — Cable bolt has the head formed in a steam hammer and the gear shift lever, above, has the ball end forged prior to completion on the gap mill. Photo, courtesy Ajax Mfg. Co.

Design Considerations

As a general rule, parts of simple design such as tie rods, tapered axles, gun barrels, drive shafts, and similar relatively plain parts should be considered as possibilities for either the gap mill or the die-rolling method. Quantity in such cases will dictate to some extent the feasibility and economy. The greatest portion of parts of a more intricate nature which can be completed or nearly completed will naturally fall into the die-rolling category where quantity makes it economically feasible, *Fig.* 35.13. The sheared flash line on such parts of course appears much the same as in ordinary forgings and must be taken into consideration.

Length and Area. Length of parts to be die rolled must fall within the present limitation of 96 inches, maximum. Cross-sectional area, also, must fall within the maximum allowable of about 23 square inches while reductions should fall within 75 per cent, maximum. Where design details not possible to roll are necessary as on the front axle ends, *Fig.* 35.8, stock can be allowed and shaped in rolling for subsequent die forging.

Flat Parts. As a rule, die rolling from flats presents the most difficulty and from rounds or squares, the least. Therefore, substantially flat parts should be avoided wherever possible for maximum economy.

Radii and Draft. Radii specified on die-rolled parts should be generous especially where reductions take place along the length of parts. The minimum edge radii which can be held in production is $1/16$-inch. Wherever possible $1/8$-inch

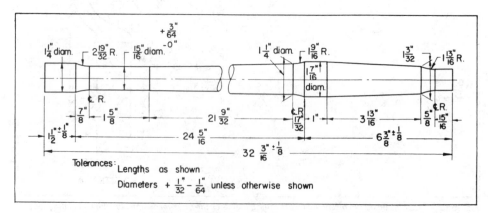

Fig. 35.15 — Automotive rear axle shaft, die rolled from 8640-H steel, illustrates tolerances held on average-diameter work. Drawing, courtesy Republic Steel Co.

minimum radius is preferred. To simplify the rolling process generous draft angles should be allowed on parts. Minimum draft angles which should be specified range from 2 to 4 degrees on a side.

Ordinarily, abrupt changes in cross-sectional design permitted by die rolling cannot be handled with the gap mill. Where such part designs are necessary, the special features must be drop forged, *Fig.* 35.14, and the shank ends can then be completed by reducing in the forging roll.

Selection of Materials

Various types of materials which can be handled in die rolling and gap mill forging naturally parallel those outlined in the previous chapters on hot upsetting and die forging. Actually some metals which offer great difficulty in die forging, such as AISI-6412, present little difficulty in die rolling. These tough steels that resist proper breaking down and working prior to finish die forging are easily rolled into blanks that conform closely to the finished shape, simplifying the final forging operation. Such steels, too, can be handled as finished or semifinished die rollings.

Forging Temperatures. Temperatures used in die rolling closely follow those employed in ordinary rolling practice. Highest temperature used is about 2100 F. Average steel parts are rolled at a temperature of about 1900 F with small parts running up to 2000 F and sometimes a little more.

Specification of Tolerances

In general, flashless blanks or parts cannot be held to close tolerances in die rolling and for obvious reasons their use as blanks does not necessitate the accuracy of finished or semifinished work. With flashless gap-mill parts, however, accuracy comparable to that of die forgings can be achieved and is satisfactory for parts which require subsequent finish machining on portions to be fitted.

In die rolling, average diameters can be held to plus $1/32$-inch and minus $1/64$-inch, *Fig.* 35.15. On large diameters, however, greater leeway is necessary and

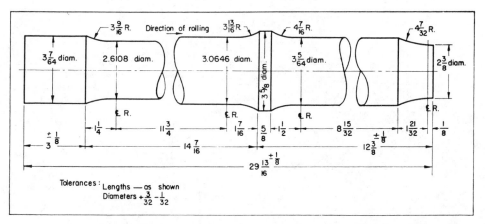

Fig. 35.16 — Rear tractor axle, die rolled from AISI 3140 steel, showing tolerances held in rolling large diameters. Drawing, courtesy Republic Steel Co.

tolerances of plus $3/32$-inch and minus $1/32$-inch are normally employed, *Fig.* 35.16. Commonly used length and width tolerances for various sections can be ascertained from *Figs.* 35.6, 35.15 and 35.16. Tolerances on flash portions follow closely those outlined under die forging and hot upsetting.

Finish quality on both die-rolled and gap-mill forgings parallels that of hot-rolled, upset and forged parts. Surface finish is suitable for most parts and as with any hot-worked material where extremely fine finish and close tolerances for fitted portions are necessary, machining operations must be employed as usual.

36

Hot Extrusion

DIRECT extrusion of metals in simple, symmetrical forms had its beginning years ago with soft, workable metals such as lead, tin, zinc, lead-tin alloys, etc. Cable sheathings, hose casings, lead pipe, wire, and similar relatively simple cross-sectional forms comprised the major production. During the intervening period of development leading to the perfection of the process and today's modern die steels, the range of metals which could be handled moved into the field of metals having higher strength and requiring higher extrusion

Fig. 36.1 — A variety of extruded aluminum cross sections showing but a few of the thousands of shapes possible. Photo, courtesy Reynolds Metals Co.

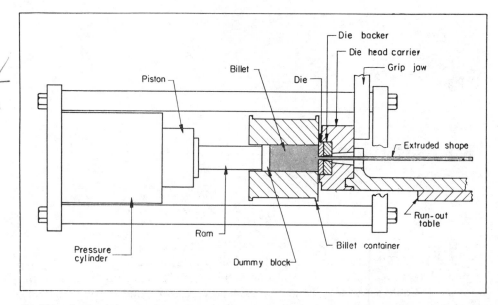

Fig. 36.2 — Sectional view showing the general setup used for direct hot extrusion of metals. Drawing, courtesy Revere Copper & Brass Inc.

temperatures. As the production technique improved it released the process from the narrow confines of simple, symmetrical sections and today shapes of almost unlimited variety can be produced in a wide selection of engineering alloys, *Fig.* 36.1.

Advantages. Offering in essence a "partial fabrication", the direct extrusion process commands its rightful niche in the panel of important production processing methods. Thorough application of the principles of designing for economical production makes it highly desirable that the advantages available through the use of extruded shapes be kept closely in mind. Conservation of valuable material and reduction in processing operations necessary to complete a unit or component makes it imperative from an economical standpoint.

The Direct Extrusion Method. Extruded shapes are normally produced by forcibly expelling, under high pressure, a hot plastic metal cylinder through a die having an opening of the desired shape, *Fig.* 36.2. Presses for these operations are usually horizontal, *Fig.* 36.3, although verticals are occasionally used. The heated metal billet is placed in the container and the hydraulic ram forces the metal to flow through the die arrangement, emerging in solidified form closely conforming to the dimensions of the die opening. In place of conventional lubricants, steel is extruded by the French Ugine-Sejournet process which utilizes a molten glass lubricant between the die and hot billet.

Scope of Process. The natural field of application for extruded shapes lies with the possibility of eliminating costly or difficult machining operations. Any part or component which may have an irregular cross section, the surfaces of which can be planes parallel to the longitudinal part axis, can be economically produced from multiple-length bars, *Fig.* 36.4. The few necessary finishing operations can be handled on production machines either before or after sawing or cutting to the necessary length for the part being made.

Further advantages of extruded shapes over and above possible production

Fig. 36.3 — Typical extruding press in operation producing plain magnesium tubing. Photo, courtesy Dow Chemical Co.

economies may be attributed to physical properties of the materials which can be extruded. Some of these considerations are: (1) Excellent machinability for subsequent processing; (2) electrical or thermal conductivity; (3) uniform density of structure; (4) acceptable surfaces without machining; (5) greater tensile strength; (6) improved ductility; (7) good toughness; (8) corrosion resistance; (9) smooth cold-worked surfaces; (10) metal color or color match; and (11) closely controlled dimensional tolerances.

Limitations. The major limitation to design in making use of an extruded shape basic stock is that of section. Unlike other methods of extrusion, shapes must have a constant cross section throughout their length and any holes, slots, grooves, etc., not parallel to the axis of the part in mind must be produced by machining or some other subsequent operation. Capacity of existing equipment limits the size of shapes which may be extruded; present maximum size of shapes in aluminum or magnesium is about 80 square inches and the shape, if solid, must fit within a 24-

inch circumscribed circle whereas, if hollow, a 20-inch circle. However, some are possible up to 39 inches diameter. In steel maximum enclosed diameter is 6.5 inches. Length is limited from 20 to about 60 feet, maximum, depending on material, size and weight. In steel, length is limited to 40 feet. In general, about 500 pounds of magnesium or aluminum is the maximum possible to extrude in one section; 375 pounds for steel. Section size limit for copper and copper-base alloys is roughly one-half that for aluminum.

Shape design in extrusions is practically unlimited. Design is not restricted to a substitute standard shape. Special sections exactly suited to a particular design can be obtained. Wall thickness throughout a cross section need not be uniform as with roll-formed or pressed shapes. Within broad limits metal may be placed as desired for strength or stiffness, or to obtain complicated contours which defy economical machining. Die cost is relatively modest, for simple shapes starting at a minimum of perhaps $150. Processing operations are extremely flexible and short runs can be made economically as well as long runs.

Dies. Direct hot extrusion requires die steels which exhibit the maximum in hot hardness and strength for economical production. The "lead-in" or bell mouth on a die for copper and copper-base alloys is on the entry side, and the bearing is on the exit side. For aluminum, zinc and magnesium, however, the bearing is at the entry side and corners are as sharp as possible with the exit side being relieved to allow for expansion of the metal as it emerges from the die.

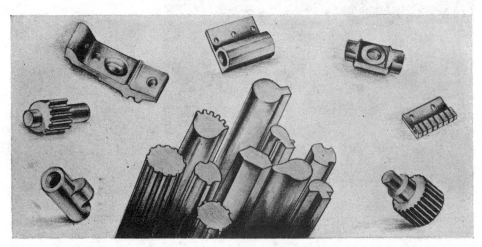

Fig. 36.4 — Sketch showing a variety of typical machine parts which are produced most economically in quantity from multiple length extruded shapes.

Aluminum and magnesium alloys are extruded as hollow shapes by using so-called "spider-web" or porthole dies wherein the metal is parted in passing over the webs supporting the center plug or mandrel and again self-welded before leaving the die exit. Copper, steel and their alloys do not bond or weld in this manner and so are seldom produced as hollow or totally enclosed shapes.

No rules or set mechanics govern the design of extrusion dies. Die production depends mainly upon trial by extrusion to determine the flow pattern to obtain equalized flow.

Fig. 36.5 — Representative of group one, these shapes are simple, symmetrical and easiest to produce. Photo, courtesy Revere Copper & Brass Inc.

Flow Pattern Effect. Variety or intricacy of cross section depends upon the metal or alloy selected, and control of the flow of metal through the die. Reduction ratio between the billet cross-sectional area and the die area also influences flow. For practical operation this ratio should never be less than 10 to 1. In order to secure uniformly high properties, however, it is desirable to secure a ratio of 15 to 1 or preferably greater.

Fig. 36.6 — A representative group of simple, partially symmetrical steel shapes with slots or serrations. Courtesy, Babcock & Wilcox Co.

Straightening. Each shape is handled as an individual problem. Dies are finished and designed to take advantage of present knowledge of particular metal flow characteristics to obviate excessive distortion. However a certain amount of twist, waviness or camber may occur, in which case straightening is employed. This is accomplished, if possible, by stretcher where the shape is symmetrical; or by hydraulic press if shapes are large and of heavy cross section as in the case of copper shapes for electrical use. Straightening by drawing with forming dies which produce no cold reduction may also be used.

Shapes of a symmetrical cross section often can only be straightened by bench and hand work. Large quantities of a single section, in any case, would warrant use of special rolls for straightening in a roll straightener.

Drawing. Ordinarily, straightening is done directly after extrusion except where reducing draws are to be made. Cold reductions achieve increased hardness, accuracy, increased tensile strength, and/or improved surface finish. All copper shapes and the higher copper content alloys must be cold-drawn to improve surfaces which are generally unsatisfactory due to the high extrusion temperature. Drawing also develops closer dimensional control and meets demands for better tolerances where required. A standard draw bench is customarily used.

Generally, copper shapes for exacting electrical applications require two cold reductions. Brass alloys, including nickel silver, which have copper contents below 58 per cent are too brittle for cold reduction and are used "as extruded". Aluminum and zinc shapes seldom are drawn.

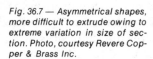

Fig. 36.7 — Asymmetrical shapes, more difficult to extrude owing to extreme variation in size of section. Photo, courtesy Revere Copper & Brass Inc.

Some drawing operations are utilized which perform no cold reduction and are termed "forming draws". A shape such as a U would be extruded as a wide V, to allow maximum die strength and life, and then would be drawn to form the desired finished section.

Design Considerations

Over and above the broad limitations outlined previously for the field of application, a number of particular design considerations affect the practicability

Fig. 36.8 — Representative tongued and grooved shapes. Designs are limited somewhat by the particular material. Photo, courtesy Revere Copper & Brass Inc.

of extrusions. To simplify the overall picture, extrusions in general can be broken down into a series of categories ranging from the simplest, most economical to the most difficult and complicated. These might be enumerated as follows:

1. Simple symmetrical shapes, *Fig.* 36.5. These are easy to extrude and are procurable in a wide variety of metals and alloys.
2. Simple, partially symmetrical shapes, *Fig.* 36.6, often with slotted or serrated surfaces. These shapes are relatively easy to produce.
3. Asymmetrical shapes, *Fig.* 36.7. These are more difficult to extrude because the metal tends to flow faster where heavier sections occur, small projections may not fill out properly, and gage is difficult to maintain.
4. Tongued or grooved shapes, *Fig.* 36.8. Where the tongue in the die has a length greater than three times its width at the narrowest point for aluminum, or greater than its width at the narrowest point for copper or copper alloys, the shape is classified as semienclosed and as a rule requires special handling consideration. As such it may become a special shape.
5. Sharp corner shapes, *Fig.* 36.9. Die and pressure limitations demand that the intersection of two surfaces at angles less than 90 degrees be avoided if sharp corners are absolutely necessary.
6. Special shapes. This group includes shapes which do not fall into the previous categories, such as the "stepped" extrusions where size of cross section is increased at various stages of the extrusion process. Also shapes such as those with long, slender tongues in which the actual shape extruded is different than that desired and the final shape is obtained by drawing or rolling.

Minimum Sections. With simple and partially symmetrical shapes the only restriction which is necessary to observe is that in relation to minimum section thickness. General data regarding section gage which can be extruded is shown in Chapter 68 along with other metal and design characteristics.

Grooves and Serrations. Details regarding minimum groove and serration size is shown in *Fig.* 36.10. V-serrations can be produced but of course the roots and crests will be rounded to 0.010-inch radius, the nearest practical approach to sharp edges.

Tongues and Grooves. A variety of plain or complicated tongued or grooved shapes can be produced. If a channel or groove is adjacent to an exterior surface and the separating leg is relatively thin in gage, such a leg will sometimes be extruded in an extended or spread condition, *Fig.* 36.11, to allow maximum strength in the die tongue. This is particularly necessary in copper or copper-base alloy extrusion. The "as extruded" leg is then formed in a die to the desired shape.

Fig. 36.9 — Sharp-corner shapes normally require surfaces which intersect at angles greater than 90 degrees. Photo, courtesy Revere Copper & Brass Inc.

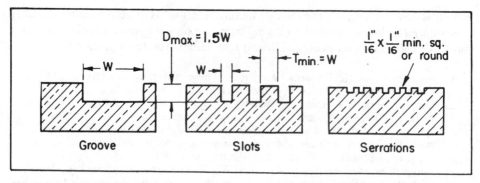

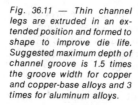

Fig. 36.10 — Design factors relative to grooves and serrations are shown above.

Fig. 36.11 — Thin channel legs are extruded in an extended position and formed to shape to improve die life. Suggested maximum depth of channel groove is 1.5 times the groove width for copper and copper-base alloys and 3 times for aluminum alloys.

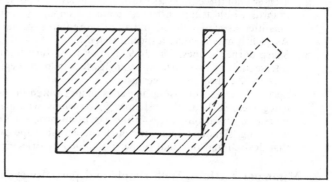

When such channels or grooves appear in multiple, the distance between channels should at least be equal to the channel width. Maximum groove or channel depth should be limited, if possible, to 1 1/2 times the width for economic reasons. With semienclosed aluminum extrusions the length or width of the groove or channel, i.e., the die tongue, should be limited to a maximum of three times the tongue neck. Occasionally this can run up to four times, but such shapes require special consideration and should be discussed with the manufacturer prior to completing the final design. Where thin legs are to be extruded integral with a relatively heavy body, the length of leg should not exceed 5 times the leg thickness.

Hollow Shapes. An almost unlimited variety of hollow shapes — simple,

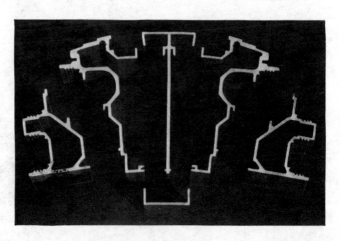

Fig. 36.12 — Group of shapes showing some unusual hollow aluminum extrusions for assembly into a complex post. Photo, courtesy Revere Copper & Brass Inc.

symmetrical, asymmetrical and special — can be readily produced, *Fig.* 36.12, in aluminum and magnesium alloys. As mentioned previously, though, copper and copper-base alloys normally cannot be handled in this manner. The nearest approach to a hollow shape in these materials is produced by extruding in an open position and die forming the section closed, *Fig.* 36.13. A minimum opening of 1/4-inch is necessary owing to springback in forming. On occasion, where flat shapes beyond the capacity of available presses are desirable, semienclosed shapes can be used to advantage in aluminum and magnesium alloys. Extruded in an enclosed or perhaps W form, such extrusions can be easily straightened into flat shapes or flat formed shapes by rolling.

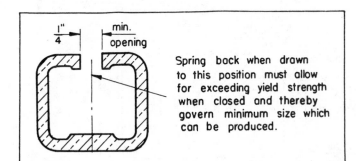

Spring back when drawn to this position must allow for exceeding yield strength when closed and thereby govern minimum size which can be produced.

Fig. 36.13 — Design indicating the nearest approach to a hollow section attainable in copper-base materials. Drawing, courtesy Revere Copper & Brass Inc.

Corners. All corners or edges considered as "sharp" are normally made with a 0.010-inch radius. Absolutely "razor sharp" corners are points of stress concentration and contribute to early die failures as well as subsequent processing difficulties. Truly sharp corners on parts should therefore be avoided wherever possible although occasionally corners can be brought to a sharp edge by the drawing operation without greatly increasing the cost. In such cases the corners are extruded to 1/32-inch radii and cold reduced to be as sharp as practicable. Recommended economical minimum radius for plain extruded shapes is 1/32-inch and in the drawn condition 1/64-inch to avoid stress concentrations and excessive costs, *Fig.* 36.14.

Combination Shapes. To obtain the structural strength of a box section, retain adequate simplicity, and use a minimum amount of metal to achieve a complicated section design, a combination of shapes can be used. Combination shapes can be produced so as to provide an assembly which interlocks to form a complete unit, *Fig.* 36.12. Tolerances can be controlled to insure a loose, sliding or drive fit for such members. Dovetails for interlocking assemblies, *Fig.* 36.15, have been standardized in a few sizes most commonly used with aluminum extrusions. Details for a 1/8-inch, 3/16-inch and 5/16-inch size are shown in *Fig.* 36.16.

Stepped Extrusions. Stepped extrusions mentioned previously, *Fig.* 36.17, have been developed for long parts such as aircraft wing spars where it is desirable to have an integral spar end cap. To date, these extrusions are possible only in aluminum. Originally, it was necessary for one surface to be common to both large and small sections. Today, however, this limitation has been overcome and it is now entirely practical to produce stepped extrusions having no surface common to both large and small sections. These extrusions must be made large at one end only, other variations are not yet possible. Sections must be simple and should have no wide, thin flanges. An allowance of $1/8$-inch for machining is usually adequate to take care of any lack of straightness or mismatch. In many cases $1/16$-inch is sufficient but, because of the highly specialized nature of these extrusions, it is preferable to consult with the supplier prior to assuming a minimum adequate allowance.

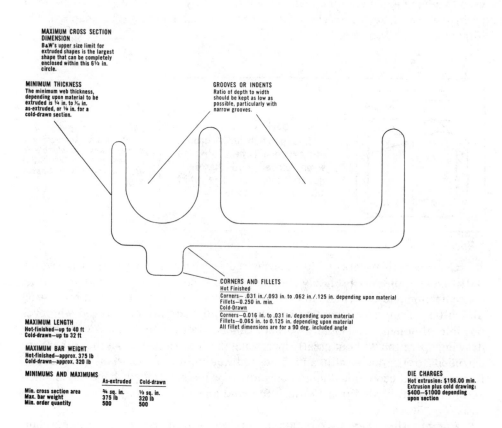

MAXIMUM CROSS SECTION DIMENSION
B&W's upper size limit for extruded shapes is the largest shape that can be completely enclosed within this 6½ in. circle.

MINIMUM THICKNESS
The minimum web thickness, depending upon material to be extruded is ¼ in. to ¾ in. as-extruded, or ⅛ in. for a cold-drawn section.

GROOVES OR INDENTS
Ratio of depth to width should be kept as low as possible, particularly with narrow grooves.

CORNERS AND FILLETS
Hot Finished
Corners— .031 in./.093 in. to .062 in./.125 in. depending upon material
Fillets—0.250 in. min.
Cold-Drawn
Corners—0.016 in. to .031 in. depending upon material
Fillets—0.065 in. to 0.125 in. depending upon material
All fillet dimensions are for a 90 deg. included angle

MAXIMUM LENGTH
Hot-finished—up to 40 ft
Cold-drawn—up to 32 ft

MAXIMUM BAR WEIGHT
Hot-finished—approx. 375 lb
Cold-drawn—approx. 320 lb

MINIMUMS AND MAXIMUMS

	As-extruded	Cold-drawn
Min. cross section area	¾ sq. in.	½ sq. in.
Max. bar weight	375 lb	320 lb
Min. order quantity	500	500

DIE CHARGES
Hot extrusion: $156.00 min.
Extrusion plus cold drawing:
$400—$1000 depending
upon section

Fig. 36.14 — Design tips for steel extrusions. Maximum radii should be used. Courtesy, Babcock & Wilcox Co.

Selection of Materials

Most commercial shapes are extruded in copper, brass, steel, aluminum, zinc, or magnesium. Copper of 99.9 per cent purity lends itself readily to shape production. A large number of copper-base alloys also can be extruded; those with the highest zinc content are the most plastic and can be produced in the most complex cross sections. Those in the range of 56 to 63 per cent copper are used the most, require the lowest extrusion pressure, and are exceptionally plastic in the temperature range from 1200 to 1550 F. Hollow shapes cannot be made in steel, copper or copper alloys. Cost of extruding zinc alloys is higher than for the brasses and consequently they are seldom used.

Copper-Base Alloys. As copper content increases, copper-base alloys become increasingly stiff and difficult to extrude. Within the pressure limitations of present equipment the complexity of section will vary markedly with copper content. Commercial copper extrusions are generally limited to relatively simple and large solid shapes. Control of lead is important at the higher temperatures and generally alpha brasses or commercial coppers of high-lead content are not commonly extruded. Large, simple sections have been made under controlled conditions with lead contents as high as 5 per cent.

In addition to these materials, extrusions have been made in phosphor bronzes, aluminum bronzes, and nickel silvers — usually those in which the combined sum of nickel and copper falls in the range of 60 to 63 per cent. Nickel-silver alloys. having higher combined percentages become extremely stiff at extrusion temperatures and can only be produced in the most simple and relatively large sections.

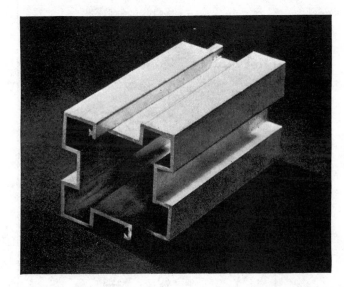

Fig. 36.15 — Combination shape which uses a standard dovetail joint for interlocking. Photo, courtesy Aluminum Company of America.

Nickel Alloys. Cupro-nickel alloys, particularly those having 20 to 30 per cent nickel content, Monel, K-Monel, and Inconel have been hot extruded but to date only simple sections such as tubes, rounds, squares, and hexagons have been successfully produced.

Lightweight Alloys. Aluminum and magnesium extruded shape production probably exceeds that of all the other materials. Most of the commercial light alloys can be extruded and owing to their favorable strength-to-weight ratio are finding increased use in portable equipment, vehicle bodies, textile machinery, and aircraft where weight savings are critical. Aluminum shapes are most often produced in 6063-T5 or T6 and 6061-T5. Hard alloys such as 7075 cost more to extrude; magnesium alloys in AZ31B and C, AZ61A, AZ80A and ZK60A.

Steel & Alloys. Steels from C-1018 through 8750 and 52100 can be extruded as can stainless grades from 302 through 430. Choice should be discussed with the producer in order to insure minimum cost.

Metal Characteristics. Because of the variations in flow characteristics,

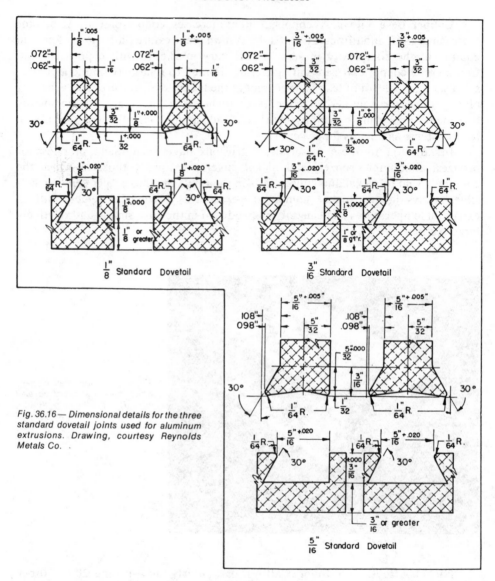

Fig. 36.16 — Dimensional details for the three standard dovetail joints used for aluminum extrusions. Drawing, courtesy Reynolds Metals Co.

reduction ratios between billet and die, and time of extrusion, a wide variation in structure is usually evident. The leading end of any extrusion differs sufficiently from the trailing end so that tensile property range in some materials may vary plus or minus 5000 psi. A minimum standard for the lead end, which would be the softest condition, may be considered as corresponding to existing ASTM standard specifications for rod.

Specification of Tolerances

Good commercial tolerances held in the general production of extruded shapes are shown in TABLE 36-I. These tolerances are generally close enough to permit use of as-extruded parts in most routine assemblies. In the event closer

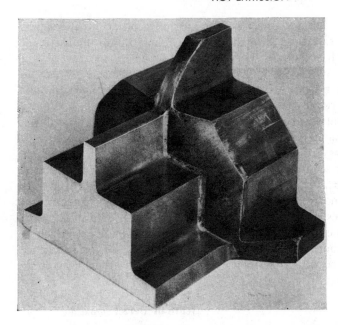

Fig. 36.17 — Section of a simple stepped extrusion showing the portion at the step. Photo, courtesy Aluminum Company of America.

tolerances are required on a specific dimension, such can be held. On payment of extras, mills will accept tolerance reductions down to one-quarter commercial.

Surface Conformance. With flat surfaces the allowable deviation is usually taken as 0.004-inch per inch of width with 0.004-inch deviation from the flat as a minimum. Allowable deviation from a specified contour is plus or minus 0.005-inch per inch for curved surfaces with 0.005-inch minimum.

Angles. Angular tolerance is 2 degrees, plus or minus, where thickness of the thinnest leg is under 0.188-inch, and $1^{1}/_{2}$ degrees for legs 0.188 and over.

TABLE 36-I

Commercial Tolerances for Extruded Shapes

(inches, plus or minus)

Dimension (Inches)	Brass (Plain Ext'd)	Muntz Metal Naval Brass (Plain Ext'd)	Mang. Bronze Nickel Silver (Plain Ext'd)	Leaded Brass Copper Naval Brass Mang. Bronze Muntz Metal (Ext'd & Drawn)	Aluminum Alloys* Magnesium Alloys	
					(Not Heat Treated)	(Heat Treated)
0.000 to 0.125	0.005	0.0075	0.010	0.003	0.007	0.010
0.126 to 0.500	0.005	0.0075	0.015	0.003	0.010	0.015
0.501 to 1.000	0.0075	0.010	0.020	0.005	0.015	0.020
1.001 to 2.000	0.010	0.015	0.025	0.0075	0.017	0.025
2.001 to 3.000	0.015	0.020	0.030	0.010	0.020	0.030
3.001 to 4.000	0.020	0.030	0.035	0.015	0.025	0.035
4.001 to 5.000	0.030	0.040	0.045	0.020	0.030	0.040
5.001 to 6.000	0.035	0.050	0.060	0.030	0.035	0.045
6.001 to 7.000	....	....	....	....	0.040	0.050
7.001 to 8.000	....	....	....	....	0.045	0.055
8.001 to 9.000	....	....	....	....	0.050	0.060
9.001 to 10.000	....	....	....	....	0.055	0.065
10.001 to 11.000	....	....	....	....	0.060	0.070
11.001 to 12.000	....	....	....	....	0.065	0.080

* Usual deviation from specified dimension where 75 per cent or more of the dimension is metal. With hollow shapes where more than 25 per cent of the dimension is space, specified tolerances must be increased from 0.004 to 0.010-inch depending on size.

Radii. Tolerances on radii are: Plus 1/64-inch for sharp corners or fillets; plus or minus 1/64-inch for any radius up through 0.187-inch; and plus or minus 10 per cent for radii of 0.188-inch and larger.

Surface Roughness. Wherever possible, for economy, surface finish as extruded should be acceptable on parts. Price penalties and restriction of suppliers may result if unnecessary requirements are placed on surface quality.

Surface quality is greatly affected by the extrudability of the material, billet heat, and speed of extrusion. Handling and straightening are rough on surface finish. Drawing reductions are usually light and do not remove all the visible scratches which may arise from heat checks and die cracks at elevated temperatures. For best surface quality on copper extrusions, two cold draws are necessary to remove normal defects due to the pressure and high temperatures needed. Surfaces sometimes can be sanded, polished or buffed to improve surface finish but as a rule such operations are not too well suited for handling in the mill and in any but large-quantity runs would be uneconomical. Typical surface roughness generally allowed with aluminum extrusions indicates the approximate depth of scratches to expect:

Section Thickness (in.)	Depth of Defect (max., in.)
Up through 0.063	0.0015
0.064 through 0.125	0.002
0.126 through 0.188	0.0025
0.189 through 0.250	0.003
0.251 and up	0.004

These scratches or surface markings are the maximum to be expected and, naturally, would be in the minority. Included in such defects are die marks, handling marks and polishing marks.

ADDITIONAL READING

[1] *A Guide to Aluminum Extrusions,* The Aluminum Association, Aluminum Extruders Council, New York, 1967.
[2] *Steel Extrusions,* Babcock & Wilcox Co., Beaver Falls, PA., 1970.

37

Cold and Impact Extrusion

EMPLOYED to a considerable extent in hot and cold forging and especially on nonferrous parts, extrusion of metals both by impact and continuous pressure takes advantage of a peculiar characteristic of metals. Plastic workability or low resistance to mass deformation, usually achieved by heating above the recrystallization range can, with some of the ductile and malleable metals, be attained *with little or no heating* through the application of tremendous impact pressures.

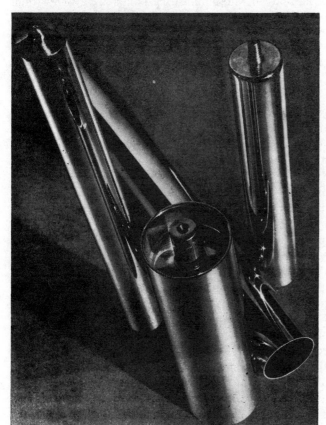

Fig. 37.1 — Some typical cylindrical impact extrusions showing special bottom design. Photo, courtesy Aluminum Company of America.

Fig. 37.2 — A group of radio shielding pots impact extruded from aluminum. Side holes require an extra punching operation. Photo, courtesy Aluminum Company of America.

Cold Extrusion. Plastic flow of metals being impact extruded is not unlike that of a heavy viscous fluid. During extrusion the metal retains its crystalline structure. Except for such metals as lead and tin whose annealing points are around room temperature, plastic flow during extrusion is the characteristic slip-plane movement of cold working. Heat generated by the impact blow exercises a profound effect and operates to reduce the extruding pressure. However, where the materials so dictate modest heating is utilized to improve processing, giving rise to the term "warm forging."

Field of Application. Although the pressures required and consequent tool duty are much more severe when compared to stamping and deep drawing, the advantages of impact extrusion are numerous. Within the size and capacity limitations of the present-day equipment, this processing method offers unique possibilities in the design and application of cups, deep shells, or tubular shaped parts, *Fig.* 37.1. The astonishing reductions which are possible by extrusion often produce with a single blow results that would otherwise require several deep drawing operations, including intermediate anneals. Consequently, fewer machines are necessary and output is greater. Common applications include cans, capsules, flashlight cases, cartridge cases, dry battery cases, tubular parts with straight or tapered walls, radio condensers and shielding pots (*Fig.* 37.2), etc. One of the most interesting applications is the impact extrusion of 10-bladed aluminum alloy turbine fans for aircraft refrigeration units which operate at 100,000 rpm.

Extrusion vs. Drawing. The point where extrusion becomes more economical than drawing depends both on the shell proportions and on shape or complexity. Rectangular or odd cross-section parts which are difficult to deep draw, in many instances, can be extruded with relative ease, *Fig.* 37.3. Again, considering ordinary round shells having plain flat bases, extrusion is economically sound when the length exceeds 1.5 times the diameter. Also, in certain cases extrusion can be used to advantage as a preliminary in deep drawing.

Special advantages of the process make it imperative for the designer to

Fig. 37.3 — Square, rectangular and odd cross section shells can be readily extruded with special bottom designs. Photo, courtesy Aluminum Company of America.

Fig. 37.4 — Group of aluminum impact extrusions showing further variations in bottom design commercially practical. Photo, courtesy Aluminum Company of America.

investigate and consider its possibilities. Impact extrusion is relegated primarily to what might be termed tubular or cup-shaped parts, and the thickness and form of the base portion need bear no relation whatever to the wall. Bases may be flat, conical, formed, or embossed, and can be of uniform thickness or may include integral lugs, bosses, depressions, or ribs, *Figs.* 37.1, 37.3 and 37.4. Wall cross section can be of almost any symmetrical shape, with or without flutes or ribs, and normally must be uniform throughout the length.

Impact Method. Essentially, the most versatile method of impact extrusion produces parts by striking a cold slug of metal, *Fig.* 37.5, approximately the shape

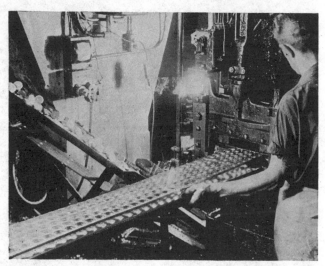

Fig. 37.5 — Slugs for impact extrusions being blanked from plate in a multiple die. Photo, courtesy Aluminum Company of America.

and size of the outer cross-sectional dimensions of the desired shell, with a high-speed impact blow. The punch used is of a size and shape required for the inside of the part to be produced while the die cavity corresponds to the external dimensions. The impact, which may be on the order of 20 to 100 tons per square inch of area depending on the ductility of the metal and the machine impact, causes the metal to act as a viscous fluid and flow up and out between the confines of the punch and die. Normally the punch is undercut above a narrow shoulder to assist in stripping. Characteristics not unlike ordinary fluids with respect to orifice effect are present and thus the wall extruded up and around a punch is neither as large in outer dimensions as the die nor as small in internal dimensions as the punch. Wall thickness therefore is always less than the die and punch clearance but varies according to the sharpness of the corners over which extrusion takes place.

Impact Presses. Impact work is usually performed on high-speed crank, eccentric or toggle presses, *Fig.* 37.6. Press speeds may range from 20 to about 60 strokes per minute, although in many cases speed is limited to 40 to 45 strokes per minute where copious quantities of coolant cannot be used to cool the tools. Actual working stroke usually occupies about one-tenth to one-fifteenth of a second. However, comparable work is done on hydraulic presses and also on newly developed high-energy impact forging presses, *Fig.* 37.7.

Ranges of Sizes. Pressures required in ordinary press extrusion, as noted previously, may range from approximately 50 tons per square inch for highly ductile, malleable materials such as pure aluminum, lead or tin to 100 tons per square inch for the lower ductility metals and alloys such as brass or copper and

Fig. 37.6 — Rectangular aluminum impact extrusions being produced on a typical high-speed mechanical press. Photo, courtesy Aluminum Company of America.

aluminum alloys. The impact method, however, reduces this requirement to roughly 20 tsi and 40 tsi, respectively. Base area of parts which can be produced, therefore, depends upon available press capacity. Lead and tin extrusions up to about 3 inches in diameter and 12 inches in length with wall thicknesses from 0.005 to 0.010-inch have been regularly produced. Aluminum parts usually range from 0.5-inch to 5 inches in diameter, overall length varying with diameter from about 4 to 10 inches, and minimum wall thicknesses also varying with size from 0.004 to 0.075-inch. However, impacts with diameters up to 26 inches and lengths to 3 feet with 2 inch walls have been produced. A great variety of rectangular and square shell designs, *Fig.* 37.3, up to approximately 3 inches have been produced but here again design, the material being worked and press capacity determine the maximum dimensions.

Hooker Extrusions. Another method of impact extrusion which makes possible the extrusion of extremely long tubular sections in the harder metals such as brass and copper is that commonly referred to as the Hooker process. This interesting modifiation of the impact method, instead of extruding the metal up and around the punch by applying pressure to a plain slug, forces it down through the die by pressure on the wall or flange of a formed slug, *Fig.* 37.8. Applications of the Hooker process, however, are rather limited and so far have found use primarily in producing small-diameter thin-wall copper, brass and aluminum seamless tubes with open or closed ends, cartridge cases, battery cases, etc.

The outstanding advantage of the Hooker process is that of producing cylindrical parts with tapered walls, *Fig.* 37.8, in but a single stroke, a feature not possible with regular impact work. Also, owing to the character of the metal movement in the Hooker process, the pressure required is less intense and, to further relieve tool duty, special preformed annealed blanks are usually employed. Unlike regular impact work, though, the base of such parts must be plain, the

Fig. 37.7 — Mesta-Ceff Press, Model HE-55 which exerts 400,000 foot pounds. Machines range from 72,000 to 1,808,000 foot pounds of energy. Courtesy, Mesta Machine Company.

forging action of the punch against the die bottom not being available and the bottom of the part remains as preformed on the blank.

Hooker Production. Production speed in extruding downward is rather fast for most parts — speeds around 125 strokes per minute being used. In some cases a separate die-trimming operation is not necessary, for a new slug forces out the previously finished part. Economical production runs usually range around 10 to 15 million pieces.

Range of Sizes. Tubular parts made by this method run from about 3/16 to 3/8-inch diameters with wall thicknesses from 0.004 to 0.010-inch. Heavier walls can be produced but generally can be made more economically by deep drawing. Lengths extruded range up to 12 inches and the maximum feasible is about 14 inches.

Cold Extrusion of Steel. Developed originally in Germany*, the cold extrusion of steel is now a reality. Basically, this method of extrusion is almost identical to the Hooker process — in extruding the metal flows *with* the direction of the punch — although, in addition, work in which the material flows *against* the

*Cold Shaping of Steel — OTS Report PB-39371.

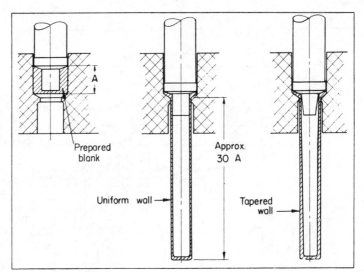

Fig. 37.8 — A modification of impact extrusion, the Hooker process forces material down through the die around the punch rather than up. Drawing, courtesy Clifford Mfg. Co.

direction of the punch as well as both *with and against* have been successfully done. Apparently, the main key to successful cold extrusion of steel is the use of a zinc phosphate coating such as Bonderizing which helps resist the tremendous forces of the operation and reduce friction to a large extent.

Cold extrusion permits the making, in one integral piece, of parts which normally are difficult to deep draw or which require a great many anneals. Extrusion pressures required range from 86 to 115 tons per square inch. Reductions as high as 70 per cent are practical, *Fig.* 37.9, strength is increased, and surfaces produced are extremely smooth and free from imperfections.

A typical cold-extruded steel part, *Fig.* 37.10 indicates the versatility of the process. Generally, a low-carbon, low-silicon, deep drawing steel of good homogeneous quality is utilized for cold extruding. This new method is now well developed and parts from a few ounces to as much as 15 pounds with diameters up to 3-1/4 inches with conventional machines and with high-energy machines, diameters up to 12 inches are commonly produced in modest to large quantities.

Design Considerations

As a rule impact extrusion should be considered primarily when a part requires à heavy bottom with a relatively thin side wall or when the heavy bottom requires integral studs, bosses, necks, or cavities inside or out, *Fig.* 37.11. Another criterion for design is cross section — parts with square, rectangular, oval, or combinations of such shapes in section being particularly suitable. When wall thickness in section must vary, including longitudinal ribs inside or out, when longitudinal wall thickness must taper, or when the ratio of length-to-diameter is high, the extrusion method should be carefully considered.

Shell Length and Walls. Generally, in impact extrusion the maximum shell length is about six times the inside diameter of cylindrical parts. Column strength of the punch is usually a practical limiting factor. In designing for a combination of several of the preceding variations mentioned, a limiting factor in bottom thickness enters into the picture, especially when wall thickness varies in cross-section.

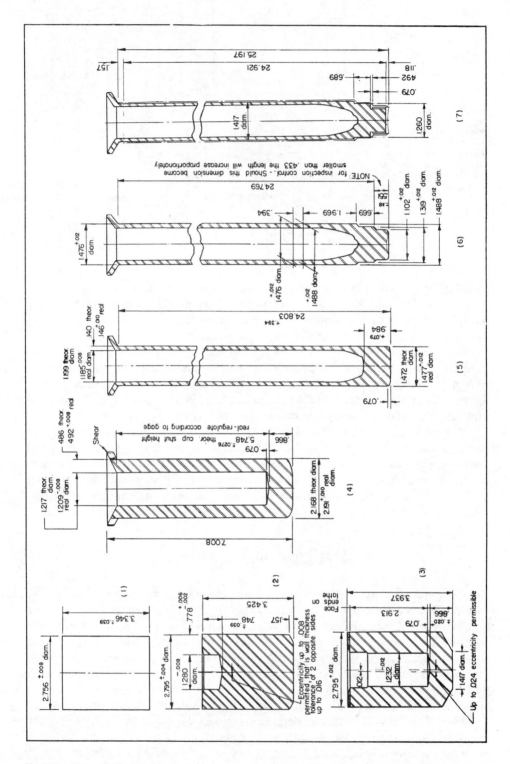

Fig. 37.9 — Flanged cylinder cold extruded in two operations (4) and (5) from the cold-formed blank shown at (3). At operation (6) the bottom is cold formed and at (7) the machined part is shown.

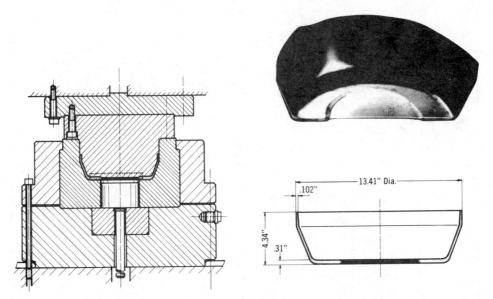

Fig. 37.10 — A typical cold extruded steel parts showing the design possibilities which the high-energy method offers. This truck wheel insert of AISI 1020 steel, weighing 16 pounds, was formed at 2250 degrees F with 400,000 foot pounds of energy on the CEFF Model HE-55.

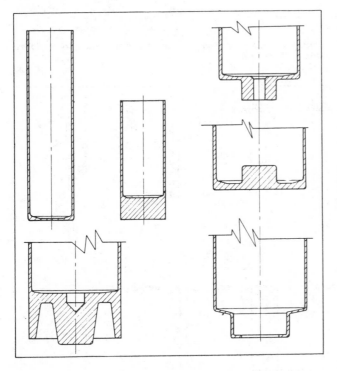

Fig. 37.11 — Typical designs well suited to impact extrusion. Drawing, courtesy Aluminum Company of America.

Sufficient metal must be provided to fill out the wall, hence it is desirable to have the finished bottom 0.010 to 0.040-inch heavier than the greatest wall, depending on production conditions. Bottoms for the Hooker method usually should be plain but can be any desired thickness, and side walls, as noted previously, can be tapered as desired within practical punch limitations.

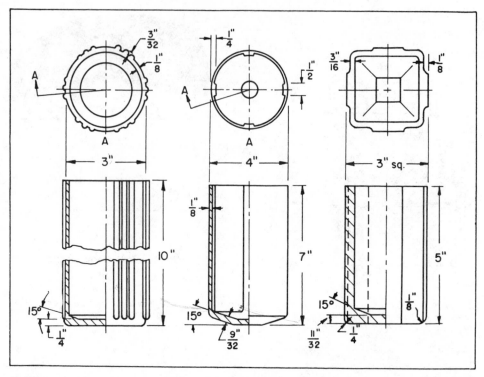

Fig. 37.12 — To assure maximum accuracy cross section of extrusion designs should be as symmetrical as possible. Drawing, courtesy Aluminum Company of America.

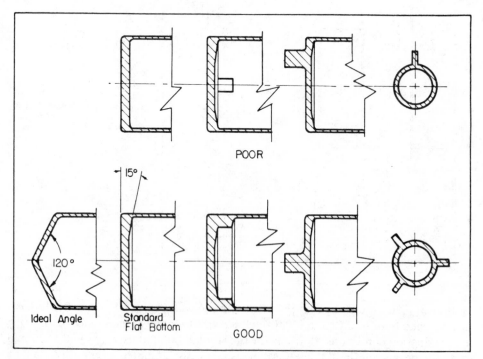

Fig. 37.13 — Heavy walls provide maximum economy in production but should be used in preference to minimum thickness only with proper consideration for metal economy. Drawing, courtesy Aluminum Company of America.

Symmetry. Owing to punch length, in most cases there is a tendency toward float during extrusion. To minimize this cause of wall variation, symmetry in design is highly desirable as well as is attention to bottom design. Cross-sectional wall design should be symmetrical as to wall variations and ribbed sections to equalize extrusion forces, *Fig.* 37.12, and, likewise, bottom design should be as symmetrical as possible to assure regular flow of metal, *Fig.* 37.12. Tapering or beveling the punch nose also assists centering of the punch and helps reduce the pressure of extrusion. In aluminum impact extrusion, a standard punch bevel of 15 degrees has been adopted for flat bottoms while a 120-degree included angle is considered the ideal for cylindrical parts, *Fig.* 37.12. Complex designs may necessitate some tooling consultation to assure satisfactory production. To assure an even, smooth flow of metal at the outer edges, corner or edge radii should be at least 0.031 to 0.062-inch and preferably greater.

 T<small>ABLE</small> 37-1 — **Practical Minimum Walls for Aluminum Impacts**
(Inches)

Outside Diameter (in.)	Alloy			
	1100	**6061**	**2014**	**7075**
1	0.010	0.018	0.031	0.036
2	0.020	0.028	0.062	0.073
3	0.030	0.038	0.093	0.110
4	0.040	0.050	0.125	0.150
5	0.050	0.075	0.155	0.180
6	0.060	0.090	0.190	0.220
7	0.075	0.107	0.220	0.250
8	0.100	0.130	0.260	0.300
9	0.110	0.143	0.300	0.345
10	0.125	0.162	0.350	0.375

Courtesy, Kaiser Aluminum & Chemical Co.

Wall Thickness. Pressures as high as 100 or more tons per square inch of bottom area have been recorded in impact production of ordinary parts. Pressures increase rapidly from a minimum with soft, pure metals and small parts to a maximum with large parts or work hardening metal or alloys. As bottom area increases, the metal is required to travel further before extrusion, hence a rectangular bottom with the same area as a round one requires more pressure. Also, more pressure is required to extrude a shell with a thin wall than one with a thick wall, It is well, therefore, to avoid specifying minimum wall thicknesses wherever possible, *Fig.* 37.13, and T<small>ABLE</small> 37-I.

Holes. Holes in projecting lugs or nozzles often can be produced or pierced during the extrusion process, or provided in the blank. Such holes, however, must be coaxial or centrally located — all other holes must be produced by a secondary operation.

Punch Angle for Hooker Extrusion. With Hooker extrusion, the flatter the punch angle and the steeper the die angle, *Fig.* 37.8, the easier the flow of metal. Good practice dictates a 20-degree angle on the punch nose and a 30 to 35-degree angle in the die. Where push-through, continuous extrusion is used these angles are usually reduced to 15 degrees each. Relation of wall thickness of a blank to wall thickness of the finished tube is usually limited to the range of 4-to-1 and 25-to-1.

Design of Steel Extrusions. Parts in steel should preferably have a symmetrical cross-section. Any changes in diameter require a radius for flow but should be 1/32-inch for maximum efficiency. Undercuts cannot be formed and tapers or draft angles should be avoided. However, punches should have a centering angle or radius. High-energy machines, however, are being used for producing unusual designs, *Fig.* 37.14.

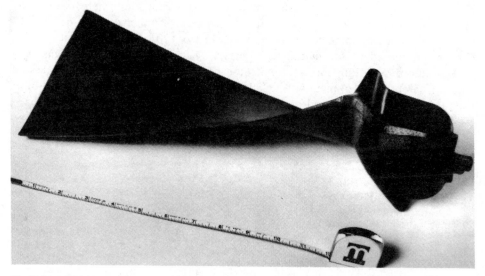

Fig. 37.14 — High-energy impact forged turbine blade. This titanium (8-aluminum; 1-molybdenum; 1-vanadium) aircraft engine fan blade was precision forged with a twist over 70° in three blows from the same die. Courtesy, Mesta Machine Company.

Selection of Materials

Cold impact extrusion applications began with tin. To date the method has also been successfully applied to parts of lead and aluminum. Zinc is usually handled with a little preheating — about 300 F — to get into the most ductile range. With the Hooker process copper and brass can be readily worked from cold slugs although it is also well adapted to producing parts from lead, tin and aluminum. Work at present is well under way in the Hooker extrusion of steels and before long parts should be commercially obtainable. Some work has been done experimentally on hot impact extrusion of small copper parts but it has not been widely commercialized.

Zinc. In general, to assure maximum workability at lowest impact extruding pressures, pure metals are necessary, especially with those metals that strain harden quickly. Purity of zinc is of much importance, a special grade 99.99 per cent pure is usually employed although an alloy with 0.6 per cent cadmium is used for dry battery cells. Zinc has the added advantage of ease of soldering, a characteristic of value in many electrical applications.

Aluminum. In impact extruding aluminum, 99.75 per cent commercially pure 2S is preferable owing to longer tool life, although 99.0 per cent pure metal is widely used. Any of the wrought aluminum alloys can be impact extruded. Alloys 1100 and 3003 are strain hardening and nonheat-treatable, so are used for moderate strength needs. Smoothest surfaces are obtained on parts produced from the 1100 alloy. Where greater strength is required, alloys 6061, 6063, 6066 and 6151 are specified. These also have better machinability. Where highest strength is imperative, alloys 2014, 7039 and 7075 are used. Tensiles from 70,000 psi to 97,900 psi can be attained by proper treatment.

In order to realize good tool life, the following percentages of reduction in area should not be exceeded:

1100 alloy	96%
6061 alloy	94%
2014 alloy	88%
7075 alloy	86%

Recommended minimum wall thickness, based on the above percentages, are shown in TABLE 37-I.

Copper and Brass. When copper is specified, a 99.9 per cent pure electrolytic grade is employed. Along with copper, 70/30 cartridge brass as well as aluminum also find use in the Hooker method. These metals show the effects of severe cold-work and consequently must be annealed both before and after extrusion if ductility is necessary in the extruded part for further working or cold forming.

Steel. Most steel cold extruded parts are produced from 1006 through 1080 carbon steels. Also, alloy steels such as 1340, 4140, 4615, 5015, 5140, 8620, 8640, and 8655 are processed.

Specification of Tolerances

As can be readily seen, conditions of processing are conducive to high accuracy during useful tool life. Wall thickness and diameter can usually be held to plus or minus 0.001 to 0.002-inch per inch with impact extrusion while bottom thickness can be held within plus or minus 0.003 to 0.005-inch depending upon design complexity. However, for maximum economy, practical tolerances will be plus or minus 0.003-inch per inch of diameter. Base, web or flange thickness will usually run plus or minus 0.012-inch for parts up to 3 inch diameter, plus or minus 0.015-inch for 3 to 5 inch diameter, plus or minus 0.020-inch for 5 to 7-1/2 inches, and plus or minus 0.030-inch for parts over 7-1/2 inches. For walls a practical rule is to expect tolerances to be plus or minus 10% of thickness. Ovality may be a factor and can run as much as 0.004-inch per inch of diameter and on critical parts may require a sizing operation. Surface finish will run from 16 to 125 rms depending on the ease with which the alloy can be extruded.

Hooker Extrusions. Parts have been produced by the Hooker process with walls as thin as 0.004-inch, accurate to plus or minus 0.0005-inch. Large parts with heavier walls, normally, should have wider tolerances. Production economy is approximately inversely proportional to the tolerance allowed on diameter. Bottoms by this process depend primarily upon the method of blank manufacture, preparation of the blank usually including a cupping and stamping operation to size the bottom.

Erosive action on the narrow edge of the die over which extrusion takes place creates severe die wear and consequently the wider the tolerances specified, the greater the die life and the lower the unit cost per piece.

Steel Cold Extrusions. In general, cold extruded steel parts will carry tolerances equivalent to those common on conventional screw machine parts. Thus, tolerances of plus or minus 0.001-inch to plus or minus 0.005-inch are typical. Normal surface finish runs from 50 to 150 rms.

Additional Reading

[1]*Aluminium Impacts Design Manual*, 1973. The Aluminum Association, New York.

38

Cold Drawing

O FFERING possibilities similar to those of the hot extrusion process, cold drawing broadens the range of application to improve product quality. The use of cold-drawn and cold-finished rounds, hexagons, squares, etc., in both ferrous and nonferrous materials is well-known and requires no comment. The use of shapes in the design of machine elements to reduce machining cost or simplify production, however, demands careful consideration and continued attention.

Process Possibilities. Although ferrous materials can be extruded hot as noted in Chapter 36, continuous shaped bars can be cold drawn to high accuracy. The process of cold drawing fills a gap in the field of shaped bars. While the variety of cross-sectional shapes that can be produced is somewhat more restricted than with extrusion, the possibilities are nevertheless sufficiently broad, *Fig.* 38.1, to offer economical advantages. By designing with the limitations of the process in mind, the economic advantages can be multiplied manyfold.

Any machine part which has an irregular cross section whose surfaces are parallel along all or a major portion of one axis are naturals to consider for production from multiple-length bars, *Fig.* 38.2. This is especially true where the amount of material necessary to remove in producing such a cross section comprises the greatest portion of the required machining. Assuming that the requirements of the design can be satisfied by a cold-drawn material and by its physical properties, the primary controlling factors which remain are the physical limitations of the process and the production quantities. In the final analysis, adapting a part to the cold drawing process often results in improved appearance, simplified design, increased strength, and longer wearing qualities.

Cold Drawing. Cold-drawn bars normally are produced from hot extruded shapes, hot-rolled bars or coils prepared especially for this purpose. The material should be of top quality inasmuch as the drawing operation involves the elongating or stretching of the material which would accentuate any surface irregularities or defects. As a rule, the order of operations is as follows: (1) Pickling; (2) washing with water and immersing in lime solution; (3) pointing; (4) cold-drawing; (5) straightening; (6) cutting to length; and (7) inspecting. Pickling is required to assure an unmarred cold-drawn finish.

Liming after washing assures complete removal of pickling acid and also serves as added die lubricant in drawing. Where bars cannot be started through the die by means of the regular pointing head, the lead end of the bars are reduced for a

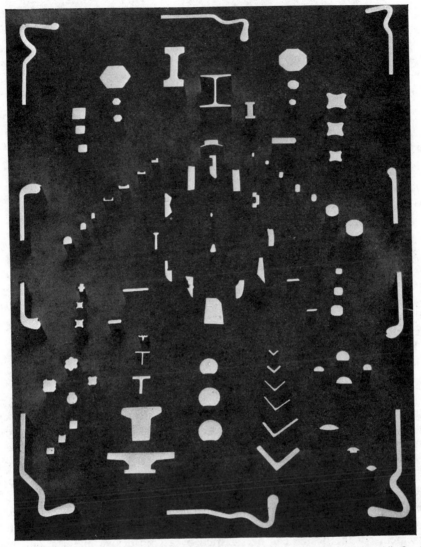

Fig. 38.1 — A variety of typical cold-drawn steel cross sections. Photo, courtesy Jones & Laughlin Steel Corp.

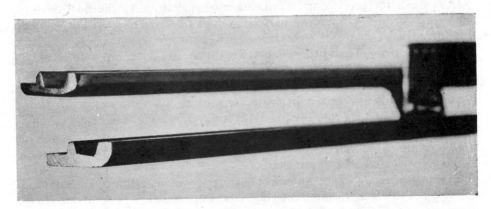

Fig. 38.2 — Finished carriage ball race for accounting machine and cold-drawn shape used.

Fig. 38.3 — Six hot rolling passes and five cold drawing passes are used to produce this special section from a standard round, left.

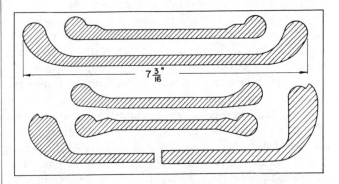

Fig. 38.4 — Special sections, hot rolled and given one cold draw to size. Wherever possible, such sections should be designed to allow symmetrical rolling. Sections can be parted after drawing to obtain two identical halves and simplify the drawing process. Drawings, courtesy Republic Steel Co.

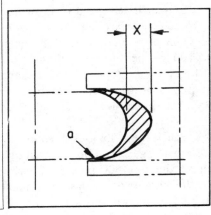

Fig. 38.5 — Turbine blade section of stainless steel. In order to hold dimension (x) to required tolerances and eliminate wrinkling or tearing of the extremely thin edge of (a) special roller drawing dies are required. Excess material drawn out at (a) is sheared to height in an extra operation.

length of 6 or 8 inches by means of rotary swaging or, with intricate sections, by overpickling in acid.

Machines Utilized. The actual drawing operation comprises the pulling or drawing of the pickled and limed hot-rolled or extruded bar through the die opening. Machines used in drawing are the well-known draw benches. These draw benches vary in size and power and may be of any convenient length, usually from 30 to 75 feet. The mechanism consists of three parts; namely, the drive, the pointing head, and the pulling head. Once fed through the die, the bar is gripped by the pulling head and the pulling head carrier is hooked to an endless block chain, which is driven by a variable-speed motor through gears, to draw the material through the die.

Multiple-Pass Sections. Special shapes are processed with a variety of stages. Multiple-pass cold drawing with intermediate annealings is carried out using standard rounds, hexagons, squares, or flats as starters, *Fig.* 38.3. As a rule it is preferable for economical reasons to use up to three cold draws only, inasmuch as too many draws may overwork certain materials. However, as many as eight to ten passes are required for some sections to produce the desired profile, *Fig.* 38.3 being a good example. Size of sections handled normally ranges from approximately 0.030-square inch weighing about 0.1-pound per foot to a maximum of 6 to 7 square inches with weight per foot of 10 to 12 pounds. The maximum width of section normally practicable runs to about 4 inches.

Cold Finished Hot-Rolled Sections. Another method comprises the cold drawing or finishing to size of sections hot-rolled to the necessary shape. Up to three hot rolling passes, and sometimes four, are used to produce the desired shape following which the bar is given one cold draw to size the section and improve surface finish. No actual shaping is done in the cold drawing operation; only a small uniform reduction is made. Only large parts are produced by this method, cross-sectional area and weight per foot being limited to about the same maximum as in multiple-pass work. Section width, however, can run up to 9 or 10 inches where required, *Fig.* 38.4.

Cold Finished Hot-Extruded Sections. These sections follow the size and weight data given in Chapter 36 covering hot-extrusion.

Special Methods. A combination of the preceding two stages or methods can also be used or, where necessary, roller dies or Turks Heads may be employed to reduce the frictional drag of solid dies. Parts such as turbine blades drawn from stainless are produced in such dies, *Fig.* 38.5.

Drawing Speed and Quantities. Speed of drawing varies widely, depending upon the equipment, bar size and section, finish desired, material analysis and physical properties, amount of reduction, etc. The range of commercial drawing runs from about 40 to 100 feet per minute, although for sections such as flats, for instance, speed might be as low as 9 fpm while for shaped wire speed might be as high as 300 fpm. Generally, drawing speed for multiple-draw shapes ranges around 15 to 25 fpm while that for shapes hot rolled and given one sizing draw, ranges around 50 to 60 fpm.

Production speeds such as these obviously require tonnage output to attain the greatest possible price advantage. A minimum quantity for small special shapes might be placed at approximately 500 pounds; where quantity can be raised to about a ton or more on the initial order with similar quantities as frequent as once a year, die costs will be but a comparatively small part of the cost per piece. Where quantity ranges from 5 to 10 or more tons, special sections may cost even less than the simplest machining operations on a standard bar.

Shapes which are hot rolled and given one sizing draw require much greater initial output. Production should range around 100 to 150 tons per rolling for economy and where 3 or 4 rollings can be made per year, roll and die costs can be reduced to insignificance.

Design Considerations

While it is preferable to begin the cold drawing operation with a hot-extruded or hot-rolled shape closely approximating the desired cross-section, the process has been developed to a point where many intricate sections can be drawn from unrelated hot-rolled rounds, squares or flats. A small hot rolling mill is sometimes used to "break down" a standard section to one more suitable for cold drawing,

Fig. 38.6 — Group of typical drawn sections which portray the general range of section variation and types of tooth forms readily drawn in production. Photo, courtesy Republic Steel Co.

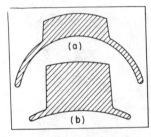

Fig. 38.7 — Special sections such as at (a) require hot rolling with a final forming draw to shape the thin extensions. Section at (b) is more economical and practical. Drawing, courtesy Republic Steel Co.

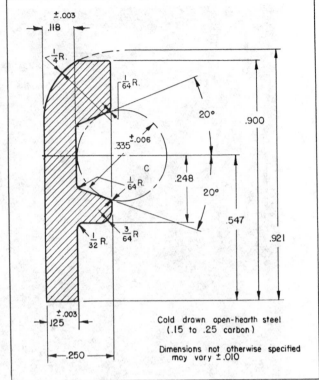

Fig. 38.8 — Typical section design used for the accounting machine carriage raceway shown in Fig. 38.2. The tolerances shown are typical of those normally held in mass production of shapes for machine parts.

Fig. 38.3. For maximum economy and prompt delivery, however, special shapes should be designed to approximate closely as possible an available hot-rolled mill section.

Reduction Limitations. There are practical limitations to the shapes which may be produced economically. Metal in all parts of a section must have sufficient strength to withstand the drawing stresses. Reduction in area and the amount of cold working is thus limited; the usual reduction in area being about 10 to 15 per cent per stage. Actual dimensional reductions or "draft" used may range all the way from $1/64$ to $1/8$-inch on diameter.

Section Thickness. Although the metal can be made to flow into extended sections of a die to a certain extent, the amount of such redistribution is limited by the strength of the bar leaving the die. Light extended sections greater than 3 times their thickness in width are extremely difficult to draw, *Fig.* 38.6, although heavier ones have been produced up to around 6 times. Pinion rod for gears ranging from the very smallest to around 1-inch diameter in mild steel, $7/8$-inch in high-carbon, and $13/16$-inch in stainless have been produced. Almost any tooth form including intermittent, spline and ratchet types can be drawn, *Fig.* 38.6.

Ordinarily, sections which are relatively wide in comparison with their thickness must be produced by hot or cold rolling or extrusion, *Fig.* 38.7, although occasionally roller dies or Turks Heads can be employed to obtain the desired results.

Undercuts. While shallow external undercuts, *Fig.* 38.8, can be produced where the depth does not exceed the width, complicated internal undercut sections are impractical. Many have been attempted but seldom are successful, *Fig.* 38.9.

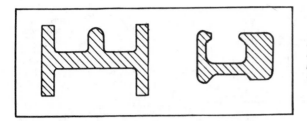

Fig. 38.9 — Undercut sections which have been attempted in cold drawing but proved impractical or uneconomical. Drawing, courtesy Republic Steel Co.

Where exacting undercuts are required such should be produced by machining, *Fig. 38.10.*

Extremely thin edges are undesirable. Edge extensions seldom can be drawn satisfactorily to a uniform length and a sharp edge. A radius is desirable and preferably a full rounded end of fair proportion in relation to the body section, *Fig. 38.11*, should be used.

Corners and Fillets. Re-entrant or inside corners in some sections cannot be obtained readily with a radius less than 1/64-inch because of excessive die wear. Where sharper corners are needed, machining may be required, and in such cases a more generous radius will extend die life considerably. In many designs an undercut can be used to achieve close fits between parts and allow desirable drawing radii where required, *Fig. 38.12.*

Generally, fillets should be specified with minimum radii of 1/64-inch on the smallest shapes with proportionally increasing radii as the size increases. External corners can, on the other hand, be sharp wherever desired.

Selection of Materials

All grades of carbon, alloy or stainless steels that can be hot-rolled or hot extruded are obtainable in cold-drawn form. The softer steels — those low in carbon and hardening alloys — yield most readily to the cold drawing operation. Consequently, the soft open-hearth steels up to 0.50 per cent carbon content are probably the most economical to draw, alloy steels follow, and stainless alloys are

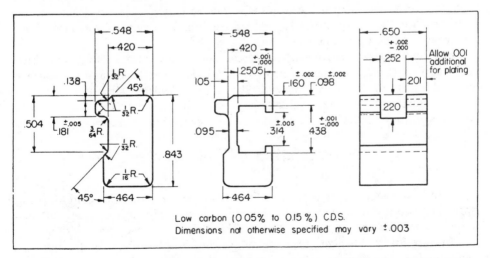

Low carbon (0.05% to 0.15%) C.D.S.
Dimensions not otherwise specified may vary ±.003

Fig. 38.10 — Accounting machine carriage stop-body blank and drawn shape used in manufacture. Undercut portion is impractical to draw.

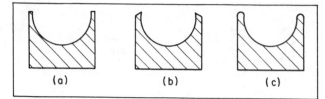

Fig. 38.11. — Extensions of section (a) may not be sufficiently strong for drawing. Section at (b) can be drawn but (c) eliminates the undesirable sharp corners and thin portions of (a) and (b). Drawings, courtesy Republic Steel Co.

Fig. 38.12 — Where close fit is necessary, a corner relief will make possible the most desirable drawing radii. Drawing, courtesy Republic Steel Co.

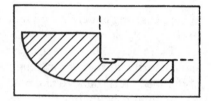

the most difficult to handle. Bessemer steels are of poor drawing quality and are undesirable for any shape requiring over one draw.

Production Cost. While a soft, ductile steel may be drawn to a comparatively simple cross section in two or three passes through progressive dies, a harder steel may require additional passes and possibly preliminary and intermediate anneals to produce the same section. Drawing cost increases and production speed decreases, generally, as the workability or ductility of the material decreases.

Machinability. Although machinability of most carbon and alloy steels is improved by cold drawing, where operations such as cutting off, drilling, tapping, or finish machining are involved, machinability is an important factor and should be kept in mind in selecting the most economical material from an overall standpoint.

Tool Steels. Regular steels are considered as those up to 0.60 carbon content, those over 0.60 carbon being classified as tool steels. Shapes can also be obtained in standard or special analysis oil or water hardening tool steels. Tool steels are especially advantageous on those machine parts requiring precise heat treatments.

Stainless Alloys. Being the most difficult to draw of all the steels, stainless varieties are restricted to use where shape can be very simple and symmetrical. Unless a roller die can be utilized, *Fig.* 38.5, sections must be relatively heavy with respect to extended portions.

Surface Quality. High surface wear resistance imparted to tooth forms which are cold drawn often obviates heat treatments. Smooth, hard, uniform surface attained in cold drawing steels provides the optimum finish desirable on the teeth of gears, ratchets, cog wheels, etc. Improvement in physical properties by cold working will often allow the use of a lower cost steel without sacrifice in strength.

Specification of Tolerances

In addition to increasing machinability and strength, the cold drawing operation also produces a smooth bright surface of excellent appearance. Paints and other finishes can be applied direct without special surface preparation.

Surface Finish. Owing to the excellent surface produced by cold drawing, the necessity for machining can often be eliminated. Thus, wherever possible the surface as drawn should be used. In addition, the improved wear resistance imparted by the cold working is largely lost if the surface is machined away. Inasmuch as cold-worked material has a tendency to distort if partially machined, parts so designed may require annealing to overcome this tendency and the additional cost of this operation should be kept in mind.

Dimensional Variation. Dimensional tolerances possible by means of cold drawing are of considerable importance inasmuch as the accuracy will eliminate most general machining operations. Whereas plus or minus $1/16$-inch is normal in the production of hot-rolled forms or shapes, commercial tolerances for cold-drawn sections range around plus or minus 0.003-inch, *Fig.* 38.10. In the largest sections and in the interests of economical production, plus or minus 0.005-inch or greater should be used, if at all possible. Noncritical dimensions should be dimensioned with wide tolerances to simplify the problem of holding the critical fits as close as possible, *Fig.* 38.8.

Gears and Small Parts. In the case of the smallest shapes which are relatively simple, critical dimensions can be held to plus or minus 0.001-inch and on occasion to a total variation of 0.001-inch. Gear, spline and ratchet tooth forms can be produced to acceptable commercial standards and often to better accuracy than with machining methods.

Lengths. Length tolerances for various types of parts are, of course, dependent upon the method by which such portions are produced. Typical tolerances practical for such parts can be found by reference to the process contemplated, turbine blades for instance, being finished by means of automatic form milling as discussed in Chapter 14.

39

Cold Heading

$\mathbf{A}$UGMENTING the other high-production processes — forging, extruding, stamping, etc. — available for the mass production of machine parts is cold heading. The many economies and advantages possible by this method of fabrication make its consideration imperative for machine parts which fall within its scope.

The Process. Cold heading is an old but in many respects a somewhat obscure process. The first automatic cold header was invented, designed and built by W. E. Ward in 1847. However, for many years these machines were relegated solely to the production of common bolts and screws and, as a consequence, were and often still are called bolt headers. Today, though, many types of parts produced on such machines bear little relation to bolts or screws, *Fig.* 39.1. In fact, many parts now

Fig. 39.1 — Typical parts made by upsetting and related operations such as extruding, piercing and trimming. In upsetting, force is applied to metal contained between punch and die to overcome elastic limit and cause defor-material. Courtesy, National Machinery Co.

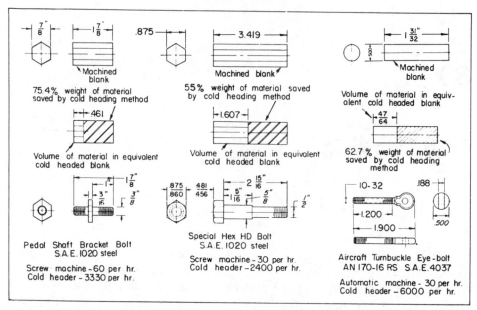

Fig. 39.2 — Comparison of material consumption and production rates with machined and cold headed parts of identical design. Drawing, courtesy American Institute of Bolt, Nut & Rivet Manufacturers.

cold headed can be produced economically in no other way. Developments and refinements in the process during recent years have been made primarily through improvements and quality of materials, better machines and ingenious die design.

Advantages. Differing from ordinary forging in that the metal — actually a wire — is upset by dies moving along its axis, as in hot upsetting, rather than normal to it, cold heading is desirable whenever practicable since the expense of heating and subsequent scaling is eliminated. In addition, a particular advantage lies in the fact that, owing to the change in grain structure produced by the cold working, parts have a greater factor of strength over those of similar dimensions which have been turned, milled, or otherwise machined. All factors being equal, when headed and machined parts which have the same tensile strength are compared, the cold-worked part is found to possess increased fatigue and shock resistance. With identical material the cold-headed part has, in addition to increased fatigue and shock resistance, a tensile strength 20 to 25 per cent greater. Materials whose strength is increased greatly by cold working — such as the various nonferrous metals and low-carbon steels — are thus considerably improved without extra operations. The smooth, compressed, toughened surfaces produced by drawing raw metal into wire are retained.

Cold heading lends itself to mass production and because it is a fast operation, a great deal of time is saved over that necessary for other methods producing a similar part. An equivalent amount of tooling for a headed part, as compared to a similar machined one, would produce a greater number of pieces, *Fig. 39.2*. Many extra operations are eliminated, and where secondary operations must be performed, parts are easily handled in large quantities by automatic setups.

Other advantages which are characteristic of headed parts are the smooth, rounded corners, built-in fillets, and definite lack of burrs. Coupled with these are often the additional ones of burnished or cold-worked finish, uniform accuracy, economy of production, and versatility of manufacture. Finished portions may be either larger or smaller than the original wire — one portion of the blank wire may be extruded down as well as another portion headed up.

Scrap Losses. Design for cold heading can effect great savings in material. A machined part must be made from bar stock, the size of which must equal the largest portion of the blank, whereas a headed part is made from wire which usually equals in diameter that of the shank on the part and is headed up to conform to the largest portion. Consequently, scrap loss in heading may be only one to three per cent while that in machining or forging may run to 75 per cent, *Fig.* 39.3.

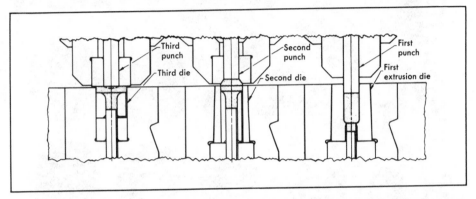

Fig. 39.3 — Cold forming tooling layout and advantage of making direct from coils in one machine. It saves raw material while producing spark plug bodies 10 times faster than previous method. Also, part is 18% stronger. Courtesy, National Machinery Co.

Machine Size Range. The scope of the cold heading or forging process ranges far beyond that ordinarily realized. The parts shown in *Fig.* 39.4 serve to illustrate the diversity in design and indicate the range of possibilities. Although special machines which handle rod over 2-inch diameter are in use, maximum diameter of wire stock which can be cold headed today in standard machines is 2-inches. For the great majority of production designs, cold heading is considered up to the 1-inch wire range; above 1-inch, hot upsetting can also be used. The final decision, however, depends upon part design and adaptability. Length of wire handled in automatic machines may run from 1/4-inch to approximately 9 or 10 inches, depending upon diameter. Length of parts produced in what are known as rod headers can be almost any reasonable figure.

Open-Die Machines. Cold heading machines are classified as solid-die or open-die headers. In open-die machines, wire is fed through two half dies which, when closed, hold the wire firmly, cut it off to length and then transfer it to heading

Fig. 39.4 — Typical cold former parts as produced on six - and seven-station machines. Courtesy, National Machinery Co.

position where punches in a horizontal ram collect and force the wire to conform to the shape of the die impression. These machines are used for parts with wide limits and those too long for solid dies. Largest open-die machine normally available is limited to 1-inch diameter wire.

Solid-Die Machines. A solid-die header or cold former has a slightly different operational sequence in that wire is fed through a quill and is cut off to length first. This piece is transferred in front of the solid die and the "first-blow" punch pushes it into the die until it strikes a knockout pin which acts as a stop. The protruding portion is then upset into the impression in the face of the die or dies by succeeding blows. After the piece is finished by one or more blows the knock-out pin ejects it. Such dies of course, make possible most exacting tolerances and the extruding operations mentioned previously.

Single-stroke headers have one die and one punch. They are used to make simple parts that can be formed in one blow. Ball headers are a variation of this type machine. Production speeds of such machines range up to 600 parts per minute.

Double-stroke headers, such as the 3/8-inch capacity machine shown in *Fig.* 39.5, have one die and two punches. Two heading blows are made on each blank in the one die. First the upsetting, gathering or coning punch contacts the workpiece. Then as the heading slide moves back from the die, the punch rocker oscillates to position the finishing punch on the centerline of the die for the second blow. This is the most popular type machine for producing screw blanks and other fasteners. Size capacities range up to 3/4 inch, and speeds vary up to 450 parts per minute. Some tubular rivet headers are double-stroke machines with mechanisms specially designed for heading and extruding the parts.

Fig. 39.5 — Double-stroke header, with capacity to handle material up to $^3/_8$ inch in diameter, has one die and two punches. First an upsetting or coning punch forms each part. Then a finishing punch is indexed into position for a second stroke. Courtesy, National Machinery Co.

Fig. 39.6 — Wire feeds through straightening rollers into a $^1/_2$-inch, four-die boltmaker. It produces 70 finished capscrews per minute. Courtesy, National Machinery Co.

Three-blow, two-die headers have two dies and three punches. Of the same basic design as double-stroke headers, these machines provide the added advantage of extruding or upsetting in the first die before double-blow heading or heading and trimming in the second die. Such machines are used to produce large-head, small-shank special fasteners and parts by combining trapped extrusion and upsetting in one simple header. They are also suitable for making stepped-diameter parts where transfer between dies would be difficult.

Progressive or transfer headers are multiple station machines with two to five dies and an identical number of punches. A simple transfer mechanism moves the workpieces from the cutter through the successive dies. Multiple upsetting blows, combined with extruding, piercing and trimming, make these versatile machines ideal for long-shank parts. Capacities range up to materials one inch in diameter, with workpiece lengths up to nine inches under the heads.

Boltmakers are three- or four-die headers that combine heading, trimming, pointing and threading in one machine, *Fig. 39.6.* They are widely used for making completely finished hexagonal head and socket head capscrews, and their versatility permits producing a wide variety of threaded special parts. Capacities range up to materials 1 1/4 inches in diameter, and production rates vary up to 300 parts per minute.

Cold nut formers are five-die machines specially designed for producing standard and special nuts. A simple transfer mechanism rotates the blanks end-for-end between successive dies. This allows working of the metal on both sides to produce well-filled, high-quality nut blanks with insignificant waste. Capacities range up to nuts one inch and larger.

Cold formers provide four, five or six dies with a variety of transfer mechanisms for increased versatility. All forming operations can be combined on such machines for making odd-shaped parts. They can be arranged to feed wire, bar or slugs, and can form metal cold or warm. Extra-large formers can handle material up to two inches in diameter or even larger.

Quantities. Generally speaking, cold heading jobs run in excess of 5000 pieces and most often over 50,000 pieces for economical production. Unless design restricts production strictly to a heading operation as with a pyramid or odd-shape upsets, an order for less than 1000 pieces would most likely be produced on a hand screw machine or lathe. Between 1000 and 5000 pieces could be handled most easily on a single-spindle Brown & Sharpe without excessive tooling costs. However, if an order runs over 25,000 pieces, say perhaps up to 40,000 or 50,000, then the cost of header dies and punches could economically offset any maintenance cost on screw machine tools. This is, of course, with due consideration to design and adaptability of the part. And, ordinarily the heading process and tooling would then be useful for much larger runs. An economical quantity run for odd-shape machine parts is usually considered as beginning at 50,000 pieces. Common types of headed parts such as bolts and other fasteners usually run upwards to the millions. This is readily conceivable inasmuch as some of the smaller screw type parts can be turned out at speeds as high as 600 per minute. Increase in diameter, complexity, or length to the maximum may slow this output speed to a few pieces per minute.

Design Considerations

A part suitable for heading is usually of simple and more or less regular contour and, if produced by machining, would require an excess of material removal, *Fig.* 39.1. Undercuts are seldom present on headed designs unless produced by a secondary operation, *Fig.* 39.7.

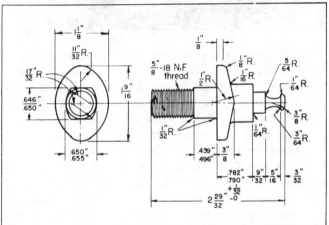

Fig. 39.7 — Elliptical collar, cold headed in two blows from ⁵/₈-inch diameter AISI 1035 steel and sheared to size, would be difficult to produce otherwise. Undercut head portion requires a secondary machining operation. Photo and drawing, courtesy Allied Products Corp.

Method of Heading. Factors which determine the method of heading are: (1) Relation of the amount of stock required for forming a head to the wire size in number of diameters, to find the number of strokes or upsetting steps necessary, and (2) the relation of the length of a blank to the wire size in number of diameters, to determine the type of die.

Upset Limitations. Experience has shown that when the length of wire to be upset into a head is less than twice the diameter of the blank, a single-stroke machine is most generally applicable. If, however, the length of wire to be upset is over twice but less than four times the diameter for the resulting blank, a double-stroke header is used. Where the length of wire to be upset is over four times the blank or wire diameter, either a three-stroke header or a larger capacity double-stroke machine or an additional reheader is necessary. In each instance, though, these figures may be exceeded if stock is well annealed and of high headability with no long, slender portions. By the same token, with hard wire of poor headability, something less than these figures must be used. Generally, the practical limit for the three-blow header is six times the wire diameter. For some regularly contoured heads which are rounded and not too thin, this figure can be upped to about 10 to 12, though only with material of good headability.

Length. The length of a part determines the type of die. With part lengths of ten wire diameters or less and relatively small size wire, a solid-die machine can be used. However, with blanks over this length ratio, an open-die header is necessary and for such cases the tool department or a producer should be consulted, especially with extremely long parts since these often must be hand-fed.

Head and Flange Diameters. As a general rule, it is well to stay within the limit of four diameters for maximum volume of upset. This sets the limit for head or flange diameters at about three to four times the shank diameter, *Fig.* 39.8, unless

Fig. 39.8 — Brake adjustment unit. Cold headed from 1/2-inch AISI 1035 steel, a flange 0.130 to 0.150-inch wide is upset to 1 1/8-inch diameter. Threaded end is extruded to size for rolling and serrations are blanked in a secondary operation. Photo, courtesy Allied Products Corp.

additional operations are included. With relatively workable materials, it is possible to produce a head or flange the final diameter of which is 4 1/2 diameters with two blows and up to 6 diameters with three blows. In a few instances, heads as much as 70 times the wire diameter have been produced by first extruding to a reduction of 80 per cent and then upsetting to 6 diameters of initial stock size. This, however, is an exception to the general rule.

Cross Sections. Unless design definitely requires it otherwise, upset portions should be specified as circular for maximum economy. These should be within the foregoing volume limitations with head thickness ranging from around 3/8 minimum to 3/4 maximum of the shank or wire diameter.

Tapers and Corners. Wherever a taper is necessary its small diameter should always be at the bottom of either die cavity, *Figs.* 39.1 and 39.4, unless special machining is justified. Sharp corners should never be specified — as large a radius as possible will extend die life immeasurably. Generous radii or tapers should be

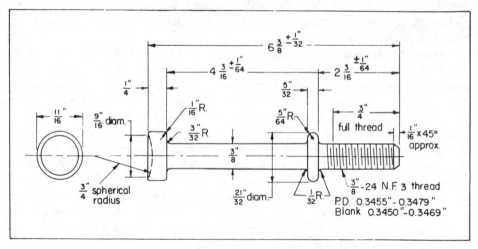

Fig. 39-9 — Necked aircraft part cold headed from AISI 1020 steel wire and flanged in a second open-die machine. Drawing, courtesy Lamson & Sessions Co.

specified for stepped diameter junctions.

Reductions. Average extrusion work with small reductions can be had with long die life. In fact, some designs in small sizes have been extruded in production with a 50 to 60 per cent reduction. This drops to around 12 to 15 per cent for $1/2$-inch wire. On heavy extrusions, however, lowered die life must often be accepted.

Special Forms. Parts having heads or portions of spherical shape can be produced easily by cold heading and shaped flanges can be blanked and shaved to size. Necked pieces, likewise, can often be produced by a double operation. The part shown in *Fig.* 39.9 was upset to form the hollow head leaving a plain shank. In a second semiautomatic open-die header, the $21/32$-inch diameter flange was formed while extruding the thread section for rolling. Many variations of this method can be utilized in design.

Threads. As can be judged from the foregoing illustrations, threads should be specified as rolled whenever possible. Thread rolling is a highly economical adjunct to the cold heading process and will effect, if specified, lowest possible cost in the manufacture of threaded parts.

Selection of Materials

In considering suitable materials, it is well to remember that those which are ultrasoft, such as pure magnesium, or ultrahard, such as hardened steels, do not make for good headed parts. Perhaps the best headability characteristics are possessed by wire material having medium hardness and excellent cold-working properties. This would imply a relative absence of free-machining elements such as sulphur, lead, phosphorous, silicon, etc., from ferrous materials. Sulphur in amounts up to 0.050 per cent has little effect, but over 0.050 per cent reduces the ability of the material to withstand cold working. Alloying elements such as manganese, nickel, chromium, molybdenum, and vanadium also reduce the capacity for cold working of ferrous materials. The various metals commonly cold worked are generally as follows in order or preference:

1. Aluminum and aluminum alloys
2. Brasses
3. Copper and copper alloys
4. Carbon steels
5. Alloy steels
6. Stainless steels

Ferrous Materials. General overall specifications for ferrous heading wire may be outlined as follows:

1. Elongation 20 to 45 per cent
2. Reduction of area 50 to 70 per cent
3. Tensile strength 70,000 to 90,000 psi
4. Wire free from seams, inclusions, etc.

Although neither the elongation nor the reduction of area alone can be considered an indication of headability, the two together are useful. If the elongation is less than that specified above, the wire is not sufficiently plastic to fill out the die impressions properly. Cracks usually form before complete upset. Similarly, if the reduction of area falls below the specifications, the wire can be headed, but only poorly, and die life decreases greatly.

Tensile strength specification is necessary not only to insure a part which fulfills requirements but also as an indication of resistance to upsetting. When the tensile falls below the minimum, the wire will be so soft that, although it can be headed, sponginess may result with attendant heading difficulties. Where the tensile strength must necessarily be over the maximum specified, special coatings must be cold-drawn onto the wire to assist the upsetting.

Common Heading Stock. The common and usually most economical heading stock is probably ferrous wire with a carbon content of 0.10 to 0.15 per cent. Most easily worked steel is that with 0.12 per cent average carbon, but this material is subject to excessive grain growth. Used widely for fasteners of all types is a 0.15 to 0.25 per cent carbon steel. Carbon steels ranging from 0.25 to 0.45 per cent are specified where maximum physical properties are to be obtained by subsequent heat treatment. Thus, AISI 1035 steel is most commonly employed for such parts.

Grain Size. Coarse-grained steels — 1 to 5 McQuaid-Ehn test — in the lower carbon ranges can be used for parts under $7/16$-inch diameter which require special physical properties after heat treatment. An oil quench, however, must be used with the coarse-grained steels. Although the coarse-grained wire in the 0.30 to 0.40 per cent carbon range usually has a better plastic flow than fine grain, the fine-grain wire — 5 to 8 McQuaid-Ehn test — must be used on parts over $7/16$-inch diameter to avoid cracking after heat treating. This is especially true with high-carbon steel parts water quenched to obtain desired properties.

Special Wires. In almost all cases, if the physical properties of a part are specified, a manufacturer can produce or obtain a wire which will be of satisfactory headability and fulfill the requirements after heat treatment. The normal run of materials specifically intended for cold heading are shown in Chapter 68.

Nonferrous Materials. The four general specifications outlined for ferrous materials apply also for nonferrous heading wire but the tensile strength need have no particular low limit. However, with extremely low-tensile strength wire, hot heading or a preheat operation must be resorted to.

Stainless Alloys. Stainless steels of the 18-8, 25-12, and the 12 to 17 per cent straight-chromium types are often used. Usual practice is to cold head the straight chromium types up to a maximum of $5/8$-inch and up to $7/16$-inch diameters with the austenitic chromium-nickel types. In excess of these diameters, hot heading must be used.

An average 302 stainless alloy will result in a 40 Rockwell C hardness on a medium upset. This alloy is hardenable only by cold working and normally can be headed to a maximum volume of $3^1/2$ diameters. Type 430, normally the most popular for heading, is not hardenable by cold working or heat treatment and normally has lower physicals. Where heat treatment is necessary, the martensitic types, such as 410, must be used but are less desirable in headability than the two previous types.

Aluminum and Magnesium Alloys. The lightweight materials include 17S, A-17S, 24S and 53S aluminum alloys. Most of these alloys are used because of special color, appearance, headability, or to reduce weight. Alloy 53S is extensively used for marine parts. Material costs are usually higher than for steel, however. The various magnesium alloys available are generally not suitable for cold heading.

Other Materials. Monel metal can be used and certain special instrument parts have been made from Invar. The high resistance of Monel to practically all corrosives makes it applicable to a wide range of industrial processing machine parts. Phosphor bronze is often cold headed into parts for marine service.

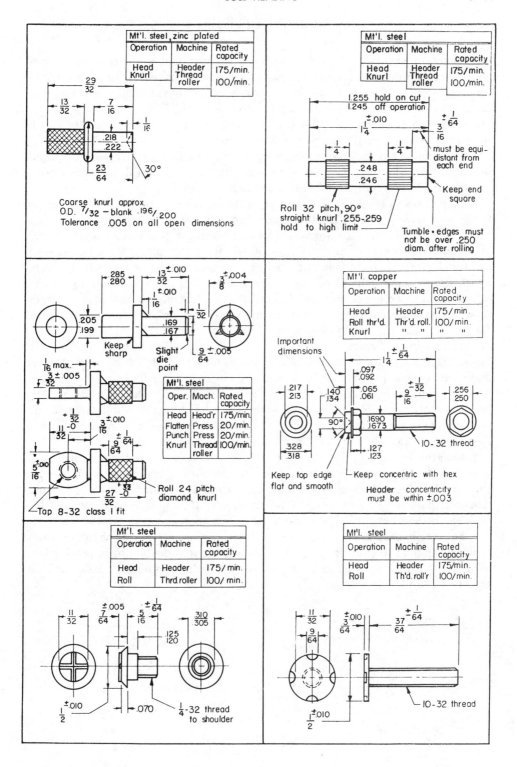

Fig. 39.10 — A group of typical machine parts headed and rolled. Typical tolerances are shown. Drawings, courtesy Lamson & Sessions Co.

Specification of Tolerances

Generally speaking, die life is inversely proportional to the tolerance limits held on parts produced. Experience has shown that limits of plus or minus 0.001-inch can be obtained readily and, in some cases, are not unreasonable. Most die-life calculations, however, are based on plus or minus 0.002-inch, a tolerance usually maintained on regular headed blanks. Reducing this to plus or minus 0.001-inch would cut average die life by 50 per cent or more.

Experiment has shown that where highly exacting limits are necessary, special honings of the dies, use of tapers, generous fillets, and special die designs can be resorted to for increasing die life to a practical length.

Tolerances vs Die Life. Based on expected die life with limits of plus or minus 0.002-inch, a minimum of 250,000 pieces can be produced using 0.10 to 0.15 per cent carbon steel, brass, or aluminum wire. This figure drops to around 125,000 pieces for steels such as AISI 4037, or 50,000 to 70,000 pieces for stainless steel or Monel wire, etc. If the tolerance can be increased to plus or minus 0.004-inch, which satisfies most general-run designs, the normal 250,000 minimum figure can be increased to well over 300,000. In a great many instances, where plus or minus 0.001-inch can be increased to plus or minus 0.005-inch or more die life as much as 60 times greater can be obtained in production.

The foregoing figures are, of necessity, based upon solid-die work. It is usually found that open-die work automatically means increased die life by several times due to use of four impressions in the die as compared to one or two in a solid die and, in addition, freedom from friction upon ejection, etc. The open-die header, however, will not hold the exacting tolerances possible with a solid-die machine and consequently is used primarily for those parts that will require further machining operations. Extrusion work cannot be done on open dies.

Length. Tolerances for the overall length of parts should be as liberal as possible. Close limits can be held on upset sections formed completely within either die but overall length should be at least plus or minus 1/32-inch, *Fig.* 39.9, or more wherever possible. Plus or minus 1/64-inch, *Fig.* 39.10, and in certain cases 0.010-inch can be had but should be considered as less economical in production.

Square Shoulders. The squareness of shoulders or collars with shank portions is normally held to plus or minus 1 degree, maximum deviation, in production. Where better tolerances are necessary drawings should so indicate to assure special attention to this requirement.

40

Thread and Form Rolling

ROLLING of screw threads and other circular forms between suitable hardened-steel dies, known and practiced for well over a hundred years, has today reached a high degree of perfection. Originally classed merely as an extremely high-production process, thread and form rolling now offers not only unusual production speed but also precision comparable to that of fine grinding, as well as wide versatility. With the introduction of cylindrical-die thread and form rolling machines during the past several decades, the economies and other practical benefits of cold rolling can now be obtained on a broad range of design applications.

Form Rolling Methods. Essentially a cold forging process, thread and form rolling produces the finished part by displacement of metal to conform with the contours of the dies employed. Two distinct methods for accomplishing this end are in use today, the flat-die and the cylindrical-die methods. In the older and more common former method, a cylindrical blank is placed between two flat hardened-steel dies — into the exposed faces of which is machined a negative impression of the desired form pattern — and as one die is moved past the other the proper profile is impressed onto the blank, *Fig.* 40.1. No material whatever is removed from the original blank. The diameter of the finished part is determined by the distance between the faces of the dies at the end of the stroke. As can be readily

Fig. 40.1 — Die arrangement used in flat-die rolling. Photo, courtesy Reed Rolled Thread Die Co.

seen, the work is formed, or forged, at a fixed cycle of penetration which is limited and controlled by the diameter of the part and the length of the dies.

In the second and most recent method of rolling, cylindrical dies are employed having negative multiple threads or a negative impression of the form pattern which is desired, *Fig.* 40.2. Driven synchronously at a predetermined speed on

Fig. 40.2 — Three-die arrangement used horizontally or vertically in cylindrical-die rolling. A center rest is used in machines with two rolls. Photo, courtesy Reed Rolled Thread Die Co.

Fig. 40.3 — Two-die thread rolling machine rolling a single-start worm of 8620 steel, ⁹/₁₆-inch diameter, 10-pitch, 1¹/₈ inches long at 8 to 10 pieces per minute. Courtesy, Landis Machine Co.

parallel shafts, the dies — either two or three — are actuated to feed in automatically to the desired finished diameter and then withdraw, *Fig.* 40.3. As with the flat-die method, the entire formed length is rolled at one time but, inasmuch as the cylindrical dies present in effect die surfaces of infinite length, the penetration cycle can be varied to suit the particular requirements of the workpiece.

Rolling and Heading. As mentioned in Chapter 39, thread rolling is undoubtedly the fastest and most efficient method of forming threads and is a valuable adjunct to the cold heading process, *Fig.* 40.4. The same is true even where contours or forms other than standard threads are concerned. Too, thread rolling is often employed in automatic screw machine work to speed up output.

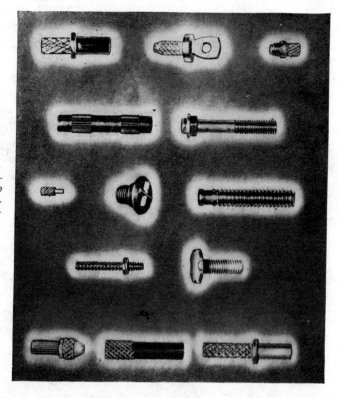

Fig. 40.4 — A group of small cold-headed parts finished on flat-die thread-rolling machines. Photo, courtesy Lamson & Sessions Co.

Range of Application. Recent advances in the technique of rolling have overcome many of the previous limitations common to thread rollers and it is now practical to roll precision threads and forms on hollow cylinders, short lengths, tapers, large diameters, etc., using any of a wide range of materials, *Fig.* 40.5. In addition, awkward or top heavy parts, *Fig.* 40.6, normally impractical for the flat-die machine, are readily handled with the triple cylindrical-die machine.

Flat-Die Machines. There are several types of machines available employing each of the rolling methods mentioned — reciprocating flat-die, rotating twin cylindrical-die and rotating triple cylindrical-die. Flat-die machines are of the inclined variety to provide automatic feed and gravity ejection, *Fig.* 40.7; horizontal machines for hand-fed work; and side-feed machines for special long work which requires a horizontal position. All are particularly well suited to the production of the more simple types of fasteners and forms.

On symmetrical threaded parts such as those up to about 1-inch in diameter, threads under 1/4-inch, designs such as gimlet point screws, knurled parts under 1-inch diameter, etc., flat-die machines excel in output speed and adaptability. Normally these machines are limited to maximum diameters of about 1.125-inch although diameters up to 1.500-inch have been produced successfully. The lower diameter limit is generally around 0.060-inch but here again threads as small as 0.006-inch have been made successfully. These should be recognized as exceptions rather than the rule.

Maximum length of rolled portion is limited primarily by the die face width on flat-die machines. This dimension runs from 0.500-inch on the smallest machine to around 8 inches on the largest machine. Die lengths which can be used range from about 1.750-inch on the smallest machine to about 15 inches on the largest.

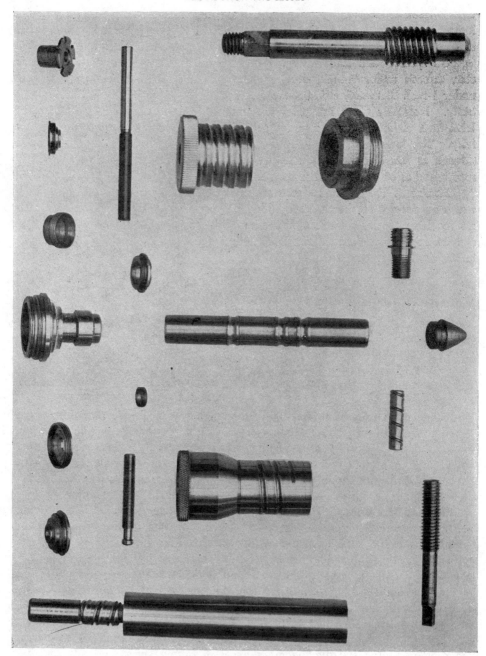

Fig. 40.5 — Group of solid and hollow machine parts with oil grooves, ball bearing grooves, short threads, multiple leads, and micrometer threads produced on cylindrical-die thread rollers.

Roll-Die Machines. Cylindrical-die machines normally are adjustable over a wider range of diameters than are flat-die machines but are not suited to the rolling of diameters under 1/4-inch, 3/8-inch being a commonly accepted lower limit. These machines are also manufactured in both horizontal and vertical types with the practical diameter which can be handled generally being about 4 inches although, in certain cases, this can be upped to 6 inches on some machines, *Fig.* 40.8. Maximum workpiece length which can be rolled in one pass ranges up to 13 1/2

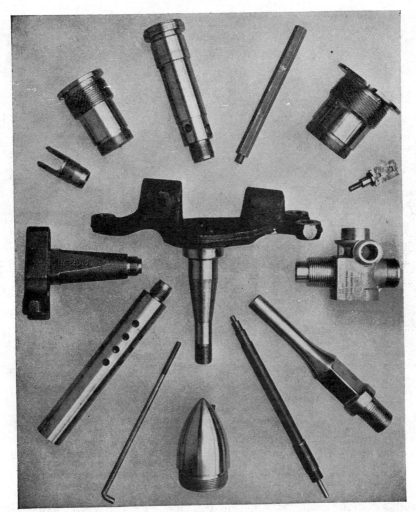

Fig. 40.6 — A representative group of awkward machine parts the threads of which were rolled on a cylindrical-die machine. Photo, courtesy Reed Rolled Thread Die Co.

inches and the rolled section is dependent upon the face width of dies which can be employed in the particular machine used. Two-roll horizontals can handle diameters of 4 inches and over, with lengths of parts varying by machine but through-feed of parts up to about 1-inch may allow lengths to 20 feet. Special machines are used for large special parts, *Fig.* 40.9.

Advantages of Rolling. As in all cold-forming operations, the product of the thread rolling machine is superior to that produced by machining. Tensile and shear strength as well as fatigue resistance of the rolled parts are increased to a marked degree. To these advantages are added those of a smooth, hard, burnished surface and the ability to produce accurate threads and forms on soft, tough and stringy materials which are impossible to machine without tearing.

Production Rate. The high speeds of which thread-rolling machines are capable make it incomparably faster than any other method suited to the production of similar designs when the material used is within the normal hardness range for rolling. Production rates for reciprocating flat-die machines range from

Fig. 40.7 — Flat-die thread roller arranged with a cold header for automatic production of machine bolts. Photo, courtesy Chrysler Corp.

30 pieces per minute with the larger machines to around 175 pieces per minute with the small automatic machines. As a rule, special parts over 0.625-inch diameter are hand-fed and output is limited normally to about 60 pieces per minute, although in cases where maximum precision is required the top rate is usually about 30 pieces. Cylindrical-die machines in some cases are slower, production rates ranging from around 10 to 40 pieces per minute with the larger, hand-fed parts but automatically fed machines can handle up to 300 parts per minute.

On occasional instances, where speed of production alone is essential and the qualities afforded by cold working can be dispensed with, hot rolling can sometimes be employed to advantage with thread rolling machines, the ease of working gained thereby allowing both greater diameter and more severe deformations than that indicated by the cold working capacity of the machines used. The added cost of heating and somewhat inferior surface finish, however, as with hot upsetting, must be offset by the advantages gained and requires careful evaluation.

Rolling Speed. Peak rolling surface speed for dies is around 314 surface feet per minute, at which point the peak load also occurs. Round dies, by virtue of constant speed of rotation, maintain the same speed of blank rotation as with flat dies while operating at about one-third less surface speed. Increased die life afforded thereby is considerable.

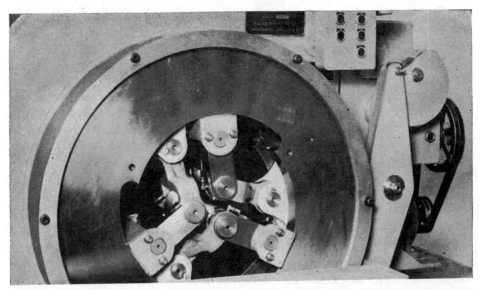

Fig. 40.8 — *A large horizontal thread and form rolling machine capable of handling diameters up to 6 inches. Photo, courtesy Reed Rolled Thread Die Co.*

Limitations. To be recognized and taken into consideration along with the advantages of quality, precision, versatility and economy of thread rolling are a number of broad limitations which normally apply to this method of processing. First, the nature of the process itself makes it impossible to produce internal threads or forms. Only in the case of some special machines used for threading extremely thin-wall tubing can internal threads be produced.

Being essentially a high-production process, thread and form rolling is seldom practical for producing small quantities and in this aspect it parallels the cold heading process. Because metal is displaced in producing a specific thread for

Fig. 40.9 — *Grob & Tesker thread roller at Saginaw Steering Gear Div., G.M.C., the world's largest, will roll ball bearing screws and splines to 12 inches diameter, Acme threads to 15 inches diameter.*

form, finished part diameter is always greater than the blank, a characteristic which is objectionable where a shank and finished diameter must be identical. An extra turning or extruding operation, however, can often correct this feature but, of course, at extra expense in many cases. Coarse-pitch threads or forms on small diameters are often difficult to produce without distortion of the blank. Too, threads or forms with steep sides and wide flats such as square threads, buttress threads, acme threads, etc., are highly susceptible to material defects and are seldom produced when the ultimate in quality is requisite. Lastly, hardness of materials used is limited to a maximum of 35 to 40 Rockwell C and beyond this point die life is too short to be practical.

Design Considerations

As can be noted from *Figs.* 40.4, 40.5 and 40.6, a wide variety of machine parts can be designed to utilize thread rolling. An almost unlimited variety of thread forms can be produced on solid and hollow parts. Special annular bearing grooves and helical oil grooves, knurls and serrations of any variety can be rolled. Single, double and triple-lead worms can be rolled, burnishing can be done, etc. Diameters are limited by the machine capacities previously enumerated. The rolled length, too, should normally fall within machine limitations. However, where forms or threads longer than the available width of dies are required, rerolling can be utilized. Continuity of such forms produced in two passes, for instance, is accomplished by overlapping, *Fig.* 40.10. Actual overall length of parts having a portion of their length rolled is relatively unrestricted — parts as long as 20 feet have been produced all threaded or with threaded or formed ends.

Diameter and Length Limitations. Limiting the application of rolling threads and forms in flat-die machines is the fact that it is impractical to make flat dies greater in length than 10 to 15 times the blank circumference and available die cavity length of the machines is limited as noted previously. Ordinarily, a minimum allowance of one-half a work turn of drop-off for starting and one and one-half turns for finishing is necessary with flat dies. Naturally, this limits the number of turns available for penetration — the rate at which the form on the dies is impressed into the blank. Dies usually must be of sufficient length to allow a minimum of four work turns of a blank in rolling and, for accurate rolling in the harder materials, a minimum of seven work turns is recommended. Length of rolled portion should not be less than one diameter to avoid the tendency of short lengths to skew in the dies.

Average penetration P per contact of a given point on a blank with the die surface cannot be less than:

$$P = D \div \left[\frac{L}{C} \right]$$

where D is the thread depth, L is the total die length and C is the pitch circle circumference, all in inches. Considering a $1/2$-13 thread rolled between two dies the total length of which is 15 inches, average penetration per contact on the blank is 0.0047-inch. While this rate is not excessive for solid blanks, it would collapse or distort hollow blanks where penetrations of 0.001-inch and less per contact often are necessary. In such cases cylindrical-die machines are employed and the number

Fig. 40.10 — Produced on a special reheader, the blank for this part was threaded in a special flat-die machine. The $^3/_4$-14 thread is 9 inches long, right, and required overlapped rolling. Photo, courtesy Allied Products Corp.

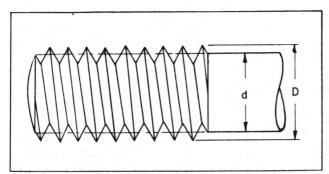

Fig. 40.11 — View of a rolled thread showing blank diameter to finished diameter relationship characteristic with rolling. Drawing, courtesy Reed Rolled Thread Die Co.

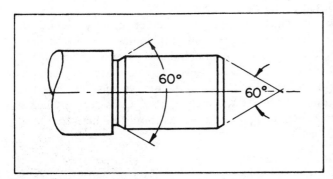

Fig. 40.12 — Where undercut necks must be used design shown should be employed for best results. Photo, courtesy Reed Rolled Thread Die Co.

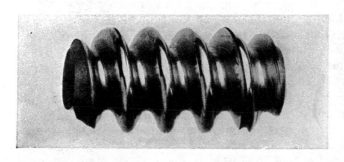

Fig. 40.13 — This special form, rolled in SAE 1020 steel, indicates the extremely high quality surface finish obtainable. Drawing, courtesy Reed Rolled Thread Die Co.

of work turns can be varied as desired to control penetration rates accurately. Too, parts shorter than one diameter in length can be readily handled in these machines because of three-point support and the possibilities of outboard steady rests.

Shoulders. Narrow or wide-shouldered work may be handled without trouble and threads ordinarily can be rolled within one-half the pitch length of a shoulder.

Blanks. Specifications for blanks are of great importance in producing accurate rolled work. Blanks must be round to start with, inasmuch as in only a few cases of noncircularity will any of the various machines correct out-of-roundness. Blanks for rolling are always smaller than the finished part, usually being within a very few thousandths of the pitch diameter of the form, *Fig.* 40.11. The volume of metal upset above the blank always approximates the material displaced in rolling but because material, hardness, finish, etc., affect the final result, the exact diameter should be established by trial. Hollow blanks may be affected by wall thickness, use of support arbors, and rate of penetration as well. The same is true of extremely thin-wall blanks mentioned in the foregoing; shell diameter will be just slightly larger than the pitch diameter but must be established by trial for accuracy.

Blank diameters should not vary along their length to assure maximum uniformity of the finished part. Ends of blanks should be beveled with a 60-degree included angle to the root diameter of the form to assure maximum die life inasmuch as this tends to minimize severe bending action on the rolling blanks. Necks adjacent to shoulders are generally unnecessary and sometimes detrimental to good rolling but where such are found desirable for design purposes, 60-degree included angles should be used as indicated in *Fig.* 40.12.

Accuracy of Blanks. Tolerances on blank diameter usually should be approximately two-thirds or less that on the final rolled diameter of the part. Where blanks can be made directly from commercial rod stock without further finishing, maximum economy can be realized. However, where accuracy such as that of Class 4 or 5 threads is desired, for example, the blanks must be centerless ground to within a total tolerance of 0.0005-inch or less. Considering Class 3 threads as an example, screw machine produced blanks should be held to a total variation of 0.001 to 0.002-inch.

Selection of Materials

Threads and forms can be rolled in any material which is sufficiently plastic to flow and to withstand the stresses of cold-working without disintegrating. As a measure of ductility for rollability, material with an elongation factor of 12% or more is desirable. However, other factors such as hardness, yield strength, the ratio of proportional limit to tensile strength, micro-structure, the degree and speed of work hardening, the modulus of elasticity, the non-metallic content of steels, and the pitch and diameter of the workpiece must be taken into account.

The yield point must be exceeded in order to accomplish a permanent deformation of the material. It is obvious that the higher the yield point, the greater the pressure that must be exerted by the machine and the tools, and the less its tendency to flow.

In general, the hardness and the elongation factor is regarded as routine indices of rollability. The hardness is easily obtained, hence it is used as a comprehensive indicator of yield point and reduction of area.

However, these two factors are not a complete index of rollability. Certain high-temperature alloys may be dead soft, show high per cent elongation and low yield, yet develop a yield point which renders cold working impossible. This would occur in materials of high work-hardening propensity on which the process of rolling itself would raise the yield point so that plastic deformation could not be completed. Such a material, for example, is Haynes Alloy #25.

Normally, any of the materials used can be rolled, including steels with hardnesses ranging up to 35 Rockwell C. Brass, bronze, nickel, nickel steel, Monel, Everdur, aluminum, silver, stainless tool steels, high-speed steels, zinc alloys, fiber, etc., have been rolled. Though ductile low-carbon cold forging materials with hardnesses up to 32 Rockwell C are most commonly and most easily rolled, it is possible to roll materials as hard as 35 to 40 Rockwell C, though of course die life is progressively lower as hardness increases. For best results with forms which require a considerable amount of metal movement in rolling or where the material is in the upper rolling hardness range, free-machining elements such as silicon, phosphorus, lead and sulphur should not be present. Elements which make for ease in machining are detrimental to good cold forging characteristics in metals and, as with cold heading, in selecting materials a compromise must be made depending primarily upon the combination of processing methods necessary to produce a specific part.

The sulfurized 1100 series steels, popular for machining, are furnished in three ranges of sulfur content as follows: 0.08 to 0.13%; 0.16 to 0.23%; and 0.24 to 0.33%. Most threads can be rolled on material in which the sulfur does not exceed 0.13%. As the sulfur content increases, the degree of cold working must be reduced until nothing more severe than the AN/UN form in material of the 0.24 to 0.33% sulfur range be recommended for rolling. This warning is because the sulfide inclusions in such material reduce its ability to withstand cold working, resulting in flaking of the rolled surfaces; and, in extreme cases, the complete disintegration of the material, particularly on splines and gears. Inclusions are elongated and broken up the same as the grains of the metal, while the metal is being rolled. They are cold worked into stringers, and the distances between them are shortened thus promoting flaking. Leaded steels are often thought to have good rollability with the idea that lead being soft and malleable will add to the cold working characteristics. The exact opposite is true for the same reason as in the case of sulfur.

Steels such as 1137, 1132, 1120, and 1117 (in which the manganese is relatively high and the sulfur relatively low) are good rolling steels resulting in good rollability, well improved physicals, and good finish, yet retaining good machineability.

The following materials are considered good for rolling: the 1300 series manganese steels; the 3100 series nickel-chromium steels; the 4000, 4100 and 4300 molybdenum steels; the 8600 nickel-chromium-moly steels; the 340 copper and brasses; the 6061 - T4 aluminums.

Hard Materials. When rolling threads in heat-treated blanks, the root of the form naturally work-hardens to a greater extent than the crests. With blanks of 34 Rockwell C hardness, the roots reach a hardness value of nearly 60C. Inasmuch as this is approximately the hardness of the dies, it can be readily seen that unless extremely low penetration rates are employed, die life would be so low as to be impractical. Naturally, production speed is sacrificed where such parts are necessary; although the quality of the product is often such as to warrant the extra cost, high-tensile aircraft fasteners being a good example where this is the case.

Specification of Tolerances

Rolled threads provide an excellent key to the tolerances available and are found practical in rolling of any type or form. Class 2 threads are extremely easy to produce and can be rolled from ordinary rod stock without further finishing. However, inasmuch as the more accurate dies have longer life, dies thread-ground to Class 3 specifications will produce Class 3 threads at a cost below that for Class 2 with inferior dies. By holding the thread depth on the dies and the blank to proper tolerances, Class 4 threads can be produced.

Cylindrical-Die Work. Cylindrical-die thread rollers will produce Class 4 and 5 threads when desired. Pitch diameters in production often can be held to within a total variation of 0.0008 to 0.001-inch from part to part, with a maximum taper or variation over the length of one thread of 0.0003-inch, while thread to shank concentricity under good conditions can be held to within a total variation of 0.0003-inch. Micrometer shafts, *Fig.* 40.5, have been rolled to plus or minus 0.0001-inch on diameter with a lead accuracy of 0.0001-inch.

Surface Finish. Rolled forms and threads have greater uniformity and better surface finish, *Fig.* 40.13, than those obtained with other methods. Comparatively speaking, surface finish of rolled threads, forms or burnished surfaces ranges from about 4 to 8 microinches. Commercial finish on ground threads is usually from 8 to 16 microinches, 60 to 65 microinches with milled threads, and 125 to 250 microinches with chased threads.

Lengths. In specifying for rolling it must be remembered that rolling usually results in lengthening of the workpiece to a certain extent. Although in most cases this effect is of no consequence, it may create out-of-limit parts where length tolerance is critical. Too, with parts such as those with rolled oil grooves or bearing grooves, the material thrown up during rolling often must be removed by means of a final centerless grinding operation and often tolerances depend primarily upon this operation.

ADDITIONAL READING

[1] *Rollability of Materials,* Landis Machine Co., Waynesboro, Pa.
[2] *Manual on How to Design Rolled Worm Threads,* Landis Machine Co., Waynesboro, Pa.

SECTION

PRODUCTION

5

PROCESSES

METAL DEPOSITION METHODS

41

Electroforming

$\mathbf{R}$ECENT developments in the electrodeposition of metals have definitely brought this somewhat obscure art into the category of precision manufacturing methods. Commonly known as electroforming when applied to the production of metallic parts by electrochemical deposition of metal, this interesting technique makes available to the design engineer the means for solving many complicated fabrication problems encountered in the design of present-day precision machines and instruments, *Fig.* 41.1.

Development. Originated over a century ago by Jacobi, the process was first used to produce electrotypes and other similar articles. Later it was utilized to

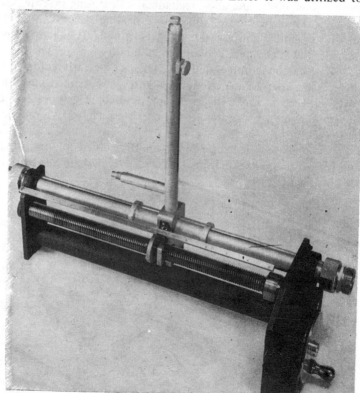

Fig. 41.1 — Microwave impedance meter, coaxial line and traveling carriage of which are electroformed telescopic tubing sections. Photo, courtesy Sperry Rand Corp.

Fig. 41.2 — A wave-meter with a micro-meter waveguide feed. The precision resona-tor of the unit is elec-troformed. Photo, courtesy Sperry Rand Corp.

make sheet copper, seamless copper tubes, etc. More recent uses include the manufacture of sound record masters, filter screens up to 400 mesh per linear inch, jointless radiators, caskets, reflectors, pipe fittings, seamless copper tanks, floats, and transformer cores. The advent of successful electroforming with iron brought into the picture the economical mass production of complicated dies such as those used in molding tires, glass and plastics as well as for embossing and die casting.

The steady trend toward more exacting tolerances in late years as well as the previously unheard of requirements presented by war-born design such as that found in radar and microwave equipment, gun sighting and computing mechanisms, electronic and mechanical calculating machines, etc., has helped to promote the utilization of the electroforming process, *Fig.* 41.2. In addition to making possible the production of parts with highly exacting repetitive dimensions, this method also makes available a quality of surface finish not possible by ordinary mass machining methods. This is especially valuable where the required surfaces are internal and irregular — often impossible to produce otherwise at any cost. Such parts might be precision telescopic tubing, round and rectangular waveguides, tee sections, tapered and twisted sections, resonators, irregular shapes, bellows, etc., *Fig.* 41.3.

The Process. Electroforming — unlike electroplating which seldom involves deposits over 0.0001-inch in thickness intended for protection or decoration —

Fig. 41.3 — Group of electroformed radar components includ-ing tapers, bends, twists, magic tee, wavemeter body, etc.

consists primarily of producing irregular mechanical parts by a process of electrodepositing metal on a master mold or matrix to a thickness consistent with the mechanical design requirements. The mold or matrix, a negative impression of the finished part, is of primary importance for naturally it determines the accuracy of the parts produced both as to dimension and surface finish.

Molds and Mandrels. A variety of methods for providing satisfactory molds or matrices are available. These can be broken down into two major groups, namely, reusable and expendable. The reusable types are used for uniform cross sections or tapers which will allow the withdrawal of the master pattern without destroying it. This is accomplished in two ways: (1) With a separating medium or (2) by relying on poor adherence of a conductive coating. Separating mediums employed with reusable steel patterns may be thin films of low-melting point metals such as tin or cadmium. Where a negative steel master can be made easily by machining, it is finished undersize, plated to size with, say, tin and the desired metal deposited to the proper thickness thereon, *Fig.* 41.4. By heating the finished electroform above

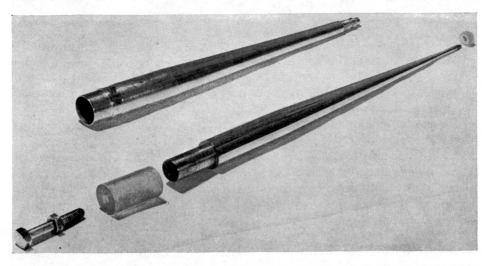

Fig. 41.4 — Reusable tapered steel pattern with masking caps and bolt. Electroformed part after finish machining operations is shown at top. Photo, courtesy Sperry Rand Corp.

the melting point of tin in a wax, oil or lead bath, the pattern can then be withdrawn. The separating medium can be used in thicknesses from one-millionth of an inch up, although 0.0002 to 0.001-inch is the usual amount mandrels are made undersize, depending upon the application. Where a complicated master pattern such as a spiral computing mechanism cam, which takes many hours to produce, is to be used, production electroforming matrices can be made by plating this master with a thin parting film and electroforming on the master the required number of negatives in iron, nickel or copper. These negatives, then, serve as matrices for producing finished parts, *Fig.* 41.5.

Poor-adherence mandrels or patterns usually are made of glass, Kovar or some other material having a relatively low coefficient of thermal expansion, *Fig.* 41.6. Surfaces of nonconductors are made conductive by spraying or dip coating with silver. This extremely thin coating becomes an integral part of the electroformed work when completed because of good adherence between the silver and the electroformed metal. Removal is effected by heating the assembly to about 600 F. Inasmuch as the silver and electroformed part expand about twice as fast as

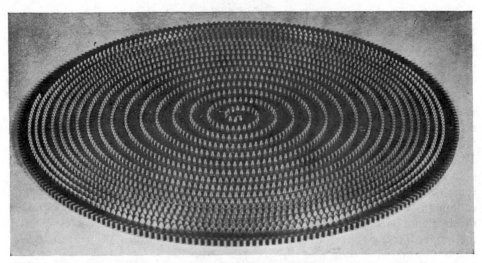

Fig. 41.5 — Electroformed spiral pin-cam of nickel-faced iron for use in computing mechanisms.

the mandrel and the silver adheres but slightly to the highly polished mandrel surface, the part comes off easily. A sprayed coating is required for low-expansion metallic patterns also, to provide a poor adherence film for easy removal. Outstanding advantage of the glass patterns or mandrels is the surface perfection obtainable by optical grinding methods.

Expendable matrices or patterns generally are used to produce parts of irregular shape, namely those whose design does not permit withdrawal of the matrix after completion — where it must change form radically to allow removal. These can be of several varieties: (1) Dissolvable metal or, (2) fusible metals or waxes. Matrices of the former type can be of commercially pure aluminum, which can be dissolved out with sodium hydroxide; cold-rolled steel or zinc, which can be dissolved out with hydrochloric acid; etc. Value of the dissolvable matrix is realized in applications where parts with extremely thin walls would be distorted by heat or where the matrix must be mechanically stronger and harder than the

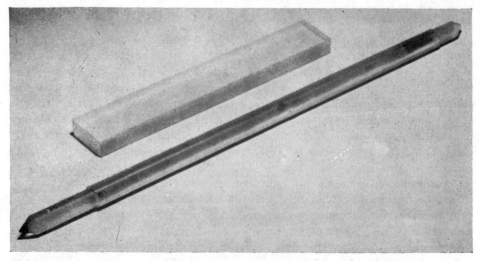

Fig. 41.6 — Typical glass mandrels for electroforming coaxial line and waveguide section used in radar. Photo, courtesy Sperry Rand Corp.

fusible metals and waxes.

Fusible metal, *Fig.* 41.7, and wax matrices or patterns are produced, as a rule, in a master die — steel, plaster, etc. — by casting. Fusible metals available can be had in a wide variety of melting points ranging from around 116.6 F to 350 F. Waxes used are rendered conductive by spraying with a metallic or graphite coating.

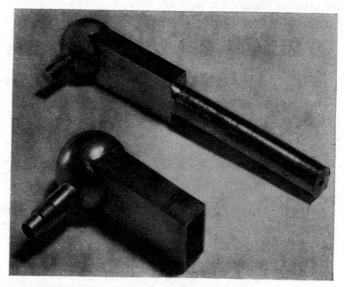

Fig. 41.7 — A special wave-guide starting section and the expendable mandrel utilized, above. Photo, courtesy Sperry Rand Corp.

Fusible-metal matrices present some problems in production and these in addition to their instability at room temperature eliminate them from consideration in applications where the maximum in accuracy is necessary. One problem is maintaining good surface finish. Another is failure of matrix to melt out completely. Traces of the matrix metal remaining inside parts after melting out must be brushed away or dissolved out in a bath, adding to the cost per part. However, since these metals and waxes can be easily and rapidly cast into matrices, fabrication of irregular machine components can be placed on a production basis most readily by this method.

Another type matrix or pattern which can be cast in a mold to produce either a conducting or a nonconducting base for electroforming is that of rubber. Rubber matrices are made conductive in much the same manner as are the waxes. Rubber matrices can be stripped and reused.

Field of Application. Electroforming is not, in the regular sense, an economical method of fabrication, but being a high-precision process, in many cases it is less expensive than conventional precision machining processes. It is in application to those designs which are either totally impossible or extremely costly to produce by other methods that electroforming finds its most useful place. In the production of three-dimensional cams, spiral computing cams, and similar intricate, critically accurate mechanical components this process opens to the designer a field of possibilities hitherto found impractical.

Metals. Metal deposited in electroforming is essentially stress free, a condition conducive to dimensional accuracy over long periods of time. Machining of electroformed parts, for the same reason, produces little or no distortion. A large selection of materials readily available for commercial plating can be utilized.

Layers or laminations of materials can be had such as copper on the inside for good conductivity and chromium on the outside for a durable, wear-resistant surface. Where expansion problems are encountered, a conducting rubber film can be bonded to a part and a metal surface electroformed over the rubber.

Wall Thickness. In production the wall thickness on irregularly shaped electroformed parts as a rule will vary considerably unless special-shape anodes are used to maintain a uniform solution throwing power. A 0.050-inch section might vary as much as 0.020-inch and sharp corners usually build up heavily, *Fig. 41.8.* As a consequence, whenever possible, no exacting limitations should be specified as to wall thickness to achieve maximum economy. Corner buildup often is an asset to design in that additional strength and rigidity is achieved at no additional cost. When a part can be machined before removing the mandrel, *Fig. 41.4,* uniformity of wall thickness can be achieved most economically, but where uniformity without machining is essential, special handling often can be used to produce the wall thicknesses desired. Maximum practical wall thickness is about 1/2-inch but there is almost no limitation to the thinness of wall.

Mass Production. Electroforming is adaptable to mass production of parts even though the making of one part is rather a matter of days more so than hours. Utilization of a multiplicity of matrices or forms and a great many plating tanks solves the problem. Production equipment is not elaborate, nor is it costly. The primary necessity for quality is control. Regardless of the size of production runs, cost per piece does not vary appreciably in electroforming — a characteristic unlike most other production methods.

Size Range of Parts. Overall size of parts generally is limited only by the size of plating tanks available. Parts from electroformed foil 0.00008-inch in thickness all the way to 28-inch diameter kettle drums have been made successfully. Size

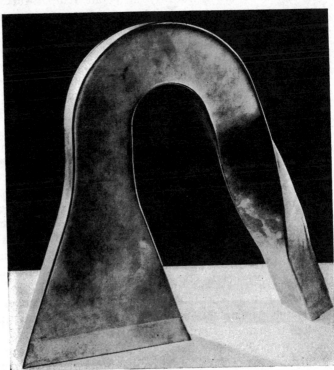

Fig. 41.8 — Electroformed feed horn for S-band radar produced by means of a fusible-metal matrix. Note heavy buildup of metal along all the corner edges.

capabilities of producers runs up to pieces 9 feet by 6 feet by 2 feet with walls to 0.200-inch thick. The important consideration in achieving a satisfactory design is that of contour. Smooth, flowing contours facilitate the electrodeposition process. Metal cannot be deposited in a recess which is deeper than it is wide owing to attraction of the deposit to the point nearest the anode. Accompanying difficulties in achieving uniform wall deposits by means of special anodes is self-evident.

Design Considerations

Although somewhat limited in design possibilities, electroforming makes possible mass reproduction of uniform and irregular shapes to both a dimensional and duplicating accuracy not possible otherwise. An outstanding case in illustration is that of microwave instruments and transmission-line components whose conducting surfaces are on the inside the apparatus. To prevent reflections and excessive absorption of energy in this skin-depth layer, it must be extremely smooth and its contour held to highly exacting tolerances. Electroforming fulfills this requirement since the surface of the matrix — in this case an external one — is easily produced by ordinary methods to the tolerances necessary.

Projections and Recesses. Delicate precision parts such as computing cams, *Fig.* 41.5, etc., which require many hours of machining and touching up to achieve the precision operation desired, can by one of the methods previously described be produced in quantity with assurance of precision duplication from part to part. The only design requirement is that the part be easily removable from the matrix. Projections, therefore, should be specified with some taper whenever possible. Recesses should be avoided in favor of shallow, smoothly rounded contours. Recesses should also be avoided in designing parts which necessitate melting out the matrix. Where recesses are specified, the width should be substantially greater than the depth.

Wall Thickness. Wall thickness deposited is limited to one dimension throughout a part, usually a maximum of 1/2-inch. As a rule, an adequate wall thickness for most applications seldom has been found to exceed 3/8-inch.

Ribs and Bosses. Solid ribs and bosses or heavy sections cannot be formed readily. Where a strengthening rib or boss is deemed necessary, these sometimes can be "grown-in", *Fig.* 41.9, inasmuch as metal can be electroformed permanently or temporarily on a base metal. This process of growing-in can also be applied to composite unit design. The electrically-heated pitotstatic tube shown in cross section in *Fig.* 41.10 is an excellent example of this method of design. The principal elements — heating spools and copper tubes — are surrounded with a lead mold or matrix formed in the desired shape of the finished tube. On this is electroformed a heavy shell of copper and a surface coat of chromium. When the matrix is melted out the unusual design shown is achieved.

Holes. A wide selection of hole sizes can be had ranging from around 0.0007-inch diameter and up. This diameter limitation of 0.0007-inch is dictated by the fact that this is the smallest size of accurately drawn wire now available. Wire, much as a mandrel, is coated with a parting compound and withdrawn from a finished electroform to produce a hole of desired size. Extremely long, small-diameter holes can be easily produced in this way. Small tubing can be utilized in a manner similar to wire. Also tubing or wire can be grown-in much the same as ribs and bosses where such design is desirable, *Fig.* 41.11.

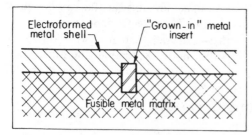

Fig. 41.9 — Where a rib or boss is deemed necessary a metal insert can be used and "grown-in" at the desired point.

Fig. 41.10 — Right — Electrically-heated pitotstatic tube for high-altitude aircraft produced by "grown-in" electroforming of the various components. Photo, courtesy Kollsman Instrument Div.

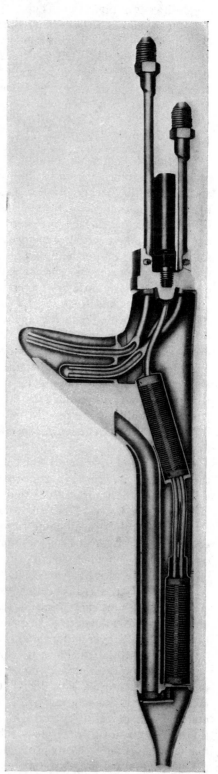

Fig. 41.11 — Electroformed heat exchanger tubes utilizing "grown-in" wires to increase surface area.

Convolutions. On parts such as the special bellows shown in *Fig.* 41.12, where ordinary methods of production cannot be applied owing to the depth of convolution, electroforming can be used successfully. In this case where smooth internal surface finish and an exact depth of convolution corresponding to some fraction of a wave length was required, an aluminum matrix or form was utilized. A special anode is necessary, however, to assure a uniform wall thickness.

Fig. 41.12 — Exacting internal dimensions are achieved by electroforming this microwave unit bellows.

Precision Assemblies. Precision gear assemblies which require an exact angular relationship, precision bores, and a high degree of concentricity often can be produced more economically by electroforming than by any other way, *Fig.* 41.5. Once a precision master is completed, any number of duplicates may be had. Should a precision gear arrangement or complicated cam require any locating work or jig work before assembly, a fixture may be produced also by electroforming. This simplifies the problem of designing for precision assembly in many cases.

Hardened Parts. Where iron is used, a part may be case hardened to a depth of 0.070 to 0.100-inch and heat treated to produce any hardness from 16 to 65 Rockwell C. Heat treating or hardening, however, often results in a loss of much of the original accuracy. Parts which must withstand considerable load can either be built up extra heavy and machined off flat, backed up with a filler of fusible metal, or supported by a steel backing plate.

Selection of Materials

Techniques for electroforming have been worked out for a wide variety of metals. These include copper, nickel, silver, gold, rhodium, iron, brass, chromium, alloys of cobalt, and nickel. Also, copper-zinc, nickel-cobalt and zinc-cadmium are used industrially. Laminated metals such as nickel-lined iron, iron-lined nickel, copper-lined iron, chromium-lined iron, chromium-lined copper, etc., can be utilized. Use of laminated design, however, involves an additional operation which naturally increases the cost of such parts.

Metals. Selection of a satisfactory metal or metals should be made on the basis of price, availability, service conditions, physical, metallurgical, and electrochemical characteristics. Most commonly used materials are copper, nickel and iron and physical design requirements can be relied upon as to which should be used. Where special conditions demand it, a laminate such as chromium or nickel might be used over one of these base metals, *Fig.* 41.13.

Fig. 41.13 — Aluminum piston for diesel engine with electroformed iron dome, 0.015-inch thick, to resist heat erosion.

Speed of Deposition. Speed of deposition occasionally can be a deciding factor in material selection. Copper can be deposited at a rate of 0.001 to 0.003-inch in 10 to 20 minutes whereas iron and nickel are deposited at about 0.001 to 0.002-inch per hour.

Characteristics. Electroformed materials possess metallurgical characteristics similar to metals that have been formed by the more common methods of rolling, drawing, forging, casting, etc. One major exception exists, however, in that the grain or crystal growth is radial rather than longitudinal. On annealing, the grain structure returns to conventional arrangement.

Electroformed iron, as deposited, Brinells about 225 to 250, has a tensile strength of about 50,000 to 55,000 psi, an elongation of 15 per cent in 2 inches, and a weight of 0.2842 pounds per cubic inch. Density is extremely high, the electroformed iron actually being less porous than glass.

Nickel can be deposited with a tensile strength ranging from 50,000 to 150,000 psi depending upon the bath used and conditions. The ordinary bath produces a tensile of 50,000 psi, 150 Brinell hardness, and an elongation of 30 per cent in 2 inches. The medium bath deposits nickel ranging from 100,000 to 120,000 psi tensile, 260 to 280 Brinell hardness and elongation of 12 to 20 per cent. The hard bath produces a deposit with a tensile ranging from 120,000 to 150,000 psi, 28 to 45 Rockwell C hardness.

Specification of Tolerances

Extreme dimensional accuracy mentioned previously as one of the outstanding assets of the electroforming process is truly just that. This refers primarily to surface finish and to variation between mass-produced parts. Unlike most processes, variation between parts can be made practically nonexistent by proper control. Dimensional duplication from part to part is easily held within 0.0001 to 0.0002-inch.

Surface Finish. Finished surface quality as well as dimensional accuracy of a part is dependent upon the accuracy of the negative metal master matrix, fusible metal, wax or rubber matrix, or dissolvable-metal mold. Surface quality of the matrix is reproduced with fidelity even to as fine as 7000 lines per inch.

Dimensions. Dimensionally, the steel mandrel or negative metal matrix and the dissolvable-metal matrix methods are superior in close-tolerance work. Parts accurate to plus or minus 0.0002-inch with surface finishes around 2 to 5 microinches can be produced. Owing to the fact that matrices of cast fusible metal, rubber and wax require a master mold, parts produced require somewhat greater tolerances. As a rule, on parts so produced, plus or minus 0.001-inch can be held. Surface finish obtainable, likewise, is not quite as good and cost is greater owing to the necessity for polishing each matrix casting before electroforming.

Wall Thickness. As previously mentioned, wall thicknesses may vary considerably in designs of irregular shape and where this can be tolerated with or without machining, maximum economy can be realized. However, with proper anode design for a specific part, light walls — say up to 0.020-inch — can be held within plus or minus 0.002-inch. Variation in heavier walls may run to plus or minus 0.005-inch.

ADDITIONAL READING

[1]"Electroformed Molds for Plastics Processing," *Plastics Design & Processing,* July 1970. p. 24.
[2]James W. Oswald, *Heavy Electrodeposition of Nickel,* The International Nickel Co., New York, N.Y., 1962.

42

Metal Spraying

SPRAYING of molten metal, commonly referred to as metallizing, like many other processes, has received considerable impetus during late years. Although relegated by many to the general maintenance and salvage field without further thought, metallizing nevertheless is fast finding many original applications along with the various other mass-production processes.

Field of Application in Basic Design. Probably the most common production application of metallizing today is that of corrosion and heat oxidation protection, *Fig.* 42.1. Of most interest, however, are those applications which give the metallizing process a place in basic design. Copper spraying can be employed to provide electrical conducting or shielding surfaces and for improvement of electrical connections as on carbon motor brushes, *Fig.* 42.2. Also, soldering surfaces can be provided on glass and other nonmetallic parts by tin spraying, *Fig.* 42.3. Perhaps the most outstanding value of the process, though, lies in the design possibilities afforded in the application of special surfaces on parts which must withstand unusual conditions. Thus, low-cost materials can be utilized for such parts and sprayed metal surfaces specified for those portions which must withstand heavy wear or corrosion or which require the unusual properties of some particular metal. Piston rods, pump rods, pistons, rams, slides, cams, motor brushes, chemical process equipment, circuit breaker bushings, etc., are among the multitude of machine parts utilizing metal spraying in original manufacture.

Metal Spraying Processes. Originally conceived by Dr. M. U. Schoop of Zurich around 1911, the process in its present state of development embraces the creation of a metallic spray by means of atomizing metal in wire or powder form with a so-called spray gun. An oxygen-gas flame melts the metal — wire fed by an air turbine or powdered metal fed from a special container by means of a hose and vibrator — and a blast of compressed air picks up and carries the molten spherical globules to the prepared surface. Upon hitting the surface, the flattened particles of metal chill almost instantly and key themselves to the base material which may be metal, wood, plaster, stone, glass, concrete, rubber, or plastics. The bond is entirely mechanical and consequently proper consideration must be given to assure a surface which affords satisfactory bonding characteristics. Ordinarily, for corrosion applications, blasting with sand or sharp angular grit roughens the surface satisfactorily. For heavy wear-resistant coatings, though, a more certain mechanical grip surface usually must be provided.

Fig. 42.1 — Above — Corrosion protection for a steel engine mount provided by sprayed 72S aluminum alloy lasts indefinitely.

Fig. 42.2 — Right — Mechanized copper-spraying equipment provides proper electrical connection for electric motor brush leads. Photo, courtesy General Electric Co.

Advantages. Experience has proved that, in the majority of cases, metal as deposited by metallizing is of superior wear-resistant quality than the parent metal. In many cases it is possible to apply much harder materials at the point where greatest wear will occur. The air-hardened characteristics of metallized deposits along with the porous nature of the deposit create ideal properties for lubricated rubbing or sliding surfaces. A good example is the redesign of a heavy bronze aircraft landing-gear strut piston to utilize a drawn-steel shell metallized on the outside with 0.045-inch thick bronze, reducing weight by one-half. The advantage of porosity in lubrication applications, however, becomes a disadvantage in certain cases when the coating is to resist the penetration of fluids.

Limitations. Ordinary sprayed metal, it is well to remember, cannot be used on those machine parts where it is desired to build up a sharp edge, or on parts where it will be subjected to point impact or bearing. Again, one of the important factors to consider in determining the economics of a metallizing application is the value of the part itself. Metallizing is relatively expensive and, therefore, is only practical on intricately machined pieces representing a large amount of labor or material costs. Spraying can be utilized to assure excellent wear or corrosion

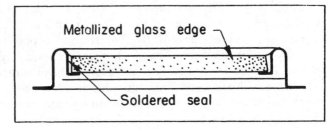

Fig. 42.3 — Pyrex glass window soldered directly to a metal frame by means of metallizing the edge, is liquid, gas and airtight. Drawing, courtesy Metco Inc.

resistance on required areas, eliminating the necessity for making an entire part from hard-to-machine, expensive or critical materials.

Hard Surfacing. Acceptable coatings of Stellite can be deposited by spraying. These coatings are especially suitable where the maximum in both wear and corrosion resistance are desired. In fact, Stellite may be deposited easily on parts of bronze, aluminum or copper — a useful application not possible by any other means. Also of interest is the fact that though powder spraying equipment is simpler than that for wire spraying and is lighter in weight, it is more difficult to handle and, with the exception of zinc, metals are more expensive in powder form than as wire, depositing efficiency is lower, and more weight of powder is necessary to obtain a specific weight of deposit. Powder guns cannot be used to deposit direct a usable finished surface of the higher-melting point metals such as steel, Colmonoy alloys, etc., so their field ordinarily is limited to easily melted metals — aluminum, zinc, etc., as well as nonmetallics such as rubber, plastics, and low-melting point vitreous materials.

Metallizing with powders of the higher-melting point metals previously mentioned, however, is successfully carried out by what is now referred to as "spray welding". Metal powder, applied to a base metal in conjunction with chromium boride crystals by means of a flame gun, is melted and permanently bonded to the base metal with a welding torch, induction heater, or controlled-atmosphere furnace. This method is particularly effective where hard, wear-resistant facings are desired which must withstand abuse, wear, impact, and have sharp edges, *Fig. 42.4*, but is limited to a final coating thickness of 0.060-inch, maximum.

Range of Sprayed Thicknesses. No definite limits exist as to the maximum practical thickness or overall areas of sprayed coatings except those imposed by the economic aspects of the application. Coatings, as a rule, can be any thickness from

Fig. 42.4 — Large machine mandrel ends with hard-faced nose section shown at left after spray welding but before bonding and at the right, after finishing.

approximately 0.002-inch upwards. Generator armatures, utilizing a sprayed copper deposit approximately 3/4-inch thick on a 6-inch diameter armature shaft, have been produced successfully. The sprayed copper surface which formed the contact area for the brushes, being air quenched naturally in the process, proved in service to be harder and more wear resistant than the original copper. Lathe beds, metallized with steel and finish ground, also have been produced, resulting in a precision bedway with excellent lubrication properties and wear resistance equal to that of hardened steel.

For many applications the ordinary type of hand-operated gun is used successfully but, where production is high and uniformity desirable, mechanized setups can be employed. Spraying of electric motor brush parts, *Fig.* 42.2, is a typical example where an automatic metal spraying machine falls just short of being completely mechanized. Metallizing of the gland zones of steam turbine shafts, with high-chromium nickel steel, a regular production setup, is shown in *Fig.* 42.5.

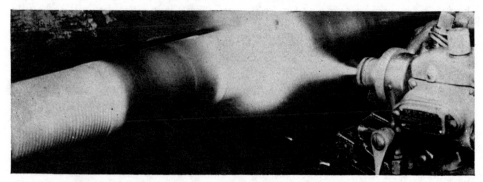

Fig. 42.5 — Metallizing the shaft of a steam turbine with high-chromium nickel steel. Photo, courtesy Metco Inc.

Design Considerations

Utilization of sprayed metals in the design of various types of parts requires attention primarily to the bonding. Most desirable method of bonding depends either upon the material or the type of application. Mechanical bonding satisfactory for plastics, corrosion-protection coatings and spray welding can be provided on parts by merely specifying a thorough sand or steel grit blast finish. No other preparatory consideration is required as with the application of heavier special surfaces which are finish machined after depositing.

Surface Preparation. Several methods of preparing surfaces for metallizing are in common use. One is by rough turning and another by grooving and knurling. Suited primarily for surfaces of rotation, the rough turning is literally a crude thread of 18 to 20 pitch approximately 0.040-inch deep, *Fig.* 42.6. The grooving method utilizes closely spaced grooves made by a tool 0.045 to 0.050-inch wide, about 0.025-inch deep, spaced about 0.015 to 0.025-inch apart and knurled to form dovetails, *Fig.* 42.7. Grooving tools are specified to produce corner radii of about 0.020-inch maximum to minimize the stresses set up, and because of this feature the grooving method is preferred over rough turning. The third or electric bonding method, a fairly recent development useful for any type of surface,

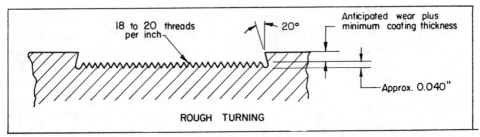

Fig. 42.6 — Design of rough threading operation used to prepare shafts for spraying. Drawing, courtesy John Nooter Boiler Works Co.

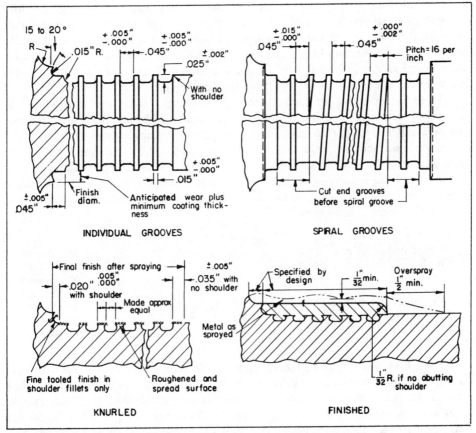

Fig. 42.7 — Design of grooving and knurling method for surface preparation to retain sprayed metal. Drawing, courtesy General Electric Co.

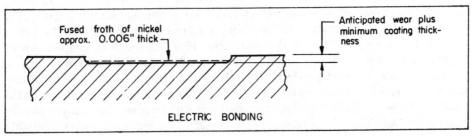

Fig. 42.8 — Typical method for preparing a surface by means of electrically fused nickel froth. Drawing, courtesy John Nooter Boiler Works Co.

provides the fastest, most economical and satisfactory preparation and, most important, does the least structural damage to the part, *Fig.* 42.8. Design for this method consists only of specifying the portion to be sprayed sufficiently undersize to suit the requirements. Ends, where undercut, should be smoothly rounded, *Fig.* 42.8, and the radius can be as desired. Bonding is not unlike a welding process, an electrode holder with six nickel electrodes is stroked across the prepared surface to produce a light, frothy, bubble-like layer of nickel about 0.006-inch in depth. Maximum mechanical bonding, especially at the point of junction, is assured by this method and is preferable in most cases.

Shoulders and Threads. Sprayed metals are much like cast metals of the same analysis, but their density is about 10 to 15 per cent less and they have little tensile strength or elasticity. Their resistance to compression, though, is good and makes them ideal for oil-impregnated bearings and bearing surfaces. It follows naturally that, except for spray welding, coatings should never be designed so as to taper out; a definite shoulder is desirable in all cases. Characteristics of sprayed metals also preclude the possibility of cutting threads in coatings.

Wear Surfaces. Where the qualities of Stellite are desired, including complete corrosion resistance, coatings should be at least 0.030-inch thick to eliminate porosity. However, coatings may be any specified thickness otherwise. Generally, where wear surfaces are required, consideration is given the maximum amount of wear anticipated and to this is added a minimum basic coat, depending upon part size, to obtain the correct depth for the spray coat. Using diameters as an indication of part size, the following minimum basic coats are normally added to the maximum expected wear figure:

Under 1-inch	0.010-inch
1 to 2-inch	0.015-inch
2 to 3-inch	0.020-inch
3 to 4-inch	0.025-inch
4 to 5-inch	0.030-inch
5 to 6-inch	0.035-inch
6-inch and over	0.040-inch

Press Fits. The foregoing figures apply to all such work except press fits. Here the minimum thickness may be 0.010-inch on the radius after machining, regardless of diameter. Tests on driver axles for locomotives designed with a 0.015-inch thick, 1.20 per cent carbon steel metallized surface in the press fit proved to be 50 per cent higher in fatigue resistance to breaking off in the wheel fit and 160 per cent greater in fatigue resistance to initiation of cracks.

Sliding Shafts. Certain applications are encountered where metallizing alone satisfies both design and cost requirements. A good example of such a case is that of a heavy boring shaft arranged to slide within a driver sleeve, a key in the sleeve operating in a long keyway in the shaft, *Fig.* 42.9. Although a combination of metals such as SAE 52100 hardened and ground steel for the sleeve and turned, ground and nitrided steel for the shaft could be used, it was found most expedient and economical to provide merely an ordinary easily machined steel sleeve (AISI 1020) and a shaft of similar material metallized with bronze. The sprayed bronze coating on the shaft not only simplifies manufacture but also makes available "oilless" lubrication, a quality not possible otherwise. Keyways for such jobs should be specified as shown in the shaft cross section.

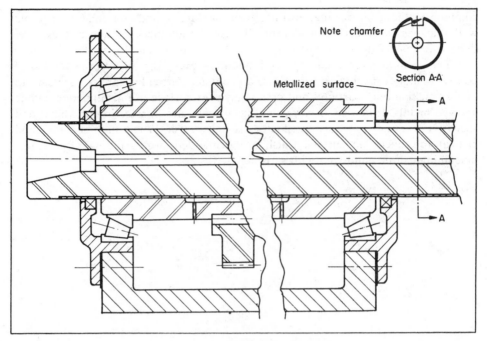

Fig. 42.9 — Sliding boring shaft, metallized with bronze, reduces cost considerably and guarantees self lubrication.

Selection of Materials

Practically every metal available in wire form, except perhaps tungsten, can be sprayed in a wire gun — brass, bronze, copper, silicon-bronze, aluminum-bronze, aluminum, magnesium, lead, tin, iron, zinc, cadmium, babbitt, nickel, carbon and alloy steels, Monel, stainless steels, Stellite alloys, molybdenum, tantalum, Nichrome, etc. High-melting point alloys, especially those available only in powder form such as those for hard facing, are readily handled by spray welding. In addition to materials suitable for powder gun spraying mentioned previously, mixtures such as powdered glass and metal can be utilized where increased bond strength and a nonporous coat are desired. Deposit efficiency is best with the wire gun and ranges from around 65 per cent for easily oxidized metals such as zinc to 90 per cent for aluminum, iron and steel.

Hardness and Porosity. Bronze coatings with a tensile strength of 30,000 psi, a compressive strength of 100,000 psi and a hardness of 85 Rockwell B may be produced. In fact, all sprayed coatings exhibit apparent hardness greater than that of the parent metal and wear much longer. Porosity of all regular sprayed bearing metals is such that they absorb approximately 10 per cent of the lubricant, reducing wear and running friction. Cases are on record where running friction has been reduced as much as 30 per cent.

Specification of Tolerances

Except in the case of protective coatings where reasonable uniformity in thickness is imperative, sprayed metal surfaces are always finished by one of the common methods — turning, milling, grinding, etc. — and consequently coating wall thickness may vary considerably. Finished tolerances available, therefore, depend primarily upon the finishing method or methods employed.

The peculiar structure of sprayed metals — excepting those spray welded of course — makes machining sometimes a problem. Particular care is necessary to avoid "picking out" particles and creating a poor surface in finishing machinable sprayed metals. Likewise, in dry grinding, special care in selection of wheels and speeds is necessary to assure satisfactory surface finish. Where high accuracy and critical tolerances on surface finish are demanded, wet grinding after rough machining can be employed with satisfactory results.

CASTING
METHODS

43

Sand Casting

Among the multitude of processes utilized in the design of machines, those of casting perhaps receive the least attention and consideration. One of the earliest processes for forming metals, casting provides a versatility and flexibility which have maintained its position throughout the years as a primary production method for machine elements.

Methods of Casting. Any metal which can be melted and poured can be cast and the size range of parts regularly produced by presently available methods is greater than that of any other process affording similar results. Generally, casting methods can be divided into those using nonmetallic molds and those using metallic molds. However, to permit a logical breakdown regarding design, these processes can best be subdivided according to the specific type of molding method used in casting, as follows:

1. Sand
2. Centrifugal
3. Permanent-mold

4. Die
5. Plaster-mold
6. Investment

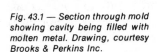
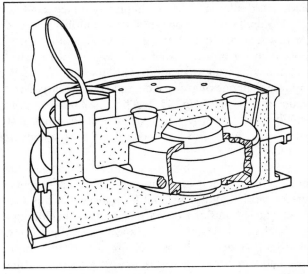

Fig. 43.1 — Section through mold showing cavity being filled with molten metal. Drawing, courtesy Brooks & Perkins Inc.

Fig. 43.2 — Coremakers giving the final touches to a large sand core prior to baking. Photo, courtesy Abex Corp.

Each of these distinct methods is covered separately along with those factors pertinent to each. As is characteristic of other methods, castings can be produced which violate some or all of the basic principles of good design but such parts are seldom economical or commercially practical. It behooves the designer, therefore, to consider the practical aspects of casting design equally as well as the production character of any other of his machine parts.

Sand Casting. Best known and still most widely applied method of casting is with sand molds. Sand casting consists basically of pouring molten metal into appropriate cavities formed in a sand mold, *Fig.* 43.1. The sand used may be natural, synthetic or artificially blended material. Along with a small amount of water and a bonding agent such as fire clay, bentonite, cereal binders or liquid binders for cohesion, the mechanically mixed sand is used to form the molds.

Molds. Two common types of sand molds are generally employed in casting — the dry sand mold and the green sand mold. In the former, the mold is dried thoroughly prior to closing and pouring while the latter is used without any preliminary drying operation. The dry sand mold is more firm and resistant to collapse than one of green sand and consequently core pieces for molds are largely made in this way. *Fig.* 43.2. The resistance of dry sand molds to contraction prohibits their use where light or intricate castings of steel are to be made. For such parts green sand molds are essential and the majority of steel castings are now made in green sand.

Shell Molding. One of the most exciting new developments in the foundry industry is the shell molding process which permits such close control of both surface finish and dimensional tolerances that it is sometimes possible to eliminate all secondary finishing operations such as machining and grinding.

Shell molding starts with a special metal pattern which is heated to about four hundred fifty degrees F. and clamped to a "dump box" containing an extremely fine sand which has been mixed with a thermosetting plastic resin. The box and pattern are inverted to bring the resin coated sand into contact with the hot pattern. After about thirty seconds the resin melts and forms a thin shell of bonded sand on the pattern. When the box and pattern are righted the excess sand drops away. After a short curing time, the shell hardens and can be stripped from the pattern by built-in ejector pins. The other half of the mold is made in the same manner and the two are "pasted" together with a resin glue. For additional mechanical support the shell is sometimes placed in a flask filled with coarse sand or shot prior to pouring.

Fig. 43.3 — Molders making the drag or bottom half of sand molds using separate metal patterns. The jolt squeezing machines shown are used largely for medium-size molds. Photo, courtesy Abex Corp.

Equipment for this process can be relatively simple or highly specialized automated installations. While pattern costs are higher than for green sand molding and the castings themselves generally cost a little more, the elimination of machining and finishing often makes for worthwhile savings in the completed component. Less pattern draft and closer tolerances often characterize shell molded castings. This process is also finding wide application in the production of cores. Shell cores have three advantages over conventional cores: baking time and core driers are eliminated, weight is saved because the cores are hollow, and extremely smooth internal passages are produced, such as in burner castings where the smooth passageway improves efficiency.

Patterns. To produce a mold for a conventional sand cast part it is necessary to make a pattern of the part. Patterns are made from wood or metal to suit a particular design with allowances to compensate for such factors as natural metal shrinkage and contraction characteristics. These and other effects, such as perhaps mold resistance, distortion, casting design, mold design, etc., which are not entirely within the range of accurate prediction generally make it necessary to adjust pattern dimensions to produce castings of the required dimensions.

Types of patterns utilized in sand casting generally can be classified as in the following:

1. Single, loose wood or metal patterns. Low in cost but only satisfactory for very small quantities. Metal better than wood but subject to same limitations — gates and risers must be cut by the molder in each mold used.
2. Gated wood or metal patterns. Satisfactory for small-run production. Wood subject to warpage and breakage.
3. Metal match-plate patterns. For large quantities of small castings. Assure more accurate reproduction than loose patterns and increased cost is compensated by increase of at least 60 per cent in production over a similar gated metal pattern.
4. Separate wood or metal cope and drag patterns. Permit good production on medium to large castings because the two parts of the molds can be produced simultaneously. Metal construction particularly suited for quantity production of medium to large size castings and to machine gang molding, *Fig.* 43.3.

In addition to the main pattern, most castings require cores, *Fig.* 43.4, to produce cavities or maintain desired contours or sections. In some cases, dry sand molds or cores are assembled to obtain an entire mold as with crankshafts and anchor chain. Thus, in addition to pattern equipment, core boxes are generally required to make the core pieces for a casting.

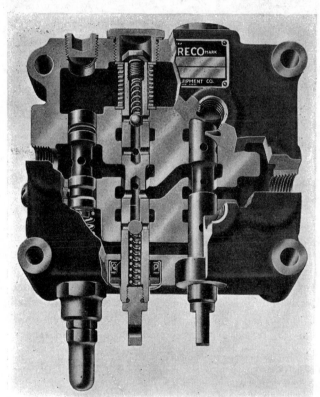

Fig. 43.4 — Cutaway view of a hydraulic control valve. Complex coring work is necessary for forming the various interconnected fluid passageways with smooth flowing contours. Photo, courtesy Hydreco Div.

Mold Making. To produce a mold, the pattern is placed within a so-called flask which is then packed with molding sand. After the sand is thoroughly packed the flask is inverted and the pattern is withdrawn leaving a cavity of the desired shape and size, *Fig.* 43.3. When both top (cope) and bottom (drag) halves of a mold are completed in this way, any cores required are placed in position and the top flask is locked over the bottom, *Fig.* 43.5. Following pouring and solidification of the metal, the sand is shaken off the casting, any cores are knocked or cleaned out

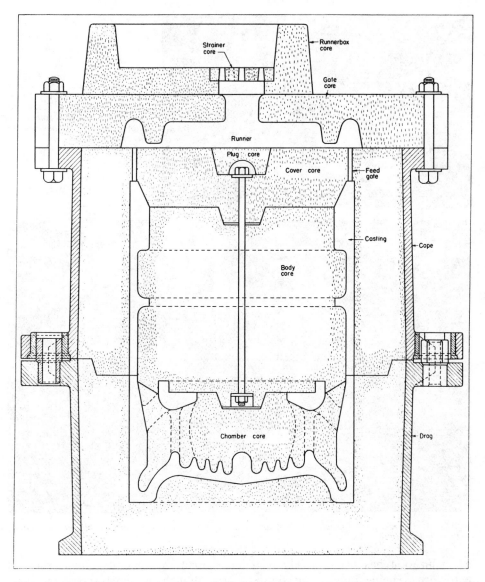

Fig. 43.5 — Cross section through assembly of diesel locomotive piston mold. Drawing, courtesy Campbell, Wyant & Cannon Foundry Co.

and the gates, runners, risers, etc., are sawed off and the castings are ground smooth, *Fig.* 43.6.

Casting Production. The wide range of production possibilities and the expendable nature of the sand utilized gives sand casting outstanding flexibility. Patterns can be readily adjusted, changed or revamped to suit design requirements without excessive cost or obsolescence of equipment. Sand casting, in addition, permits the production of castings of a size and intricacy otherwise impractical, *Fig.* 43.7.

By hand methods one or a few parts may be readily cast at reasonable cost. Through the use of mechanized equipment for rapid, automatic production of molds, mass production of castings has been highly successful. The degree of

Fig. 43.6 — Above — Sawing off the gates, runners, risers, sprues, etc., from a Hastelloy Alloy B valve casting. Blind risers and gating used can be readily seen. Photo, courtesy Haynes Stellite Div., Cabot Corp.

Fig. 43.7 — Unusual giant cast steel saddle support for a main suspension bridge cable. Photo, courtesy Bethlehem Steel Corp.

mechanization has been largely dependent upon the quantity. Items such as automotive engine blocks and similar parts, though extremely complex, have reached a high degree of mechanization. In many cases, where quantities warrant the expense, almost completely automatic continuous production of sand castings has been successful utilizing the shell mold process.

Design Considerations

The final cost of a casting involves far more than merely the cost of the metal. It is the ultimate cost of the machine part itself that is of most importance. Maximum economy in the final analysis may evolve from a casting design of greater rather than lesser weight, more machining rather than less, etc. Design which recognizes the inherent limitations present in casting will invariably help to assure maximum soundness and low scrap or rework. In selection of the material to

Fig. 43.8 — Casting design, above, having two undercuts requiring dry sand cores, was improved greatly by redesign shown below to simplify molding and casting entirely in a green sand mold. Draft is all in one direction and less snagging is required. A cost reduction of 10.6 cents per casting resulted despite an increase of 0.7-pound in weight. Photos, courtesy Warner Automotive Parts Div., Borg-Warner Corp.

be cast, the foundry characteristics of individual metals must be kept in mind. Each requires patterns which are made to suit and are not generally interchangeable. Foundry craftsmanship and know-how can seldom substitute economically for inadequate design.

Design Attributes. An ideal design, both from the standpoint of foundry practice and economy, should be possible to cast in green sand with a two-part mold. Where necessary, green sand cores should be usable and preferably the mold itself should provide for portions ordinarily cored. Size and awkwardness should be minimized to permit use of the smallest possible flask or where large machine parts are concerned, the smallest pit space. Ordinarily, gating and heading problems should be left to the foundry, but in all cases design should recognize the need for proper gating and heading positions on a casting, the need for free flow of metal to all parts of the mold, and the need for risers to feed the heavier sections continuously during solidification to compensate for shrinkage.

Patterns. Although the designer of castings would seldom be involved in pattern work, it is useful to keep in mind this requirement of the process. Designers should recognize the fact that although the removal of a casting from a sand mold brings in no withdrawal problems, the making of the sand mold by withdrawal of the pattern after ramming certainly does. Design to facilitate production of the sand molds will result in highly worthwhile economies, *Fig.* 43.8.

Contraction. Shrinkage allowances must be made on patterns to counteract the contraction in size as the metal cools. The amount of shrinkage is dependent upon the design of the casting, type of metal used, solidification temperature, and mold resistance. All portions of a casting may not shrink uniformly as a result of these factors and, being a volumetric contraction, it is difficult to predict accurately. Consequently, determination of the proper amount to use is a result of experience and should be discussed with the producing foundry. Shrinkage values normally employed are shown in TABLE 43-1.

TABLE 43-1 — **Pattern Shrinkage Allowance**
(inches per foot)

Metal	Shrinkage*
Aluminum alloys	1/10 to 9/32
Beryllium copper	1/8
Copper alloys	3/16 to 7/32
Everdur	3/16
Gray irons	1/8
Hastelloy alloys	1/4
Magnesium alloys	1/8 to 11/64
Malleable irons	1/16 to 3/16
Meehanite	1/10 to 9/32
Nickel and nickel alloys	1/4
Steel	1/8 to 1/4
White irons	3/16 to 1/4

* Smaller value applies generally to large or cored castings of intricate design. Larger value applies to small to medium simple casting designs with unrestrained shrinkage.

Draft. All surfaces perpendicular to the parting plane of the mold, i.e., in the direction in which the pattern or patterns are withdrawn, require draft to permit easy withdrawal. The usual amount for general-purpose work is 1/8-inch per foot or 1/2-degree per side. If the surface is not to be machined and the amount of slope does not interefere with its function, greater amounts of draft — one degree or more — may simplify molding and casting. With high-production molding equipment where pattern drawing is mechanized, less draft is necessary. Normally, in such cases 1/16-inch per foot is satisfactory.

Interior features of a pattern ordinarily require greater draft than exterior portions. On high-production matchplate work about 2 degrees is necessary although in cases this can be reduced to one degree. Greater drafts are preferable if design permits. Unless the drawing specifies, draft may be divided by the patternmaker, subtracting half and adding half. However, it is usually preferable to add the draft material but *not* to incorporate in a drawing any dimensions which include the added thickness or length.

Parting Lines. Molds, in general, are most conveniently parted in a flat plane. Casting design, therefore, is most economical if the part can be molded with a flat rather than a curved parting plane, *Fig.* 43.3. Although high-production matchplate equipment permits some deviation from flat parting planes without serious mold preparation problems, it should be recognized that these irregular matchplates are far more costly than the use of a standard universal plate.

In addition to striving for simple design which permits molding with a straight parting line, wherever possible only a single parting in the mold should be required. The fewer partings in the mold, the less the work required to produce it. Pattern equipment is less complicated and the part can be more readily adapted to machine molding. Thus, three-part mold designs should be avoided by proper consideration during design.

Coring. All interior features of a cast design which are not possible to form in the mold with the pattern proper must be produced by means of coring, *Fig.* 43.9.

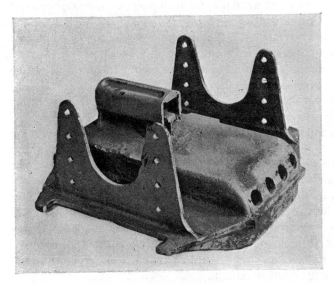

Fig. 43.9 — Floating draft sill center brace for a freight car. This steel casting which weighs approximately 425 pounds and measures 26 by 30 by 12 inches high, is a good example of a difficult coring and molding job requiring careful location of gates and risers. This design was considered superior to numerous composite designs tried in an effort towards simplification. Photo, courtesy Symington-Gould Corp.

Cores, in general, are costly to produce and set and wherever possible should be eliminated. Because of the lower cost, green sand cores are preferable where cores cannot be eliminated. Because of the low strength of green sand cores, internal features produced with these cores cannot exceed their diameter or widest dimension in length or depth. Also, green sand cores require considerably greater draft than dry sand cores, which often can be made without any draft allowance.

Where cores are necessary, it should be noted that they require adequate support in the mold. Core prints — sections added to the pattern to leave an imprint in the mold in which the corepiece can be placed — are required to provide this feature and the core extension which fits into the core print naturally will leave a hole in the casting wall, *Fig.* 43.1. The vertical print surfaces should have at least 2 degrees of draft and may have as high as 15 degrees. In general, the length of print should equal its diameter or width. Prints must be so placed that the core weight is uniformly supported and so that any shifting, floating or sagging is prevented. Where prints are not practical, metal chaplets for core support can be and are used, *Fig.* 43.10, but it is usually preferable to avoid the necessity for their use by proper design, especially with pressure castings. Stub cores also should be avoided.

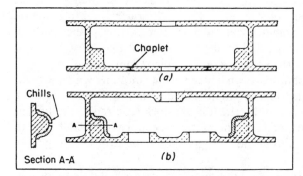

Fig. 43.10 — Cross section of a magnesium sand casting originally designed as at (a) which required chaplets for proper support of the sand core. Redesign at (b) eliminates the need for chaplets by using two spaced core print holes. Holes are strengthened by beads and are larger to facilitate escape of gases and shakeout of chill pieces. Drawing, courtesy American Magnesium Corp.

Core gases produced in casting are usually vented through the core prints as the simplest exit. Where a print is not available for such purposes one usually must be added. Besides the problem of venting core gases from a cored pocket there is the additional requirement that it must be possible to remove the core sand readily after solidification. Large core support prints not only facilitate setting and anchoring but also greatly simplify cleanout of the core sand, a feature which often becomes an acute foundry problem, *Fig.* 43.11.

Locating Surfaces. Stable checking surfaces should be indicated on any casting drawing to facilitate ready dimensional check of the part. Pattern shop, foundry, machine shop if required, and inspection departments should all be able to check a casting from the same points. Wherever possible such surfaces should be located on the same side of the parting line away from the possible influence of shift of the cope, drag or a core. Several widely spaced surfaces which have no finish allowance and which are to be held to fairly close limits serve best for locational checking.

The requirements of locating, fixturizing and laying out for machining should also be kept in mind. If good locating points for checking are not available, it may prove that the part will be difficult to hold or lay out. In many cases special pads for locating, laying out or fixturizing a part should be added. The slight additional expense may prove to be a production economy.

Machining Allowances. Many castings — especially shell molded castings — can be put to use in the "as-cast" condition; that is, without any final machining or finishing other than the cleaning and sandblasting treatment usually employed. Where design is such that the normal tolerances obtainable in sand or shell casting are satisfactory, maximum economy dictates use in the as-cast state. However, where more exacting dimensional accuracy is required, machining must normally be used.

To permit machining to flatness only, as with surface grinding, where some unmachined portions are satisfactory, a very minimum of additional metal need be allowed. If machining to a full flat or formed surface without imperfections is desirable, a greater amount of metal must be allowed. The allowance utilized for a particular casting depends upon the metal, the size and shape, the position of the

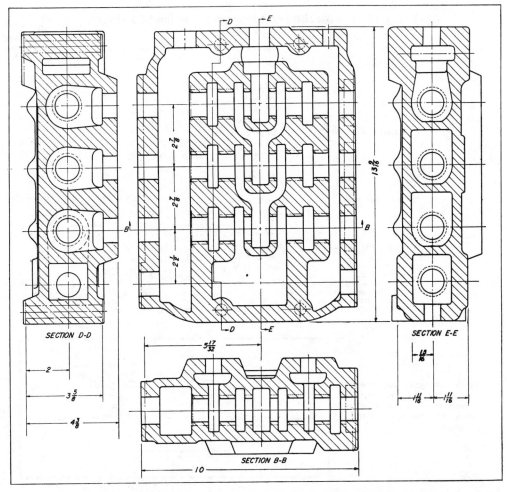

Fig. 43.11 — Drawing for a hydraulic valve casting. Complex coring such as this requires the ultimate in foundry skill. Complete removal of core sand from passages often presents a problem where hydraulic service is concerned. Drawing, courtesy Hydreco Div.

particular surface in the mold, warpage or surface variations, machining method, and degree of surface perfection desired, *Fig.* 43.12. Nonferrous metals, such as brass, are less porous and have a thinner surface skin than ferrous metals and consequently permit the production of a fine finish with less metal removal. It is generally desirable to have surfaces which are to be machined cast in the drag side of the mold. For those to be cast in the cope, extra allowance should be made because of unavoidable surface impurities which rise to the top during pouring. In TABLE 43-II are given the usual finish allowances utilized with various materials and casting sizes.

Section Uniformity. All cast metals require proper feeding of the molten metal into the mold to insure maximum soundness. As the molten metal solidifies, a contraction in volume takes place, requiring additional molten metal to continuously flow in to fill the voids. Risers on the casting are added by the foundry at the most desirable positions, to perform this duty, *Fig.* 43.13. In order that both the risers and the gates can adaquately feed a solidifying casting, it must be recognized that the section sizes and proportions of the casting must be designed so

TABLE 43-II — **Machining Allowances for Sand Castings***

(inches)

Metal	Casting Size	Finish Allowance
Cast Irons	Up to 12 in.	$\frac{3}{32}$
	13 to 24 in.	$\frac{1}{8}$
	25 to 42 in.	$\frac{3}{16}$
	43 to 60 in.	$\frac{1}{4}$
	61 to 80 in.	$\frac{5}{16}$
	81 to 120 in.	$\frac{3}{8}$
Cast Steels†	Up to 12 in.	$\frac{1}{8}$
	13 to 24 in.	$\frac{3}{16}$
	25 to 42 in.	$\frac{5}{16}$
	43 to 60 in.	$\frac{3}{8}$
	61 to 80 in.	$\frac{7}{16}$
	81 to 120 in.	$\frac{1}{2}$
Malleable Irons	Up to 8 in.	$\frac{1}{16}$
	9 to 12 in.	$\frac{3}{32}$
	13 to 24 in.	$\frac{1}{8}$
	25 to 36 in.	$\frac{3}{16}$
Nonferrous Metals	Up to 12 in.	$\frac{1}{16}$
	13 to 24 in.	$\frac{1}{8}$
	25 to 36 in.	$\frac{5}{32}$

* Courtesy American Foundrymen's Society.
† Allowance ranges up to 1 inch. Values are for ordinary finishes on the drag side of the mold where distortion is an unlikely factor. For surfaces to be cast in the cope allowances often must be doubled.

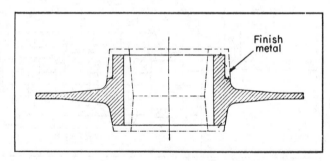

Fig. 43.12 — Typical part showing the effect of machining allowances on the design of the casting. Drawing, courtesy Hills-McCanna Co.

as to permit the required flow of metal.

As a rule, metal solidifies as a gradually increasing envelope or wall within the mold, starting at the mold surfaces and progressing inward in all directions. Because the thinnest sections generally solidify most rapidly, it can be seen that any very great variation in section thicknesses would result in the creation of severe contraction stresses in the soft and perhaps yet molten heavier adjacent sections. The ideal design should be so proportioned that sections farthest from the feed heads solidify first, with solidifying action progresssing to the feed head and risers, which are the last to cool.

It can be seen that severe variations in wall thickness often can result in castings having heavy intermediate sections which cannot be fed. If thin sections "freeze" before heavy interior portions, shrinkage defects are likely to occur in these so-called hot spots. This is especially true in steel castings where shrinkage is most severe, TABLE 43-I. Where substantially uniform section thickness is not practical throughout a casting, as is often the case, the design should take into consideration the problem of feeding and solidification sequence and provide adequate sections leading to the heavier portions or allow progressive increase in section toward the feeders.

Where adequate risers can be applied, however, no arbitrary limits on

Fig. 43.13 — A magnesium alloy nose section after removal from sand mold. Sprues, gates, runners, and risers have not yet been removed. Heavy risers and chilling are necessary to assure sound metal in such a complex casting by equalizing cooling between heavy and thin sections.

reasonable variations in section thickness need be considered, *Fig.* 43.13. As a rule, sections remote from risers may vary plus or minus 20 per cent without recourse to special foundry technique. Somewhat greater variations can often be handled through the proper application of chills or other special handling, *Fig.* 43.10. It has been found that ferrous chills of sufficient size will effect a solidification or freezing rate in steel approximately $2^1/2$ times that in stand only. Thus a heavy section away from a riser can be properly fed through a thinner section providing the section is not over $2^1/2$ times the thickness of the connecting section. Chills generally cannot be used in iron casting, but the foregoing relationships with most metals other than steel would not differ greatly.

Section Thickness. Wherever applicable, good practice dictates use of the minimum section thicknesses compatible with the design. However, the lightest wall which can be successfully cast depends upon the design intricacy, size and metal to be used. The normal minimum section recommended for various metals is shown in TABLE 43-III. Section thickness must be proportioned to the area as well

TABLE 43-III — **Minimum Sections for Sand Castings**
(inches)

Metal	Section
Aluminum alloys	$\frac{3}{16}$
Copper alloys	$\frac{3}{32}$
Gray irons	$\frac{1}{8}$
Magnesium alloys	$\frac{5}{32}$
Malleable irons	$\frac{1}{8}$
Steels	$\frac{1}{4}$*
White irons	$\frac{1}{8}$

* For sections longer than 12 inches, minimum thickness should be increased gradually to 1 inch at 150 inches length, tapering off to $1\frac{3}{8}$ inch at 300 inches.

as to the adjoining sections as previously noted.

Wall thicknesses down to 3/16-inch can be poured in steel where size is relatively small and 1/4-inch walls are cast regularly on small and medium-size parts. For highly stressed or pressuretight parts, however, wall thickness must be about double these values. With large castings, walls of 9/16-inch are required to assure free flow of metal in the mold. Sections up to as much as 4 feet thick have been cast in steel.

Gray iron can generally be cast into sections as light as 1/8-inch and remain gray. High-tensile cast iron cast into sections under 1/4-inch, however, many become hard and white at corners and edges.

Malleable iron can be cast in sections as thin as 1/8-inch and less in part sizes up to about 2 inches. Generally, however, sections from 3/16-inch to 1/4-inch which are properly ribbed for strength have been found most practical. Sections of 1/2 to 3/4-inch are most practical for castings of medium size. Large, flat areas will tend to sag in the mold and consequently 3/16-inch sections are restricted to areas up to approximately 8 by 8 inches. Larger areas require progressively greater thicknesses to suit. The maximum section practical in malleable is about 4 inches.

With the nonferrous materials, section varies with the alloy, size, service pressure, etc. In brasses and bronzes the minimum section is generally set at 3/32-inch to avoid the excessive temperatures necessary to pour lighter walls. In aluminum sand-casting alloys, 3/16-inch minimum wall is required for the majority of small and medium-size castings although walls as light as 1/8-inch can be poured where size is not excessive. The usual minimum wall thickness in magnesium alloys is 5/32-inch although under favorable conditions sections as light as 1/8-inch can be

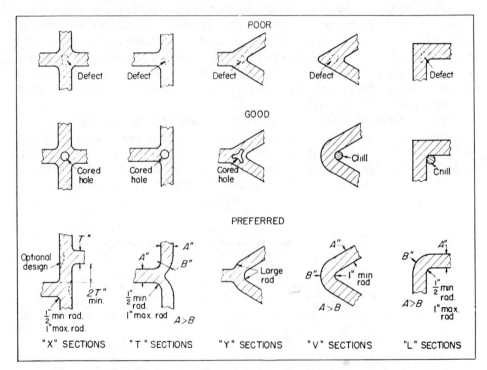

Fig. 43.14 — Design of unfed junctions in steel castings are particularly important. Heavy sections where walls join are critical regarding adequate feeding to eliminate shrinkage voids. Drawing, courtesy Steel Founders' Society of America.

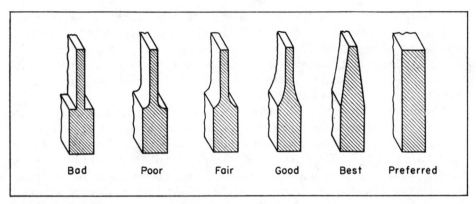

| Bad | Poor | Fair | Good | Best | Preferred |

Fig. 43.15 — Recommended design for blending of unequal sections to be joined. Drawing, courtesy Steel Founders' Society of America.

cast. Thin sections are extremely difficult to cast in large areas, especially if the area is in a horizontal position while pouring. Sections in excess of 2 inches can be cast in magnesium with good mechanical properties.

Section Junctions. Adjoining or connecting sections of a part are generally far more important and critical in a casting than are the section proportions. Where features of a design come together to create a heavy section, a hot spot, as previously mentioned, is normally created. If proper feeding can be assured, only the solid contractional stresses need be considered. However, if away from the feeders the same problem as with section variation occurs — shrinkage voids within the larger and slower cooling area. Because steel is the most critical in this respect it can usually be assumed that a design satisfactory for steel will be generally satisfactory for other metals, *Fig.* 43.14.

When member junctions cannot be fed directly, it is preferable to maintain substantially uniform walls throughout. For instance, groups of strengthening ribs should not converge into a single "X" intersection, *Fig.* 43.14. All ribs should be staggered or displaced so as to maintain a normal wall between junctions. Where very heavy section junctions are necessary, coring can sometimes be used to maintain uniformity, *Fig.* 43.14, although this practice increases the cost.

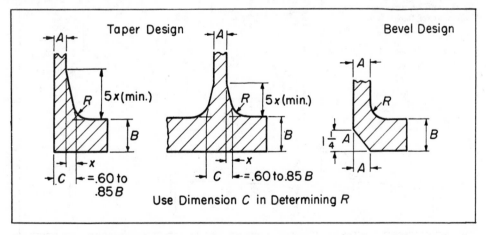

Fig. 43.16 — Recommended design of junctions for malleable iron where one section is over 1.66 greater than the other. Variations in adjoining walls over 2.5 times are not advisable. Drawing, courtesy Malleable Founders' Society.

Where sections of differing thickness must be joined it is generally advantageous to blend in the lighter section by a gradual taper, *Fig.* 43.15. In steels where the variation exceeds 33 1/3 per cent, a 15-degree taper is recommended. For smaller variations a radius can be used satisfactorily. Recommendations for malleable iron junctions are shown in *Fig.* 43.16, and those for aluminum or magnesium alloys are shown in *Fig.* 43.17.

Corners and Fillets. Sharp corners exhibit considerable lag in cooling and are potential points of high stress with subsequent rupture being the frequent result. Adequate fillets at all section junctions are therefore a "must" for strength and soundness, *Fig.* 43.18. The fillet radius should generally be equal to the section thickness and, in joining unequal sections, it should be based on the mean dimension of the two walls. Seldom should the radius be greater than the section — too great a radius may develop a hot spot of metal. Suggested radii of fillets and corners for aluminum or magnesium alloys are shown in *Fig.* 43.17, and for steel in *Fig.* 43.14.

Bosses and Lugs. Design features such as large bosses and lugs usually constitute heavy metal sections which demand adequate feeding for soundness, *Fig.* 43.19. In general, raised bosses or pads should be eliminated for ease in molding or should be blended into the surface to improve the casting, *Fig.* 43.20. Where bosses interfere with the removal of the pattern from the mold, metal should be added to permit withdrawal and eliminate the need for loose pattern pieces, *Fig.* 43.21. The primary consideration should be that the pattern can be readily withdrawn, *Fig.* 43.22. Where cores can be eliminated, design appearance and cost usually are

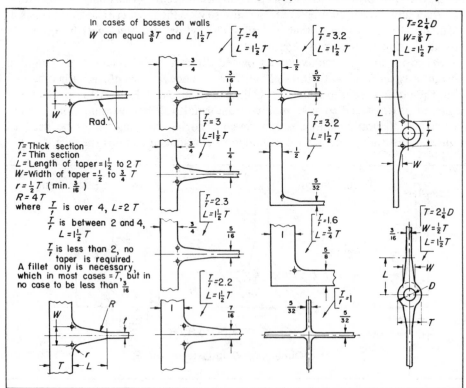

Fig. 43.17 — *Design recommendations for section junctions in magnesium alloy sand castings. Drawing, courtesy Hills-McCanna Co.*

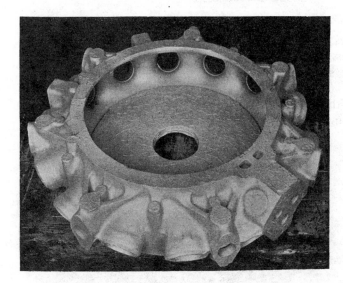

Fig. 43.18 — Magnesium alloy supercharger front housing showing junction design, fillets and smooth flowing curves.

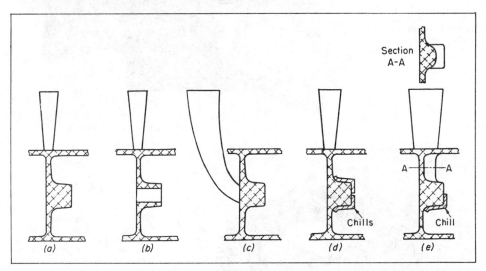

Fig. 43.19 — Heavy round boss in a thin wall (a) is inaccessible to riser and will develop microshrinkage. Coring at (b) helps but is not always possible; horn risers (c) produce sound metal but are costly even if feasible; iron chills (d) reduce defects but design with feed channel as at (e) is best. Drawing, courtesy American Magnesium Corp.

Fig. 43.20 — Bosses and pads which increase metal thickness should be blended into the casting wall as shown at (b) rather than as at (a) to improve metal flow. Drawing, courtesy Meehanite Metal Corp.

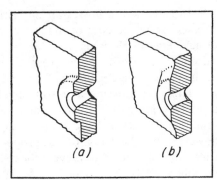

improved.

Large isolated bosses, such as those used as hubs in gear blanks, wheels, pulleys, and other sections, cannot be poured where the connecting spokes or webs are slender. Uniform cross sections should be used in the spokes and rim with coring to reduce the internal hub section as well as the joint between the spokes and the rim. Spoke designs with an S curve or a dished solid web on circular rimmed parts will allow the cooling stresses to be relieved without hot tearing.

Fig. 43.21 — Magnesium sand casting showing a variety of bosses. The large boss, top center, and those in narrow web section are designed to be fed directly and those on top, either side, are carried on into the wall to permit withdrawal of the pattern. Photo, courtesy Permanente Metals Corp.

Fig. 43.22 — Magnesium sand casting showing raised pads on slanted surfaces properly designed for easy pattern withdrawal. Photo, courtesy Permanente Metals Corp.

In cases where bosses are in critical locations subject to severe operational stresses, as in the case of many aircraft engine castings, the use of typical boss design, *Fig. 43.23*, is undesirable. Actually, it has been found that in such cases raised bosses, *Fig. 43.24*, whether of round or odd shape, obviate all fatigue failure at the critical edge junctions formed by machined surfaces.

Ribs. Reinforcing and stiffening of cast parts in the past has largely been sought by means of added ribbing. Ribs also have been found to be useful in retaining flatness in large areas during the casting operation. It has been proved, however, that the actual design proportions of ribbing and of rib spacings must be carefully determined to prevent the resulting part from being weaker than one with unribbed walls. Ribs can be used to greatly increase the section stiffness by controlling the rib spacings. However, a slender rib adds little to the moment of inertia which determines the relative stiffness, but increases the distance to the extreme fiber and actually can decrease the section modulus, hence, the strength.

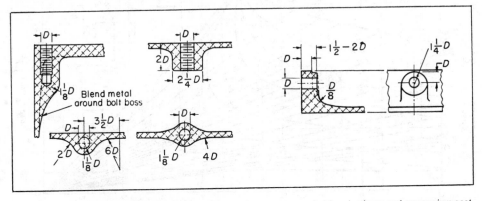

Fig. 43.23 — Typical flange and other bolt-hole boss designs recommended for aluminum and magnesium castings. Drawing, courtesy American Magnesium Corp.

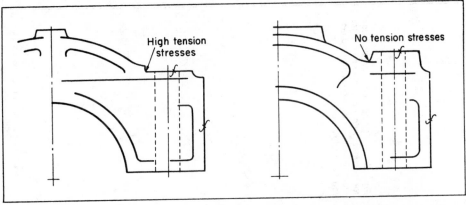

Fig. 43.24 — Design with raised bosses at points of high tension stress prevents fatigue failures which arise from machined spotface edge.

As rib height increases beyond the section thickness to which the rib is attached, stresses at the rib junction increase rapidly to a point where the ribbed design actually becomes weaker under repeated stresses than an unribbed section, especially where the rib width to rib spacing ratio is large. From the charts shown in *Fig.* 43.25, the design value of well proportioned ribs is evident. Thick, short ribs prove to have greater overall effectiveness and at the same time also can be cast with much greater ease. Generous fillets at the rib junctions, *Fig.* 43.26, improve design considerably. The rib designs shown in *Fig.* 43.27 have been found to be a distinct improvement in highly stressed structures such as aircraft engine components.

Conventional types of ribs should be governed by the rules regarding uniformity of section including the staggering of ribbing to avoid any possibility of massive junctions. Where such ribs are required by design they should be flat topped or beaded, rather than round topped, to reduce outer fiber stresses and minimize cracking, *Fig.* 43.28. The usual draft angle should be applied to ribs, especially for those in a vertical position in the mold and the base should be

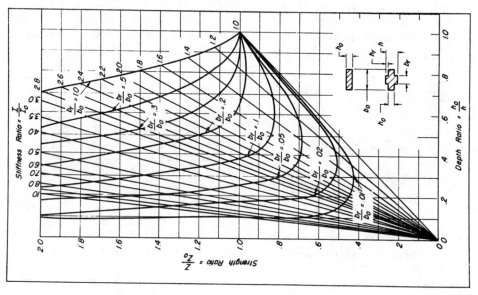

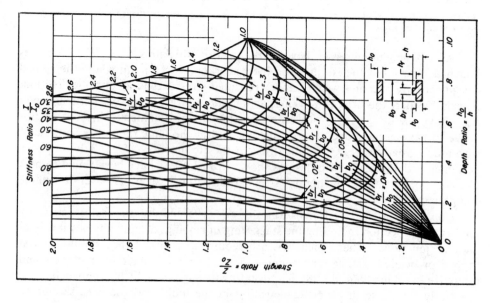

Fig. 43.25 — Strength and stiffness ratios for sections with ribs on one or both sides. Dimension b_o represents rib spacing for multiple ribs.

From Randich, MACHINE DESIGN, Sept. 1949.

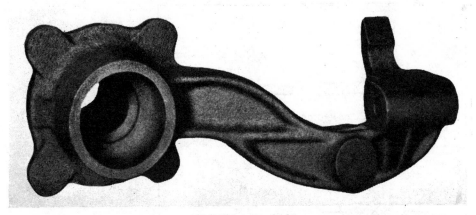

Fig. 43.26 — Rear axle support and spring bracket cast in CWC 74 steel alloy showing flowing ribs and generous fillets.

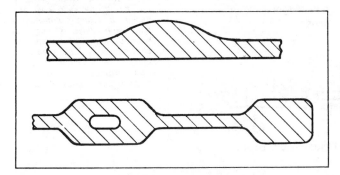

Fig. 43.27 — Low flowing wide ribs in aluminum and magnesium aircraft engine structures provide stronger, stiffer, lighter design easier to cast and clean.

blended-in, as previously noted. Bead-top ribs are more efficient than plain ones but increase cost considerably owing to the coring necessary. Bead design for malleable iron is shown in *Fig.* 43.28. Generally, designing to place ribs in compression rather than tension is most satisfactory and helps eliminate fatigue cracking.

Beads. Beaded design is often desirable to secure stiffness at the edge of holes, particularly in light sections. Such beads, however, should not interfere with drawing of the pattern from the mold or cost may increase considerably. Beads at the parting line and those on the upper or lower mold surfaces are most easily handled. Bead design for aluminum, magnesium and for malleable is shown in *Fig.* 43.28.

Holes. Small holes are normally not cored unless necessary for proper maintenance of wall section; in most materials drilling is the usual procedure but in malleable iron, holes which do not exceed the metal thickness can be cold punched if desired. Medium and large-size holes are readily cored and may often be useful for internal core support or core sand cleanout. Core diameter for such holes should not be less than one-half the section thickness and, as a rule, the smallest cored hole which can be cast satisfactorily is approximately 1/4-inch. Medium or large-size holes which require closer tolerances than are obtainable in coring must be bored or otherwise finished to size and require the addition of proper machining allowances. If sufficient draft is allowed and holes are equal to or preferably greater than the section, green sand cores can be used with maximum economy of production.

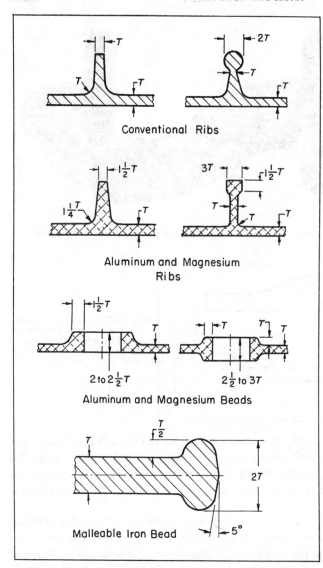

Conventional Ribs

Aluminum and Magnesium Ribs

Aluminum and Magnesium Beads

Malleable Iron Bead

Fig. 43.28 — Conventional design for ribs to be sand cast and design for casting in aluminum or magnesium alloys. All edges or corners should be rounded. Also shown are typical bead-edge designs recommended for sand-cast malleable irons and aluminum or magnesium alloys.

Pockets and Recesses. Deep pockets and recesses which complicate molding, prevent ready withdrawal from the mold, or require special coring should be eliminated unless such features are a distinct requirement. Undercut recesses should be avoided if at all possible and pockets should be of clean design with plenty of draft for easy drawing. Sharply projecting and overhanging features should be eliminated from cored sections inasmuch as they are easily broken off or washed out in pouring.

Inserts. Cast-in parts or inserts can be employed in the design of many castings. Such inserts required special foundry technique generally complicating the job of producing a sound casting and should never be utilized or called for unless no other design approach is possible. Only a mechanical bond is developed when metal is cast around an insert which may be cold or preheated to avoid shrink cracking during cooling. Consequently, to develop a good mechanical bond, special knurling, grooving or other keying features must be used on the insert piece.

Inserts of steel, cast iron or bronze are often used with magnesium castings to provide increased hardness, bearing strength or wear resistance at specific places. Steel or copper tubing can be used for ducts — oil passage in crankshafts, for instance — and steel wires or ribbons can be used to obtain shapes not possible to cast. Pole pieces of laminated steel have also been cast into magnesium housings. Copper-base inserts for magnesium or aluminum usually require a chromium plated or sprayed iron finish to prevent undesirable alloying of the metals.

Welded Design. Occasionally a specific design is found to be incompatible with practical production requirements. Although innumerable parts have been produced at considerable cost regardless of complex design, in most cases a more practical design solution is possible. One such approach to simplified production of sound steel castings is by means of welding. One difficult part can be designed as several more easily cast pieces. Regardless of the fact that final cost does not always prove to be lower, the result in sound castings which require little or no repair of defects is important in dollars-and-cents.

One such design is shown in *Fig.* 43.29. This 22,800-pound low-carbon alloy-steel casting redesigned after considerable difficulty in production and service, proved to require no rework and eliminated actual losses on unit sales. The result was two easily cast pieces in place of one impractical piece.

Cast-iron parts in a similar manner may be designed for welding or brazing. If desirable, even dissimilar metals may be welded to cast iron parts. For successful results, very careful procedures must be followed in the welding and brazing operations.

Selection of Materials

In general, almost any of the metals used in machine structures can be selected for use in a cast part on the basis of physical properties suited to the stresses and environmental conditions involved. It is necessary, however, to observe the major restrictions on size and wall thickness permissible.

Steels. Size of sand castings varies considerably with the material used. Steel castings may range in weight from but a few ounces to many tons, the largest on record being about 230 tons. Steel castings as much as 70 feet long and 10 feet wide with intricate structural design are cast regularly.

Irons. In cast irons, machine parts weighing from a fraction of an ounce to twenty tons or more are regularly produced. Likewise, in malleable iron tiny parts weighing but a fraction of an ounce are cast and the major portion of castings range up to 25 pounds. Malleable castings, however, in the hundreds of pounds have been used.

Aluminum Alloys. Aluminum sand castings usually range from a few ounces to about 500 pounds although castings weighing as much as 7000 pounds have been made.

Copper-Base Alloys. These alloys are available in a wide range of compositions and offer many desirable characteristics for various machine elements. The common copper-base sand casting alloys are shown in Chapter 68 along with general data on their field of uses.

Magnesium Alloys. Magnesium sand castings up to 10 feet in length with weight up to 1000 pounds have been produced with presently available equipment.

Fig. 43.29 — Ponderous steam turbine cylinder casting and valve chest seat as redesigned for simpler casting by utilizing a welded joint. Photos, courtesy Allis-Chalmers Mfg. Co.

Those alloys commonly employed are shown along with major characteristics in Chapter 68.

Malleable Iron. The high machinability of malleable iron along with its freedom from casting strains and unique combination of high strength, ductility and shock resistance makes it an excellent material for many machine parts. Its freedom from the majority of casting design limitations regarding intricacy and section variation often permits the production of parts otherwise uneconomical to produce.

Specification of Materials. In designating cast materials, the ASTM specifications or the minimum mechanical physical properties to which the casting must conform should be used rather than chemical compositions. Due to differences in raw materials, foundry practices, etc., different manufacturers may use varying methods for obtaining the same desired physicals. Thus with cast irons, if properties are critical, these must be specified in order to obtain the proper iron available from the wide variety of types produced. The term "cast iron" is nonspecific and therefore is only a general classification. Also, the terms "semisteel" and "high-test cast iron" are generally obsolete and should not be used.

Specification of Tolerances

In a single casting, the metal shrinkage ordinarily varies over the casting and in different sections anywhere from about $1/32$-inch to as much as $5/8$-inch from the dimensions which were provided for in the pattern. Exact reproducible tolerances for castings are for the most part a result of experience with many differing designs. Close tolerances, especially, always cannot be accurately predicted and may require some trial runs with subsequent pattern adjustments to obtain suitable results.

Generally speaking, sand castings can be produced to tolerances which are approximately one-half the maximum shrinkage allowance for the metal being cast, TABLE 43-I. Actually, the contraction of heavy castings of large size is less per foot than for small castings with light sections. Typical production tolerances applicable to sand castings produced in various materials are shown in TABLE 43-IV.

TABLE 43-IV — **Typical Tolerances for Sand Castings**

(inches per foot, plus or minus)

Metal	Tolerance*
Aluminum alloys	1/32
Beryllium copper	1/16
Cast irons	3/64
Cast steels	1/16
Copper alloys	3/32
Magnesium alloys	1/32
Malleable irons	1/32

* Tolerance indicated generally for all dimensions, including wall thickness, core shift, etc. For dimensions under 12 inches, tolerance indicated is minimum.

Surface finish on sand castings naturally is a result of the sand mix used in making the molds. The larger the part the greater the possibility for surface imperfections to creep in. The general run of sand castings range in surface roughness from about 250 to 1000 microinches, rms, and today some types of castings are being produced with surface roughness under 250 mircoinches with dimensional tolerances of plus or minus 0.005-inch per foot. These, however, are the result of shell-molding techniques and such refinement of surface and dimension should not be demanded unless the expense involved warrants the recourse to this process.

Additional Reading

[1] *Ferrous Casting Design,* Central Foundry Div., General Motors Corp., Saginaw, Mich.

[2] "Ductile Iron: An Overview of Where and How It's Used," John Schuyten, *Metal Progress,* November, 1975.

44

Permanent- Mold Casting

A S DEMAND for quality castings in production quantities increased, the attractive possibilities of metal molds brought about the development of the permanent-mold process. Though not as flexible regarding design as sand casting, the use of metal-mold casting made possible the continuous production of quantities of castings from a single mold as compared to batch production from individual sand molds. Basically, the process differs from sand casting only in the mold itself but certain inherent limitations beyond those characteristic in sand casting are present.

Metal Molds and Cores. In permanent-mold casting both metal molds and cores are used, the metal being poured into the mold cavity with the usual gravity head as in sand casting, *Fig.* 44.1. Molds are normally made of dense cast iron or Meehanite, large cores also of cast iron and small or collapsible cores of alloy steel. All necessary sprues, runners, gates, and risers must be machined into the mold and the mold cavity itself is made with the usual metal shrinkage allowances, *Fig.* 44.2. The mold is usually composed of one, two or more parts which may swing or slide for rapid operation. Whereas in sand casting the longest dimension is always placed in a horizontal position, with permanent-mold casting the longest dimension of a part is normally placed in a vertical position, *Fig.* 44.3.

Semipermanent-Mold Casting. Where metal cores are employed, it is necessary to withdraw the cores before final complete solidification has taken place to avoid shrink cracking. If castings are large or intricate, sand cores are normally necessary, especially where metal cores would be impossible to withdraw, *Fig.* 44.4. In such cases, the process is generally termed semipermanent-mold casting and those portions of the parts produced by the sand cores duplicate sand casting results. Because of the usual vertical parting line, cores must be provided with projections on the core prints to permit insertion into one-half of the mold prior to closing, *Fig.* 44.5.

Low-Pressure Permanent-Mold. Low-pressure permanent-mold casting is different from conventional PM casting in that it has fully automatic pouring and utilizes gas pressure in lieu of gravity feed of the metal. A view of the process is shown in *Fig.* 44.6. From 5 to 15 psi of air or nitrogen pressure is applied to the surface of molten metal. This fills the die cavity without turbulence and molds are heated and/or cooled so as to promote uniform cooling of the metal. On cooling, the mold opens with the casting on the moving half ejecting the casting as the mold

Fig. 44.1 — Typical arrangement of metal dies for permanent-mold casting. A magnesium casting is being removed. Photo, courtesy Dow Chemical Co.

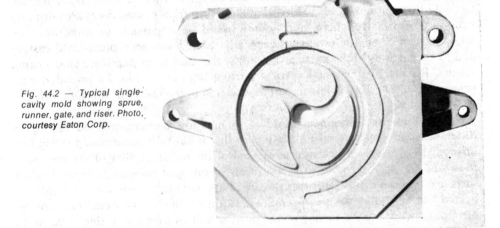

Fig. 44.2 — Typical single-cavity mold showing sprue, runner, gate, and riser. Photo, courtesy Eaton Corp.

Fig. 44.3 — Pouring permanent-mold castings from 600-pound furnaces. Photo, courtesy Wellman Bronze & Aluminum Co.

moves upward. It produces accurate, dense castings with modest die costs.

Production Quantities. Wherever quantities are in the range of 500 pieces or more, permanent-mold casting becomes competitive in cost with sand casting, and if design is simple, runs as small as 200 pieces are often economical. Production runs of 1000 pieces or greater will generally produce a favorable cost difference. High rates of production are possible and multiple-cavity dies with as many as 16 cavities can be used, *Figs.* 44.5 and 44.7. In casting gray iron in multiple molds as many as 50,000 castings per cavity are common with small parts. With larger parts

Fig. 44.4 — Gray iron permanent-mold casting for a valve body used for high-pressure oil systems showing intricate sand coring. Photo, courtesy Eaton Corp.

Fig. 44.5 — Permanent mold for the casting shown in Fig. 44.4 with cores in position. Core hangers can be seen at the top end of the sand core.

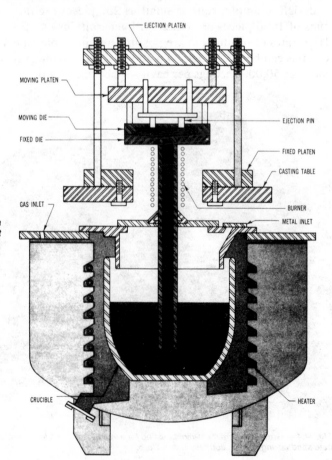

Fig. 44.6 — Sectional view of a typical low-pressure permanent mold casting machine.[1]

Fig. 44.7 — Typical multiple-cavity mold for producing iron valve bodies. Photo, courtesy Eaton Corp.

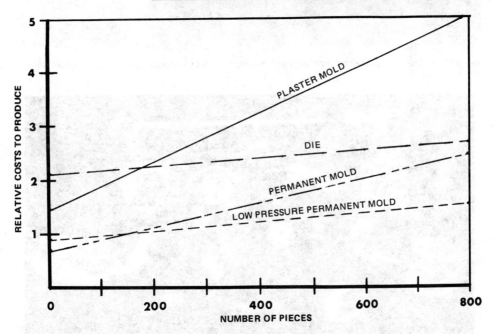

Fig. 44.8 — Cost comparison of various casting systems. At approximately 40,000 pieces die casting crosses and becomes more economical than LPPM casting. Chart, courtesy Western Electric Engineer.[2]

of gray iron, weighing from 12 to 15 pounds, single-cavity molds normally yield 2000 to 3000 pieces per mold on an average. Up to 100,000 parts per cavity or more is not uncommon with nonferrous metals, magnesium providing the longest die life. Low-pressure permanent-mold casting is economical for quantities up to 40,000 pieces, *Fig.* 44.8.

Production Speed. Small nonferrous parts of simple shape have been produced in multiple-cavity molds at speeds as high as 100 pieces per hour or more. Single-cavity molds may produce from 20 to 40 or more per hour. Exact production rate depends upon size, complexity and time required for the metal to solidify. Gray-iron parts produced in automatic molding machines require approximately two to six minutes apiece. Productionwise, permanent-mold casting rates midway between

Fig. 44.9 — Group of permanent-mold copper-base alloy castings. Intricate design and excellent surface achieved is apparent.

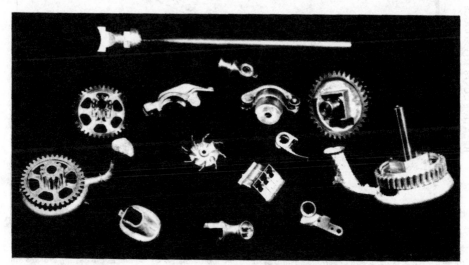

Fig. 44.10 — Group of aluminum-bronze permanent-mold castings. Two gears are shown both with and without sprue and runner attached. Two parts have steel shaft inserts. Photo, courtesy McGill Mfg. Co.

sand casting and die casting.

Sizes. Commonly, nonferrous permanent and semipermanent-mold castings may range in weight from approximately an ounce to about 100 pounds but permanent-mold castings up to 500 pounds and semipermanent-mold parts over 500 pounds have been made. Size may range up to 4 feet in diameter and four to six feet in length. Gray-iron parts may range in weight from about 8 ounces to 15 pounds.

General Advantages. Over and above the advantages in production speed and cost, permanent-mold castings offer a much smoother surface finish and accuracy than sand castings. Metallurgically, the permanent-mold chill effect produces finer

TABLE 44-I[2]

CASTING PROPERTY	DIE	PERM-ANENT MOLD	Low PRESSURE PERMANENT MOLD	SHELL	SAND	PLASTER	INVEST-MENT
Strength	B	A	A	B	B	B	B
Structural Density	C	A	A	B	C	B	B
Reproducibility	A	B	A	B	C	B	B
Pressure Tightness	C	A	A	B	C	B	B
Cost per Piece[b]	A	B	A	B	C	D	D
Production Rate	A	B	A	C	C	D	D
Flexibility as to Alloys	C	B	A	B	A	B	B
Tolerances	A	B	A	C	C	A	A
Design Flexibility	C	B	B	B	A	A	A
Size Limitation	B	B	B	B	A	B	C
Time to Obtain Tooling	B	B	B	B	A	B	B
Pattern or Mold Cost[c]	C	B	B	B	A	B	B

COMPARISON OF ALUMINUM CASTING METHODS[a]

[a]Ratings indicate relative advantage, A being best.
[b]In the case of multiple cavities in permanent or sand molds, that process may produce the most economical casting.
[c]The cost of die casting dies plus tooling for machining may be less than the same over-all cost for sand or permanent mold, machined castings. SG castings require mill/drill type machining and very little set-up time with minimal tooling.

grained metal with maximum strength, uniformity and soundness. Strength of castings generally is from 10 to 20 per cent greater than when sand cast. Less allowance for machining is required and variation from part to part is almost negligible. An overall comparison of permanent-mold casting with sand, die and plaster-mold casting can be gained from the ratings of TABLE 44-I.

Dimensional accuracy of permanent-mold casting is much better than with sand casting and many parts can be used in the as-cast condition or with only a minimum amount of finishing, *Fig.* 44.9. In addition, parts with inserts may be readily cast in nonferrous metals to provide complex parts largely complete when removed from the molds, *Fig.* 44.10.

Design Considerations

The problem of design for permanent-mold casting closely parallels the basic procedures found in successful sand casting design. Good design, however, is even more important inasmuch as the slight pattern corrections possible with sand casting are not easily incorporated into a finished permanent mold. Widely varied designs are practicable, the primary factor to be considered being that, in operation, the mold sections leave the major portions of the cast surface of a part at angles of 45 to 90 degrees to obviate mold abrasion and facilitate extraction. Secondly, design should permit the use of a straight parting line, which is *always* desirable. For economy and ease in production, design should be modified wherever necessary to permit a straight parting line in the mold.

Section Variations and Corners. Variations in section thickness should be avoided wherever possible as is normally the case in sand casting. Adjoining light and heavy sections should be gradually blended together. Sharp-corners should never be used; adequate fillets aid metal flow and help prevent shrinkage or cracking. Isolated heavy sections or bosses should be avoided unless such features occur at the parting line, *Fig.* 44.11. This is especially true of portions isolated by thin sections. It is imperative that adequate feed of metal to all sections be possible and, where necessary, heavy ribs should be added to assure full flow into bosses and other features from the risers, *Fig.* 44.12.

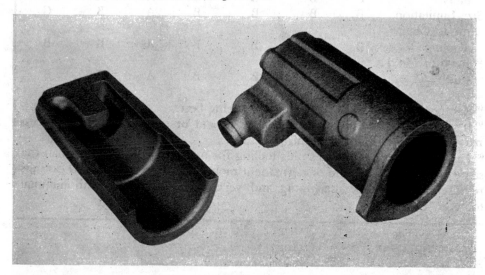

Fig. 44.11 — Gray iron hydraulic cylinder casting produced in permanent molds. Internal and external bosses occur at parting line and wall uniformity is maintained by flutes. Photo, courtesy Eaton Corp.

Section Thickness. In general, the lightest wall thickness that can be readily cast in permanent molds is 1/8-inch. Under special conditions and where size of area of wall is small, this can often be reduced to 3/32-inch with copper-base and some aluminum alloys. Light sections should never be so located that shrinkage will create cracking. The thinnest wall feasible in gray iron is about 3/16-inch. In magnesium, wall thickness of 5/32-inch, minimum, is preferred in sections over 2 inches wide and 3/16-inch for sections over 4 inches in width or 30 square inches in area. In magnesium alloys, sections from 1/8-inch to more than 2 inches in thickness

Fig. 44.12 — Blank gray iron casting for a worm gear. Ribs assure proper flow to heavy hub section and improve design.

may be cast pressuretight and to full strength. Pressuretightness may at times require some deviation from the normal minimum metal removal allowances, however, *Fig.* 44.13.

Cored Holes. Designs requiring holes not in the direction of mold parting require loose pieces, sand cores, or retractable metal cores. These requirements add to mold cost and decrease production rate.

For cored parts in semipermanent-mold casting, the minimum size of core is dependent upon its relationship to the section adjacent. Core diameter should not be less than one-half the section thickness and seldom, if ever, can be less than $1/4$-inch on the smallest parts. Slender cores are always a problem and should always be avoided.

Retractable metal cores, permanent or movable, are used to provide holes, steps, shoulders, recesses or other features, such as uniform wall thickness, where these features are not so intricate as to prevent withdrawal of the core from outside the mold. Complex coring is well to avoid and core pulling attachments on most permanent-mold machines require that movable cores be normal to the direction of mold parting.

Metal cores under $1/4$-inch in diameter are impractical but $1/4$-inch holes can, in some cases, be successfully cored in magnesium and aluminum alloys. For most average-size parts, minimum core size should be held to $3/8$-inch if possible to assure satisfactory life under the high-temperature, erosion and shrinkage stresses. Minimum core size which has been found practical with copper-base alloys is $3/8$-inch. Length of cored hole should be at least two and not over six times the core diameter depending upon the diameter, TABLE 44-II.

TABLE 44-II — **Permanent-Mold Metal Core Sizes**
(inches)

Diameter (d)	Length* (max.)
$\frac{1}{4}$ to $\frac{3}{8}$	2d
$\frac{3}{8}$ to $\frac{5}{8}$	4d
$\frac{5}{8}$ or over	6d

* Depth of cored hole

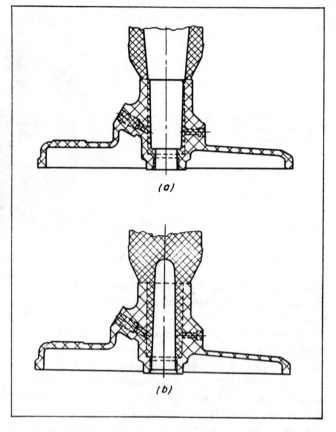

Fig. 44.13 — Heavy sections of this aluminum alloy permanent-mold casting as originally designed (a) were impossible to feed properly with resultant shrinkage and leakage. By permitting additional machining as at (b) adequate feed to the various sections is assured with a simpler core producing sound, pressure-tight metal. Photo, courtesy Permold Co.

Metal cores permit greater production speed, produce better surface quality and are much less expensive than sand cores. Consequently, all undercuts should be avoided to permit use of metal cores. Undercuts create casting difficulties, require more mold parts, result in more wear on moving parts, and require wider tolerances. Where an undercut must be utilized or is unavoidable in a cored section it should be kept free from complications with wide access openings to permit withdrawal of a collapsible core. Though more expensive than plain steel cores, collapsible steel cores are much preferred to sand cores.

Draft. Positive draft must be provided on all surfaces which do not have a natural draft to permit easy withdrawal from the mold halves. Draft requirements vary owing to the tendency for the metal to shrink away from the outer surfaces of the mold cavity and against the inner faces. Draft on outer surfaces is thus normally about 3 to 4 degrees while that for inner surfaces and cored holes should be 4 to 7 degrees. The minimum draft angle for all outer surfaces is generally set at 2 degrees with the inner at 3 but should not be used unless required and approved by the caster. Actual minimum angle may be increased or decreased by the shrinkage characteristics of the particular part, an item which can only be determined by experience. In special cases, draft as low as 1 degree per side is not unusual. Less draft is permissible on retractable metal cores, the preferred being about $1\,^{1}/_{2}$ to $2\,^{1}/_{2}$ degrees per side and the minimum, $^{1}/_{2}$ to 1 degree.

Ribs. Design of webs and ribs for strength or stiffness should closely follow the precepts outlined in Chapter 43 covering sand casting. The tendency to over-rib castings should be avoided and to assure stiffness in the production casting, the limitations on proportions of ribs should be closely observed. In addition, this practice also simplifies the problem of casting sound, accurate parts.

Machining Allowances. Permanent-mold castings require machining of fitted surfaces much as do sand castings. However, much less metal allowance normally is required owing to closer accuracy and better surface finish. Approximately 1/32 to 1/16-inch of metal is allowed on average-size castings and this may increase to as much as 1/8-inch with the largest. With gray iron the allowance ranges from 0.040 to approximately 3/32-inch per surface. Design may affect the actual metal allowance somewhat in cases where adequate metal flow is imperative, *Fig. 44.13*, and especially where leaktight walls are necessary in nonferrous metals.

Identification Markings. Where identifying letters or numerals are to be included in the mold, such features should be kept away from positions which interfere with or affect the casting structure or mold operation. Preferably, lettering or numerals should be raised and must be 3/16-inch in height or greater. Smaller letters rapidly become illegible owing to wear. Any points required for locating the casting for machining or for checking should be kept widely spaced and it should be determined by consulting the caster that such positions will not interfere with the location of gates and risers.

Inserts. Use of cast-in inserts is often of considerable value in design. Permanent-mold castings can be made with a wide variety of insert designs. Heating elements, fasteners, shafts, and similar features are common. Ordinarily, as in other methods of casting, inserts require some mechanical anchoring, *Fig. 44.14*. Knurls, grooves, keys or special lugs are generally employed.

Designers must recognize the shrinkage stresses which may result with inserts. Sufficient metal must be provided around inserts to prevent cracking from contraction of the metal on cooling. In some cases, inserts may be preheated to prevent problems. Other inserts, such as rings, may be slotted where contraction stresses must be held to a minimum.

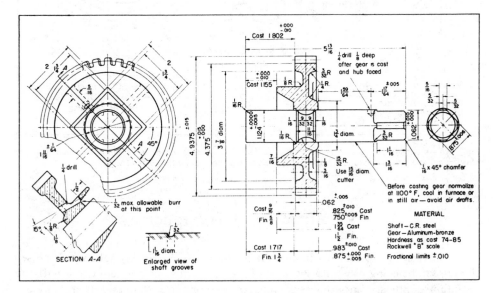

Fig. 44.14 — Aluminum-bronze worm gear permanent-mold cast onto a steel shaft. Grooves and key cuts are used to retain the shaft against rotation. Cast teeth require only a light hobbing cut.

Selection of Materials

Permanent-mold casting has been most widely applied to aluminum and magnesium alloys but gray iron and copper-base alloys are gradually gaining favor in this direction.

Aluminum Alloys. Widely employed in permanent-mold castings, these alloys provide numerous excellent characteristics and are eminently suited to this method. Aluminum alloys normally used for permanent-mold castings are covered in Chapter 68 along with pertinent data regarding their principal characteristics and uses. Parts up to 500 pounds have been cast.

Copper-base alloys for casting are given in Chapter 68. Aluminum bronze however, is most often used for permanent-mold jobs owing to better fluidity and casting characteristics, *Fig.* 44.15. Strong, sound and hard castings are secured but the high shrinkage (2 per cent) is hard on dies and cores.

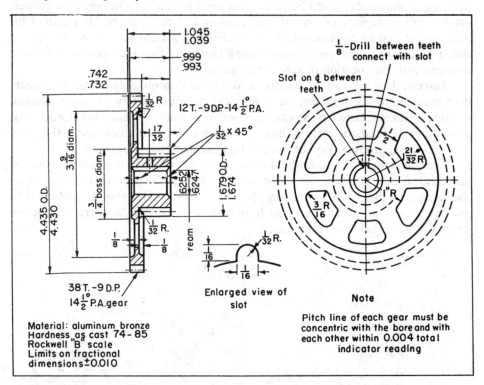

Fig. 44.15 — Permanent-mold cast aluminum-bronze gear for a stoker transmission. Teeth are finished by a light shaving cut and only the bore and side faces are machined. Photo, courtesy McGill Mfg. Co.

Specification of Tolerances

Dimensional tolerances in permanent-mold work can be held much closer than in sand casting. Weight tolerance of plus or minus one per cent can usually be maintained. Surface smoothness is often equal to or better than that of pressure die castings.

For most aluminum and magnesium alloy parts produced, solid-die tolerances are figured as plus or minus 0.010-inch for the first inch or less, plus 0.0015-inch for each additional inch. With copper-base alloy parts tolerances are figured as plus or minus 0.010-inch for the first inch or less, plus 0.005-inch for each additional inch. Those dimensions occurring across the moving parts of a mold cannot be held to quite as close an accuracy as those in one of the machined cavities. Consequently an additional 0.010-inch is allowed for such dimensions.

On the general run of parts, tolerances on dimensions cast in the fixed cavity may run plus or minus 0.020-inch and across the parting line, plus or minus 0.030-inch. On average-size castings which are simple in design it is often practical to hold plus 0.010-inch and minus 0.020-inch on dimensions across the parting line. In the fixed-cavity dimensions, this may be reduced to plus or minus 0.010-inch, *Fig.* 44.14, without serious difficulty.

Metal shrinkage may exert considerable effect upon the accuracy of cored holes. Generally, however, straight cored holes may be held to plus 0.005-inch and minus 0.010-inch on diameter.

REFERENCES

[1] Eugene E. Sprow, "Low-Pressure Casting for High-Performance Parts," *Machine Design*, April 5, 1973.
[2] John V. Milos and Paul A. Yeisley, "Manufacturing Aluminum Castings and Extrusions for Use in SG System Submarine Cable Repeaters," *Western Electric Engineer*, July, 1975.

45

Centrifugal Casting

Obscure among methods of casting metals, the centrifugal processes in recent years have made considerable progress. Though a routine production technique for more than a quarter of a century, centrifugal casting as a modern process has evolved largely in the last 25 years owing to increasingly stringent design requirements. Seeking both improved production possibilities as well as lower cost means for manufacture of unusual machine parts resulted in recourse to centrifugal methods of casting to gain some of the obvious advantages of the static casting process while eliminating some of its shortcomings.

Centrifugal Casting Methods. As can be surmised, feeding of the molten metal in centrifugal casting is achieved through centrifugal force produced with a revolving mold. Thus, in effect, it is basically a pressure casting method as against static casting which is carried out at atmospheric pressure with the height of the gates and risers providing the only means of pressure.

In centrifugal casting of metals, generally impurities such as dirt, sand and slag, owing to lower specific gravity, are forced by differential force to the inside

Fig. 45.1 — Chromium-molybdenum alloy-steel cylinder barrel casting shown as centrifugally cast, rough machined and finished. Photo, courtesy American Cast Iron Pipe Co.

surface of the spinning mold along with any gas or air pockets. Elimination of these inclusions results in a more sound cast metal with uniform strength throughout, greater density, and decidedly better physical properties. For this reason, the process has found wide usage for pressuretight parts and components subject to high loads and stresses, *Fig.* 45.1. Physical properties for the most part compare favorably with those of forgings and in many cases permit the economical production of parts from metals which are extremely difficult or impossible to forge. In addition, the method permits the satisfactory production of unusual designs not economical by static casting methods.

True Centrifugal Casting. Best known and simplest of the centrifugal techniques is generally referred to as true centrifugal casting. In this method the mold is spun about its own axis and the inside of the casting is formed by centrifugal force acting on the metal, *Fig.* 45.2. No core is required and the interior cavity becomes a true cylinder, regardless of the exterior shape, whose diameter is determined by the volume of metal poured. The external surface need not be plain cylindrical in shape but may be modified by flanges, bands, flutes, or projections; the major requirement is that the shape be symmetrical about its central axis to avoid unbalance, *Fig.* 45.3.

With this method molds may be rotated about their own axes either in a horizontal or a vertical position. The inner surface with horizontal rotation always becomes a true cylinder but in vertical rotation the force of gravity becomes

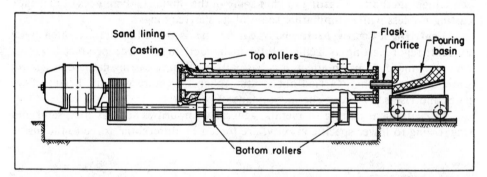

Fig. 45.2 — Horizontal type true centrifugal casting machine for long parts. Sand lining is inserted in vertical position by packing around a short plug which is raised progressively. Drawing, courtesy American Cast Iron Pipe Co.

Fig. 45.3 — Fluted tubes centrifugally cast into 16-foot lengths for use in making electric motor stators. As-cast tube is 4³/₄ inches in diameter and has a bore of 4¹/₁₆ inches. The OD tolerance is plus or minus 0.015-inch while that for the ID is plus or minus 0.08-inch. Photo, courtesy American Cast Iron Pipe Co.

effective and a paraboloid form may be produced. Where length is moderate, however, this effect is only minor. Because of the excessive speed which would be required to force metal to rise along vertical mold walls evenly where taper is undesirable, part length can seldom exceed twice the diameter.

Horizontal-Axis Castings. In general, horizontal-axis casting is chosen where part length exceeds the diameter. It is well adapted for the manufacture of short, *Fig.* 45.4, and long tubular parts such as pipe, tubing, cylinder barrels, hollow

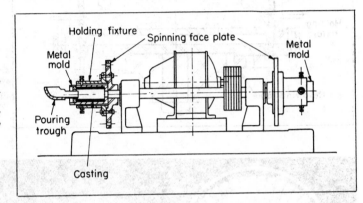

Fig. 45.4 — True centrifugal casting machine with two molds. Drawing, courtesy American Cast Iron Pipe Co.

drive shafting, sleeves, bearings, liners, rolls, shells, etc. Outside diameters of parts which can normally be produced range from approximately 2 inches to 50 inches in lengths up to 347 inches, maximum. The largest true centrifugal casting poured to date required a charge of 58,935 pounds of metal and measured $54^3/4$ OD by $48^7/16$ ID by 342 inches long ($54^1/8$ by 51 by 302 inches finished). The maximum wall thickness poured normally may range up to about 5 inches.

Vertical-Axis Castings. With vertical-axis casting, parts which have diameters exceeding their length are the usual production, *Fig.* 45.5. Diameters may range up to approximately 6 feet and lengths generally to several feet.

Semicentrifugal Casting. Almost identical to true centrifugal casting, the term semicentrifugal is applied to the process where cores are employed to modify the casting or bore produced. The vertical method is always used and stepped bores, noncylindrical bores or other cored features of wide variety in designs can be had with good production, *Fig.* 45.6. As can be readily seen, parts with small-diameter bores may have metal at the bore which is not as good as that at the periphery.

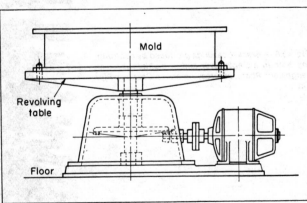

Fig. 45.5 — Floor type vertical centrifugal casting machine for large-diameter parts. Drawing, courtesy American Cast Iron Pipe Co.

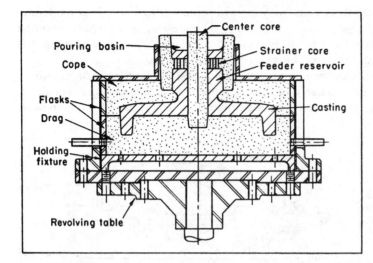

Center core

Pouring basin

Cope

Strainer core

Feeder reservoir

Flasks

Drag

Casting

Holding
fixture

Revolving table

Fig. 45.6 — Semi-
centrifugal vertical casting setup for producing a
cored flywheel. Drawing,
courtesy American Cast
Iron Pipe Co.

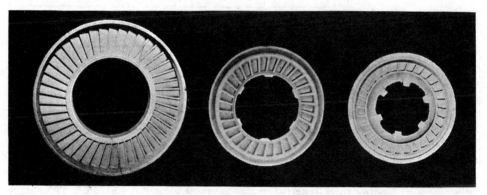

Fig. 45.7 — Group of turbosupercharger diaphragms produced by the semicentrifugal method in dry sand cores.
Photo, courtesy Lebanon Steel Foundry Co.

Fig. 45.8 — Bracket castings produced by centrifuging both in a single layer and in multiple layer arrangement. Photo, courtesy American Cast Iron Pipe
Co.

Where bores are sufficiently large, superior castings can be produced which are extremely difficult or impossible to make by other methods, *Fig.* 45.7.

Centrifuge Casting. The vertical centrifugal casting process is also used to produce nonsymmetrical castings which cannot be rotated about their own centerline, thus becoming in reality a method of pressure casting. Centrifugal force by this method is utilized to aid in forcing molten metal to fill a group of molds arranged in a single layer, or groups of molds located about a central sprue in multiple layers, *Fig.* 45.8.

The problems involved with centrifuged castings are largely those found in sand casting. Design of such parts therefore closely follows sand casting procedures. The primary advantages in centrifuging are improved metal quality, greatly increased yield and improved production by virtue of the radial-feed and stack methods, *Fig.* 45.9. In one instance, with 12 automatic vertical machines and 21 men, stacked molds having 144 cavities produced a maximum output of 28,800 castings per hour.

Molds. A variety of mold materials is used in centrifugal casting. Sand is probably the most common, especially for single pieces and large parts. Where variations in OD and flanges are necessary the cylindrical sand mold is cut away to form the desired shape. The poor thermal conductivity of sand molds, however, makes them undesirable for certain nonferrous alloys subject to segregation defects.

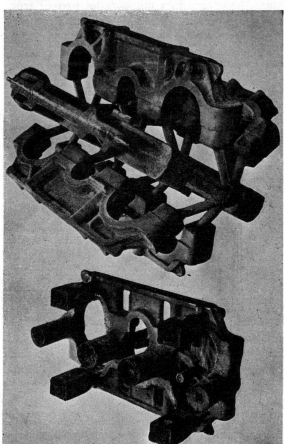

Fig. 45.9 — Centrifuged triple bearing bracket, top, and previous static casting showing greatly improved yield and decreased cleanup effected by designing for centrifugal casting. Photo, courtesy Falk Corp.

Where production quantity justifies, permanent molds are widely used. These may be of metal or graphite. Cast-iron molds are used with bronzes and brasses while steel or cast iron can be used with aluminum alloys. Graphite molds have extremely high conductivity and are exceptionally well suited for parts which require maximum soundness in heavy sections. Graphite molds are suitable only for small or medium-size runs, the average being about 100 pieces per mold. With metal molds the average run is approximately 500 pieces.

Design Considerations

Centrifugal castings, as with most production parts, may require serious consideration of the process limitations in order to obtain maximum benefit. Again, a sand casting may require some redesign to adapt it to the centrifugal method in order to obtain more uniformly sound castings. True centrifugal casting is especially worth consideration, of course, where circular or substantially circular parts are in view. Some machinery parts may be cast to finished or substantially complete form, *Fig.* 45.10.

Over and above the use of centrifuging previously mentioned, in some cases, small quantities of parts which ordinarily would be forged can be successfully cast by centrifugal methods with steel quality closely approximating that in forging, *Fig.* 45.11. Design of centrifuged castings, however, does not parallel that of true centrifugal parts. With centrifuged castings, practice should follow that in sand casting to assure best results even though pressure feeding is available.

Flanges. A variety of moderate flange designs are practical in design, *Figs.* 45.1 and 45.2. With cylindrical parts it is preferable to limit the use of external flanges to a single end of a part. Flanges, however, may be located at either end or

Fig. 45.10 — Centrifugally cast bevel gear having a net weight of 316 pounds. Photo, courtesy Falk Corp.

Fig. 45.11 — Group of large cast-steel connecting rods produced by centrifuging. Photo, courtesy Falk Corp.

in the central area. The casting of flanges at two ends has been done but difficulties arise in contraction of the metal.

Fillets and Radii. In true centrifugal castings the need for radii adjoining flanges and other design features is not as pronounced as in sand castings but a definite radius should always be specified to avoid the problems of maintaining sharp edges in production. Sharp corners, as in other casting processes, are detrimental to mold life.

External Contours. The external shape of true centrifugal castings may be almost any symmetrical shape — multisided, fluted, etc. It is important, however, that the mold be balanced where large parts are concerned and this factor should be borne in mind for such parts. Small parts, however, where the unbalance is less severe have been made with external bosses located in a variety of positions and with a variety of heights.

Internal Contours. Where internal contours of true centrifugal castings require flanges or other protrusions, it is necessary to increase wall thickness to suit and finish machine to the desired design, *Fig.* 45.12. Semicentrifugal castings may be cast with internal contours largely as desired, *Fig.* 45.13, through the use of coring.

Wall Thickness. With true centrifugal castings, wall thickness may range from approximately 1/4-inch, minimum, to 5 inches, maximum. Steel tubes under 10 feet in length and 2 feet in diameter have been cast with 1/4-inch walls. As a rule, however, wall thickness must be greater, especially where machining is

Fig. 45.12 — Rough centrifugal stainless casting with the rough machined and finish machined ring for aircraft jet engine. Photo, courtesy Cooper Alloy Foundry Co.

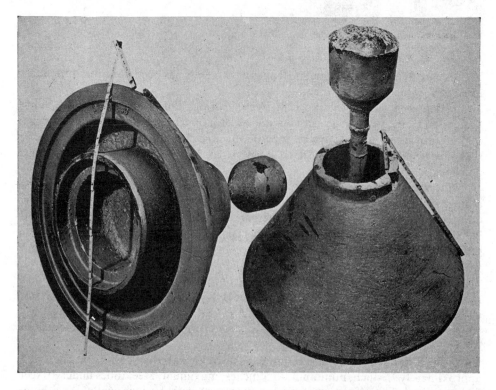

Fig. 45.13 — Tapered steel head semicentrifugally cast in sand with internal coring to produce the desired steps in the bore. Photo, courtesy Falk Corp.

contemplated. With centrifugal castings, thinner walls can be cast satisfactorily than in static sand casting.

Cored Holes. Recommended minimum size of holes or central bores produced is 1 inch. Whenever holes of smaller size are desired, drilling after casting should be employed. If cored holes are desired in solid ends, these should be located as near the axis of rotation as possible to insure proper feeding and progressive shrinkage. Where a hole does not run clear through a part, production will be facilitated by casting a through hole and plugging to suit after machining, *Fig. 45.14.*

Machining Allowances. Metal allowance for machining varies with the part design and method of casting. In sand molds the internal allowance may range from 3/16-inch to 1 1/2 inches per side depending upon the metal and the size. Ring blanks with heavy walls tend to shrink at the center of the inner bore and require an extra allowance. With permanent molds only 1/4-inch, average per side, is required. Outside allowance ranges from 5/16 to 5/8-inch on diameter. Where the surface is to be machined to a perfect finish, more metal is desirable than for cases where merely a cleanup is required. Where no fit is contemplated, as-cast finish may be satisfactory.

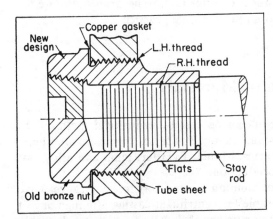

Fig. 45.14 — Bronze condenser stay rod nut as centrifugally cast at lower cost than the original design machined from bar stock. Drawing shows composite design using hollow casting. Drawing, courtesy Allis-Chalmers Mfg. Co.

Where steel parts are concerned, metal allowance for machining the inside surface for press fits and general purposes is 1/8-inch per side, minimum. The average should generally be about 20 per cent of the wall thickness per side. For sliding and running fits, 30 per cent of the wall should be allowed with 1/4-inch being a minimum per side. For external surfaces the average metal allowance is 1/8 to 3/16-inch with 1/4-inch being a maximum per side even on walls up to 5 inches thick.

Spokes and Rims. In the design of spoked wheels, gears or pulleys the problem of proper spoke size and design is important, as it also is in sand casting. Sufficient spoke area or thickness is of paramount importance to assure proper feed of metal to the rim section without premature freezing. It has been determined by extensive tests[1] that the ratio of spoke cross-sectional area to rim volume (that portion fed by one spoke) should be at least 1 to 10. The ratio of spoke thickness to rim volume, in the same design, should be at least 1 to 13.

Draft. Allowance in permanent molds for draft may sometimes be needed. With cast-iron molds under 10 inches in diameter, draft requirements increase, as diameter decreases, to as much as 1/8-inch per foot with the smaller diameters. If this amount can be doubled, mold life can be extended. Above 10 inches no draft is necessary. Sand molds for true centrifugal castings require no draft but those for centrifuged parts, of course, follow usual sand casting practice.

Selection of Materials

All alloys which can be sand cast can be cast by centrifugal methods. In general, carbon steel, alloy steels, stainless steels, brasses, bronzes, aluminum alloys, aluminum and manganese bronzes, Monel, Everdur, plain and alloy cast irons, and copper have been cast in production.

Desirable Metals. Those alloys of aluminum with short freezing ranges have been found to be generally most desirable for centrifugal casting although any of the alloys which can be otherwise cast can be handled. Alloys which are desirable to avoid wherever possible are those containing a high phosphorus content. Phosphorus causes a high degree of "wetability" and this characteristic makes removal from the mold difficult. Also desirable to avoid are high lead (20 per cent or more) alloys which have a tendency to centrifuge and, also, alloys which are "hot-short", a characteristic creating cracking.

Improvement in Properties. Most metals show a distinct improvement in mechanical properties when centrifugally cast. Properties of castings made in sand molds normally are about 10 per cent higher than those of equivalent static castings. Those made in "chill" permanent molds exhibit properties 30 to 40 per cent better considering tensile strength, elongation, reduction in area, and hardness. Yield strength, however, is much less affected.

Properties from Centrifuging. In centrifuged castings, the quality of metal is normally not quite equal to that in true centrifugal castings inasmuch as it is difficult to obtain the desired directional solidification. Added material normally is allowed at the central portions to assure best possible quality, *Fig. 45.9*. Again if large-diameter castings require small-diameter hub sections, the center portion may show distinctly lower physicals than the rim portions.

Specification of Tolerances

In the production of cast parts by the centrifugal method, the tolerances typical in sand casting largely prevail. Most parts require machining and consequently actual casting tolerances assume rather minor importance.

Surface Roughness. Surface roughness of parts formed against sand may range upward from 250 microinches, rms, as is typical with sand castings, *Fig.* 45.15.

Fig. 45.15 — Idler wheel hubs cast in horizontal-axis machines with sand molds and cores. Excellent finish is visible. Photo, courtesy American Cast Iron Pipe Co.

Those made in permanent molds usually are under 250 microinches in roughness.

General Tolerances. Metal shrinkage affects dimensional tolerances as in sand casting. Actual reproducible dimensional accuracy therefore may vary considerably. For most parts and especially semicentrifugal and centrifuged castings, the tolerances should follow those covered in Chapter 43. Typical tolerances applicable range from plus or minus $1/16$-inch to $1/8$-inch. Some true centrifugal castings can be held to plus or minus $1/32$-inch on outside diameters and uniformity of wall thickness when desired.

Once the foundry procedures have been properly adjusted and experimental testing completed, it is possible to produce some parts to even better tolerances, *Fig.* 45.3. In some instances, tolerances of plus or minus $1/64$-inch on external dimensions thus may be realized. Owing to variation in charging weight, however, inside diameter tolerances on true centrifugal castings ordinarily will be plus or minus $1/16$-inch or greater. As has been indicated, wall thickness with vertical true centrifugal castings is normally greater at the bottom than the top depending upon the speed at which the mold is rotated.

REFERENCES

1 "Centrifugal Casting of Aluminum Alloy Wheels in Sand Molds" — Northcott & Lee. *Journal of the Institute of Metals,* London, Vol. 71, 1945.
2 "Centrifugal Casting of Copper Alloy Wheels in Sand Molds" — Northcott & Lee, *Journal of the Institute of Metals,* London, Vol. 73, Part 8, 1946-47.

46

Die Casting

HISTORY of the development of modern die casting as a production process can be traced back more than a century to the first attempts at casting lead bullets before the American Revolution. The first recorded attempt at pressure casting is found in patent papers, dated 1849, covering a crude machine designed for the casting of lead alloy type. It was not until some twenty years later, however, that a machine for making general-purpose castings appeared.

In 1907 the first "gooseneck" hot-chamber die casting machine was patented by Van Wagner. Air pressure was used in this machine to force the metal into the closed die, affording the first step toward uniformity of the product. As machines were improved, the die casting of lead and tin alloys was superseded by the use of zinc alloys from about 1912. Aluminum as a die casting alloy was introduced in 1914 but received little attention until the late twenties. The die casting of copper-base alloys entered successful production in the early thirties and adoption of these alloys as well as those of aluminum resulted in the final perfection of the cold-chamber die casting process. In the past few years magnesium alloys have gained importance in this field and now rank next to the aluminum alloys. Most recently, work has been progressing with the die casting of both steels and stainless steels in

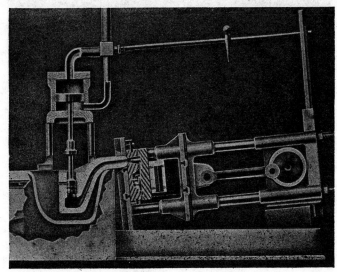

Fig. 46.1 — Hot-chamber die casting machine sectioned to show method of injection.

46-01

modified machines. However, economic production is not yet attainable.

Die Casting Methods. As mentioned, there are basically two methods for producing die castings. These have been termed *hot-chamber* and *cold-chamber* to indicate their primary difference. In general, the hot-chamber method is suited primarily for the casting of alloys having melting temperatures in the neighborhood of 1400 F or less and which do not have an affinity for iron. The cold-chamber machine can be used as a substitute for the hot-chamber machine but is especially designed for handling alloys having higher melting temperatures approaching 1800 to 1900 F, for alloys which have an affinity for iron such as aluminum, and for the casting of parts which require the highest possible density.

Hot-Chamber Machine. Earliest of the methods of die casting so far developed, hot-chamber casting includes all machines wherein the injection cylinder is immersed in a continuously heated pot of molten metal. Molten metal which enters the gooseneck or injection chamber from the pot is forced into the closed dies by direct air pressure, or by a ram which may be actuated by mechanical, pneumatic, or hydraulic means, *Fig.* 46.1.

The direct air or gooseneck machines, used to some extent for casting aluminum alloys, is limited severely in the pressure which can be applied to force the metal into the dies. Maximum air pressure on the metal is about 500 to 600 psi and this pressure is insufficient to exert any great amount of squeeze action for obtaining good density. Production rate on these machines varies according to the casting size but does not often exceed 300 shots per hour.

Plunger hot-chamber machines, *Fig.* 46.1, generally are available in capacities from 125 to over 1000 tons, and are used for casting alloys of zinc, lead, antimony, and tin. Production rate, depending upon the size of the machine, ranges up to 750 shots per hour or more.

Cold-Chamber Machine. Operation of the dies in cold-chamber casting is

Fig. 46.2 — Cold-chamber die casting machine shown in production of large electric cleaner castings. Photo, courtesy Hydraulic Press Mfg. Co.

identical to that found in the hot-chamber method. Major differences between the
two methods are in the means for injecting the metal and in the pressures used. In
cold-chamber machines, *Fig.* 46.2, the pot of melted metal is separate from the
machine and the required amount is ladled into the plunger chamber, *Fig.* 46.3,
and from thence squeezed into the die cavities. The latest machines available now
offer automatic ladling of the metal into the chamber and automatic removal of
the castings.

Capacities generally available for cold-chamber casting may range from about
200 to over 4500 tons although some modern machines are larger. In most cases,
however, increased pressure on the metal is achieved by using a smaller diameter
plunger and as a consequence, the volume of metal cast is reduced accordingly.

Cold-chamber machines are normally produced both with horizontal and
vertical injection cylinders. In the horizontal machine the metal is injected from a
horizontal cylinder, *Fig.* 46.3, whereas in the vertical machine the injecting
cylinder is upright, *Fig.* 46.4. Vertical machines available are capable of rates
ranging from 60 to 300 shots or more per hour depending upon the size of the
casting being made. Horizontal-injection models available are capable of rates up
to approximately 400 shots per hour or more. Completely automated die casting
machines perform all operations including trimming on parts weighing up to 10
pounds (zinc) and fitting an area of 25 inches by 36 inches, maximum.

Sizes of Castings. Practical range of casting size and weight depends greatly
upon the available production casting equipment. Generally, minimum casting
weight in zinc or copper-base alloys is something less than one-half ounce. Zinc die
castings weighing but 0.000004-pound (250,000 pieces per pound) are regularly

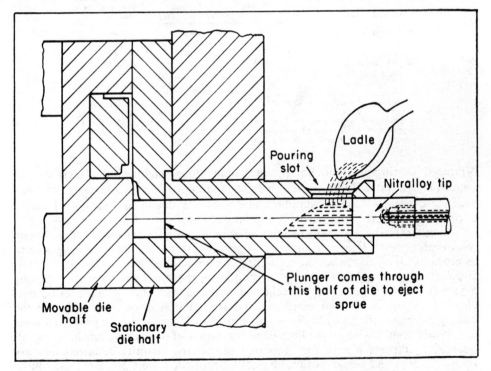

Fig. 46.3 — Cross section through dies and injection plunger of typical horizontal cold- chamber machine. Drawing, courtesy Reed Prentice Corp.

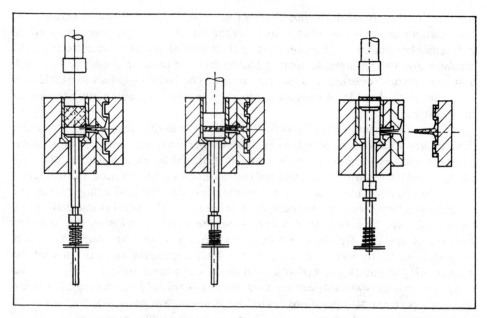

Fig. 46.4 — Cross sections showing casting sequence with typical vertical injection cold-chamber machine. Drawing, courtesy Hydropress Inc.

TABLE 46-I — **Maximum Weight and Volume of Diecastings***

Material	Weight (lb. max)	Volume (cu. in., max)	Projected Area (sq. in., max)
Magnesium alloys	20	300	625
Aluminum alloys	75	750	625
Zinc alloys	70	240	800
Copper-base alloys	14†	50	250

*Typical values for standard equipment. In special cases and with special machines these values have been exceeded.

†Standard machines are capable of injecting up to 30 pounds of copper-base alloys but practical considerations such as hand ladling and heat dispersal with such large volumes often make their use uneconomical and sometimes impractical. Certain manufacturers do not consider that more than 14 pounds in copper-base alloys can be economically die cast until improved die steels are available.

produced. Castings in aluminum or magnesium alloys weighing but a fraction of an ounce are also possible. Maximum weights and approximate volumes of castings which can be produced on standard machines are shown in TABLE 46-I which covers the common categories of die-casting alloys.

In general, size of average production diecastings seldom exceeds 24 inches on any one dimension. However, the free die space on standard machines ranges up to as much as 50 by 65 inches on cold-chamber units and 29 by 40 inches for hot-chamber units. Height of castings is limited by the maximum free die opening available which ranges up to 30 inches on cold-chamber and 18 1/2 inches on hot-chamber machines. Special machines and special dies have made possible the casting of parts up to 84 inches in length and 1200 square inches in projected area.

Production. Owing to the high-capacity output of die casting machines, small quantities of parts are seldom economical. Where quantity requirements are relatively high and design is suitable for the process, lowest possible cost can be achieved. Practical production quantities which will justify use of die casting usually range from 1000 to 5000 pieces upward, depending upon the particular design of the part to be cast and the metal to be utilized.

Production cost is also tied in closely with the metal to be cast. Useful die life in casting varies with the melting point of the metal. Die life with copper-base alloys, which are cast at 1750 to 1900 F, ordinarily is expected to run from 5000 to 25,000 fillings. With alloys having melting points not exceeding 1650 F, die life is better and from 10,000 to 50,000 shots can be expected. At these temperatures, die life of 50,000 pieces or more is generally obtainable, especially with thin-wall parts. Core life with the former dies is lower, ranging from 6000 to 20,000 shots generally.

Die life when casting aluminum and magnesium alloys is much better, the lower melting point — 1065 to 1165 F — permitting 50,000 to 500,000 fillings. Zinc alloys are normally cast at 750 to 800 F and die life at this temperature is maximum — one-half to five million or more shots are possible.

Production is also determined to a degree by the type of die. Depending upon the type of part and its size, multiple-cavity dies can be used. From two to 30 or more cavities can be filled at one shot.

Design Considerations

To achieve the maximum benefit from die casting methods, it is necessary to observe certain rules which avoid costly or impossible die designs. Because permanent steel dies are used, it is obvious that parts must follow many of the same characteristic design limitations found in permanent-mold casting.

Design possibilities with die casting are almost unlimited, parts produced by these methods since their inception are too numerous to mention. Typical of

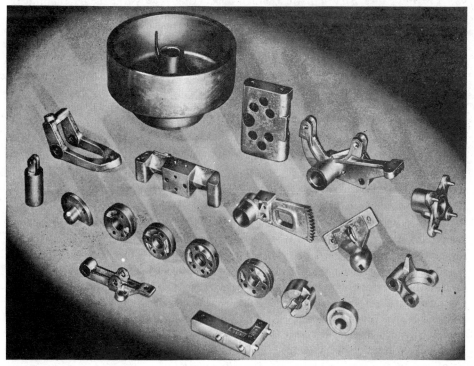

Fig. 46.5 — Group of typical zinc diecastings. These parts form the mechanism for a commercial automatic record player. Photo, courtesy New Jersey Zinc Co.

present-day use of die-cast components are the parts shown in *Fig.* 46.5. As can be readily recognized, lowest possible die cost and consequently part cost are obtainable by adhering to maximum simplicity in design. The greater the quantity to be produced, however, the greater are the opportunities to utilize complex die designs to cost advantage in eliminating machining and other subsequent production operations, *Fig.* 46.6.

Fig. 46.6 — Group of aluminum and magnesium diecastings showing unusually intricate designs possible. Photo, courtesy Litemetal Di-Cast Inc.

Parting Lines. To eliminate as much die sinking as possible the parting line or place where the die halves separate should be established on a single plane parallel to the surface of greatest length. This also permits ejection of the part over the shortest possible distance, a feature which facilitates production of satisfactory castings. Die partings should never be such that a feather edge is left on one half.

Establishment of the proper plane of parting may often depend upon design features, the die casting machine used, die operation, tolerances, metal soundness, etc. It is desirable to visualize the final design result to determine the problems which may result from the normal flash which occurs at the parting line.

Seldom should the parting be established at sharp-cornered or complex contours; it is much simpler to trim flash if it is of smooth uniform outline. It is especially important to avoid flash lines on flat or plain surfaces; evidences after trimming are difficult to remove. Frequently a bead or rounded edge can be used at the parting line and such design makes flash removal a simple, neat operation, *Fig.* 46.7. Only where a part can be completed in one half of a die, or where final machining operations can be utilized for flash removal should sharp-edge partings be considered.

Where close tolerances are desirable without recourse to additional finishing operations, the parting plane may be critical. It is often desirable to locate close-tolerance portions so as to be produced in the solid portion of one half of the die rather than between the two halves. An additional consideration is necessary with pieces cast into the lower half of the die on a vertical-injection machine; where strength and density are primary considerations, the parting line normally would be placed to avoid trapped air and porosity, *Fig.* 46.8.

When costly die slides and movable cores can be avoided by use of irregular or

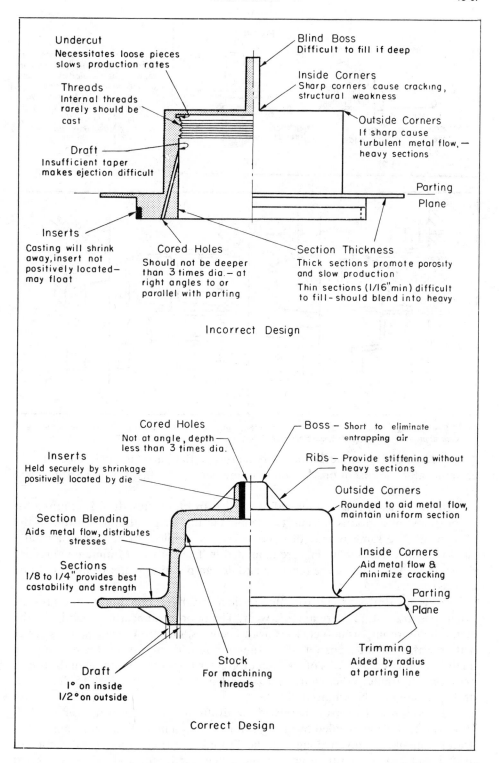

Incorrect Design

Undercut
Necessitates loose pieces
slows production rates

Threads
Internal threads
rarely should be
cast

Draft
Insufficient taper
makes ejection difficult

Inserts
Casting will shrink
away, insert not
positively located—
may float

Cored Holes
Should not be deeper
than 3 times dia.— at
right angles to or
parallel with parting

Blind Boss
Difficult to fill if deep

Inside Corners
Sharp corners cause cracking,
structural weakness

Outside Corners
If sharp cause
turbulent metal flow,—
heavy sections

Parting
Plane

Section Thickness
Thick sections promote porosity
and slow production
Thin sections (1/16"min) difficult
to fill—should blend into heavy

Correct Design

Cored Holes
Not at angle, depth
less than 3 times dia.

Inserts
Held securely by shrinkage
positively located by die

Section Blending
Aids metal flow, distributes
stresses

Sections
1/8 to 1/4" provides best
castability and strength

Draft
1° on inside
1/2° on outside

Stock
For machining
threads

Boss – Short to eliminate
entrapping air

Ribs – Provide stiffening without
heavy sections

Outside Corners
Rounded to aid metal flow,
maintain uniform section

Inside Corners
Aid metal flow &
minimize cracking

Parting
Plane

Trimming
Aided by radius
at parting line

Fig. 46.7 — Original design for a part and redesign showing important design factors to observe. Drawing, courtesy Litemetal Di-Cast Inc.

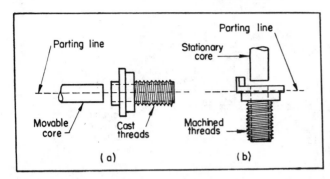

Fig. 46.8 — Movable core is necessary with the threaded part (a) whereas a stationary core can be used with (b). In vertical casting, (a) would be favored owing to denser metal as air would be entrapped with die (b). Drawing, courtesy Milwaukee Die Casting Co.

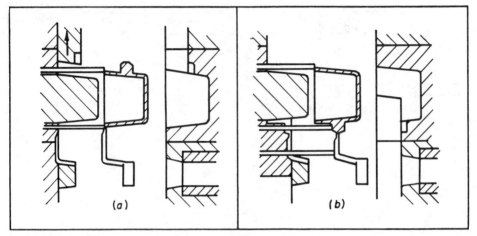

Fig. 46.9 — A change in design and parting line for this casting eliminates a slide (a) and yet permits production of a central projection (b). Drawing, courtesy Harvill Mfg. Co.

"joggled" parting lines, deviation from the rule of flat partings can often be highly economical in die cost or production time and perhaps both. Such a case is shown in *Fig.* 46.9.

Draft. To assure ready removal of castings from the dies, draft is required for all portions substantially normal to the parting line or parallel to the line of die movement. The same is true of cores or slides. Typical draft values normally employed for various materials are tabulated in TABLE 46-II. Minimum values are shown in the table but where function and design permit, greater drafts should be allowed.

Generous drafts extend die life considerably by eliminating to the greatest extent possible natural core and die wear. The poorer the bearing qualities of the metal the more important generous draft becomes. Generous draft permits easier withdrawal, eliminates possible distortion in removal and assures the best surface quality obtainable. For ribs of 1/8-inch or greater thickness, a minimum draft of 3 degrees is desirable. Where narrower ribs are necessary, as much as 15 degrees of draft per side may be required.

General practice is to merely show draft on drawings wherever needed. However, it is recommended by most producers that a note showing the amount of draft permissible be given along with the desired direction. Where the direction of draft is not critical, latitude in placement may facilitate production. Suggested drawing indication is from *minimum* allowable hole size, *Fig.* 46.10, unless design factors dictate a maximum hole limitation.

TABLE 46-II — **Approximate Draft Values for Diecastings**

(Minimum, inches per inch of length per side)

Alloy	Side Walls*	Cores†	
		Depth	Diameter
Magnesium	0.0025	0.003‡	0.0005
Aluminum	0.008 to 0.010	0.006‡	0.001
Zinc	0.002 to 0.005	0.003	0.003
Copper-base	0.015 to 0.020	0.020	0.015

* Greatest draft possible should always be used to reduce die wear, especially where metal shrinkage would interfere in withdrawal.

† For cored holes through hubs and bosses where metal shrinkage may be severe, a draft of 3 to 5 degrees or more should be allowed. Depth of hole or core length values apply per side and are minimum for cores under 1 inch in diameter. Values for diameter apply per inch of diameter over 1 inch and are added to the basic value for core length. The same depths and drafts apply to noncircular holes, dimensions being based on width or length.

‡ Holes under ¼-inch require greater draft; for instance, ⅛ to ¼-inch holes should have 0.008-inch for aluminum and 0.004-inch for magnesium per inch of length per side, minimum, and 1/10 to ⅛-inch diameters, 0.010-inch for aluminum and 0.005-inch for magnesium per inch of length per side, minimum. Holes of maximum depth require at least 0.025-inch per inch of depth in magnesium and double the amount for aluminum.

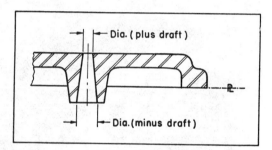

Fig. 46.10 — Method for indicating direction in which draft is to be used. Drawing, courtesy Newton-New Haven Co.

Wall Sections. As is generally true in the design of all castings, wall sections should be as uniform as possible. Uniformity of wall thickness is extremely important and should be adhered to closely in order to avoid undue shrinkage variations and warpage. Where walls must vary somewhat, gradual blending should be used to make satisfactory transitions. A good rule is to make the length of tapered section greater than four times the difference in wall thicknesses. To permit the most rapid cooling and, hence, casting cycle the lightest wall feasible is generally recommended. In general, the walls of diecastings should fall within the practical range of ¹/₁₆-inch to ⁵/₁₆-inch. Extremely heavy walls ordinary result in voids and porosity. The limits of wall dimensions generally practicable in the die casting of various alloys are shown in TABLE 46-III.

TABLE 46-III — **Section Limits for Diecastings**

(inches)

Alloy	Wall Section Thickness		
	Min Small Parts	Min Large Parts	Maximum
Magnesium	0.031*	0.062	0.500†
Aluminum	0.040	0.080	0.500†
Zinc	0.015	0.050	0.500‡
Copper-base	0.031	0.062	0.500

* Where area to be cast does not exceed 10 sq in. Usual minimum practical wall is set at 0.040-inch. Practical minimum section thickness for walls can be roughly estimated as approximately one-thousandth the projected area to be cast.

† Sections greater than 0.150-inch are rarely required.

‡ Sections up to 1½ inches have been cast and are satisfactory where some voids are permissible. However, walls over ½-inch are rarely economical.

Where concentrations of metal such as bosses or heavy ribs are utilized, the unequal shrinkage compared with that of adjacent thin walls results in some shrinkage on the wall surface. These shrink marks or "shadow-hollows" may be undesirable and in cases where the metal volume cannot be reduced, the shrinkage can often be masked by low ribbing or low relief designs. As a rule, shrink marks seldom occur with walls over 0.100-inch in thickness.

Large, plain flat walls should be avoided wherever a perfect, unblemished area is desired. To eliminate the cost of rejects from natural imperfections in such surfaces, it is recommended that they be crowned, curved, or broken up by beads, steps or low relief.

Coring. To assist in attaining uniformity of section as well as to provide holes, coring is widely employed, *Fig.* 46.11. Use of cores as "metal savers" and to provide uniform walls in a casting assures maximum economy in design and eliminates heavy portions which are inherently subject to shrink porosity which reduces overall strength.

Fig. 46.11 — Outboard motor cylinder block cold-chamber die cast from aluminum with steel cylinder wall inserts. Photo, courtesy Hydraulic Press Mfg. Co.

Fixed cores which form an integral part of the die halves and that have axes parallel to the die motion are least expensive and produce the most accurate results. It is possible, however, to utilize movable corepieces which slide on or within the dies. Large, deep holes should almost invariably be cored for maximum economy, but small holes can often be drilled or punched with equal or better economy. Holes may be designed in almost any relationship, *Fig.* 46.12, but where multiple-cavity dies are used, the number of sides from which cores can be applied is distinctly limited.

For maximum core life, the diameter-to-length ratio should be kept small. Large cores withstand the heat and mechanical abuse much longer than small ones. Generally, circular or noncircular cored blind holes can be designed to have depths up to five times the diameter in zinc, lead or tin-alloy castings and up to three times the diameter in aluminum, magnesium or copper-base alloys. These depths are based upon the minimum draft allowable, TABLE 46-II, and apply to minimum-size holes. As size increases, the practical length increases gradually and

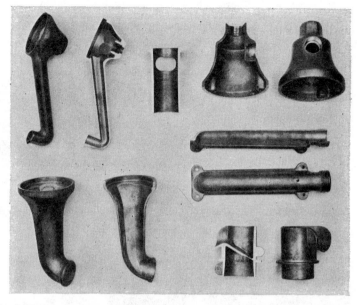

Fig. 46.12 — Group of copper base alloy parts showing deep recesses and intricate coring obtained by means of punches. Photo, courtesy Titan Metal Mfg. Co.

holes from $1/4$ to $1/2$-inch and over can be made from six to eight diameters in length. With holes over $1/4$-inch where a more generous draft can be used, say 0.025-inch or more per inch per side, lengths of cored hole up to ten times diameter and greater can be produced. Where the core can be supported on both ends these increased depths are more easily and economically obtainable.

With adequate intermediate support even longer holes have been produced. For instance, in a die-cast magnesium manifold in which a series of outlet ports permitted support from side cores, a hole 0.218 to 0.299-inch in diameter by 10.750 inches long has been successfully cast in production. The approximate minimum size of core which can be used with various materials is shown in TABLE 46-IV.

TABLE 46-IV — **Minimum Core Diameters**

(inches)

Alloy	Diameter
Magnesium	$\frac{3}{32}$
Aluminum	$\frac{3}{32}$
Zinc	$\frac{3}{32}$ *
Copper-base	$\frac{3}{16}$

* On very small parts, holes of $\frac{1}{64}$-inch diameter are regularly produced.

Cores which intersect or form a joint in a part are feasible, *Fig.* 46.13, but often may form a flash which cannot be readily removed. Collapsible core-pieces can be used to produce special holes not otherwise possible to cast but are seldom economical.

Corner Radii and Fillets. Wherever possible, outside corner radii of parts should match internal fillet radii to maintain uniformity of section. If sharp outside corners must be used, a minimum radius of 0.015-inch should be specified for economy. Because it is difficult to sink a sharp corner into a die (other than those produced at the parting line), it is recommended that such corners be rounded to a

Fig. 46.13 — Cold-chamber aluminum carburetor die-casting shows intricate intersecting coring utilized. Photo, courtesy Hydraulic Press Mfg. Co.

Fig. 46.14 — Gasoline engine cylinder head showing thin, closely spaced ribs as cast requiring no machining.

$^1/_{16}$-inch radius and that the radius never be less than $^1/_{32}$-inch. The top edges of ribs and webs should be generous and a full radius is desirable wherever possible, *Fig.* 46.14.

Sharp internal corners in diecastings (external corners in dies) are detrimental. Fillets facilitate the flow of metal in the dies and lengthen the die life considerably. To avoid overlarge fillets which may create shrink voids, a general formula has been recommended: Make the radius approximately equal to one-third the sum of the two adjacent wall thicknesses. With magnesium diecastings, however, an exception exists and generally fillet radii should be equal to the sections joined.

Bosses. Since in diecasting design the lightest possible walls are desirable, it is generally necessary to provide bosses or hubs at points in the part where fasteners are to be used or mating parts positioned, *Fig.* 76.15. Minimum practical wall sections should be used and where additional strength or rigidity is desired, reinforcing ribs can be employed together with bosses. Such ribs also facilitate metal flow to the bosses. Wherever possible, all bosses should be on the same side of the casting to aid adherence to one die half for ejection.

Where bosses must be inclined at an angle to the parting line or major surface of a part, design should be such as to allow forming by the solid die. If this

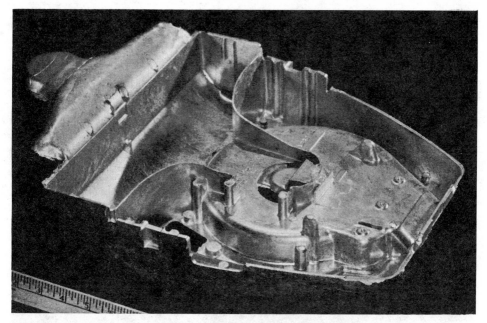

Fig. 46.15 — Intricate vacuum cleaner aluminum casting weighing about five pounds. Thin walls, bosses, drill-points, and ejector locations are easily discernible. Photo, courtesy Hydraulic Press Mfg. Co.

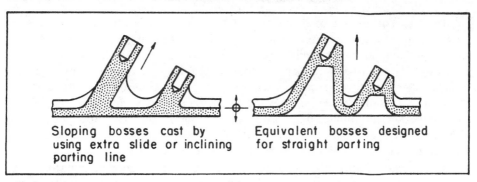

Sloping bosses cast by using extra slide or inclining parting line

Equivalent bosses designed for straight parting

Fig. 46.16 — Design modification for bosses at an angle to parting line. Drawing, courtesy Aluminum Co. of America.

procedure creates a boss with a large base, the major portion should be hollowed out to maintain a uniform wall throughout, *Fig.* 76.16.

To accommodate fasteners, bosses may be tapped from a cored hole. Boss and hole location will be accurate and the usual countersink for tapping can be incorporated with the core. Where cores are not advisable or practical, it is possible to provide drillpoints where such are deemed desirable.

Recesses. All design innovations which will eliminate any recessed or undercut portions in the direction of die or core travel should be explored thoroughly prior to use of such details. Undercut and recessed portions can be formed only with expensive dies and as a result of complexity, useful die life is also shorter. Though recesses can be formed, economy is best served by design to eliminate them wherever possible.

Where external undercuts are at right angles to the parting line, a stationary core in the die can often be used by judicious selection of a parting line. Where the recess interferes with parting or falls at right angles to the line of die travel, a slide or core is usually required, *Fig.* 46.9. Internal undercuts are most difficult and normally necessitate the use of loose pieces, "knockouts" or collapsible cores. These should be avoided if at all possible and ordinarily are feasible only where quantity permits the additional die cost.

The most economical method for obtaining internal reliefs or grooves without resorting to expensive collapsing punches is by use of inserts. Die-cast rings of proper design can be slipped onto the punch and cast into place, *Fig.* 46.17. Owing to the chill effect from the punch, the insert does not adhere and can be machined out in a simple operation.

Fig. 46.17 — Section through a flight instrument housing shows insert of same alloy as housing die cast in place, left. Right, assembly shows smooth internal groove obtained by machining out the insert.

Inserts. The use of cast-in inserts extends also to a variety of machined, formed, stamped, or cast pieces which are to become an integral part of a design. An adequate amount of material must be provided around the insert for support and design should usually be such that solidification contraction of the metal exerts the desired locking force. Heavy knurls, holes, grooves, slots, or other grip surfaces are desirable. Studs should be designed to have a shoulder surface slightly above the cast metal to eliminate edge flash from entering the threads and the possibility of pulling out under load. Wherever possible a flange shoulder should be used on the stud to permit sealing off the metal and to add strength.

Often, where coring is impossible, special diecastings, tubes, etc., may be cast-in to achieve the desired design result. Where inserts are to be used, care in the selection of material combinations must be exercised to prevent galvanic action.

Threads. Plain studs may often be cast onto a part for fastening purposes. Generally, these should be left plain and riveted or gripped with a fastener which requires no threads. Threaded cast studs are generally weak. External threads under 3/4-inch as a rule must be machine threaded, cast threads rarely being economical. Generally, where threads are coarser than 24 pitch and over 3/4-inch in pitch diameter, casting is feasible. However, where parts are small, threads up to 40 pitch have been successfully cast in zinc alloys. Where external threads can be cast split across a parting line, cost is not increased but *is* where a loose piece must be inserted and removed from each casting to form a flashless thread.

Where threads need not be over the full diameter, casting is highly economical and two or three flat sides can be employed, *Fig.* 46.18. The sides are normally produced with draft; flash removal is a simple operation and the resulting surface generally does not contact the mating thread.

Markings. Wherever any type of markings are desired on diecastings the specification should be for raised design. Depth of markings should not exceed their

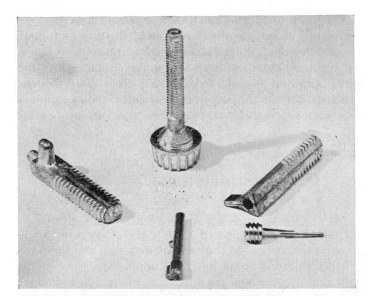

Fig. 46.18 — Small threaded parts designed with flats to simplify casting. Photo, courtesy Gries Reproducer Corp.

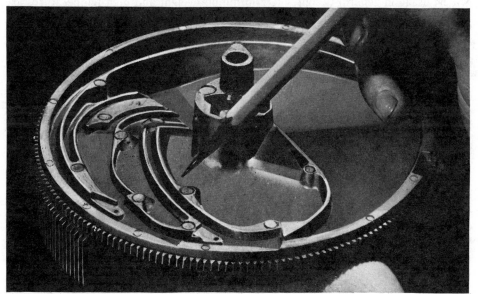

Fig. 46.19 — Complex record changer drive cam die cast in zinc alloy showing ejector spots employed. Photo, courtesy New Jersey Zinc Co.

face width, a ratio of 1 to 1.5 in depth to face width being an economical maximum. Depressed lettering on the casting requires exceedingly expensive die work. Lettering, of course, should occur only on surfaces parallel to that of the parting line to avoid special slides for ejection. Draft angle for lettering or markings should be 20 degrees or greater.

Ejector Locations. A variety of means are used to permit rapid, positive ejection of castings from the die cavities. Most commonly used are ejector pins, *Fig. 46.9.* Although the designer will seldom know where ejectors are to be used, it should be kept in mind that wherever such pins occur in the die face a flash will result. Any surface which must be free from such marks should be so indicated. Occasionally, special ejector lugs are cast onto a part to permit removal without marking or deforming. The result of ejectors can be seen in *Fig.* 46.19.

Pressure-Mold Castings. The design of pressure-mold diecastings follows closely that of ordinary commercial castings except that uniformity of section is imperative. Design must avoid any details which induce undesired turbulence and general design must be as simple as possible. Conditions such as sharp corners, raised lettering, cast-in inserts, cores in the path of the gate entry, and cast threads except opposite the gate are not acceptable. Maximum section is 3/16-inch, except in special cases.

Selection of Materials

Today, die casting permits the use of a fairly wide range of basic metals and their alloys. As die materials have been improved, the higher melting point metals have become more and more economical to utilize. Although lead and tin alloys cast well, they seldom find application in machines and appliances.

Zinc alloys. Modern zinc alloys, which are die cast at 750 to 800 F, find the widest usage in general design where minimum weight and high-temperatures are not important factors. Strength of zinc alloys parallels that of the other die-casting alloys excepting copper-base alloys. Zinc alloys can be readily finished by almost any commercial method. Use of zinc-alloy castings at temperatures above 300 F is not recommended.

Aluminum and Magnesium Alloys. Owing to higher melting temperatures, aluminum, magnesium and copper-base alloys are more expensive to cast than zinc alloys, generally in this sequence. Aluminum and magnesium alloys are very stable and most often are selected for low-weight applications, for cases where temperatures over 300 F are encountered, or for corrosion resistance.

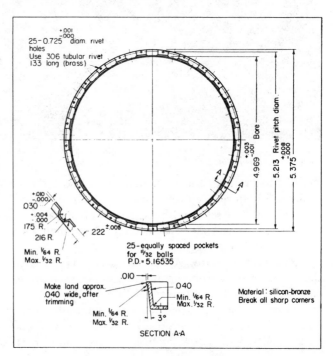

Fig. 46.20 — Unusual ball retainer for an antrifriction bearing die cast in silicon-bronze alloy. Drawing, courtesy McGill Mfg. Co.

Copper-Base Alloys. Copper-base alloys offer maximum-strength parts but have the adverse characteristic of a high melting point. Die life is short and surface quality of castings is not as good as with the other alloys. Owing to the fact that less overall heat is dispersed through the die, castings with average or light walls are most economical to cast, *Fig.* 46.20. Copper-base alloys are utilized primarily for castings where strength, hardness, or corrosion resistance are important in design. Ordinary die-casting brasses or bronzes afford tensile strengths ranging up to 85,000 psi. The aluminum bronzes increase this range to around 105,000 psi. The silicon bronzes, though lower in physicals, have the highest castability and are used where thin sections must be filled.

Comparative data in TABLE 46-V are presented in order to show the general factors for selection.

TABLE 46-V — Selector for Die Casting Alloys*

Selection Factor	Aluminum Alloys ASTM Nos. 360, 380, 413 and 518	Copper Alloys (Brass) ASTM Nos. 858,878,879	Magnesium Alloys ASTM No. AZ91B	Zinc Alloys ASTM Nos. AG40A. AC41A
Mechanical Properties				
Tensile strength	2	1 (strongest)	3	2
Impact strength	3	1 (toughest)	3	2
Elongation	3	1 (most ductile)	3	2
Dimensional stability	2	1 (most stable)	3	3a
Resistance to cold flow	2	1 (most resistant)	2	3
Brinell hardness	3	1 (hardest)	4	2
Physical Constants				
Electrical conductivity	1 (highest)	2	3	2
Thermal conductivity	1 (highest)	2	3	2
Melting pointb	2	3 (highest)	2	1 (lowest)
Weight, per cu. in.	2	4	1 (lightest)	3
Casting Characteristics				
Ease, speed of casting	2	3	2	1 (easiest)
Maximum feasible size	1	2	1	1
Complexity of shape	1	2	1	1
Dimensional accuracy	2	3	2	1 (most accurate)
Minimum section thickness ..	2	3	2	1 (thinnest)
Surface smoothness	2	3	2	1 (smoothest)
Cost				
Die costc	2	3	2	1 (lowest)
Cost per pound	2	4	1	3
Production cost	2	3	2	1 (lowest)
Machining cost	2	3	1	1
Finishing costd	3	2	3	1 (lowest)
Cost per piecee	2	3	2	1 (lowest)
Use				
Extent of application	1 (most used)	4	3	2

*Courtesy New Jersey Zinc Co.

aThrough the use of a low temperature annealing treatment, Alloy No. AG40A can be made virtually stable in dimensions.

bA low melting point is favorable in reducing die cost and upkeep and facilitates casting.

cDies for casting the low melting point alloys are least expensive and have longest life.

dIncludes polishing and buffing expense as well as ease of applying all types of commercial finishes, both electrodeposited and organic.

eBased on die, material and fuel costs, production speed and machining and finishing costs.

Specification of Tolerances

Generally, die casting offers closer dimensional tolerances than any other production casting process producing comparable results. As is true with any design, tolerances should be held to a minimum *only* on the dimensions which so

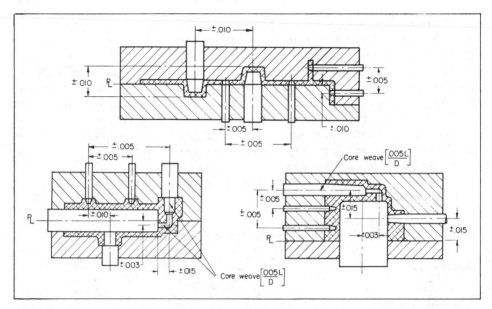

Fig. 46.21 — Diagrammatic sketch showing average production tolerances that can be maintained with aluminum die casting. Tolerances for zinc are about half these values and for copper-base alloys, about double. Drawing, courtesy Harvill Mfg. Co.

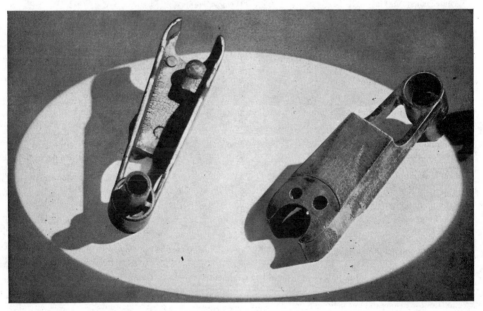

Fig. 46.22 — Copper-base casting showing effect of heat on die. Veins produced from cracking are visible.

require. Longer die life and greater economy result where generous tolerances are possible.

Noncritical Tolerances. On fractional dimensions which are not critical, a total tolerance of plus or minus 0.010-inch is generally assured on lengths up to about 6 or 7 inches or 0.0015-inch per inch over 7 inches and need not be indicated.

Commercial Tolerances. Closest tolerances can be held on portions which are formed within solid portions of a die. Dimensions across the die parting line are normally about double. Average commercial tolerances for die castings of various materials are shown in TABLE 46-VI. In *Fig.* 46.21, tolerances are depicted for various commonly encountered die arrangements. The tolerance on flatness of surfaces is generally obtained by the formula 0.0015x, where x is the longest dimension of the flat surface, in inches.

TABLE 46-VI — Diecasting Tolerances*
(Minimum, plus or minus)

Alloy	One Dimension (inch per inch)	Minimum Total (inch)
Magnesium	0.0015	0.002
Aluminum	0.0015	0.002
Zinc	0.001	0.0025
Copper-base	0.003	0.005

*Tolerances apply to dimensions formed within cavities of solid dies. Basic tolerance for the first inch is double the amounts given. Dimensions across die parting require an additional tolerance of plus 0.003 to 0.010-inch. Dimensions affected by mismatching of opposite halves require 0.010-inch added to the plus tolerance. Those affected by moving cores require 0.015-inch added to the plus tolerance.

Flash Trimming. As a rule, the flash on castings is removed in a trimming die. Commercial trimming tolerance is generally considered within 0.010-inch of the true edge.

Warpage. Some warpage occurs in die casting both on ejection from the die and on cooling. Where heat-affected metals such as magnesium are cast, warpage ordinarily does not exceed 0.005-inch per linear inch or more than 0.015-inch total in any dimension of 8 inches with a minimum to be expected of 0.005-inch.

Surface Quality. Smoothness of surfaces on die castings depends greatly upon the metal cast and especially on its melting temperature. Low-melting metals such as zinc produce castings with excellent surfaces which require little and, in some cases, no special finishing. Up through the aluminum and magnesium alloys, surface finish or roughness to be expected in commercial runs of castings ranges from 40 to approximately 100 microinches, rms. To assure the smoothest surface on all exterior portions of a casting, it is recommended that a draft of at least 5 degrees be specified.

With the higher melting point metals, surface roughness is somewhat greater owing to heat checking of the die surfaces under the enormous pressures utilized. With new dies copper-base die castings have surfaces equal to or better than that found in permanent-mold casting. However, as dies are used and heat checking results in fine cracks in the die surfaces, the metal is forced under pressure into the cracks and results in corresponding raised veins on the casting surfaces, *Fig.* 46.22. Ordinarily these veins can be removed with an average amount of polishing.

Dimensioning. Owing to the complexity created by shrinkage, critical or particular points or faces should be separately dimensioned from one datum line to permit minimum tolerances to be specified. Long dimensions, of course, require greater tolerances than short ones.

Mismatch. Dimensions which terminate in opposite die parts are affected by die and core wear which creates mismatch on occasion owing to the forces involved in casting. Generally, this is not serious but wherever possible about 0.015-inch should be allowed. Thus, all diameters on one corepiece will be concentric with

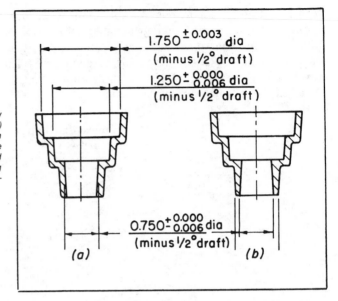

Fig. 46.23 — Control of coring by direction of draft permits part (b) to be produced with 0.750-inch hole concentric with OD. Core becomes part of solid die and eliminates core shift with long core of design (a). Drawing, courtesy Newton-New Haven Co.

each other but not necessarily with any outside diameters formed by the die cavities. Careful selection of coring by the direction of draft allowed on holes will eliminate any undesirable problems, *Fig.* 46.23.

ADDITIONAL READING

[1] J. T. Bennett, "Die Casting Stand-ins," *Machine Design*, July 24, 1975.
[2] D. C. Herrschaft, "Thin Wall Zinc Die Castings," A.S.M.E. paper 73DE15, 12 pp., April 9, 1973, Philadelphia.

47

Plaster-Mold Casting

SIMILAR in many general aspects to permanent-mold casting, the plaster-mold process represents a further step in the search for the ideal mold medium. Used many centuries ago in fine arts for producing bronze statuary, plaster as a mold material for the production of castings on a commercial scale has only recently come into use. Developed largely during the past forty years, plaster-mold castings, *Fig.* 47.1, are now produced regularly in production quantities by at

Fig. 47.1 — *A group of typical nonferrous plaster-mold castings produced in quantity. Photo, courtesy Universal Castings Corp.*

least three slightly differing techniques.

Plaster-Mold Process. In general, the various methods of plaster-mold casting are similar. The plaster, also known as gypsum or calcium sulphate, is mixed dry with other elements such as talc, sand, asbestos, and sodium silicate. To this mix is added a controlled amount of water to provide the desired permeability in the mold. The slurry which results is heated and delivered through a hose to the flasks, all surfaces of which have been sprayed with a parting compound. The plaster slurry readily fills in and around the most minute details in the highly polished brass patterns. Following filling, the molds are subjected to a short period of vibration and the slurry sets in five to ten minutes.

Molds. Molds are extracted from the flask with a vacuum head, following which drying is completed in a continuous oven, *Fig.* 47.2. Copes and drags are

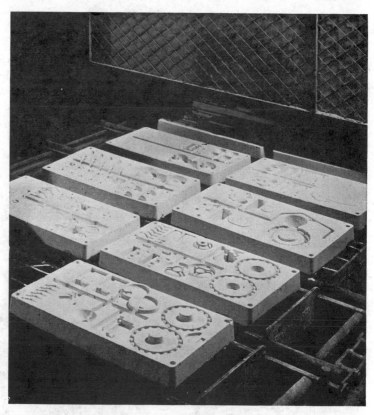

Fig. 47.2 — Finished plaster molds emerging from the continuous drying oven. Photo, courtesy Universal Castings Corp.

then assembled, with cores when required, and the castings poured. Upon solidification, the plaster is broken away and any cores used are washed out with a high-pressure jet of water.

Advantages of Plaster-Mold Casting. The most outstanding advantage in use of plaster as a mold material is that it can be varied in thermal capacity from an excellent insulator to a mild chill. Completely dehydrated plaster has no chill action and thus slow cooling of the metal along with the yielding character of the material prevents stress concentrations that might otherwise develop. Plaster molds permit accurate control of shrinkage, distortion or warpage being negligible. The surfaces produced are extremely smooth and the finest details are sharp and undistorted, *Fig.* 47.1. The self-venting and dry inert character of the mold creates

Fig. 47.3 — Flasks on the production line showing various patterns assembled into each for economical runs in the same metal. Photo, courtesy Universal Castings Corp.

little or no agitation of the metal, eliminating gas porosity and nonuniform density.

Major limitation of plaster-mold casting, however, is that only nonferrous metals may be cast. Generally, those with melting points over 1900 F cannot as yet be cast; the maximum temperature molds can withstand is about 2200 F. Experiments, however, are being carried on to adapt the process for other metals and in some cases bronze alloys with melting points as high as 2400 F have been successfully handled.

Patterns. Tooling cost or pattern cost is relatively low. The average pattern generally costs approximately $150 to $200. As the patterns are subjected only to the soft plaster being run over them, there is little abrasion, pressure or shock and consequently pattern life is practically unlimited. As changes are desired, it is usually possible to alter original patterns at low cost.

Patterns are made in split style, for cope and drag molds, and are mounted on

Fig. 47.4 — Upper left is a turbine for a torque converter mass produced. Upper right is a 10-inch aircraft wheel in which a metal saving of two pounds was made. Lower part is a 6-inch, 2-stage shrouded impeller which required critical balance accuracy for operation at 1500 fps peripheral speed and smooth internal surfaces. Vanes of impeller are held within 0.005-inch. Photo, courtesy Eclipse-Pioneer Div., Bendix.

precision machined unit strips in a standard flask, *Fig.* 47.3. Thus, numerous separate patterns can be arranged for casting simultaneously, reducing cost to a minimum. The patterns are accurately constructed and finished. Made from engraver's brass, patterns are held within a maximum variation of plus or minus 0.003-inch and are given a high polish. Because the plaster mold must be easily removable from the flask, pattern design must follow that found desirable for any casting method.

Size of Parts. Size of castings produced is of considerable importance owing to the need for a standardized production line. The standard flask size is 12 by 18 by $3^1/2$ to 4 inches in overall depth. This size is most common and permits the casting of parts singly or in multiple from the smallest desired to the largest single piece which can be molded allowing a reasonable margin of plaster over the pattern.

Straight partings are preferable and the maximum depth of pattern from the center parting line is $1^1/2$ inches or a total, for a symmetrical part, of about 3 inches. Total depth may also depend upon the parting line. Where nonsymmetrical parts with other than a straight center parting are necessary, the greatest depth from one side only of the parting line is 2 inches. Parts weighing up to 15 pounds have been produced in these standard molds.

Parts too large for a standard mold are handled in oversize molds on a special semiautomatic production line. Oversize flasks used in present operations have been standardized as 10 by 18 by $2^1/2$ inches, 12 by 21 by 6 inches, and 24 by 36 by 12 inches. In special operations, machine parts up to 35 inches in diameter and

Fig. 47.5 — Group of machine parts showing gear teeth, slots, vanes, ratchets, and cams. Photo, courtesy Universal Castings Corp.

weighing over 200 pounds have been successfully cast up to the present time.

Production. Use of plaster-mold castings is relatively unlimited by quantity requirements. Volumes running from a few hundred pieces per year to several million pieces can be produced equally well. The unit strip flask system and standardized metals make small runs and even single parts for experiment feasible. Depending upon the design complexity, production per pattern per week on the unit strip plan runs about 150 to 250 pieces and with full plate of patterns this is increased to 300 to 450 pieces per pattern per week.

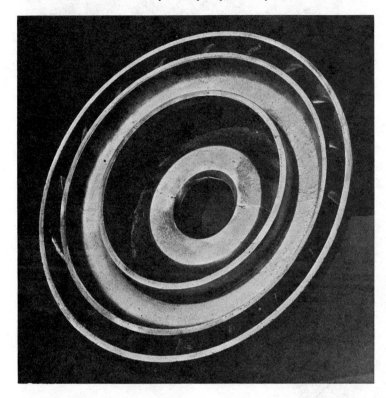

Fig. 47.6 — Cast aluminum runner for automotive torque converter with intricate passages and sharp-edge blades. Photo, courtesy Universal Castings Corp.

Design Considerations

Most important feature of the plaster-mold casting process is the practicability of producing parts to exacting dimensions so that machining, finishing, balancing, etc., are almost totally eliminated, *Fig.* 47.4. Parts having internal splines, ratchet teeth, stops, blind gear teeth, airfoil-section blades, integral rivets, etc., which are ordinarily difficult and expensive to machine, can be readily produced, *Fig.* 47.5. Gears may be cast to usual machining accuracy so as to necessitate only a boring operation.

Draft. As with most casting methods, draft is required but with plaster-mold casting it is needed in order to facilitate withdrawal of the plaster mold from the patterns. Outside surfaces usually require a 1/2-degree draft angle and inside surfaces, 1/2 to 3 degrees. Holes or cavities formed by loose cores require no draft but integral core sections require from 1/2 to 1 degree per side.

Cores. Several pieces can be consolidated into one part by means of separate core pieces but never by loose pieces on the pattern. Undercuts and intricate passages are readily produced, *Fig.* 47.6. Holes under ¹/₂-inch diameter normally can be drilled more economically than cored, *Fig.* 47.7. Drill spots, however, can

Fig. 47.7 — Several parts of odd shape in which the holes have been cored. Photo, courtesy Universal Castings Corp.

Fig. 47.8 — Machine parts on which identifying markings can be seen. Depressed lettering is most economical owing to ease in stamping markings into brass patterns. Photo, courtesy Universal Castings Corp.

be placed on the pattern to eliminate fixtures. Cores for special coarse-pitch thread forms, grooves, etc., are readily produced and present little difficulty in casting.

Wall Thickness. Plaster, being a poor conductor of heat, is ideal for producing thin, sound walls. Where the area does not exceed 2 square inches, a wall of 0.040-inch can be cast. For larger areas the practical minimum wall thickness increases, for 4 to 6 square inches a thickness of 0.062-inch being minimum and for areas up to 30 square inches the minimum is generally set at 0.093-inch. Tapering sections are readily producible and blade sections can be cast with knife-sharp edges.

Stock Allowances. Round, square or odd-shape holes may be cast with only a minimum of stock for machining. Ordinarily where machining is required for any reason, $1/32$-inch metal allowance is sufficient but less can be allowed on small parts and cored holes, especially where broaching is to be used.

Markings and Inserts. Markings of almost any nature can be clearly reproduced. Lettering may be either raised or depressed as desired, *Fig.* 47.8. Additional features such as steel or other inserts may be cast-in with the usual considerations for galvanic action and for sufficient strength to resist cracking of the casting on cooling.

Other Design Features. Typical design features such as ribs, bosses, lugs, corner radii, fillets, etc., should follow general practice as outlined in previous chapters covering other casting methods. Like diecasting, plaster-mold casting permits the production of numerous intricate design features to an accuracy that requires little or no machining, *Fig.* 47.9.

Fig. 47.9 — Group of intricate parts with all design features completed by plaster-mold casting. Little machining is required before assembly. Photo, courtesy Universal Castings Corp.

Selection of Materials

To effect maximum economy and satisfactory castings, metals suitable for plaster-mold casting have been developed and standardized. Lower scrap and efficient grouping of production patterns is achieved by specifying one of the standard alloys the characteristics of which are given in Chapter 68.

In addition to these standard alloys almost any other castable metal in the nonferrous group can be cast if necessary, although at higher cost. Pure copper can be cast and beryllium-copper alloys produce excellent castings which have high physical properties and which will harden without warpage or relative dimensional change. Beryllium-copper alloys offer tensile strengths up to 170,000 psi and hardnesses to Rockwell C46. Others possible include silicon-bronzes, nickel bronzes, nickel-copper, tin bronzes, and silicon-aluminum-bronze.

Specification of Tolerances

In general, a tolerance of plus or minus 0.005-inch per inch of dimension can be held in production. Where the dimension crosses the parting line the tolerance is increased to plus or minus 0.010-inch per inch. In specific portions of a casting, however, more exacting tolerances are practical. For instance, overall tolerances of plus or minus 0.005-inch for portions within one half of a mold and plus or minus 0.010-inch where the dimension crosses the parting line are being commercially produced today.

Hole Tolerances. Cored dimensions within a casting can also be held to plus or minus 0.005-inch per inch. Center-to-center dimensions of holes produced by separate cores can be held to the same limits. Tolerances on integral core portions must take into consideration the draft angle which is required.

Flatness Limits. Flatness of surface which can be obtained depends to some extent upon the casting design. As a rule, areas of roughly one square inch are readily held flat within 0.005-inch. Larger surfaces, say approximately six square inches, may sustain a warpage ranging from 0.007 to 0.015-inch depending upon design intricacy.

Roughness. Surface roughness in production can be consistently held to 30 microinches, rms, or better. The velvet-smooth surfaces produced are eminently suited for any parts which are to be electroplated.

48

Investment Casting

L AST but not least in the group of major casting methods is that best termed as investment casting. An ancient process, it was known and used by the Chinese prior to the year 50 B.C. Benvenuto Cellini used the process to create his celebrated "Perseus with the Head of Medusa". Modern history of investment casting, however, dates from shortly after the turn of the century when Dr. W. H. Taggart interested the dental profession in the advantages of the process in producing metal dentures.

Developments and improvements in technique which followed led to widespread usage of the process by the jewelry industry in the middle thirties. Value of the process for solving many problems in the production of intricate

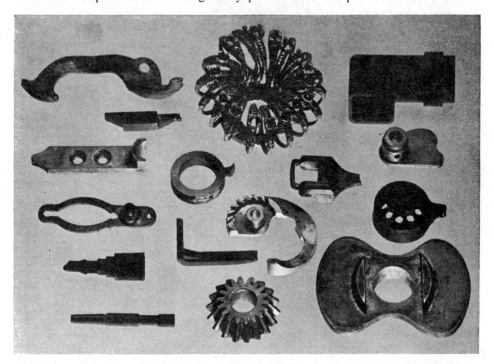

Fig. 48.1 — A "tree" of investment-cast rings and a variety of machine parts produced by the same means using centrifugal feeding of the metal.

machine parts first gained recognition during World War II and from the resulting industrial experience, *Fig.* 48.1, the investment casting process has become one of the present basic production methods.

Advantages of Investment Casting. Contrary to popular opinion, the major factor which influenced application of the process from the start was its outstanding fidelity of reproduction rather than any actual dimensional accuracy or repetitive accuracy. Accuracy of the process regarding relatively small parts is high but high primarily in comparison to other methods of casting and *not* to precision machining. Main advantages of the process, therefore, may be summed up as: (1) Ability to produce extremely intricate or compound shapes which are impossible, unreasonable or more costly by other means; (2) practicability of casting to size, with a minimum of finishing required, parts from alloys not readily machinable in production; and (3) the possibility of producing small parts from metals with melting points beyond the limitations of die casting. A group of typical machine parts which exemplify some of these factors is shown in *Fig.* 48.2.

Fig. 48.2 — Group of investment castings with intricate shapes typical of those ideally suited to this method.

Special Process Characteristics. The process of investment casting does not resemble closely any of the other casting methods. The only resemblance is to permanent-mold casting in that the longest dimension of a cast part is normally placed in a vertical position and mold venting is generally a problem. However, vacuum casting eliminates the problem and insures dense, nonporous metal. The major difference is that the molds are one-piece and cannot be examined before pouring.

Casting Process. In general the process of investment casting consists of the following basic steps: (1) Forming the expendable pattern, (2) gating and assembling patterns, (3) investing the pattern, (4) melting out of patterns and drying the investment mold, (5) pouring the castings, and (6) removing and cleaning up the castings. Production of accurate expendable patterns is all-important in order to obtain the necessary results.

Molds. Master molds for production of patterns in quantity are made of rubber, low melting point metals or steel. Rubber and soft-metal master molds are

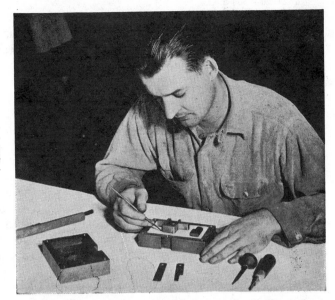

Fig. 48.3 — Master pattern is embedded in plaster of Paris in one-half of the master-mold flask to pour the soft-metal mold for the opposite half. Photo, courtesy International Nickel Co. Inc.

made, respectively, either by vulcanizing or by casting against a master pattern of the part desired. The master pattern used is generally made with an allowance for shrinkages which occur and held to tolerances one-third those desired in the final casting, *Fig.* 48.3. Steel master molds are generally sunk directly from the drawing specifications.

Use of rubber master molds is more or less confined to the casting of jewelry. Soft-metal master molds are generally found satisfactory for parts without too many complications or close tolerances. They are generally preferred, however, for parts having contours or shapes which are difficult to generate or reproduce, as with turbine blades, etc. Steel master molds are most satisfactory for long runs and for parts requiring the maximum in dimensional accuracy or having small cores,

Fig. 48.4 — Master soft-metal mold with corepiece and wax patterns being produced on typical injection press. Photo, courtesy International Nickel Co. Inc.

thin slots and intricate, involved shapes.

Expendable Patterns. Although the process is sometimes termed the "lost-wax" method, this is a misnomer today inasmuch as wax is not the only pattern material utilized. Numerous materials have been employed for patterns but today the materials in common use are waxes, plastics and frozen mercury. Properties of each are different, require different handling and are suited for different applications.

Waxes are probably most commonly employed in production, being simplest to handle, *Fig. 48.4*. Polystyrene plastic patterns, produced by regular injection molding, are harder than wax patterns, withstand more abuse and can be held to very close tolerances. The plastic cost is about one-third that of waxes. Plastic patterns are generally best suited for long runs owing to greater expense in steel master molds and injection equipment.

Most recent pattern material is mercury which is poured into dies and frozen below minus 40 F. Advantages are claimed in improved accuracy and greater size of part possible. More intricate parts may be cast by assembling a number of frozen mercury pattern pieces.

Frozen mercury patterns, like those in waxes, are readily bonded to each other to produce complex assemblies or to sprues for casting. The maximum number of patterns practicable are mounted on a single sprue for casting simultaneously, *Fig. 48.5*, and minimum individual pattern size thus permits obvious savings.

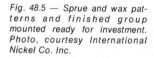

Fig. 48.5 — Sprue and wax patterns and finished group mounted ready for investment. Photo, courtesy International Nickel Co. Inc.

Investments. Type of investment utilized in making the final mold for casting depends upon the metal to be cast. That used for nonferrous alloys generally consists of silica with plaster of Paris as a binder. Ferrous alloys which react with plaster generally require other investment materials such as a refractory and a suitable binder which also will resist the higher temperatures involved.

In preparing high-temperature investments, a primary investing dip coating is given the patterns, *Fig. 48.6*. Typical is one of sodium silicate, a wetting agent and silica which is finally dusted over with a somewhat coarser grade of silica. After attaching the flask to the bottom plate on which the patterns are mounted, the secondary investment (typical is one of silica, grog of ground fire brick, and a mix of alcohol and tetraethylsilicate plus one-half old investment) is poured around the patterns. This is jolted during the short setting period and after hardening the patterns are melted out. Following a final thorough preheat the metal is poured, *Fig. 48.7*.

With mercury patterns a clay such as Kaolin is often used. This is deposited in layers normally to $1/8$-inch in thickness, by dipping, although up to $1/2$-inch has

Fig. 48.6 — Dipped and dusted patterns are dehumidified thoroughly before pouring the secondary investment. Photo, courtesy Haynes Div., Cabot Corp.

Fig. 48.7 — Pouring a group of investment castings by the common static method. Photo, courtesy Allis-Chalmers Mfg. Co.

been used. After setting at low temperature the frozen mercury pattern is allowed to melt out or is flushed out with warm mercury. The remaining investment shell is then fired to produce a hard, permeable ceramic mold which is surrounded with dry sand, preheated and the casting poured.

Casting. Investment castings are poured by static, centrifugal, air pressure, or vacuum methods. High-temperature alloys and small charges (up to about 15 pounds) are processed in electric-arc furnaces, especially where air pressure is used. Larger charges, static pouring, vacuum casting and centrifugal casting are done largely in induction furnaces or gas-fired crucibles. Advantages of centrifugal pouring, as originally practiced in dental work, is gaining favor owing to improved metallurgical character of the cast metal.

Size of Castings. Generally speaking, investment casting is best suited for the production of reasonably small parts. Maximum dimensions and weight of parts are limited in production by the flasks available and furnace capacities. The largest cylindrical mold usually used is 10 inches in diameter by 20 inches long and the largest box flask is 9 by 12 by 15 inches.

Maximum capacity of pressure pouring available is about 15 pounds. Induction furnaces up to 150 pounds capacity are used for static pouring. Maximum weight of a part or parts poured generally does not exceed 50 to 70 per cent of the furnace charge owing to the size of the sprue necessary, *Fig.* 48.8. About the largest pressure castings ordinarily produced in quantity range around

Fig. 48.8 — Gated pattern for a rotor disk showing size of sprues used. Other patterns include a conveyor crank and turbine blades. Photo, courtesy Allis-Chalmers Mfg. Co.

3 1/2 pounds and for static pouring about 15 pounds with a maximum dimension of approximately 6 to 8 inches. Parts from 18 to 24 inches long and weights up to almost 100 pounds have been poured statically. Smallness of size is practically unlimited — one of the smallest parts produced to date weighs less than 0.002-pound.

Production quantities. Investment casting offers no competitive advantage for large quantities of parts which can be produced in several setups on automatic machinery. Where the quantity is small, however, the lower cost of tooling can often make the casting more economical. One specific example of this nature indicated that the point of equal cost per piece was reached at 5000 pieces. The limit, of course, varies with the particular part design. Production runs may thus range from as few as six pieces to as many as 100,000 pieces or more. Economical production runs for average machine parts generally range from 500 to 5000 pieces. To amortize the higher costs involved in casting with plastic patterns, production runs of 5000 to 10,000 pieces are required with greater quantities resulting in distinct savings over wax patterns.

Design Considerations

In general, any relatively small part which would require a fair amount of machining to produce is a possible candidate for investment casting. Where overall cost of parts, machining, fasteners, and assembly can be eliminated by one investment casting requiring but a small amount of finishing, highly economical results can be obtained, *Fig.* 48.9. Cost savings per piece as high as 95 per cent have

Fig. 48.9 — Beryllium-copper calculator variable cam with a maximum dimension of about 2 inches as investment cast. Original cam cost $30.00 to assemble from stampings by brazing while casting cost approximately $2.00. Photo, courtesy Arwood Corp.

been recorded in numerous cases. Many otherwise expensive operations can often be eliminated. Typical of these are precision operations such as drilling, threading, tapping, tapering, grinding, etc., usually necessary for assembly purposes. The casting in *Fig.* 48.10 eliminated a shrink fit.

Ideal Design. The ideal part for this method of production weighs in the range from approximately one ounce to one pound and has uniform sections within the range of 0.035-inch minimum, to 1/4-inch, maximum. Good design for casting in

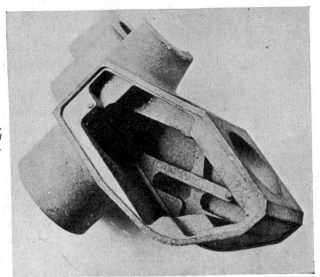

Fig. 48.10 — This binocular housing, cast in magnesium, formerly required a shrink fit assembly of two parts. Photo, courtesy Arwood Corp.

general applies in the case of detail design for investment casting and must be adhered to for good results.

Sections and Thickness. Though the preceding specifications indicate that uniform sections from 0.035 to 1/4-inch are ideal, castings with sections from 0.020 to 3/8-inch and occasionally 1/2-inch are regularly produced. Sections as light as 0.015-inch are possible where they are continuations of tapered sections and do not exceed 1/4-inch in length. It is unwise to demand, unless absolutely necessary, any continuous section less than 0.025-inch, except over very limited areas. For large areas, wall thickness should never be less than 0.040 to 0.050-inch. Areas up to approximately 12 square inches may be cast in sections as thin as 0.040-inch, minimum, with reasonably accurate flatness where adequately supported or ribbed.

Heaviest section possible to produce is about 1 inch. However, heavy sections should be avoided if possible and ribbing substituted to obtain the desired stiffness and strength. Due regard should be given to rib proportions to avoid loss of strength.

Uniformity of section is ideal but, of course, varying sections such as in turbine blading are feasible where desired, *Fig.* 48.11. Wherever sections vary, however, the transition should be gradual and the effects of uneven cooling between sections of varying thickness should be carefully considered.

Edges. Thin edges are difficult to produce in wax without distortion or handling difficulties. The minimum practical edge thickness is generally considered to be 0.012 to 0.015-inch provided there is no abrupt section change and a gradual taper of 20 degrees included angle, minimum, is used. Thinner edges have been cast, i.e., a trailing edge of 0.005-inch on a blade section of 0.015-inch, but are not practical as a rule, especially in the hard alloys.

Edge radii of 0.005-inch can be held at the crests of serrations having a fine pitch and wide included angles. In casting serrations some difficulty is encountered with small globules of metal which often form at the roots of sharp or small included angles.

Draft. Investment castings do *not* require draft on outer walls or cores regardless of the fact that the expendable patterns are formed in permanent type

Fig. 48.11 — Group of typical castings showing widely varying sections and range in size and complexity of design. Photo, courtesy International Nickel Co. Inc.

dies. Occasionally, however, it is found advantageous to use a draft of about $1/2$-degree to permit proper removal of the patterns. Those surfaces which should have no draft should be indicated to permit use of draft elsewhere when necessary for production.

Fillets. Generous fillets are advantageous both functionally and in producing a stronger pattern. Minimum fillet radii should be 0.015-inch and whenever possible at least $1/16$ to $1/8$-inch is preferable.

Bosses and Studs. Bosses used for fastening are generally preferred to studs but integrally cast studs can be used if desired, *Fig.* 48.12. Integral studs are more economical than inserted types.

Fig. 48.12 — Cast machine side frame showing numerous cast studs and shaped bosses. Photo, courtesy International Nickel Co. Inc.

Fig. 48.13 — Aircraft clamp piece used for quick disconnect on engine, cast in 410 stainless steel, showing a variety of holes and complex shapes. Drawing details are shown on page 48-11. Photo, courtesy Haynes Div., Cabot Corp.

Holes. Coring can be used to advantage in many cases and can be of any size or shape if sufficiently large, *Fig.* 48.13. General limitation on cored holes is that the diameter be not less than one-half the section thickness. For instance, in high-temperature metals the smallest diameter hole generally considered feasible is 0.050-inch and in the soft nonferrous alloys, about 0.020-inch.

Generally, through holes can be cast to depths about three times their diameter. Where circumstances so permit, however, smaller holes and holes with l/d ratios over 10 can be produced by means of special ceramic cores. For example, a hole $1/64$-inch in diameter with a depth of $3/4$-inch has been successfully produced.

Holes to be cored should, if at all possible, be through holes to permit adequate core support from both ends. Blind holes should be avoided. It is difficult to cast blind holes to depths greater than twice the hole diameter, especially in high-temperature alloys. A good rule is to restrict blind holes to a maximum depth of $1 1/4$ times the diameter. Undercuts, again, are possible but are best avoided for economy.

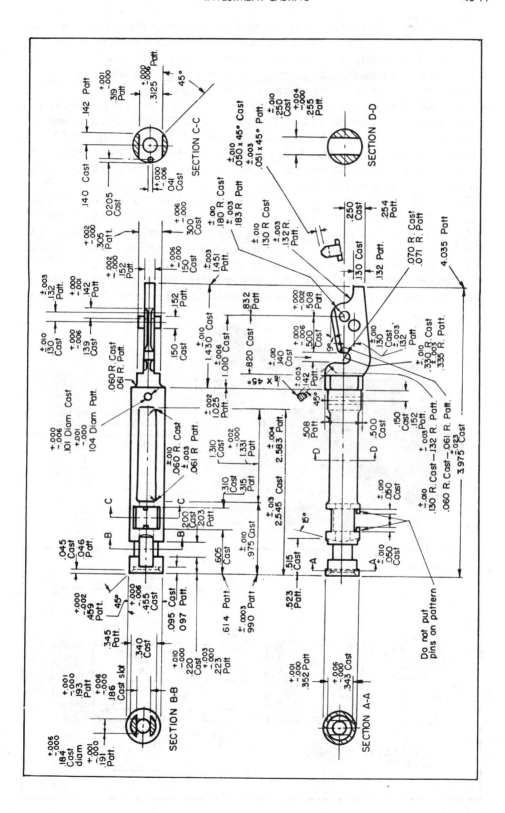

Machining Allowances. Those surfaces which require closer tolerances than this method provides must be machined and an allowance for this must be made. Finish stock generally allowed ranges from 0.010-inch on small parts to a maximum of 0.040-inch on large ones. Holes to be tapped should always have a preliminary reaming or boring operation to assure best possible threads.

Threads. Barring special unmachinable alloys, size or form of thread, it is generally more economical to cut or grind threads. Either external or internal threads can be produced but the resulting accuracy seldom exceeds that of Class 1, *Fig.* 48.14. Class 2 threads have been produced with plastic patterns. Owing to the tendency for small globules of metal to adhere to the roots of threads it is generally necessary to remove them by machining.

External threads smaller than 1/4-inch and pitches finer than 20 in nonferrous alloys or 14 to 16 in ferrous alloys are seldom feasible. Threads longer than 1/2-inch must have the pattern pitch-compensated for shrinkage. Where close "machining type" tolerances are desired it is not recommended that threads or gears be cast, unless circumstances permit casting tolerances, *Fig.* 48.15.

Parting Lines and Gating. A visible parting line is generally found on investment castings and in design this fact should be kept in mind. Areas that must

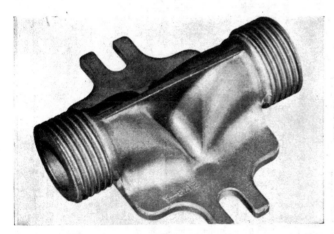

Fig. 48.14 — Cast fitting with Acme threaded end connections used as cast. Photo, courtesy Haynes Div., Cabot Corp.

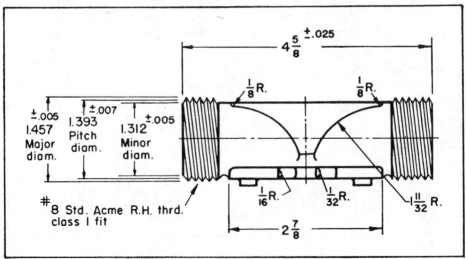

Fig. 48.15 — Investment-cast Hastelloy C alloy gears for a steel mill pickling tank drive. Photo, courtesy Haynes Div., Cabot Corp.

be kept free of parting lines should be indicated with the thought in mind that restrictions should permit use of the simplest possible two-piece pattern die.

Gates, of course, must be used to feed the casting and also must be removed. Gating areas necessary for a particular casting cannot always be predicted and, hence, the best practice in design is to indicate those surface areas where gates would be undesirable.

Selection of Materials

Practically all metals which can be melted are being cast by the investment process. It not only provides economy for unusual designs in the more common metals but makes possible many otherwise impossible machine parts in metals not machinable to any great degree. All standard cast materials are readily available as are many special alloys such as those of cobalt, nickel and chromium, and vacuum refined copper and pure iron.

Physical properties of investment-cast metals are generally equal to the mean between the transverse and longitudinal values for rolled bars of the same metal or alloy. On an average they are about 10 per cent greater than the properties of sand-cast metal. Heat-treatable metals respond well without warpage. Any of the tool steels may be used and are often found advantageous in cases where hardened parts must be used. Nitriding steels have been cast but their as-cast hardness is about 400 Brinell and little machining can be done without annealing. Corrosion and wear-resistant alloys such as Stellite or Vitallium, which cannot be machined or forged, are readily cast and offer tensile strengths to 110,000 psi at ordinary temperatures. At 1500 F tensile strength of 65,000 psi is typical with these metals.

Heat-treated aluminum-bronze provides an approximate tensile strength of 85,000 psi with a Rockwell hardness of B-90. Monel "S" or "H" provides a tensile of 120,000 psi and a hardness of Rockwell C-32. Heat treated stainless Type 420 offers the same tensile strength with slightly different properties. Beryllium-copper on the other hand offers tensile strengths up to 170,000 psi in heat-treated form. In the steels, heat-treated AISI 3140 gives a tensile strength up to 220,500 psi.

Owing to the close control of melt possible, some special castings of closely controlled composition have been produced by this method. Alloy gray iron pistons, cylinder sleeves, compressor valve sleeves, and other parts have been cast by these methods.

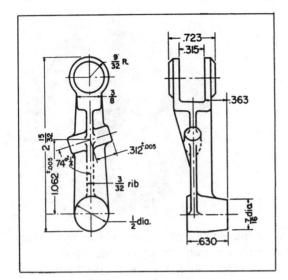

Fig. 48.16 — Aerial camera crank investment-cast in Type 410 stainless steel.

Specification of Tolerances

The investment casting process has often been termed "precision casting," a name which has been misleading at times. Actually, considering dimensions of more than 5 inches in length, the as-cast tolerances are not substantially better than those found in careful permanent-mold casting or even in present-day top-quality sand casting. Actual dimensional tolerances on the *average* are specified as plus or minus 0.005-inch per inch for dimensions over 1 inch and plus or minus 0.003-inch per inch on those under 1 inch. The effect of size on tolerances is easily seen.

Dimensional Variation. Some degree of difference is present in casting ferrous and nonferrous alloys when considering tolerances. *Minimum* general tolerances are usually set at plus or minus 0.002-inch per inch for nonferrous alloys and plus or minus 0.004-inch per inch for ferrous. Occasionally, dimensions of 1/4-inch or less can be held to plus or minus 0.0005-inch and many parts have been produced with cores, thin slots and shapes held to tolerances of plus or minus 0.001-inch. The problem is not holding the desired accuracy on necessary portions of the part but in many cases is that of holding too high an accuracy on unnecessary portions. Lowest production cost is achieved where tolerances can be specified to plus or minus 0.010-inch.

Where a gate is removed by grinding, the tolerance ordinarily can be no closer than plus or minus 0.010-inch over the gate area. For a critical dimension across the parting line, it is usually desirable to allow 0.001-inch total in addition to the nominal tolerance per inch allowed.

Warpage. Some allowance must be made for possible warpage and it is not always practical to maintain long, thin sections perfectly straight. Parallel accuracy of sections such as those in an H-shape are difficult to keep in true parallel. Design, if practicable, should provide ribbing to assist in retaining such faces against warpage.

Surface Quality. Large areas are not always possible to cast free from imperfections. Areas which must be free from defects should be so marked. Cost of parts is affected directly by the degree to which minor imperfections are permissible. Surface roughness generally can be held to 70 to 80 microinches, rms, without added expense. With plastic patterns surfaces with roughness from 10 to 20 microinches, rms, have been produced.

Angular Variations. Angular accuracy is often difficult to maintain, but, as a rule, a tolerance on angles of plus or minus 1/2-degree can be held. It is practical, however, on some parts to hold this to plus or minus 1/4-degree, *Fig.* 48.16. With precision steel pattern molds, gears with plus or minus 0.001-inch on tooth dimensions and plus or minus 10 minutes angular tolerance have been made.

MOLDING
METHODS

49

Powder Metallurgy

$\mathbf{F}$OR lack of a better, more concise term for the pressing and sintering of metal powders into machine parts of all varieties, the generally recognized one, "powder metallurgy", is normally employed. Although not always considered as such, powder metallurgy is a production process and recent developments in technique have advanced it to a point where it now must be ranked alongside the many others readily available to the designer.

Historically, powder metallurgy antedates most other methods, the first actual recorded experiments being those of Wollaston who in 1829 produced the first ductile platinum by reducing ammonium chloro-platinate, cold pressing and sintering. Continuous research led to the first practical commercial use in producing tungsten wire in 1916. The first porous metal bearings were marketed just following World War I, but in the ensuing years improvements and new applications developed at a rather slow pace. With the outbreak of World War II, however, the vast potentialities of powder metallurgy began to be realized for its value not only as a method of fabricating parts whose physical characteristics are impossible to produce otherwise, but a large-volume mass-production process having excellent speed and material economy.

Field of Application. Successful applications of parts produced from metal powders to date are rather impressive. Beyond the well-known oilless or self-lubricating bearings and like parts, there lies a tremendous field of machine parts which, with the particular characteristics available through powder metallurgy, can be redesigned to achieve greatly improved performance, longer service life, simplified design and manufacture, lower cost per part, etc. Representative of these are external and internal gears, external and internal ratchets, levers, sliding blocks, cams of all varieties, clutch friction facings, internal and external splines, rollers, guides, permanent magnets, piston rings, bushings, turbine blades, small clutches, spacers, etc., *Fig.* 49.1.

Metal Structure. A variety of metallurgical methods can be utilized to produce end products from metal powders as well as from combinations of metallic, non-metallic, and intermetallic powders. Several general classes of metal-powder structure can be set up: The first, in which consolidation during sintering is primarily a particle-to-particle cohesion or contact fusion of particles containing a melting constituent; the second, in which one of the powders used acts as a melting medium, bonding or cementing together a high melting-point constituent; and a third, in

Fig. 49.1 — Group of representative precision machine parts produced by powder metallurgy. The tremendous range of design possibilities can be seen from this wide selection of production parts. Photo, courtesy American Powdered Metals, Division of the Barden Corp.

which consolidation of a fairly high melting-point metal is achieved as in the first category but, lacking high density, the compact is impregnated with low melting-point metal.

Steps in Production. Various steps in the procedure which may be employed in producing parts are generally as follows: (1) Selection of the powder or powders best suited for the part being designed as well as for the most rapid production; (2) wet or dry mixing of powders where more than one powder is to be used; (3) pressing in suitable dies; (4) low-temperature, short-time sintering usually referred to as presintering for increasing strength of fragile parts, removing lubricants, etc.; (5)

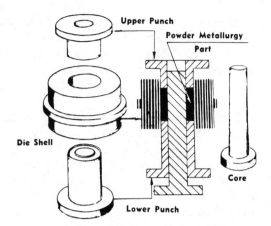

Fig. 49.2 — Simple punch and die arrangement for a cylindrical part. Telescoping punches are used for more complex parts. Courtesy, Metal Powder Industries Federation.

Typical Die and Components

machining or otherwise forming of presintered parts; (6) sintering green compacts or presintered parts to obtain the desired mechanical properties such as proper density, hardness, strength, conductivity, etc.; (7) impregnating operation for low-density sintered compacts, usually by dipping in molten metal so as to fill all pores or by allowing the impregnant to melt and fill the pores during sintering; (8) coining or sizing operation, cold or hot, when necessary to attain more exacting dimensional tolerances and also improve properties; and (9) a hot-pressing operation to replace the usual pressing and sintering.

Presses. Most commonly used commercially is the plain cold pressing and sintering cycle. Presses may be mechanical, hydraulic or a combination of both. Generally, small parts which can be made at high speed with relatively low pressures are best produced in mechanical automatic presses with single or multiple-cavity dies. Such presses for average parts usually are built in pressure capacities ranging to 100 or 150 tons although sizes to 1500 tons are available. Units over 75 to 150-ton capacity and ranging up to as much as 5000 tons are designed as combination mechanical-hydraulic or fully hydraulic. Presses are generally of the single-punch type or of the rotary multiple-punch type. Single-punch presses are either of the single-action type which compresses with the top punch only, or of the double-action type which employs movement of the lower punch simultaneously with the upper to obtain more uniform compacting and automatic ejection. As many as two upper telescoping punch movements and three or four lower punch telescoping movements are used along with side core movements for complex part designs, *Fig.* 49.2.

Fully automatic mechanical high-speed rotary presses may utilize multiple dies ranging from 6 to 35 sets. Thus, production may reach as many as 1000 pieces per minute, although part design is somewhat limited and pressures must be low. Overall, average production proceeds at a rate of 500 to 50,000 pieces per hour. However, largest parts produced on big hydraulic presses may require one-half minute apiece. Where parts are of large size or when maximum uniformity in density is essential and maximum speed is not so necessary, hydraulic presses are employed.

Compacting Pressures. Pressing or compacting of powder-metal parts generally requires anywhere from 5 to around 100 tons pressure per square inch.

Common garden-variety parts are produced with 20 to 50 tons per square inch pressure. The broad limitations posed by pressing area on size of parts which can be handled is readily apparent. It is interesting to observe, however, that the final density of a powder metallurgy product is not determined by the pressure under which it is cold-pressed or briquetted. Rate of pressure application, particle size, type of material, sintering time and temperature, occluded gas, etc., also have their effect upon final density and size.

Press Stroke Limitations. Press stroke also presents certain broad limitations as to part size. The compression ratio between the volume of powder in a die before and after pressing is dependent upon loading weight, particle size, form and composition, metal hardness, and pressure utilized. With most common metals and alloys this ratio is usually 3-to-1, but may vary from 5-to-1 up to 10-to-1 with fine powders and 2-to-1 up to 4-to-1 with medium-size powders. Coarse particles are seldom used except in special cases. Combined with the compression ratio is the additional general limitation that diameter-to-length ratio of parts be restricted to a maximum of 3-to-1.

Size of Parts. Single and double-action presses up to about 85 to 300-ton capacity, which are widely available, will handle parts up to about 4 inches in diameter with 6 to 8 inches of die fill. Generally, however, larger presses to 1000-ton capacity are capable of handling sizes up to 10 inches and more in diameter with up to 8 inches of die fill. Cylindrical bearings up to 18 inches in diameter weighing some 233 pounds have been produced in quantity and, where economical, this can be increased to around 36 inches in diameter. Bronze plates 20 inches by 30 inches by 1-inch thickness also have been produced.

Production Quantities. Owing to the speed of the operations in producing parts from powders, economical quantities are necessarily high. Volumes ordinarily must run from 20,000 to 50,000 pieces before a cost advantage can be obtained. In a fair number of instances, economical production of large parts may start at as low as 5000 pieces. Where numerous highly complex machining operations can be eliminated, lots as small as 500 pieces have proved advantageous.

Where few parts are required, these should be machined from cored or solid bar stock if design permits. High cost of powders, from 8 cents to about $1.00 per pound, and normal die costs, although not usually excessive — about $150 for small, simple parts to perhaps $1800 or more for large or complicated ones — naturally preclude the production of but few parts much the same as in the other high-production processes. Ordinarily an average die will produce 50,000 pieces before wear necessitates refitting or replacement.

General Advantages. To properly assess the advantages which give powder metallurgy its greatest value as a production method, it should *not* be considered generally as a competitor of other methods. Competition on an even basis with other methods may even be unwise; powder metallurgy should yield a distinct contribution and should be considered primarily for parts which either cannot be produced by other methods at all or which cannot be produced economically, *Fig.* 49.3. In general, a few of the specific applications are: (1) Production of solid ingots or parts from highly refractory metals such as tungsten, tantalum, columbium, and molybdenum which cannot be fused commercially in available furnaces; (2) parts from several materials of divergent characteristics which do not alloy in the molten state, including metallics, intermetallics and nonmetallics, that

Fig. 49.3 — P/M parts of copper-steel composition perform under heavy duty conditions in a hydrostatic transmission. Powder metallurgy here eliminates major machining operations and major costs, yet close tolerances are held — piston bores within 0.0007-inch, straightness and parallelism of the bores within a total eccentricity of 0.001-inch, hole to hole bore position within $^1/_2$-degree. Photo, courtesy Metal Powder Industries Federation.

combine to give a product which retains each material's desired characteristics; (3) precise-dimension parts so hard or brittle as to preclude any other means of shaping; (4) parts with characteristics unobtainable by other methods, such as controlled porosity, extremely high density, frictional characteristics, controlled inductance, etc.; (5) laminated parts with more intimate binding than possible by conventional means; (6) precision parts which can be produced more economically by powder metallurgy than by any other method primarily by elimination of further finishing operations; and (7) preforms for hot forging with little or no waste.

Hot Pressing. Where the ordinary procedure for handling powder metal parts does not result in a satisfactory density or pressures required are extremely high, hot pressing or hot densification can often be utilized. In this method, the powders are heated in the dies and pressure applied to form the part. The hot pressing method yields high and nearly ideal density (to 98%) and greater strength at relatively low pressures. Carbide parts up to 100 square inches in cross section, greatest dimension of which can be 18 inches, with a length of 8 inches have been produced by hot pressing, especially parts too large for regular cold pressing and sintering and thin-wall parts that tend to go out-of-round. Long carbide bars are normally extruded.

Coining. Ordinarily, to obtain desirable density and precision, parts of materials other than carbides are coined or sized after sintering but naturally this adds to the cost. Compacts up to 10 square inches in cross section are readily coined in hydraulic presses at a rate of about 4 to 6 pieces per minute with hand feeding and up to 10 per minute with automatic feeding. Generally dies will produce 100,000 to 200,000 pieces before refinishing is necessary and may produce up to 1,000,000 pieces before replacement.

Design Considerations

As with other production processes, economically sound design of powder-metal parts demands consideration and adherence to certain inherent limitations present. It must be remembered that, normally, metal powders lack plastic flow when subjected to pressure and those with a high-flow factor — that is, slow flow — make even filling of a die difficult and give rise to bridging, especially with narrow sections, with resultant wide variations in density. Consequently, limitations of form are inherent in design of powder-metal parts; it is impossible to press powder into re-entrant angles, sharp corners or undercuts, *Fig.* 49.4. In general, the overall design of parts should be as symmetrical as possible to promote uniform density in the powder structure.

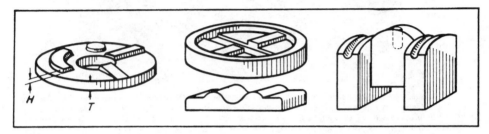

Fig. 49.4 — Axial projections, complex shapes, slots, grooves, blind holes, and recesses are easily obtainable so long as such design details can be produced by the action of one punch or two opposed punches. Where H is greater than T/4 telescoping punches are required. Drawing, courtesy Moraine Products Div., General Motors Corp.

Fig. 49.5 — Disc driver hub for Singer sewing machine produced from a sintered copper-steel alloy. Photo, courtesy Dixon Sintaloy, Inc.

Undercuts. Reliefs, undercuts, etc., on either the inside or outside of a part are beyond the capabilities of conventional equipment, and today the technology is available for production of such components, *Fig.* 49.5. Ordinarily pressure cannot be applied from the sides but only from the top and bottom of a die so such features which cannot be built into the upper and lower punches cannot be employed. This eliminates internal or external threads as well as holes or bosses normal to the axis of pressing and such features, if required, must be produced by a secondary machining operation. However, through use of the Olivetti process, many unusual parts can be produced, *Fig.* 49.6.

Section Thickness. Cross-sectional wall thickness of parts, owing to the method of pressing, must be uniform. Certain minimum limitations are recommended regarding wall thickness, TABLE 49-I. Wall thickness should be as great as practicable and seldom less than 0.062-inch. In general practice, a

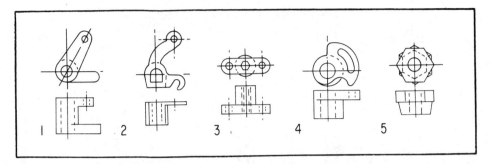

Fig. 49.6 — (1) Offset lever — with upper and lower projections in different directions — can be made on split-die equipment in one pressing, eliminating substantial machining. With conventional P/M tooling methods, part could not be withdrawn from the die. (2) Design of overhanging sections can be implemented in P/M when using split-die techniques and equipment. With conventional equipment, these sections would crack and break off during normal pressing motions. Also, the tangent areas where the overhang meets the main body are permissible with the split-die method. (3) In this case, there is a problem where the flange overlaps the hub. The part could not be ejected from conventional tooling. With Olivetti tooling, such a configuration presents no problem. (4) Cam combines an overhang problem with a knife-edge tangent hub and flange. A flat at this area would be required with conventional tooling. The Olivetti technology makes a smooth tangential contour possible. (5) Angular shaft configuration with an overlapping hex-shaped nut could not be withdrawn from presently designed tools. Split-die equipment provides a solution. (6) Ordinarily, cantilevered offset cam could be made in P/M only as a two-piece assembly requiring costly fastening operations. With split-die technology, it is made in one piece, assuring economy and precise orientation. (7) View "a" shows a back angle which could not be obtained in P/M on conventional equipment without secondary machining operations. (8) A split pinion without teeth in the upper or lower section is shown. With conventional P/M, the overhang would require secondary machining and increase cost. (9) Two-level part could not be tooled in conventional equipment. If pressed as shown in view "a", the part would not be extracted from the tooling; if pressed in the direction of view "b", with the long rectangular piece in the lower tool, the area between the two legs could not be plunged from the upper punch. (10) Root diameter of the outer gear of this combination cluster gear is very close to the outside diameter of the smaller toothed gear. This would necessitate a very thin tool member if conventionally tooled — at high compacting pressures such thin tool sections break down. The Olivetti system allows very close sections — or touching sections — without any risk of tool breakdown. Drawings, courtesy Dixon Sintaloy, Inc.

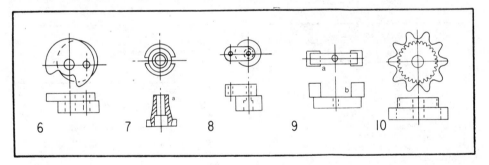

minimum wall of 0.050-inch is recommended for cylindrical sections up to 0.750-inch and proportionally greater as diameter increases.

Owing to the almost negligible flow of the powders, large and abrupt changes in section or thickness as well as nonsymmetrical cross sections should be avoided to guard against variations during sintering, *Fig.* 49.7.

Although flat sections as thin as 0.015 to 0.020-inch have been pressed successfully, coining is usually necessary to obtain the required density. Recommended practice is to limit flat sections to a minimum of 0.032-inch wherever possible. Thin flanges as well as extremely large-diameter flanges on parts also should be avoided. The diameter of flange on any part over 3/4-inch long should not exceed 1.5 times the part O.D., *Fig.* 49.8.

Corners and Fillets. Sharp corners at points of intersection are detrimental, a minimum radius fillet of 0.010 to 0.015-inch should always be specified on internal

PRODUCTION PROCESSES

T<small>ABLE</small> 49-I — **Wall Thickness Limitations** *
(inches)

Wall Thickness (min)	Length (max)	Outside Diameter (max)
0.032	0.500	0.500
0.040	0.625	0.750
0.045	0.750	1.000
0.050	0.875	1.125
0.055	1.000	1.250
0.060	1.125	1.375
0.065	1.250	1.500
0.070	1.375	1.625
0.075	1.500	1.750
0.080	1.625	1.875
0.085	1.750	2.000
0.090	1.875	2.250
0.095	2.000	2.500
0.100	2.500	2.750
0.110	2.750	3.000
0.120	3.000	3.250
0.130	3.250	3.500
0.140	3.500	3.750
0.150	3.500	4.000

* Courtesy Keystone Carbon Co.

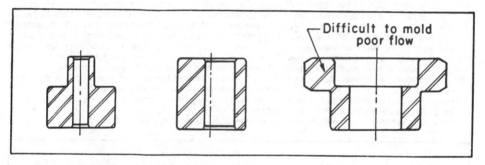

Fig. 49.7 — Limitations on large, abrupt changes in cross-sectional area are based mainly upon nonuniform changes in sintering which may cause considerable warpage. Parts such as these should be avoided wherever wall variations are greater than 2 to 1.

Fig. 49.8 — Sharp corners on flange intersections should be avoided to obviate weakness and production problems. Drawing, courtesy Presmet Corp.

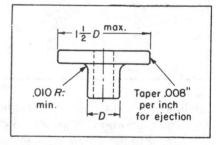

corners, *Fig.* 49.8. Again, however too large a fillet also causes some difficulty.

All corners should be beveled 0.010 to 0.015-inch minimum, to avoid formation of flash or fins but bevels should not be too large to obviate punches with fragile, feather edges. Accepted standard bevels are given in T<small>ABLE</small> 49-II.

Spherical Parts. Parts with a radius should have a small land at the edge to avoid fragile punch characteristics, *Fig.* 49.9. It is well to remember, though, such curved designs as *Fig.* 49.9 are particularly difficult to handle inasmuch as they require considerable loose powder and plastic flow. Where spherical portions are employed, an approximate design is preferable to increase tool life and simplify tool design, *Fig.* 49.10.

TABLE 49-11 — Standard Edge Bevels *
(inches)

Wall Thickness	Bevel (45°)
0.020 to 0.030	0.008 to 0.010
0.030 to 0.040	0.010 to 0.012
0.040 to 0.060	0.012 to 0.015
0.060 to 0.080	0.015 to 0.018
0.080 to 0.110	0.018 to 0.021
0.110 to 0.130	0.021 to 0.024
0.130 to 0.160	0.024 to 0.027
0.160 and up	0.027 to 0.031

* Courtesy Keystone Carbon Co.

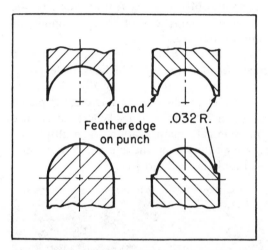

Fig. 49.9 — Parts pressed with a radius on the punch should be designed to have a small land to avoid premature tool breakage. Drawing, courtesy Presmet Corp.

Fig. 49.10 — An approximate design for spherical surfaces permits simplified tooling and extends tool life considerably. Drawing, courtesy Keystone Carbon Co.

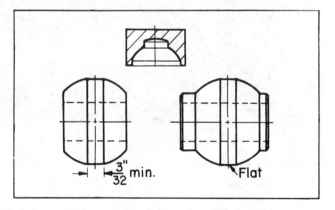

Cross-Sectional Shape. Parts with radial projections and recessions, *Fig. 49.11*, must be carefully designed to avoid bridging of the powder fill across narrow splines which gives rise to density variation, weakness and ejection difficulties. Too, parts having narrow splines may result in early die breakage owing to the high pressures involved and consequently the designer should always strive to assure maximum possible punch sections for greatest die life. In addition, whenever part tolerances permit, taper of cavity side walls from 0.0005-inch per inch and up should be allowed to reduce ejection difficulties.

Radial projections and variations that do *not* include variation in length can

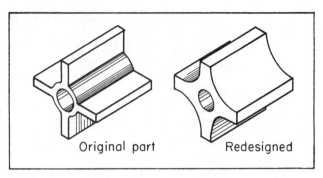

Fig. 49.11 — Narrow radial splines as well as slots should be avoided to guard against poor density and weak punch design. Drawing, courtesy Presmet Corp.

Original part　　　　　Redesigned

generally be produced without creating any problems. These features increase the surface contact with the die, however, and should be kept to a minimum for adequate die life and also to assure uniform density.

Projections. As a general rule, parts in which axial variations — bosses, projections, recesses, etc. — are not greater than 1/4 the body thickness or length of the part are practical and considered "naturals" for this process, *Fig.* 49.12. The primary reason for this is that such parts can be produced by a single upper punch with sufficiently uniform density to satisfy general requirements. Where variation is greater than 1/4, *Fig.* 49.12, the mechanical compensation of multiple or telescoping compound punches is necessary to maintain the average compression ratio of 3-to-1. Cost, of course, for such tools is much greater.

Fig. 49.12 — Gunsight piece made of bronze with projections greater than section thickness.

Slots and Holes. Slots, grooves, blind holes, and other irregularities are easily obtainable, *Fig.* 49.13, but wherever possible the sides should have a draft allowance of approximately 0.008-inch per inch, *Fig.* 49.8. The minimum hole permissible is 1/8-inch with lengths up to 1/2-inch maximum. For each 1/4-inch increase in length the ID should be increased at least 1/32-inch. The limit on length of hole is generally 3 1/2 inches. Holes *not* in the direction of pressing generally cannot be produced.

Hole depth can be held the same as in drilling but a smaller included angle at the bottom should be allowed, say 90 degrees, or a rounded punch tip, *Fig.* 49.14. Also, a draft allowance of 0.001-inch per inch, minimum, in such cases should be

Fig. 49.13 — Slots, grooves, blind holes, keys, etc., are particularly economical in relation to similar characteristics produced by other processing methods. Photo, courtesy Aluminum Company of America.

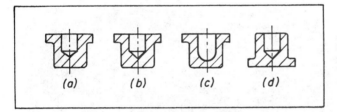

(a)   (b)   (c)   (d)

Fig. 49.14 — Minor design alterations, (b) or (c), permit economical production of blind holes ordinarily drilled. Design (d) is preferred for lower cost.

allowed whenever possible. Although blind hole designs such as that of *Fig.* 49.14*b* and *c* have been produced with special tooling, it is preferred for economy to keep the flange on the blind end as shown at *d* to permit ejection with ordinary tooling.

Tapered holes are also easy to produce but the small end should be made cylindrical for a length of at least 1/32-inch to prevent the top punch from contacting the taper on the core rod of the die setup, *Fig.* 49.15.

Stepped Shoulders. Where stepped shoulders are contemplated, a minimum step width of 1/32-inch can usually be produced but preferably each step should be at least 1/8-inch greater in diameter than the body or preceding step, *Fig.* 49.16. The same limitation holds for counterbores. A step of at least 0.100-inch should be allowed between the root diameter of gears and a hub diameter or major diameter in the case of cluster gears.

Part Length. Practical lengths which can be produced, as previously indicated, depend upon the press stroke available and the compression ratio of the material. In general, production has been confined to parts with not over 8 square inches of section area and 3 inches in length in the direction of pressing. Parts 12 inches in diameter by 9 inches in length, however, have been made although the

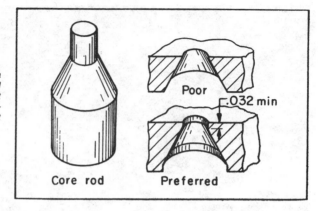

Fig. 49.15 — Tapered holes require a short cylindrical land at the small end approximately as shown to protect the punch. Land at large end is optional. Drawing, courtesy Presmet Corp.

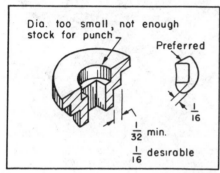

Fig. 49.16 — Circular steps should be at least 1/32-inch or greater; the internal counterbore shown for the square hole would have to be increased or produced by machining afterward. Drawing, courtesy Presmet Corp.

Fig. 49.17 — Arcing tip for a large air blast circuit breaker, made from copper-tungsten powders. Thread is machined after sintering. Photo, courtesy P. R. Mallory & Co. Inc.

press stroke for such parts is great and not generally available.

Length of parts should be comparable to the cross-sectional area; overlong sections exhibit low density at central positions or at the bottom (when produced with only one punch). Ordinarily, with cylindrical sections, it is well to avoid a length-to-diameter ratio greater than 3-to-1.

Gears. Undercut gear teeth which are difficult to machine can be readily produced by pressing. Extremely small gears such as those used in clocks are pressed in quantities effecting savings of 60 to 90 per cent over that found with other methods. Helical gears are also practical today. Limitations, however, are that pitch diameters under 5/16-inch are not practical. Helix angles should not exceed 24 degrees. Center holes are preferably round but in some cases flatted, square or odd-shaped center holes can be produced.

Inserts and Threads. Inserts are generally not pressed-in during processing. However, such parts can be pressed into place after sintering. Since threads create re-entrant angles, these features cannot be pressed but must be machined after completion of the sintering operation, *Fig.* 49.17.

Hot Pressing. Parts which can be hot pressed are generally free from many of the design restrictions normally encountered with ordinary cold pressing and sintering. However, inasmuch as this is a relatively new method of processing and not widely available, designs should be discussed with the producer before completion.

Compound Parts. Compound parts can be produced readily by utilizing the copper impregnating operation to braze two or more sintered-steel pieces together to form one high-strength complex part not economically possible by other means.

Machining. After sintering, undercuts and other features not practical to press can be machined by regular methods. However, to effect maximum economy, designs should be possible to complete in the dies. The more additional machining operations necessary, the less favorable will be the economies, *Fig.* 49.18.

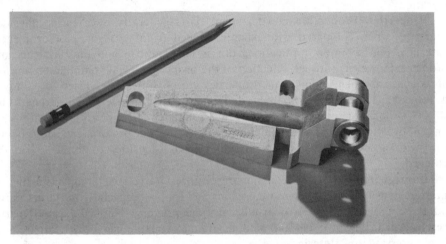

Fig. 49.18 — Copper infiltrated P/M steel door hinge for a commercial van is considerably more economical when compared to one made by conventional methods. Photo, courtesy Metal Powder Industries Federation.

Carbide Parts. For carbide bushings a wall thickness approximately 15 to 25 per cent of the diameter usually is most economical. Cost of making thin-wall parts is greater than the saving in material. On thin strips a proportion of length-to-thickness between 8-to-1 and 12-to-1 is usually most economical. Beyond this point cost of manufacture increases rapidly. Presintered parts can be machined into a wide variety of shapes and forms prior to final sintering.

Identification Markings. Names, directions and markings can be incorporated into the dies whenever such are on the face of either punch. Broad, strong letters are desirable to assure good legibility and for economy should be raised on the finished part. Depth of lettering from 1/64 to 1/32-inch is generally employed with characters about 1/16-inch high or greater. Line width of 1/64-inch or greater with either a round cross section or flat face with 30-degree sides are usually used.

Selection of Materials

A wide variety of materials is available in powder form for use in producing powder parts. Most commonly used are probably bronze, tin, brass, copper, iron, iron alloys, steel, carbides, and aluminum. Beryllium, stainless steel, nickel, copper-nickel alloys, tungsten, molybdenum, tantalum, manganese, zinc, and in fact almost any of a wide variety of the common alloys are available. In addition, of course, are the nonmetallics such as silicon and graphite. Unique in powder-metal processing, however, is the fact that parts may be produced from metals or combinations of metals, not possible to alloy, melt or otherwise work into useable form.

Particle Size. Powders having particles over 100 mesh or 150 microns in diameter are rarely used. Generally, the range of particle size is always below 150 microns and upward of 50 per cent of the powder will be below 45 microns. Powders behave differently, even when of the same material, so uniformity of mix is essential for obtaining uniform characteristics in parts. Hard powders are difficult to press while soft powders are much easier to handle. Powders which work harden also are more difficult to press and may require an annealing operation.

Properties. Production powder-metal parts generally have lower physical properties when compared with parts produced from wrought, cast or forged materials. For common applications this is no handicap and the physicals are sufficient. However, through selection of the proper material combination and/or subsequent treatment, equivalent properties are possible. The variety of compositions and specifications available today is virtually limitless. Tensile strengths to 200,000 psi or more can be attained by heat treating. Where densities to within 97 per cent by coining is not adequate, forging of the powder preform can and is being done with no waste of material. By this means, densities to 99.98% can be achieved.

Impregnated Parts. Impregnating porous presintered steel compacts achieves optimum density with low-pressure pressing and is ideal, especially for larger parts and those too complicated to produce in one pressing. Naturally, small parts present a handling problem and seldom offer reasonable economy compared to simple pressing and sintering.

Applications. Aluminum has been used widely for bearings. Its limitations on use for machine parts have been eliminated and today, a great variety of parts can be made to advantage, *Fig.* 49.13. Actually, aluminum yields high-density parts with lower compacting pressures than those required by other metals. Tensile

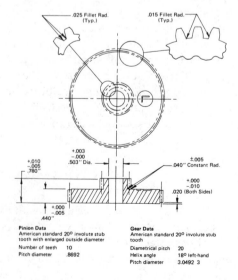

Fig. 49.19 — Automotive worm gear and pinion showing typical tolerances on production parts. Gear teeth have no machining marks, hence result in quieter operation. Material is nickel steel heat treated and tempered to 130,000 psi tensile. Drawing, courtesy Hoeganaes Corp.

Pinion Data
American standard 20° involute stub tooth with enlarged outside diameter

| Number of teeth | 10 |
| Pitch diameter | .8692 |

Gear Data
American standard 20° involute stub tooth

Diametrical pitch	20
Helix angle	18° left-hand
Pitch diameter	3.0492 3

strengths to as high as 48,100 psi are achieved with sintered 201 AB alloy. Beryllium and titanium are used for specialty parts; the various brasses and bronzes for bearings; filters and screens in addition to a wide variety of machine parts; copper and copper alloys for corrosion-resistant parts, filters, screens, friction disks, etc.; iron in combination with aluminum, nickel and cobalt for magnets and pure for machine parts, magnetic cores, radio and electronic parts, etc.; heavy metal for radium containers, in rotors of gyroscopes and as counterweights in machines.

Heat-Treated Parts. The various irons and steels can usually be handled much the same as ordinary parts as far as heat treating, case hardening, hardening, tempering, and drawing are concerned. It is well to remember, though, some of the accuracy inherent in the process may be sacrificed on such treatment. Copper-impregnated iron and steel parts also can be precipitation hardened.

Controlled Porosity. Porosity of the various materials can be accurately controlled in production and can usually be varied from about 5 to 50 per cent.

Specification of Tolerances

Inasmuch as the economy factor of powder metallurgy is directly influenced by the tolerances to be met, these should be as liberal as possible to extend normal die life to the maximum. Ordinarily, on parts merely pressed and sintered, it is possible to hold radial or side-to-side dimensions to plus or minus 0.0015 to 0.002-inch per inch, providing sections are fairly uniform and shape not too complicated, *Fig.* 49.19. Length tolerances — along axis of pressing — on small pieces can be held to plus or minus 0.002-inch, plus or minus 0.005-inch up to about $1^{1}/_{2}$ inches of length, plus or minus 0.0075-inch up to about 3 inches, and plus or minus 0.010-inch on longer parts. Flange thicknesses can also be held to these length tolerances. Standard concentricity tolerances for sleeves and bearings is normally 0.003-inch total indicator reading below $1^{1}/_{2}$-inch bore, 0.004-inch to 3-inch bore and 0.006-inch over 3-inch bore.

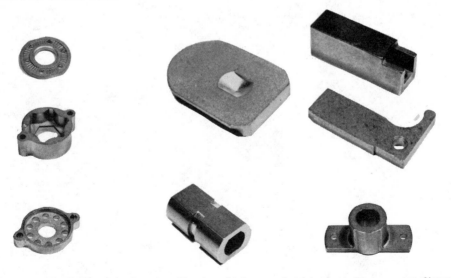

Fig. 49.20 — Fine finish and detail can be achieved as with these small stainless parts. Photo, courtesy Sintered Specialties Div., Panoramic Corp.

Sizing Tolerances. Closer tolerances can be held by sizing or coining at additional expense. Plus or minus 0.0005-inch per inch normal to the axis of pressing can be held. Concentricity tolerances can be reduced to 0.002-inch total and occasionally to 0.001-inch. Length of parts can be held to similarly closer tolerances also by a coining operation.

Economical Limits. For maximum economy, parts should be specified to finish tolerances available from the sintering operation or greater, if at all possible, since machining of powder metal parts, though sometimes performed, is *not* recommended. On secondary operations such as drilling and tapping of holes, threading, addition of features impossible to press, etc., the ordinary machining tolerances apply.

Gears. Both spur and helical gears can be held to excellent limits; tolerances on OD or pitch diameter are generally plus or minus 0.001-inch with ID plus or minus 0.0005-inch. Concentricity between ID and OD seldom varies more than 0.002 total indicator reading. Gear thickness generally can be held to plus or minus 0.005-inch and those under 1/4-inch can be held to plus or minus 0.003-inch.

Finish. Surface finish tolerances of parts in the as-sintered form should be accepted wherever possible to achieve maximum economy. Finish in the as-sintered form is unusually smooth and shiny, *Fig.* 49.20, and excepting those cases where impregnants are used, some degree of porosity usually exists but is not detrimental to operation or service. In fact, the value of this characteristic for lubrication purposes is perhaps the best-known feature of the powder metallurgy process. Where coining is employed as a final operation surface finish is excellent and porosity is hardly discernible.

ADDITIONAL READING

[1]C. L. Downey and H. A. Kuhn, "Designing P/M preforms for forging axisymmetric parts," *International Journal of Powder Metallurgy & Powder Technology*, vol. 11, no. 4, October 1975.
[2]B. L. Ferguson, S. K. Suh and A. Lawley, "Impact behavior of P/M steel forgings," *International Journal of Powder Metallurgy & Powder Technology*, vol. 11, no. 4, October 1975.
[3]James Robertson, "Use of Powdered Metals for Short Runs," presented at the ASME Design Engineering Conference & Show, Philadelphia, Pa., April 9-12, 1973, New York: ASME, Publication 73-DE-25.
[4]*Creating/ with Metal Powders*, Riverton, NJ: Hoeganaes Corporation.
[5]*Powder Metallurgy Design Guidebook*, Princeton, NJ: Metal Powder Industries Federation.
[6]*An Engineering Handbook on Powder Metallurgy*, Toledo, Ohio: The Bunting Brass and Bronze Company.

50

Plastics Molding

Charles E. Chastain

AMONG the basic engineering materials widely utilized and accepted to-day, plastics are rapidly approaching a dominant position. The volume of plastics used today in the U.S. already exceeds all non-ferrous metals on a cubic inch basis and will, by 1985 it is estimated, also exceed that of ferrous metals.

Several important factors account for this explosive growth during the past twenty years. Plastics can be easily fabricated on a high volume basis to very accurate tolerances and usually require no additional machining or finishing. They are typically strong, tough, lightweight, corrosion resistant and durable. But by varying the chemical formulation of the plastic, properties can be varied from brittle to impact resistant, clear to opaque, electrically conductive to insulating, weather resistant to photodegradable, high-temperature rigidity to low-temperature flexibility, and can be economically molded into almost any shape at a cost usually less than competing materials.

However, plastics should not be considered simply as a cheap substitute for metal but as an engineering material in its own right, *Fig.* 50.1, with its own unique combination of properties, limitations and processing requirements. The designer must understand that plastics are better than metals in some respects and worse in other respects. Plastics are suitable for applications not only where they provide similar properties to competing materials at a lower cost, but more importantly where they provide properties not obtainable with any other material.

Types of Plastics. The multitude of types of plastics now available for use in design can be broken down generally into two basic categories: (1) Thermoplastic and (2) thermosetting. The thermoplastics, as the name indicates, are heat-softening. After heating, forming and cooling, they can be remelted again. On the other hand, thermosetting plastics, or thermosets, are heat-hardening and, once having been molded under heat and pressure, the synthetic polymerizes into a dense, hard infusible mass that cannot be remelted again.

Forms of Plastics. For convenience, plastics may be subdivided into several basic forms according to the means by which parts are produced. Generally, these are: (1) Molding, (2) laminating and (3) casting. Each of these methods of making finished plastics parts consists of several or more variations which permit a wide selection and extend the practical possibilities in design. Casting has been utilized

Charles E. Chastain is President, Plastics Technical Concepts, Inc., Chicago.

Fig. 50.1 — Injection molded phenolic reactor turbine wheel which replaces multipiece cast aluminum unit in all model lines. Courtesy Ford Motor Co.

mostly for tools, dies, models, patterns and sample parts with limited applications possible for production machine parts.

Molding Methods. The various useful molding method subdivisions and their area of materials or design application are generally as follows:

1. Compression . . . . Thermosetting; powder or premix with paper, fiber or cloth fillers.
2. Transfer . . . . Thermosetting; complicated designs with cores, inserts, etc.
3. Injection . . . . Thermoplastics; powder form; fast molding of intricate parts. Thermosetting; complicated designs with cores, inserts, etc.
4. Extrusion . . . . Thermoplastics; tubes, rods, sheets, shapes, etc.
5. Cold . . . . Refractory and nonrefractory; resistance to high heat and arcs.
6. High-pressure laminated . . . . Thermosetting; paper, glass fiber and cloth filled phenolics, high strength, electrical properties, deep draws.
7. Low-pressure laminated . . . . . Thermosetting; wood, paper, glass fiber, or cloth-filled phenolic or polyester resins for large size products.

Compression Molding. In compression molding a thermoset plastic, normally as powder, preform or premix, is loaded directly into the hot die cavity, which is held at temperatures ranging from 290° to 350°F. Pressure from 2000 to 20,000 psi is applied, held for a curing period, and then the finished part is ejected. Time for polymerization or curing depends upon the design, small thin-wall pieces requiring as little as 20 seconds.

During compression molding, the "force" or compressing member of the open mold applies the molding pressure directly on the unmelted plastic material, forcing it to melt and flow, filling the cavity as the mold closes, *Fig. 50.2.* Pressure is maintained until the hot plastic "sets" at which time the mold is opened and the hot part ejected.

Transfer Molding. Transfer molding was developed largely to avoid the disadvantages found with compression molding. Unlike compression molding, the mold is closed without loading plastics material into the cavity. Instead, powder, pellet or preform material is loaded into a separate cylinder and, under heat and pressure of approximately 15,000 psi, is transferred or injected into the closed dies, *Fig. 50.3.* Curing time required after transfer is usually less than that of compression molding curing time. However, transfer molding has more waste since the runners, sprue, and transfer cylinder residue cannot be remelted and reused.

Employed for producing thermosetting plastics, transfer molding permits the use of intricate inserts and slender cores not practicable with the compression method. In plunger-transfer molding, *Fig.* 50.4, the transfer pot is built into the mold on the parting line, reducing the amount of wastage. A more recent development, screw transfer molding, brings the transfer method very close to injection molding as used for thermoplastics. The plastic material is "plasticated" in shot quantities of from 15 to 1000 g. and forced by ram or screw into a preheated closed die. Good flowing materials with a long flow period are desirable.

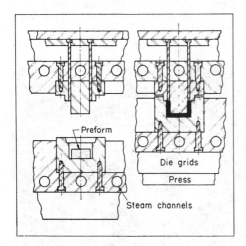
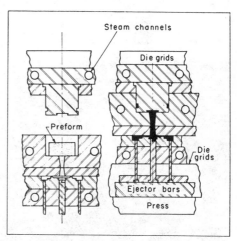

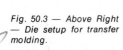

Fig. 50.2 — Above — Typical arrangement of dies for compression molding.

Fig. 50.3 — Above Right — Die setup for transfer molding.

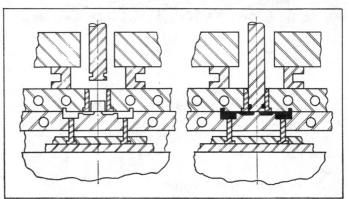

Fig. 50.4 — Right — Duplex or plunger arrangement for transfer molding.

Injection Molding. This is the most successful and widely used method for molding thermoplastics. The molding process begins with granulated material fed from a hopper, in measured amounts, to the heating chamber where it attains a fluid consistency under temperatures from 250° to 500°F. The softened plastic is then automatically injected into a chilled mold by means of a screw or plunger, *Fig.* 50.5, under pressures from 10,000 to 30,000 psi. After a short cooling period, the dies open and eject the completed solidified piece or group of pieces. Reaction injection molding of two metered urethane components produces complex structural foam parts with excellent physical properties, *Fig.* 50.6. Notable features which have helped make injection molding of thermoplastics popular are: (1) Less material loss — gates and runners can be reused; (2) fewer finishing operations — flash characteristic in compression molding is eliminated; and (3) high thermal

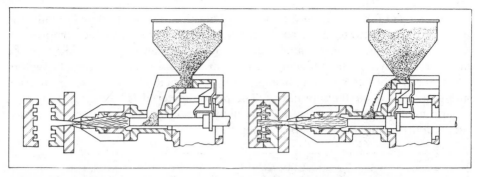

Fig. 50.5 — Typical setup for injection molding of thermoplastics. Drawing, courtesy Amos Molded Plastics Div.

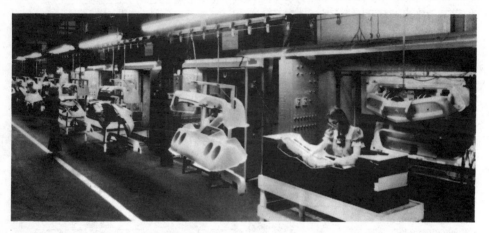

Fig. 50.6 — Reaction injection molding line that can produce over one million bumper size parts per year. Shown are front and rear fascia components for current model cars. Courtesy, Davidson Rubber Co., Inc., Div., McCord Corp.

efficiency — the heating chamber and mold remain at constant temperatures.

Recently, injection molding has also been adapted to mold thermosets by preheating and mixing the thermoset powder in the injection molding machine, then injecting the softened material directly into a hot thermoset mold in a manner similar to transfer molding. The melt is more uniform than compression or transfer molding and parts will usually cure faster, although with higher stress levels.

Extrusion Molding. Molding of thermoplastics by continuous extrusion is not unlike ordinary extrusion methods. Granular material is fed by hopper into a heating cylinder through which it is forced by a screw into a heated shape die. Continuous lengths are produced at an average rate of 500 to 1000 feet per hour, the hot material cooling and hardening on the take-off belts.

Die requirements are relatively inexpensive and equipment is much simpler than for the other methods of forming plastics. Generally, thermoplastic materials are the only ones extruded.

Blow Molding. This process is extensively used for making bottles and other hollow plastics parts having relatively thin walls. To blow mold a part, an extruder first extrudes a hollow tube (parison) in a downward direction, where it is captured at the proper time between two halves of a shaped mold. This closes the top and bottom of the tube. Air is then blown into the soft parison, expanding it until it uniformly contacts the inside contours of the cold mold and solidifies. Then the mold automatically opens, the part is ejected and a new cycle begins.

Polyethylene is by far the most commonly used plastic for blow-molded articles, but use of PVC, polypropylene and polystyrene is rapidly increasing. Bottles, watering cans, display fruit and hollow toys are typical blow-molded products.

Rotational Molding. Some hollow parts having heavy walls and/or complex shapes may be produced more economically by rotational molding. Here, a premeasured amount of plastic in powder or liquid form is placed in the bottom half of a cold mold, after which the mold is closed and rotated continuously about its vertical and horizontal axes to distribute the material uniformly over the inner mold surfaces. The rotating mold then passes through a heated oven where the plastic is fused or cured. While continuing to rotate, the mold subsequently passes into a cooling chamber where it is quickly cooled. This solidifies the plastic inside into a uniformly thick, hollow shell whose outside surface and contour faithfully duplicate those of the inner mold surface. The mold opens, the finished plastic part is removed and the cycle is repeated.

Hollow balls, squeeze toys, tanks, hobbyhorses, heater ducts and auto armrest skins are commonly made by rotational molding. Polyethylene and PVC account for almost all the plastics now processed this way.

Cold Molding. In the cold molding of parts[1], the powdered material is compressed cold to the desired shape, the method closely paralleling that for powder metals. The compressed forms are then baked in ovens at a temperature of 225° to 500°F. for a period of 12 to 14 hours depending upon size. Suitable plastics for cold molding are: (1) Nonrefractory types made with bitumen or phenolic resin base, and (2) refractory types made with inorganic binder compositions. Refractory materials receive in addition to the regular bake a special high-humidity baking treatment.

Laminated Molding. Laminated plastics are available in simple stock forms such as sheet, rod and tube which may be punched, sawed or machined as desired. However, where other forms are desired, laminated molding is utilized. In this method, layers of impregnated paper, woven fibers or cloth are located in a mold much like an ordinary compression mold, *Fig. 50.7*, subjected to pressure and curing temperature. Mold cycles are generally slow and may range from minutes to as much as several hours.

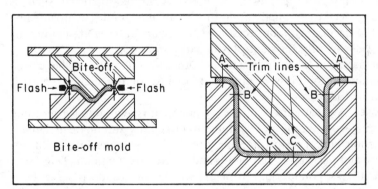

Fig. 50.7 — Typical molding arrangements for laminated materials. Drawing, courtesy General Electric Co.

Where the parallel laminated structure is undesirable in parts for strength purposes, a variation of the laminate structure known as molded macerate is employed. Here, by shredding or chopping the cloth or paper, better flow is produced and a distorted, nonuniform, interlocked formation is obtained.

Although they show lower mechanical properties than straight laminates, the molded macerates have a more uniform overall strength. Like laminates, they are not suited to fast molding cycles.

In addition to high-pressure laminated molding is that of low-pressure laminated molding which utilizes primarily thermosetting resins capable of void-free polymerization at pressures under 500 psi. Plies of resin-coated or resin-impregnated wood, paper, cloth, or glass fiber are formed over a die by pressure applied through a flexible diaphragm, *Fig.* 50.8, with heat to set the resin. Besides steam, other methods such as the vacuum bag, *Fig.* 50.9, and blanket are used to actuate the diaphragm.

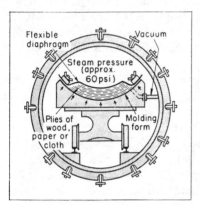

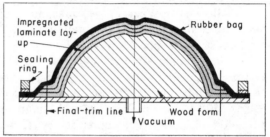

Fig. 50.8 — Left — Low-pressure laminated molding in autoclave with steam.

Fig. 50.9 — Vacuum-bag process for molding laminates.

Advantages of low-pressure laminated molding are found in the practically unlimited size of part in area, easy positioning of inserts or preformed shapes to provide assemblies or undercuts, low cost of production molds and equipment, thin sections of unusual strength and lightness, ease of forming hollow or complex shapes, and wide choice of material. Disadvantages lie in smooth, mold-finished surface on only one side of part and accuracy required in preliminary build-up of layers owing to poor flow characteristics of the materials.

Another important method of producing molded laminates is by the hand lay-up technique. Here layers of polyester or epoxy saturated material such as fiberglass cloth are placed over a male mold or into a female mold, compacted by hand and allowed to harden without pressure. After removal, they are among the strongest and most chemically resistant plastic parts produced.

Although canvas, nylon fabric, asbestos, and other fibrous materials have been used as reinforcements, fiberglass is preferred because of the high strength it imparts to the finished laminate.

Molding Presses. A great variety of press equipment is available for molding plastics parts, ranging from simple manual types to large fully automatic machines. Common compression and transfer molding presses today range in capacity to 4000 tons and more and parts up to 20 pounds and greater are common. Television cabinets from 12 to 35 pounds and from 15 to 18 inches deep have been compression molded on larger presses. One of the largest in use today exerts 4000 tons molding pressure to make possible considerably increased sizes, *Fig.* 50.10.

Injection molding machines, *Fig.* 50.11, in common usage generally range up to 800 ounces (G.P. Polystyrene), or 1520 cu. in. shot capacity. However, there are in use today standard machines with a great variety of sizes extending up to 4570

Fig. 50.10 — Down-acting compression molding press for both die try-out and production of fiberglass reinforced body panels exerts 4000 tons pressure. Courtesy, K. R. Wilson, Inc.

Fig. 50.11 — Injection molding machine with hydromechanical mold clamp which exerts 1000 tons pressure with an ejection pressure of 25 tons. Courtesy, Cincinnati Milacron Co.

tons with 1175 ounces shot capacity.

The weight capacity of any machine is, of course, dependent upon the specific gravity of the particular plastic. For instance, a 22-ounce acetate rated machine would be 16-ounce rated with polystyrene. Parts which exceed these weight figures are generally termed large-size injection-molded parts.

Production Speeds. Ordinarily, injection molding presses can operate at speeds up to 30 cycles per minute, but the average is about 6 cycles per minute. The average for compression molding is about 2 cycles per minute.

Production speeds are highly dependent on the type of plastics used, part shape, wall thickness and tolerances required. Multi-cavity molds are used to increase production rates, but they always run at a slower cycle than a single-cavity mold. Thus an eight-cavity mold will usually produce only 5 to 7 times the number of parts per minute of a single cavity.

Design Considerations

The first overall consideration in determining the most desirable method of molding, as well as that for practical molding in general, is accepted to be the piece design. Part detail design for molding thus becomes an outstandingly important consideration.

Wall Thickness. Of paramount importance in design of molded parts, wall thickness should be maintained as uniform as possible. Abrupt changes should be avoided. Nonuniform walls result in nonuniform shrinkage, warpage, high stress concentrations, production difficulties and higher material costs, *Fig.* 50.12. Whenever practicable, heavy sections in molded parts should be cored out to attain a uniform wall.

Fig. 50.12 — Group of typical precision parts produced on injection molding equipment with from 6 to 70-ounce injection capacity. Courtesy, Cincinnati Milacron Co.

Minimum desirable wall thickness in compression molding is generally accepted as being 0.062-inch with depth to 2 inches, and increases with size. Walls as light as 0.031-inch are molded occasionally.

For injection-molded parts wall thickness should seldom be less than 0.050-inch unless travel of material is short and pressures high. Under any circumstances 0.015-inch is about minimum. Preferable wall ranges from 0.070 to 0.125-inch.

In general, laminates such as high strength glass-fabric parts are limited regarding practical wall thickness. Finished walls can range from 0.020 to 0.250-inch or more, but the lightest wall compatible with the design being, of course, most economical. Uniformity of thickness is a must, *Fig.* 50.13.

Draft. An important consideration in design is facility in ejection of parts from the mold after solidifying. Taper or draft is required on all surfaces normal to the parting plane of the mold. Interior walls, ribs, webs, recesses, and cavities should have a taper of at least one degree. Production costs increase sharply when taper is not included on outside wall design although it is possible to mold straight sides.

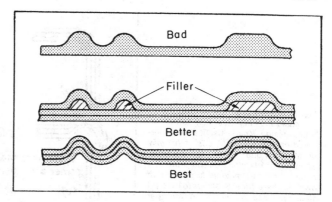

Fig. 50.13 — Comparison of wall variations in parallel laminates. Glass fiber and premolded filler sections can be used but are expensive. All added wall thickness requires special hand fitting of the laminate.

Taper per side as low as 1/4-degree can be used, but average draft runs from 1/2 to 5 degrees.

Generally, practical draft with short walls can be figured at 0.005-inch per inch, minimum, but with longer lengths this can be decreased to as low as 0.001-inch per inch taper, per side. Cold-molded items require a minimum draft of 3 to 5 degrees as do deep barriers and bosses on all parts.

Paper, cloth or glass-fabric laminated molded parts require the most liberal draft possible to permit economical production.

Radii and Fillets. Generous fillets and radii on parts not only improve flow of the material in the mold but also add as much as 50 per cent to the strength. Fillets are preferred to sharp corners in all cases except at the parting line, *Fig.* 50.12. All parts should be designed with radii of 0.020-inch, minimum, and for cold-molded parts, 1/16-inch, minimum. Stress concentration studies indicate a minimum fillet radius equal to one-fourth the wall thickness should be observed for good service.

Owing to the character of laminated molding, generous corner radii at all points are essential. Sharp or square corners are difficult to produce and involve either special cutting of the laminations before pressing to prevent rupture or machining after molding, *Fig.* 50.14. Flowing lines materially aid the molding process.

Holes. Certain slenderness ratios have been established for molded work regarding the core pins necessary for producing holes. These ratios establish the practical limitations on hole proportions. In compression molding, blind holes of 1/16-inch diameter or less should not exceed their diameter in length. Larger blind holes are limited to a depth of about twice their diameter. Deeper holes must, as a rule, be through holes. Limitations for intricate molded pieces are given in TABLE 50-I, with accompanying *Fig.* 50.15, and provide a reasonable factor of safety. Where holes are to be used for driving self-tapping screws, the blind hole should have a one-degree taper per side. Depth of hole should accommodate at least a 3/16-inch length of screw.

Through holes, according to the foregoing ratios, can thus be made up to four times the diameter by two core pins which meet, the larger extending the greater length. One pin, however, should be at least 0.015 to 0.030-inch greater than the other to accommodate possible misalignment between mold halves. Where through-hole core pins can be supported in both mold halves, ratios up to five-to-one have been found practical.

Transfer and injection molding materials permit much greater hole depths. Commonly, ratios of three or four-to-one and occasionally six-to-one can be used.

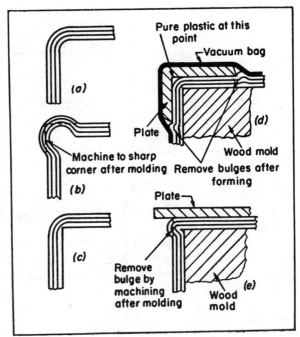

Fig. 50.14 — Possible corner designs for laminated molding. Rounded corner, (a) is best design. If part must fit a sharp-cornered mating assembly, design (b) may be used although corner is weakened. Shown at (c) is effect of sharp internal corner. Laminate material is pulled thin over the sharp edge. Sketches (d) and (e) show methods of producing stronger sharp internal corner. These techniques increase costs and produce weak points if bulges are removed. Drawing, courtesy Laminated Plastics Inc.

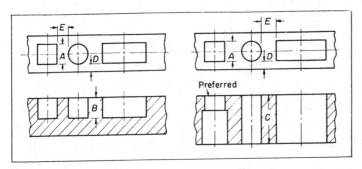

Fig. 50.15 — Design of common compression-molded holes. Values for various dimensions are shown in TABLE 50-I.

With core pins supported in both mold halves, ratios up to twelve-to-one have been employed.

Poorly located holes may result in increased shrinkage and cracking. Holes under 3/32-inch diameter should be at least one diameter from the edge or wall of a part and never less than 1/16-inch from another hole. For larger holes lesser ratios are permissible as indicated in TABLE 50-I. In addition, holes should, if possible, be located parallel to the direction of mold movement to eliminate special molds and attachments. Square hole core pins tend to wear at the corners and where high accuracy is necessary special reinforced edges may be required on the core pins. Deep, narrow slots can often be made practical by including a heavier center shank portion for strengthening the core pin, *Fig.* 50.16.

Through holes in laminated-molded parts are most often drilled after molding. Where parts are to be fastened with rivets, size of rivets is limited to 5/16-inch owing to the possibility of fracturing the laminate in upsetting larger sizes. Where countersunk rivets are used, laminate thickness should provide at least 1/32-inch of cylindrical hole between the bottom of the countersink and the opposite face. In the event breakthrough of the countersink is not avoidable, a plain

Table 50-1 — Limitations on Small Cored Holes
(inches, see *Fig.* 50.15)

Hole Diameter A (Min)	Blind Depth B (Max)	Thru Depth C (Max)	Side Walls D (Min)		Hole Spacings E (Min)	
			Hot Molded	Cold Molded	Hot Molded	Cold Molded
$\frac{1}{16}$	$\frac{1}{16}$	$\frac{1}{8}$	$\frac{1}{16}$	$\frac{3}{32}$	$\frac{5}{64}$	$\frac{7}{64}$
$\frac{5}{64}$	$\frac{3}{32}$	$\frac{3}{16}$	$\frac{1}{16}$	$\frac{3}{32}$	$\frac{3}{32}$	$\frac{1}{8}$
$\frac{3}{32}$	$\frac{1}{8}$	$\frac{1}{4}$	$\frac{1}{16}$	$\frac{3}{32}$	$\frac{3}{32}$	$\frac{1}{8}$
$\frac{7}{64}$	$\frac{5}{32}$	$\frac{5}{16}$	$\frac{3}{32}$	$\frac{1}{8}$	$\frac{1}{8}$	$\frac{5}{32}$
$\frac{1}{8}$	$\frac{3}{16}$	$\frac{3}{8}$	$\frac{3}{32}$	$\frac{1}{8}$	$\frac{1}{8}$	$\frac{5}{32}$
$\frac{5}{32}$	$\frac{1}{4}$	$\frac{1}{2}$	$\frac{3}{32}$	$\frac{1}{8}$	$\frac{1}{8}$	$\frac{3}{16}$
$\frac{3}{16}$	$\frac{5}{16}$	$\frac{5}{8}$	$\frac{1}{8}$	$\frac{1}{8}$	$\frac{1}{8}$	$\frac{7}{32}$
$\frac{7}{32}$	$\frac{3}{8}$	$\frac{3}{4}$	$\frac{1}{8}$	$\frac{3}{16}$	$\frac{3}{32}$	$\frac{1}{4}$
$\frac{1}{4}$	$\frac{7}{16}$	$\frac{7}{8}$	$\frac{1}{8}$	$\frac{3}{16}$	$\frac{3}{16}$	$\frac{9}{32}$
$\frac{5}{16}$	$\frac{9}{16}$	$1\frac{1}{4}$	$\frac{3}{32}$	$\frac{1}{4}$	$\frac{1}{4}$	$\frac{11}{32}$
$\frac{3}{8}$	$\frac{11}{16}$	$1\frac{3}{8}$	$\frac{5}{32}$	$\frac{1}{4}$	$\frac{1}{4}$	$\frac{11}{32}$
$\frac{7}{16}$	$\frac{13}{16}$	$1\frac{5}{8}$	$\frac{3}{16}$	$\frac{3}{8}$	$\frac{3}{8}$	$\frac{1}{2}$
$\frac{1}{2}$	$\frac{15}{16}$	$1\frac{7}{8}$	$\frac{3}{16}$	$\frac{3}{8}$	$\frac{3}{8}$	$\frac{1}{2}$

hole should be used with a countersink washer.

Undercuts and Recesses. Whenever an undercut portion is used the problem of removal from the mold must be considered. Undercuts which necessitate split molds increase part cost considerably and, generally, internal undercuts are impractical. Internal recesses should be extended to the parting line to permit the part to be drawn out of the cavity or off the plunger. Internal and external recesses can sometimes be used to form a side hole in this manner, *Fig.* 50.17.

Bosses. Essential for mounting and assembly, bosses should be located whenever practicable in corners. Three bosses are preferable for mounting purposes. If four are necessary, a grinding operation is usually required to bring the top surfaces into planity. All mounting bosses should project 0.015-inch or more above all other projections in the same area.

High bosses which must be formed in the upper mold half should be avoided. High bosses have a tendency to trap gas and may be weak. Small bosses should not exceed two diameters in height. All large bosses should be hollowed to maintain wall uniformity. Wherever possible, especially on high bosses, a draft of 5 degrees on all sides is recommended along with generous fillets.

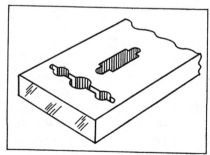

Fig. 50.16 — To strengthen deep, narrow-slot core pins, a round center portion is recommended.

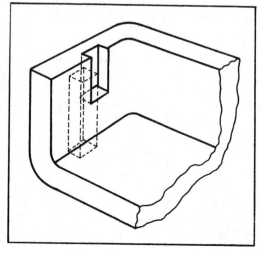

Fig. 50.17 — Both external and internal recesses should be extended to the parting line to simplify molding. In this design a side hole is produced. Drawings, courtesy General Electric Co.

Bosses, much as variations in wall thickness, are difficult to produce in laminated molding and should be eliminated in favor of raised or dished portions with uniform walls, *Fig.* 50.13. Filled thermoset parts, however, are not subject to the same limitations and follow more closely the requirements for regular compression moldings, *Fig.* 50.18.

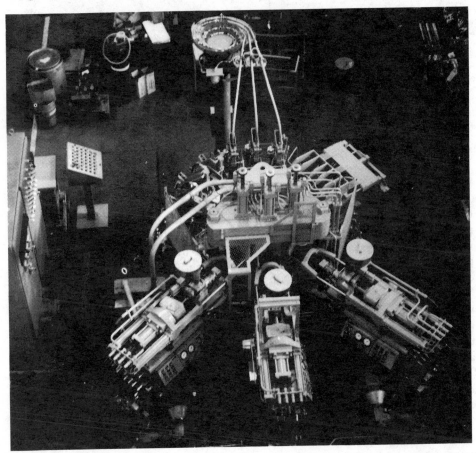

Fig. 50.18 — Asbestos-filled phenolic compound is used to mold a copper shell and aluminum bushing into a commutator assembly by automatic screw-transfer molding in this rotary six-station triple-head machine. Courtesy, Stokes Div., Pennwalt Corp.

Ribs. Large, unsupported surfaces should be ribbed for strength and to resist warpage. In addition to improving strength, ribs act as runners or feeders to assist material flow to isolated bosses, pads or projections. Ribs require generous draft and should have at least 5 degrees taper per side, *Fig.* 50.19. Rib pattern over large areas should be staggered and consideration should be given the possibility of hollowing the ribs to maintain a uniform section. If rib sections are too heavy, shrink marks may appear opposite. The rib proportion in *Fig.* 50.19 will largely eliminate any possibility of shrink marks if the wall relationship of 2R is maintained.

With laminated-molded pieces, ribs may be molded into the design but generally more intricate molds are necessary and are expensive.

Threads. As a rule, holes molded with threads are not always satisfactory if the

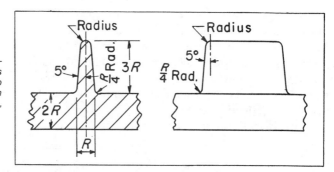

Fig. 50.19 — Recommended proportions for ribs. Rib spacings should not be closer than the rib base width dimension utilized in multiribbed designs. Drawing, courtesy General Electric Co.

thread must engage metal, especially where frequent disassembly is necessary. Also, owing to shrinkage, the pitch of molded threads of fair length may not fit standard metal threads but will match molded mating threads, *Fig.* 50.20.

Fig. 50.20 — Noryl 20 per cent glass reinforced resin plastic water softener valve outperforms metal and reduced cost by 25 per cent. Dimensional stability permits molded threads for inlet/outlet and tank connections. Courtesy, General Electric Co.

Generally, female threads over 5/16-inch diameter and coarser than 32 pitch are molded. Tapping of threads over 1/4-inch is impractical in phenolics. Small diameter threads, especially under 3/16-inch, are usually more economical to mold as plain holes and tap. Tapped threads range from 68 to 75 per cent full thread and in small sizes the best possible fit is Class 2 with 68 per cent full thread. Tapped holes should have a molded-in counterbore or countersink to prevent chipping in tapping and assembly. Wall thickness of three diameters around tapped holes is recommended for adequate strength.

Internal threads may be molded integral and external threads may be molded in two sections with flash and parting lines removed by a sizing die after molding. Threads in blind holes with limited space or thin walls should be molded. Standard threads should always be specified. Threads should terminate abruptly in a shoulder of major diameter; feather edges are undesirable on mold parts. Likewise, threads should start at least 0.031-inch in from the face normal to the thread axis to avoid fragile feather edges, *Fig.* 50.21. Strongest possible molded threads result from use of macerated or flock-filled materials.

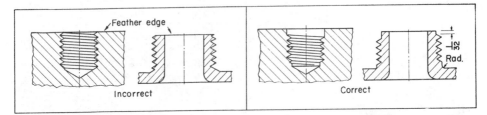

Fig. 50.21 — Feather edges on threads should be avoided in molding. Drawing, courtesy General Electric Co.

Fig. 50. 22 — Group of Valox Thermoplastic polyester resin parts. Dimensional stability and heat resistance makes this material excellent for electrical parts with inserts, gears, etc. Courtesy, General Electric Co.

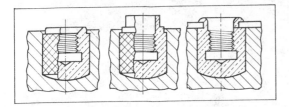 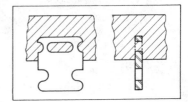

Fig. 50.23 — Typical female threaded inserts for bolting and riveting to metal attachments.

Fig. 50.24 — Stamped insert lug showing punched hole for anchorage.

Inserts. A wide variety of inserts are employed for purposes of fastening, mounting, electrical connection, etc., *Fig.* 50.22. If frequent disassembly is a thread requirement, the best design approach is to employ metal inserts. Threaded inserts, *Fig.* 50.23, require adequate thread length for assembly, sufficient strength to resist crushing in molding and proportions to assure suitable anchorage. If used, well-rounded diamond knurls for anchoring are recommended, although grooving or punching can be used, *Fig.* 50.24. Female through inserts should be avoided and means for preventing molding material from entering the threads is necessary with male or female pieces.

Three types of male inserts shown in *Fig.* 50.25 exemplify good and poor

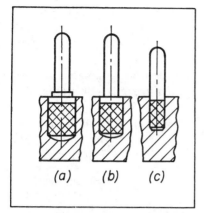

Fig. 50.25 — Male inserts with double seal (a) are best, single shoulder (b) provides reasonably adequate compound seal, shoulderless (c) requires added finishing operation at increased cost.

Fig. 50.26 — Large hub inserts in nylon mixer drive gears. Knurled grip surface can be seen. Photo, courtesy Chicago Molded Products Corp.

design. For adequate anchorage, it is desirable to have an embedded length equal to two diameters and the coarse diamond knurl should terminate about 1/32-inch from the exposed shoulder to permit molding the insert in with about 0.005 to 0.010-inch projecting above the adjoining molded area. All inserts should be normal to the parting line of the mold if possible, for maximum economy.

Sufficient material is required around inserts to withstand the shrinkage on cooling without cracking. In any case, walls under 3/32-inch are seldom recommended and should be avoided. These **minimum** thicknesses of material around the insert are suggested:

Insert Material	Material Thickness (minimum)
steel	equal to outer maximum of insert
brass	0.9 times outer radius of insert
aluminum	0.8 times outer radius of insert

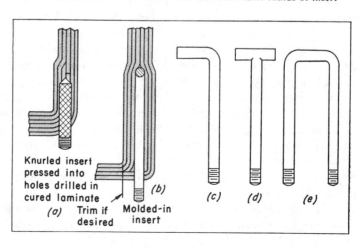

Fig. 50.27 — Knurled type inserts (a) are pressed into drilled holes after curing laminated molded part. Molded-in type at (b) are of typical designs shown at (c), (d), and (e). Drawing, courtesy Laminated Plastics Inc.

Knurled insert pressed into holes drilled in cured laminate

(a) Trim if desired

(b) Molded-in insert

(c)   (d)   (e)

The majority of inserts are made of brass or steel, *Fig.* 50.26. Many types are standardized and available from stock. However, for special designs other materials such as porcelain, woven wire mesh, laminated phenolic materials, glass fiber, Mycalex, etc., are widely utilized.

Where flanges are undesirable for holding or securing laminated molded parts, inserts can often be used to advantage also. If high strength is desirable, T or U-bolt type inserts are preferred, *Fig.* 50.27.

Shrinkage. Molded parts are usually 0.000 to 0.050-inch per inch shorter than the mold after cooling. Some warpage also may occur. From the cold mold to the cold piece average shrinkage for *some* plastics is approximately as follows:

	Shrinkage (in./in.)
Polyvinyl Chloride (Rigid)	0.001 to 0.004
Phenolic, (G.P. Grades)	0.003 to 0.009
ABS Polymer	0.003 to 0.008
Modified Acrylic	0.004 to 0.008
Cellulose acetate	0.002 to 0.005
Glass Reinforced Lexan	0.007
Methyl methacrylate	0.003 to 0.008

Identifications and Designs. For identity markings or designs with machined die cavities, raised letters could be employed, *Fig.* 50.12, and where dies are produced by hubbing, depressed letters should be utilized. Letters which project 0.003-inch above the surface are legible and about 0.010-inch is used as a desirable maximum value for lettering up to 1/8-inch. Where lettering or other markings are over 0.030-inch high, draft and fillets should be used. Intricate designs which project about 0.003-inch can be produced by photographing and etching. Costs for various raised or depressed lettering compare as follows:

	Machined	Hubbed
Raised on molded surface	Low	High
Raised on depressed pad	Medium	High
Depressed on molded surface	High	Low
Photoengraved lettering	Low	Low

Sharp stamps will print or roll markings onto a surface without heat. Deep branding to 0.006 or 0.008-inch depth, however, is accomplished with heated dies.

Finishing. Letters to be filled with paint should be sharp edged from 0.005 to 0.030-wide, half as deep as wide, and with rounded bottom corners, *Fig.* 50.28. Where painting is necessary, adherence can be improved by sand-blasting the part or the mold surface. Plastics can be coated with copper, chromium, silver, gold, and other metals by electroplating, metal spraying or vacuum metallizing.

Ejection. Ordinarily, parts must be removed after molding by means of positive ejection pins, *Figs.* 50.2, 50.3 and 50.4. Such pins must bear on solid, firm portions which will not easily permit deformations. On a design, it is well to indicate where such pins can be utilized without affecting the function of the finished part. As a rule, ejection pins leave a small indentation, and an allowance either up or down of 0.015-inch should be permitted. High ribs and barriers require knockout pins beneath and preferably a circular boss should be provided if practicable.

Parting Lines. Needless to say, molding is simplified and cost is held to the minimum by adhering to straight partings. Design of parts should be modified

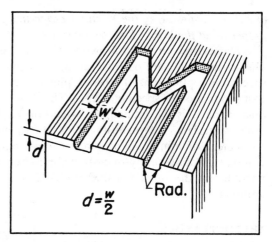

Fig. 50.28 — Typical design for letters to be filled with paint.

wherever possible to permit mold parting in one plane. This parting, for economy, should be so located that flash can be removed with a minimum of finishing. A bead or sharp edge simplifies removal.

Postforming. Laminates postformed hot preferably should have generous radii throughout. Typical forming radii vary from 3/32-inch for 1/16-inch sheet to 2 1/2 inches for 3/8-inch sheet. Minimum radii for single and compound curvatures are generally set at two times the sheet thickness. Some drawing is possible and 1/16-inch sheet, for instance, can be drawn to a depth of 1-1/2 inches for a 3-inch spherical cup. Deep-drawn or complicated shapes having an ultimate elongation of 10 per cent lengthwise, 12 per cent crosswise, and 4 per cent diagonally are possible.

Selection of Materials

Selection of the proper plastic for any particular product application is important. The product designer must answer the key question — what does the part do? What is expected of it in terms of the forces it must bear, the temperature range it must endure or the time of operation?

Once the designer can answer such questions he can then establish ranges for such material parameters as degree of stiffness or flexibility; tensile and impact strength; electrical properties; gas permeability; long term creep properties and so on. There will be an "essential value," a value that the material must possess to even be considered for use in the part. A "desirable value" for the parameter must also be established by taking into account uncertainties in calculations and tests, and safety factors. Finally, there is a range above which the value of the parameter is unimportant.

As an example, the designer may find that his part must be strong enough to withstand a compressive force of 6000 psi (essential), although he would feel more comfortable with a value of 8000 psi (desirable), while a value above 10,000 psi would be unimportant on the basis of strength alone.

If the designer also includes categories for the engineering factors of production (quantity, speed, machinery); design limitations due to materials, methods and economics; and material cost/lb, cost/cu.in. or cost/unit strength, he can make a knowledgeable choice of material. *The material to be used is simply that one that satisfies all of the essential properties plus the greatest number of desirable properties at the lowest final unit cost.*

Properties. In the following, the various important properties of typical plastics materials are outlined along with some of their typical applications. Few successful applications of molded plastics can be made only on the basis of lower cost production or speed of output. The major factor influencing the use of molded plastics parts is the attainment of certain physical properties not available with other materials. Adaptability of design to molding follows in importance and attainment of minimum production cost, output speed, uniformity of product, elimination of assembly operations, etc., become highly valuable added features possible by thorough consideration of all the advantages and limitations of molding.

THERMOPLASTICS

POLYETHYLENE

Trade Names
Poly-eth — Petrothene — Polythene — Alathon — Hi-Fax — Fortiflex — Dylan.

Properties
An excellent combination of toughness, good physical and electrical strengths, lightweight, easy processing conditions and low cost. Good flexibility — Anti-stick surface, resistant to many chemicals. Easy heat sealable — Low water, vapor permeability — Films can breathe. Low heat resistance (120°F) — Wax-like surface — Easily stress and flex-cracked — Process sensitive — Difficult to hold tolerances — Cold flows under load.

Applications
Squeeze bottles — Chemical Tanks — Toys — Garbage cans — Film bags — Wrapping — Disposable gloves — Coating paper: milk cartons, boxes, news wrapping — Wire and cable insulation — Bottle closures — Irrigation pipe.

Other Notes
High density: Linear — High crystallinity — Stiff.
Low density: Branched — Low crystallinity — Flexible. Cross-linking is possible.
Low heat degradation — Permits unlimited use of regrind without loss of properties.

POLYPROPYLENE

Trade Names
Poly-pro — Marlex — Petrothene — Pro-fax — Escon — Avison.

Properties
Heat resistance to 250°F — Very low specific gravity — Floats on water — Fairly hard, rigid and scratch resistant compared to polyethylene — Good impact strength above 0°F — Good chemical resistance — Excellent resistance to stress or flex cracks — Electroplatable.

Applications
Picnic plates — Cups — Hospital dinnerware — Coffeemaker parts — Vaporizer housing — Washing machine pumps — Rope — Indoor-outdoor rugs — Living hinge — Packaging films (.010-.021" thick, .080 land) — Auto parts: battery cases, kick panels, fan shrouds, heater ducts, fender fillers, bucket seat backs.

Other Notes
Normally does not stress-crack. Fillers: Talc, Glass. Usually molded in slightly warm molds (100-150°F) — Requires higher pressures to mold than polyethylene.

POLYSTYRENE

Trade Names
Styron — Dylene — Lustrex — Bakelite — Fostarene — Ampacet — Polystyrol.

← Fucking Thermosets

Properties
Low price — Hard — Rigid — High shine — Easy to process — Metallic sounding — Low moisture absorption — Odorless, tasteless — Scrap easily reclaimed — Easily surface decorated — Limited heat resistance — Limited solvent resistance — Resists many acids, Alkali — Poor outdoor stability — Low shrinkage during molding — Many modifications possible to make foams, rubbers and copolymers.

Applications
Model builders kits — Toys — Novelties — Bath tile — Display packaging — Appliance housings — Glassware — Foam: Ice buckets, shipping cartons, thermal insulation and furniture components — Camera cases — Luggage — Pipe — Refrigerator Liners — Lighting panels.

Other Notes
Expandable foams. (Gas - Chemical) — Parts — Slabs — Strands. SAN (A copolymer with higher strengths — Higher costs — Harder to process) — Impact types (Modified by mechanical means — Butadiene — PVC) — Styrene — Butadiene elastomers (Swim fins — Rubber bands — Tennis shoes — Very low shrink — Difficult to mold and eject) — Styrene also a part of ABS Terpolymers.

POLYVINYL CHLORIDE

Trade Names
Exon — Geon — Pliovac — Marvinol — Plaskon — Tygon — Vyram — Vinylite.

Properties
Unplasticized PVC is hard — Tough — Very resistant to weather, moisture and chemicals — **Self-**extinguishing. Plasticized PVC is softer — More flexible — Less resistant to all environments **but** much easier to process — Both have excellent electricals.

Applications
Unplasticized: Pumps — Impellers — Pipe — Siding — Window channels — Molding trim. Plasticized vinyls: Seat covers, garden hose, floor tile, dolls, hobby horses, wire insulation, blister packaging, film, raincoats, strip coatings and bottles.

Other Notes
Very difficult to process unplasticized PVC because temperature of processing causes heat degradation. PVC also contains plastisols, lubricants, stabilizers, and fillers.

POLYVINYLIDENE CHLORIDE

Trade Names
Saran — Geon.

Properties
High tensile strength (10,000 psi, but when film is oriented — 45,000 psi) — High clarity — Very low gas permeability — Excellent chemical resistance — Excellent film and fiber former. Quick solidification in molding thick sections. Molds at high speeds — Strong welds.

Applications
Saran wrap — Fibers for outdoor furniture — Bristles — Doll hair — Upholstery — Pipe linings — Chemical resistant pipe and fittings.

POLYVINYL ACETATE

Trade Names
Polyco — Plymul — Elvacet — Gelva — Vinylseal.

Properties
Easily emulsified in water — Excellent adhesive properties — Very soft surfaces — Subject to easy cold-flow.

Applications
Wood glues (Elmer's type) — Plaster sealer — Latex paints — Hair sprays — Textile finish for feel (soft or hard hand) — Felt & straw body.

Other Notes
Latex paints are either styrene-butadiene (Kem-tone) or PVAc emulsions (fast growing) — Outdoor latex is PVAc. Often polyvinyl alcohol is added to glue and paint to increase adhesion — Polyvinyl pyrrolidone in hair spray.

POLYVINYL ALCOHOL

Trade Names
Lemac — Reynolon — Elvanol

Properties
Excellent resistance to hydrocarbons — Very good adhesive — Impervious to most dry gases — can be dissolved by water or alcohols.

Applications
Water soluble bags — Mold releases — Adhesion improvers to PVAc emulsions and PVB films.

Other Notes
Polyvinyl alcohol is made by hydrolizing polyvinyl-acetate — Degree of hydrolization determines degree of water solubility — 80% OH is soluble in cold water.

POLYVINYL FLUORIDE

Trade Names
Tedlar (PVF)

Properties
Extremely good weather resistance — Very difficult to process — Heat degraded.

Applications
Weather resistant coating for siding — Metal.

TETRAFLUORO ETHYLENE

Trade Names
Teflon — Halon — Fluon (TFE)

Properties
Very high heat resistance (450° to 550°F) — Excellent anti-stick surface — Good impact at -110°F — Best chemical resistance of TP especially at 400°F — Lowest coefficient of friction of plastics — Unmatched resistance to heat aging — Nonburning — Stable physical properties up to 450°F — High specific gravity (2.2). Poor creep resistance.

Applications
Bearings — Bushings — Slides — Where temperature, vacuum or FDA will not allow use of standard lubricants — No-stick fry pans — Bakery pans — Molds — Dies — Pumps — Impellers — Valves — Tubing — Lab ware — Gaskets — Seals — Packing — Pipe tape — Electrical insulation.

Other Notes
Not processed as other T.P. — Has no true melt point — Must be sintered or solvent carried, then fused in oven to achieve homogenuity. Fillers can greatly alter properties (Carbon-metal). C-F bond is

the most stable organic bond. Almost zero water absorption. Increased load pressure (10,000 psi) — Raises thermal degradation resistance to 700°-800°F.

CHLOROTRIFLOURO ETHYLENE

Trade Names
Kel-F — Plaskon — (CTFE)

Properties
Has most of the same properties as TFE but can be processed on conventional equipment. Excellent resistance to heat, chemicals, moisture, impact and abrasion — Zero moisture absorption. Better dimensional stability than TFE — Nonburning — Has antistick properties — Low coefficient of friction — Low gas permeability.

Applications
Bearings — Bushings — Slides (molded types) — Chemical pumps — Impellers — Valves — Tubing — Seals — Gaskets — Packing — Electrical insulation — Closures — Lab stoppers — Films for medical packages.

Other Notes
Can be obtained with 25% plasticizer — Or fillers. Very corrosive during molding — Requires stainless steel molds. Often needs stress-relieving after molding.

POLYVINYLIDENE FLUORIDE

Trade Names
Kynar (PVF$_2$)

Properties
A compromise between PE and TFE — Heat resistance of 300°F with easy processing — Compared to other fluoroplastics, PVF$_2$ has higher tensile, better abrasion resistance, is tougher, stronger and has reduced cold flow. High shrinkage (.025"/in.) — High molding stresses.

Applications
Films for packaging — Extruded tubing and pipe — Electrical wire and cable insulation — Heat shrinkable tubing — Porous pen points — Pumps — Valves — Seals — Metal coating.

Other Notes
PVF$_2$ is the only fluoroplastic being extruded solid. Requires 1 to 3 hours for stress relieving at 290°F.

FLUORINATED ETHYLENE-PROPYLENE

Trade Names
Teflon (FEP)

Properties
Similar to TFE but very easy to process. Excellent resistance to heat, chemicals, weather and impact. Nonstick surface — Abrasion resistant — Nonburning — High electrical strengths — Low coefficient of friction.

Applications
In general the same applications as TFE but where fast molding in production will offset the higher cost of FEP resins. Wire insulation — Pumps — Valves — Bearings — Metal coatings — Tubing — Tapes — Films.

Other Notes
Actually a co-polymer of F.E. and F.P. — Easiest fluoroplastic to process.

CELLULOSE NITRATE

Trade Names
Celluoid — Tenite CN — Mitron — Pyralin — Herculoid

Properties
Great toughness — Best of thermoplastics — Very inflammable — Explosive — Not weather resistant — Age embrittles, discolors — Low moisture absorption — Dimensionally stable — Very easy to mold from sheets with low heats.

Applications
Billiard balls — Washable collars — Cuffs — Ivory-like piano keys — Patent leather — First transparent unbreakable film — Decorative box covers — Float toys — Cellulose lacquers — Duco cement.

Other Notes
14% nitrogen in CN. 13% in guncotton — Camphor best plasticizer — Cellophane, rayon are regenerated cellulose — Cellulose is a natural polymer, not plastic.

CELLULOSE ACETATE

Trade Names
Tenite CA — Celanese — Vuepak — Plastacele — Lumarith — Hercocel CA — Kodapak.

Properties
Very tough — Hard surface — Easy to clean — Permeable to water vapor, but not water — Good weather resistance — Brilliant transparent colors — Easily moldable compared to cellulose nitrate — Poor solvent resistance — Age embrittles — Poor dimensional stability — Absorbs moisture.

Applications
Safety photographic film — Instrument dials — Point or purchase display signs — Toys — Film for pre-packaged display goods — Eye glass frames — Sunglasses — Toothbrushes — Combs — Lacquer — Dope — Nail polish — Blisters — Recording tape.

Other Notes
Requires plasticizer for molding — Generally odorless, tasteless — Cellulose acetate (2 OH's substituted) — Triacetate (3 OH's).

CELLULOSE PROPIONATE

Trade Names
Tenite CP — Forticel — Celanese.

Properties
Very tough — Hard surface — Easy to clean — Needs less plasticizer than other cellulosics — Good weather resistance — Short molding cycles — Compatible with many plasticizers, each different — No residual smell or taste — Best balance of properties of cellulosics.

Applications
Pen and pencil barrels — Tooth brushes — Auto steering wheels — Knobs — Blisters — Telephone housings — Transistor radios — Eye glass frames (large use) — Screwdrivers — Lipstick and compact cases.

Other Notes
Usually 5 propionate — 1 acetate substituted.

CELLULOSE BUTYRATE

Trade Names
Tenite CAB — Celanese CAB.

Properties
Very tough — Hard surface — Easy to clean — Compatible with high boiling plasticizers — Best weather resistance of cellulosics — Molds at lower pressure but higher softening point — Brilliant transparent colors — Best dimensional stability — Best moisture and chemical resistance of cellulosics.

Applications
Outdoor lighted signs — Tail light lens — Steering wheels — Storm windows — Window shades — Eye glasses — Blister packaging — Extruded pipe — Handles — Flashlight reflectors — Metalizable.

Other Notes
Usually 2 butyrate — 1 acetate substituted.

ETHYL CELLULOSE

Trade Names
Ethocel — Hercocelle — Campco.

Properties
Very high shock resistance — Tough — Low melt temperature — Not flammable — Excellent low temp shock resistance — No odor or taste — Compatible with many plasticizers — High surface shine — Hard — Lowest specific gravity of cellulosics.

Applications
Luggage — Football and baseball helmets — Flashlight cases — Display boxes — Cigar holders — Bowling duck pins — Arm rest molding — Blister packaging — Hot melt strip coatings.

Other Notes
2.2-2.4 Ethyl substituted — Never 3. Very soft flow material.

NYLON

Trade Names
Zytel — Nypel — Plaskon — Fosta.

Properties
Excellent toughness and wear resistance — Anti-frictional properties — Self extinguishing — Good chemical resistance and electricals — Hygroscopic (2.5% — poor dimensional stability) — High molding shrinkage — .010-.015"/in. — Many different nylons and additives.

Applications
Gears — Pumps — Valves — Bushings — Screws — Nuts — Zippers — Chemical ware and tubing — Rollers — Slides — Blister packaging — Brush bristles — Carpeting — Hosiery — Clothes — Textile fibers.

Other Notes
Polymerization of amino acid (nylon 6-11-12) or diamine with dibasic acid (nylon 6,6-6,10). Must be dry (0.3%) to process but normally absorbs up to 2.5% in use. Relative moisture absorption: Nylon 6>6,6 6,10>11>12. Tartan and Astroturf are Nylon "grass". Fillers: Fiberglass — Moly — TFE — Nucleated.

A B S

Trade Names
Royalite — Kralastic — Cyclolac — Lustran — Abson — Tybrene.

Properties
Unusually high rigidity with high impact — Excellent embossing properties — Electroplatable, best of thermoplastics — Fair chemical resistance — Variable rigidity/flexibility ratio — Cold formed with metal tooling — Resistant to staining — Good lacquer bond.

Applications
Suitcases — Tote boxes — Appliance housings — Household pipe — High heels — Telephone cases — Auto instrument cluster — Kick panels — Grill work — Headlight housings — Crash pad and arm rest covers — Football helmets — Office machine housings.

Other Notes
ABS is a terpolymer of Acrylonitrile-Butadiene-Styrene. Butadiene increases flexibility. Serious moisture pickup on pellets. Plating requires stress-free parts, special formulation. Quick freeze gives short injection cycle.

ACRYLIC

Trade Names
Plexiglas — Lucite — Acrylite — Implex — Bavick — Polycast — Perspex.

Properties
Optical clarity — Brilliant colors — Outstanding weather resistance — Excellent electricals — nontracking — Good wear resistance — Hard shiny surface — Scratches easily — Fair chemical resistance — Bends light around corners.

Applications
Tail light lenses and reflectors — Aircraft windows — Skylights — Contact lenses — Outdoor lighted signs — Imbedments — Knobs — Buttons — Jewelry boxes — Plastic safety glasses — Outdoor coatings — Artist paints — Dentures — Illuminated dials — Medallions — Models.

Other Notes
Problems with moisture pickup on pellets. Solvent welded — Clear joints. Coefficient of thermal expansion 10 x 8 glass. Modified with PVC for high impact. 92% white light passed — 5 years loss — 1%.

ACETAL

Trade Names
Delrin — Celcon (modified)

Properties
Engineering plastic — Strong — Tough — Stiff — High resistance to heat and fatigue — Highest chemical resistance — No solvent at RT — Highest abrasion resistance — Very low creep — High dimensional stability — Low moisture absorption — Ductile — Wide range of colors.

Applications
Bearings — Bushings — Rollers — Gears — Water pump impellers — Case — Zippers — Auto instrument cluster — Office machine housings — Levers — Cams — Hardware — Handles — Casters — Lock parts.

Other Notes
Fast freeze-off — Large vents. Nontracking — Burns slow, soot free. Hot welding or hot plate joining. Surface treatment possible. Molecule packs into tight crystals. Low tendency to stress crack, holds self-tapping screws like metal.

POLYCARBONATE

Trade Names
Lexan — Merlon.

Properties
Engineering plastic — Direct replacement of Zinc — Transparent — Superior toughness — Rigid —
Excellent outdoor and dimensional stability — Good optical clarity — Very low creep under load —
Physiologically inert — Heat resistance: 250°-300°F — Self extinguishing — Very stable under
moisture — Low, predictable shrinkage gives accurate moldings.

Applications
Street lamp housings — Lenses — Windows — Office machine housings — Drill housings — One piece
sunglasses — Pump impellers — Appliance housings — Auto bumpers — Plastic safety glasses.

Other Notes
Very hygroscopic — pellets must be dry to 0.5% or strengths will drop drastically — 4 hrs. at 170°F.
Design exactly like zinc with tighter tolerances. Use very generous fillets and radii. Cheaper than most
metals on cu. in. basis.

POLYALLOMERS

Trade Names
Tenite PA

Properties
Excellent hinge flex life — Better than PP — Unusually high stiffness and impact resistance — Better
abrasion resistance than PE or PP — Better flow and stress crack resistance than PE — Short molding
cycles — Very high crystallinity — Broad processing temp and pressure — Better than PE or PP —
Low density (0.90) — Resists flex blush — Used −40° to 210°F under no-load conditions.

Applications
Notebook covers — Bowling ball bags — Luggage — Typewriter cases — Tackle boxes — Welting for
auto door and window — Open mesh hardware sacks.

Other Notes
Not a blend or normal copolymer. Allomer: Same crystal form — Different chemical form.

POLYPHENYLENE OXIDE

Trade Names
GE PPO — Noryl (modified PPO).

Properties
Excellent engineering plastic — Very tough — Wide temperature range (−275°F to +375°F) —
Superior dimensional stability even in water — Best resistance to creep of any thermoplastic — Pellets
not hygroscopic, drying not usually needed — Very high resistance to acids, bases, steam. Requires
high pressures and temperature to mold.

Applications
Many electrical parts and housings — Hot water pumps and impellers — Utensils and containers for
steam sterilizing — FDA approved food pouches for oven cooking.

Other Notes
Noryl is modified PPO for much easier processing — Has much better flow — Rapid freeze-off — But
at some sacrifice of heat resistance (+220°F) and lower physical strengths. PPO will crosslink if too
hot in molding.

IONOMER

Trade Names
Surlin.

Properties
Excellent transparency — 80-92% transmission — Very tough and abrasion resistant even at −300°F — Very low mold shrinkage (0.002-0.006"/in.) — Permits deep drawing in thermoforming — Films are strong, stiff, tough — Easy tear propagation in films, hard to start — Resistant to all common solvents at R.T. — Easily heat sealed — Fast thermoforming.

Applications
Blister packaging — Flexible packaging — Blow bottles — Safety shields — Coating Nylon, PP, PES for heat sealing. Multiwall bags — Closures.

Other Notes
High melt strength — Ionic bonds thermally reversible — Processed like polyethylene except at higher temperatures.

POLY METHYL PENTENE

Trade Names
TPX

Properties
Excellent transparency — 90% transmission — Very high heat resistance (300°F to 430°F) — Lowest specific gravity — (0.83) — Good chemical resistance — Very low water absorp. — Highly crystalline — Sharp melting point (464° F) — Superior electricals, especially at 300°-400°F.

Applications
Medical bottles — Packages — Containers — Pipe and tubing for transparent chemical use — Disposable syringes — Clear cooking pouches — Light fixtures — Cosmetic containers.

Other Notes
Similar to P P in most ways except higher heat resistance. Electricals similar to TFE but easier processing.

POLYSULFONE

Trade Names
Udel

Properties
Good transparency — Engineering plastic — Tough — High mechanical strength — Heat resistance to +345°F — Resists creep or cold flow — Hard, shiny surfaces — Excellent electricals at +345°F — Self-extinguishing — Resistant to mineral acid, alkalii.

Applications
Power tool housings — Computer parts — Switches — Dishwasher tubs — Appliance housings — Pipe — Steam sterilizable utensils — Bottles — Film.

Other Notes
Excellent resistance to high temperature insulation cut-thru. Little heat degradation on reprocessing, must be dry. Requires primer for painting — Can be electroplated.

POLYIMIDE

Trade Names
Vespel

Properties
Very high heat resistance — 500°F continuous, 900°F short. Lowest coefficient of thermal expansion of any plastic — Approaches that of aluminum. Low friction — Very wear resistant — No melting point — Excellent electrical and physical strengths.

Applications
Bushings — Bearings — Slides — Protective coatings for metal — Component parts for aircraft and space hardware.

Other Notes
Processes only by sintering or solution coating. Prepreg fiberglass also used for structural parts.

THERMOSETS

PHENOLIC

Trade Names
Bakelite — Durez — Catalin — Resinox — Textolite.

Properties
Strong — Rigid — Tough — Heat resistant. Very low cost (pound or volume). Brittle in thin sections. Extremely good electrical properties — Good dimensional stability — Low cold flow. Molded colors: brown or black. Excellent chemical resistance — Scratch resistant — Very good adhesive.

Applications
Electrical insulation — Plugs — Sockets — Knobs — Appliance handles — Bases — Housings — Binder for masonite — Formica — Grinding wheels — Varnish coatings — Plywood adhesives — Washing machine agitators — Pump impellers — Radio cabinets — Auto parts — Silent gears — Bearings.

Other Notes
Volatile by-products during cure. One-step and two-step resins. Fillers are wood flour, asbestos, or mineral fillers. Cast phenolic: Billiard balls, marbelized castings.

UREA

Trade Names
Beetle — Styplast — Plaskon.

Properties
Very good electrical insulation — Excellent surface hardness — Bright colors — No taste or odor — Weather and sunlight resistant — Very low cost when filled — Naturally static-free — Excellent textile finish — Superior wood adhesive.

Applications
Electrical switches — Plugs — Devices — Bottle closures — Buttons — Buckles — Control knobs — Cosmetic containers — Camera cases — Appliance housings — Plywood adhesive — "Chipboard" binder — Permanent press fabrics — "Drip-dry".

Other Notes
Volatile byproducts during cure. Fillers: wood floor, glass fibers and cotton. Faster set and cure than melamine, but with lower flow. Amino resin class.

MELAMINE

Trade Names
Cymel — Plaskon — Plenco — Melantine — Melmac.

Properties
Very hard surface — High gloss — Scratch and stain resistant — Odorless & tasteless — Bright colors — Printed overlay — Tough — Steam cleanable — Excellent moisture, chemical resistance — Naturally static free — Good textile finish — Good wood glue — Excellent electrical properties.

Applications
Dinnerware — Plates — Cups — Handles. Surface layer for decorative laminates. Buttons — Bottle closures — Shaver housing — Plywood adhesives — "Drip-dry" textile finishes — Wool treatment (preventing shrinkage).

Other Notes
Volatile byproducts during cure. Dinnerware is alpha-cellulose filled. General purpose is wood-flour filled.

ALKYD

Trade Names
Glaskyd — Plaskon

Properties
Excellent arc and heat resistance — Hard — Tough — Dimensionally stable — Bright colors — Cures without volatiles — Fast molding cycles (15-30 sec.) — Low molding pressure (500 psi) — Very free flowing during molding.

Applications
Electrical insulation — Plugs — Bases — Auto distributor caps — Encapsulation — Resistors — Capacitors — Switch gear — Circuit breakers — Coil bobbins — T-V tuner — Paints (large usage).

Other Notes
Alcohol + Acid = Al-cid (alkyd). Mostly premix (gunk) molded. No "breathing" necessary. Wide range of fillers.

DIALLYL PHTHALATE

Trade Names
Dapon — Polydap — Durez.

Properties
Exceptional dimensional stability — Excellent high heat and chemical resistance — Very hard — Tough — Low moisture absorption — No volatiles during molding — Superior shelf stability — Low pressures for molding — Full range of colors — Outstanding electrical properties, especially under high humidity conditions.

Applications
Electrical insulation — Multiprong plugs — Close tolerance electrical sockets — Printed circuit boards — Decorative laminates — Wood surfacing — Molded coils — Resistors — Capacitors — Impregnant for wood.

Other Notes
No post-mold deformation. Very low shrinkage for polyester (0.006").

POLYESTER

Trade Names
Polylite — Koppers — Selectron — Marco — Hetron — Vibrin — Dapon.

Properties
Excellent resistance to heat and chemicals — Excellent electrical properties — No volatiles during cure — Liquid at R.T. — Unlimited colors: Transparent to opaque. Very high gloss finishes — Hard — Can be cured in low cost molds — Rigid or flexible — Fast or slow cure — High mechanical strengths if reinforced.

Applications
Sports car bodies — Boats — Housings — Bath tubs — Chemical tanks — Hoods — Ducts — Fishing

poles — Synthetic marble — Imbedments — Modern art — Protective coating — Decorative panels — Skylites — Awnings — Electrical insulating — Potting — Reinforced plastics mechanical parts.

Other Notes
Matched metal molding — Spray-up, hand layup — Filament winding — Fibers: (Dacron) — Film: (Mylar), (saturated esters) — Styrene monomer common diluent.

EPOXY

Trade Names
High Strength — Epon — Araldite — Bakelite — Epi-rez.

Properties
Outstanding toughness and chemical resistance — Excellent adhesive ability — Outstanding electrical properties — Liquid without diluents or solvents — Can be solidified without heat or pressure — Properties variable with modifiers (wide range) — No volatiles during cure — Very low shrinkage (0.001"/in.) — Very dimensionally stable. High mechanical strengths.

Applications
Encapsulation of electrical components — Metal-forming dies, jigs, fixtures, molds. All purpose adhesives — Protective coatings — Wear resistant — Tanks and coatings for chemical resistance.

Other Notes
Can be liquid cast in low cost molds. A great variety of possible formulations using 20 resins, 30 curing agents, 20 fillers, and 40 modifiers.

POLYURETHANE

Trade Names
Adiprene — Solithane — Baker (thermoset) — Estane — Texin — Roylar (thermoplastic).

Properties
Great wear resistance — Tear resistance — High resilience — Rubber-like — Good impact — Excellent resistance to chemicals — Excellent foam former (flexible-rigid) — Good electrical properties — High coefficient of friction — Poor hysteresis — Poor heat conductor.

Applications
Auto lubrication seals — Truck wheels — Auto crash padding — Seat cushions — Mattress foam — Furniture cushions — Rigid foam furniture components — Synthetic leather — Electrical potting of coils — Coatings — Seamless flooring.

Other Notes
Polyester foams: Stiffer — Slow recovery. Polyether foams: Softer — Fast recovery. One shot — Two shot — Moisture cure.

SILICONE

Trade Names
Silastic — GE.

Properties
Excellent high and low temp flexibility — Rubber-like elastomer — Good electricals at high temperature — Water repellent — Little viscosity change with temperature — Stable properties over wide temp (−70° to +450°F) — Curable without heat or pressure.

Applications
High temp seals — Gaskets — Hose — Grommets — Potting electrical components — lubricant for extreme temperatures — Water repellent paints — Heat resistant paints — Anti-foamers — Mold releases — Adhesives — Medical use: Heart valve, veins, prosthetics.

Other Notes
Organic-inorganic polymer. Fluids — Soft rubbers — Brittle solids. First stainless steel blade coating.

Specification of Tolerances

Several items must be considered in arriving at the practical tolerances for molded plastics parts: (1) Molding variations, (2) shrinkage, (3) warpage, and (4) aging.

Molding Variations. As a general rule, tolerances should be as generous as possible to allow for temperature change effects in mold and material. Shrinkage of the molded part must be allowed for when designing the mold. Recommended minimum tolerances are as follows:

Method	Position of Surface	Tolerance (in./in.)
Compression Molding	Perpendicular to parting line	0.008 to 0.010
	Parallel to parting line	0.005
Transfer or Injection		
Molding	Perpendicular to parting line	0.005
	Parallel to parting line	0.002

It is, of course, obvious that the closer the tolerances, the higher the production costs in dies, tooling and molding control. Too, all the various factors which accumulate to increase the molding variations must be considered.

Finish. The surface quality of molded parts depends largely upon the mold surface utilized. Standards for plastics, as yet, have not been established for guidance in setting design specification. At least one manufacturer has set up a series of surface standards for appearance only and these are as follows:

As Molded Finish:
1. Highly polished gloss.
2. Semigloss, commercial.
3. Low-cost for experimental parts.
4. Fine-grain parallel lines.
5. Modification of item 4 with deeper markings.
6. Satin, from abrasive air blast of molds.

Mechanically Applied Finish:
1. Dull, grained parallel lines.
2. High polish, glossy.
3. Low-cost buff, fair gloss.

Often, finish is important and glare from glossy surfaces is undesirable. Too, some standard is desirable to assure duplication of finish in some cases and to simplify the specification of surface finish.

Better finish as well as efficient flash removal is obtained by Wheelabrating plastics parts with a blast of nonabrasive pellets such as crushed apricot pits or crushed walnut shells. These materials do an excellent job of flash removal without marking the molded surfaces, eliminating approximately 95 per cent of all hand finishing operations.

REFERENCES

[1]R. S. Krolick and M. Ryan, "If Your Run is Over 700 to 8000 Parts, Consider Cold Press Molding", *Modern Plastics*, June 1972, p 86.

ADDITIONAL READING

[1]R. M. Bell, "Characteristics and uses of fiber reinforced molding compounds for precision parts, "*ASME Design Engineering Conference & Show Proceedings*, ASME 70-DE-63, May 11-14, 1970.

[2]Martin Friedland, "Cold forming of plastics," *ASME Design Engineering Conference & Show Proceedings*, ASME 70-DE-35, May 11-14, 1970.

[3]B. F. Hantz, "Cold-molding materials may solve your plastics selection problem," American Insulator Corp.

[4]American Can Company, "High-speed plastic molding," *Mechanical Engineering*, p. 51, April 1975.

[5]General Electric Co., *Design Tips, Lexan,* Pittsfield, Mass: Plastics Dept., General Electric Co.

[6]*Standards and Practices of Plastics Custom Molders,* New York: The Society of the Plastics Industry, Inc.

[7]*Injection Molding*, New York: Society of the Plastics Industry, Inc., 1975.

[8]T. Zehev and I. Klein, *Plasticating Extrusion*, New York: Van Nostrand Reinhold Co., 1970.

[9]J. D. Ferry, *Viscoelastic Properties of Polymers,* New York: Wiley-Interscience, 1970.

[10]*An Introduction to Extrusion*, Committee on Plastics Education, New York: Society of the Plastics Industry, Inc., 1968.

[11]*Molding Thermoset Materials*, New York: Society of the Plastics Industry, Inc., 1970.

51

Rubber Molding

LIKE the plastics, rubber and rubberlike synthetics provide almost unlimited possibilities as component parts of machine assemblies. Again, it is a case where specific properties, largely unattainable with any other materials, *Fig.* 51.1, are a design necessity and some knowledge of rubbers, their advantages and limitations, and their manufacture into parts is invaluable.

General Design. As a rule, it is possible to design or redesign a rubber part to approach closely the ideal conditions for economical production without affecting

Fig. 51.1 — Variety of precision molded rubber products requiring special properties of resilience, tensile strength and low permanent set. Courtesy, International Packings Corp.

or destroying the intended functional characteristics. However, the satisfactory designing of a rubber machine part, *Fig.* 51.2, requires a fair understanding of some of the properties of these materials, the problems of molding and of mold making. As with plastics materials, the particular compound to be employed has some effect on the method of manufacture which can be most readily used. Consequently, it is necessary to consider the various methods of molding which can be utilized and their individual design advantages and/or limitations.

Rubbers. The rubber materials which can be utilized for molded parts now range from natural stocks from various areas of the world to a full array of rubberlike synthetics. The crude or synthetic-base material is compounded with

Fig. 51.2 — Group of pump gears, impellers and a boat propeller made from oil and water-resistant Buna N synthetic rubbers. Photo, courtesy Goodyear Tire & Rubber Co.

curing agents, antioxidants, accelerators, lubricants, etc., and is thoroughly kneaded on mixing mills. It emerges as large slabs for storage pending subsequent manufacturing operations. This "green" or uncured compounded rubber can vary from a soft gummy state to a hard leathery condition. In no case is it liquid.

The nature of this uncured stock largely determines just how it will have to be shaped or prepared to fit a mold cavity for proper flow. A soft compound can be roughly shaped and laid in or adjacent to the mold cavity, whereas a hard stock requires careful tailoring and direct placement in the cavity.

Preliminary Processing. To obtain uncured stock of controlled size and shape for molding, therefore, the compounded material is processed by one of several preliminary operations: (1) Slabbing, (2) calendering, or (3) extruding (termed "tubing" in the rubber field). In slabbing, the compounded rubber is loaded into a mill, consisting of two large steel rolls which operate at slightly different surface speeds, which kneads and heats the compound and reduces its plasticity. After sufficient milling, the rubber is cut off in slabs, the gage of the stock being determined by the roll spacing. Cutting templates for "slabbing off" pieces to proper contour for mold loading are often employed.

The calender prepares the rubber compound in thin sheets which can be held to close tolerances. In calendering, *Fig.* 51.3, a ribbon of rubber is fed from a warmup mill to the calender rolls, from which it emerges as a sheet or strip. Rubberizing or "frictioning" of fabric is also done in a calender.

Extruding or tubing is carried out in a screw type extruder, shape and size of the stock being determined by the contour of the die. After cooling in a water tank, the extruded raw stock is cut to the necessary lengths for molding.

Fig. 51.3 — Sheeting out rubber on a calender. Photo, courtesy Goodyear Tire & Rubber Co.

Molding Methods. Oldest and still most widely used method for producing finished rubber parts is simple compression molding using mechanical presses, *Figs.* 51.4 and 51.5. With some of the newer methods of manufacture now in use, however, there are five practicable molding processes. These are (1) Compression molding, (2) transfer-injection molding, (3) full-injection molding, (4) extrusion molding, and (5) form-dip molding.

Compression Molding. This method of molding consists in placing a piece or pieces of prepared stock in the heated mold cavity, bringing the halves of the mold together under pressure of 500 to 2500 psi, and curing, *Fig.* 51.6. Heat for curing is supplied by the platens of the press utilized. Depending upon size, it is possible to mold from one to as many as 360 pieces per mold.

Fig. 51.4 — Production line compression molding geared to high volume produces some 2 million parts per day at International Packings Corp.

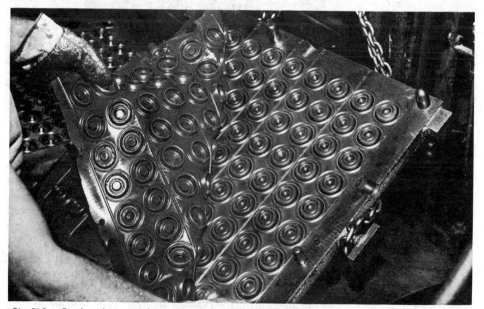

Fig. 51.5 — Section of compression mold and view showing removal of cured parts. Photo courtesy Goodyear Tire & Rubber Co.

Fig. 51.6 — World's largest high-pressure rubber molding press capable of compression and transfer molding shown with one half of a pump liner weighing 300 pounds. Platens measure 6 by 9 feet and deliver 2500 psi. Courtesy, Gates Rubber Co.

Transfer-Injection Molding. Now coming into wide use, transfer-injection molding permits the use of a single piece of prepared compound. Intricately shaped parts can be molded with improved efficiency over the compression method. The prepared piece is placed in a charging cavity in the mold and forced at high pressure through runners or channels into the mold cavity, *Fig.* 51.7. Usually the mold is opened and closed by hydraulic pressure, a separate plunger being used for injection.

Full-Injection Molding. With this recent development in molding methods, an extrusion head is used as an integral part of the molding machine. Lengths of extruded compounded rubber stock are fed directly into the extruder which, in turn, injects or forces the material into the mold cavity or cavities. These units, *Figs.* 51.8 and 51.9, are entirely automatic in operation with the exception of stripping the finished parts from the mold.

Cavities are laid out so that the stock is injected into a central canal which branches into two or more feeders and thence to the cavities. High pressure and turbulence developed during injection result in high temperature, reducing materially the necessary curing period. Resultant savings in molding justifies the use of this equipment in many instances.

Extrusion Molding. A wide variety of uniform cross-section parts can be extruded rapidly to the desired shape. Very intricate sections are practicable. As mentioned, screw type feed is employed for forcing the stock through the die, *Fig.* 51.10. Unlike the other molding methods, extrusion does not permit curing during the cycle. After removal from the water cooling tank, extruded stock is cured under temperature and pressure by means of steam and after cooling, is cut to the necessary lengths.

As a rule, extruded parts up to about 6 inches diameter are produced. Sections under 3/16-inch diameter can be extruded to lengths of about 500 feet, maximum. A tube of 6-inch diameter and 1/8-inch wall, however, can only be produced in lengths to about 50 feet. Extruded shapes require fair volume production for economy inasmuch as at least 100 pounds of stock is required to put an extruder into operation.

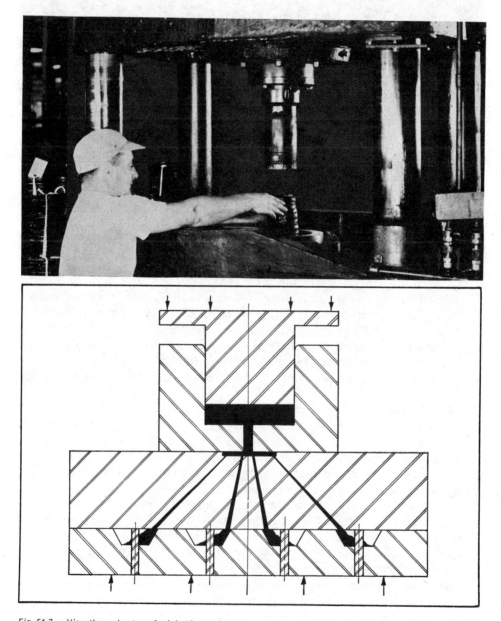

Fig. 51.7 — View through a transfer-injection mold, below, and operator shown loading press. Photo and drawing, courtesy Goodyear Tire & Rubber Co.

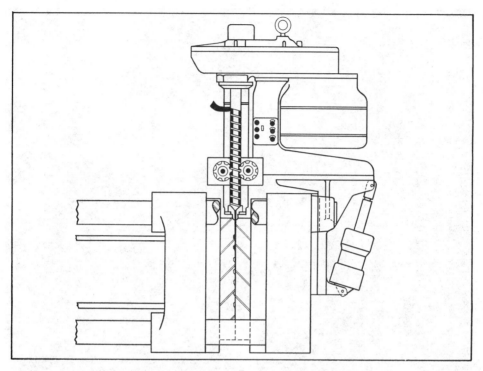

Fig. 51.8 — Screw type injection molding press and cross section through extrusion head and dies. Courtesy Goodyear Tire & Rubber Co.

Fig. 51.9 — Rotary injection molding machine for high volumes. Shot size is adjustable and a different mold can be used in each of the four stations. Unit injects up to 85 cu. in. shot. Courtesy, Stokes Div., Pennwalt Corp.

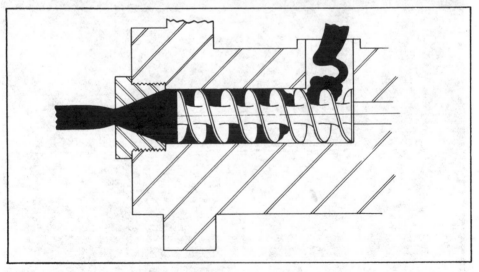

Fig. 51.10 — Extruder producing a roll covering section and view through typical extrusion head and die. Photo and drawing, courtesy Goodyear Tire & Rubber Co.

Form-Dip Molding. The form-dipped process starts with the dipping of an aluminum mandrel into a bath of a coagulating agent, then into the liquid rubber latex. The latex coagulates on the mandrel in a thick or thin rubber coating, depending on the time in the bath. After a heat cure, the rubber coating is removed from the mandrel. This completes the process, but normally, the dipping mandrel will be recycled for additional production. Recycling can be done indefinitely, which presents one of the economies of the process; you can get hundreds or thousands of parts from limited tooling.

The configuration of the mandrel must be such that the latex part can be stripped off. Therefore, the open end of the part must be large enough to slip over the largest cross section of the mandrel. As a guide, the largest cross-sectional area should be no more than six times that of the smallest cross-sectional area. Typical facilities allow making dipped rubber parts as large as 4 feet long by 12 inches in diameter, or as small as 1/4 inch in diameter by 1 inch long.

Production Considerations. The final design of a molded rubber part normally represents a compromise between the desired end result and the economically feasible end result. Also important is the production quantity to be required. Lowest production cost will dictate one method for but few parts whereas another may be feasible for large quantity output.

Design Considerations

Molded rubber parts require design techniques somewhat different from those ordinarily found desirable with other methods of production. Four major points which must be kept in mind for economical design are:

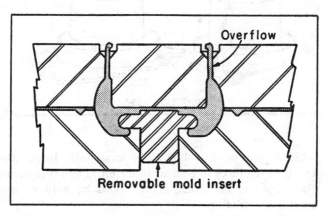

Fig. 51.11 — Cross section through dies showing flash and overflow. Part is inflation cup cap for milking machine. Drawing, courtesy Gates Rubber Co.

1. Can a mold be made to produce the part economically?
2. Can the mold be operated economically in production?
3. Can the piece be finished economically?
4. Can the flash be removed without affecting the designed function of the part?

With these general factors under scrutiny at all times, minimum production cost and maximum satisfaction can be reasonably assured.

Parting Lines. As in all molding methods, placement of the mold parting or partings is extremely important not only for assuring simplest possible mold design and operation, but also for simplifying flash removal and finishing. Any particular location on a part which, for design purposes, should be free from flash should be so indicated on the drawing. Because nonfills result in rejects, molded rubber parts invariably have overflow flash and molds are designed to accommodate this condition, *Fig.* 51.11. Circular flash, such as in *Fig.* 51.11, is readily removable automatically and cheaply. The part at *a* in *Fig.* 51.12 is difficult to trim whereas the one molded vertically as at *b* is much more economical.

Edges. Sharp or feather edges should be avoided in design inasmuch as they are difficult to mold and to trim accurately. The flat edge as shown at *b* in *Fig.* 51.13 is not only easier to mold and trim but is better looking. A sharp edge is difficult to hold to close tolerances and such mold portions wear to a rounded edge

Fig. 51.12 — Chicken picker machine finger molded horizontally at (a) and vertically at (b). Soft rubber makes possible mold at (b) and lower cost flash removal. Drawing, courtesy Gates Rubber Co.

which causes poor sealing and excess flash. In addition, a sharp edge generally creates a tendency to trap air in the mold and results in incomplete filling out. The flat edge, on the other hand, could be held to normal production molding tolerances indefinitely.

Corners. Rounded edges or corners must be used judiciously. Oftentimes, the use of a generously rounded edge makes it necessary to place the parting and flash line at a point which creates removal difficulties. Rounded edges, therefore, should be avoided at or near the parting. The sketch at *a* in *Fig.* 51.14 shows the flash line as dictated by a round edge. Unless the trim allowance on flash is generous, difficulty is encountered in removal operations. Where flash trim must be held to a few thousandths, the design with a square corner as at *b* in *Fig.* 51.14 is preferable.

Where edges are not at or near the parting line of the mold, rounded corners are the preferred design. As great a radius on corners, *Fig.* 51.15a, as possible will help improve compound flow in the mold and eliminate air trapping, characteristics generally present in molding sharp corners, *Fig.* 51.15*b*. This is particularly true in molding soft compounds.

Holes. Where holes through parts are along the major parting line necessary for molding, mandrels or core pieces are utilized. These must be so located that both molding of the piece and locating and removal of the mandrels are practicable, *Fig.* 51.16. Wherever possible, the diameter-to-length ratio of holes

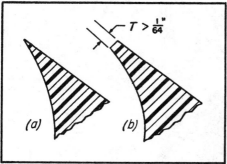

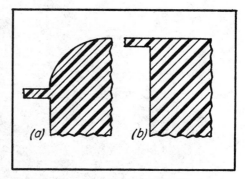

Fig. 51.13 — Flat edge (b) is more economical to produce than sharp edge shown at (a), which cannot be held in continuous production. Drawings, courtesy Sirvene Div., Chicago Rawhide Mfg. Co.

Fig. 51.14 — Rounded edges at the parting (a) make flash removal expensive whereas square edges (b) can be removed economically.

should be kept low. Pressures in molding are high and if mandrels or cores are too slender, support pins are necessary to keep the core centered. Such supports, however, leave holes which require filling.

Where holes are required for attaching molded rubber parts to other members, adequate spacing is necessary. Holes should be spaced generously from each other and also as far from edges as possible to eliminate tearing.

Fig. 51.15 — Rounded corners improve design wherever the flash line is not involved, (a), right. Drawing, courtesy Sirvene Div., Chicago Rawhide Mfg. Co.

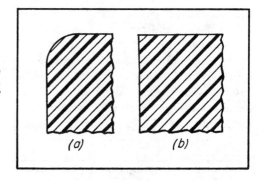

Undercuts. Re-entrant curves and deep undercuts create problems in production molding. The undercut shown at *a* in *Fig.* 51.17 would be difficult to remove from the mold owing to the extreme difference in diameters on the core or mandrel pin. Furthermore, dimensions as at the small diameter of *Fig.* 51.17a, are difficult to maintain because of extreme distortion and set during removal. To eliminate these problems and to avoid possible tearing of the parts at the inner sharp corner, a design modification such as shown at *b, Fig.* 51.17, is recommended. This part would be much stronger and much more economical to produce.

Fig. 51.16 — Aircraft fuel cell fitting which required a split cavity mold with a loose and permanent insert.

Undercuts can be formed with solid mandrels or cores, as in *Fig.* 51.11, only where an elastic stock is used. Where compounds having low-stretch properties are required, cores or mandrels of the collapsing type must be used. Molding is much slower in such instances and consequently more expensive. Thus, when a rubber over 80 durometer hardness is specified, undercuts should be avoided for economy. A design such as that shown at *b, Fig.* 51.12, could not be produced successfully in a horizontal mold using a low tear resistance stock.

Deep narrow undercuts should be avoided, not only because of difficulty in removal from the mold but because of the weakness of the mold projection needed to form the undercut. Mold repair costs are excessive with such designs. Depth of undercut should never exceed twice the width, *Fig.* 51.18, and width should be at least 1/16-inch and preferably greater.

Convolutions. Parts with convolutions having very thin walls and sharp corners as at *a* in *Fig.* 51.19 should be avoided whenever possible. Wall thickness with convolutions should be at least 0.050-inch or greater and as great a radius as possible should be used on the corners, *Fig.* 51.19*b*. Elimination of the sharp corners will improve flex life considerably and sufficient wall thickness allows for some shift of the mold mandrel commonly encountered.

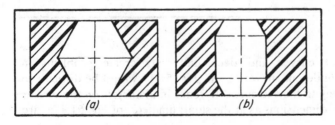

Fig. 51.17 — Undercut design (a) creates problems of tearing and distortion which are avoided by design (b). Drawing, courtesy Sirvene Div., Chicago Rawhide Mfg. Co.

Metal Inserts. Where inserted pieces are to be completely embedded in rubber, *Fig.* 51.20a, it is difficult to assure positive location. As a rule, rubber flow in the mold varies and exact position of full floating inserts is impossible to predict. Thus, where inserts are employed the parts should be designed so that the mold assures positive positioning, *Fig.* 51.21.

It is well to remember that inserts are subjected to pressure during molding, sometimes as high as 10,000 psi, and must be capable of withstanding such a load without deformation. Inserts must resist collapse and be such that displacement in molding is not possible.

Where rubber is to be molded into a metal shell of cylindrical or cup design, it is well to remember that parts having the rubber flush with the metal as at *b* in *Fig.* 51.20 are difficult to trim and clean. The rubber should be extended beyond the

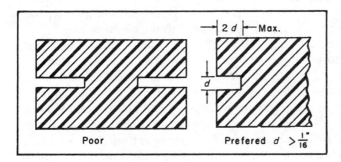

Fig. 51.18 — Narrow, deep under-cuts are poor design. Proportions should never exceed those indicated. Drawings, courtesy Sirvene Div., Chicago Rawhide Mfg. Co.

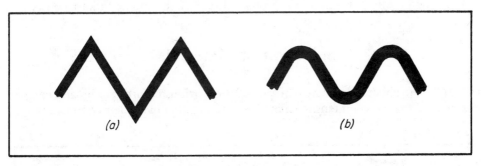

Fig. 51.19 — Sharp corners (a) on convolutions are detrimental to flex life and design (b) is preferable.

shell in all instances as at *c* in *Fig.* 51.20 to overcome these objections. In addition, if a perfectly flat end face or faces are desirable, the extension makes sufficient compound available for grinding. End faces are usually concave after molding, largely because of shrinkage, and grinding is often desirable.

For best adhesion of rubber to metal inserts or attachments the metal surface should be smooth. Corrugations, serrations, knurling, etc., decrease the effective adhesion and should be avoided. Rubber to metal adhesion strength of 250 psi can normally be expected under average manufacturing conditions. Where rubber-metal parts are being designed for stressed applications, other important factors must be considered which are beyond the scope of production design.

Where rubber parts are fastened to metal it is essential that chafing does not occur at the areas of contact. All flexing or distortion should occur in the main

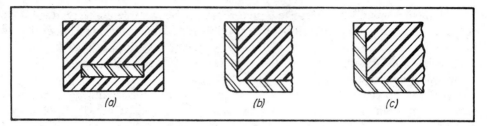

Fig. 51.20 — Totally embedded metal inserts (a) are difficult to position. Design (b) is difficult to trim and (c) is preferred for economy.

body of the rubber. A general rule of thumb to follow in designing for deflection is to provide a column of rubber at least four times the extent of movement. Thus, for 1/2-inch of flexing, a column of rubber not less than 2 inches should be provided.

Draft. Where rubber of 90 to 95 durometer hardness is to be molded, it is good practice to allow a slight draft on all surfaces normal to the mold parting.

Form-Dipped Parts. The drawing, *Fig.* 51.22 shows a simple rubber item that can be produced by the form-dipped process. It also gives the terms that are commonly used to describe a form-dipped item. The configuration and size of the item determine whether the tooling will be machined, sand-cast or fabricated from aluminum. Tooling should be designed with a starting point for entering the latex bath. This can be a radiused or a pointed end. Sharp edges produce a thin area in the rubber item. The single wall thickness of an item can range from .006" to .125".

Often a minor design change will enable an item to be form-dipped rather than compression molded, resulting in substantial savings in time and tooling costs.

Selection of Materials

Natural rubber compounds are made from various natural crudes and can be formulated to give good tensile strength and elongation over their complete range

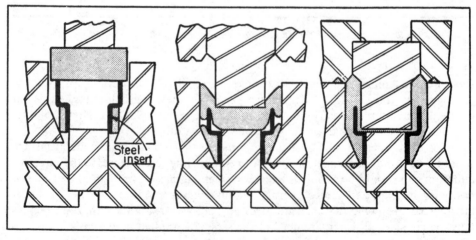

Fig. 51.21 — Embedded inserts should be designed to be positioned by the dies positively. This oil well swab cup has a stamped steel insert located by the center pin. Drawing, courtesy Gates Rubber Co.

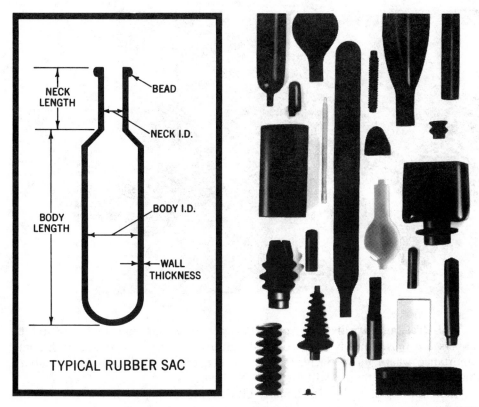

Fig. 51.22 — Drawing showing typical form-dipped design nomenclature and typical parts. Courtesy, Oak Rubber Co.

of hardness. For resilience or elasticity, natural rubbers are generally unexcelled by any of the synthetics so far developed.

Oil-Soluble Synthetics. The elastomer GRS or Buna S is an all-purpose synthetic used for the bulk of present synthetic rubber products. Relatively inexpensive, it has fairly good mechanical properties with the exception of tensile strength which is about one-tenth and resiliency which is about two-thirds that of natural rubber.

Fig. 51.23 — Hycar nitrile rubber vacuum valve for Bendix automatic washer. Valve is closed by vacuum and opened by water pressure. Low compression set or cold flow, low water absorption and resistance to chemicals make this material an excellent choice for this application. Photo, courtesy B.F. Goodrich Co.

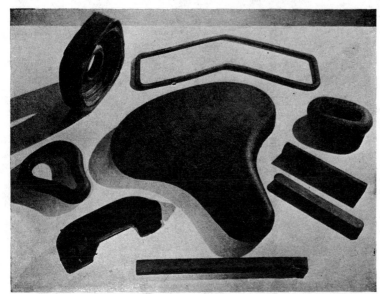

Fig. 51.24 — Group of molded cellular parts including a motor mount, vibration damper, motorcycle seat, and various machine gaskets. Photo, courtesy Sponge Rubber Products Co.

Oil-Resistant Synthetics. Apart from the foregoing so-called oil-soluble rubbers are those known as the oil-resistant group. These latter are of special importance where most machine parts are concerned. Each type has peculiar characteristics and no one will meet all requirements.

Thiokol identifies the organic polysulfide types having a fairly wide range of durometer hardnesses. These compounds are practically unaffected by gasoline, fuel oils, kerosene, lubricating oils, dilute acids, or alkalis. Benzol causes some swelling. Thiokol FA offers poor cold flow properties. However, the newer ST type eliminates this problem and offers excellent low-temperature flexibility. Flexibility is retained to minus 65 F.

Buna N includes several types which vary only slightly in characteristics. Excellent resistance to mineral oils, animal oils, fats, vegetable oils, some solvents, gases and water are typical. Resiliency is similar to that of Buna S. Benzol, toluol and carbontetrachloride cause considerable swelling but no deterioration. Resistance to high temperatures (225 F), compression set and abrasion are excellent also. Flexibility to minus 65 F may be had with some types. Compounds may be produced ranging from extremely soft to bone hard and the material will not flow even under extreme pressures, *Fig.* 51.23.

Neoprenes are higher swelling elastomers than Thiokol or Buna N. The compounds are flameproof and offer excellent resistance to temperatures up to 250 F. Stable under severe flexing and resistant to sunlight or ozone, these elastomers may be used with gasoline, oils, air, natural or manufactured gases, weak acids, and alkalis. Flexibility to minus 40 F and occasionally lower temperatures may be had. Type GN is somewhat more resilient than type E but is a little more difficult to process. Special type FR can be compounded for operating conditions down to minus 70 F or lower. Tensile strength and water resistance of FR, however, is lower and swelling in oils and solvents slightly higher than that of the other two types.

Silicone rubber synthetics have poor common properties such as tensile strength, elongation at break, and resistance to abrasion compared to natural and other synthetics. They do have, however, the important property of resistance to extreme heat, the useful range being from minus 100 F to 500 F. They retain flexibility, resiliency and hardness for long periods, have low compression set and favorable electrical properties. The material has no tendency to stick or adhere to metallic surfaces. It does not deteriorate in oil but does in gasoline. Cellular silicone rubber, Cohrlastic, is also available for use in vibration dampers, fairing, seals, gaskets, etc., *Fig.* 51.24.

Rubber Specifications. In the design of any rubber part it is necessary to specify the desired compound properly. ASTM specifications should be used for adequate interpretation of requirements to the parts manufacturer. Mechanical rubber specifications may be found in ASTM Specification D735-48T and those for cellular rubber from ASTM Specification D1056-49T.

TABLE 51-I — **Characteristics of Rubber and Synthetics**

Material	Desig-nation (Gov't)	Tensile Strength Pure Gum	Tensile Strength Black Rein-forced	Hardness (Durom-eter)	Mixing Eff. (%)	Extrud-ing (%)	Calen-dering (%)	Cohesion	Molding
Natural	.....	3000	4500	any	100	100	100	Excellent	Excellent
Buna S	GRS	400	3000	any	85	90	90	Fair	Fair
Buna N	GRA	600	3500	any	50	50	50	Poor	Excellent
Butyl	GRI	3000	3000	28 to 85	90	90	100	Good	Good
Thiokol	GRP	300	1500	35 to 90	75	75	50	Poor	Good
Neoprene	GRM	3500	3500	25 to 90	100	75	90	Good	Good

In the case of cellular rubber parts, *Fig.* 51.24, density or compression should be specified. Also, it should be determined whether a mechanical skin is desired on cellular rubber parts. Color must be noted or the common black or gray will be assumed. Any special finish must also be noted on the drawings. Average machined-mold finish is commercial and any other usually results in increased expense. General comparative characteristics of rubbers are given in TABLE 51-I.

Specification of Tolerances

In the design of molded parts, adequate consideration should be given to dimensional tolerances inasmuch as these directly affect cost. A good minimum dimensional tolerance for small parts is plus or minus 0.010-inch although plus or minus 0.005-inch can be held commercially, *Fig.* 51.25.

Production Economy. The effect of tolerances on production economy can be shown most effectively by means of a typical production example. If, for instance, a 1-inch diameter by 1/4-inch thick disk was to be molded to plus or minus 0.005-inch on diameter and thickness, a 100-cavity mold could be used. Were this tolerance reduced to plus or minus 0.003-inch, however, a smaller mold with 16 to 25 cavities of much greater cost and lower productivity would be necessary to hold the closer tolerances and parts cost would be greatly increased.

Production Tolerances. Applicable tolerances generally must vary according to size, and the usually accepted tolerance on dimensions *not* affected by flash, or within the machined portions of the mold, is plus or minus 1/2 per cent for precision components. However, plus or minus 1 per cent is commercially acceptable. The commercial tolerance on dimensions perpendicular to the flash plane or parting line are as follows:

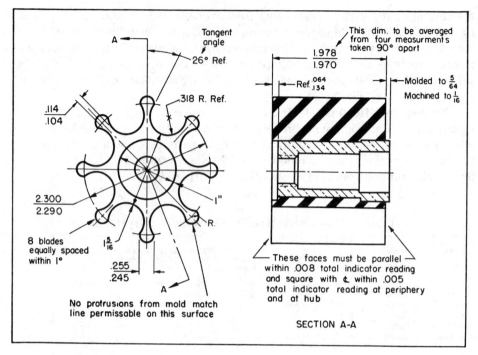

Fig. 51.25 — Synthetic rubber rotor for a combination vacuum-pressure pump.

0 to ¹/₂-in. incl.	±0.015-in.
¹/₂ to 1 in. incl.	±0.018-in.
1 to 3 in. incl.	±0.025-in.
4 to 8 in. incl.	±0.050-in.

The allowable variations in wall thickness as indicated by runout on cylindrical or circular parts are generally as follows:

0 to 2¹/₂-in. diams.	0.020-in. total ind. reading
2¹/₂ to 5-in. diams.	0.031-in. total ind. reading
5 to 10-in. diams.	0.062-in. total ind. reading

Commercial tolerances on dimensions of special extruded shapes are:

0 to ¹/₄-in. incl.	±0.015-in.
¹/₄ to ¹/₂-in. incl.	±0.031-in.
¹/₂ to 1 in. incl.	±0.046-in.
1 to 2 in. incl.	±0.062-in.

Shrinkage. After a rubber part is removed from the mold, shrinkage takes place on cooling to the extent of ¹/₂ to 4 per cent, depending upon the rubber stock employed and the shape of the part. These variations make it expensive to obtain closer tolerances on parts than those indicated in the foregoing. However, it can be done but at higher cost.

Hot Distortion. Where a part is to be made from a soft grade of material and stretching or contracting is needed to remove the finished pieces from the cavities.

an additional allowance must be made for hot distortion.

Compression Rates. Where shear or compression rates are specified, a tolerance of plus or minus 15 per cent is required to allow for variations in rubber modulus, metal dimensions, and testing discrepancies.

Cellular Parts. Typical tolerances on dimensions of cellular rubber molded parts or special shapes as suggested by ASTM are as follows:

Sponge (Open cells)

Thickness (inches)

$1/4$ and under	$\pm 1/32$-in.
$1/4$ to 3	$\pm 1/16$-in.

Length or width (inches)

$1/4$ and under	$\pm 1/32$-in.
$1/4$ to 3 incl.	$\pm 1/16$-in.
3 to 6 incl.	$\pm 1/8$-in.
6 and over	$\pm 1/4$-in.

Expanded (Closed cells)

Thickness (inches)

$1/8$ to $1/2$ incl.	$\pm 1/16$-in.
$1/2$ to $1 1/2$ incl.	$\pm 3/32$-in.
$1 1/2$ to 3 incl.	$\pm 1/8$-in.

Length or width (inches)

6 and under	$\pm 1/4$-in.
6 to 12	$\pm 3/8$-in.
12 and over	$\pm 3\%$

Form-Dipped Parts. Tolerances can generally be held to plus or minus 0.004-inch and, on some selected parts less.

ADDITIONAL READING

[1] Harry L. Boyce, "Designing Molded Rubber Parts," *Mechanical Engineering*, September, 1970.

52

Ceramics Molding

$\mathbf{A}$MONG the earliest objects fashioned by man, ceramic products have withstood the test of centuries and still find application in numerous fields. Exceptional properties of ceramics have increased their use and their availability and have brought about the development of highly specialized technical ceramics suited to unusual industrial applications, *Fig.* 52.1. Properly chosen and suitably designed, technical ceramics can fill a real need in the design of industrial equipment.

Failure to consider the use of ceramics in mechanical equipment generally stems from the rather meager knowledge of these materials and lack of data with which to design such parts to assure not only better performance but lower production costs.

Types of Ceramics. Traditionally, ceramic materials have been composed largely of naturally occurring clays, alone or in admixture with various amounts of quartz, feldspar and other nonmetallic minerals. The term ceramic often implies, in addition, the field of articles made from glass but in this discussion glasses will not be considered owing to the distinct difference in manufacture.

High-Clay Ceramics. In all probability, parts produced from compositions consisting predominantly of clays account for the largest tonnage of manufactured ceramics. For convenience, these may be said to have a clay content in excess of 50 per cent. These ceramics are characterized by high shrinkage — approximately 25 per cent by volume — during drying and firing and, therefore, exhibit the widest size variation. To a large extent, the production tolerances are a direct function of the clay content. Parts produced include almost all structural clay products, clay-base refractories, most chemical porcelains and stoneware, electrical and mechanical porcelains.

Low-Clay Ceramics. This classification includes steatite and other low-loss dielectrics, special and super-refractories, and special porcelains. Steatite normally carries over 80 per cent talc (hydrated magnesium silicate) which is bonded with ceramic fluxes. Shrinkage is low and close tolerances can be held.

Clay-Free Ceramics. This small but nevertheless important group of ceramics has compositions entirely devoid of clay. Included in the group are the pure oxide types and the so-called cermets or ceramals which have properties between metals and ceramics.

A combination of pure oxide ceramic and a metal constituent, both of which

Fig. 52.1 — A variety of intricate ceramic parts designed for industrial, electronic, and machine applications. Photo, courtesy American Lava Corp.

have good high-temperature properties, the so-called metal-ceramics are being developed to meet the present-day need for a material capable of withstanding operating temperatures to 2400 F or more. The metal constituent is employed to provide thermal conductivity and shock resistance while the ceramic provides resistance to deformation under high stress at high temperatures.

Production Methods. Second most important effect on the tolerances to which a ceramic part can be manufactured is produced by the method of forming involved. Various methods are used for forming ceramic bodies prior to firing and, consequently, practical production design must take cognizance of not only the tolerance effect of the method required but also the general range and limitations imposed. Categorically speaking the various forming methods are either of the wet or the dry process type, *Fig.* 52.2.

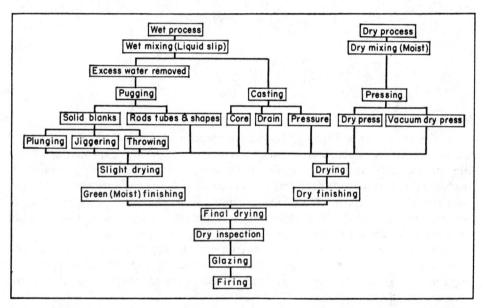

Fig. 52.2 — Chart of ceramics manufacturing processes and procedures utilized today. Drawing, courtesy Locke Inc.

Ceramic to be formed by one of the wet methods is given the desired plasticity through the use of a suitable quantity of plastic clay together with sufficient water to provide a relatively soft, plastic mass. When the physical requirements of the fired material make it impossible to secure the required plasticity by use of clay and water, the mix may be plasticized by means of organic binders and plasticizing agents. With low-clay and clay-free ceramics, however, the dry process is often used, semimoist granular powder having 2 to 12 per cent water by weight being compressed in dies to the desired form. Ceramics to be molded by casting are made fluid by the addition of defloculants such as sodium silicate, sodium carbonate, or solutions of salts which yield hydroxyl iron and do not ordinarily contain any more water than those prepared for ordinary wet methods.

Plastic Wet Processes. Oldest and probably most diversified general process, plastic wet forming consists of five methods of preliminary shaping — extrusion, throwing, jiggering, pressing, and casting. Method of preliminary forming to be used depends upon the part design features and the economical manufacturing considerations.

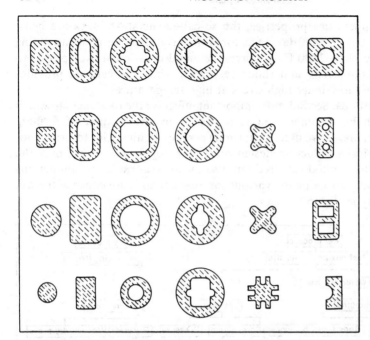

Fig. 52.3 — Group of cross sections showing possibilities of extrusion for uniform sections. Drawing, courtesy Locke Inc.

After preliminary forming, the molded pieces are subjected to a drying period under controlled heat and humidity. Degree of drying depends upon the dimensional accuracy desired in the finished part. Any final shaping required is performed at this stage and, if glazing is necessary, the piece is dipped or sprayed with a mixture of glazing materials suspended in a relatively high percentage of water and again dried. To develop the final physical, electrical and mechanical characteristics the formed pieces are fired in a kiln at approximately 2200 F.

, *Extrusion.* Most frequently used extruder is the auger or screw type, usually fitted with a vacuum chamber for removal of occluded gases. Often referred to as pugging, extrusion may be done with or without a mandrel or core to shape sections of almost any variety, *Fig.* 52.3.

Although this method is limited to pieces of uniform cross section, machining prior to firing may be used to produce the desired external shape, *Fig.* 52.4. While internal machining can be done, it is expensive and is better avoided. Extrusions from 0.030-inch to 18 inches outside diameter and 60 inches length can be produced. Labor and tool costs are moderate.

Throwing. A cut blank of plastic material is placed on a revolving plaster disk or table and while turning is shaped roughly by hand with the aid of tools, *Fig.* 52.5.

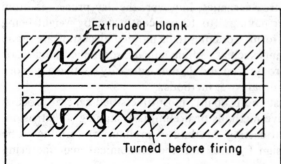

Extruded blank

Turned before firing

Fig. 52.4 — Extruded hollow blanks and outline showing design machined before firing. Drawing, courtesy Westinghouse Electric Corp.

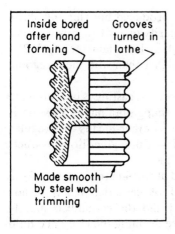

Inside bored after hand forming

Grooves turned in lathe

Made smooth by steel wool trimming

Fig. 52.5 — Left — A typical hand thrown and turned mechanical porcelain part.

Fig. 52.6 — Right — A porcelain fitting which is jiggered in a plaster mold and trimmed or turned.

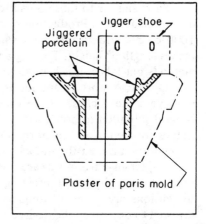

Jiggered porcelain

Jigger shoe

Plaster of paris mold

Fig. 52.7 — Below Left — A typical porcelain part hot pressed in a plaster mold with a rotating tool which is heated and oiled.

Fig. 52.8 — Below Right — Special shape cast partly with a core and partly without a core where it is impractical. Drawings, courtesy Westinghouse Electric Corp.

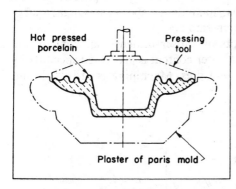

Hot pressed porcelain

Pressing tool

Plaster of paris mold

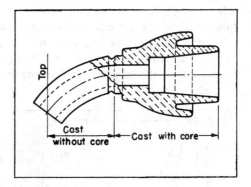

Top

Cast without core

Cast with core

Hollow blanks with several bore sizes, tapered bores, solid pieces, and round hollow shapes with one closed end are thus produced. External contours and bores are finished ordinarily by turning prior to firing.

Parts up to about 8 inches maximum diameter, 24 inches maximum length or height, and 1/2-inch minimum bore can be made by this method. Output is low with this particular method, tools are inexpensive but as a general rule labor cost is high.

Jiggering. With this method, treated material is placed in a revolving plaster-of-Paris mold and worked into the mold contours simultaneously by hand and by means of a profile tool, *Fig.* 52.6. These partly dried shapes are then trimmed and machined if necessary after further drying. Bowl-shape parts, cylindrical parts with tapered round bores at one or both ends, and bow shapes with deep skirts can be produced. Sizes ordinarily practicable range up to 22 inches diameter and 10 inches length. Production rate is only moderate, mold and tool expense is moderate, but labor cost is high.

Pressing. In this method a plaster-of-Paris mold is also used, the material being roughly worked to the desired shape. The roughly shaped material, in the mold, is then placed in a suitable press and with an oily, hot, rotating tool or punch is compressed to desired shape, *Fig.* 52.7. On removal any finishing or machining operations necessary are performed.

Parts similar to those possible with jiggering are produced but ribs and skirts cannot be as deep. Production rate is rapid and ordinarily quantities must be over 1000 pieces. Practicable sizes range up to about 14 inches diameter and $12^1/_2$ inches length or height. Tooling and setup costs are higher than with the foregoing methods but labor cost is moderate.

Casting. A more recent development of the wet process, casting is probably the most versatile because of the variations in size and shape obtainable. Casting is generally limited to production of parts not possible by other means — irregular shapes not having an axis of rotation, and large parts — although hollow products of revolution are easily made by this method, *Fig.* 52.8.

Plaster-of-Paris molds are used, the porous mold absorbing the moisture from the mix or "slip" and creating a deposited shell of ceramic. Where inside dimensions are of no consequence, no core is used and the ceramic is merely allowed to deposit to the desired wall thickness before draining the mold. To hold specific internal dimensions and avoid boring, cores are employed to permit build-up of ceramic to the desired shape and wall thickness.

Generally, parts up to 80 inches height and 36 inches diameter are readily obtainable although sections up to 45 inches diameter have been produced. As a rule, weight is limited to about 1400 pounds, maximum. Labor cost is high, a large number of molds is required, and production rate is slow.

Vacuum Extrusion and Pressing. To obtain closer dimensional control and impervious structure, ceramic with reduced water content is processed specially by evacuation of the mix before forming in steel dies. Machining is ordinarily not needed.

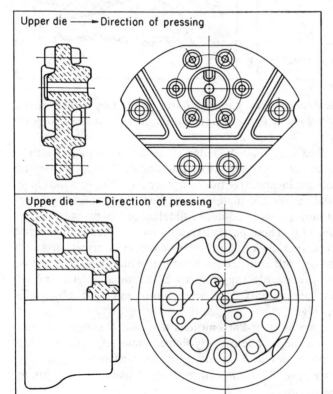

Upper die ——→ Direction of pressing

Upper die ——→ Direction of pressing

Fig. 52.9 — Complicated part pressed in evacuated dies. Charge is weighed to insure accuracy.

Fig. 52.10 — Porcelain electrical part designed for dry pressing in metal dies. Drawing, courtesy Westinghouse Electric Corp.

Best suited for simple shapes and pieces, this method is economical where large quantities are the order — say 5000 pieces or more. Extrusion is limited to outside diameters up to 1.625 inches, lengths to 10 inches and inside diameters up to 0.140-inch. Pressing, *Fig.* 52.9, is limited to a maximum height of 7 inches, maximum length of 12 inches, maximum width of 10 inches, and projected areas up to 120 square inches. Tool and setup costs are high and labor cost is moderate.

Dry Process. Differing from the plastic process of forming in that only a semimoist granular powder is used, the dry process is subject to much less dimensional variation in manufacture. Pressing in metal dies is the only method of forming used. Typical parts possible are: Flat or ribbed plates of any form or curvature with or without holes, slots, recesses, and barriers; U and other shapes; intricate box shapes with ribbed partitions and side holes or slots. Machining or other finishing is seldom required, *Fig.* 52.10.

An economical method for producing large quantities at moderate cost, dry pressing is most generally limited to relatively small parts in the neighborhood of 2 to 4 inches. Maximum length or width possible is 14 inches, height is limited to about 7 inches and projected area to 196 square inches.

High-clay ceramics generally result in a porous structure with this method but the low and clay-free types can be produced with excellent results. The close dimensional control possible makes the process advantageous with the latter groups. Dry pressing is well adapted to high production, and automatic presses may be used.

Design Considerations

Structurally, the problem of designing with ceramics resolves itself into one of design to compensate for or avoid involving the undesirable lack of ductility and low tensile strength. The best design procedure is to take all possible advantage of the high compressive strength of these materials and avoid as much as possible all loading under tension. Pronounced weakness at notches makes it imperative to avoid sharp change in section thickness or direction.

Wall Thickness. Where hollow extrusions are designed, the wall thickness may range from 10 to 20 per cent of the outside diameter, but should be held to 1 3/4 inches, maximum, if possible. Rectangular or square shapes require liberal internal fillets of at least 1/2-inch radius. The cross section of such parts should be held as uniform as possible to aid drying. Thus, concentric external radii should be used to match the internal fillets, and wall thickness should be held to 1 1/4 inches or less.

Where heavy walls are necessary for external machining on hollow extruded sections, wall thickness should be minimized to avoid high losses due to differential in shrinkage of finished items, *Fig.* 52.11. Sections in any case should be kept as uniform as possible without abrupt variations. As a general rule, sections for wet pressing, should not exceed 1 inch.

Fillets and Radii. Along with uniform section thickness, generous fillets should be employed to reduce stresses and aid the flow of material in forming. Fillets at bosses, shoulders, webs, ribs, depressions, or counterbore should depend upon the height of these features. For those features formed by the mold or die, the minimum radius should be equal to the depth or height of the adjoining feature for dimensions of 1/4-inch or less. With sizes from 1/4-inch to 2 inches, the minimum radius should be 1/4-inch and for those over 2 inches the radius should be at least

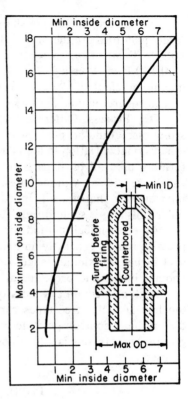

Fig. 52.11 — View of part machined inside and out after extrusion. Chart indicates desirable proportion of inside to outside diameters for extrusion. Holes as small as 0.010-inch and outside diameters of 0.030-inch have been produced. Drawing, courtesy Westinghouse Electric Corp.

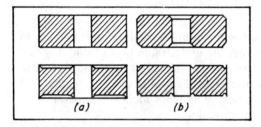

Fig. 52.12 — Sharp edges on pressed parts (a) are less desirable than bevel edges (b) owing to better appearance and less fin trouble after firing. Drawing, courtesy American Lava Corp.

one-eighth the depth or height.

For parts to be machined, smaller fillet radii are practicable. On dimensions up to 3/32-inch the radius can be equal to the height or depth, from 3/32 to 3/4-inch values the minimum radius should be 3/32-inch, and for deeper or higher features it should again equal at least one-eighth the dimension.

Corners or edges should be rounded off to improve appearance and obviate chipping. Minimum radius practicable is 1/16-inch and larger values are preferable. Beveled or chamfered edges are generally desirable on dry pressed parts. A standard chamfer is generally accepted at 1/32-inch by 45 degrees. The bevel not only reduces die wear but simplifies fin removal after firing, *Fig.* 52.12.

Ribs. Protruding ribs or fins should be spaced apart at least as much as they are wide or deep. A full radius on the top of ribs is advantageous. Where walls are relatively light, ribs should be employed for strengthening and to prevent distortion, *Fig.* 52.13.

Draft. Some draft is required on parts for most processes used. In forming, both molds and cores must be withdrawn and at least 1/64-inch per inch should be provided although 1/32-inch per inch per side is more desirable. Large cores should

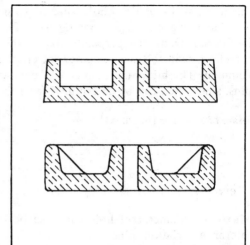

Fig. 52.13 — Dry pressed part showing improved design with strengthening rib. Drawing, courtesy Westinghouse Electric Corp.

Fig. 52.14 — Metal die setup for pressing parts. Outer walls and corepiece should have no draft as shown. Drawings, courtesy American Lava Corp.

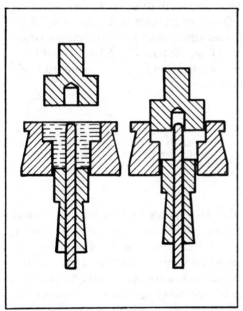

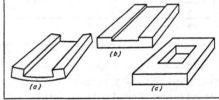

Fig. 25.15 — Deep groove in part (a) results in warpage. Shallow groove design (b) is better but a recessed design (c) is most desirable alternate.

have not less than 2 degrees per side. For dry pressing, plain outside surfaces and through holes should be made without draft, *Fig.* 52.14.

Grooves. Deep grooves which run over the length of a part tend to cause warping. To avoid this condition grooves should be kept shallow and not over one-half the piece thickness if possible. A recess is preferable wherever practicable in lieu of a channel shape, *Fig.* 52.15.

Holes. In general, holes should be 1/16-inch diameter or over for economy, although holes as small as 0.010-inch are produced in some parts. Wall thickness between holes should be as substantial as possible. The same is true of walls between holes and edges. Thin walls in dies are fragile and may be expensive to maintain. Wherever practicable, walls between holes should equal at least one-half the hole diameter and should never be less than 1/32-inch in any case. Thinner walls are sometimes possible but it is inadvisable to have walls under one-third the hole diameter. Walls between holes and edges should never be less than 1/8-inch.

The firing shrinkage along the axis of extrusion or in the direction of pressing is different from that perpendicular to the axis or direction of pressing. Holes, therefore, tend to become somewhat elliptical, as much as 2 to 3 per cent being common. The minimum diameter of clearance hole to permit assembly of bolts, pins, studs, etc., can be determined from the following specifications:

Diam of Fastener (in., max)	**Clearance** (in., min)
1/16 to 5/32	0.006
3/16 to 3/8	0.008
13/32 to 5/8	0.010
21/32 to 1	0.012
over 1	0.016

Maximum diameter of hole can then be found by applying the expected total tolerance variation, plus.

Shrinkage also affects hole spacings, and generally it is undesirable to attempt to have round holes match the hole spacings in a metal attachment. One or all of the holes in the ceramic part should be elongated to provide ready assembly. Two holes should be elongated in the direction of maximum shrink by an amount equal to the total tolerance on the distance between hole centers, *Fig.* 52.16. If only one of two holes is to be elongated, the amount of elongation should be equal to twice the tolerance on the distance between hole centers.

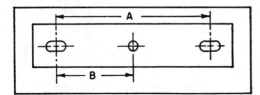

Fig. 52.16 — Holes should be elongated purposely to overcome the effect of firing shrinkage. One hole, "B", requires twice the elongation needed with two, "A"

Threads. It is impractical to mold threaded holes in ceramics, especially porcelains. With dry pressed clay-free ceramics, however, holes may be tapped. Coarse tapped threads are preferable and resist crumbling found with fine pitches. Smallest tapped hole which can be made satisfactorily is a number 3-48. Smaller sizes are not recommended. The number of threads in contact should be kept at a minimum owing to shrinkage. It is seldom necessary to use over 12 threads engagement and the ideal fit is obtained with about eight threads engagement in a 12-thread hole. For shallow holes, a screw two-thirds to three-quarters of the tapped depth is adequate. Where screws must pass through a part, the thickness and pitch should be selected so that six or seven threads are used, the minimum being four and the maximum, eight. Holes should be counterbored before tapping to reduce losses from cracking.

External threads may be cut sharp in clay-free ceramics but are more durable if an extended pitch can be used, *Fig.* 52.17. Fillets and radii for threads, as with grooves and slots, are recommended to add to the strength and to resist corner breakage. Threading is most economical if full length, *Fig.* 52.17b. To produce only a portion of thread is more expensive, *Fig.* 52.17c and to thread to a shoulder or undercut more expensive, *Fig.* 52.17d. Contact on external threads should be from six to twelve threads. Metal parts should be proportioned to keep contact under twelve.

Fig. 52.17 — *Extended-pitch threads (a) are preferable. Full length (b) is more economical than (c) or (d). Drawing, courtesy American Lava Corp.*

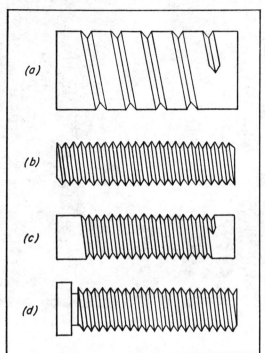

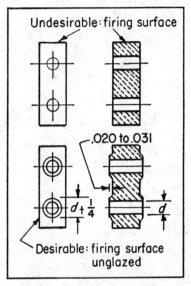

Fig. 52.18 — *Where ground surfaces are required for fit to metal parts, reduced boss area is preferred design, above. Drawing, courtesy Westinghouse Electric Corp.*

Machining. Parts which must be otherwise machined before firing require some consideration. External fillets should have at least 1/8-inch radii if possible. Deep undercuts should be avoided. Since internal machining is extremely expensive it should not be used. In turning external shapes the final piece must be capable of resisting the loads without breaking. With solid bodies, the ratio between the length and root or minimum turned diameter should not be greater than 6 or 7 to 1.

Where parts must fit flat surfaces after firing, grinding is usually necessary. Maximum economy is possible by employing a group of bosses to reduce the area to be finished. Bosses around holes should be about 1/4-inch greater in diameter than the hole. Boss height may be about 0.020-inch but 1/32-inch is preferable, *Fig.* 52.18.

Almost any required accuracy can be had by grinding but stock removal is expensive. However, numerous grinding operations are regularly employed in mechanical ceramic and stoneware parts where their properties of corrosion resistance, hardness, etc., are imperative, *Fig.* 52.19.

Firing Surface. To permit solid resting of parts during firing in the kiln, a firing surface should be left unglazed, *Fig.* 52.18. Long flat pieces should be fired on one long side to reduce warping.

Ordinarily tubular and bushing parts with a height not exceeding six times the diameter of the base are sufficiently steady for foot firing but longer parts must be hung. Hanging is expensive and, generally, pieces with up to 160 cubic inches of volume per square inch of supporting area can be fired in this manner. Designs which require lengths longer than fifty or sixty inches should be avoided.

Fig. 52.19 — Centrifugal pump impeller being finish ground to close machine tolerances. Photo, courtesy U.S. Stoneware Co.

Fig. 52.20 — Group of mechanical parts including special textile machine thread guides and other wear-resistant inserts. Photo, courtesy American Lava Corp.

Lettering. Block lettering with raised characters on a depressed surface is preferred for dry pressed parts. All others are generally marked with ceramic ink on unglazed portions.

Selection of Materials

Ceramic materials are generally chosen for mechanical or industrial equipment because of corrosion, heat or wear resistance, *Fig.* 52.20, or for electrical properties. Design to obtain the advantage of these properties must take into consideration the relatively low strength and brittle character of these materials. The general physical properties of some ceramics are listed in Chapter 68.

Specification of Tolerances

With porcelain parts the combined drying and firing shrinkage runs from one-seventh to one-eighth, kiln shrinkage alone being one-tenth to one-twelfth. With such conditions the problem of tolerances is critical, especially since shrinkage is not always uniform.

Commercial Tolerances for Porcelains. It has been established that a standard tolerance between glazed and unglazed surfaces of plus or minus 3 per cent or 0.030-inch per inch or fraction thereof will permit practical commercial manufacture with reasonable cost for wet process porcelain. With dry process porcelains, the practicable limits are somewhat less than those for the wet process — about 2 per cent, or plus or minus 0.020-inch per inch. Minimum single tolerance on dimensions less than 1 inch with the former is 0.030-inch, plus or minus, and with the latter is 0.020-inch.

Critical Tolerances for Porcelains. Where close or so-called critical tolerances must be specified, cost is increased but generally a tolerance of 2 per cent, plus or minus, with a minimum of 0.020-inch in the case of wet process parts and 1 1/2 per cent, plus or minus, with a minimum of 0.015-inch for dry process parts can be obtained. Dimensions between unglazed surfaces of parts can generally be held to about one-half these figures if necessary.

Die Wear. For dry process pressed parts, dimensions determined primarily by the upper and lower die positions or those affected by side wear are subject to the pressures used and variations in amount of fill. Such dimensions should have tolerances increased by 50 to 100 per cent, if possible.

Tolerances for Clay-Free Ceramics. Standard practice with dry process low-clay or clay-free technical ceramics such as steatite is to hold plus or minus 1 per cent on critical dimensions over unglazed surfaces with a minimum of plus or minus 0.005-inch on general dimensions and plus or minus 0.010-inch on thickness dimensions. Over glazed surfaces the limits are plus or minus one per cent plus 0.010-inch, with a minimum of plus or minus 0.015-inch. Where thickness is in excess of one inch, the recommended tolerance is plus or minus 3 per cent. All noncritical dimensions should carry the standard critical plus 0.015-inch.

On hole spacings, a tolerance of plus or minus 0.005-inch between centers measuring under 1/2-inch and plus or minus one per cent on larger dimensions can be held. On unglazed holes standard tolerances are plus or minus 0.005-inch for

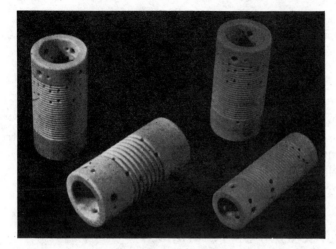

Fig. 52.21 — Group of electronic
coil forms showing numerous
plain and tapped holes. Photo,
courtesy Henry L. Crowley & Co.

holes up to 1/2-inch and plus or minus one per cent for 1/2-inch and larger holes. For glazed holes these tolerances are increased to plus or minus 0.012-inch for diameters up to 0.600-inch and plus or minus 2 per cent on holes 0.600-inch and greater. Screw threads, *Fig.* 52.21, conform generally to American Standard Class 1 fits.

FABRICATING
METHODS

53

Arc Welding

$\mathbf{A}$RC welding is one of several fusion processes for joining metals, and the process most widely used in production and construction whether manual or mechanized. By the application of intense heat, metal at the joint between two parts is melted and caused to intermix — directly or, more commonly, with an intermediate molten filler metal. Upon cooling and solidification, a metallurgical bond results. Since the joining is by intermixture of the substance of one part with the substance of the other part, with or without an intermediate of like substance, the final weldment has the potential for exhibiting at the joint the same strength properties as the metal of the parts. This is in sharp contrast to nonfusion processes of joining — such as soldering, brazing, or adhesive bonding — in which the mechanical and physical properties of the base materials cannot be duplicated at the joint.

In arc welding, the intense heat needed to melt metal is produced by an electric arc. The arc is formed between the work to be welded and an electrode that is manually or mechanically moved along the joint (or the work may be moved under a stationary electrode). The electrode may be a carbon or tungsten rod, the sole purpose of which is to carry the current and sustain the electric arc between its tip and the workpiece. Or, it may be a specially prepared rod or wire that not only conducts the current and sustains the arc but also melts and supplies filler metal to

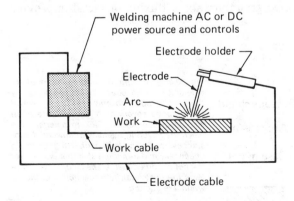

Fig. 53.1 — The basic arc-welding circuit.

Omer Blodgett is Design Consultant for The Lincoln Electric Co., Cleveland, Ohio.

the joint. If the electrode is a carbon or tungsten rod and the joint requires added metal for fill, that metal is supplied by a separately applied filler-metal rod or wire. Most welding in the manufacture of steel products where filler metal is required, however, is accomplished with the second type of electrodes — those that supply filler metal as well as providing the conductor for carrying electric current.

Basic Welding Circuit. The basic arc-welding circuit is illustrated in *Fig.* 53.1. An ac or dc power source, fitted with whatever controls may be needed, is connected by a ground cable to the workpiece and by a "hot" cable to an electrode holder of some type, which makes electrical contact with the welding electrode. When the circuit is energized and the electrode tip touched to the grounded workpiece, and then withdrawn and held close to the spot of contact, an arc is created across the gap. The arc produces a temperature of about 6500°F at the tip of the electrode, a temperature more than adequate for melting most metals. The heat produced melts the base metal in the vicinity of the arc and any filler metal supplied by the electrode or by a separately introduced rod or wire. A common pool of molten metal is produced, called a "crater." This crater solidifies behind the electrode as it is moved along the joint being welded. The result is a fusion bond and the metallurgical unification of the workpieces.

Arc Shielding. Use of the heat of an electric arc to join metals, however, requires more than the moving of the electrode in respect to the weld joint. Metals at high temperatures are reactive chemically with the main constituents of air — oxygen and nitrogen. Should the metal in the molten pool come in contact with air, oxides and nitrides would be formed, which upon solidification of the molten pool would destroy the strength properties of the weld joint. For this reason, the various arc-welding processes provide some means for covering the arc and the molten pool with a protective shield of gas, vapor, or slag. This is referred to as arc shielding, and such shielding may be accomplished by various techniques, such as the use of a vapor-generating covering on filler-metal-type electrodes, the covering of the arc and molten pool with a separately applied inert gas or a granular flux, or the use of materials within the core of tubular electrodes that generate shielding vapors.

Whatever the shielding method, the intent is to provide a blanket of gas, vapor, or slag that prevents or minimizes contact of the molten metal with air. The shielding method also affects the stability and other characteristics of the arc. When the shielding is produced by an electrode covering, by electrode core substances, or by separately applied granular flux, a fluxing or metal-improving

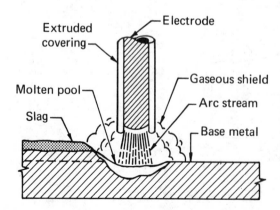

Fig. 53.2 — How the arc and molten pool are shielded by a gaseous blanket developed by the vaporization and chemical breakdown of the extruded covering on the electrode when stick-electrode welding. Fluxing material in the electrode covering reacts with unwanted substances in the molten pool, tying them up chemically and forming a slag that crusts over the hot solidified metal. The slag, in turn, protects the hot metal from reaction with the air while it is cooling.

function is usually also provided. Thus, the core materials in a flux-cored electrode may supply a deoxidizing function as well as a shielding function, and in submerged-arc welding the granular flux applied to the joint ahead of the arc may add alloying elements to the molten pool as well as shielding it and the arc.

Fig. 53.2 illustrates the shielding of the welding arc and molten pool with a covered "stick" electrode — the type of electrode used in most manual arc welding. The extruded covering on the filler metal rod, under the heat of the arc, generates a gaseous shield that prevents air from contacting the molten metal. It also supplies ingredients that react with deleterious substances on the metals, such as oxides and salts, and ties these substances up chemically in a slag that, being lighter than the weld metal, arises to the top of the pool and crusts over the newly solidified metal. This slag, even after solidification, has a protective function; it minimizes contact of the very hot solidified metal with air until the temperature lowers to a point where reaction of the metal with air is lessened.

While the main function of the arc is to supply heat, it has other functions that are important to the success of arc-welding processes. It can be adjusted or controlled to transfer molten metal from the electrode to the work, to remove surface films, and to bring about complex gas-slag-metal reactions and various metallurgical changes.

Overcoming Current Limitations. The objective in commercial welding is to get the job done as fast as possible so as to lessen the time costs of skilled workers. One way to speed the welding process would be to raise the current — use a higher amperage — since the faster electrical energy can be induced in the weld joint, the faster will be the welding rate.

With manual stick-electrode welding, however, there is a practical limit to the current. The covered electrodes are from 9 to 18-in. long, and, if the current is raised too high, electrical resistance heating within the unused length of electrode will become so great that the covering overheats and "breaks down" — the covering ingredients react with each other or oxidize and do not function properly at the arc. Also, the hot core wire increases the melt-off rate and the arc characteristics change. The mechanics of stick-electrode welding is such that electrical contact with the electrode cannot be made immediately above the arc — a technique that would circumvent much of the resistance heating.

Not until semiautomatic guns and automatic welding heads, which are fed by continuous electrode wires, were developed was there a way of solving the resistance-heating problem and, thus, making feasible the use of high currents to speed the welding process. In such guns and heads, electrical contact with the electrode is made close to the arc. The length between the tip of the electrode and the point of electrical contact is, then, inadequate for enough resistance heating to take place to overheat the electrode in advance of the arc, even with currents two or three times those usable with stick-electrode welding.

This solving of the "point-of-contact" problem and circumventing the effects of resistance heating in the electrode was a breakthrough that substantially lowered welding costs and increased the use of arc welding in industrial metals joining. In fact, through the ingenuity of welding equipment manufacturers, the resistance-heating effect has been put to work constructively in a technique known as long-stickout welding. Here, the length of electrode between the point of electrical contact in the welding gun or head and the arc is adjusted so that resistance heating almost — but not quite — overheats the protruding electrode. Thus, when a point on the electrode reaches the arc, the metal at that point is about ready to melt. Thus, less arc heat is required to melt it — and, because of this, still higher welding speeds are possible.

Selecting A Welding Process

In selecting a process for production welding, a primary consideration is the ability of the process to give the required quality at the lowest cost. Here, the cost factor must include not only the operating cost, but also the amortization of the capital costs of equipment over the job and those jobs that may reasonably be expected to follow. Thus, a fabricator may have a job of running straight fillets where fully mechanized submerged-arc equipment would be the ultimate solution to low-cost, quality welding. But, unless the job involves a great amount of welding, or there is assurance that it will be followed by other jobs of a similar nature, the lowest cost to the fabricator — taking into account the feasibility for amortization — might be with a hand-held semiautomatic gun or even with the stick-electrode process. The job is, thus, the starting point in the selection of a welding process.

The job defines many things. It defines the metals to be joined, the number of pieces to be welded, the total length of welds, the type of joints, the preparations, the quality required, the assembly-positioning requirements — to name but a few. Process selection, thus, considers first the needs of the job at hand. Unless the job will be adequately repetitive or subsequent jobs will be similar in requirements, however, it may be desirable to select a process that is not optimum but which will be most efficient.

First, the job tells what metals are to be welded. If the metals are steel, a wide range of processes may be considered. If the metals are aluminum and magnesium, the available processes will be gas tungsten-arc and gas metal-arc. These processes and stick-electrode welding may also be chosen if maximum weld properties are required in high-alloy steels or corrosion-resistant materials.

If the metals to be welded are the carbon or low-alloy steels, selection becomes more difficult. More processes are applicable, and varying degrees of mechanization are possible. The welding is likely to be heavy or involve many feet — or even miles — of weld, with varying types of joints. Furthermore, there may be precise requirements for such joints in terms of fill, follow, freeze, or penetration.

Taking into account the fact that process selection is almost self-determining when the metals to be welded are aluminum, magnesium, copper, titanium, or the other nonferrous metals — or that the stick electrode is almost certainly to be the choice when welding steel where the immediate cost is of no importance, the following text will be concerned with choosing the optimum process for the *production* welding of steels. Here, intangibles no longer are effective, and only technology and pure economics prevail.

In the *production* welding of steel, mechanization immediately enters the picture. It can't be ignored, since some forms of mechanization are even applicable to tack welding. Here, also, stick-electrode welding is ruled out, since it can't be competitive with the mechanized processes.

All of the mechanized arc-welding processes have been developed as a means of reducing welding costs. Each of these processes has certain capabilities and limitations, and selecting the best process for a particular job can be a difficult, if not confusing, task. Unless the user makes a correct selection, however, he may lose many of the benefits to be gained in moving to mechanization and, thus, not achieve his objective of welds that meet the application requirements at the lowest possible cost.

There are varying degrees of mechanization with all of the mechanized processes — from semiautomatic hand-held equipment to huge mill-type welding installations. Thus, one needs a "common denominator" when processes are to be compared. Semiautomatic equipment possibly provides the best common denominator. Anything added to semimechanization is extraneous to the process capability *per se* and merely amounts to "putting the semiautomatic process on wheels."

With semimechanization and the capabilities of the various processes subject to semimechanization in mind, a four-step procedure can be used in deciding which is the preferred process for the particular production welding job.

The Four-Step Procedure. The four steps involved in process selection are:

1. First, the joint to be welded is analyzed in terms of its requirements.
2. Next, the joint requirements are matched with the capabilities of available processes. One or more of the processes are selected for further examination.
3. A check list of variables is used to determine the capability of the surviving processes to meet the particular shop situation.
4. Finally, the proposed process or processes indicated as most efficient are reviewed with an informed representative of the equipment manufacturer for verification of suitability and for acquisition of subsidiary information having bearing on production economics.

This four-step approach is not mathematically precise, since it requires judgments. However, these *are made in the proper sequence* — which is most important for arriving at correct results. Also, on vital points, such as the joint factors of greatest significance, the judgments are usually clear-cut, with the possibility of error being negligible.

The main problem in applying this approach will be in assessing the capability of various processes to supply the needs of the joint. This will necessitate review of manufacturers' literature, or the directing of pertinent questions to equipment makers — and, to be sure that no process is omitted from consideration, data on each must be available.

STEP I — Analysis of Joint Requirements. When economy in welding is a prime objective, the needs of any joint can be expressed in four terms:

1. Fast-fill ... meaning high deposition rate.
2. Fast-freeze ... meaning the joint is out-of-position — overhead or vertical.
3. Fast-follow ... being synonymous with high arc speed and very small welds.
4. Penetration ... meaning the depth of penetration into the base metal.

Fast-fill is an obvious requirement when a large amount of weld metal is needed to fill the joint. Only by a high deposition rate can a heavy weld bead be laid down in minimum arc time. Fast-fill becomes a minor consideration when the weld is small.

Fast-freeze implies out-of-position — the need for quick solidification of the molten crater. The joint may additionally be a fill-type or penetration-type, but fast freezing is of paramount importance. Not all semiautomatic processes are capable of being used on fast-freeze joints.

Fast-follow implies the ability of the molten metal to follow the arc at rapid travel speed, giving continuous, well-shaped beads, without "skips" or islands. The crater may be said to have low surface tension; it wets and washes in the joint readily. Fast-follow is especially desirable on relatively small single-pass welds, such as those used in joining sheet metal. It is rarely a requirement of a large or multiple-pass weld; with these, the arc speed automatically will be optimum when the deposition rate is sufficient.

Penetration varies with the joint. With some, it must be deep to provide adequate mixing of the weld and base metal, and with others it must be limited to prevent burnthrough or cracking.

Any joint can be categorized in terms of its needs for these four factors. Examination of the joint detail is all that is required for such categorization.

Fig. 53.3 shows examples of fill, follow, freeze, and penetration joints. Note that in the case of the first three types, the adjective "fast" describes what is wanted for maximum efficiency. In the case of penetration, however, either deep or shallow penetration may be desirable, depending upon the joint and the application requirements of the assembly.

In (a) of *Fig.* 53.3, a substantial volume of weld metal is required for fill. The process that will give fill most rapidly while meeting other requirements would be given high priority in the four-step selection procedure. In (b), the weld-metal requirements are minimal, and the process that would permit the fastest arc speed would get consideration. In (c), only a process that would permit out-of-position welding would be useful. Here, submerged-arc would be ruled out. But, in (d), deep penetration is indicated as necessary, in which case submerged-arc would definitely be considered.

The examples in *Fig.* 53.3 represent extremes. Although freeze joints are immediately recognizable, many of the joints encountered are not easily categorized in terms of fill, follow, and penetration. Various combinations of these three characteristics may be needed in the joint. Where there is a question, a judgment must be made as to the "weight" of each characteristic in the joint requirements.

In analyzing joint requirements, it is virtually impossible for the experienced engineer or welding supervisor to make an error that would substantially affect the selection. The three factors are "weighted" — not quantitatively defined. As long as the order of significance of the three factors is assessed correctly, the information needed for matching up joint requirements with process capabilities has been obtained.

STEP II — Matching Joint Requirements With Processes. The equipment manufacturer's literature usually will give information on the ability of various processes to deliver the requirements of the joint. It is virtually impossible to be misled at this point, since the deposition-rate and arc-speed characteristics of each process can be clearly defined. Penetration is related to the current density of the process, but also affected by the thickness of the plate to be welded. In-plant experience with various processes will also be helpful, but care must be exercised that personal prejudices do not lead to destruction of the objectivity of the analysis. Once the capabilities of various processes are known, they can be matched with the joint requirements and tentative selections made.

Fill

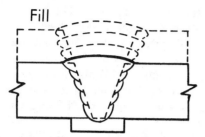

(a) Substantial volume of
weld metal required.

Follow

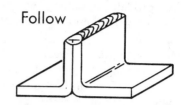

(b) Minimal weld metal required.
Usually, only the edges or
surfaces of plates being
joined must be melted together.

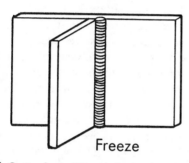

Freeze

(c) Out-of-position joint, where
control of molten crater
is important.

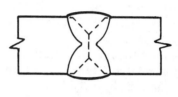

Penetration

(d) Joint requires deep penetration
into base metal (maximum
penetration). Bridging a gap
(minimum penetration) may be
required if fitup is poor.

Fig. 53.3 — Examples of fill, follow, freeze, and penetration joints.

If the joint in question is the one illustrated by (a) in *Fig.* 53.3, it must be rated 75% fill, 20% penetration, and 5% follow. Fill is obviously all-important; penetration is nominal; and follow is unimportant. No matter what percentages are assigned, as long as they are in the proper order, one has a starting point for matching them with a welding process. The predominant need for fill suggests, possibly, submerged-arc or self-shielded flux-cored welding with a fast-fill electrode. The nominal need for penetration suggests that the latter might be more desirable, since submerged-arc with its deeper penetration could lead to burnthrough.

STEP III — The Check List. Considerations other than the joint itself have bearing on selection decisions. Many of these will be peculiar to the job or the welding shop. They can be of overriding importance, however, and they offer the means of eliminating alternate possibilities. When organized in a check-list form, these factors can be considered one-by-one, with the assurance that none is overlooked. Some of the main items to be included on the check list are:

Volume of Production. It must be adequate to justify the cost of the process equipment. Or, if the work volume for one application is not great enough, another application may be found to help defray investment costs.

Weld Specifications. A process under consideration may be ruled out because it does not provide the weld properties specified by the code governing the work. The specified properties may be debatable as far as defining weld quality, but the code still prevails.

Operator Skill. Replacement of manual welding by a mechanical process may require a training program. The operators may develop skill with one process more rapidly than with another. Training costs, just as equipment costs, require an adequate volume of work for their amortization.

Auxiliary Equipment. Every process will have its recommended power sources and other items of auxiliary equipment. If a process makes use of existing auxiliary equipment, the initial cost in changing to that process can be substantially reduced.

Accessory Equipment. The availability and the cost of necessary accessory equipment — chipping hammers, deslagging tools, flux lay-down and pickup equipment, exhaust systems, et cetera — should be taken into account.

Base-Metal Condition. Rust, oil, fitup of the joint, weldability of the steel, and other conditions must be considered. Any of these factors could limit the usefulness of a particular process — or give an alternative process a distinct edge.

Arc Visibility. With applications where there is a problem of following irregular seams, open-arc processes are advantageous. On the other hand, when there is no difficulty in correct placement of the weld bead, there are decided "operator comfort" benefits with the submerged-arc process; no headshield is required and the heat radiated from the arc is substantially reduced.

Fixturing Requirements. A change to semiautomatic process usually requires some fixturing if the ultimate economy is to be realized. The adaptability of a process to fixturing and welding positioners is a consideration to take into account, and this can only be done by realistically appraising the equipment.

Production Bottlenecks. If the process reduces unit fabrication cost, but creates a production bottleneck, its value may be completely lost. For example, highly complicated equipment that requires frequent servicing by skilled technicians may lead to expensive delays and excessive over-all costs.

The completed check list should contain every factor known to affect the economics of the operation. Some may be peculiar to the weld shop or the particular welding job. Other items might include:

Protection Requirements	Range of Weld Sizes	Weld Cleaning Costs
Application Flexibility	Seam Length	Housekeeping Costs
Setup Time Requirements	Ability to Follow Seams	Initial Equipment Cost
Cleanliness Requirements		

Each of these items must be evaluated realistically, recognizing the peculiarities of the application as well as those of the process and of the equipment used with it.

Insofar as possible, human prejudices should not enter the selection process; otherwise, objectivity will be lost. At every point, when all other things are equal, the guiding criterion should be welding cost.

STEP IV — Review of the Application by Manufacturer's Representative. This step in a systematic approach to process selection may seem redundant. However, a basic thesis is that at every step the talents of those who know best should be utilized. Thus, the check list to be used is tailored by the user to his individual situation. The user, and he alone, knows his situation best. It is also true that the manufacturer of the selected equipment knows its capabilities best — will be able to clear up any questions, supply up-to-date information about the process, point out pitfalls in its use, and give practical application tips.

If the analysis of the factors involved has been correct, and the exploration of possible processes thorough, it is almost axiomatic that the equipment manufacturer's representative will confirm the wisdom of the selection. If something important has been overlooked, however, he will be in a position to point it out and make recommendations. This final review before installation of the equipment is to bring to the job all the information bearing on successful application as well as to verify the decision.

The Shielded Metal-Arc Process

The shielded metal-arc process — commonly called "stick-electrode" welding or "manual" welding — is the most widely used of the various arc-welding processes. It is characterized by application versatility and flexibility and relative simplicity in the equipment. It is the process used by the small welding shop, by the home mechanic, by the farmer for repair of equipment — as well as a process having extensive application in industrial fabrication, structural steel erection, weldment manufacture, and other commercial metals joining. Arc welding to persons only casually acquainted with welding usually means shielded metal-arc welding.

With this process, an electric arc is struck between the electrically grounded work and a 9 to 18-in. length of covered metal rod — the electrode. The electrode is clamped in an electrode holder, which is joined by a cable to the power source. The welder grips the insulated handle of the electrode holder and maneuvers the tip of the electrode in respect to the weld joint. When he touches the tip of electrode against the work, and then withdraws it to establish the arc, the welding circuit is completed. The heat of the arc melts base metal in the immediate area, the electrode's metal core, and any metal particles that may be in the electrode's covering. It also melts, vaporizes, or breaks down chemically nonmetallic substances incorporated in the covering for arc shielding, metal-protection, or metal-conditioning purposes. The mixing of molten base metal and filler metal from the electrode provides the coalescence required to effect joining.

As welding progresses, the covered rod becomes shorter and shorter. Finally, the welding must be stopped to remove the stub and replace it with a new electrode. This periodic changing of electrodes is one of the major disadvantages of the process in production welding. It decreases the operating factor, or the percent of the welder's time spent in the actual operation of laying weld beads.

Another disadvantage of shielded metal-arc welding is the limitation placed on the current that can be used. High amperages, such as those used with semiautomatic guns or automatic welding heads, are impractical because of the long (and varying) length of electrode between the arc and the point of electrical contact in the jaws of the electrode holder. The welding current is limited by the resistance heating of the electrode. The electrode temperature must not exceed the "break down" temperature of the covering. If the temperature is too high the covering chemicals react with each other or with air and therefore do not function properly at the arc. Coverings with organics break down at lower temperatures than mineral or low hydrogen type coverings.

The versatility of the process, however — plus the simplicity of equipment — are viewed by many users whose work would permit some degree of mechanized welding — as overriding its inherent disadvantages. This point of view was formerly well taken, but now that semiautomatic self-shielded flux-cored arc welding has been developed to a similar (or even superior) degree of versatility and flexibility in welding, there is less justification for adhering to stick-electrode welding in steel fabrication and erection wherever substantial amounts of weld metal must be placed. In fact, the replacement of shielded metal-arc welding with semiautomatic processes has been a primary means by which steel fabricators and erectors have met cost-price squeezes in their welding operations.

Principles of Operation. Welding begins when the arc is struck between the work and the tip of the electrode. The heat of the arc melts the electrode and the surface of the work near the arc. Tiny globules of molten metal form on the tip of the electrode and transfer through the arc into the molten weld "pool" or "puddle" on the work surface.

The transfer through the arc stream is brought about by electrical and magnetic forces. Movement of the arc along the work (or movement of the work under the arc) accomplishes progressive melting and mixing of molten metal, followed by solidification, and, thus, the unification of parts.

It would be possible to clamp a bare mild-steel electrode into the electrode holder and fuse-join two steel parts. The resulting weld would lack ductility and soundness if judged by the present standards of weld quality. The weld metal so deposited would contain oxides and nitrides resulting from reaction of the molten metal with oxygen and nitrogen of the atmosphere. An essential feature of the electrode used in the shielded metal-arc process is a covering or coating, applied to the core metal by extrusion or dipping, that contains ingredients to shield the arc and protect the hot metal from chemical reaction with constituents of the atmosphere.

The shielding ingredients have various functions. One is to shield the arc — provide a dense, impenetrable envelope of vapor or gas around the arc and the molten metal to prevent the pickup of oxygen and nitrogen and the chemical formation of oxides and nitrides in the weld puddle. Another is to provide scavengers and deoxidizers to refine the weld metal. A third is to produce a slag coating over molten globules of metal during their transfer through the arc stream and a slag blanket over the molten puddle and the newly solidified weld. *Fig.* 53.4 illustrates the decomposition of an electrode covering and the manner in which the

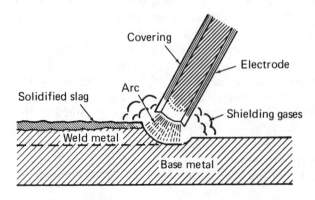

Fig. 53.4 — Schematic representation of shielded metal-arc welding. Gases generated by the decomposition and vaporization of materials in the electrode covering — including vaporized slag — provide a dense shield around the arc stream and over the molten puddle. Molten and solidified slag above the newly formed weld metal protects it from the atmosphere while it is hot enough to be chemically reactive with oxygen and nitrogen.

arc stream and weld metal are shielded from the air.

Another function of the shield is to provide the ionization needed for ac welding. With alternating current, the arc goes out 120 times a second. For it to be reignited each time it goes out, an electrically conductive path must be maintained in the arc stream. Potassium compounds in the electrode covering provide ionized gaseous particles that remain ionized during the fraction of a second that the arc is extinguished with ac cycle reversal. An electrical path for reignition of the arc is thus maintained.

A solid wire core is the main source of filler metal in electrodes for the shielded metal-arc process. However, the so-called iron-powder electrodes also supply filler metal from iron powder contained in the electrode covering or within a tubular core wire. Iron powder in the covering increases the efficiency of use of the arc heat and thus the deposition rate. With thickly covered iron-powder electrodes, it is possible to drag the electrode over the joint without the electrode freezing to the work or shorting out. Even though the heavy covering makes contact with the work, the electrical path through the contained powder particles is not adequate in conductivity to short the arc, and any resistance heating that occurs supplements the heat of the arc in melting the electrode. Because heavily-covered iron-powder electrodes can be dragged along the joint, less skill is required in their use.

Some electrodes for the shielded metal-arc process are fabricated with a tubular wire that contains alloying materials in the core. These are used in producing high-alloy deposits. Just as the conventional electrodes, they have an extruded or dipped covering.

Power Source. Shielded metal-arc welding requires relatively low currents (10 to 500 amp) and voltages (17 to 45), depending on the type and size electrode used. The current may be either ac or dc; thus, the power source may be either ac or dc or a combination ac/dc welder. For most work a variable voltage power source is preferred, since it is difficult for the welder to hold a constant arc length. With the variable voltage source and the machine set to give a steep volt-ampere curve, the voltage increases or decreases with variations in the arc length to maintain the current fairly constant. The equipment compensates for the inability of the operator to hold an exact arc length, and he is able to obtain a uniform deposition rate.

In some welding, however, it may be desirable for the welder to have control over the deposition rate — such as when depositing root passes in joints with varying fitup or in out-of-position work. In these cases variable voltage performance with a flatter voltage-amperage curve is desirable, so that the welder can decrease the deposition rate by increasing his arc length or increase it by shortening the arc length.

Fig. 53.5 illustrates typical volt-ampere curves possible with a variable voltage power source. The change from one type of voltage-ampere curve to another is made by changing the open-circuit voltage and current settings of the machine.

The fact that the shielded metal-arc process can be used with so many electrode types and sizes — in all positions — on a great variety of materials — and with flexibility in operator control makes it the most versatile of all welding processes. These advantages are enhanced further by the low cost of equipment. The total advantages of the process, however, must be weighed against the cost of per foot of weld when a process is to be selected for a particular job. Shielded

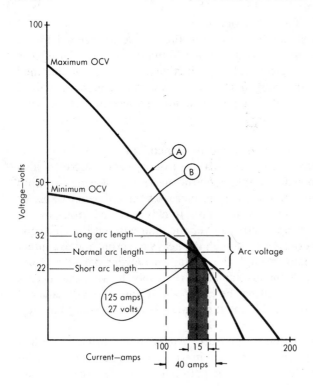

Fig. 53.5 — Typical volt-ampere curves possible with a variable voltage power source. The steep curve (A) allows minimum current change. The flatter curve (B) permits the welder to control current by changing the length of the arc.

metal-arc welding is a well recognized way of getting the job done, but too faithful adherence to it often leads to getting the job done at excessive welding costs.

The Submerged-Arc Process

Submerged-arc welding differs from other arc-welding processes in that a blanket of fusible, granular material — commonly called flux — is used for shielding the arc and the molten metal. The arc is struck between the workpiece and a bare wire electrode, the tip of which is submerged in the flux. Since the arc is completely covered by the flux, it is not visible and the weld is run without the flash, spatter, and sparks that characterize the open-arc processes. The nature of the flux is such that very little smoke or visible fumes are developed.

The Mechanics of Flux Shielding. The process is either semiautomatic or full-automatic, with electrode fed mechanically to the welding gun, head, or heads. In semiautomatic welding, the welder moves the gun, usually equipped with a flux-feeding device, along the joint. Flux feed may be by gravity flow through a nozzle concentric with the electrode from a small hopper atop the gun, or it may be through a concentric nozzle tube-connected to an air-pressurized flux tank. Flux may also be applied in advance of the welding operation or ahead of the arc from a hopper run along the joint. In fully automatic submerged-arc welding, flux is fed continuously to the joint ahead of or concentric to the arc, and full-automatic installations are commonly equipped with vacuum systems to pick up the unfused flux left by the welding head or heads for cleaning and reuse.

During welding, the heat of the arc melts some of the flux along with the tip of the electrode as illustrated in *Fig.* 53.6. The tip of the electrode and the welding

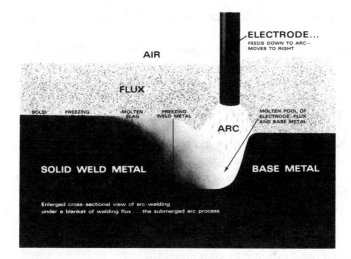

ELECTRODE...
FEEDS DOWN TO ARC—
MOVES TO RIGHT

AIR

FLUX

SOLID   FREEZING   MOLTEN   FREEZING   MOLTEN POOL OF
                           SLAG   WELD METAL   ELECTRODE-FLUX
                                                        AND BASE METAL

ARC

SOLID WELD METAL   BASE METAL

Enlarged cross-sectional view of arc-welding
under a blanket of welding flux . . . the submerged arc process

Fig. 53.6 — The mechanics of the submerged-arc process. The arc and the molten weld metal are buried in the layer or flux, which protects the weld metal from contamination and concentrates the heat into the joint. The molten flux arises through the pool, deoxidizing and cleansing the molten metal, and forms a protective slag over the newly deposited weld.

zone are always surrounded and shielded by molten flux, surmounted by a layer of unfused flux. The electrode is held a short distance above the workpiece, and the arc is between the electrode and the workpiece. As the electrode progresses along the joint, the lighter molten flux rises above the molten metal in the form of a slag. The weld metal, having a higher melting (freezing) point, solidifies while the slag above it is still molten. The slag then freezes over the newly solidified weld metal, continuing to protect the metal from contamination while it is very hot and reactive with atmospheric oxygen and nitrogen. Upon cooling and removal of any unmelted flux for reuse, the slag is readily peeled from the weld.

There are two general types of submerged-arc fluxes, bonded and fused. In the bonded fluxes, the finely ground chemicals are mixed, treated with a bonding agent and manufactured into a granular aggregate. The deoxidizers are incorporated in the flux. The fused fluxes are a form of glass resulting from fusing the various chemicals and then grinding the glass to a granular form. Fluxes are available that add alloying elements to the weld metal, enabling alloy weld metal to be made with mild steel electrode.

Advantages of the Process. High currents can be used in submerged-arc welding and extremely high heat developed. Because the current is applied to the electrode a short distance above its tip, relatively high amperages can be used on small diameter electrodes. This results in extremely high current densities in relatively small cross sections of electrode. Currents as high as 600 amperes can be carried on electrodes as small as 5/64-in., giving a density in the order of 100,000 amperes per square inch — six to ten times that carried on stick electrodes.

Because of the high current density, the melt-off rate is much higher for a given electrode diameter than with stick-electrode welding. The melt-off rate is affected by the electrode material, the flux, type of current, polarity, and length of wire beyond the point of electrical contact in the gun or head.

The insulating blanket of flux above the arc prevents rapid escape of heat and concentrates it in the welding zone. Not only are the electrode and base metal melted rapidly, but the fusion is deep into the base metal. The deep penetration allows the use of small welding grooves, thus minimizing the amount of filler metal per foot of joint and permitting fast welding speeds. Fast welding, in turn, minimizes the total heat input into the assembly and, thus, tends to prevent problems of heat distortion. Even relatively thick joints can be welded in one pass by submerged-arc.

Welds made under the protective layer of flux have good ductility and impact resistance and uniformity in bead appearance. Mechanical properties at least equal to those of the base metal are consistently obtained. In single-pass welds, the fused base material is large compared to the amount of filler metal used. Thus, in such welds the base metal may greatly influence the chemical and mechanical properties of the weld. For this reason, it is sometimes unnecessary to use electrodes of the same composition as the base metal for welding many of the low-alloy steels. The chemical composition and properties of multipass welds are less affected by the base metal and depend to a greater extent on the composition of the electrode, the activity of the flux, and the welding conditions.

Through regulation of current, voltage, and travel speed, the operator can exercise close control over penetration to provide any depth ranging from deep and narrow with high-crown reinforcement, to wide, nearly flat beads with shallow penetration. Beads with deep penetration may contain on the order of 70% melted base metal, while shallow beads may contain as little as 10% base metal. In some instances, the deep-penetration properties of submerged-arc can be used to eliminate or reduce the expense of edge preparation.

Multiple electrodes may be used, two side by side or two or more in tandem, to cover a large surface area or to increase welding speed. If shallow penetration is desired with multiple electrodes, one electrode can be grounded through the other (instead of through the workpiece) so that the arc does not penetrate deeply.

Deposition rates are high — up to ten times those of stick-electrode welding. *Fig.* 53.7 shows approximate deposition rates for various submerged-arc arrangements, with comparable deposition rates for manual welding with covered electrodes.

Submerged-arc welding may be done with either dc or ac power. Direct current gives better control of bead shape, penetration, and welding speed, and arc starting is easier with it. Bead shape is usually best with dc reverse polarity (electrode positive), which also provides maximum penetration. Highest deposition rates and minimum penetration are obtained with dc straight polarity. Alternating current minimizes arc blow and gives penetration between that of dcrp and dcsp.

Applications and Economies. With proper selection of equipment, submerged-arc is widely applicable to the welding requirements of industry. It can be used with all types of joints, and permits welding a full range of carbon and low-alloy steels, from 16-gage sheet to the thickest plate. It is also applicable to some high-alloy, heat-treated, and stainless steels, and is a favored process for rebuilding and hardsurfacing. Any degree of mechanization can be used — from the hand-held semiautomatic gun to boom or track-carried and fixture-held multiple welding heads.

The high quality of submerged-arc welds, the high deposition rates, the deep penetration, the adaptability of the process to full mechanization, and the comfort characteristics (no glare, sparks, spatter, smoke, or excessive heat radiation) make it a preferred process in steel fabrication. It is used extensively in ship and barge building, railroad car building, pipe manufacture, and in fabricating structural beams, girders, and columns where long welds are required. Automatic submerged-arc installations are also key features of the welding areas of plants turning out mass-produced assemblies joined with repetitive short welds.

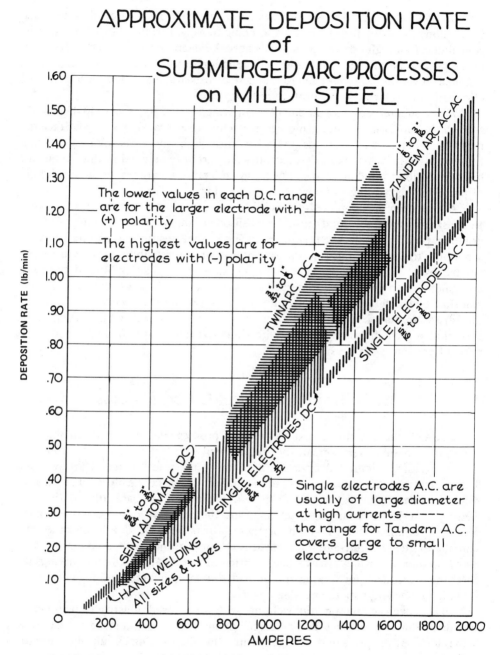

Fig. 53.7 — Approximate deposition rates of various submerged-arc arrangements, compared with the deposition rates of stick-electrode welding.

The high deposition rates attained with submerged-arc are chiefly responsible for the economies achieved with the process. The cost reductions when changing from the manual shielded metal-arc process to submerged-arc are frequently dramatic. Thus, a hand-held submerged-arc gun with mechanized travel may reduce welding costs more than 50%; with fully automatic multiarc equipment, it is not unusual for costs to be but 10% of those with stick-electrode welding.

Other factors than deposition rates enter into the lowering of welding costs. Continuous electrode feed from coils, ranging in weight from 60 to 1,000 pounds, contributes to a high operating factor. Where the deep-penetration characteristics of the process permit the elimination or reduction of joint preparation, expense is lessened. After the weld has been run, cleaning costs are minimized, because of the elimination of spatter by the protective flux.

When submerged-arc equipment is operated properly, the weld beads are smooth and uniform, so that grinding or machining are rarely required. Since the rapid heat input of the process minimizes distortion, the costs for straightening finished assemblies are reduced, especially if a carefully planned welding sequence has been followed. Submerged-arc welding, in fact, often allows the premachining of parts, further adding to fabrication cost savings.

A limitation of submerged-arc welding is that imposed by the force of gravity. In most instances, the joint must be positioned flat or horizontal to hold the granular flux. To deal with this problem, weldment positioners are used to turn assemblies to present joints flat or horizontal — or the assemblies may be turned or rotated by a crane. Substantial capital investments in positioning and fixturing equipment in order to use submerged-arc welding to the fullest extent, and thus gain full advantage of the deposition rate, have proved their worth in numerous industries. Special fixturing and tooling have been developed for the retention of flux and molten metal in some applications, so that "three-o'clock" and even vertical-up welding is possible.

The Self-Shielded Flux-Cored Process

The self-shielded flux-cored arc-welding process is an outgrowth of shielded metal-arc welding. The versatility and maneuverability of stick electrodes in manual welding stimulated efforts to mechanize the shielded metal-arc process. The thought was that if some way could be found for putting an electrode with self-shielding characteristics in coil form and feeding it mechanically to the arc, welding time lost in changing electrodes and the material loss as electrode stubs would be eliminated. The result of these efforts was the development of the semiautomatic and full-automatic processes for welding with continuous flux-cored tubular electrode "wires." Such fabricated wires (*Fig.* 53.8) contain in their cores the ingredients for fluxing and deoxidizing molten metal and for generating shielding gases and vapors and slag coverings.

In essence, semiautomatic welding with flux-cored electrodes is manual shielded metal-arc welding with an electrode many feet long instead of just a few inches long. By the press of the trigger completing the welding circuit, the operator activates the mechanism that feeds the electrode to the arc. He uses a gun instead of an electrode holder, but it is similarly light in weight and easy to maneuver. The only other major difference is that the weld metal of the electrode surrounds the shielding and fluxing chemicals, rather than being surrounded by them.

Full-automatic welding with self-shielded flux-cored electrodes is one step further in mechanization — the removal of direct manual manipulation in the utilization of the open-arc process.

One reason for incorporating the flux inside a tubular wire is to make feasible the coiling of electrode; outside coverings such as used on stick electrodes are

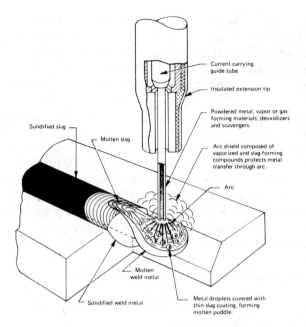

Current-carrying guide tube

Insulated extension tip

Powdered metal, vapor or gas-forming materials, deoxidizers and scavengers.

Arc shield composed of vaporized and slag-forming compounds protects metal transfer through arc.

Arc

Solidified slag

Molten slag

Molten weld metal

Solidified weld metal

Metal droplets covered with thin slag coating, forming molten puddle.

Fig. 53.8 — Principles of the self-shielded flux-cored arc-welding process. The electrode may be viewed as an "inside-out" construction of the stick electrode used in shielded metal-arc welding. Putting the shield-generating materials inside the electrode allows the coiling of long, continuous lengths of electrode and gives an outside conductive sheath for carrying the welding current from a point close to the arc.

brittle and would not permit coiling. The "inside-out" construction of the fabricated electrodes also solved the problem of how to make continuous electric contact at a point in the welding gun close to the arc.

One of the limitations of the stick electrode is the long and varying length of electrode between the point of electrical contact in the electrode holder and the electrode tip. This limits the current that can be used because of electrical resistance heating. High currents — capable of giving high deposition rates — in passing through an electrode length more than a few inches long would develop enough resistance heating to overheat and damage the covering. But when electrical contact can be made close to the arc, as with the inside-out construction of tubular electrodes, relatively high currents can be used even with small-diameter electrode wires.

The inside-out construction of self-shielded electrode, thus, brought to manually manipulated welding the possibility of using higher-amperage currents than feasible with stick-electrode welding. As a result, much higher deposition rates are possible with the hand-held semiautomatic gun than with the hand-held stick-electrode holder.

Higher deposition rates, plus automatic electrode feed and elimination of lost time for changing electrodes have resulted in substantial production economies wherever the semiautomatic process has been used to replace stick-electrode welding. Decreases in welding costs as great as 50% have been common, and in some production welding deposition rates have been increased as much as 400%.

The intent behind the development of self-shielded flux-cored electrode welding was to mechanize and increase the efficiency of manual welding. The semiautomatic use of the process does just that — serves as a direct replacement for stick-electrode welding. The full-automatic use of the process, on the other hand, competes with other fully automatic processes and is used in production where it gives the desired performance characteristics and weld properties, while eliminating problems associated with flux or gas handling. Although the full-automatic process is important to a few industries, the semiautomatic version has the wider application possibilities. In fact, semiautomatic self-shielded flux-cored welding has potential for substantially reducing welding costs when working with steel wherever stick-electrode welding is used to deposit other than minor volumes of weld metal. This has been proved to be true in maintenance and repair welding, as well as in production work.

Advantages of the Process. The advantages of the self-shielded flux-cored arc-welding process may be summarized as follows:

1.   When compared with stick-electrode welding, gives deposition rates up to four times as great, often decreasing welding costs by as much as 50 to 75%.

2.   Eliminates the need for flux-handling and recovery equipment, as in submerged-arc welding, or for gas and gas storage, piping, and metering equipment, as in gas-shielded mechanized welding. The semiautomatic process is applicable where other mechanized processes would be too unwieldy.

3.   Has tolerance for elements in steel that normally cause weld cracking when stick-electrode or one of the other mechanized welding processes are used. Produces crack-free welds in medium-carbon steel, using normal welding procedures.

4.   Under normal conditions, eliminates the problems of moisture pickup and storage that occur with low-hydrogen electrodes.

5.   Eliminates stub losses and the time that would be required for changing electrodes with the stick-electrode process.

6.   Eliminates the need for wind shelters required with gas-shielded welding in field erection; permits fans and fast air-flow ventilation systems to be used for worker comfort in the shop.

7.   Enables "one-process," and even "one-process, one-electrode," operation in some shop and field applications. This, in turn, simplifies operator training, qualification, and supervision; equipment selection and maintenance; and the logistics of applying men, materials, and equipment to the job efficiently.

8.   Enables the application of the long-stickout principle to enhance deposition rates, while allowing easy operator control of penetration.

9.   Permits more seams to be welded in one pass, saving welding time and the time that otherwise would be consumed in between-pass cleaning.

10.   Is adaptable to a variety of products; permits continuous operation at one welding station, even though a variety of assemblies with widely different joint requirements are run through it.

11.   Provides the fast filling of gouged-out voids often required when making repairs to weldments or steel castings.

12.   Gives the speed of mechanized welding in close quarters; reaches into spots inaccessible by other semiautomatic processes.

13.   Provides mechanized welding where mechanized welding was formerly impossible, such as in the joining of a beam web to a column in building erection.

14.   Enables the bridging of gaps in fitup by operator control of the penetration without reducing quality of the weld. Minimizes repair, rework, and rejects.

The Gas-Shielded Arc-Welding Processes

As noted in the preceding, the shielded metal-arc process (stick-electrode) and self-shielded flux-cored electrode process depend in part on gases generated by the heat of the arc to provide arc and puddle shielding. In contrast, the gas-shielded arc-welding processes use either bare or flux-cored filler metal and gas from an external source for shielding. The gas is impinged upon the work from a nozzle that surrounds the electrode. It may be an inert gas — such as argon or helium — or carbon dioxide (CO_2), a much cheaper gas that is suitable for use in the welding of steels. Mixtures of the inert gases, oxygen, and carbon dioxide also are used to produce special arc characteristics.

There are three basic gas-shielded arc-welding processes that have broad application in industry. They are the gas-shielded flux-cored process, the gas tungsten-arc (TIG) process, and the gas metal-arc (MIG) process.

The Gas-Shielded Flux-cored Process. The gas-shielded flux-cored process may be looked upon as a hybrid between self-shielded flux-cored arc welding and gas metal-arc welding. Tubular electrode wire is used (*Fig.* 53.9), as in the self-

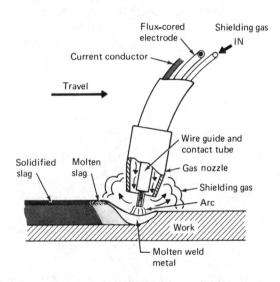

Fig. 53.9 — Principles of the gas-shielded flux-cored process. Gas from an external source is used for the shielding; the core ingredients are for fluxing and metal-conditioning purposes.

shielded process, but the ingredients in its core are for fluxing, deoxidizing, scavenging, and sometimes alloying additions, rather than for these functions plus the generation of protective vapors. In this respect, the process has similarities to the self-shielded flux-cored electrode process, and the tubular electrodes used are classified by the American Welding Society along with electrodes used in the self-shielded process. On the other hand, the process is similar to gas metal-arc welding in that a gas is separately applied to act as arc shield.

The guns and welding heads for semiautomatic and full-automatic welding with the gas-shielded process are of necessity more complex than those used in self-shielded flux-cored welding. Passages must be included for the flow of gases. If the gun is water-cooled, additional passages are required for this purpose.

The gas-shielded flux-cored process is used for welding mild and low-alloy steels. It gives high deposition rates, high deposition efficiencies, and high operating factors. Radiographic-quality welds are easily produced, and the weld metal with mild and low-alloy steels has good ductility and toughness. The process is adaptable to a wide variety of joints and gives the capability for all-position welding.

Gas Metal-Arc Welding. Gas metal-arc welding, popularly known as MIG welding, uses a continuous electrode for filler metal and an externally supplied gas or gas mixture for shielding. The shielding gas — helium, argon, carbon dioxide, or mixtures thereof — protects the molten metal from reacting with constituents of the atmosphere. Although the gas shield is effective in shielding the molten metal from the air, deoxidizers are usually added as alloys in the electrode. Sometimes light coatings are applied to the electrode for arc stabilizing or other purposes. Lubricating films may also be applied to increase the electrode feeding efficiency in semiautomatic welding equipment. Reactive gases may be included in the gas mixture for arc-conditioning functions. *Fig.* 53.10 illustrates the method of which shielding gas and continuous electrode are supplied to the welding arc.

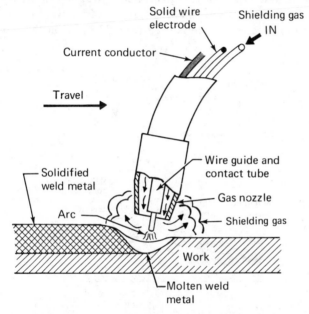

Fig. 53.10 — Principle of the gas metal-arc process. Continuous solid-wire electrode is fed to the gas-shielded arc.

MIG welding may be used with all of the major commercial metals, including carbon, alloy, and stainless steels and aluminum, magnesium, copper, iron, titanium, and zirconium. It is a preferred process for the welding of aluminum, magnesium, copper, and many of the alloys of these reactive metals. Most of the irons and steels can be satisfactorily joined by MIG welding, including the carbon-free irons, the low-carbon and low-alloy steels, the high-strength quenched and tempered steels, the chromium irons and steels, the high-nickel steels, and some of the so-called superalloy steels. With these various materials, the welding techniques and procedures may vary widely. Thus, carbon dioxide or argon-oxygen mixtures are suitable for arc shielding when welding the low-carbon and low-alloy steels, whereas pure inert gas may be essential when welding highly alloyed steels. Copper and many of its alloys and the stainless steels are successfully welded by the process.

Welding is either semiautomatic, using a handheld gun to which electrode is fed automatically, or full-automatic equipment is used. Metal transfer with the MIG process is by one of two methods: "spray-arc" or short circuiting. With spray-arc, drops of molten metal detach from the electrode and move through the arc column to the work. With the short-circuiting technique — often referred to as short-arc welding — metal is transferred to the work when the molten tip of the electrode contacts the molten puddle.

MIG welding is a dc weld process; ac current is seldom, if ever, used. Most MIG welding is done with reverse polarity (DCRP). Weld penetration is deeper with reverse polarity than it is with straight polarity. MIG welding is seldom done with straight polarity, because of arc-instability and spatter problems that make straight polarity undesirable for most applications.

The gas metal-arc process can be used for spot welding to replace either riveting, electrical resistance, or TIG spot welding. It has proved applicable for spot welding where TIG spot welding is not suitable, such as in the joining of aluminum and rimmed steel. Fitup and cleanliness requirements are not as exacting as with TIG spot welding, and the MIG process may be applied to thicker materials.

Gas Tungsten-Arc Welding. The AWS definition of gas tungsten-arc (TIG) welding is "an arc welding process wherein coalescence is produced by heating with an arc between a tungsten electrode and the work." A filler metal may or may not be used. Shielding is obtained with a gas or a gas mixture. *Fig.* 53.11 shows the principles of the process.

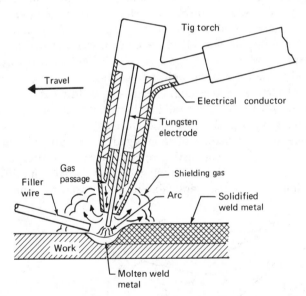

Fig. 53.11 — Principles of the gas tungsten-arc process. If filler metal is required, it is fed into the pool from a separate filler rod.

Essentially, the nonconsumable tungsten electrode is a "torch" — a heating device. Under the protective gas shield, metals to be joined may be heated above their melting points so that material from one part coalesces with material from the other part. Upon solidification of the molten area, unification occurs. Pressure may be used when the edges to be joined are approaching the molten state to assist coalescence. Welding in this manner requires no filler metal.

If the work is too heavy for the mere fusing of abutting edges, and if groove joints or reinforcements, such as fillets, are required, filler metal must be added. This is supplied by a filler rod, manually or mechanically fed into the weld puddle. Both the tip of the nonconsumable tungsten electrode and the tip of the filler rod are kept under the protective gas shield as welding progresses.

Materials weldable by the TIG process are most grades of carbon, alloy, and stainless steels; aluminum and most of its alloys; magnesium and most of its alloys; copper and various brasses and bronzes; high-temperature alloys of various types; numerous hard-surfacing alloys; and such metals as titanium, zirconium, gold, and silver. The process is especially adapted for welding thin materials where the requirements for quality and finish are exacting. It is one of the few processes that are satisfactory for welding such tiny and thin-walled objects as transistor cases, instrument diaphragms, and delicate expansion bellows.

The gases employed in the TIG process are argon and helium and mixtures of the two. Filler metals are available for a wide variety of metals and alloys, and these are often similar, although not necessarily identical to the metals being joined.

Specialized Arc-Welding Processes

The joining of metals may be accomplished by methods other than those utilizing electrical energy. Oxyacetylene, friction, explosive, and thermit welding are examples. There are also various methods of welding that make use of electrical energy that are not "arc" welding. Examples are the variations of electrical-resistance welding, ultrasonic, electron-beam, and electrodeposition welding.

There are also several processes that may appropriately be labeled "arc" welding that do not fall in the categories described in the preceding text. In some cases, however, these specialized arc-welding processes may be looked upon as modifications of basic welding processes.

Electroslag Welding. Electroslag welding is an adaptation of the submerged-arc welding process for joining thick materials in the vertical position. *Fig.* 53.12 is a diagrammatic sketch of the electroslag process. It will be noted that, whereas some of the principles of submerged-arc welding are applied, in other respects the process resembles a casting operation. Here, a square butt joint in heavy plate is illustrated, but the electroslag process — with modifications in equipment and technique — is also applicable to T joints, corner joints, girth seams in heavy-wall

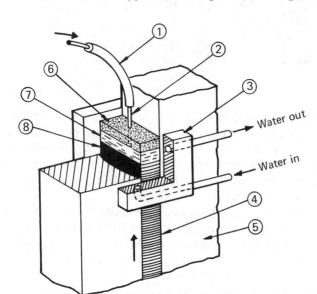

Water out

Water in

Fig. 53.12 — Schematic sketch of electroslag welding. (1) electrode guide tube, (2) electrode, (3) water-cooled copper shoes, (4) finished weld, (5) base metal, (6) molten slag, (7) molten weld metal, (8) solidified weld metal.

cylinders, and other joints. The process is suited best for materials at least 1-inch in thickness, and can be used with multiple electrodes on materials up to 10-inches thick without excessive difficulties.

As illustrated by the open, square butt joint in *Fig.* 53.12, the assembly is positioned for the vertical deposition of weld metal. A starting pad at the bottom of the joint prevents the fall-out of the initially deposited weld metal, and, since it is penetrated, assures a full weld at this point. Welding is started at the bottom and progresses upward. Water-cooled dams, which may be looked upon as molds, are placed on each side of the joint. These dams are moved upward as the weld-metal deposition progresses. The joint is filled in one "pass" — a single upward progression — of one or more consumable electrodes. The electrode or electrodes may be oscillated across the joint if the width of the joint makes this desirable.

At the start of the operation, a layer of flux is placed in the bottom of the joint and an arc is struck between the electrode (or electrodes) and the work. The arc melts the slag, forming a molten layer, which subsequently acts as an electrolytic heating medium. The arc is then quenched or shorted-out by this molten conductive layer. Heat for melting the electrode and the base metal subsequently results from the electrical resistance heating of the electrode section extending from the contact tube and from the resistance heating within the molten slag layer. As the electrode (or electrodes) is consumed, the welding head (or heads) and the cooling dams move upward.

In conventional practice, the weld deposit usually contains about 1/3 melted base metal and 2/3 electrode metal — which means that the base metal substantially contributes to the chemical composition of the weld metal. Flux consumption is low, since the molten flux and the unmelted flux above it "ride" above the progressing weld.

The flux used has a degree of electrical conductivity and low viscosity in the molten condition and a high vaporization temperature. The consumable electrodes may be either solid wire or tubular wire filled with metal powders. Alloying elements may be incorporated into the weld by each of these electrodes.

Weld quality with the electroslag process is generally excellent, due to the protective action of the heavy slag layer. Sometimes, however, the copper dams are provided with orifices just above the slag layer, through which a protective gas — argon or carbon dioxide — is introduced to flush out the air above the weld and, thus, give additional assurance against oxidation. Such provisions are sometimes considered worthwhile when welding highly alloyed steels or steels that contain easily oxidizing elements.

The electroslag process has various advantages, including no need for special edge preparations, a desirable sequence of cooling that places the outside surfaces of weld metal under compressive rather than tensile stresses, and a relative freedom from porosity problems. Special equipment, however, is required, and few users of welding have enough volume of welding subject to application of the process to warrant the cost of such equipment.

Electrogas Welding. Electrogas welding is very similar to electroslag welding in that the equipment is similar and the joint is in the vertical position. As the name implies, the shielding is by carbon dioxide or an inert gas. A thin layer of slag, supplied by the flux-cored electrode, covers the molten metal, and the heat is supplied by an arc rather than by resistance heating as in the electroslag process.

Stud Arc Welding. Stud arc welding is a variation of the shielded metal-arc process that is widely used for attaching studs, screws, pins, and similar fasteners to a larger workpiece. The stud (or small part), itself — often plus a ceramic ferrule at its tip — is the arc-welding electrode during the brief period of time required for studding.

In operation, the stud is held in a portable pistol-shaped tool called a stud gun and positioned by the operator over the spot where it is to be weld-attached. At a press of the trigger, current flows through the stud, which is lifted slightly, creating an arc. After a very short arcing period, the stud is then plunged down into the molten pool created on the base plate, the gun withdrawn from it, and the ceramic ferrule — if one has been used — removed. The timing is controlled automatically and the stud is welded onto the workpiece in less than a second. The fundamentals of the process are illustrated in *Fig.* 53.13.

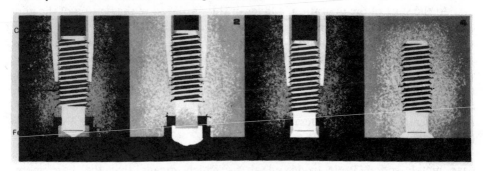

Fig. 53.13 — Principles of stud welding, using a ceramic ferrule to shield the pool. (A) The stud with ceramic ferrule is grasped by the chuck of the gun and positioned for welding. (B) The trigger is pressed, the stud is lifted, and the arc is created. (C) With the brief arcing period completed, the stud is plunged into the molten pool on the base plate. (D) The gun is withdrawn from the welded stud and the ferrule removed.

Stud welding is a well developed and widely used art. Information needed for its application with the carbon and alloy steels, stainless steels, aluminum, and other nonferrous metals may be found in the *AWS Welding Handbook*, Sixth Edition, Section 2.

Plasma-Arc Welding. Plasma-arc (or plasma-torch) welding is one of the newer welding processes, which is used industrially, frequently as a substitute for the gas tungsten-arc process. In some applications, it offers greater welding speeds, better weld quality, and less sensitivity to process variables than the conventional processes it replaces. With the plasma torch, temperatures as high as 60,000°F are developed and, theoretically, temperatures as high as 200,000°F are possible.

The heat in plasma-arc welding originates in an arc, but this arc is not diffused as is an ordinary welding arc. Instead, it is constricted by being forced through a relatively small orifice. The "orifice" or plasma gas (see *Fig.* 53.14) may be supplemented by an auxiliary source of shielding gas.

"Orifice" gas refers to the gas that is directed into the torch to surround the electrode. It becomes ionized in the arc to form the plasma and emerges from the orifice in the torch nozzle as a plasma jet. If a shielding gas is used, it is directed onto the workpiece from an outer shielding ring.

The workpiece may or may not be part of the electrical circuit. In the "transferred-arc" system, the workpiece is a part of the circuit, as in other arc-welding processes. The arc "transfers" from the electrode through the orifice to the work. In the "nontransferred" system, the constricting nozzle surrounding the electrode acts as an electrical terminal, and the arc is struck between it and the

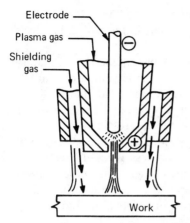

Electrode

Plasma gas

Shielding gas

Work

Fig. 53.14 — Diagrammatic sketch of the plasma-arc torch.

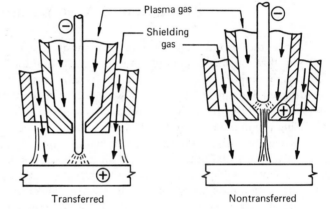

Fig. 53.15 — Transferred and nontransferred arcs.

Plasma gas

Shielding gas

Transferred

Nontransferred

electrode tip; the plasma gas then carries the heat to the workpiece. *Fig.* 53.15 illustrates transferred and nontransferred arcs.

The advantages gained by using a constricted-arc process over the gas tungsten-arc process include greater energy concentration, improved arc stability, higher welding speeds, and lower width-to-depth ratio for a given penetration. "Keyhole" welding — or penetrating completely through the workpiece — is possible.

Atomic-Hydrogen Welding. The atomic-hydrogen process of arc welding may be regarded as a forerunner of gas-shielded and plasma-torch arc welding. Although largely displaced by other processes that require less skill and are less costly, it is still preferred in some manual operations where close control of heat input is required.

In the atomic-hydrogen process, an arc is established between two tungsten electrodes in a stream of hydrogen gas, using ac current. As the gas passes through the arc, molecular hydrogen is disassociated into atomic hydrogen under the intense heat. When the stream of hydrogen atoms strikes the workpiece, the environmental temperature is then at a level where recombining into molecules is possible. As a result of the recombining, the heat of disassociation absorbed in the arc is liberated, supplying the heat needed for fusing the base metal and any filler metal that may be introduced.

The atomic-hydrogen process depends on an arc, but is really a heating torch. The arc supplies the heat through the intermediate of the molecular-disassociation, atom-recombination mechanism.

The hydrogen gas, however, does more than provide the mechanism for heat transfer. Before entering the arc, it acts as a shield and a coolant to keep the tungsten electrodes from overheating. At the weld puddle, the gas acts as a shield. Since hydrogen is a powerful reducing agent, any rust in the weld area is reduced to iron, and no oxide can form or exist in the hydrogen atmosphere until it is removed. Weld metal, however, can absorb hydrogen, with unfavorable metallurgical effects. For this reason, the process gives difficulties with steels containing sulfur or selenium, since hydrogen reacts with these elements to form hydrogen sulfide or hydrogen selenide gases. These are almost insoluble in molten metal and either bubble out of the weld pool vigorously or become entrapped in the solidifying metal resulting in porosity.

Fixtures and Manipulators

Once the production welding operations reach a stage of mechanization beyond that of the handheld and manually manipulated semiautomatic gun, fixtures, manipulators, and weldment positioners become important items of welding equipment. In fact, even with the hand-held semiautomatic gun some device for rotating the weldment or mechanically positioning it may be essential for realizing the maximum potential of semiautomatic welding.

When the full-automatic welding head is used, a fixturing device becomes a necessity. It may be nothing more than a support for holding the material to be welded while a self-propelled tractor carries the welding head over the joint, or it may be a complex mechanism, electrically or electronically controlled, and capable of such functions as "sensing" and self-correction. In some industries, such as the automotive, the weldment fixtures may cost many thousands of dollars.

Types of Fixtures. The objective in fixturing is to take maximum advantage of mechanized welding's ability to lower welding costs and increase productivity. The fixtures are usually devices that present the work or welding equipment in a position that assures maximum welding speed. Often these devices also systematize and control the flow of material to and from the welding equipment.

Welding fixtures fall into two broad classifications — those that act on the parts being welded, and those that act on the welding equipment. The type of action may be supporting, clamping, grounding, imparting movement, to name only a few. As an example, a welding fixture for the work may clamp the parts together and support them so that the weld can be made in the flat position. This fixture may hold the work stationary — in which case the welding equipment moves (or is traveled) over it — or may be mechanized and move the work under the stationary welding equipment. Some typical work-moving types of fixtures are rotating positioners, power turning rolls, lathes, and mill-type setups, as illustrated in *Fig. 53.16.*

Fixtures that act on the welding equipment — some types are called welding manipulators — similarly may provide a stationary support only or both support and move (travel) the welding equipment. The welding equipment may be a welding head or a welding gun, or may include all of the equipment in the welding circuit, including the power sources and even the flux-recovery unit.

Fig. 53.16 — Some typical work-moving welding fixtures are the (a) rotating positioner, (b) power turning rolls, (c) lathe, and (d) continuous mill.

The fixtures for holding the head stationary range from a simple arm and mounting bracket to the very complex installations common in continuous-mill welding. The types of fixtures that both support and travel the welding equipment include the beam-with-carriage, the self-propelled tractor, the spud welder, and a variety of boom-type welding manipulators (*Fig.* 53.17). Variations in the boom-type manipulators include column-stationary and boom-travel variable and both boom and column movable with variable travel.

Welding fixtures may also perform other functions that are not directly related to the actual welding operation. Thus, the fixturing may be an integral part of a conveyor system or may also position the work for a machining operation, such as drilling or grinding. Sometimes the designer even tries to incorporate part-forming into the welding fixture system. This is not recommended, however, since experience has proved that divorcing the welding fixture from all forming operations leads to more trouble-free operation.

Fig. 53.17 — These pictures illustrate four different types of welding equipment fixtures; (a) beam-with-carriage, (b) self-propelled tractor, (c) spud welder, and (d) boom-type manipulator with variable travel of both boom and column.

Considerations in Fixture Selection. Elaborate welding equipment or work fixturing usually requires tremendous production volume of a single or reasonably similar parts. If part volume does not exist, the next thing to consider is substantial footage of welding that lends itself to mechanization. In this case, fixture adaptability, simplicity, and cost often are the important fixture-selection factors. This may mean combining a variety of work-supporting or holding devices with a simple beam-with-carriage fixture, as in *Fig.* 53.18. Even a small self-propelled tractor — if it can be classified as a welding fixture — mounting a semi-automatic welding gun may produce the best answer for a given job, since it features fixturing simplicity and represents a nominal equipment investment.

Joint geometry also plays an important role in the selection of a welding fixture. For example, circumferential welds are made with the welding head stationary and the work moving (*Fig.* 53.19). Welds in a straight line usually require a fixture that travels the head along the joint with a stationary work fixture

Fig. 53.18 — The welding fixture here combines a work-holding device and a beam-with-carriage support for the welding head. The elevator doors being fabricated are held in perfect alignment as a submerged arc butt weld is made at 150 inches per minute.

Fig. 53.19 — Making a circumferential weld with the welding head stationary and the work moving. Here the submerged-arc head is supported by a simple pedestal arrangement.

providing clamping to maintain part alignment and fitup.

Welding Head Stationary. A stationary welding head calls for work movement and invariably leads to more complex work fixturing. In fact, the work fixturing for a stationary head can be the ultimate in automation (*Fig.* 53.20). Some fixtures have been designed to have the weight of the workpiece dropping into the fixture activate its positioning and clamping and initiate the welding cycle, thus, even eliminating the simple act of the welding operator pushing a start button.

Sometimes the head mounting of head-stationary fixturing will allow movement up and down or to one side to allow passage of the workpiece into and out of the fixture. It can also permit adjustment of the head position as welding progresses. This is still considered a head-stationary fixture, since welding speed is controlled by the movement of the work under the head.

Although a stationary welding head fixture is often considered the simplest type of fixturing, when it is incorporated as part of a continuous mill welder it

Fig. 53.20 — The work fixturing for a stationary head can be the ultimate in automation, as this fixture for welding spring pads on axle housings in an automotive plant.

Fig. 53.21 — The welding head is stationary in this mill-type fixture for pipe welding. The pipe is formed from plate; then continuously fed into the mill and welded as it passes under the head.

becomes part of one of the most complex welding fixtures (*Fig.* 53.21). In a mill welding fixture, the work is continuously fed under the stationary welding head, usually at high speeds. Often, the work is formed just ahead of the welding fixture and machined immediately after welding. Naturally, this complicates the basic work-fixturing operations of supporting, clamping, grounding, and part movement.

The expanding use of multiarc welding — where three, four, and even six welding heads are installed on the same welding fixture — is increasing the number of head-stationary welding installations. The shear weight and bulk of the welding equipment needed for these multihead installations tends to become excessive for a standard beam-with-carriage fixture or boom-type welding manipulator. The stationary head also eliminates the problem of finding ways to travel the control system, power sources, flux supply and recovery system, consumable electrode, and welding cables that must be solved with head-movable welding.

Welding Head Movable. With head-moving welding fixtures, the work remains stationary, and the welding head travels along the joint. Head movement may result from the travel of a carriage on a beam or rail, from a traveling boom mounted on a stationary column, from the travel of the entire manipulator riding on a track in the floor (*Fig.* 53.22), or from the travel of a tractor riding the surface of the workpiece (*Fig.* 53.23).

Fig. 53.22 — This boom-type welding manipulator travels everything along the joint being welded — equipment, flux, flux recovery system, wire-electrode, power sources, and controls.

Fig. 53.23 — This self-propelled tractor riding on the workpiece simplifies fixturing. In this case both the welding equipment and workpiece are traveling, but in opposite directions to keep the weld puddle in the flat position. Note the spirit level to maintain this position.

Fig. 53.24 — The beam-with-carriage fixture usually offers head travel in straight lines parallel to the beam. The length of the weld that can be made without stopping is limited by the length of the beam.

Fig. 53.25 — A boom-type welding manipulator can service a number of welding stations. If the loading, unloading, and welding cycles are properly in sequence, the welding operating factor can be substantially increased.

The beam-with-carriage-type welding-equipment fixture usually offers head travel in straight lines that run parallel to and on one side of the support for the beam (*Fig.* 53.24). The length of the weld that can be made without stopping is limited by the length of the beam.

A boom-type welding manipulator equipped for both boom and column travel can swing the welding head through a 360° arc. It is capable of welding in a straight line in any direction away from the column to a distance somewhat less than the length of the boom and can service welding stations located anywhere within this area (*Fig.* 53.25). It duplicates the performance of a beam-with-carriage-type fixture when the boom is held stationary and the entire manipulator is traveled on the floor tracks. In addition, it can weld on either side of the column supporting the boom. Weld length is limited by the length of the tracks on which the manipulator carriage is traveling. Both the beam-with-carriage and the boom-type welding manipulator can serve as welding-head-stationary fixtures.

Considerations in Weldment Design

A weldment design program starts with a recognition of a need. The need may be for improving an existing product or for building an entirely new product, using the most advanced design and fabrication techniques. In any event, many factors must be taken into account before a design is finalized. These considerations involve asking numerous questions and doing considerable research into the various areas of marketing, engineering, and production.

Analysis of Present Design. Insofar as possible, when designing an entirely new machine or structural unit, an attempt should be made to gain information about market requirements the new product is aimed to meet. If a new machine is to replace an older model, the good points and the deficiencies of the predecessor machine should be understood. The following questions can help in reaching the proper decision:

1. What do customers and the sales force say about the older machine?
2. What has been its history of failures? What features should be retained or added?
3. What suggestions for improvements have been made?
4. Was the old model overdesigned?

Determination of Load Conditions. The work the machine is intended to do, or the forces that a structural assembly must sustain, and the conditions of service might cause overload should be ascertained. From such information, the load on individual members can be determined. As a starting point for calculating loading, the designer may find one or more of the following methods useful:

1. From the motor horsepower and speed, determine the torque in inch-pounds on a shaft or revolving part.
2. Calculate the force in pounds on machine members created by the dead weight of parts.
3. Calculate the load on members of a hoist or lift truck back from the load required to tilt the machine.
4. Use the maximum strength of critical cables on a shovel or ditch digger that have proved satisfactory in service, to work back to the loads on machine members.
5. Consider the force required to shear a critical pin as an indication of maximum loading on the machine.
6. If a satisfactory starting point cannot be found, design for an assumed load and adjust from experience and tests; at least the design will be well proportioned.

Major Design Factors. In developing a design, the designer should think constantly about how decisions will affect production, manufacturing costs, performance, appearance and customer acceptance. Many factors far removed from engineering considerations, per se, become major design factors. Some of these are listed below, along with other relevant rules:

1. The design should satisfy strength and stiffness requirements. Overdesigning is a waste of materials and runs up production and shipping costs.

2. The safety factor may be unrealistically high as indicated by past experience.

3. Good appearance has value, but only in areas that are exposed to view. The print could specify the welds that are critical in respect to appearance.

4. Deep and symmetrical sections resist bending efficiently.

5. Welding the ends of beams rigid to supports increases strength and stiffness.

6. The proper use of stiffeners will provide rigidity at minimum weight of material.

7. Use closed tubular sections or diagonal bracing for torsion resistance. A closed tubular section may be many times better than an open section.

8. Specify nonpremium grades of steel wherever possible. Higher carbon and alloy steels require preheating, and frequently postheating, which are added cost items.

9. Use standard rolled sections wherever possible. Use standard plate and bar sizes for their economy and availability.

10. Provide maintenance accessibility in the design; do not bury a bearing support or other critical wear point in a closed-box weldment.

Layout. To the designer familiar only with castings, the laying out of a weldment for production may seem complex because of the many choices possible. Variety in the possibilities for layout, however is one of the advantages of welded design; opportunities for savings are presented. Certain general pointers for effective layout may be set forth:

1. Design for easy handling of materials and for inexpensive tooling.

2. Check with the shop for ideas that can contribute to cost savings.

3. Check the tolerances with the shop. The shop may not be able to hold them, and close tolerances and fits may serve no useful purpose.

4. Plan the layout to minimize the number of pieces. This will reduce assembly time and the amount of welding.

5. Lay out parts so as to minimize scrap.

6. If possible, modify the shape and size of scrap cutouts, so that such material may be used for pads, stiffeners, gear blanks and other parts.

7. If a standard rolled section is not available or suitable, consider forming the desired section from blanks flame-cut from plate. It is also possible to use long flat bar stock welded together, or to place a special order for a rolled-to-shape section.

8. In making heavy rings, consider the cutting of nesting segments from plate to eliminate excessive scrap.

Plate Preparation. Flame cutting, shearing, sawing, punch-press blanking, nibbling, and lathe cutoff are methods for cutting blanks from stock material. The decision relating to method will depend on the equipment available and relative costs. It will be influenced, however, by the quality of the edge for fitup and whether the method also provides a bevel in the case of groove joints. Experience has suggested the following pointers:

1.   Dimensioning of the blank may require stock allowance for subsequent edge preparation.

2.   Not all welds are continuous. This must be borne in mind when proposing to prepare the edge and cut the blank simultaneously.

3.   Select the type of cut that can be made in one pass. For single-bevel or single-V plate preparation, a single torch cut is used; for double-bevel or double-V, a multiple-torch.

4.   When a plate planer is available, weld metal costs can be reduced with thick plate by making J or U-groove preparations.

5.   Consider arc gouging, flame gouging, or chipping for back-pass preparations.

Forming and Special Sections. Forming can greatly reduce the cost of a weldment by eliminating welds and, often, machining operations. Thickness of materials, over-all dimensions, production volume, tolerances, and cost influence the choice of forming methods. The suggestions below should be useful in making decisions:

1.   Create a corner by bending or forming rather than by welding from two pieces.

2.   Roll a ring instead of cutting from plate in order to effect possible savings.

3.   Form round or square tubes or rings instead of buying commercial tubing if savings could be effected.

4.   Put bends in flat plate to increase stiffness.

5.   Use press indentations in plate to act as ribs, instead of using stiffeners to reduce vibration.

6.   Use corrugated sheet for extra stiffness.

7.   The design problem and cost of manufacture may be simplified by incorporating a steel casting or forging into the weldment for a complicated section.

8.   Use a small amount of hardsurfacing alloy applied by welding, rather than using expensive material throughout the section.

Welded Joint Design. The type of joint should be selected primarily on the basis of load requirements. Once the type is selected, however, variables in design and layout can substantially affect costs. Generally, the following rules apply:

1.   Select the joint requiring the least amount of weld filler metal.

2.   Where possible, eliminate bevel joints by using automatic submerged-arc welding, which has a deep-penetration arc characteristic.

3.   Use minimum root opening and included angle in order to reduce the amount of filler metal required.

4.   On thick plate, use double-grooves instead of single-groove joints to reduce the amount of weld metal.

5.   Use a single weld where possible to join three parts.

6.   Minimize the convexity of fillet welds; a 45° flat fillet, very slightly convex, is the most economical and reliable shape.

7.   Avoid joints that create extremely deep grooves.

8.   Design the joint for easy accessibility for welding.

Size and Amount of Weld. Overwelding is a common error of both design and production. Control begins with design, but must be carried throughout the assembly and welding operations. The following are basic guides:

1.   Be sure to use the proper amount of welding — not too much and not too

little. Excessive weld size is costly.

2. Specify that only the needed amount of weld should be deposited. The allowable limits used by the designer include the safety factor.

3. The leg size of fillet welds is especially important, since the amount of weld required increases as the square of the increase in leg size.

4. For equivalent strength, longer fillet welds having a smaller leg size are less costly than heavy intermittent welds.

5. Sometimes, especially under light-load or no-load conditions, cost can be reduced by using intermittent fillet welds in place of a continuous weld of the same leg size.

6. To derive maximum advantage from automatic welding, it may be better to convert several short welds into one continuous weld.

7. Place the weld in the section with the least thickness, and determine the weld size according to the thinner plate.

8. Place the weld on the shortest seam. If there is a cutout section, place the welded seam at the cutout in order to save on the length of welding. On the other hand, in automatic welding it may be better to place the joint away from the cutout area to permit the making of one continuous weld.

9. Stiffeners or diaphragms do not need much welding; reduce the weld leg size or length of weld if possible.

10. Keeping the amount of welding to a minimum reduces distortion and internal stress and, therefore, the need and cost for stress relieving and straightening.

Use of Subassemblies. In visualizing assembly procedures, the designer should break the weldment down into subassemblies in several ways to determine which will offer cost savings. The following are points to note:

1. Subassemblies spread the work out; more men can work on the job simultaneously.

2. Usually, subassemblies provide better access for welding.

3. The possibility of distortion or residual stresses in the finished weldment is reduced when the weldment is built from subassemblies.

4. Machining to close tolerances before welding into the final assembly is permitted. If necessary, stress relief of certain sections can be performed before welding into the final assembly.

5. Leak testing of compartments or chambers and painting before welding into the final assembly are permitted.

6. In-process inspection (before the job has progressed too far to rectify errors) is facilitated.

Use of Jigs, Fixtures, and Positioners. Jigs, fixtures, and welding positioners should be used to decrease fabrication time. In planning assemblies and subassemblies, the designer should decide if the jig is simply to aid in assembly and tacking or whether the entire welding operation is to be done in the jig. The considerations listed below are significant:

1. The jig must provide the rigidity necessary to hold the dimensions of the weldment.

2. Tooling must provide easy locating points and be easy to load and unload.

3. Camber can be built into the tool for control of distortion.

4. The operating factor can be increased by using two jigs, so that a helper can load one while the work in the other is being welded.

5. Welding positioners maximize the amount of welding in the flat downhand position, allowing use of larger electrodes and automatic welding.

Assembly. The assembly operations affect the quality of the welds and fabrication costs. Even though the designer may not have control of all the factors entering into good assembly procedures, he should be aware of the following:

1. Clean work — parts free from oil, rust, and dirt — reduces trouble.
2. Poor fitup can be costly.
3. A joint can be preset or prebent to offset expected distortion.
4. Strongbacks are valuable for holding materials in alignment.
5. When possible, it is desirable to break the weldment into natural sections, so that the welding of each can be balanced about its own neutral axis.
6. Welding the more flexible sections first facilitates any straightening that might be required before final assembly.

Welding Procedures. Although the designer may have little control of welding procedures, he is concerned with what goes on in the shop. Adherence to the following guidelines will help to effect the success of the weldment design:

1. Welding helpers and good fixtures and handling equipment improve the operating factor.
2. Backup bars increase the speed of welding on the first pass when groove joints are being welded.
3. The use of low-hydrogen electrodes eliminates or reduces preheat requirements.
4. The welding machine and cable should be large enough for the job.
5. The electrode holder should permit the use of high welding currents without overheating.
6. Weld in the flat downhand position whenever possible.
7. Weld sheet metal 45° downhill.
8. If plates are not too thick, consider the possibility of welding from one side only.
9. With automatic welding, position fillets so as to obtain maximum penetration into the root of the joint: flat plate, 30° from horizontal; vertical plate, 60° from horizontal.
10. Most reinforcements of a weld are unnecessary for a full-strength joint.
11. Use a procedure that eliminates arc blow.
12. Use optimum welding current and speed for best welding performance. If appearance is not critical and no distortion is being experienced, usual speed frequently can be exceeded.
13. Use the recommended current and polarity.
14. Consider the use of straight polarity (electrode negative) or long stickout with automatic welding to increase melt-off rate.
15. On small fillet welds, a small-diameter electrode may deposit the weld faster by not overwelding.

Distortion control. Distortion is affected by many factors of design and shop practice. Some points on shop procedures to control distortion of which the designer should be aware include:

1. High-deposition electrodes, automatic welding, and high welding currents tend to reduce the possibility of distortion.
2. The least amount of weld metal, deposited with as few passes as possible, is desirable.

3. Welding should progress toward the unrestrained portion of the member, but backstepping may be practical as welding progresses.

4. Welds should be balanced about the neutral axis of the member.

5. On multipass double-V joints, it may be advisable to weld alternately on both sides of the plate.

6. Avoid excessive prestressing members by forcing alignment to get better fitup of the parts.

7. Joints that may have the greatest contraction on cooling should be welded first.

Cleaning and Inspection. The designer, by his specifications, has some effect on cleaning and inspection costs. He also should recognize the following shop practices that affect these costs:

1. Industry now accepts as-welded joints that have uniform appearance as finished; therefore, do not grind the surface of the weld smooth or flush unless required for another reason. This is a very costly operation and usually exceeds the cost of welding.

2. Cleaning time may be reduced by use of ironpowder electrodes and automatic welding, which minimize spatter and roughness of surface.

3. Spatter films can be applied to the joint to reduce spatter sticking to the plate. Some electrodes and processes produce little or no spatter.

4. Sometimes a slightly reduced welding speed or a lower welding current will minimize weld faults. Lower repair costs may result in lower over-all costs.

5. Overzealous inspection can run up welding costs; it is possible to be unreasonably strict with inspection.

6. Inspection should check for overwelding, which can be costly, and can also contribute to distortion.

The Design of Welded Joints

The loads in a welded steel design are transferred from one member to another through welds placed in weld joints. Both the type of joint and the type of weld are specified by the designer.

Fig. 53.26 shows the joint and weld types. Specifying a joint does not by itself describe the type of weld to be used. Thus, ten types of welds are shown for making a butt joint. Although all but two welds are illustrated with butt joints here, some may be used with other types of joints. Thus, a single-bevel weld may also be used in a T or corner joint (*Fig.* 53.27), and a single-V weld may be used in a corner, T, or butt joint.

Fillet-Welded Joints. The fillet weld, requiring no groove preparation, is one of the most commonly used welds. Corner welds are also widely used in machine design. Various corner arrangements are illustrated in *Fig.* 53.28. The corner-to-corner joint, as in *Fig.* 53.28(a), is difficult to assemble because neither plate can be supported by the other. A small electrode with low welding current must be used so that the first welding pass does not burn through. The joint requires a large amount of metal. The corner joint shown in *Fig.* 53.28(b) is easy to assemble, does not easily burn through, and requires just half the amount of the weld metal as the joint in *Fig.* 53.28(a). However, by using half the weld size, but placing two welds, one outside and the other inside, as in *Fig.* 53.28(c), it is possible to obtain the same total throat as with the first weld. Only half the weld metal need be used.

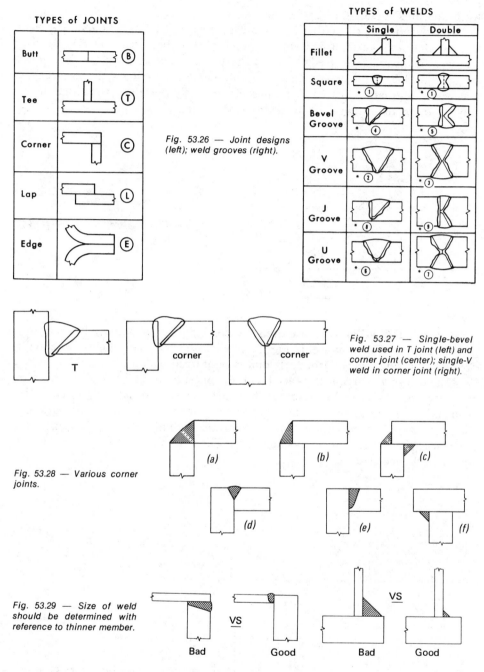

TYPES of JOINTS

Butt	(B)
Tee	(T)
Corner	(C)
Lap	(L)
Edge	(E)

TYPES of WELDS

	Single	Double
Fillet		
Square	* ①	* ①
Bevel Groove	* ④	* ⑤
V Groove	* ②	* ③
J Groove	* ⑧	* ⑨
U Groove	* ⑥	* ⑦

Fig. 53.26 — Joint designs (left); weld grooves (right).

Fig. 53.27 — Single-bevel weld used in T joint (left) and corner joint (center); single-V weld in corner joint (right).

T   corner   corner

Fig. 53.28 — Various corner joints.

(a)   (b)   (c)

(d)   (e)   (f)

Fig. 53.29 — Size of weld should be determined with reference to thinner member.

Bad   VS   Good   VS   Bad   Good

With thick plates, a partial-penetration groove joint, as in *Fig.* 53.28(d) is often used. This requires beveling. For a deeper joint, a J preparation, as in *Fig.* 53.28(e), may be used in preference to a bevel. The fillet weld in *Fig.* 53.28(f) is out of sight and makes a neat and economical corner.

The size of the weld should always be designed with reference to the size of the thinner member. The joint cannot be made any stronger by using the thicker member for the weld size, and much more weld metal will be required, as illustrated in *Fig.* 53.29.

In the United States, a fillet weld is measured by the leg size of the largest right triangle that may be inscribed within the cross-sectional area (*Fig.* 53.30). The throat, a better index to strength, is the shortest distance between the root of the joint and the face of the diagrammatical weld. As *Fig.* 53.30 shows, the leg size used may be shorter than the actual leg of the weld. With convex fillets, the actual throat may be longer than the throat of the inscribed triangle.

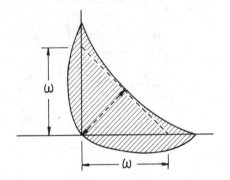

Fig. 53.30 — Leg size, w, of a fillet weld.

Groove and Fillet Combinations. A combination of a partial-penetration groove weld and a fillet weld (*Fig.* 53.31) is used for many joints. The AWS prequalified, single-bevel groove T joint is reinforced with a fillet weld.

The designer is frequently faced with the question of whether to use fillet or groove welds (*Fig.* 53.32). Here cost becomes a major consideration. The fillet welds in *Fig.* 53.32(a) are easy to apply and require no special plate preparation. They can be made using large-diameter electrodes with high welding currents, and, as a consequence, the deposition rate is high. The cost of the welds increases as the square of the leg size.

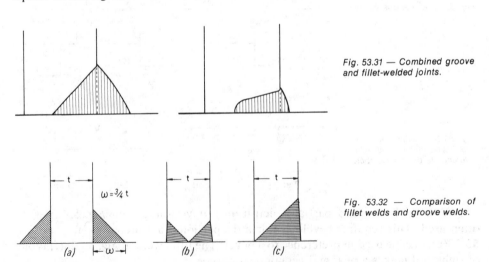

Fig. 53.31 — Combined groove and fillet-welded joints.

Fig. 53.32 — Comparison of fillet welds and groove welds.

In comparison, the double-bevel groove weld in *Fig.* 53.32(b), has about one-half the weld area of the fillet welds. However, it requires extra preparation and the use of smaller-diameter electrodes with lower welding currents to place the initial pass without burning through. As plate thickness increases, this initial low-deposition region becomes a less important factor, and the higher cost factor decreases in significance. The construction of a curve based on the best possible determination of the actual cost of welding, cutting, and assembling, such as illustrated in *Fig.* 53.33, is a possible technique for deciding at what point in plate thickness the double-bevel groove weld becomes less costly. The point of intersection of the fillet curve with a groove-weld curve is the point of interest. The accuracy of this device is dependent on the accuracy of the cost data used in constructing the curves.

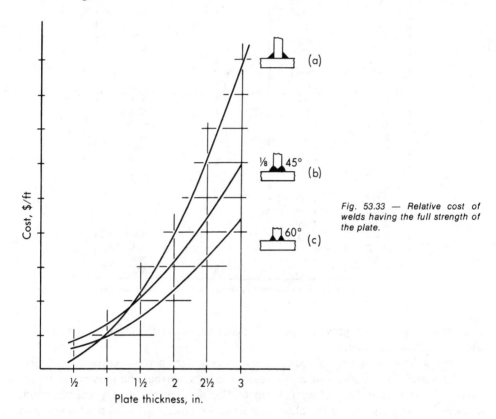

Fig. 53.33 — Relative cost of welds having the full strength of the plate.

Referring to *Fig.* 53.32(c), it will be noted that the single-bevel groove weld requires about the same amount of weld metal as the fillet welds deposited in *Fig.* 53.32(a). Thus, there is no apparent economic advantage. There are some disadvantages, though. The single-bevel joint requires bevel preparation and initially a lower deposition rate at the root of the joint. From a design standpoint, however, it offers a direct transfer of force through the joint, which means that it is probably better under fatigue loading. Although the illustrated full-strength fillet weld, having leg sizes equal to three-quarters the plate thickness, would be sufficient, some codes have lower allowable limits for fillet welds and may require a leg size equal to the plate thickness. In this case, the cost of the fillet-welded joint may exceed the cost of the single-bevel groove in thicker plates. Also, if the joint is so positioned that the weld can be made in the flat position, a single-bevel groove weld would be less expensive than if fillet welds were specified.

Sizing of Fillets. TABLE 35-I gives the sizing of fillet welds for rigidity

TABLE 35-I — Rule-of-Thumb Fillet-Weld Sizes
Where the Strength of the Weld Metal
Matches the Plate

Plate thickness (t)	Strength design	Rigidity design	
	Full-strength weld (ω = 3/4 t)	50% of full-strength weld (ω = 3/8 t)	33% of full-strength weld (ω = 1/4 t)
Less than 1/4	1/8	1/8*	1/8*
1/4	3/16	3/16*	3/16*
5/16	1/4	3/16*	3/16*
3/8	5/16	3/16*	3/16*
7/16	3/8	3/16	3/16*
1/2	3/8	3/16	3/16*
9/16	7/16	1/4	1/4*
5/8	1/2	1/4	1/4*
3/4	9/16	5/16	1/4*
7/8	5/8	3/8	5/16*
1	3/4	3/8	5/16*
1-1/8	7/8	7/16	5/16
1-1/4	1	1/2	5/16
1-3/8	1	1/2	3/8
1-1/2	1-1/8	9/16	3/8
1-5/8	1-1/4	5/8	7/16
1-3/4	1-3/8	3/4	7/16
2	1-1/2	3/4	1/2
2-1/8	1-5/8	7/8	9/16
2-1/4	1-3/4	7/8	9/16
2-3/8	1-3/4	1	5/8
2-1/2	1-7/8	1	5/8
2-5/8	2	1	3/4
2-3/4	2	1	3/4
3	2-1/4	1-1/8	3/4

* These values have been adjusted to comply with AWS-recommended minimums.

designs at various strengths and plate thicknesses, where the strength of the weld metal matches the plate.

In machine design work, where the primary design requirement is rigidity, members are often made with extra heavy sections, so that the movement under load will be within very close tolerances. Because of the heavy construction, stresses are very low. Often the allowable stress in tension for mild steel is given as 20,000 psi, yet the welded machine base or frame may have a working stress of only 2000 to 4000 psi. The question arises as how to determine the weld sizes for these types of rigidity designs.

It is not very practical to calculate the stresses resulting in a weldment when the unit is loaded within a predetermined dimensional tolerance and then use these stresses to determine the forces that must be transferred through the connecting welds. A very practical method, however, is to design the weld for the thinner plate, making it sufficient to carry one-third to one-half the carrying capacity of the plate. This means that if the plate were stressed to one-third to one-half its usual value, the weld would be sufficient. Most rigidity designs are stressed much below these values; however, any reduction in weld size below one-third the full-strength value would give a weld too small an appearance for general acceptance.

Groove Joints. *Fig.* 53.34 indicates that the root opening (R) is the separation

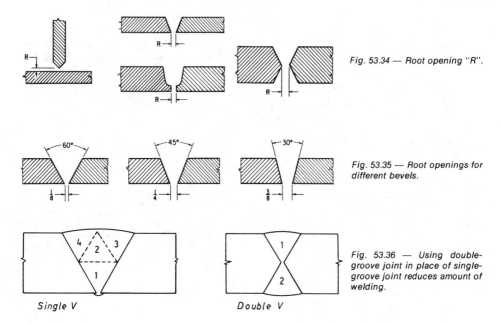

Fig. 53.34 — Root opening "R".

Fig. 53.35 — Root openings for different bevels.

Fig. 53.36 — Using double-groove joint in place of single-groove joint reduces amount of welding.

Single V   Double V

between the members to be joined.

A root opening is used for electrode accessibility to the base or root of the joint. The smaller the angle of the bevel, the larger the root opening must be to get good fusion at the root.

If the root opening is too small, root fusion is more difficult to obtain, and smaller electrodes must be used, thus slowing down the welding process.

If the root opening is too large, weld quality does not suffer but more weld metal is required; this increases welding cost and will tend to increase distortion.

Fig. 53.35 indicates how the root opening must be increased as the included angle of the bevel is decreased. Backup strips are used on larger root openings. All three preparations are acceptable; all are conducive to good welding procedure and good weld quality. Selection, therefore, is usually based on cost.

Root opening and joint preparation will directly affect weld cost (pounds of metal required), and choice should be made with this in mind. Joint preparation involves the work required on plate edges prior to welding and includes beveling and providing a root face.

Using a double-groove joint in preference to a single-groove (*Fig.* 53.36) cuts in half the amount of welding. This reduces distortion and makes possible alternating the weld passes on each side of the joint, again reducing distortion.

In *Fig.* 53.37(a), if the bevel or gap is too small, the weld will bridge the gap leaving slag at the root. Excessive back-gouging is then required.

Fig. 53.37(b) shows how proper joint preparation and procedure will produce good root fusion and will minimize back-gouging.

In *Fig.* 53.37(c), a large root opening will result in burnthrough. Spacer strip may be used, in which case the joint must be back-gouged.

Backup strips are commonly used when all welding must be done from one side, or when the root opening is excessive. Backup strips, shown in *Fig.* 53.38(a), (b), and (c), are generally left in place and become an integral part of the joint.

Spacer strips may be used especially in the case of double-V joints to prevent

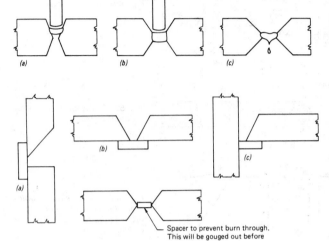

Fig. 53.37 — (a) If the gap is too small, the weld will bridge the gap, leaving slag at the root; (b) a proper joint preparation; (c) a root opening too large will result in burnthrough.

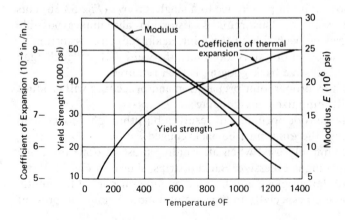

Fig. 53.38 — Backup strips — (a), (b), and (c) — are used when all welding is done from one side or when the root opening is excessive, a spacer to prevent burn-through (d) will be gouged out before welding the second side.

Spacer to prevent burn through.
This will be gouged out before
welding second side

burnthrough. The spacer in *Fig.* 53.38(d) to prevent burnthrough will be gouged out before welding the second side.

Weldment Distortion

Distortion is one of several major problems that occur with welding processes. Distortion in a weldment results from the nonuniform expansion and contraction of the weld metal and adjacent base metal during the heating and cooling cycle of the welding process. During such a cycle, many factors affect shrinkage of the metal and make accurate predictions of distortion difficult. Physical and mechanical properties, upon which calculations must in part be based, change as heat is applied. For example, as the temperature of the weld area increases, yield strength, modulus of elasticity, and thermal conductivity of steel plate decrease, and coefficient of thermal expansion and specific heat increase (*Fig.* 53.39). These changes, in turn, affect heat flow and uniformity of heat distribution. Thus, these variables make a precise calculation of what happens during heating and cooling

Fig. 53.39 — Changes in the properties of steel with increases in temperature complicate analysis of what happens during the welding cycle — and, thus, understanding of the factors contributing to weldment distortion.

difficult. Even if the calculation were simple, of greater value in the design phase and in the shop is a practical understanding of causes of distortion, effects of shrinkage in various types of welded assemblies, and methods to control shrinkage and to use shrinkage forces to advantage.

Distortion. In a welded joint, expansion and contraction forces act on the weld metal and on the base metal. As the weld metal solidifies and fuses with the base metal, it is in its maximum expanded state — it occupies the greatest possible volume as a solid. On cooling, it attempts to contract to the volume it would normally occupy at the lower temperature, but it is restrained from doing so by the adjacent base metal. Stresses develop within the weld, finally reaching the yield strength of the weld metal. At this point, the weld stretches, or yields, and thins out, thus adjusting to the volume requirements of the lower temperature. But only those stresses that exceed the yield strength of the weld metal are relieved by this accommodation. By the time the weld reaches room temperature — assuming complete restraint of the base metal so that it cannot move — the weld will contain locked-in tensile stresses approximately equal to the yield strength of the metal. If the restraints (clamps that hold the workpiece, or an opposing shrinkage force) are removed, the locked-in stresses are partially relieved as they cause the base metal to move, thus distorting the weldment.

Shrinkage in the base metal adjacent to the weld adds to the stresses that lead to distortion. During welding, the base metal adjacent to the weld is heated almost to its melting point. The temperature of the base metal a few inches from the weld is substantially lower. This large temperature differential causes nonuniform expansion followed by base metal movement, or metal displacement, if the parts being joined are restrained. As the arc passes on down the joint, the base metal cools and shrinks just like the weld metal. If the surrounding metal restrains the heated base metal from contracting normally, internal stresses develop. These, in combination with the stresses developed in the weld metal, increase the tendency of the weldment to distort.

The volume of adjacent base metal that contributes to distortion can be controlled somewhat by welding procedures. Higher welding speeds, for example, reduce the size of the adjacent base metal zone that shrinks along with the weld.

Controlled expansion and contraction is applied usefully in flame-straightening or flame-shrinking of a plate or weldment. For example, to shrink the center portion of a distorted plate, the flame from a torch is directed on a small, centrally located area. The area heats up rapidly and must expand. But the surrounding plate, which is cooler, prevents the spot from expanding along the plane of the plate. The only alternative is for the spot to expand in thickness. In essence, the plate thickens where the heat is applied. Upon cooling, it tends to contract uniformly in all directions. When carefully done, spot heating produces shrinkage that is effective in correcting distortion caused by previous heating and cooling cycles.

Shrinkage of a weld causes various types of distortion and dimensional changes. A butt weld between two pieces of plate, by shrinking transversely, changes the width of the assembly, as in *Fig.* 53.40(a). It also causes angular distortion, as in *Fig.* 53.40(b). Here, the greater amount of weld metal and heat at the top of the joint produces greater shrinkage at the upper surface, causing the edges of the plate to lift. Longitudinal shrinkage of the same weld would tend to deform the joined plate, as shown in *Fig.* 53.40(c).

BUTT WELDS

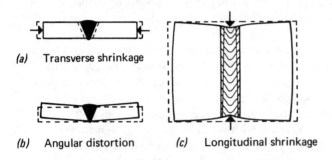

(a)   Transverse shrinkage

(b)   Angular distortion   *(c)*   Longitudinal shrinkage

FILLET WELDS

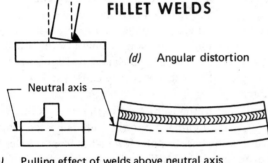

(d)   Angular distortion

Fig. 53.40 — How welds tend to distort and cause dimensional changes in assemblies.

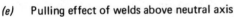

(e)   Pulling effect of welds above neutral axis

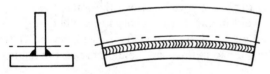

(f)   Pulling effect of welds below neutral axis

Angular distortion is also a problem with fillets, as illustrated in *Fig.* 53.40(d). If fillet welds in a T-shaped assembly are above the neutral axis (center of gravity) of the assembly, the ends of the member tend to bend upward, as in *Fig.* 53.40(e). If the welds are below the neutral axis, the ends bend down, as in *Fig.* 53.40(f).

Shrinkage Control. If distortion in a weldment is to be prevented or minimized, methods must be used both in design and in the shop to overcome the effects of the heating and cooling cycle. Shrinkage cannot be prevented, but it can be controlled. Several practical ways can be used to minimize distortion caused by shrinkage:

1. Do not overweld. The more metal placed in a joint, the greater the shrinkage forces. Correctly sizing a weld for the service requirements of the joint not only minimizes distortion, it saves weld metal and time. The amount of weld metal in a fillet can be minimized by use of a flat or slightly convex bead, and in a butt joint by proper edge preparation and fitup. Only the effective throat, dimension T in *Fig.* 53.41(a), in a conventional fillet can be used in calculating the design strength of the weld. The excess weld metal in a highly convex bead does not increase the allowable strength in code work, but it does increase shrinkage forces.

Proper edge preparation and fitup of butt welds, *Fig.* 53.41(b), help to force

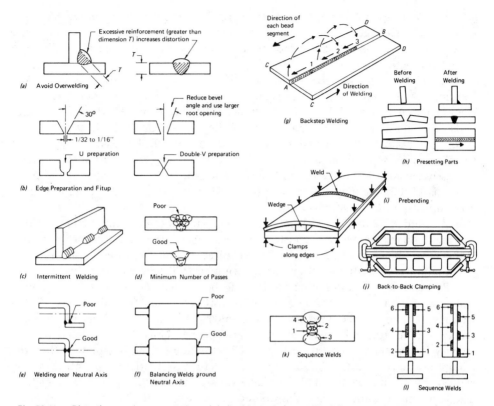

Fig. 53.41 — Distortion can be prevented or minimized by techniques that defeat — or use constructively — the effects of the heating and cooling cycle.

the use of minimum amounts of weld metal. For maximum economy, the plates should be spaced from 1/32 to 1/16 in. apart. A bevel of 30 degrees on each side provides proper fusion at the root of the weld, yet requires minimal weld metal. In relatively thick plates, the angle of bevel can be decreased if the root opening is increased, or a J or U preparation can be used to decrease the amount of weld metal used in the joint. A double-V joint requires about one-half the weld metal of a single-V joint in the same plate thickness.

In general, if distortion is not a problem, select the most economical joint. If distortion is a problem, select either a joint in which the weld stresses balance each other or a joint requiring the least amount of weld metal.

2. Use intermittent welding. Another way to minimize weld metal is to use intermittent rather than continuous welds where possible, as in *Fig.* 53.41(c). For attaching stiffeners to plate, for example, intermittent welds can reduce the weld metal by as much as 75%, yet provide the needed strength.

3. Use as few weld passes as possible. Fewer passes with large electrodes, *Fig.* 53.41(d), are preferable to a greater number of passes with small electrodes when transverse distortion could be a problem. Shrinkage caused by each pass tends to be cumulative, thereby increasing total shrinkage when many passes are used.

4. Place welds near the neutral axis. Distortion is minimized by providing a smaller leverage for the shrinkage forces to pull the plates out of alignment. *Fig.* 53.41(e) is illustrative. Both design of the weldment and welding sequence can be used effectively to control distortion.

5. *Balance welds around the neutral axis.* This practice, shown in *Fig.* 53.41(f), offsets one shrinkage force with another to effectively minimize distortion of the weldment. Here, too, design of the assembly and proper sequence of welding are important factors.

6. *Use backstep welding.* In the backstep technique, the general progression of welding may be, say, from left to right, but each bead segment is deposited from right to left as in *Fig.* 53.41 (g). As each bead segment is placed, the heated edges expand, which temporarily separates the plates at B. But as the heat moves out across the plate to C, expansion along outer edges CD brings the plates back together. This separation is most pronounced as the first bead is laid. With successive beads, the plates expand less and less because of the restraint of prior welds. Backstepping may not be effective in all applications, and it cannot be used economically in automatic welding.

7. *Anticipate the shrinkage forces.* Placing parts out of position before welding can make shrinkage perform constructive work. Several assemblies, preset in this manner, are shown in *Fig.* 53.41(h). The required amount of preset for shrinkage to pull the plates into alignment can be determined from a few trial welds.

Prebending or prespringing the parts to be welded, *Fig.* 53.41(i), is a simple example of the use of opposing mechanical forces to counteract distortion due to welding. The top of the weld groove — which will contain the bulk of the weld metal — is lengthened when the plates are sprung. Thus the completed weld is slightly longer than it would be if it had been made on the flat plate. When the clamps are released after welding, the plates return to the flat shape, allowing the weld to relieve its longitudinal shrinkage stresses by shortening to a straight line. The two actions coincide, and the welded plates assume the desired flatness.

Another common practice for balancing shrinkage forces is to position identical weldments back to back, *Fig.* 53.41(j), clamping them tightly together. The welds are completed on both assemblies and allowed to cool before the clamps are released. Prebending can be combined with this method by inserting wedges at suitable positions between the parts before clamping.

In heavy weldments, particularly, the rigidity of the members and their arrangement relative to each other may provide the balancing forces needed. If these natural balancing forces are not present, it is necessary to use other means to counteract the shrinkage forces in the weld metal. This can be accomplished by balancing one shrinkage force against another or by creating an opposing force through the fixturing. The opposing forces may be: other shrinkage forces; restraining forces imposed by clamps, jigs, or fixtures; restraining forces arising from the arrangement of members in the assembly; or the force from the sag in a member due to gravity.

8. *Plan the welding sequence.* A well-planned welding sequence involves placing weld metal at different points about the assembly so that, as the structure shrinks in one place, it counteracts the shrinkage forces of welds already made. An example of this is welding alternately on both sides of the neutral axis in making a butt weld, as in *Fig.* 53.41(k). Another example, in a fillet weld, consists of making intermittent welds according to the sequences shown in *Fig.* 53.41(l). In these examples, the shrinkage in weld No. 1 is balanced by the shrinkage in weld No. 2, and so on.

Clamps, jigs, and fixtures that lock parts into a desired position and hold them until welding is finished are probably the most widely used means for controlling distortion in small assemblies or components. It was mentioned earlier in this section that the restraining force provided by clamps increases internal stresses in the weldment until the yield point of the weld metal is reached. For typical welds on low-carbon plate, this stress level would approximate 45,000 psi. One might expect this stress to cause considerable movement or distortion after the welded part is removed from the jig or clamps. This does not occur, however, since the strain (unit contraction) from this stress is very low compared to the amount of movement that would occur if no restraint were used during welding.

9. Remove shrinkage forces after welding. Peening is one way to counteract the shrinkage forces of a weld bead as it cools. Essentially, peening the bead stretches it and makes it thinner, thus relieving (by plastic deformation) the stresses induced by contraction as the metal cools. But this method must be used with care. For example, a root bead should never be peened, because of the danger of either concealing a crack or causing one. Generally, peening is not permitted on the final pass, because of the possibility of covering a crack and interfering with inspection, and because of the undesirable work-hardening effect. Thus, the utility of the technique is limited, even though there have been instances where between-pass peening proved to be the only solution for a distortion or cracking problem. Before peening is used on a job, engineering approval should be obtained.

Another method for removing shrinkage forces is by stress relief — controlled heating of the weldment to an elevated temperature, followed by controlled cooling. Sometimes two identical weldments are clamped back to back, welded, and then stress-relieved while being held in this straight condition. The residual stresses that would tend to distort the weldments are thus removed.

10. Minimize welding time. Since complex cycles of heating and cooling take place during welding, and since time is required for heat transmission, the time factor affects distortion. In general, it is desirable to finish the weld quickly, before a large volume of surrounding metal heats up and expands. The welding process used, type and size of electrode, welding current, and speed of travel, thus, affect the degree of shrinkage and distortion of a weldment. The use of iron-powder manual electrodes or mechanized welding equipment reduces welding time and the amount of metal affected by heat and, consequently, distortion.

Preheating and Stress Relieving

In some welding operations, it is necessary to apply heat to the assembly before starting the welding. In others, a postheat — or application of heat after welding — is needed to relieve the internal stresses that have been developed. With certain weldments, heat may also be applied between welding passes to maintain a required temperature. Each of these applications of heat has a bearing on the quality of weld or the integrity of the finished weldment, and, in code work, control of temperature before, during, and after welding may be rigidly specified.

Preheating — When and Why. Preheating is used for one of the following reasons: (1) To reduce shrinkage stresses in the weld and adjacent base metal — especially important with highly restrained joints; (2) To provide a slower rate of cooling through the critical temperature range (about 1600°F to 1330°F), preventing excessive hardening and lowering ductility of both the weld and heat-affected area of the base plate; and (3) To provide a slower rate of cooling through the 400°F range, allowing more time for any hydrogen that is present to diffuse away from the weld and adjacent plate to avoid underbead cracking.

As suggested, a main purpose of preheat is to slow down the cooling rate — to allow more "Time at Temperature," as illustrated in *Fig.* 53.42. Thus, the **amount** of heat in the weld area as well as the temperature is important. A thick plate could be preheated to a specified temperature in a localized area and the heating be ineffective because of rapid heat transfer, the reduction of heat in the welding area, and, thus, no marked effect on slowing the cooling rate. Having a thin surface area at a preheat temperature is not enough if there is a mass of cold metal beneath it into which the heat can rapidly transfer.

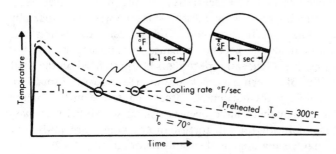

Fig. 53.42 — A main purpose of preheat is to slow down the cooling rate. As the insets show, there is a greater temperature drop in one second at a given temperature (T_1) when the initial temperature of the plate is 70°F than when the initial temperature is 300°F. In other words, the cooling rate (°F/sec) is slower when preheat is used.

Because of the heat-absorption capacity of a thick plate, the heat-affected zone and the weld metal after cooling may be in a highly quenched condition unless sufficient preheat is provided. What really matters is how long the weld metal and adjacent base metal is maintained in a certain temperature range during the cooling period. This, in turn, depends on the amount of heat in the assembly and the heat transfer properties of the material and its configuration. Without adequate preheat, the cooling could be rapid and intolerably high hardness and brittleness could occur in the weld or adjacent area.

Welding at low ambient temperatures or on steel brought in from outside storage on cold winter days greatly increases the need for preheat. It is true that preheating rids the joint of moisture, but preheating is usually not specified for that purpose.

The Amount of Preheat Required. The amount of preheat required for any application depends on such factors as base metal chemistry, plate thickness, restraint and rigidity of the members, and heat input of the process. Unfortunately, there is no method for metering the **amount** of heat put into an assembly by a preheat torch. The best shop approach for estimating the preheat input is a measure of the temperature at the welding area by temperature-indicating crayon marks or pellets. These give approximate measures of temperature at the spots where they are placed, which measurements are taken as indices to the heat input and are correlated with thickness of metal and chemistry of metal in tables specifying minimum preheat temperatures. Thus, temperature is the gage to preheat inputs, and preheating to specified temperatures is the practical method of obtaining the amount of preheat needed to control the cooling rate after welding.

There are various guides for use in estimating preheat temperatures, including the recommendation of the suppliers of special steels. No guide, however, can be completely and universally applicable because of the varying factors of rigidity and restraint in assemblies. Recommendations are, thus, presented as "minimum preheat recommendations," and they should be accepted as such. However, the quenched and tempered steels can be damaged if the preheat is too high and special precautions must be taken.

The American Welding Society and the American Institute of Steel Construction have established minimum preheat and interpass temperature requirements for common weldable steels, as shown in TABLE 53-II. While material thickness, ranges of metal chemistry, and the welding process are taken into account in the minimum requirements, some adjustments may be needed for specific steel chemistry, welding heat input, joint geometry, and other factors.

Generally, the higher the carbon content of a steel, the lower the critical cooling rate and the greater the necessity for preheating and using low-hydrogen electrodes.

Methods of Preheating. The method of preheating depends on the thickness of the plate, the size of the weldment, and the heating equipment available. In the production welding of small assemblies, preheating in a furnace is the most satisfactory method. Another satisfactory method is torch heating, using natural gas premixed with compressed air. This produces a hot flame and burns clean. Torches can be connected to convenient gas and compressed-air outlets around the shop. Acetylene, propane and oil torches can also be used. On large weldments, banks of heating torches may be used to bring the material up to temperature quickly and uniformly.

Electrical strip heaters are used on longitudinal and girth seams on plate up to 2 in. thick. The heaters are clamped to the plate parallel to the joint and about 6 in. from the seam. After the plate reaches the proper preheat temperature, the heaters may remain in place to add heat if necessary to maintain the proper interpass temperature. Other means of preheating are induction heating — often used on

TABLE 53-II — **Minimum Preheat and Interpass Temperature.**
AWS D1.1-Rev. 1-73, Table 4.2.[1,2]
(Degrees F)

		Welding Process			
Thickness of Thickest Part of Point of Welding — Inches	Shielded Metal-Arc Welding with other than Low-Hydrogen Electrodes	Shielded Metal-Arc Welding with Low-Hydrogen Electrodes; Submerged Arc Welding; Gas Metal-Arc Welding; or Flux Cored Arc Welding		Shielded Metal-Arc Welding with Low-Hydrogen Electrodes; Submerged-Arc Welding with Carbon or Alloy Steel Wire, Neutral Flux; Gas Metal-Arc Welding; or Flux Cored Arc Welding	Submerged Arc Welding with Carbon Steel Wire, Alloy Flux
	ASTM A36,[4] A53 Gr. B, A106, A131, A139, A375, A381 Gr. Y35, A500, A501, A516, A524, A529, A570 Gr. D and E, A573; API 5L Gr. B; ABS Gr. A, B, C, CS, D, E, R	ASTM A36, A106, A131 A139, A242 Weldable Grade A375, A381 Gr. Y35, A441, A516, A524, A529, A537 Gr. A and B, A570 Gr. D and E, A572 Gr. 42, 45, 50, A573, A588, A618, API 5L Gr. B and 5LX Gr. 42; ABS Gr. A, B, C, CS, D, E, R, AH, DH, EH	ASTM A572 Grades 55, 60 and 65	ASTM A514, A517	ASTM A514, A517
To 3/4, incl.	None[3]	None[3]	70	50	50
Over 3/4 to 1-1/2, incl.	150	70	150	125	200
Over 1-1/2 to 2-1/2, incl.	225	150	225	175	300
Over 2-1/2	300	225	300	225	400

1 Welding shall not be done when the ambient temperature is lower than zero F. When the base metal is below the temperature listed for the welding process being used and the thickness of material being welded, it shall be preheated (except as otherwise provided) in such manner that the surfaces of the parts on which weld metal is being deposited are at or above the specified minimum temperature for a distance equal to the thickness of the part being welded, but not less than 3 in., both laterally and in advance of the welding. Preheat and interpass temperatures must be sufficient to prevent crack formation. Temperature above the minimum shown may be required for highly restrained welds. For quenched and tempered steel the maximum preheat and interpass temperature shall not exceed 400°F for thickness up to 1-1/2 in., inclusive, and 450°F for greater thicknesses. Heat input when welding quenched and tempered steel shall not exceed the steel producer's recommendation.

2 In joints involving combinations of base metals, preheat shall be as specified for the higher strength steel being welded.

3 When the base metal temperature is below 32°F, preheat the base metal to at least 70°F and maintain this minimum temperature during welding.

4 Only low-hydrogen electrodes shall be used for welding A36 steel more than 1 inch thick for bridges.

piping — and radiant heating.

High accuracy is not required in preheating carbon steels. Although it is important that the work be heated to a minimum temperature, no harm is done if this temperature is exceeded by 100°F. This is not true, however, for quenched and tempered steels, since welding on an overheated plate may cause damage in the heat-affected zone. For this reason the temperature should be measured as accurately as possible with such steels.

Temperature-indicating crayons and pellets are available for a wide range of temperatures. A crayon mark for a given temperature on the work will melt suddenly when the work reaches that temperature. Two crayon marks, one for the lower limit and one for the upper limit of temperature, show clearly when the work is heated to the desired temperature range.

Several types of portable pyrometers are available for measuring surface temperature. Properly used, these instruments are sufficiently accurate, but must be periodically calibrated to insure reliability.

Thermocouples may be attached to the work and used to measure temperature. Thermocouples, of course, are the temperature-sensing devices in various types of ovens used for preheating small assemblies.

Interpass Temperatures. Usually a steel that requires preheating to a specified temperature also must be kept at this temperature between weld passes. With many weldments, the heat input during welding is adequate to maintain the interpass temperature. On a massive weldment, it is not likely that the heat input of the welding process will be sufficient to maintain the required interpass temperature. If this is the case, torch heating between passes may be required.

Once an assembly has been preheated and the welding begun, it is desirable to finish the welding as soon as possible so as to avoid the need for interpass heating.

Since the purpose of preheating is to reduce the quench rate, it logically follows that the same slow cooling should be accorded all passes. This can only be accomplished by maintaining an interpass temperature which is at least equal to the preheat temperature. If this is not done, each individual bead will be subjected to the same high quench rate as the first bead of a non-preheated assembly.

Stress Relief. Stress relieving is defined as heating to a suitable temperature (for steel, below the critical); holding long enough to reduce residual stresses; and then cooling slowly enough to minimize the development of new residual stresses. Stress relieving should not be confused with normalizing or annealing, which are done at higher temperatures.

The ASME Code requires certain pressure vessels and power piping to be stress relieved. This is to reduce internal stress. Other weldments such as machine-tool bases are stress relieved to attain dimensional stability after machining.

Heating and cooling must be done slowly and uniformly; uneven cooling could nullify much of the value of the heat treatment or even cause additional stresses in the weldment. In general, the greater the difference between maximum and minimum thickness of the component parts, the slower should be the rate of temperature change. If the ratio of maximum to minimum thickness of the component parts is less than 4 to 1, heating and cooling rates should not exceed 400°F/hour divided by the thickness in inches of the thickest section. However, the heating and cooling rates should not exceed 400°F/hour. If the ratio of the thicknesses of the component parts vary more, rates should be reduced accordingly. For example, with complex structures containing members of widely varying thicknesses, heating and cooling rates should be such that the maximum temperature difference of sections of the same weldment should not exceed 75°F. Temperatures of critical sections can be monitored using thermocouples mounted on the weldment.

The stress-relief range for most carbon steels is 1100 to 1200°F, and the soaking time is usually one hour per inch of thickness. * For the low-alloy chrome-molybdenum steels, with the chromium in the range of 1/2 to 2-1/4% and the molybdenum up to 1%, the stress-relief range is 1250 to 1300°F for one hour. Some of the higher alloy steels require more soaking time. For example, E410, E502, and E505 weld metal is stress relieved at 1550 to 1600°F for two hours, and E430 at 1400 to 1450°F for four hours.

Local stress relieving can be done on girth joints of pipe and pressure vessels when required by codes. The same precautions are necessary as for furnace heating: slow heating, time-at-temperature, and slow cooling.

---

*AWS proposes to say "per inch of weld thickness, when the specified stress-relief is for dimensional stability, the holding time shall be one hour per inch of thickness of the thicker part." However, sometimes ASME Sect. VIII refers also to the "thinner of two adjacent butt-welded plates."

54

Spot Welding

$\mathbf{A}$MONG the metal fabricating processes which comprise the joining of several parts by means of welding, one of the simplest and most economical is resistance welding. In resistance welding processes the fusing temperature is generated solely through the resistance offered to the passage of an electric current by the parts to be welded. Differing from other methods of welding in that no additional material is required in the welding operation, resistance welding methods utilize the application of mechanical pressure to forge the electrically heated portions together. Effect of the pressure is to refine grain structure and produce a weld with physical properties equal in most cases to the parent metal and occasionally superior.

Four Processes. Generally, resistance welding can be divided into four principal types or classifications: (1) Spot welding, (2) seam welding, (3) projection welding, and (4) butt or flash welding. Certain of these processes may be subdivided and these subdivisions will be discussed under each specific

Fig. 54.1 — Aircraft tail boom. Redesigning for spot welding saved 200 manhours per ship and reduced the weight of the boom by 3 pounds.

classification. Succeeding chapters will cover each of the latter three major classifications separately in order to deal properly with their specifically different aspects.

Major Advantages. The major characteristic of all resistance welding processes is high speed of operation. Individual welds are made in seconds or fractions of a second. Thus, these methods are ideally suited for large-quantity production. They are, however, with proper design considerations, also applicable for short-run work with equally satisfactory results and economy. Actual cost comparisons show substantial savings when applied to designs to be produced in quantities as low as a thousand pieces and less, *Fig.* 54.1.

Spot Welding. In all probability spot welding is the most widely used of the resistance welding processes. Equipment may be quite simple and inexpensive; a single-phase transformer, a pair of electrodes and a means for applying pressure to the electrodes comprising the basic machine. High current at low potential, from the transformer secondary, is localized by the electrodes and under the applied pressure creates the weld spot.

Types of Machines. Types of spot welds can be readily divided into the categories of single and multiple welds. These can then be further divided into direct and indirect welds. Single type spot welding machines, wherein but one weld is made during one operation of the machine, are built in two general types — portable or gun machines and stationary or pedestal.

Stationary Single-Spot Welders. Stationary machines are available in two general styles: The rocker-arm or walking beam type and the press type. Spot welders have been classified and standardized by the Resistance Welder

Photo, courtesy Precision Welder & Machine Co.

Fig. 54.2 — Standard RWMA Size 3 air-operated rocker-arm spot welder. At lower front is current regulator and at top right front is air pressure regulator and gage indicating welding pressure. At the rear of the machine is shown the automatic sequencing control. Photo, courtesy Teledyne Precision-Cincinnati.

Manufacturers' Association into specific sizes, kva ratings and throat depths, as shown in TABLE 54-I.

TABLE 54-1 — Specifications for Standard Welders

Rocker-Arm Spot Welders		
Size No.	Rating (kva)	Throat Depth (in.)
1	15	12–18–24
2	30	12–24–30
3	50	12–24–36
Press-Type Spot Welders		
Size No.	Rating (kva)	Throat Depth (in.)
1	50–75	18–24–36
2	100–150	18–24–36
3	150–250	18–24–36
4	300–400–500	18–24–36

Rocker-arm machines have a rocking upper electrode arm which is actuated by foot, air or fully automatic motor-driven cam action. These machines are highly flexible, offer maximum clearance around the welding horns and are adaptable to a wide range of work. Usually limited to the lighter gages, standard machines seldom exceed 50 kva, *Fig.* 54.2, although manufacturers offer equipment ranging up to 75 kva. Two thicknesses from 0.016 to 0.032-inch can be handled on a 20 kva welder and 0.016 to 0.125-inch on the 75 kva welder. Rocker-arm welders have throat depths ranging up to 48 inches commonly and occasionally to as much as 72 inches. Welding speed usually may range to about 180 spots per minute but special machines increase this figure as high as 350.

Over 50 kva, machines normally are of the press type which carry the upper electrode in a slide, *Fig.* 54.3. These machines are available in sizes for welding medium up to the heaviest gage metals than can be handled. Press type welders are air operated, hydraulically operated or motor driven, hydraulic operation usually being used for the larger sizes and capacities.

Available ratings in press type welders range to as high as 800 kva with throat depths up to 60 inches. A 75 kva welder will handle two 0.125-inch pieces while a 400 kva welder permits the welding of two thicknesses of 0.500-inch sheet. Speeds of press type welders average from 12 to 180 spots per minute, the higher speeds being accomplished on the lighter gages.

Bench welders are produced in the same general types as the other stationary machines and are designed for light precision work such as that in instrument, radio, electrical, and similar fields. These machines generally range from 0.5 to 30 kva in capacity and occasionally as high as 75 kva. Speeds usually range from 10 to 200 or 300 spots per minute.

Portable Single-Spot Welders. Where the work is too bulkly to take to the welding machines, portable machines are usually used. These permit the rapid assembly of a few or a great many parts held in a fixture, *Fig.* 54.4. Portability, however, is limited; heavy bulky cables connect the gun to the transformer and these cables must be as short as possible; not over 10 feet and preferably less than six. Portable welders or guns are of widely varying designs and are built to accommodate an equally wide variety of designs for low, moderate and mass-production quantities of parts.

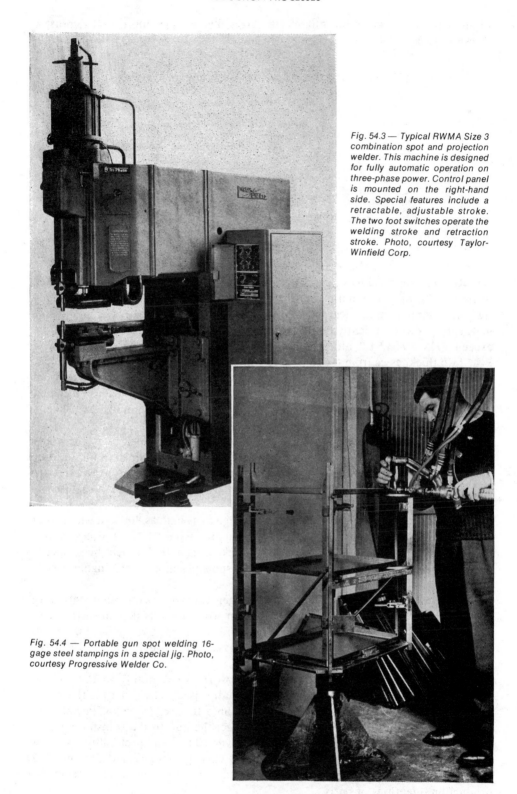

Fig. 54.3 — Typical RWMA Size 3 combination spot and projection welder. This machine is designed for fully automatic operation on three-phase power. Control panel is mounted on the right-hand side. Special features include a retractable, adjustable stroke. The two foot switches operate the welding stroke and retraction stroke. Photo, courtesy Taylor-Winfield Corp.

Fig. 54.4 — Portable gun spot welding 16-gage steel stampings in a special jig. Photo, courtesy Progressive Welder Co.

Portable guns can be classified as pinch guns, expansion guns and bar or lever type guns. Pinch guns are made with one stationary jaw and one movable jaw in the following styles: (1) Common C type; (2) revolving jaw gun similar to the C type, except one or both jaws can be revolved to permit positioning clearance; (3) scissors or rocker-arm type; (4) clam-shell type, similar to the scissors gun except that both jaws open wide to permit positioning over a wide section; and (5) universal type, a special scissors gun which permits 360-degree rotation of the gun proper on the horizontal axis and about 170 degrees on the vertical axis.

Expansion welding guns, primarily for use with welding fixtures, are of two types; short-circuit and single-cable guns. The short-circuit gun merely conducts current from the upper electrode of a fixture to the lower electrode of the fixture in which the work is located, both electrodes being connected to the transformer secondary. These guns are made to suit the work and are either air or hydraulically operated. Single-cable guns are similar in operation to the short-circuit type except that the lower electrode is energized. The other side of the transformer secondary is connected to the gun by means of the single cable.

Bar or lever type guns are for use with fixture work and may be manually, pneumatically or hydraulically operated. The workpiece is positioned on a stationary fixture electrode which is connected to the transformer secondary. The other side of the transformer is connected to a copper-alloy bar which encircles the fixture. Of the short-circuit type, the gun is merely inserted between the bar and the work. With the point of the gun properly positioned, pressure may be applied manually or otherwise and the welding cycle started by pressing a switch.

Fig. 54.5 — Pneumatic ultra speed multiple-gun welder for welding back panels to outer shells of refrigerator cabinets, left, at the rate of 100 cabinets per hour, with 120 spots per cabinet. The machine contains three transformers with double series weld circuits making four welds per transformer. The guns are sequenced by means of a geneva-motion operated sequencing valve. The work is loaded in a convenient position, then the lower platen of the machine elevates to welding position. Photo, courtesy Taylor-Winfield Corp.

Capacity of portable welding guns ranges from about 15 to 400 kva and some to 500 kva. Welding speeds may be as high as 100 spots per minute and as low as a few per minute.

Multiple-Spot Welders. Two or more pairs of electrodes cannot properly distribute current to each weld unless some means for insuring equal distribution is used. Although two or four spots are often produced by means of a series arrangement on a single circuit, *Fig.* 54.5, multiple spots are normally accomplished in these machines by means of a separate secondary transformer circuit to each pair of electrodes. Normally custom-built and fully automatic, multiple-spot machines are usually quite expensive and may have from two to 100 or more electrodes. The tremendous savings possible, however, justify such special machines where large-quantity production of standard parts is contemplated, *Fig.* 54.6.

Types of Welds. Special types of spot welding processes are known as series, indirect and invisible welding. In series welding, *Fig.* 54.7, two welds can be made at once without any markings, indentations or discoloration appearing on one side of the assembly. Welding with all the pressure on one side of the work permits the use of spot welding on large tanks and bodies where otherwise an unduly long throat would be necessary. Space between welds must be at least 2 inches on the lightest gages and up to 6 or 8 inches on the heavier gages.

Indirect welding, *Fig.* 54.8, is used on parts which prevent proper passage of the current due to design. Invisible welds are specially controlled so as to obviate all effects of heat and shrinkage. In some cases the electrode produces a bulge which is subsequently dressed off and in others indirect introduction of the current

Photo, courtesy National Electric Welding Machines Co.

Fig. 54.6 — Multiple-transformer spot welder for welding the header that goes over the door of an automobile. Both right and left-hand headers are welded on this special machine. Photo, courtesy Resistance Welder Corp.

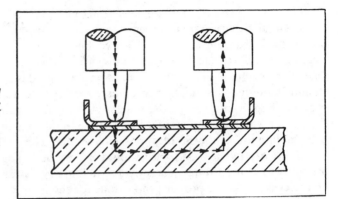

Fig. 54.7 — Assembly showing series-welded joint. Drawing, courtesy Resistance Welder Manufacturers Association.

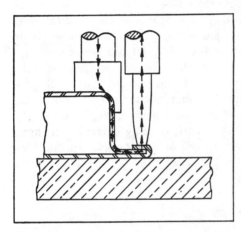

Fig. 54.8 — Spot-welded assembly showing a typical joint design for an indirect weld. Drawing, courtesy Resistance Welder Manufacturers Association.

is used to distribute current in the one part and minimize markings.

Applicable Range of Gages. The range of spot welding applications is extremely wide and can be used for parts made from foils as thin as 0.001-inch, such as in electronic tube elements, to those made from thicker sections such as those required in the machine-tool, automobile, aviation, appliance and similar machine building industries. On materials of low conductivity — less than 20 per cent of standard copper — use of pulsation welding is preferable where gage is greater than 1/8 to 3/32-inch. With this feature, which involves repeated application of welding power under maintained electrode pressure, electrode life is brought into the practical range and welding of parts such as 1-inch to 1-inch, 1/2-inch to 21/2-inch, a 11/2-inch piece between two 1-inch pieces, or a 1/2-inch piece to a piece any thickness up to 4 inches, is made practicable. Single-impulse welding is usually considered limited to spot welding on gages up to 1/8-inch in steels, 5/32-inch for aluminum and 1/8-inch for magnesium.

Design Considerations

Differences between good and poor design of parts for resistance welding are often so small as to appear trivial, but to obtain good welds and also obtain maximum economy, proper design practices must be closely followed. Careful consideration of all environmental conditions should be made before final selection of one specific resistance welding method.

General Design Aspects. Heavy materials require a great amount of heat and power and as a rule can be fabricated best by arc or gas welding. Resistance welding, therefore, can be applied most advantageously to thin-metal fabrication. One of the primary advantages of resistance welding is speed, and thus assemblies should be designed so as to permit use of as simple a machine as possible. It is specially important to avoid the necessity for intricate current carrying members in the welder which complicate feeding parts into the machine. Design should always contemplate use of the simplest possible welding electrodes.

Joint Strength. In no case should a part be changed from one method of assembly to spot welding merely by a switch in processes. The entire design should be reviewed to assure proper proportions for the joints and adequate accessibility. This is specially true if joint strength is of importance. Strength of individual spots can be readily obtained and multiple-spot joints can be computed provided spot spacing meets minimum requirements. Pertinent factors regarding spot welds in AISI 1010 low-carbon steel, the most readily welded material, are shown in TABLE 54-II. In certain cases it is necessary to develop a joint strength closely approximating that of the parent metal sheet. To accomplish this, additional rows of spots can be used until the sum of the spot diameters equals the width of the sheet. Recommended spacings must be used and intermediate rows are usually staggered. Unless spots are spaced far enough apart, welding current will shunt through previous spots and produce weak welds. Minimum spot spacing for

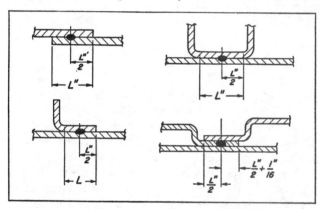

Fig. 54.9 — Sketch of parts showing overlap and spot spacings for various types of joints. Drawing, courtesy Lockheed Aircraft Corp.

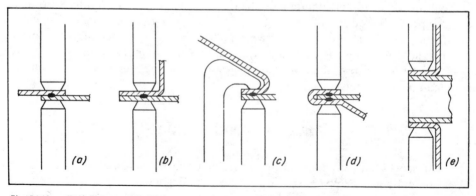

Fig. 54.10 — Typical spot-welded joints and typical electrodes required to produce them. Drawing, courtesy Resistance Welder Manufacturers Association.

TABLE 54-II*

Single-Impulse Welding Low-Carbon Steel[1]

Data Common To All Classes of Spot Welds					Setup for Best-Quality, Class A Welds					Setup for Medium-Quality, Class B Welds					Setup for Good-Quality, Class C Welds				
Thickness of Work Pieces	Electrode[2]		Min. Weld Sp.[3] c/1 to c/1	Min. Over-lap L	Weld Time	Electrode Force	Welding Current	Nugget Dia.	Av. Ten. Shear Str. ±14%	Weld Time	Electrode Force	Welding Current	Nugget Dia.	Av. Ten. Shear Str. ±17%	Weld Time	Electrode Force	Welding Current	Nugget Dia.	Av. Ten. Shear Str. ±20%
	Min. D	Max. d																	
(in.)	(in.)	(in.)	(in.)	(in.)	(c)[4]	(lb)	(amp)	(in.)	(lb)	(c)[4]	(lb)	(amp)	(in.)	(lb)	(c)[4]	(lb)	(amp)	(in.)	(lb)
0.010	3/8	1/8	1/4	3/8	4	200	4000	0.13	235	5	130	3700	0.12	200	15	65	3000	0.11	160
0.021	3/8	3/16	3/8	7/16	6	300	6100	0.17	530	10	200	5100	0.16	460	22	100	3800	0.14	390
0.031	3/8	3/16	1/2	7/16	8	400	8000	0.21	980	15	275	6300	0.20	850	29	135	4700	0.18	790
0.040	1/2	1/4	3/4	1/2	10	500	9200	0.23	1350	21	360	7500	0.22	1230	38	180	5600	0.21	1180
0.050	1/2	1/4	7/8	9/16	12	650	10300	0.25	1820	24	410	8000	0.23	1700	42	205	6100	0.22	1600
0.062	1/2	1/4	1 1/16	5/8	14	800	11600	0.27	2350	29	500	9000	0.26	2150	48	250	6800	0.25	2050
0.078	5/8	5/16	1 3/8	11/16	21	1100	13300	0.31	3225	36	650	10400	0.30	3025	58	325	7900	0.28	2900
0.094	5/8	5/16	1 5/8	3/4	25	1300	14700	0.34	4100	44	790	11400	0.33	3900	66	390	8800	0.31	3750
0.109	5/8	3/8	1 13/16	13/16	29	1600	16100	0.37	5300	50	960	12200	0.36	5050	72	480	9500	0.35	4850
0.125	5/8	3/8	2	7/8	30	1800	17500	0.40	6900	60	1140	12900	0.39	6500	78	570	10000	0.37	6150

* Courtesy Taylor-Winfield Corp.

[1] Low-carbon steel is hot rolled, pickled, and lightly oiled with an ultimate strength of 42,000 to 45,000 psi similar to SAE 1005—SAE 1010. Surface of steel lightly oiled but free from grease, scale or dirt.

[2] Electrode material is RWMA Class 2 alloy.

[3] Minimum weld spacing is that distance for which no increase in welding current is necessary to compensate for the shunted current effect of adjacent welds. Spot spacing to be used as a guide only. Spacing should be determined according to stress requirements. Greater spacings should be used whenever allowed by structural considerations. When spotwelding sheets of unequal thickness, the thinner sheet determines the edge distance, spot spacings, strength, etc.

[4] Weld time is given in cycles.

aluminum and magnesium is usually considered as equal to 8 times the gage, 16 times being preferable.

Spot Size. Welding electrode tip size directly affects the size and consequently shear strength of a weld, and proper tip area is a function of material gage. A satisfactory formula for determining the electrode tip diameter for thin sheets is: $d = 0.1 + 2t$ where t is the material thickness, and $d = \sqrt{t}$ for thick material. Then, for every sheet thickness with its proper weld nugget there is a minimum distance which is required from the edges of the weld nugget to the edge of the part. Proper overlap of joints, *Fig.* 54.9, is given in TABLE 54-II along with the other pertinent information. Too small an overlap not only produces a poor and sometimes weak joint, but often necessitates special and more costly electrodes. Such joints should be modified if at all possible to conform with minimum acceptable standards. Minimum distance from the spot centerline to the edge of the sheet in aluminum or magnesium is 4 times gage, 6 times being preferred.

Joint Design. Proper joint design is often the key to economical and satisfactory spot welded assemblies. In *Fig.* 54.10 joint (*a*) shows a typical lap which can be produced on any machine. Flange joint (*b*) is also easily accessible. Should flange (*b*) be turned in, difficulty may be encountered in getting the work into and out of the machine, particularly if the leg is high. In such cases, a special machine is often required. Joint (*c*) is sometimes encountered and normally is undesirable. With sufficient room for the upper electrode, however, it can generally be made with a portable gun. Joint (*d*) is often used but should be avoided. A considerable portion of the current passes through the outer sheet and fails to pass through the joint. Flanged stamping spot-welded to a tube, shown at (*e*) can be made if the tube is sufficiently strong to resist collapse. For best results the sleeve should make a slip fit.

Accessibility. Good and poor design practices in regard to adequate accessibility are shown in *Fig.* 54.11. From these sketches an idea as to simplifying joint assembly and achieving maximum production economy can be gained. Typical standard electrode tips and operations are shown in *Fig.* 54.12. Generally, tubular or hollow shapes with a minimum internal clearance of 2 inches can be handled to produce a lap seam along the length. Lengths at this minimum internal clearance are limited with parts having one closed end but double lengths can be had with open ended parts as follows:

Internal Clearance	Length (max)		Gage (max)			
	One end	Open ends	Alum.	Mag.	Stainless	L.C.Steel
2 in.	9	18	0.025	0.025	0.035	0.035
3 in.	18	36	0.032	0.032	0.050	0.050
4 in.	20	40	0.040	0.040	0.072	0.072
10 in.	40	80	0.064	0.064	0.090	0.090
12 in.	45	90	0.064	0.064	0.090	0.090
18 in.	45	90	0.064	0.064	0.090	0.090

While these values are not fixed limitations they do permit use of standard available equipment and eliminate recourse to special or little used machines.

Metal Thickness. It is not important that two pieces to be welded have the same thickness. A piece of 16-gage sheet will weld to a 6-inch plate as readily as to another 16-gage sheet. It is desirable, however, to maintain a maximum ratio between two gages of 3 to 1 for best results in any metal.

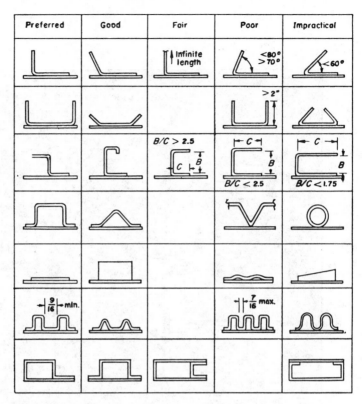

Fig. 54.11 — Chart of joints showing good and poor practices in regard to accessibility. Drawing, courtesy P. R. Mallory & Co.

Multiple thicknesses of low-carbon or corrosion-resistant steels may be spot welded together subject to a number of restrictions enumerated in the following:

1. When all sheets are of equal gage (preferred) any number of thicknesses may be welded provided the overall thickness of the stack does not exceed one inch.
2. When outside sheets are heaviest, gages in the pile-up need not be equal, but the maximum thickness ratio between any two thicknesses must not exceed 3 to 1. Up to eight thicknesses may be satisfactorily welded in this manner but overall thickness of the stack should not exceed one inch.
3. When the thinnest piece or pieces are on the outside (not necessarily equal), the maximum ratio between any two sheets must not exceed 2 to 1. Up to seven thicknesses may be satisfactorily welded but maximum thickness should not exceed one inch.

In aluminum or magnesium alloys these restrictions are somewhat more stringent:

1. When all sheets are of equal gage (preferred) up to three thicknesses may be welded provided the overall thickness of the stack does not exceed 5/16-inch for aluminum or 1/4-inch for magnesium.
2. When sheets are of unequal gage, up to three thicknesses may be welded provided the outside sheets are equal, thickness ratio between any two sheets does not exceed 2 to 1 and the total thickness of the stack does not exceed 5/16-inch for aluminum or 1/4-inch for magnesium. If thicknesses do not vary over 25 per cent they may be welded in any position.

In order to complete the design of a spot-welded assembly properly, complete data regarding the joints must be given. To supply such information to the manufacturing division, use of standard welding symbols developed by the

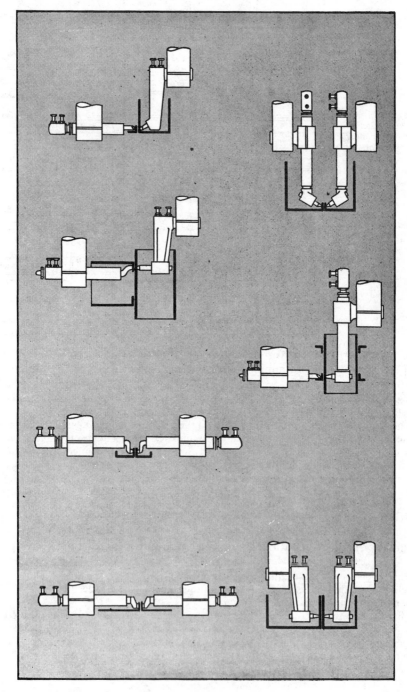

Fig. 54.12 — Typical standard electrode tips and operations. Drawing, courtesy P. R. Mallory & Co.

American Welding Society is recommended. *Fig.* 54.13 summarizes all this information and in use it can be simplified or modified to meet individual requirements.

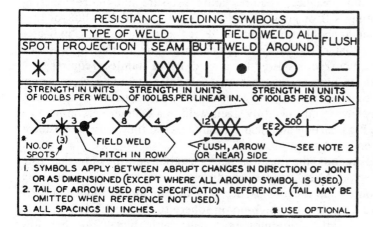

Fig. 54.13 — Legend showing symbols for use in specifying resistance-welded joints on drawings. Drawing, courtesy Resistance Welder Manufacturers Association.

Selection of Materials

Ordinary low-carbon steel with from 0.05 to 0.15 per cent carbon is the most readily weldable material and the weldability of all other materials is judged therefrom. Low-carbon steel has sufficiently high electrical resistance, a wide plastic range and offers the fewest metallurgical problems. As the carbon content is raised, embrittlement of the weld zone enters the picture and this becomes critical in the range of 0.25 to 0.35 per cent carbon. Up to 0.50 to 0.60-carbon content, satisfactory heat treatments may be applied automatically in the machine itself. Beyond this point, postheat is of little value and the end use of the product must be considered for subsequent annealing and heat treatment.

TABLE 54-III — Weldability Ratings of Various Metals*

Metal or Alloy	Rating
Low-Carbon Steel (SAE 1010)	A-100
Stainless Steel (18-8)	B-75
Aluminum (2S)	C-250
Aluminum (52S)	C-300
Magnesium (Dow M)	B-290
Magnesium (Dow J)	B-230
Nickel	B-130
Nickel Alloys (Monel, etc.)	B-65
Nickel Silver (30% Ni)	B-125
Nickel Silver (10% Ni)	B-100
Copper	F-350
Silver	F-350
Red Brass (80% Cu)	E-150
Yellow Brass (65% Cu)	C-150
Phosphor Bronze (95% Cu)	C-120
Silicon Bronze (up to 3% Si)	B-80
Aluminum Bronze (8% Al)	C-80

*A = excellent, B = good, C = fair, D = poor, E = extremely poor, and F = impractical. Dash figure indicates current.

Weldability. In TABLE 54-III are shown in the relative weldability ratings of metals most commonly spot welded. The letter prefix indicates the general weldability, everything considered in relation to low-carbon steel (rated A or excellent). Stainless, for instance, requires more pressure and more exacting control, hence its rating is B or good. Stainless is readily weldable, but for best results the carbon content should not exceed 0.08 per cent. The rating C indicates only fair weldability, that of D indicates poor. The figure in the rating index indicates the amount of current required for a weld, hence stainless requires only 75 per cent as much current as low-carbon. Materials rated E are extremely difficult to weld and those indicated F are commercially impractical or impossible.

Most proprietary steels and low-alloy steels weld in a manner very much like low-carbon but, like high-carbon and heat-treatable steels, they must be postheat treated to eliminate embrittlement.

In the nonferrous metals, the plastic range is smaller and they are therefore more difficult to weld. Also, surface conditions must be more accurately controlled. Some nonferrous metals, particularly aluminum, introduce further difficulties because of a tendency to pick-up or adhere to the electrodes.

Most of the wrought alloys of aluminum are weldable, but due to their low electrical resistance, more power is required. Hence aluminum requires $2^{1}/_2$ to 3 times as much current as low-carbon steel, TABLE 54-III. Aluminum has a short plastic range and control must be exacting. Oxide coating is undesirable and hence all aluminum alloys must be cleaned just prior to welding by pickling or wire brushing. Brass, lead, bronze, and magnesium likewise require careful cleaning.

The weldability of other metals can be judged roughly by their electrical and thermal conductivity. The higher the conductivity, the greater the difficulty in welding. Thus silver and copper are difficult to weld largely because no satisfactory electrode material is available; copper will weld readily to the electrode itself.

Dissimilar metals can be spot welded. However, many factors enter into such combinations and no specific rules can be made. The chart in TABLE 54-IV shows most of the common metals that can be spot welded with conventional production methods.

TABLE 54-IV — Spot Welding Metal Combinations and Weldabilities

Metals	Al	St. St.	Br.	Cu	Gal. Fe	St.	Pb.	Mon.	Ni	Nich.	Tin Pl.	Zn	Ph. Br.	Ni Sil	Te. Pl.
Aluminum	B	E	D	E	C	D	E	D	D	D	C	C	C	F	C
Stainless Steel	F	A	E	E	B	A	F	C	C	C	B	F	D	D	B
Brass	D	E	C	D	D	D	F	C	C	C	D	E	C	C	D
Copper	E	E	D	F	E	E	E	D	D	D	E	E	C	C	E
Galvanized Iron	C	B	D	E	B	B	D	C	C	C	B	C	D	E	B
Steel	D	A	D	E	B	A	E	C	C	C	B	F	C	D	A
Lead	E	F	F	E	D	E	C	E	E	E		C	E	E	D
Monel	D	C	C	D	C	C	E	A	B	B	C	F	C	B	C
Nickel	D	C	C	D	C	C	E	B	A	B	C	F	C	B	C
Nichrome	D	C	C	D	C	C	E	B	B	A	C	F	D	B	C
Tin Plate	C	B	D	E	B	B		C	C	C	C	C	D	D	C
Zinc	C	F	E	E	C	F	C	F	F	F	C	C	D	F	C
Phosphor Bronze	C	D	C	C	D	C	E	C	C	D	D	D	B	B	D
Nickel Silver	F	D	C	C	E	D	E	B	B	B	D	F	B	A	D
Terneplate	C	B	D	E	B	A	D	C	C	C	C	C	C	D	B

*A=excellent, B=good, C=fair, D=poor, E=very poor, and F=impractical.

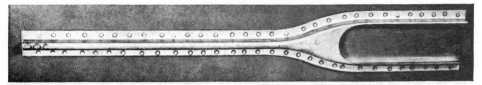

Fig. 54.14 — Stainless steel cockpit emergency release frame assembly exhibiting typical weld spots.

Fig. 54.15 — Spot welding in aircraft offers better production speed, strength, rigidity, and aerodynamic smoothness. Photo, courtesy Boeing Aircraft Co.

In aircraft parts spot welding of bare 24S or 75S is not acceptable. Bare to clad metal is acceptable providing assemblies can be anodized or chromodized and primed afterward. Spot welded magnesium parts are acceptable only if of a nonstructural or noncritical nature.

Effect of Coatings. The greatest hindrance to successful welding is scale. To obtain best results materials should be specified as free from scale, at least in the weld zone. Some coated materials can be satisfactorily welded, others not. Terne plate can be welded, though not as readily as uncoated steel. Zinc plate welds readily but the zinc coating is destroyed in the weld zone. Tin plate can be welded without injuring the coating. Coatings such as Bonderizing, Parkerizing, Browning, or Brunofixed are impossible to weld. Sherardized sheet can be spot welded but slight rusting occurs on the weld surfaces. From a production point of view, Sherardized and electrozinc surface coatings are recommended when rustproofing of components must be used prior to spot welding.

Specification of Tolerances

Generally, resistance spot welding has little or no effect on part tolerances, the actual critical dimensions of most components being dependent upon some prior method of forming. Dimensions over an assembly of two or more pieces is merely that of the sum of the various part thicknesses. Tolerances on the final assembled and welded stack, of course, are additive, being comprised of all the individual part variations. Spot welding adds no metal; actually, on cooling, the weld shrinks somewhat and along with the pressure during welding leaves a slightly hollow spot at the point of contact on each side, *Fig.* 54.14.

Where absolutely smooth, undiscolored finish is necessary, one of the previously mentioned techniques can be employed for obtaining such surfaces. For economy the as-welded surfaces should be acceptable, *Fig.* 54.15.

Tolerances on the fit of assembled parts to be spot welded should be as close as practicable. Poorly formed, buckled or warped parts often cause shunting of the current away from the weld. Poorly fitted parts which require the welding machine to accomplish a pressing operation are uneconomical to spot weld in production.

Use of fixtures to locate an assembly of parts to be welded offers wide possibilities in the manufacture of highly accurate assemblies. Assembled tolerances for spot welded assemblies should be no closer than warranted by the design contemplated in order to simplify the design of the fixtures necessary. Spot spacings are normally held within plus or minus $1/8$-inch of spacings designated.

55

Seam Welding

IN PRINCIPLE, seam welding is accomplished in the same manner as single-spot welding. Combined use of heat, generated by resistance of the interfaces of the metal to the passage of an electric current, and applied pressure to forge together the fused metal produces each weld. Application of pressure and current in seam welding, however, is by means of rotating wheel-like electrodes, *Fig.* 55.1.

Basic Process. Unlike spot welding where the electrodes are opened and closed to produce each weld, work to be seam welded is passed between the wheels which remain in contact during the entire operation. When the circuit is closed, the current produces a weld at the point of pressure between the wheels. The wheels continue to rotate and after a preset off-time interval, another surge of current produces the next weld, and so on.

When the work is passed between the electrodes so that the line of weld is at right angles to the throat of the welder, the process is usually termed *transverse* or

Fig. 55.1 — Refrigerator evaporator being fabricated by seam welding. Photo, courtesy General Electric Co.

Fig. 55.2 — Circumferential seam welding of a large flange to a jet engine cone assembly. Photo, courtesy General Electric Co.

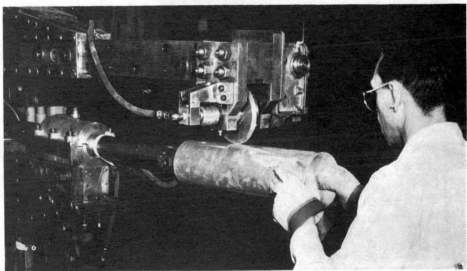

Fig. 55.3 — Longitudinal seam welding of aluminum aircraft duct. Tube is close to minimum diameter. Photo, courtesy Lockheed Aircraft Corp.

circumferential seam welding, *Fig.* 55.2. However, when the work is processed so that the weld seam is parallel to or into the throat of the machine, it is normally termed *longitudinal* seam welding, *Fig.* 55.3.

Types of Welds. Several types of seam welds can be produced and selection naturally depends upon the application. Normal practice in seam welding is to produce *pressuretight* joints by adjusting wheel speed and off-time to obtain a succession of overlapping spot welds, *Fig.* 55.4. Actual fusion of the metal must overlap and the required number of spots per inch varies with metal thickness, TABLES 55-I and 55-II. Occasionally it is desirable to have a spaced series of spots

Table 55-1†
Seam Welding Low-Carbon Steel With Interrupted Current[1]

Data Common To All Welding Speeds								Pressure-Tight Joint																Nonpressure-Tight Joint Roll-Spot Welding				
	Electrode[2]			Electrode Force		Min. Overlap L		Maximum Welding Speeds Recommended					Average Welding Speeds Recommended					Minimum Welding Speeds Recommended								Welding Speeds (In. per min)		
Thickness of work pieces	Min. d	Nor. d	Min. D	Min.	Normal	Use with min. d	Use with nor. d	Heat Time	Cool Time	Welding Speed	Welds per Inch	Welding Current	Heat Time	Cool Time	Welding Speed	Welds per Inch	Welding Current	Heat Time	Cool Time	Welding Speed	Welds per Inch	Welding Current	Heat Time	Cool Time	Max.	Avg.	Min.
(in.)	(in.)	(in.)	(in.)	(lb)	(lb)	(in.)	(in.)	(c)[3]	(c)	(In. per min)		(amp)	(c)	(c)	(In. per min)		(amp)	(c)	(c)	(In. per min)		(amp)	(c)	(c)			
0.010	1/8	3/16	3/8	400	400	1/4	3/8	1	1	116	15.5	11,500	2	1	80	15	8,000	2	3	49	14.5	7,500	2	*	480	200	100
0.021	5/32	3/16	3/8	460	550	5/16	7/16	2	1	107	11	12,500	2	2	75	12	11,000	3	3	45	13	9,000	2	*	425	185	90
0.031	3/16	1/4	1/2	530	700	5/16	1/2	2	1	103	11.5	15,000	3	2	72	10	13,000	2	4	42	14	12,000	3	*	370	180	85
0.040	3/16	1/4	1/2	600	900	3/8	1/2	2	2	98	9	18,300	3	3	67	9	15,000	2	4	39	15.5	13,500	3	*	310	170	80
0.050	3/16	1/4	1/2	700	1050	7/16	9/16	2	2	95	9.5	20,000	4	3	65	8	16,500	4	4	37	12	14,000	4	*	250	160	75
0.062	3/16	3/8	1/2	750	1200	7/16	5/8	3	1	91	10	21,000	4	4	63	7	17,500	4	4	36	12.5	15,400	4	*	190	150	70
0.078	1/4	3/8	5/8	980	1500	1/2	11/16	3	1	85	10.5	22,000	6	5	55	6	19,000	6	6	30	10	16,000	5	*	170	140	60
0.094	1/4	3/8	3/4	1010	1700	1/2	3/4	4	2	80	7.5	23,000	7	6	50	5.5	20,000	6	6	27	11	17,000	6	*	160	120	55
0.109	3/16	1/2	3/4	1120	1950	9/16	13/16	4	2	75	8	25,000	9	6	48	5	21,000	6	6	25	12	18,500	6	*	150	100	50
0.125	1/8	1/2	3/4	1240	2200	5/8	7/8	4	2	70	8.5	27,500	11	7	45	4.5	22,000	6	6	23	13	21,000	6	*	140	90	45

† Courtesy Taylor-Winfield Corp.

[1] Low-carbon steel is hot rolled, pickled, and lightly oiled with an ultimate strength of 42,000 to 45,000 psi similar to SAE 1005—SAE 1010. Surface of steel lightly oiled but free from grease, scale or dirt. Radius crowned electrodes work satisfactorily in seam-welding low-carbon steels also.

[2] Electrode material is RWMA Class 2 alloy.

[3] Heat and cool times are given in cycles of 60-cycle frequency.

TABLE 55-II — Seam Welding Stainless with Interrupted Current*

Thickness In.	USS Gauge	Elec. Width d Min. In. (Note 10)	D Min. Radius In.	D Min. In.	Elec. Force Min. lbs.	Elec. Force Normal lbs.	Min. Contact. Overlap L In. (Note 4)	MAX Heat Time Cycles	MAX Cool Time Cycles	MAX Weld Speed In/Min	MAX Welds per Inch	MAX Amperes	AVG Heat Time Cycles	AVG Cool Time Cycles	AVG Weld Speed In/Min	AVG Welds per Inch	AVG Amperes	MIN Heat Time Cycles	MIN Cool Time Cycles	MIN Weld Speed In/Min	MIN Welds per Inch	MIN Amperes	NP Heat Time Cycles	NP Cool Time	NP Max In/Min	NP Avg In/Min	NP Min In/Min
.006	38	1/8	1-1/2	3/16	300	300	1/4	1	1	90	20	6,300	2	1	60	20	4,000	2	2	44	20	3,600	2	*	180	120	80
.008	35	3/16	1-1/2	3/16	330	350	1/4	1	1	90	20	6,400	2	1	60	18	4,600	2	3	40	18	4,150	2	*	180	120	80
.010	32	3/16	1-1/2	3/16	375	400	1/4	1	1	90	20	6,600	3	2	60	16	5,000	3	2	44	15	4,500	3	*	180	120	80
.012	30	3/16	1-1/2	1/4	425	450	5/16	2	2	80	15	6,800	3	2	55	15	5,600	3	3	40	15	5,050	3	*	160	110	80
.014	29	7/32	1-1/2	1/4	470	500	5/16	2	2	80	15	7,200	3	2	55	14	6,200	3	3	42	14	5,600	3	*	160	110	80
.016	28	7/32	1-1/2	1/4	550	600	5/16	2	2	80	15	7,600	3	2	55	14	6,700	3	3	43	14	6,000	3	*	160	110	80
.019	26	1/4	1-1/2	3/4	600	650	5/16	2	2	80	15	8,300	3	2	55	13	7,300	3	3	44	13	6,500	4	*	160	110	80
.021	25	1/4	1-1/2	1/4	650	700	3/8	2	2	80	15	9,000	3	2	55	13	7,900	3	3	44	13	7,100	4	*	160	110	80
.025	24	1/4	3	3/8	775	850	7/16	2	2	75	12	12,000	3	2	50	12	9,200	3	5	38	12	8,300	5	*	150	100	78
.031	22	1/4	3	3/8	900	1000	7/16	2	2	75	12	13,500	3	3	50	12	10,600	3	5	38	12	9,600	5	*	150	100	75
.040	20	1/4	3	3/8	1050	1300	1/2	3	3	70	10	14,500	3	4	47	11	13,000	3	6	36	11	12,300	6	*	140	94	70
.050	18	5/16	3	1/2	1200	1600	5/8	3	3	65	11	15,500	4	4	45	10	14,200	4	6	35	10	13,500	7	*	130	90	65
.062	16	5/16	3	1/2	1400	1850	5/8	3	3	60	12	16,000	4	5	40	10	15,100	4	6	36	10	14,500	8	*	120	80	60
.070	15	5/16	3	5/8	1600	2150	11/16	4	3	58	9	16,400	4	5	40	9	15,900	4	8	34	9	15,700	10	*	116	80	55
.078	14	3/8	3	5/8	1700	2300	11/16	4	3	56	9	16,900	4	6	38	9	16,500	4	9	31	9	16,500	10	*	112	80	50
.094	13	7/16	3	3/4	1800	2550	3/4	5	3	55	8	17,000	5	5	40	9	16,600	5	8	31	9	16,600	12	*	110	80	40
.109	12	1/2	3	3/4	1900	2950	13/16	5	3	53	8.5	17,300	5	7	38	8	16,800	5	12	27	8	16,800	14	*	106	76	30
.125	11	1/2	3	7/8	2000	3300	7/8	5	4	50	8	18,000	6	6	38	8	17,000	6	13	24	8	17,000	15	*	100	76	25

(Column groups: **DATA COMMON TO ALL WELDING SPEEDS** — Thickness, USS Gauge, Electrode Width and Shape, Electrode Force, Min. Contacting Overlap. **PRESSURE TIGHT JOINT** — Recommended Maximum / Average / Minimum Welding Speeds, each with Heat Time, Cool Time (Note 5), Welding Speed In. per Min., Welds per Inch, Welding Current Amperes. **NON-PRESSURE TIGHT JOINT (ROLL SPOT WELDING)** — Heat Time, Cool Time, Welding Speeds In. per Min. Max/Avg/Min.)

NOTES:

1 Types of steel - 301, 302, 303, 304, 308, 309, 310, 316, 317, 321, 347, and 349.

2 Electrode material is RWMA Class 3 alloy.

3 Surface of steel is free from grease, scale or dirt.

5 Heat and cool times are indicated in cycles of 60 cycle frequency.

6 Electrode force does not provide for force to press ill-fitting parts together.

7 Use maximum electrode force at maximum speed and when welding cold worked material of 150,000 PSI or above ultimate tensile strength.

8 For large assemblies minimum contacting overlap should be increased by 30% for ease of guiding.

9 Large assemblies should be welded at the slower speeds for ease of handling and guiding.

10 For best appearance use radius faced electrodes. Average application can be welded with flat face electrodes.

* Adjust the cool time to give the desired spot spacing.

Courtesy Taylor-Winfield Corp.

for joints not necessarily pressuretight. Known as *roll-spot* welding, *Fig.* 55.4, such a series of closely spaced spots are produced by adjusting wheel speed and off-time so that individual welds are produced as the work is fed between the wheels. This type of seam weld can be made at great speed, certain machines producing for instance 600 spots per minute spaced 1/2-inch apart on 22-gage mild steel. A third type of joint, known as the *mash-seam* welded joint, is produced by reducing the normal joint overlap to approximately 1 1/2 times the gage so that the two pieces are forged into each other as the welding progresses, *Fig.* 55.4. The resultant joint thickness is slightly greater than one sheet thickness and closely resembles a butt joint. Limitations of this joint are requirements for tack or spot welding, clamping,

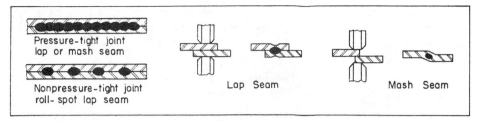

Fig. 55.4 — Types of seam welded joints utilized. Drawing, courtesy Taylor-Winfield Corp.

Fig. 55.5 — Roll-spot welding a reinforced doubler to aluminum channel. Photo, courtesy North American Rockwell Inc.

or holding in a jig to retain part dimensions and difficulty in obtaining satisfactory joints in any but the lighter gages.

Types of Machines. Standard or special machines to produce any of the types of joints by the foregoing seam-welding methods may be classified generally as: (1) Circular, (2) longitudinal, or (3) universal. Thus, the machine is classified primarily by the method whereby the work is fed into the machine as discussed previously.

Circular Seam Welders. In all probability, the most widely used machine, the circular seam welder, is designed to produce transverse or circumferential seam welds. Because the work is fed into the machine at right angles to the welder throat or arms, length of continuous seam is almost unlimited, *Fig.* 55.5. As can be seen, these machines are used primarily for making pressuretight and roll-spot circular or cylindrical end seams, *Fig.* 55.6, as well as single or multiple long straight or curved seams.

Standard RWMA machine sizes are shown in Table 55-III. However, machines up to 500 kva with throat depths to as much as 60 inches are produced.

Longitudinal Seam Welders. For producing longitudinal seams, the longitudinal welder operates with the work being fed into the machine throat, *Fig.* 55.7. These machines are adapted primarily for making side or straight seams on cylindrical or tubular work of varying cross section. Generally, this particular seam welding method is limited to rather short, straight seams — as a rule approximately 6 inches less than the available throat depth of the particular machine, TABLE 55-III.

TABLE 55-III — **Specifications for Standard Seam Welders**

Size No.	Rating (kva)	Throat Depth (in.)
1	50–75	18–24–30
2	100–150–200	18–30–42
3	250–400	24–36–48

Fig. 55.6 — Seamwelding a flexible section for the jet tailpipe of a bomber. Photo, courtesy North American Rockwell Inc.

Universal Seam Welder. Welders which combine the basic features of both longitudinal and circular machines are generally termed universal seam welders. These are designed so as to permit the upper welding head to be swiveled 90-degrees to either position and have a replaceable lower arm with welding units to match. Welds of either the circular or the longitudinal type can be made readily.

Special Machines. In addition to the regular run of standard machines, a variety of special or semispecial units is widely used. These include: (1) Machines with traveling wheels to allow the work to be clamped stationary while the seam is completed; (2) machines with a traveling platen to allow the work to be passed under the roll to complete the seams; (3) machines with multiple rolls to complete more than one seam simultaneously, *Fig.* 55.8; (4) machines designed for portable seam welding of unusual parts, *Fig.* 55.9; and (5) machines specially arranged to facilitate difficult operations in mass production, *Fig.* 55.10.

Machine Drives and Production. In conventional machines the wheel or roll

Fig. 55.7 — Longitudinal seam welder showing roll arrangement. Photo, courtesy Lockheed Aircraft Corp.

Fig. 55.8 — Dual-wheel seam welder for fabricating refrigerator evaporators. Operation with a shuttle table is automatic. Photo, courtesy Progressive Welder Co.

drive may take one of several forms: The upper or lower rolls may be direct gear driven or powered by a friction drive. The upper roll may be positively gear driven with the lower roll idling or, in the case of circular work, the lower may be driven with upper roll idling. Friction drive with knurled drive rolls is the only method that permits power application to both rolls and therefore is most widely used. Where space requirements do not permit driving the lower roll, only the upper is driven. Tin plate, terne plate, galvanized steel, and other coated steels should be welded on knurl-driven machines inasmuch as the knurled drive rolls continuously break up the surface of the welding wheels thereby removing pieces of the coating picked up during passage over the sheet.

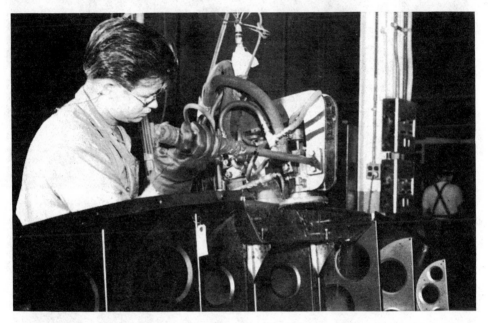

Fig. 55.9 — Special portable seam welder used to join the pan and half shell of aircraft drop tank. Photo, courtesy Lockheed Aircraft Corp.

Speed of passage of the work through the wheels normally ranges from about 12 to 360 inches per minute, although higher speeds up to about 50 feet per minute are used with continuous application of current.

Gage Limitations. Seam welding is a fast and economical method of joining sheet material together for gas or liquidtight joints but unlike spot welding is mainly limited to the lighter gages. The general range of thicknesses normally considered in the practical seam welding category is from about 0.10-inch to 1/8-inch. Although it is possible to seam weld gages to as thin as 0.002 to 0.003-inch, this is seldom done since it requires special equipment. Standard available machines produced by some manufacturers are capable of welding two sheets of 3/16-inch thickness but this top limit is seldom exceeded because above this gage production equipment becomes too costly and output too slow.

Design Considerations

Seam welding, like spot welding, is most readily applied and successfully used

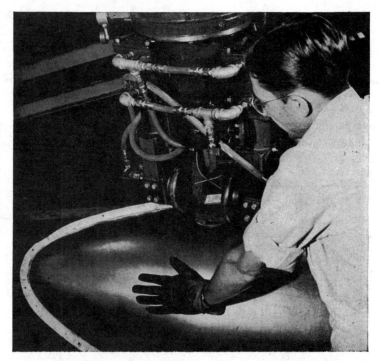

Fig. 55.10 — Special revolving head machine for welding a filler well to a gas tank shell. Head revolves on vertical axis to produce a circular seam weld completely surrounding the filler cap. Photo, courtesy Lockheed Aircraft Corp.

if the various aspects and limitations incident to the process are considered. By proper design practices maximum economy likewise will be assured.

Gage. As previously noted, seam welding is somewhat limited as to the range of sheet thicknesses which can be welded. Generally two low-carbon sheets from 0.010-inch to 1/8-inch define the maximum and minimum limits. In stainless, two sheets from about 0.006-inch up to 1/8-inch can be handled on regular machines.

Continuous Seams. Pressuretight joints require a specific minimum number of spots per inch to obtain continuous fusion. Thus approximately 12 per inch are required for 0.010-inch stock ranging down to about 5 for 1/8-inch material. For a given material and thickness combination there is a definite maximum welding speed determined by the maximum possible rate of heat input which is limited by surface burning of the material and rate at which the welds solidify under wheel pressure. Maximum speeds and welds per inch are given in TABLES 55-I and 55-II.

Joints. Also given are the recommended joint overlaps for various gages of low-carbon steel. Parts to be lapped or flanged and welded require sufficient overlap or flange width to prevent burning of the edges or spattering. Two overlap dimensions are shown in the table but the larger is preferred and should be employed wherever possible. The smaller dimension only provides reasonably good results and usually it is necessary to use fixtures to assure centering of the weld on the narrow flange. For large assemblies the contacting overlap should be increased by 30 per cent to insure sound joints and facilitate guiding. Details for stainless are given in TABLE 55-II.

Proper overlap and fitup is of primary importance in seam welding. Where design or features of an assembly are such as to present difficulty in controlling the overlap in the welding operation, the necessary lap should be fixed by a few preliminary spot welds. Fitup of joints to be welded should be good and seams should require no electrode force to create a good, close fit during the weld.

Circular seam welds are particularly critical in this respect and wherever possible a press fit or similar means for insuring intimate contact in the welding zone should be used where best quality welds are required.

Wheel Clearances. Where circular welds, *Fig.* 55.6, are concerned, the smallest inside diameter which can be handled is about 4 inches. This allows a suitable electrode diameter and reasonable service life. However, material gage has some influence on ID and generally on the lightest gages, diameters down to 2 inches can be handled, *Fig.* 55.11 and TABLE 55-IV. By proper selection of wheel diameters and width, satisfactory work clearance and desirable heat balance can be attained.

TABLE 55-IV — **Electrode and Work Diameters**
(inches, see *Fig.* 913)

Thickness (each piece)	D_1 (min)	D_2 (min)	D_3	W_2	W_3
0.010	2	$1\frac{3}{4}$	12	$\frac{1}{8}$	$\frac{1}{4}$
0.020	3	$2\frac{3}{4}$	12	$\frac{1}{8}$	$\frac{1}{4}$
0.040	4	$3\frac{3}{4}$	11	$\frac{3}{16}$	$\frac{1}{4}$
0.062	6	$5\frac{3}{4}$	12	$\frac{3}{16}$	$\frac{3}{8}$
0.078	8	$7\frac{3}{4}$	8	$\frac{3}{16}$	$\frac{3}{8}$
0.125	12	10	10	$\frac{5}{16}$	$\frac{3}{8}$

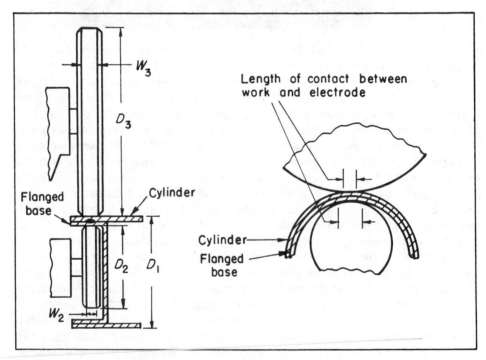

Fig. 55.11 — Recommended combinations of electrodes and work diameters. Dimensions are shown in Table 55-IV accompanying. Drawing, courtesy Taylor-Winfield Corp.

Smaller diameters can be welded, *Fig.* 55.12, especially where solid inserts eliminate crushing, but most small-diameter parts require special equipment and technique.

Cross-Sectional Shapes. Where cross sections of circular welds are other than round, the minimum internal radius allowable at corners is 2 inches. Preferable

Fig. 55.12 — Seam welding a tube for an automotive shock absorber. Photo, courtesy General Electric Co.

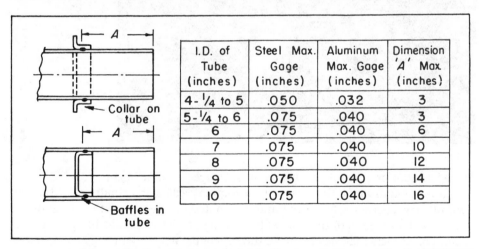

I.D. of Tube (inches)	Steel Max. Gage (inches)	Aluminum Max. Gage (inches)	Dimension 'A' Max. (inches)
4-¼ to 5	.050	.032	3
5-¼ to 6	.075	.040	3
6	.075	.040	6
7	.075	.040	10
8	.075	.040	12
9	.075	.040	14
10	.075	.040	16

Fig. 55.13 — Limitations of circular seam welds relative to tube diameter, material gage and location in collar and baffle arrangements. Drawing, courtesy Lockheed Aircraft Corp.

radius is 3 inches on steel and 6 inches for aluminum. On large square sections, for instance, use of 2-inch radius corners slows down the straight welds to the speed required in traversing the small corner radius and in such cases it is preferred to design in as large a radius as possible to speed output. Position of such welds in relation to the part length also is limited as is the material gage, *Fig.* 55.13.

Design of End Caps. Design of the joint for dished or flat ends as well as side or centered joints must be such as to allow access to the rolls, *Fig.* 55.14. Where part design does not allow roll clearance, special electrodes can sometimes be used,

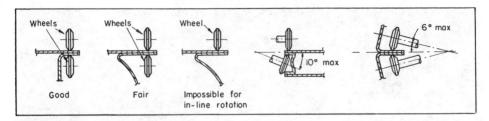

Fig. 55.14 — Design of joints for dished or flat ends to allow access to rolls. Drawing, courtesy Lockheed Aircraft Corp.

Fig. 55.15 — Special hollow electrode developed for seam welding a special hydraulic reservoir. Photo, courtesy Lockheed Aircraft Corp.

Fig. 55.15, but production must warrant such special attachments.

Seam Length on Hollow Parts. Longitudinal seams on cylindrical or tubular parts of varying cross section are limited, as noted previously, by the throat of the welder available. Length of seam which can be produced should be restricted to about 6 inches less than the available throat length. Where both ends are open the possible seam length can be doubled. As a rule, outstanding seam welds on tubular parts prove the most economical design for small quantities while overlapped seam welds are more economical in large quantities.

External Shape Seams. In welding flanged parts such as tanks and similar parts where the weld must turn along an external radius, *Fig.* 55.16, the minimum external radii advisable are as follows:

22 through 18 gage (0.030-0.050-in.) .... 2 in.
16 gage (0.062-in.) .... 2¹/₄ in.
14 gage (0.078-in.) .... 2¹/₂ in.
13 through 11 gage (0.094-0.125) .... 3 in.

The minimum radius in aluminum is 3 inches. On concave radii where the sides are normal to the flange, the inside radius should be no less than 10 inches in aluminum and 6 inches in steel. When the sides form part of a circle with radius not exceeding 4 inches, the welding radius can be reduced to a minimum of 6 inches in aluminum and 3 inches in steel, *Fig.* 55.17, because of greater wheel clearance over the seam.

Good Design. The good and poor design practices shown in the preceding chapter on spot welding apply largely in seam welding also. Typical wheel diameters may range from a minimum of 2 inches to as much as 24 inches with widths shown in Tables 55-I and 55-II. Joint design should allow for use of standard setups wherever possible.

Specifications. In completing the design of a seam-welded part, much as in spot welding, it is essential that complete data regarding the joints be included. In supplying such essential information, use of the recommended standard welding symbols developed by the American Welding Society is considered good practice. These symbols are summarized in the legend shown in Chapter 54.

Wheel Limitations. Width of weld at the centerline between the work pieces ranges from 1.5 to 3 times the gage of the metal being welded. The higher ratio of weld width to sheet thickness occurs in welding the thinner gages and the ratio decreases with increase in gage. The practical width of wheel face is limited to a minimum of about 1/8 to 3/16-inch by excessive wear; hence the higher ratios with the lighter gages.

Where obstructions which break up the continuity of the welded seam must be present in a part, these can often be straddled by the use of notched wheels. If the obstruction is on one side only, a single notched wheel and a round wheel can be

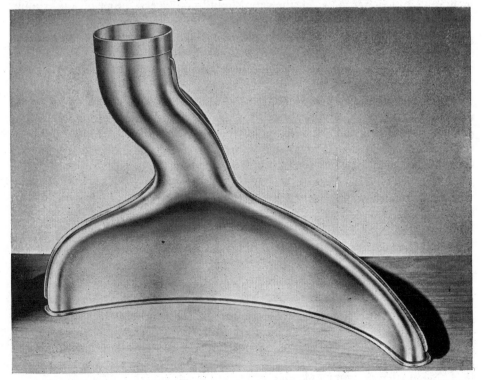

Fig. 55.16 — Airplane propeller deicing hub duct drop hammered from 0.018-inch stainless showing unusual contoured seam welded flanged edges. Photo, courtesy B. H. Aircraft Co. Inc.

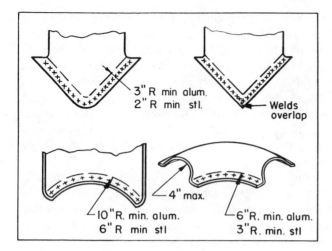

Fig. 55.17 — External radii on flanged parts are limited to the minimum values shown. Where internal radii on flanged parts are used, larger radii are necessary to permit access to the wheel. Drawing, courtesy Lockheed Aircraft Corp.

used or in the case of obstructions on both sides, two geared notched wheels are used. Start of the weld is made with the notch close to the obstruction and the seam is made by rolling away from the obstruction with the wheel. The obvious production problems created by such designs indicate the desirability for avoiding them whenever possible to do so.

Gage Ratio. Although gage of sheets to be seam welded preferably should be equal, such need not be the case. Generally, for average production in aluminum, unequal gages should not exceed a ratio of 2 to 1 with a maximum pileup of 0.128-inch total. In low-carbon steel, the ratio should not exceed 3 to 1 or a maximum pileup of 0.150-inch. Likewise, in stainless the ratio should not exceed 3 to 1 but pileup may range to 0.186-inch, maximum. In welding three thicknesses (maximum) the following restrictions apply: (1) Similar alloy combinations, (2) a maximum thickness ratio of 2 to 1, and (3) applications on which indentation can be permitted on both sides.

Seam Thickness. Where thickness through a finished weld must be a minimum, mash-seam welding is used. A slight overlap of 1 to 1 1/2 times the gage is provided and the pieces must be held by clamps or by prior spot welding. Spot weld tacking is preferred and for proper holding will vary from 1 1/2-inch spacings on 0.020-inch stock to 3-inch spacings on 0.078-inch stock. From mash welding the resulting seam is about 10 to 25 per cent thicker than the original gage. Mash-seam welding is limited to low-carbon and coated steels of 14-gage (0.078-inch) or less and should not be used unless absolutely necessary.

Selection of Materials

Generally speaking, the data covering selection of materials for spot welding, included in the previous chapter, also applies in seam welding. Materials that can be welded include carbon steels, stainless steels, zinc and other coated steels, nickel alloys, weldable bronzes and aluminum. Low-carbon steel should be free from any rust, scale, paint, grease or oil. Also, if possible, the material should not have any coating where highest quality welds are desired although steels with zinc, tin or lead-coated surfaces can be welded.

Comparative weldability of various metals covered in the previous chapter also applies as does the weldability of dissimilar metals.

Specification of Tolerances

As with spot welding, seam welding has little effect on part tolerances, specific working dimensions of welded parts being dependent upon the method used in fabricating the various components. Where precise locational tolerances are necessary, fixtures can be employed, *Fig.* 55.9, or the parts can be tack welded prior to seam welding.

With any properly made seam weld the wheels leave depressions which normally cannot be prevented. Where marring of the surface must be prevented, series welds using a wide wheel or shunting bar can be employed as with spot welding.

56

Projection Welding

A N OUTGROWTH of simple resistance spot welding, the method termed projection welding has a definite place in the production picture. Most important, perhaps, is the fact that with projection welding two or more parts can be permanently joined most nearly as dictated by design; the major limitations of spot welding are not present.

Basic Process. With projection welding, current flow and electrical heating to create individual weld spots or fused buttons are concentrated at specific points by means of projections on the parts rather than by the machine electrodes. By means

Fig. 56.1 — Close-up of the projection welding of fan units for large electric motor. Six welds are made at each index and 25 fans per hour produced. Photo, courtesy General Electric Co.

of these projections on one or both parts to be welded, relative location of weld spots on the parts can be accurately determined beforehand thus eliminating any possible error in manufacture, *Fig. 56.1*. Much faster than spot welding, this method allows the simultaneous creation of multiple welds; as few as one or as many as thirty or more spots can be made at a single impulse, depending upon the power available. Also, a large number of welds can be accurately placed in a limited area without experiencing the characteristic shunting difficulties generally found in spot welding.

Advantages. Ordinarily, projection welding requires two or three times the electrode pressure of spot welding and an average operation requires up to twice the current. It does, however, make possible the resistance welding of parts which are either impossible or impractical to spot weld conventionally. Not only is projection welding adaptable to flat or regular parts of sheet, plate, tubing, etc., as is welding, it also can be used to join irregular shapes and thick parts such as forgings, heavy stampings, complex machined pieces, and wire, *Fig. 56.2*.

Projection welding electrodes or, in some cases, dies normally have a large-area contact with the parts to be welded. This allows introduction of high currents without injuring the part surface and results in long electrode die life.

Projection Welding Machines. Because greater pressure is required in projection welding than in spot welding, considerably more powerful machines are used. Owing to the need for the heads to come together smoothly, uniformly and

Fig. 56.2 — Machined collar piece and wire forms, projection welded into one unit. Photo, courtesy General Electric Co.

parallel, press type welders constructed to afford maximum rigidity in the arms are used, *Fig. 56.3*. Low-inertia movement of the welding head is desirable to assure steady application of pressure as the projections collapse during the weld cycle. The lower electrode holder is usually made adjustable to accommodate a range of part sizes and in some cases is designed to accommodate special high-production parts, *Fig. 56.4*. Complex assemblies of parts are often handled in totally special-purpose machines, *Fig. 56.5*.

Standard projection welders conform to the general specifications outlined by the Resistance Welder Manufacturers Association, TABLE 56-I. Electrode or platen area ranges from 6 by 6 inches to 12 by 12 inches, both upper and lower. Spacing between platens is adjustable on all sizes and can be varied from 4 inches to 12 inches with the upper platen in the down position. Ram drives may be pneumatic, hydraulic or mechanical. Machines available actually range up to 60 inches in throat depth and offer capacities up to 800 kva or more.

Fig. 56.3 — Typical press type projection welder. Photo, courtesy Resistance Welder Corp.

Design Considerations

Careful analysis of the design of parts to be projection welded as well as cognizance of the various factors relating to proper projection size and shape can result in highly satisfactory, economical parts, *Fig.* 56.6. Theoretically, projections may be of an almost unlimited variety of forms. In practical application, however, certain types have been found most satisfactory and economical and hence are recommended for use wherever possible.

Heat Balance. In projection design it is always desirable to have a point or line contact from which the weld may grow. Ordinarily, the weld grows from the center outward and consequently the most satsifactory projection is the round or button type. To obtain the vitally needed heat balance — equal distribution of heat between the two parts being welded — it is often necessary to change to rib or elongated projections on extremely heavy sections.

Projection Designs. Generally, for welding sheet ranging in thickness from 24 to 13-gage, the button or spherical projection is used. Suggested standards for dimensioning such projections are shown in *Fig.* 56.7 and TABLE 56-II. Stock ranging from 12 to 5-gage in thickness is most satisfactorily welded by means of the conical type projection shown in *Fig.* 56.7. Dimensions suggested for satisfactory cone type projections are given in TABLE 56-III. Projection sizes actually may be varied considerably to suit design requirements as indicated in *Fig.* 56.8. Under no circumstances should partially sheared plugs or semipunchings be used inasmuch as such projections ordinarily will not withstand the initial pressure prior to application of the current without collapse.

Projections which have been set up as standard by AWS which produce a fused weld zone roughly equivalent to that of spot welds in the same gage sheet are shown

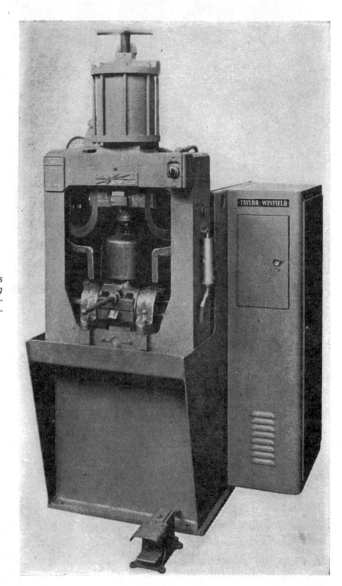

Fig. 56.4 — Air-operated press type projection welder for making annular welds on high-production parts. Photo, courtesy Taylor-Winfield Corp.

Fig. 56.5 — Special projection welder for welding typewriter body assemblies. Photo, courtesy Wean United, Inc.

TABLE 56-I — Specifications for Standard Projection Welders
(press type)

Size No.	Rating (kva)	Throat Depth (in.)	Platen Size (in.)	Stroke (in.)
1	30–50–75	12–18–30	6 x 6	2
2	100–150	12–18–30	7 x 7	3
3	150–250	12–18–30	9 x 9	4
4	300–400–500	12–18–30	12 x 12	4

Fig. 56.6 — Door interlock, with two pins and a boss to be welded to $1/8$-inch steel strip. Redesign for projection welding saved 35 per cent over plug welding formerly used. Photo, courtesy Westinghouse Electric Corp.

TABLE 56-II — **Button Projections**
(inches, see *Fig.* 56.7)

Gage	A	B	C	D	E	R
13	0.0937	0.075	0.050	0.180	0.065	1/64
14	0.0781	0.075	0.050	0.180	0.065	1/64
15	0.0703	0.075	0.045	0.172	0.055	1/64
16	0.0625	0.060	0.045	0.172	0.050	1/64
17	0.0562	0.055	0.040	0.156	0.045	1/64
18	0.050	0.050	0.040	0.156	0.040	1/64
19	0.0437	0.050	0.040	0.125	0.035	1/64
20	0.0375	0.050	0.035	0.125	0.035	1/64
21	0.0344	0.050	0.030	0.125	0.030	1/64
22	0.0312	0.050	0.030	0.125	0.030	1/64
23	0.0281	0.050	0.025	0.109	0.025	1/64
24	0.025	0.050	0.025	0.109	0.025	1/64

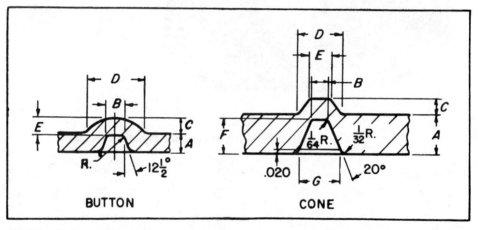

BUTTON **CONE**

Fig. 56.7 — Design of button projections usually used for embossing or stamping into 13 to 24-gage stock and cone projections usually used for 5 to 12-gage stock. Dimensions are given in Tables 56-II and 56-III. Drawings, courtesy P. R. Mallory & Co., Inc.

TABLE 56-III — **Cone Projections**
(inches, see *Fig.* 56.7)

Gage	A	B	C	D	E	F	G
5	0.218	0.080	0.075	0.210	0.100	0.200	0.220
6	0.2031	0.080	0.070	0.203	0.094	0.182	0.206
7	0.1875	0.080	0.070	0.203	0.094	0.166	0.185
8	0.1718	0.080	0.060	0.190	0.080	0.138	0.166
9	0.1562	0.080	0.060	0.172	0.080	0.122	0.155
10	0.1406	0.080	0.060	0.172	0.080	0.110	0.145
11	0.125	0.080	0.055	0.172	0.080	0.100	0.138
12	0.1093	0.080	0.055	0.172	0.080	0.090	0.131

in *Fig.* 56.9. Dimensions for these button type projections are given in TABLE 56-IV along with the expected shear strength for various tensile strengths in low-carbon and stainless steels.

As a rule, for welding fairly heavy pieces to very thin sheet under 24-gage the annular upset as produced by a prick punch in the thin sheet will serve as a suitable projection. Actually, though projection welding of sheets under 24-gage has been

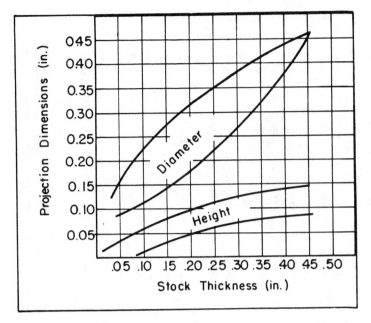

Fig. 56.8 — Range of projection dimensions normally found satisfactory in production. Designs may be safely varied within these ranges. Chart, courtesy P. R. Mallory & Co. Inc.

done, the use of the common projections formed in sheet under 0.015-inch thick is not recommended because of insufficient strength for the operation. For successful welds in sheet under 0.030-inch, it is recommended that long, narrow projections be used. The *D* dimension across such welds can be about 0.093-inch with the length about 0.250-inch. Another which has been found successful is an annular design about 0.110-inch in diameter, 0.015-inch high, 0.070-inch pitch diameter, with a 0.030-inch flat at the center.*

Formed Parts. When projections are made on formed or shaped pieces, the mating contact surface for which is not flat, it is usually desirable to use elongated projections, *Fig.* 56.10. Elongated projections can also be used to provide for welding corners of sheet, *Fig.* 56.11*a*. In some cases the sheet ends can be sheared to leave projections of suitable shape, *Fig.* 56.11*b*.

Ring Projections. Projections for gas and watertight joints are normally made with ring or annular embossed projections, *Fig.* 56.12. Annular projections can also

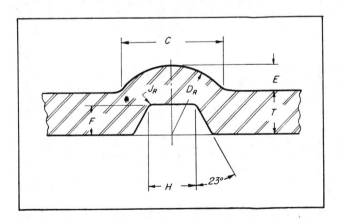

Fig. 56.9 — Design of standard AWS projection employed for setting up standard weld strengths shown in Table 56-IV on page 56-08.

*The Projection Welding of 0.010 and 0.030-inch Steel Sheet — Nippes, Gerken and Maciora, *The Welding Journal,* Sept. 1950, Pp 441-s.

TABLE 56-IV — AWS Standard Projections and Characteristics of Welds
(see Fig. 56.9)

T (gage, in.)	C (diam, in.)	E (height, in.)	D (rad, in.)	H (diam, in.)	F (depth, in.)	J (rad, in.)	Shear Strength in Steels† (lb, min, single proj.)			Fused Diam (min in.)
							to 70,000 psi	70,000 to 150,000 psi	150,000 psi up	
0.010	0.055	0.015	0.033	0.035	0.015	0.005	130	180	250	0.112
0.012	0.055	0.015	0.033	0.035	0.015	0.005	170	220	330	0.112
0.014	0.055	0.015	0.033	0.035	0.015	0.005	200	280	380	0.112
0.016	0.067	0.017	0.042	0.039	0.020	0.005	240	330	450	0.112
0.021	0.067	0.017	0.042	0.039	0.020	0.005	320	440	600	0.140
0.025	0.081	0.020	0.050	0.044	0.025	0.005	450	600	820	0.140
0.031	0.094	0.022	0.062	0.050	0.030	0.005	635	850	1100	0.169
0.034	0.094	0.022	0.062	0.050	0.030	0.005	790	1000	1300	0.169
0.044	0.119	0.028	0.078	0.062	0.035	0.005	920	1300	2000	0.169
0.050	0.119	0.028	0.078	0.062	0.035	0.005	1350	1700	2400	0.225
0.062	0.156	0.035	0.105	0.081	0.043	0.005	1950	2250	3400	0.225
0.071	0.156	0.035	0.105	0.081	0.043	0.005	2300	2800	4200	0.281
0.078	0.187	0.041	0.128	0.104	0.055	0.010	2700	3200	4800	0.281
0.094	0.218	0.048	0.148	0.115	0.065	0.010	3450	4000	6100	0.281
0.109	0.250	0.054	0.172	0.137	0.075	0.015	4150	5000	7000	0.338
0.125	0.281	0.060	0.193	0.154	0.085	0.015	4800	5700	8000	0.338
0.140	0.312	0.066	0.217	0.172	0.096	0.015	6000	....	....	0.437
0.156	0.343	0.072	0.243	0.191	0.107	0.015	7500	....	....	0.500
0.171	0.375	0.078	0.265	0.210	0.118	0.015	8500	....	....	0.562
0.187	0.406	0.085	0.285	0.229	0.130	0.015	10,000	....	....	0.562
0.203	0.437	0.091	0.308	0.240	0.143	0.020	12,000	....	....	0.625
0.250	0.531	0.110	0.375	0.285	0.175	0.025	15,000	....	....	0.687

† Values shown for two thicknesses only, the lighter gage being the critical. Shop welds will vary and machine settings should be made to approximately 20 per cent greater values to obviate any welds below strengths shown.

be used advantageously for attaching large circular parts or bosses as well as a variety of other similar designs.

Thickness Differences. To assure creation of proper fusion temperature at precisely the same instant in both of two parts being welded, it is desirable that projections for parts of unequal thickness be made in the heavier piece. A projection in the thin piece would burn off before the heavy piece was brought up to fusion temperature. It has been found, however, that where the ratio of thicknesses is greater than 3 to 1, it is usually more practical to place the projection in the thinner of the two pieces, regardless. Where formation of projections is only practical in the lighter part, heat balance can sometimes be attained by means of hard electrode material of low conductivity to back up the thin part. Another factor concerns unlike metals being joined; projections should always be formed in the part made of the metal with the higher electrical conductivity, barring other unusual conditions.

Fig. 56.10 — Elongated projections as shown on this refrigeration machine ring and bracket are recommended for welding pieces with curved surfaces. Photo, courtesy General Electric Co.

Solid Projections. It is well to remember that the formation of projections such as those in *Fig.* 56.7 results in a reduction in area through the welded portion which, in some designs, may be undesirable. In many cases projections can be machined, cast, forged or headed during production of the part to avoid the stamped projection. Integral projections, as these are called, can be of an almost unlimited variety. Machined projections may be of the annular, pyramid, elongated or spherical types, or of special design. Some common annular types are shown in *Fig.* 56.13. Special annular projections or bevel forms are used to weld tubing, the type in *Fig.* 56.14*a* being preferred where minimum flash is desirable inside. Where tubing is to be welded to a flat surface and minimum flash is desirable outside, the bevel is usually placed on the inside, *Fig.* 56.14*b*. Solid parts are similarly handled, *Fig.* 56.15. Where parts are not turned, the pyramid or modified pyramid projections frequently can be used, *Fig.* 56.16, to good production advantage as can the elongated types.

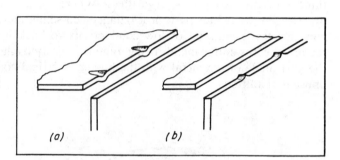

Fig. 56.11 — Types of projections which can be used effectively for end welding corners.

(a)   (b)

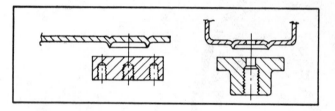

Fig. 56.12 — Annular pressed projections are used for watertight welds and attaching bosses.

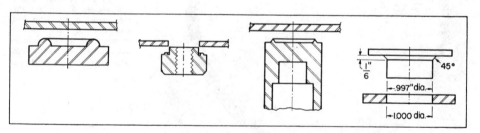

Fig. 56.13 — Some common annular types of integral projection designs which can be machined or formed.

Flashless Projections. On accurate, heavy assemblies in which no flash is desired between the mating parts after welding, a special type of coined projection has been used. Typical projection designs for 1/4-inch and 1/2-inch plate are shown in *Fig.* 56.17. The volume of the depression surrounding the projection is made approximately 70 per cent of the volume of the spherical portion above the surface of the plate.

A variety of projections — annular, spherical, elongated, pyramidal, dome, and multiple — can be produced by coining, forging or heading. Typical varieties are shown in *Fig.* 56.18 which contains a representative group of projection welding fasteners.

Multiple Projections. For welding to sheet 1/8-inch and under, multiple

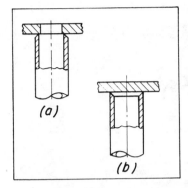

(a)

(b)

Fig. 56.14 — Left — Bevels are used in welding tubular sections to flats. The external bevel (a) results in minimum flash inside and (b) minimum flash outside.

Fig. 56.15 — Right — Ring projection welded assembly showing machined projection. Very little flash results.

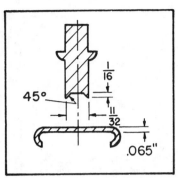

Fig. 56.16 — Modified pyramid projections milled into 7/8 by 2 by 4-inch steel block for a strapping tool.

projections provide optimum results with minimum current, pressure and time. These projections located radially as far from the part axis as possible and equally spaced offer the best results. In using multiple projections, however, it is well to keep in mind the problems encountered in assuring good contact on all projections with groups over three. Obviously with up to 3 projections, slight variations in height would have no deleterious effect on the operation. With more than three, however, projections must be held to very close tolerances on height or difficulty is experienced in obtaining simultaneous collapse. Such conditions are well to avoid.

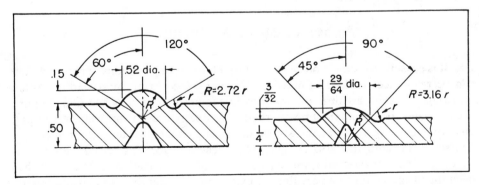

Fig. 56.17 — Special type of projection used for heavy plate where no flash and close fit of parts is desired. Volume of depression is made about 70 per cent of the volume of projection above the plate. Drawing, courtesy Swift Electric Welder Co.

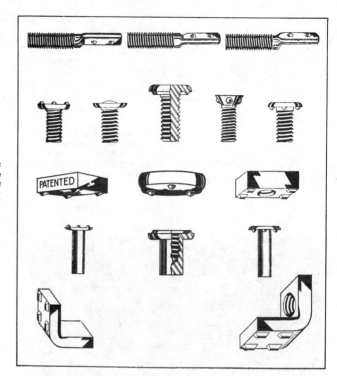

Fig. 56.18 — Typical projections normally produced by forging are exemplified in these standard projection welding fasteners. Drawings, courtesy Ohio Nut & Bolt Co.

Where multiple welds are utilized with two sheet thickness which do not vary more than 3 to 1, distance between projections S, *Fig.* 56.19, should never be less than two times the projection diameter D, (*Fig.* 56.7). Actually, to produce separate welds without lapping of the fused zones, minimum weld spacings recommended for spot welds, TABLE 54-II, should be employed. Overlap of straight welded portions O, *Fig.* 56.19, should be at least equal to $2.4D$ or preferably greater, and so designed that the weld button is centered in the overlap.

Electrode contact on parts should extend at least 5 times the projection diameter in all directions from the weld to assure uniform heat distribution. To simplify the welding of contoured parts, therefore, the area immediately surrounding a projection should be kept free as possible from excessively complex interference or contours.

Selection of Materials

It is possible to projection weld all the metals or combinations of metals which can be resistance welded, as noted in Chapter 54 covering spot welding. Some materials, however, are difficult or impossible to upset in welding and some require preheat and postheat treatments rendering their use in projection welding unsatisfactory. Again, some metals are not strong enough to support a projection. Brasses in general and also copper are undesirable for this reason. Projection welding of copper-base alloys is considered generally unreliable. Aluminum has been found successful to a very limited extent only — mainly with extruded parts. Very thin steel sheet (under 24-gage) is ordinarily more easily spot welded than projection welded.

Economical Materials. Most satisfactory materials for projection welding are, generally:

    Low-carbon steels (carbon 0.20 max)
    Stainless steels (types 309, 310, 316, 317, 321, 347 and
        349, nonhardenable, carbon 0.15 max)
    Copper-silicon and copper-tin alloys
    Nickel and nickel alloys
    Monel

Any of these materials can be used together, dissimilar combinations being readily weldable. Projections, of course, should be made in the metal with the highest conductivity.

Optimum welds on all materials are obtained with parts bright finished, that is, clean surfaces without scale, foreign materials or plating. Where assemblies are too large or complex for plating after welding, parts can be plated separately prior to welding. Zinc plating (electrogalvanizing) is the best and does not materially affect the welding operation. As in spot or seam welding, such finishes as Parkerizing, black oil finish, etc., should not be used inasmuch as they are poor electrical conductors. Where parts plated or to be plated on one side are to be projection welded, the effect of the operation on the plating or subsequent plating operation should be investigated to preclude difficulties in production.

Specification of Tolerances

Owing to the fact that all welds in projection welding are made simultaneously, the process lends itself to fixturizing. Too, many parts can be welded accurately by means of only the positioning afforded by the electrode platens. Assembly tolerances obtained are directly dependent upon the fixtures used and to assure minimum part cost, widest possible tolerances should be

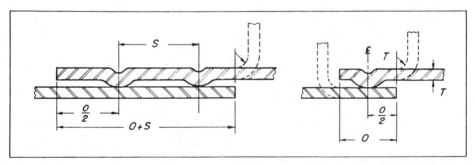

Fig. 56.19 — Standard AWS arrangement of weld laps and weld positioning. Overlap must be flat and not include any radii from forming.

allowed. Accurate relative location of parts in assembly can also be assured by means of semipunchings in one part which fit into mating holes in the other. Such details can usually be produced simultaneously with the projections.

In materials up to and including 0.050-inch thickness, the projection diameter, *D*, **Fig.** 56.7, is usually held to a tolerance of plus or minus 0.003-inch and in heavier materials, to plus or minus 0.007-inch. The height *C* is normally held to a tolerance of plus or minus 0.002-inch in gages up to and including 0.050-inch while plus or minus 0.005-inch is allowed on heavier gages.

Regardless of conditions, the strength of production welds will vary to some extent. Consequently, specifications and machine settings should be at least 20 per cent greater than the minimum strength values shown in TABLE 56-IV to eliminate the possibility of poor below-strength welds.

57

Butt Welding

FOURTH and last in the series of major resistance welding processes is butt welding. Consisting primarily of welding end-to-end or abutting pieces of metal, butt welding differs from spot, seam and projection welding in the fundamental procedure utilized to obtain the heat required for welding. Parts are clamped in current-carrying dies, one of which is powered to bring the two pieces together. By either joint resistance or arcing and flashing, the ends of the parts are brought to a temperature within the plastic range of the metal, allowing pressure from the moving die to upset and forge the parts into a single unit, *Fig.* 57.1.

Field of Application. Butt welding permits the welding together of rods, bars, sheets, tubes, stampings, forgings, channels, angles, and an almost limitless variety of complex parts so designed as to allow satisfactory handling by this method. Practically all the metals capable of being spot, seam or projection welded can be readily butt welded.

Welding Process. Unlike the other resistance methods, in butt welding the welding occurs simultaneously over the entire area of contact providing an

Fig. 57.1 — Flash welding trailer siderails into continuous lengths. Photo, courtesy Fruehauf Trailer Co.

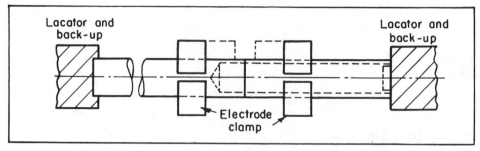

Fig. 57.2 — Typical arrangement of parts to be welded by the upset or flash-butt method and a front view into a typical flash welder. Courtesy, Wean United Inc.

extremely uniform, homogeneous joint having a strength factor approximately equal to that of the parent metal itself. The uniform heating and cooling throughout the weld zone minimizes trapped, localized stresses and the operation is completed with such speed that portions beyond the weld zone are not heated sufficiently to cause warpage, scaling or heat oxidation. As with the other resistance methods, no metal is added and chemical composition remains virtually unchanged. Material losses due to upsetting action are usually small.

Butt Welding Methods. Butt welding normally can be subdivided into two major classes: (1) Upset-butt welding, and (2) flash-butt welding. While flash-butt welding is probably the most versatile and widely used, each method has specific attributes which afford a broad overall range of application.

Upset-Butt Welders. With upset welders the full upset pressure is first applied to the work to be joined after clamping in the electrodes, *Fig.* 57.2. Following the application of pressure, current is passed through the part for a sufficient length of time to heat the portions to be welded. Welding temperature is achieved by resistance only and the applied pressure upsets or forges the heated portions together. Current is cut off as the forging action is completed.

A variety of upset welders is available ranging from hand-operated units in which both clamping and upsetting are hand actuated, to mass-production machines in which current application and upsetting action are fully automatic. Upsetting force may be provided by pneumatic or hydraulic power or mechanically through gears and cams. For upset-butt welding small sections and copper wire or rod, a spring operated pushup mechanism is often used for quick and automatic forging pressure. In some cases pressure is applied by means of weights and levers.

Flash-Butt Welders. With flash welders the parts to be joined are clamped in the electrodes much the same as in upset welders but with the ends of the parts not quite touching. Current is applied simultaneously with a low upsetting pressure and as the metals touch, arcing and flashing take place creating considerably greater heat than by contact resistance only. As the flashing continues, vapor explosions expel slag and other impurities and finally when the metal has reached welding temperature the upsetting pressure is suddenly increased, extinguishing the arc and forging the pieces into a strong weld.

Fig. 57.3 — Synchro-Matic flash welder specially designed for automatically welding typewriter frames. Photo, courtesy Thompson Electric Welder Co.

Flash welders follow closely the styles available in upset welders, in fact, both classes of welds may be made on the same machine by changing the method of operation. While manually operated machines are available they are limited to small cross sections and general work, automatic power-operated machines being preferable for production runs and more consistent quality, *Fig.* 57.3. Spring, gear and cam, pneumatic, or hydraulic drives are normally provided as with upset machines. The standard range of machine capacities as set up by the Resistance Welder Manufacturers' Association is listed in TABLE 57-I. Generally, machines with capacities up to 1200 kva are available. Special machines of a much wider range in capacity, however, have been manufactured and are extensively used, *Fig.* 57.4.

Percussive Welders. Also confined to the production of butt-welded joints, percussive welding is a rather specialized method principally used for butt welding small-diameter wires. With these machines, an intense electrical discharge occurs simultaneously or just prior to the application of pressure to the parts in the form of a hammer blow. Heating is accomplished by the playing of the arc across the metal

Fig. 57.4 — Formed and drawn automotive rear axle halves showing welder loaded and in the flashing operation. Photo, courtesy Chrysler Corp.

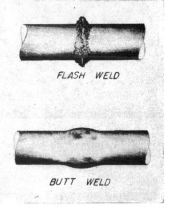

FLASH WELD

BUTT WELD

Fig. 57.5 — Typical flash-butt and upset-butt welds.

faces, resistance of the metal being of little consequence. Because of this, materials impossible to weld by upset or flash methods, such as copper to stainless for instance, can be readily welded. Fusion, being confined to the surface of the parts, results in little or no flash. Areas to be welded must be concentrated as in round or square sections, thin or wide faces being impractical. Total area must not be greater than 1/2-square inch but most applications are mainly with wire parts under 1/8-inch in diameter.

Use of Welders. The primary advantage of flash over upset welders is their lower current demand. Where large areas are to be welded, the upset method is apt to require excessive amounts of current. As a consequence, upset welding is utilized primarily for welding small and medium-size cross sections, whereas the flash method permits the joining of section areas of 80 or more square inches in low-carbon steel. Flash-butt welding, however, is extremely versatile and need not be confined to large sections. Usual machine capacities make possible the welding of areas from 0.10 to 28 square inches. However, areas as small as 0.0015 square inches can be welded and machine capacities may range from 1/8-kva on up. Wire as small as 0.004-inch in diameter has been upset-butt welded in production.

Selection of Process. Although either process may be optional on certain designs, generally such is not the case and actually there is no clear line of demarcation between the two methods. In some instances the material may determine whether an upset or a flash-butt weld is to be used; again it may be the design of the particular pieces to be welded.

TABLE 57-I — RWMA Specifications for Standard Butt Welders

Size No.	Rating (kva)	Upset Pressure (lb, max)	Platen Opening (in.)	Platen Size (width x length, in.)
		Upset-Butt Welders		
000	2	12	$\frac{9}{16}$	1 x 1⅜
00	5	70	$\frac{1}{2}$	1½ x 3
0	10	120	$\frac{7}{8}$	1¾ x 4
		Flash-Butt Welders		
1	20	2250	2	12 x 12
2	50	4500	2½	16 x 16
3	100	11,500	3	20 x 20
4	150	19,600	4	24 x 24
5	250	38,000	5	28 x 28

The upset-butt weld is primarily adapted for welding ferrous materials which can be joined at temperatures below the melting point. Where the flash spatter and ragged upset characteristic of flash welding is undesirable or impractical to remove, the upset-butt weld often proves more satisfactory, although the upset is usually greater, *Fig.* 57.5. The flash method, however, is satisfactory for welding many nonferrous metals and alloys not possible to handle by means of the upset-butt method. If complete removal of the excess metal is necessary, it is generally much less expensive and less difficult to remove the flash from a flash weld than it is to form an equivalent upset weld.

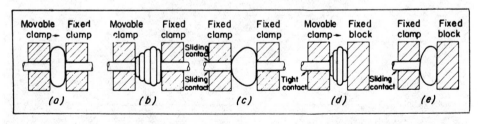

Fig. 57.6 — Upsets produced on butt welder: (a) Single, (b) multiple pushup, (c) large sliding pushup, (d) multiple end, (e) large sliding end. Drawing, courtesy Thompson Electric Welder Co.

Electric Upsetting. As mentioned in Chapter 34 covering hot upsetting operations, the upset-butt welder can be used to produce gathers or upsets on bar stock, much like the electric gathering machine. Electric upsetting to increase the stock diameter at specific points can be applied to low-carbon steel, aluminum,

Fig. 57.7 — Upset welder dies showing a flat blanked part riveted to bar by means of electric upsetting on each side of a hole.

stainless steels, bronze, brass, Monel, and like materials. Size of upset or volume of gather is dependent upon the die movement of the machine. By proper control of the current, location of the workpiece, selection of the die materials for

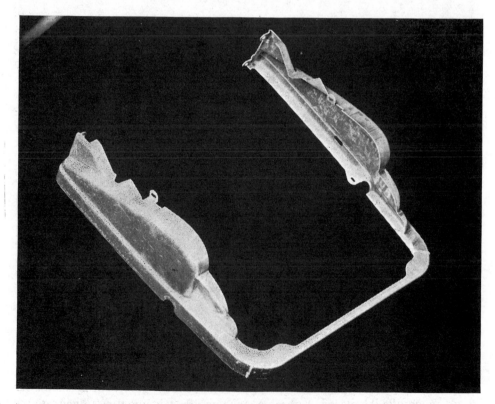

Fig. 57.8 — Stamped typewriter mask composed of three sections flash welded into a single unit on the machine shown in Fig. 57.3. Photo, courtesy Thompson Electric Welder Co.

conductivity, etc., uniform annular upsets, eccentric upsets, multiple upsets, spherical, or narrow upsets can be produced, *Fig.* 57.6. Special shapes may be obtained by the use of special dies designed to confine the material to the desired contour during upsetting. Electric riveting can also be done and in many instances upsets can be employed for fastening parts to rod or tubing economically, *Fig.* 57.7.

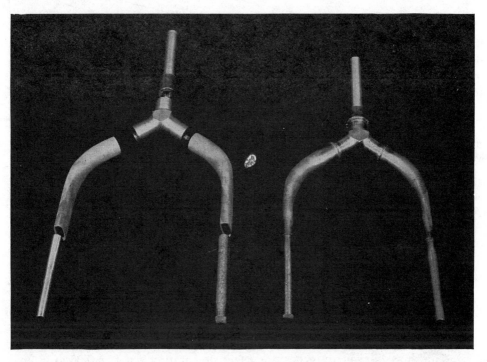

Fig. 57.9 — Aircraft link assembly unit comprised of shaped bar, formed tubes, machined forgings, and turned barstock. Welding flash is readily visible on completed assembly. Photo, courtesy McDonnell Douglas Corp.

Design Considerations

Owing to the wide range of application afforded by butt welding methods, an unusual degree of flexibility in design is made possible. Fabrication of tubular parts, combination tubular and solid fittings, combination sections, and composite parts made up of stampings, castings and forgings or arrangements of several in similar as well as dissimilar metals is readily accomplished by these methods, *Figs.* 57.8 and 57.9. Also possible are combinations of electric upsetting and butt welding to complete a part, *Fig.* 57.10. Utilization of butt welding to obtain forged assemblies too complex as a unit but with individual parts well within the practical range is also an excellent means for cost reduction, *Fig.* 57.11.

In order to obtain full advantage of the many economies afforded by butt welding, proper design of component parts is essential. Careful adherence to the limitations normally present and attention to those design requirements which are imperative for satisfactory welds will yield highly efficient results.

Upset-Butt Limits. As previously noted, the upset-butt method is primarily confined to the welding of small to medium sections in ferrous materials. Owing to

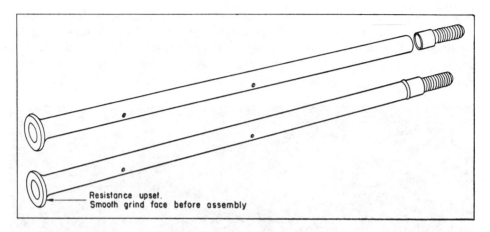

Fig. 57.10 — Flash welded tubular aircraft strut on which the end is electric upset to form a flange.

greater accuracy necessary in preparing upset-butt welded surfaces to assure equalized contact resistance, greater current requirements, larger upset produced, and somewhat lower weld strength, flash welding is preferred and more widely used. However, the advantages which may be important in design are mainly that the flash spatter encountered with flash welds is not present and the upset or bulge produced is smooth and symmetrical with little or no extrusion of the metal, *Fig.* 57.5.

Flash-Butt Areas. Cross-sectional areas for flash welding may range to approximately 30 square inches commonly. Some machines available increase this to around 80 square inches and in special-purpose machines this has been increased to as much as 240 square inches. Joint face preparation is not critical as in upset welding inasmuch as the arcing and flashing action burns away the irregularities to produce normal mating cross sections, *Fig.* 57.12.

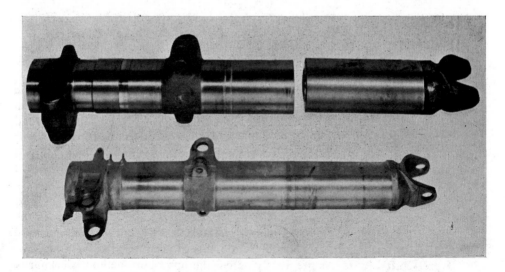

Fig. 57.11 — Aircraft shock absorber cylinder forged in two pieces and flash-butt welded to simplify forging.

Heat Balance. As in other resistance welding methods, heat balance is of extreme importance in design of butt-welded joints. The cross-section, area of contact and contour should be made identical or as nearly identical as possible on both pieces. Exceptions to this rule have been successfully handled but, in general,

TABLE 57-II — Common Ranges of Flash and Upset Welding Pressures

Material	Thickness (in.)	Upset Pressure (psi, min)	(psi, max)
Aluminum			
Soft alloys	Up to ⅝	4500	5500
High strength alloys			20,000
Copper			
Plain	Up to ¾	5000	7500
Steels			
AISI C1020	Up to ⅜	6000	8000
AISI C1020	⅜ to 1	7000	10,000
AISI C1020	1 to 2	9000	11,000
AISI C1020	2 to 4	10,000	12,000
AISI 2340	Up to 1	10,000	12,000
AISI 3140	Up to 1	10,000	12,000
AISI 4140	Up to 1¼	11,000	18,000
AISI 6145	Up to 1¼	11,000	18,000
Stainless alloys	Up to 2	13,000	25,000
High-temperature alloys			35,000

the ratio of weld areas should not exceed 5 to 4. The mass of metal should be balanced so as to permit easy gripping and accurate alignment in welding. Good and poor design practices in regard to heat balance are shown in *Fig.* 57.13. Multiple flash-butt welds on the same part are practicable where the elements are of the same thickness and variation in areas not greater than 5 to 1.

Fig. 57.12 — *Drop forgings flash welded together to form a component for a weldment. Photo, courtesy Lukens Steel Co.*

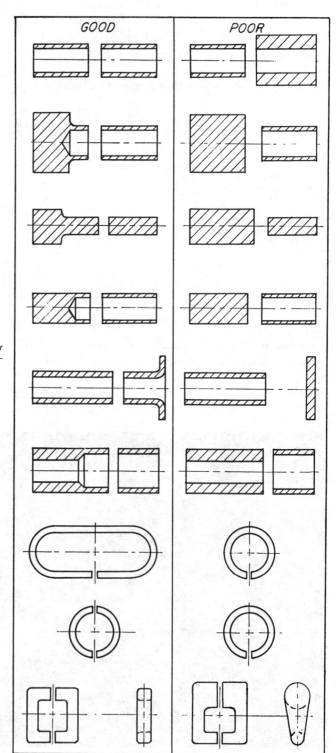

Fig. 57.13 — Good and poor design practices regarding butt-welded joints.

Sheet and Tubing. To assure good production design, satisfactory welds and economy, certain limitations on length of joint with flat sheets and diameter of tubing must be observed. In *Fig.* 57.14, the recommended maximum joint length for flat sheets is shown and in *Fig.* 57.15 is that for tubing. Other sizes can be run but ordinarily incur excessive problems in production or result in only fair or poor joints.

TABLE 57-III — **Weld Allowances for Flash Welding Steels** *

(material lost, inches)

Dimension	Tubes and Sheets	Solid Bars
0.010	0.060	
0.020	0.115	
0.030	0.175	
0.040	0.230	
0.050	0.280	0.050
0.060	0.330	
0.070	0.385	
0.080	0.435	
0.090	0.475	
0.100	0.520	0.082
0.110	0.570	
0.120	0.610	
0.130	0.650	
0.140	0.700	
0.150	0.730	0.120
0.160	0.770	
0.170	0.800	
0.180	0.840	
0.190	0.870	
0.200	0.900	0.150
0.250	0.010	0.180
0.300	0.120	0.210
0.350	1.210	0.250
0.400	1.290	0.285
0.450	1.350	0.320
0.500	1.410	0.350
0.550	1.465	0.390
0.600	1.505	0.425
0.650	1.555	0.450
0.700	1.610	0.480
0.800	1.675	0.540
0.900	1.730	0.600
1.000	1.800	0.660
1.050		0.690
1.100		0.720
1.150		0.750
1.200		0.780
1.250		0.810
1.300		0.840
1.400		0.900
1.500		0.960
1.600		1.020
1.700		1.080
1.800		1.140
1.900		1.200
2.000		1.260

* Courtesy, Resistance Welder Manufacturers' Association. Data based on welding without preheat two pieces of similar material.

Clamping Pressures. Design of parts to be butt welded should always take into consideration the clamping pressures required. The cross section must be sufficiently strong to withstand the pressure without distortion or crushing and to assure good electrical contact within the clamps. Common practice is to provide a clamping pressure 2¹/2 to 3 times the upsetting force. Coefficient of friction between the jaws and the workpiece may also be of importance. For instance, the clamping pressure with stainless steels must be approximately 40 to 50 per cent higher than for mild steels owing to the lower coefficient of friction present. Upset pressure for upset-butt welding ranges from 2500 to 8000 pounds per square inch of weld area and for flash-butt welding, from 5000 to 25,000 or more, depending upon the materials. Pressures commonly used for various materials are shown in TABLE 57-II, the lower values generally being employed with upset-butt work.

Upset Losses. An allowance must be made in design to provide for the flash or upset losses in forging together the two heated parts. Total allowance ranges from one to six times the stock thickness and generally the ratio of the amount of metal loss to the gage thickness decreases as the gage increases. The final pushup allowance is normally about one-fourth to one-third the total loss. In TABLE 57-III the amounts generally encountered in average production practice are tabulated and apply where ratio of maximum to minimum sectional dimension does not exceed 1.5 to 1. To be considered also along with upset loss is length of section necessary to accommodate the clamp. Usual clamp lengths are shown in TABLE 57-IV.

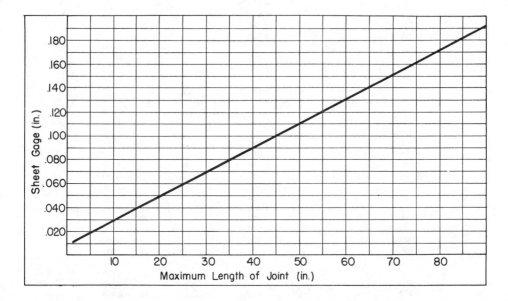

Fig. 57.14 — Chart giving RWMA standard relationship between sheet thickness and maximum length of joint.

Angle Joints. In welding corners or angle joints, it is well to avoid compact sections such as bars, squares, rounds, etc., where the angle between the parts is less than 90 degrees. To assure a satisfactory joint with good fusion at the outer corner, the angle should seldom be less than 150 degrees. Flat sections, however, can be welded satisfactorily to an included corner angle of 90 degrees if the length of joint is at least 20 times the stock thickness. Stampings or other formed sections of light gages are readily handled also, *Fig.* 57.8.

End Preparation. For joining two solid parts, preparation of a matching extension on one of the pieces is recommended, *Fig.* 57.13. The length of extension to be used on some plain diameters may be found from TABLES 57-III and 57-IV. With solid sections larger than 1/4-inch in OD, a small bevel is desirable on one of the pieces, *Fig.* 57.16, to aid in starting the flash. Very large sections may be beveled to a point at the angles shown. Solid ends should be plain, i.e., without holes. Where turned parts have centers of appreciable size or depth in the ends to be welded, these should be removed to avoid poor fusion. Tubular extensions or joints should be beveled where wall gage is over 3/16-inch, *Fig.* 57.17. If the

TABLE 57-IV — **Clamp Widths for Flash Welding**†

Dimension	Width with Locator	Width without Locator
0.250	0.375	1.00
0.312	0.375	1.00
0.375	0.375	1.50
0.500	0.375	1.75
0.750	0.500	2.00
1.000	0.750	2.50
1.50	1.000	3.00
2.00	1.25	*
2.50	1.75	*
3.00	2.00	*
3.50	2.25	*
4.00	2.50	*
4.50	2.75	*
5.00	2.75	*
5.50	3.00	*
6.00	3.25	*
6.50	3.50	*
7.00	3.75	*
7.50	4.00	*
8.00	4.25	*
8.50	4.50	*
9.00	4.75	*
9.50	5.00	*
10.00		*

† Courtesy Resistance Welder Manufacturers' Association.
* Not recommended without use of backups.

variation in wall thicknesses exceeds 15 per cent, it may be advisable to consider a ring type projection weld instead, inasmuch as the wall variation for flash welding should be well below 15 per cent and preferably the areas should be equal. With flat sheets of 3/16-inch gage or greater, the face of at least one sheet should be beveled as shown in *Fig.* 57.18 to faciliate welding. The bevels and face of parts designed for flash welding need not be precise or extremely flat and with good surface contact as with upset parts. Cast, forged or machined bevels and faces are equally suitable.

Selection of Materials

All the various materials which are readily weldable by other resistance methods can be handled in flash welding. Upset-butt welding, as mentioned previously, is largely confined to ferrous materials. Upset welding of copper, copper alloys, magnesium and aluminum is not considered reliable and in some instances is impractical except with small-diameter parts. Aluminum alloys 2S, 3S, 53S and 61S are generally weldable by the flash-butt method but for mechanical reasons must be in sections greater than 0.050-inch thick. Of the brasses, high-

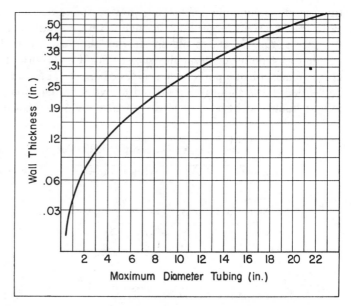

Fig. 57.15 — Chart showing
RWMA standard relationship
between wall thickness and
maximum diameter of tubing
practicable.

leaded alloys may give brittle welds and, in general, brasses with high zinc content weld better than those with high copper content. As noted in the previous chapters on resistance welding, weldability can in part be judged by the conductivity of the metal involved. Those high in electrical conductivity require much greater currents. Thus current required for aluminum is approximately 100,000 amperes per square inch whereas that for mild steel is only 20,000 to 30,000.

Weldability of Steels. The weldability of steels is considered in great degree from the number of steps required to insure a good weld. Low-carbon steels are readily weldable without any further treatment. As carbon and alloy content increases it becomes imperative to preheat or postheat or both to get maximum weld soundness. All types of steels are readily welded, though some combinations may require special procedures. Ease with which the various groups are forged can be judged on the basis of the following classifications:

Low forging strength steels (upset pressure 10,000 psi) AISI 1020 1112, 9115, etc.
Medium forging strength steels (upset pressure 15,000 psi) AISI 1045, 1065, 1335,
    3135, 4130, 4140, etc.
High forging strength steels (upset pressure 25,000 psi) AISI 4340, 4640, stainless iron
    (12 per cent chromium), stainless steel (18-8 cutlery), high-speed steel, etc.
Extrahigh forging strength steels (upset pressure 35,000 psi) All steels with extra high
    compressive strength at elevated temperatures.

Surfaces. Ordinarily, with upset welding, rolled, forged or cast parts must be machined to assure clean, smooth and well fitting joints. For flash welding however, no special face preparation is necessary. Electrode contact surfaces on rolled stock, stampings or grit cleaned forgings are satisfactory. Rust, scale, grease, or other deposits should be avoided as in other resistance work.

Metal Combinations. Dissimilar metals are most easily welded by the flash process and require less critical preparation. Even refractory type metals such as tungsten, molybdenum, and tantalum can be welded to other materials such as iron-base alloys, copper-base alloys and nickel alloys. Flash welding produces joints with the greatest strength and uniformity.

Percussive welding, although severely restricted as to size and complexity, makes possible the butt welding of round or tubular sections together or to flat surfaces in otherwise unweldable metal combinations. Steel can be welded to aluminum or magnesium, silver to copper, Stellite to ferrous or nonferrous materials, copper to aluminum, stainlees steels or other corrosion-resistant materials to ferrous or nonferrous materials and, where necessary, conventional metal combinations on which flash is not acceptable.

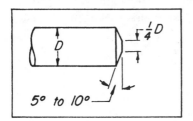

Fig. 57.16 — RWMA standard bevel design recommended for solid pieces over ¹/₄-inch in diameter.

Fig. 57.17 — RWMA standard tube-end preparation recommended for tubing over ³/₁₆-inch in wall thickness.

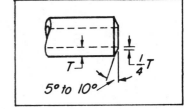

Fig. 57.18 — RWMA standard end preparation recommended for sheet over ³/₁₆-inch in gage thickness.

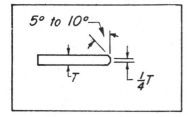

Specification of Tolerances

Dimensional accuracy with butt welding can be unusually close and, as a rule, it is better with flash than upset work. Final accuracy of parts produced, however, will depend upon the accuracy of the parts prior to welding. Where close tolerances are necessary, the locating points and surfaces on the pieces and on the welder dies must be of equal or greater accuracy than that desired on the final weldment.

Offset Tolerances. Diameter tolerances affect directly the offset or misalignment of the welded parts. Tubular or round ends, for instance, held to plus or minus 0.005-inch would result in a final possible misalignment or offset of 0.010-inch. In general, parts under 1-inch in diameter prepared with a tolerance of plus or minus 0.002-inch can be held to a concentricity of plus or minus 0.005-inch. With a tolerance of plus or minus 0.010-inch on a 6-inch diameter by ³/₄-inch wall tube, a concentricity of plus or minus 0.030-inch can be held. Very small parts centerless ground to close limits are often practical to hold to total runout of 0.004-inch, maximum.

Length Tolerances. Length or overall dimension of a welded assembly is also influenced by the preset tolerances of the individual parts. If welded without locators, the final length tolerance after welding will be equal to the sum of the tolerances before welding. If backup locators are used, the final length tolerance will normally be less than that of one piece; the accuracy being largely set by the locator spacings. For best results, length tolerance specified on parts to be welded should be kept to plus or minus 0.010-inch or better. Final length tolerance after welding in such cases will be less than plus or minus 0.010-inch. Flash welded parts with two joints to take into consideration, under the same preliminary tolerances, can be held to plus or minus 0.031-inch or better on overall length. Where greater variations in production lengths are acceptable, of course wider preliminary tolerances can be used with a definite economy.

58

Brazing

THE MANIFOLD advantages of brazing, both designwise and economically, have been recognized for many years. Furnace brazing on a mass production scale was first introduced commercially about 1930. Developments in the succeeding years have broadened the range of application and expanded the basic brazing alloys available to such an extent that today almost any metal or metals can be brazed by any one of four basic methods, *Fig.* 58.1.

Basic Process. Generally, the term brazing is applied to that group of welding processes in which metal components are joined together by means of a nonferrous metal or alloy bonding medium having a melting point over 800 F but less than that of the parent metals or alloys. In the heating operation the brazing medium melts, wetting the surfaces to be joined and creeping into and through each joint by capillary attraction. A limited amount of alloying with the parent metal takes place and any slight excess of brazing medium usually forms a neat fillet at the outer extremity of each joint. Brazed joints, properly made, have a resultant strength somewhere between that of the cast brazing metal and that of the base metal.

Advantages. Adaptable to the joining of a wide variety of ferrous and nonferrous metals or alloys as well as certain refractory metals, brazing in many cases offers the only satisfactory answer to the problem of economical design of complex assemblies, *Fig.* 58.2. Further opportunities lie in the joining of different types of separate components — screw machine parts, stampings, cold-headed parts, tubular sections, forgings, and machined pieces most easily produced by

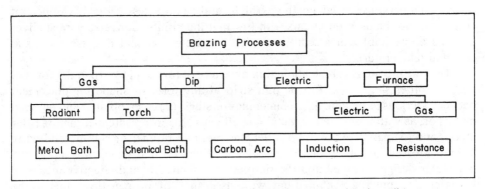

Fig. 58.1 — General breakdown of basic brazing processes and their various offshoots.

specific separate methods — into finished assemblies not practical to complete without recourse to a formidable number of more expensive manufacturing operations, *Fig.* 58.3. Of equal importance is the fact that a decidedly higher rate of production per unit of labor invariably results with parts properly designed for the brazing process.

Brazing Mediums. A wide variety of brazing mediums are available and selection depends mainly upon the application and parent metals involved. In addition to the time-honored copper and brasses with which the brazing process started, mediums now include a wide selection of silver brazing (sometimes called

Fig. 58.2 — A group of furnace-brazed assemblies indicating the complexity of design possible. Photo, courtesy Westinghouse Electric Corp.

solders) alloys as well as those for aluminum and magnesium, the most recently developed. A general cross section of these brazing mediums is listed in TABLE 58-II along with some of their major properties and characteristics.

Both silver and copper brazing are widely employed. In general, silver brazing is less expensive than copper brazing inasmuch as somewhat less work is involved, furnace temperatures need not be as high, and fits are less critical. Copper, of course, cannot be used on metals with low melting points. Conversely most silver alloys are unsuitable for assemblies which require subsequent heat treatments at elevated temperatures.

Brazing mediums can be applied in a variety of forms: (1) Wire in the form of rings or other shapes; (2) foil or thin strip of any form or shape for placement between sections to be joined; (3) punchings or slugs; (4) electroplated and sprayed metals or, as with aluminum, a metal coated with an appropriate medium or filler metal; and (5) pastes of copper powder, lacquer and thinner or commercial paste alloys. Most desirable form of brazing medium normally depends upon the particular design involved and the simplest, most suitable and effective means of inserting or applying the medium with assurance of proper flow during the operation.

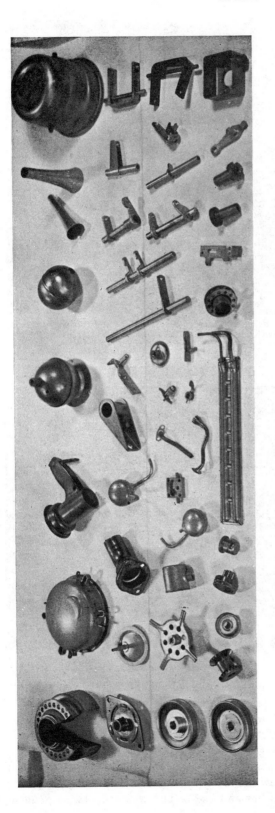

Fig. 58.3 — Groups of typical furnace-brazed parts produced in large quantities. Photo, courtesy Electric Furnace Co.

Fig. 58.4 — The machine for bronze brazing the rear fan assembly to the seat post and to the crankcase housing (six joints all together) on 26-inch and 27-inch boy's and girl's bikes. The torches and flow meters are shown at the two preheat stations. Photo, courtesy Norbell Corp.

Fig. 58.5 — Removing a fixture of copper-brazed parts from a salt-bath furnace. Photo, courtesy A. F. Holden Co.

Brazing Methods. In its present state of development some seven separate methods of completing the brazing operation are in use. The method by which parts to be brazed are heated to melt and flow the brazing medium depends greatly upon factors such as the particular design, its complexity, quantity to be produced, etc.

Gas Brazing. Preferential heating of assemblies at the portions to be brazed is accomplished by means of torch or gas-radiant heating. Gas brazing is used for soft solders, phos-copper, aluminum, low fuming bronze and silver brazing alloys. By avoidance of heating entire assemblies and use of fast wire feeders, subsequent substantial cost savings are often important, *Fig.* 58.4. In the brazing of many assemblies, gas torches — oxyacetylene, oxyhydrogen, oxycommercial or commercial gas — either hand operated or machine mounted can be used to advantage. Automated machines are most economical.

Dip Brazing. Brazing of parts by dipping is carried out by either of two methods. In metal-bath brazing, the fluxed parts are immersed in a molten bath of brazing medium. Parts are suspended in the bath to the required depth and held until they attain bath temperature, after which immersion in a cleaning solution removes the flux. Only relatively small parts are handled in this manner for obvious reasons. Chemical-bath or salt-bath brazing is adapted to larger parts and consists of immersing the parts to be brazed in a molten salt solution, *Fig.* 58.5.

Fig. 58.6 — Compressor valve plates and components parts, including brazing wires, salt-bath copper brazed to avoid decarburization of the material. Photo, courtesy A. F. Holden Co.

Salt-bath brazing has the advantage of heating parts rapidly and simultaneously, providing complete protection against decarburization. In fact, parts made of suitable steel may be silver brazed and carburized simultaneously, although the case is usually only 0.001 to 0.005-inch thick. Copper brazing in salt baths is especially advantageous with medium-carbon, alloy and high-carbon steels inasmuch as the decarburization sometimes encountered with atmosphere furnaces is not present, *Fig.* 58.6. In addition, no fluxing of the parts is necessary with salt-bath copper brazing.

Electric Brazing. Three methods of electric brazing utilized are: Arc, induction and resistance. Arc brazing consists of heating the joint by means of an electric arc formed between the base metal and an electrode or between two electrodes.

Induction brazing offers an advantage in that only shallow annealing of the parts is produced with little oxidation and distortion. Preferential heating of only the local braze area, as with gas brazing, reduces the heating requirements and makes possible rapid joining of complex assemblies, *Fig.* 58.7. However, special heating coils are normally required for each job, *Fig.* 58.8.

In resistance brazing, the third of the electric methods, heat is generated by passing an electric current through the parts to be brazed. Regular resistance welding machines can be used or the current can be applied through a set of carbon blocks, *Fig.* 58.9. Resistance welders offer accurate timing and very high-speed brazing on small assemblies. With either resistance method, heating is performed under pressure maintained until the metal solidifies; clean, neat joints are obtained with little heat effect on the parts.

Fig. 58.7 — Farm machine sprocket assembly designed for induction brazing with silver alloy. Photo, courtesy Induction Heating Corp.

Fig. 58.8 — Induction brazing equipment showing parts in place in induction coils. Photo, courtesy Induction Heating Corp.

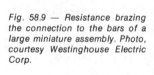

Fig. 58.9 — Resistance brazing the connection to the bars of a large miniature assembly. Photo, courtesy Westinghouse Electric Corp.

Furnace Brazing. Both electric and gas fired brazing furnaces in box, wire-mesh belt and roller hearth types are used, *Fig.* 58.10. The primary advantages of furnace brazing are low costs, flexibility allowing the brazing of light and heavy sections, extremely high production speed, capacity to handle any brazing medium, no need for fluxing except with brass (or other alloys containing zinc) and aluminum, and no limit to the number of joints which can be brazed simultaneously. Perhaps most important is the factor of multiple-joint brazing; all joints on an assembly can be completed at once regardless of their inaccessibility, nonsymmetrical shape, depth, or number, *Fig.* 58.11. Heat treatments such as normalizing, carburizing or hardening of ferrous materials can be made by proper control of the brazing cycle and furnace atmosphere.

Fig. 58.10 — Discharge end of a wire-mesh belt furnace in production brazing of automobile generator pulleys. Photos, courtesy Electric Furnace Co.

Fig. 58.11 — Below — Controlled atmosphere electric and fuel fired furnaces permit the simultaneous brazing of almost any number of joints in complex units such as these domes for hermetically sealed compressors.

Design Considerations

The first consideration in the design of brazed parts is that of the type of joint to be specified. Some factors of joint design depend upon the brazing medium to be utilized. Copper brazing demands extremely close fits and silver alloy brazing allows slight clearances but in no case does high strength emanate from filling large spaces or from large fillets.

Strength of Joints. The strongest, most satisfactory and economical joints are obtained by using small clearances. Shear strength of joints, in general, averages from 25,000 to 40,000 psi depending upon the joint clearance. Strength of copper-brazed joints in steel assemblies normally range from 27,000 to 33,000 psi with proper clearances. However, this strength may drop to as low as 22,000 with a considerable gap. Silver-brazed joints will often test with a minimum shear strength of 35,000 psi. Where design and production is rigidly controlled such as with primary components for aircraft, tests actually show a mean shear strength of approximately 43,000 psi. Overall values on such parts range from 38,000 to 52,000 psi.

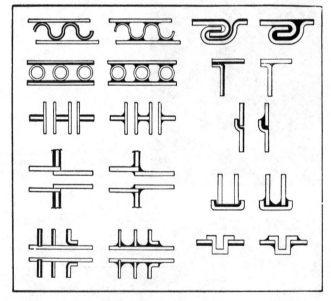

Fig. 58.12 — Typical lap joints utilized in designing brazed assemblies, below. Typical joints with braze-clad sheet, before and after brazing, are shown at the right.

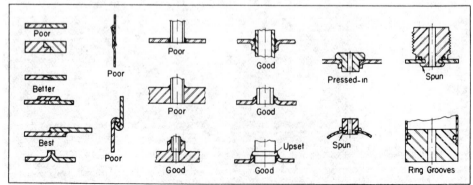

Types of Joints. The three fundamental types of joints which can be utilized in designing brazed assemblies are the butt, scarf, and lap or shear. Butt joints are the least desirable and, if at all possible, should be avoided. Where there is no alternative, butt joints should be carefully designed to assure a close fit and good flatness over the joint surface. Scarf joints are not much more satisfactory but can be used if the scarf angle is fairly acute. Stepped scarfs are also acceptable.

Joint Design. Lap joints are the most useful and satisfactory and are the only acceptable joint for primary aircraft and similar components. The lap can be made whatever length is necessary to maintain adequate strength and factor of safety. In general, laps in copper, brass and other nonferrous alloys with a depth approximately three times the gage will produce a joint as strong or stronger than the parent metal. With metals of different gage the lap depth is generally based on the thinner piece. Where dissimilar metals or critical parts are to be joined it may be desirable to compute the required amount of overlap as follows by equating the strengths on a total or unit length basis:

$$Sb = bls/f \text{ or } l = Sf/s$$

where S = the strength of the weakest member (tensile strength x thickness) in psi; l = the depth of lap in inches; b = the circumference, length, or unit length of the lap in inches; f = the factor of safety; and s = the unit shear strength of the brazing alloy used. A unit shear strength of 25,000 psi is normally used for common brazing alloys and the factor of safety may range from 2 to 5. For low stresses and static loadings, a value of 2 is satisfactory but for primary structures involving dynamic stresses, a higher value should be specified to suit the conditions. Either flat or tubular lap joints may be calculated with the preceding formula and the load must be equally distributed over the entire braze area.

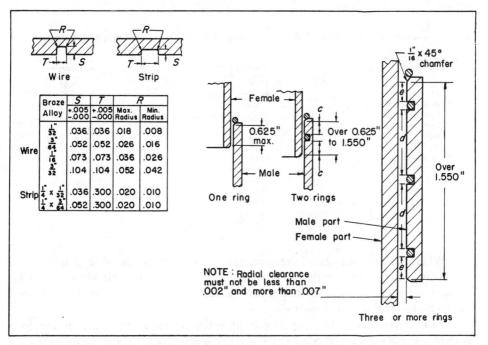

Fig. 58.13 — Lap joint design for primary structures showing groove design for wire and ribbon. Spacings for maximum capillary rise are important. Drawing, courtesy Beech Aircraft Corp.

Typical lap type joints generally utilized in brazing design are shown in *Fig.* 58.12. For maximum economy in production, joint depth or overlap should be no greater than necessary. Generally, it should not be required that the brazing metal from one preplaced ring, slug or paste deposit penetrate or creep between more than 5/8-inch of overlap where silver-brazed steels are involved. Copper brazing of steel is not as restricted; copper at furnace temperatures of 2000 to 2100 F is highly

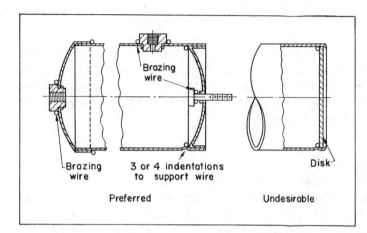

Brazing wire

Brazing wire

3 or 4 indentations to support wire

Disk

Preferred

Undesirable

Fig. 58.14 — End head design for cylindrical parts with area lap is preferable. Drawing, courtesy Electric Furnace Co.

fluid and penetrates more deeply. Combinations with high melting point silver alloys as well as copper which are mutually soluble to a high degree — for instance, copper on nickel or silver alloy on copper — usually are restricted to depths of penetration under 1/4-inch.

TABLE 58-I — Typical Average Joint Clearances Used with Brazing
(inches)

Material	Brazing Alloy	Clearance
Aluminum	Brazing sheet	0.000§
Ferrous metals	Copper	0.000
Steel	Copper	0.0005 to 0.0015
Stainless steel	Copper and silver	0.0015
Ferrous metals	Silver	0.0015 to 0.002
Brass	Silver	0.002 to 0.003
Nickel	Silver	0.002 to 0.003
Monel	Silver	0.002 to 0.003
Aluminum	Aluminum	0.002 to 0.008†
Aluminum	Aluminum	0.008 to 0.025*
Ferrous alloys	Brass (60-40)	0.030
Copper alloys	Brass (60-40)	0.030

§ Line contact.   † Short laps.   * Long laps.

Where design requires long overlaps, it is well to provide a number of reservoirs of brazing metal to assure sound joints. Design criteria adopted for critical primary aircraft structures are shown in *Fig.* 58.13. Braze-clad sheet obviates this problem of penetration.

Joint Clearances. The lap depth plus the average clearance in the joint provides the data necessary to calculate the amount of brazing metal required. Because the strength of brazed assemblies is directly affected by the joint clearance,

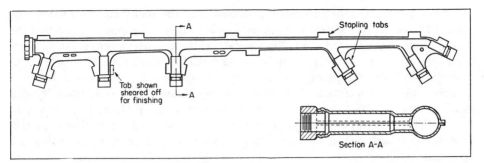

Fig. 58.15 — Tubular aluminum conduit fitting stamped in halves with tabs which are stapled together for the brazing operation. Tabs are sheared off after operation is completed. Drawing, courtesy Electric Furnace Co.

particular attention must be paid this detail. Copper gives the highest strength joints when the fit is from zero to minus 0.002-inch (interference) and maximum penetration also results within these limits. The practical limits generally followed in design of copper-brazed steel parts is to use a maximum press fit of approximately 0.001-inch per inch of diameter and for the opposite extreme, a maximum gap of 0.0005-inch per inch of diameter.

Satisfactory joint clearances for silver alloys are somewhat larger than those for copper and can range from 0.002 to 0.007-inch, the preferable clearance being approximately 0.001 to 0.003-inch. Phos-copper alloy allows a fit clearance up to 0.003-inch but the strength of such joints is relatively low. With aluminum, press fits must be avoided. Clearances of 0.006 to 0.010-inch are suitable for lap joints under 1/4-inch in length; proportionally greater clearances up to 0.025-inch are required for laps of greater length. With braze-clad sheet, line contact is desirable. A general summary is shown in TABLE 58-I.

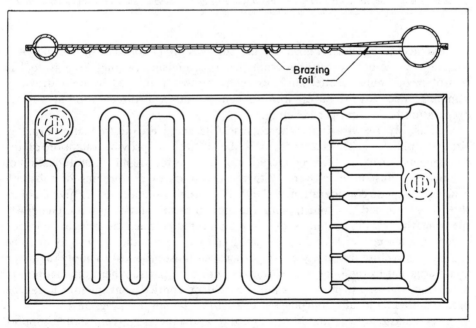

Fig. 58.16 — Steel evaporator is assembled with foil and all four edges are folded over to retain alignment. Unit is brazed in the flat under vacuum and later formed into the conventional U-shape. Drawing, courtesy Electric Furnace Co.

These general fits for parts are predicated upon the assumption that the metals involved have approximately equal coefficients of thermal expansion and that the parts are uniformly heated at a fairly low rate. Consequently, where conditions are otherwise it will be necessary to modify the actual part tolerances to assure a fit which will fall within the foregoing maximum ranges. For instance, in the design of such parts as tanks, differential in expansion may dictate the most desirable type of joint. Solid steel disk types, *Fig.* 58.14, will not braze properly and the dished or inverted type head which can be pressed into place with adequate flange contact provides the best solution. Such heads remain in proper contact during the operation.

Vertical Joints. Wherever joints are to be brazed in a vertical position it is well to provide a small chamfer on the edge adjacent to the braze or faying surface, *Fig.* 58.13. A chamfer in the horizontal position, however, will often obviate capillarity — hence should be avoided.

Fig. 58.17 — Hollow aircraft turret torque tube has heavy ends and gear-mount silver brazed in a salt bath. The finished part withstands a 2000 inch-pound torque test. Photo, courtesy Voss Bros. Mfg. Co.

Surface Roughness. Finish on all surfaces to be brazed should be of a roughness value not over 100 microinches, rms. Finishes ranging from 40 to 100 rms offer optimum results. Surfaces such as those produced by cold drawing, stamping, deep drawing, or ironing can be utilized direct without further preparation besides proper cleaning.

Assembly for Brazing. Where close tolerances are desired after brazing, either extremely small clearances must be utilized to provide retention of the components during the brazing operation or fixtures for alignment and location of components must be employed. With the exception of dip brazing or salt-bath brazing, adequate consideration of the flow of brazing metal must be given. Joints should be designed to avoid flowing the molten metal against gravity. Generally speaking, the brazing medium should not be located between the faces to be joined except in a few cases where the parts are free to move together by gravity or to be squeezed into close proximity by weighting. It is most practical to apply the braze metal adjacent to the joint and in contact with both parts. Where the medium in wire, slug, foil or other form cannot be held or applied conveniently it can be preassembled by means of grooves prepared for the purpose, *Fig.* 58.13.

Brazed assemblies should be designed to allow the use of very simple support fixtures. Jigs or fixtures to clamp or hold assembled parts together should be avoided wherever possible because of high maintenance, etc. The press fits possible

TABLE 58-II — Typical Commercial Brazing Alloys and Their Characteristics

Alloy	Composition								Temperature Range† (F)	Remarks
	Cu	Ag	Zn	Cd	Mn	Sn	P	Ni		
Copper	99+	..	..	..	..	..	..	..	1981	No flux required ordinarily. Steels having alloying elements of manganese, silicon, vanadium and aluminum in excess of 2% total require fluxing.
Brass	60	..	40	..	..	..	..	..	1600–1750	Requires flux. Used for brazing steel assemblies. Average strength equal to that of copper. For joining cast iron to steel. For torch brazing iron.
071 (Handy and Harman)	85	7	..	..	..	8	..	..	1735–1805	For use where high temperature, low silver contact is wanted for subsequent heat treatment.
603 (Handy and Harman)	30	60	..	..	..	10	..	..	1095–1325	Used on electronic equipment where zinc and cadmium are undesirable.
Phos-Copper (Westinghouse)	93	..	..	..	..	..	7	..	1304–1382	For copper, brass or bronze. No flux necessary with copper. Cannot be used with steels.
720 BT* (Handy and Harman)	28	72	..	..	..	..	..	..	1435	Silver-copper eutectic.
501 ETX (Handy and Harman)	34	50	16	..	..	..	..	..	1275–1425	A general-purpose alloy for joining ferrous, nonferrous and dissimilar metals or alloys. Ductile, has high strength and good corrosion resistance.
Sil-Fos (Handy and Harman)	80	15	..	..	..	..	5	..	1185–1300	For copper, brass, bronze and other nonferrous materials only. Not suitable for ferrous metals.
Easy-Flo* (Handy and Harman)	15.5	50	16.5	18	..	..	..	..	1160–1175	Lowest melting point alloy for high-strength joints. For all ferrous and nonferrous materials and dissimilar metals. For cast iron.
Easy-Flo 3 (Handy and Harman)	15.5	50	15.5	16	..	..	..	3	1195–1270	Used for building up ferrous and nonferrous materials, bridging and fileting. Also for low-temperature applications on Stellites, carbides, tungsten and refractory alloys. Good corrosion resistance.
655 (Handy and Harman)	28	65	..	..	5	..	..	2	1385–1445	For high-temperature applications on Stellites, carbides, tungsten, and refractory alloys.
852 (Handy and Harman)	..	85	..	..	15	..	..	..	1750–1765	For applications involving materials with high strength at elevated temperatures.
Easy-Flo 45* (Handy and Harman)	15	45	16	24	..	..	..	..	1125–1145	Lowest melting point alloy. Similar to original Easy-Flo. For ferrous, nonferrous or dissimilar metals.
Easy-Flo 35 (Handy and Harman)	26	35	21	18	..	..	..	..	1125–1295	Used for building up, bridging and fileting.
Braze 051* (Handy and Harman)	58	5	37	..	..	..	..	..	1575–1600	For use on steel where subsequent heat treatment is necessary.
SS (Handy and Harman)	30	40	28	..	..	..	..	2	1240–1435	For low-temperature applications on Stellites, carbides, tungsten, and refractory alloys.
Alcoa No. 716	(Brazing filler medium)								970–1085	For brazing 2S, 3S, 4S, 52S, 53S, 61S or 63S alloys. In wire or shim form. 53S and 61S are brazed at 1060 to 1090 F.
Brazing Sheet No. 11 or 12 (Alcoa)	Alcoa 3S alloy core coated on one side (#11) or two sides (#12) with brazing filler								1100—1140‡	General-purpose alloy sheet for forming into parts. Less distortion than Nos. 1 and 2.
Brazing Sheet No. 21 or 22 (Alcoa)	Heat treatable alloy core coated on one side (#21) or two sides (#22) with coating of brazing filler								1100—1120‡	Same heat treating practices as for 61S. Can be quenched direct from brazing furnace. General-purpose, high strength.
Nicrobraz (Wall Colmonoy)	1850–2050									Corrosion-resistant medium for stainless series 300 and 400, Inconel, Monel, alloy steels, tool steels, carbon steels, and special stainless steels. Same procedure as for copper brazing. Strength at 2000 F equal to parent metal.

* Eutectic Alloys. Melt quickly and completely within a narrow temperature range at once. Wide temperature range alloys have more sluggish flow and should be used where fit is poor or where bridging or fileting is desired.
† Lower figure is the melting point and upper is the flow point at which all of the alloy is molten.
‡ Operating range when brazing with these materials.

with copper brazing offer the most economical alternate to fixtures; fixtures are seldom necessary inasmuch as plain press fits, light scratch knurl press fits, and similar innovations can be incorporated into a design to provide whatever location or retention is necessary. Fixtures are not always practical with brazed aluminum work and, consequently, various methods of retaining are employed.

Lock seaming, crimping, flaring, staking, spinning and like fastening methods can be utilized to obviate fixtures. Aluminum parts, for instance, have been stamped with tabs which were stapled together for the brazing operation and later sheared off, *Fig.* 58.15. The evaporator trap in *Fig.* 58.16 is interesting in that a

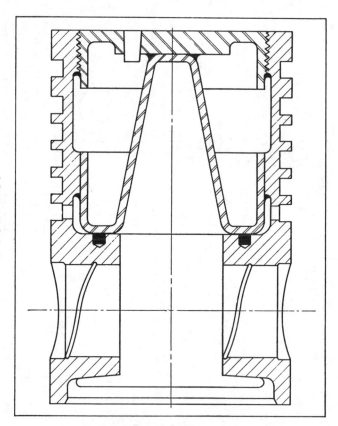

Fig. 58.18 — Cast iron sodium-cooled piston is silver brazed at the four points indicated to an iron head and steel container. Treatment of the parts in a molten salt bath at 850 F to remove the surface graphite permits successful brazing of cast iron. Drawing, courtesy Kolene Corp.

vacuum is used to promote close contact between the two parts as the foil medium melts. Tanks ordinarily should be vented and likewise any parts fitted into blind holes or recesses require some venting to avoid blowing apart during heating.

Variation in Mass. Generally, it is most economical to design parts for furnace brazing which have a fairly uniform wall thickness and no great differences in mass. Too great a difference is reflected in unequal heating with resulting unequal expansion problems during the operation. With due consideration of the extra heat consumed with heavier components, some designs nevertheless are completely justified.

Parts of aluminum may range from 0.006-inch to a maximum of 1/2-inch in thickness and occasionally this maximum can be increased. Hollow aircraft turret torque tubes have been salt-bath brazed with heavy sections at each end and at intermediate points, *Fig.* 58.17. Brazed cast-iron pistons, specially designed for sodium cooling, also have been produced successfully, *Fig.* 58.18, with special treatment to prepare the cast iron properly.

Selection of Materials

Almost any of the common engineering materials such as steel, stainless steel, alloy steel, aluminum, copper, copper alloys, cast iron, nickel, and nickel alloys can be brazed. In addition to the general comments covered in the foregoing, reference regarding specific materials and brazing alloys is made in TABLE 58-II.

Copper brazing pastes available in various compositions permit the application of the medium by means of special applicators, guns, etc. In consideration of the cost of forming and applying wire or other brazing forms, paste mediums offer considerable dollars-and-cents production savings.

Aluminum alloys are available clad on one or both sides with brazing alloy. Today, brass, bronze, copper, beryllium-copper, Invar, Monel, nickel, nickel silver, and precious metals and steels are available clad on one or both sides with silver brazing alloy. These sheet materials can be formed, stamped or blanked as desired and assembled for brazing without any preplacement of brazing medium. The advantage where blind or hidden joints and hollow parts are required is obvious.

Air hardening steels may be brazed and hardened in one operation and carburizing steels may be carburized during the brazing process. The medium and higher carbon steels, however, which require heat treatments such as this incur additional expense owing to closer control and increased handling.

Actually, the smoothness of surfaces seems to have little or no significant effect on the strength of brazed joints. The joint spacing generally proves to be the critical design factor and requires proper preliminary consideration. Rough ground, fine ground or lapped surfaces result in joints of substantially equivalent strength and, as a consequence, the most economical method can be used. Sand blasted surfaces should be avoided inasmuch as they produce joints having low strength. Ultimate tensile strength has been found to be approximately one-half that developed by ground or lapped joint surfaces.

Tests[1] have shown that, in general, the strength of the brazed joint depends upon the joint spacing. For maximum strength assemblies the following factors are emphasized for observance:

1. Minimum joint spacings
2. Capillary feeding of brazing metal
3. Oxygen-free copper or copper-base alloy brazing metal
4. No sand blasted or otherwise contaminated joint surfaces
5. Furnace atmospheres of high purity.

Observance of these factors produced steel joints more than 50 per cent higher in strength than those commercially encountered. Tensile tests on chromium steel specimens showed ultimate tensile strengths with oxygen-free copper brazing metal of approximately 125,000 psi, which was equivalent to those found with special copper-alloy brazing metals. However, where strength at elevated temperatures is desirable, the alloys are preferable. Brazing alloys with up to 10 per cent alloying elements possess greater strength. In tests[1] at 930F these alloyed brazing metals produced joints with 40 per cent greater strength and have withstood the severe conditions on steam turbine rotors where the blades were retained by brazing.

Specification of Tolerances

Final assembled tolerances on brazed assemblies are dependent primarily upon the methods utilized in producing the various components. Where parts are finish machined after brazing, the desired tolerances can be held as required. However, where assemblies are to be completed by brazing and no further finishing is to be done, attention of the method of assembly for brazing will insure satisfactory commercial accuracy.

REFERENCES

[1]"A Study of Furnace Brazing Applied to 12% Chromium Low-Carbon Steel" — by T. H. Gray, *Transactions* of American Society for Metals, Vol. XXXIX, 1947, page 453.

ADDITIONAL READING

[1]*Aluminum Brazing Handbook,* The Aluminum Association, New York, 1974.
[2]*Aluminum Soldering Handbook,* The Aluminum Association, New York, 1974.
[3]*Possibilities in Production Brazing,* Handy & Harman, New York, 1973.

SECTION

PRODUCTION

9

PROCESSES

TREATING
METHODS

59

Heat Treating Steels

Norman N. Brown

$\mathbf{A}$LTHOUGH the selection and specification of heat treatments is seldom considered too seriously as a vital part in the overall production-design picture, these factors are important and can often have a tremendous effect on final costs. Because it would be amiss to overlook this area in design, it should be covered in detail as it affects design. However, owing to the tremendous scope of the subject only the most common treatments for steels can be covered; those specialized and more limited in scope of necessity are omitted.

Conditioning Processes. In a large number of instances, some type of treatment is usually necessary to condition steels either before or after processing of any kind. Processes for conditioning metals therefore assume an important place in the design procedure. True, the treatment of metal for desirable properties may have no effect whatsoever on the actual problem of mechanical or functional design; but, again, it may have a considerable effect. The designer should at least be aware of the general problem in order to cope intelligently with it prior to actual release for production.

Heat Treatments. As a rule, all heat treatments accomplish one of two purposes : (1) Soften the metal or (2) harden the metal. These results can be achieved by differing means and with different metals or designs, specific treatments are most economical or most satisfactory.

Softening Treatments. In general, there are the following most commonly used methods or treatments designed for softening or altering the structure of the metal:

1. Annealing
2. Normalizing
3. Stress Relieving.

These treatments, as contrasted with hardening methods, are of the nonquenching type, only controlled heating and cooling being utilized.

Annealing. This treatment consists of heating above the Ac_3 or critical temperature of the metal followed by controlled furnace cooling to below the Ar_1 or transformation-end temperature. It is used to condition the metal so it is most

---

Norman N. Brown is Vice President of Wheelock Lovejoy & Co., Div., Alco Standard Corp.

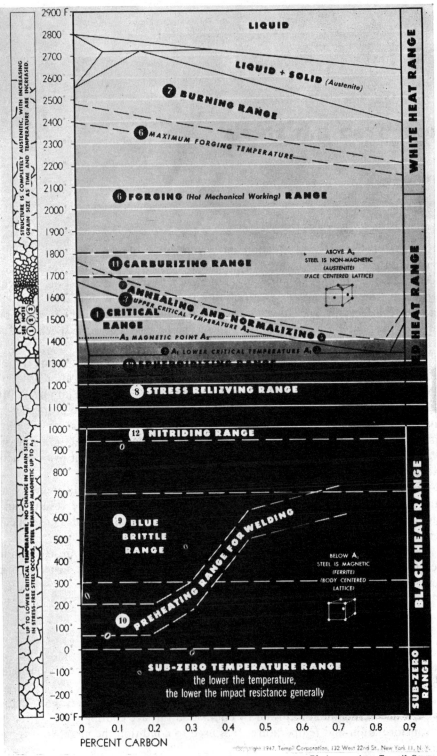

Fig. 59.1 — Overall chart showing the general treatment picture for steels. Terms shown in the diagram at the left, page 59-04, are briefly as follows:

1. CRITICAL RANGE. In this range steels undergo internal atomic changes which radically affect the properties of the material.

2. LOWER CRITICAL TEMPERATURE (A_1). Termed Ac_1 on heating, Ar_1 on cooling. Below Ac_1 structure ordinarily consists of ferrite and pearlite (see definitions below). On heating through Ac_1, the solid ferrite and the solid pearlite begin to dissolve in each other to form austenite (see below) which is nonmagnetic. This dissolving action continues on heating through the critical range until the solid solution is complete at the upper critical temperature.

3. UPPER CRITICAL TEMPERATURE (A). Termed Ac_3 on heating, Ar on cooling. Above this temperature the structure consists of austenite which coarsens with increasing time and temperature. Upper critical temperature is lowered as carbon increases to 0.85 per cent (eutectoid point).

4. ANNEALING consists of heating steels to slightly above Ac_3, holding for austenite to form, then slowly cooling in order to produce small grain size, softness, good ductility and other properties. On cooling slowly the austenite transforms to ferrite and pearlite.

5. NORMALIZING is similar to annealing except that cooling is done in still air. On cooling, austenite transforms giving somewhat higher strength and hardness and slightly less ductility than in annealing.

6. FORGING RANGE extends to several hundred degrees above the upper critical temperature.

7. BURNING RANGE is above the forging range. Burned steel cannot be cured except by remelting.

8. STRESS RELIEVING consists of heating to a range definitely below the lower critical temperature, holding for one hour or more per inch of thickness, then slowly cooling. Purpose is to allow the steel to relieve itself of locked-up stresses.

9. BLUE BRITTLE RANGE occurs approximately at 300 to 700 F, in which range steels are more brittle than above or below this range. Peening or working of steel should not be done in this range.

10. PREHEATING FOR WELDING is carried out to prevent crack formation.

11. CARBURIZING consists of dissolving carbon into surface of steel by heating to above critical range in presence of carburizing compounds.

12. NITRIDING consists of heating certain special steels to about 1000 F for long periods in the presence of ammonia gas. Nitrogen is absorbed into the surface to produce extremely hard "skins."

13. SPHEROIDIZING consists of heating to just below the critical range to put the cementite constituent of pearlite in globular form. This produces softness and in many cases good machinability.

FERRITE is practically pure iron (in plain-carbon steels) existing below the lower critical temperature. It is magnetic and has very slight solid solubility for carbon.

AUSTENITE is the nonmagnetic form of iron and has the power to dissolve carbon and alloying elements.

CEMENTITE or iron carbide is a chemical compound of iron and carbon, Fe_3C.

PEARLITE is a mechanical mixture of ferrite and cementite.

EUTECTOID STEEL contains approximately 0.85 per cent carbon.

MARTENSITE is an extremely hard constituent of steels, formed by the rapid transformation of austenite.

FLAKING occurs in many alloy steels and is a defect characterized by localized microcracking and "flake-like" fracturing. It is usually attributed to hydrogen bursts. Cure consists of cycle cooling to at least 600 F before air cooling.

OPEN OR RIMMING STEEL is not completely deoxidized and the ingot solidifies with blowholes.

KILLED STEEL has been deoxidized at least sufficiently to solidify without appreciable gas evolution.

SIMPLE RULE: Brinell hardness divided by two, times 1000, equals approximate tensile strength in pounds per square inch.

suitable for machining and other cold finishing processes. The chart in *Fig.* 59.1 shows graphically the various temperatures required with different steels. Annealing for machinability is needed with most steels other than those specifically designed to be free-machining, especially with steels over 0.35 to 0.40 per cent carbon.

General rules for suitable annealing treatments for all steels to obtain optimum machinability are difficult to specify. In many cases, however, certain procedures have proved desirable. These might be outlined for the particular type of processing as in the following:

1. *Rough metal removal.* Where removal of metal requires little accuracy or fine finish, the softest obtainable condition of coarsely spheroidized microstructure, having a hardness as low as 80 Rockwell B, offers least resistance. Somewhat higher hardnesses with accompanying partially spheroidized lamellar or a coarse lamellar structure offer better chip formation.
2. *Broaching.* Medium to fine lamellar microstructures with hardnesses of 91 to 102 Rockwell B have proved well suited to broaching operations.
3. *General-purpose machining.* A medium to coarse lamellar structure is preferred for general-purpose machining operations. Structures showing up to 50 per cent spheroidization are good and hardnesses generally range from 87 to 98 Rockwell B, depending upon the type of steel and structure.
4. *Cold working.* Anneals for cold forming, upsetting or heading operations may vary considerably. Depending upon the amount of working required, the metal may range from fairly high hardnesses to dead soft. Maximum working, of course, requires maximum softness.

Normalizing. Parts which have been hot worked and allowed to cool in air such as found with forging, rolling, upsetting, etc., usually have nonuniform structures, grain size, and hardness. To eliminate distortion, achieve uniform structure, uniform response to heat treatment, etc., a normalizing treatment is used. As the term indicates it is principally designed to achieve a "norm" or uniformity. It consists of heating to at least 100 F above Ac_3 and cooling in still air to room temperature.

Stress Relieving. Cold working or work done at any temperature below about 1425 F induces internal stresses which, unless eliminated, frequently create difficulties in subsequent processing operations. Stress relieving is virtually a subcritical anneal, the steel being heated to a temperature below Ac_1 and allowed to cool slowly in air. It is particularly applicable after straightening and machining operations and before or after heat treatments to obviate distortion.

Hardening Treatments. To improve the physical properties of medium and higher carbon content steels, as well as alloy steels, over that of low-carbon common garden-variety steels, heat treatments to increase hardness are utilized. Hardening is accomplished in several ways and steps:

1. By heating to austenitizing temperature (above Ac_3) and quenching or continuously cooling to room temperature in air, oil, water, molten chemical or metallic salts, or other liquid or gaseous media.
2. By heating and interrupting the quench.
3. By conventional or interrupted quenching and tempering.

As contrasted with softening treatments, hardening methods are generally of the quenching type and the effectiveness of the quench in producing hardness is a direct function of the carbon content of the steel, *Fig.* 59.2. The austenitic structure

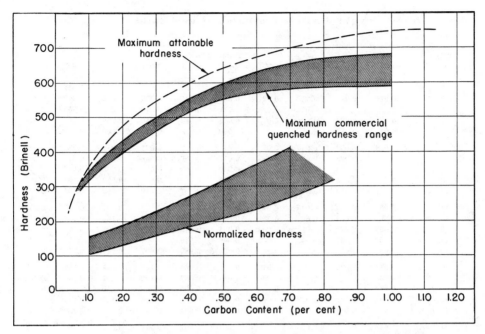

Fig. 59.2 — Approximate relationship between carbon content and maximum hardness attainable and normalized hardness. Size, alloy content and rate of cooling all affect final hardness and hence values are only relative.

formed above Ac_3, *Fig.* 59.1, is transformed into the desired martensitic structure by rapid quenching, which suppresses the formation of the softer ferritic and cementitic forms. To permit milder oil quenching and less distortion, alloying elements are added to steels to slow down the transformation and thus control the depth and severity of the hardening.

Depending upon the type of steel, differing treatments are employed to achieve the most desirable type of hardened condition. Those commonly used are generally:

1. Case hardening.
2. Surface hardening.
3. Through hardening.

Case Hardening. Hardening of a surface layer of a steel part, by essentially the addition of carbon or nitrogen, may be accomplished by one of several different methods. These case hardening processes are:

1. Carburizing.
2. Cyaniding.
3. Carbonitriding.
4. Nitriding.

Carburizing. This is a process whereby carbon is added to the outer layer of a steel part, within limitations, and the piece is quench hardened. Generally, the steels employed are initially low in carbon content. The methods of adding carbon in carburizing can be classified as pack, gas and liquid. Amount and depth of case is generally a function of temperature and time, TABLE 59-I. Greatest degree of control is found with gas carburizing and maximum depth also is provided by the

TABLE 59-I — Carburizing Case Depths *
(inches)

Time (hours)	Temperature (F)									
	1400	1450	1500	1550	1600	1650	1700	1750	1800	1850
1......	0.008	0.010	0.012	0.015	0.018	0.021	0.025	0.029	0.034	0.040
2......	0.011	0.014	0.017	0.021	0.025	0.030	0.035	0.041	0.048	0.056
3......	0.014	0.017	0.021	0.025	0.031	0.037	0.043	0.051	0.059	0.069
4......	0.016	0.020	0.024	0.029	0.035	0.042	0.050	0.059	0.069	0.079
5......	0.018	0.022	0.027	0.033	0.040	0.047	0.056	0.066	0.077	0.089
6......	0.019	0.024	0.030	0.036	0.043	0.052	0.061	0.072	0.084	0.097
7......	0.021	0.026	0.032	0.039	0.047	0.056	0.066	0.078	0.091	0.105
8......	0.022	0.028	0.034	0.041	0.050	0.060	0.071	0.083	0.097	0.112
9......	0.024	0.029	0.036	0.044	0.053	0.063	0.075	0.088	0.103	0.119
10......	0.025	0.031	0.038	0.046	0.056	0.067	0.079	0.093	0.108	0.126
11......	0.026	0.033	0.040	0.048	0.059	0.070	0.083	0.097	0.113	0.132
12......	0.027	0.034	0.042	0.051	0.061	0.073	0.087	0.102	0.119	0.138
13......	0.028	0.035	0.043	0.053	0.064	0.076	0.090	0.106	0.123	0.143
14......	0.029	0.037	0.045	0.055	0.066	0.079	0.094	0.110	0.128	0.149
15......	0.031	0.039	0.047	0.057	0.068	0.082	0.097	0.114	0.133	0.154
16......	0.032	0.039	0.048	0.059	0.071	0.084	0.100	0.117	0.137	0.159
17......	0.033	0.040	0.050	0.060	0.073	0.087	0.103	0.121	0.141	0.164
18......	0.033	0.042	0.051	0.062	0.075	0.090	0.106	0.125	0.145	0.169
19......	0.034	0.043	0.053	0.064	0.077	0.092	0.109	0.128	0.149	0.173
20......	0.035	0.044	0.054	0.066	0.079	0.094	0.112	0.131	0.153	0.178
21......	0.036	0.045	0.055	0.067	0.081	0.097	0.114	0.134	0.157	0.182
22......	0.037	0.046	0.056	0.069	0.083	0.099	0.117	0.138	0.161	0.186
23......	0.038	0.047	0.058	0.070	0.085	0.101	0.120	0.141	0.164	0.190
24......	0.039	0.048	0.059	0.072	0.086	0.103	0.122	0.144	0.168	0.195
25......	0.039	0.049	0.060	0.073	0.088	0.106	0.125	0.147	0.171	0.199
26......	0.040	0.050	0.061	0.075	0.090	0.108	0.127	0.150	0.175	0.203
27......	0.041	0.051	0.063	0.076	0.092	0.110	0.130	0.153	0.178	0.206
28......	0.042	0.052	0.064	0.078	0.094	0.112	0.132	0.155	0.181	0.210
29......	0.042	0.053	0.065	0.079	0.095	0.114	0.134	0.158	0.185	0.214
30......	0.043	0.054	0.066	0.080	0.097	0.116	0.137	0.161	0.188	0.217

* From F. E. Harris, *Metal Progress*, August, 1943.

gas-diffusion method. Deep cases of 0.300 to 0.400-inch are practicable by this diffusion process.

Cyaniding. The cyaniding method provides a superficial case of good wear resistance and high hardness. Generally, this consists of heating the parts in a molten bath of cyanide salts and following with quenching. Usual immersion periods are from one-half to one hour. Nitrogen produced by decomposition of the bath produces iron nitrides in the case and these impart the high hardness.

Carbonitriding. A combination of nitrogen and carbon is added to the surface by heating in a suitable gaseous atmosphere and quenching. A modified carburizing process, this method effects lower costs where volumes are large and creates less distortion than carburizing. However, case thickness produced is limited and seldom exceeds 0.030-inch.

Nitriding. This process consists of adding nitrogen to the surface by heating, in a suitable gas atmosphere, alloy steels capable of assimilating the nitrogen. No quenching is required, the hardness being produced by the nitrides formed from the combination of nitrogen, iron and alloying elements. Special Nitralloy steels are generally used. With chromium-aluminum-molybdenum steels a 50-hour cycle will produce a 0.015-inch case of which 0.005 to 0.007-inch exceeds 900 Vickers in hardness. A 100-hour cycle will produce a case of 0.030-inch with 0.010 to 0.012-inch in excess of 900 Vickers hardness. With large sections, normalizing prior to

Fig. 59.3 — Take-off drive shaft of AISI 4140 steel flame hardened to 59 Rockwell C on the spline. Photo, courtesy Cincinnati Milacron Co.

Fig. 59.4 — Automatic flame hardener used in treating the splined shaft shown in Fig. 59.3. Photo, courtesy Cincinnati Milacron Co.

heat treating is recommended, the heat treating being necessary for best nitriding results.

Surface Hardening. Where moderately hard surfaces are satisfactory, rapid surface hardening of medium-carbon steels can be highly economical. This can be accomplished by several methods, namely, flame or induction hardening. Rapid heating of the steel by flames or high-frequency current is followed by quenching. Advantage besides speed is extremely low distortion on symmetrical parts.

Flame Hardening. Where extremely high hardnesses are not necessary, superficial hardening of the surface of medium and occasionally high-carbon steels may be accomplished by flame hardening, *Fig.* 59.3. This method merely consists of rapidly heating the surface by means of a torch or torches and quenching by spray or immersion. Apparatus may be simple or elaborate, *Fig.* 59.4. Parts flame hardened may or may not have a previous heat treatment. Many parts, however, are first conventionally heat treated.

Induction Hardening. In a manner similar to that of flame hardening, induction hardening produces a superficial surface effect, *Fig.* 59.5. Electrical frequencies from 1000 to 500,000 cycles are employed and, for heating, special fitted induction coils are necessary. For simple or uniform parts, coils are no problem but where intricate or unusual parts are concerned, induction coils may prove difficult.

Through Hardening. To obtain throughout a given part the optimum in mechanical properties — adequate strength together with the greatest toughness

Fig. 59.5 — Salisbury axle induction hardened full length on Tocco machine. Surface is hardened to 55 Rockwell C 1/8-inch deep and subsequently drawn back to 43 to 47. Use of AISI 1033 with flame hardening instead of AISI 4140 and conventional treatment reduced material costs, permitted machining prior to hardening, and cut machining time by 75 per cent. Photo, courtesy Ohio Crankshaft Co.

possible at that strength — through hardening is normally employed. This usually consists of heating to some point above the critical temperature and quenching. In the interests of toughness, tempering or drawing back the as-quenched hardness is used immediately following the quench. Alloys steels are often necessary to obtain the desired hardening and, especially where sections vary widely, some compromise may be desirable to avoid brittle-hard sections.

Tempering. To achieve the degree of toughness desired in a piece after quenching, tempering is generally desirable. The operation consists primarily in reheating a part, after hardening, to a low or moderately high temperature and cooling. This serves to "draw back" in varying amounts on the hardness, *Fig.* 59.6, but must normally be done to achieve a degree of toughness or plasticity necessary to resist fatigue and cracking. A general rule-of-thumb is that a part must be held at drawing heat for one hour per inch of cross section to insure thorough heating.

Quenching. A number of quenching media are employed for through hardening depending upon the part, the steel used and the design. The common ones are: (1) Water, (2) oil, and (3) air. Water is generally used on plain-carbon steels that require fast cooling to attain top hardness. Quenching is severe and if parts are intricate or if toughness is desirable, oil should be used.

The oil quench is probably most widely used, especially with alloy steels. The

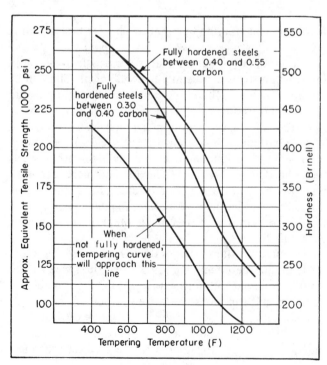

Fig. 59.6 — Typical curves showing relation between hardness and tempering temperatures (SAE Handbook).

greater sensitivity of alloy steels to hardening necessitates this milder quench and, being slower, the quench induces less distortion. Cooling in air blast is used with highly alloyed (tool) steels. Minimum hardening distortion with maximum hardness makes these steels ideal where the low distortion can be justified economically.

Interrupted Quenching. Generally applied to medium and high-carbon steels, these processes are designed to reduce distortion and cracking in the hardening treatments. Three interrupted-quench treatments used are: (1) Martempering, (2) austempering and (3) isothermal. In martempering the operation is designed to equalize temperature gradients within parts before the structure transformation progresses very far. A part is quenched into a molten salt, or a high flash point oil, at a temperature approximately the *Ms* point (martensitic transformation starts) of the steel (300 to 500 F) and held until the entire mass has reached bath temperature. This is followed by cooling to room temperature in air.

Parts with substantial differences in mass or complicated design are particularly subject to growth, shrinkage and twisting distortion. With steels of high hardenability, cracking may take place. Generally, these factors are caused by the unequal stresses created during rapid austenite-martensite transformation (see discussion under hardening treatments) occurring at different times in different portions of a part. Martempering obviates to a great extent these factors. About 3 to 4 inches is the usual maximum section which can be fully hardened by this quenching method.

Austempering is somewhat similar to martempering in purpose but has found most use with plain-carbon steels and with alloy steels in the lower alloy content range. The quench is generally a molten salt or molten low melting alloy and the piece is quenched direct from heat treating temperature. However, by holding the bath at 400 to 700 F the martensitic structure which forms at lower temperatures is avoided and only bainite forms. The process is particularly suited for relatively light sections, generally less than $1/2$-inch and seldom over 1 inch in thickness. An additional asset is found in the unusual toughness of parts which are austempered. Hardnesses range from 48 to 58 Rockwell C or about the same as a piece conventionally tempered in the range of 450 to 650 F but accompanying toughness is much greater.

Isothermal quenching consists of cooling in an agitated salt bath, at 400 to 600 F, until temperature is equalized and transferring to a bath at a higher heat to draw the part. This treatment is generally used for obtaining physicals similar to those from austempering but in larger sizes. Sections up to 2 inches can generally be handled. Diagrammatic pictures of these quenching processes are shown in *Fig.* 59.7.

Design Considerations

In general, the most important consideration in the design of parts to be heat treated is the effect of the design configuration on the success of the treatment in production. Extremely intricate parts naturally are potentially troublesome to handle as are parts with widely varying sections or masses, *Fig.* 59.8. The problem, therefore, becomes one of selecting a steel and a treatment compatible with the necessary design, which will offer adequate results. The results, of course, would be in terms of strength, toughness, wear resistance, and maximum retention of

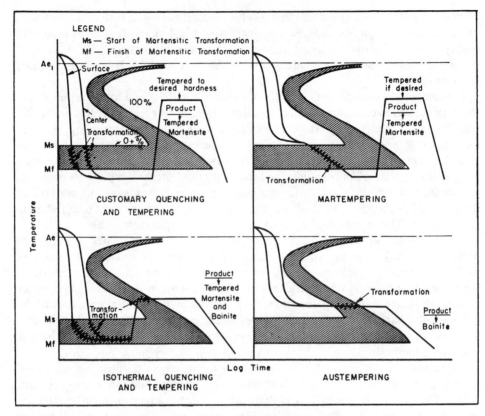

Fig. 59.7 — Diagrammatic charts showing the comparative characteristics of the various tempering processes. Drawings, courtesy U.S. Steel Corp.

accuracy. The final, but not least, consideration is that of reasonable or satisfactory cost.

Service Requirements. A practical approach to selection and specification of the most desirable heat treatment would be by means of service properties. These may be subdivided into several categories as follows:

1. Hard. Resistance to wear and indentation.
2. Tough. Resistance to torque, impact, fatigue or pull.
3. Hard and tough. Any combination of hardness and toughness requirements.

Service requirements for a specific machine part or parts will generally indicate the category and also help determine the most likely steel to use.

Hardness. In the interests of hardness, generally one of the following treatments is employed: (1) Case hardening, (2) surface hardening, or (3) through hardening. Case hardening is perhaps the most common for obtaining hardness only. Generally, low-carbon steels with carbon contents under 0.25 per cent are used and where distortion is a problem the alloy grades are preferable. Alloy grades also result in a more uniform and generally harder case with sections over 3 or 4 inches than do plain-carbon grades.

Hardness and wear resistance of the nitrided case are superior, and the cyanided, carbonitrided and carburized cases follow in descending order. The chart of *Fig.* 59.9 gives an idea as to the general ranges possible with the various

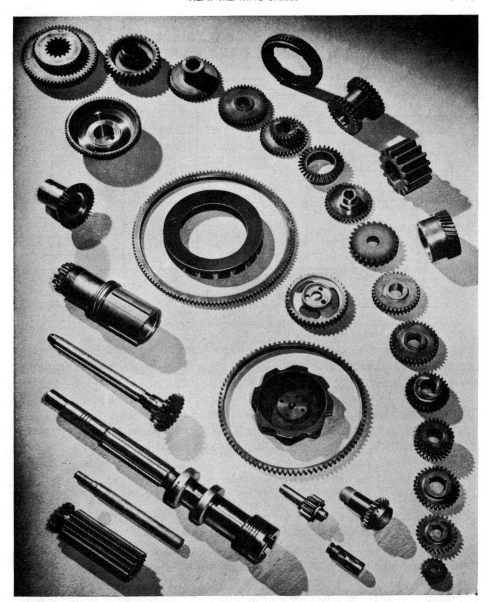

Fig. 59.8 — Group of flame hardened machine parts exhibiting widely varying sections and intricate design. Photo, courtesy Cincinnati Milacron Co.

available case hardening processes.

Depth of case required must be adequately considered. Where excessive shock and pressures are negligible, depth of case depends upon the distortion that may occur and on such factors as grinding, wear permissible, etc. The process to use and specifications to call for must therefore be predicated to a great extent upon the part design in the interests of obviating distortion. Drawing specifications may be selected from the chart of case hardening treatments shown in *Fig.* 59.10 on the basis of the general design of the part in question. In average cases of pack carburizing an allowance of 0.015 to 0.025-inch is usually made for grinding to take care of distortion. Where distortion is greater, it is usually more economical to

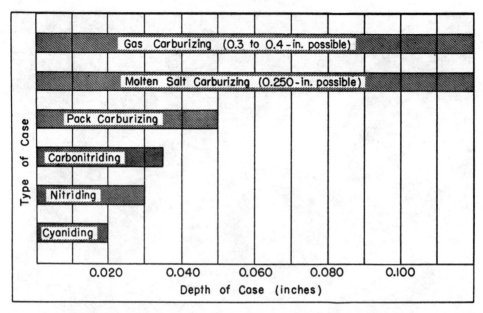

Fig. 59.9 — General ranges in case depth possible by different methods of case hardening.

select some other process or combination which results in less distortion. A typical allowance provided for liquid carburizing is about 0.005 to 0.010-inch although in many instances, distortion can be held to 0.001-inch.

Where hardnesses over 60 Rockwell C are not imperative and values under 55 to 60 are satisfactory, flame or induction hardening can be used, *Fig.* 59.11. For wear-resistant hardness, plain-carbon steels of 0.40 to 0.50 per cent carbon such as AISI 1045 or 1050 are satisfactory. Generally, the drawing specification should read:

> Machine.
> Stress relieve (where part is intricate or considerable machining is involved).
> Flame (or induction) harden designated surfaces.
> Temper (where induced stresses may result in undesired distortion;
>   drawing at 400 F reduces stresses 40 to 50 per cent).

With parts where high hardness cannot be obtained without undue distortion, through-hardening is often used. Plain high-carbon steels over 0.60 per cent carbon (usually AISI 1080 or 1090) and oil and water-hardening alloy or tool steels employed also afford greater depth of hardness. In the interests of hardness only, the use of through-hardening is limited and may be less economical.

Toughness. Through-hardening is generally the major treatment for obtaining toughness. In the interests of maximum toughness only, hardnesses under 300 Brinell are generally specified and alloy steels in the medium-carbon range are employed. Occasionally a low-carbon alloy can be used to advantage.

Most of these steels can be machined in heat-treated condition under 300 Brinell although machinability varies with the analysis and hardness. Generally, machinability is in the following order with the best machining grades first:

1. 4640	6. 5140
2. 4145	7. 6145
3. 8642	8. 2345
4. 4340	9. 3250
5. 3140	10. 1045

Type of Steels AISI	Anticipated Distortion According to Design			
	Negligible Distortion	Low Distortion	Medium Distortion	Considerable Distortion
	Symmetrical Chunky Simple			Slender Long, narrow Irregular
Low-carbon 1020 1112 1117 12614 etc. (all sizes)			Machine Carbonitride Harden Grind	
Low-carbon 1020 1112 1117 12614 etc. (under 3-in.)	Machine Carburize (or cyanide) Harden Grind	Rough machine Normalize Finish machine Carburize (or cyanide) Harden Grind		
Low-carbon alloys 4615 3115 8620 5120 etc. (large sizes)	Machine Carburize (or cyanide) Harden Grind	Machine Carburize (or cyanide) Harden Grind	Machine Stress relieve Carburize (or cyanide) Harden Grind — Rough machine Normalize Finish machine Carburize (or cyanide) Harden Grind	Machine Carbonitride Harden Grind — Rough machine Normalize Finish machine Carbonitride Harden Grind — Rough machine Normalize Finish machine Carburize Martemper Grind — Rough machine Carburize Straighten Finish machine Stress relieve Martemper Grind
Medium-carbon alloys 4140 4640 etc. Nitralloy				Machine Stress relieve Nitride — Preheat-treat Machine Stress relieve Nitride — Anneal Rough machine Heat treat Finish machine Stress relieve Nitride — Preheat-treat Rough machine Stress relieve Finish machine Nitride

Fig. 59.10 — Selector for case hardening specifications. Treatments are arranged to permit lowest production cost compatible with minimum possible distortion.

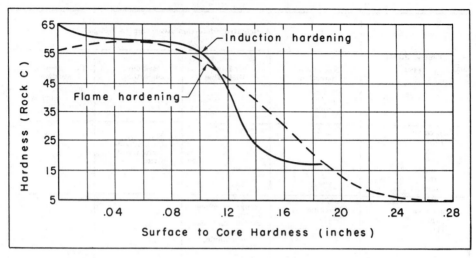

Fig. 59.11 — Graph showing typical hardness curves for flame and induction hardening.

The water-quenching plain-carbon steels distort more than the oil-hardening types and specifications must take this factor into consideration. If parts are long and slender, straightening may be necessary. Such parts should normally have the following specification:

> Anneal
> Heat treat
> Straighten
> Stress relieve
> Machine

If the machining involved is considerable, it is desirable to first rough machine to eliminate any great amount of machining after hardening. The specification would be:

> Anneal
> Rough machine
> Heat treat
> Finish machine

Where the highest range of hardnesses are concerned and the part is of fairly massive shape, distortion would be small and the steps could be:

> Anneal
> Machine
> Heat treat
> Grind

Within the alloy steel grades there is comparatively little difference in physical properties between the medium (5140, 4145, 8642, etc.) and rich (4340, 3250, etc.) alloys under 300 Brinell hardness. The rich alloys, however, exhibit somewhat better impact strength at the same hardnesses up to 300 Brinell.

Hardness-Toughness. In the hardness-toughness category a wide variety of combinations are possible. Almost always, the combination is a matter of compromise between the two physical properties. As hardness increases beyond

300 Brinell, toughness or resistance to impact and fatigue decrease to the point where at maximum hardness this property all but disappears.

Case hardened parts require steels with proper core properties, especially those to be selectively case hardened. Where high compressive loads are encountered, a fairly high core hardness is beneficial in supporting the case. However, if great toughness is the design requirement, a low core hardness may be preferable. Steels tabulated as to both case and core hardening properties are given in TABLE 59-II. Generally, in order to assure adequate toughness, the case thickness should not exceed more than one-sixth the section thickness. The relationship of case thickness relative to the service expected is shown in *Fig. 59.12*.

TABLE 59-II — **AISI Classification of Steel Cases and Cores for Hardenability**

			Hardenability Classification						
——Low——			————Medium————				——High——		
Case*									
1000	1100			2500	4300		1300	4100	8600
				3300	4800		2300	4600	8700
					9300		4000	5100	
Core									
4017	4608		1320	3120	4621	8620	2517	4820	
4023	4615		2317	4032	4812	8622	3310	9310	
4024	4617		2512	4119	4815	8720	3316	9315	
4027	8615		2515	4317	5115	9420	4320	9317	
4028	8617		3115	4620	5120		4817		

* As original carbon content has no effect on case hardening, last two digits are omitted.

In the range of directly hardenable alloy steels, the breakdown as outlined in AISI Standards is extremely helpful in specification. Five groups are set up as follows under mean carbon content in per cent ("Hardenability" relates to the depth and distribution of hardness):

1. *Carbon* 0.30 *to* 0.37. For heat-treated parts requiring moderate strength and great toughness. Frequently used for water-quenched parts of moderate section size or oil-quenched parts of small size. Typical parts are connecting rods, axle shafts, studs, steering knuckles, etc.
   Low hardenability — AISI 1330, 1335, 4037, 4042, 4130, 5130, 5132, 8630
   Medium hardenability — AISI 2330, 3130, 3135, 4137, 5135, 8632, 8635, 8637, 8735, 9437
2. *Carbon* 0.40 *to* 0.42. For heat-treated parts requiring higher strength and good toughness. Used for medium and large-size parts, low and medium-hardenability steels being generally employed for axle shafts, propeller shafts and high-hardenability steel for particularly large axles, shafts and aircraft parts. Generally oil quenched.
   Low hardenability — AISI 1340, 4047, 5140, 9440
   Medium hardenability — AISI 2340, 3140, 3141, 4053, 4063, 4140, 4640, 8640, 8641, 8642, 8740, 8742, 9442
   High hardenability — AISI 4340, 9840
3. *Carbon* 0.45 *to* 0.50. For heat-treated parts requiring fairly high hardness and strength along with moderate toughness. Parts are usually gears and similar stressed machine components.
   Low hardenability — AISI 5045, 5046, 5145, 9747, 9763
   Medium hardenability — AISI 2345, 3145, 3150, 4145, 5147, 5150, 8645, 8647, 8650, 8745, 8747, 8750, 9445. 9845
   High hardenability — AISI 4150, 9850

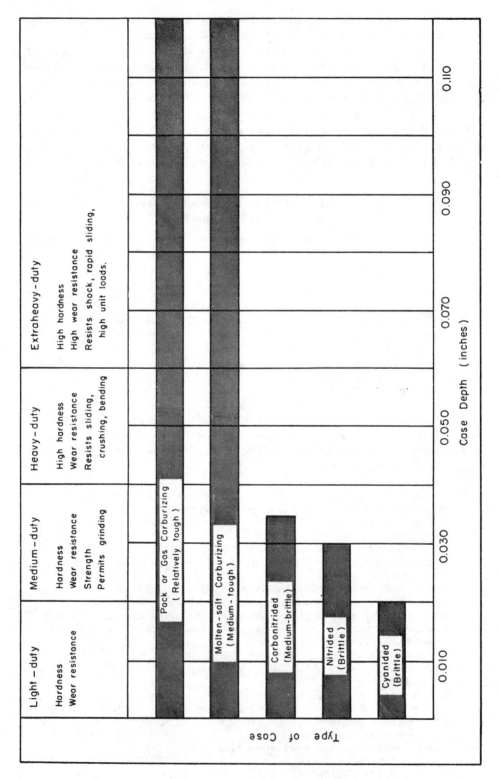

Fig. 59.12 — Chart showing relationship of case thickness to overall service conditions of hard-tough parts.

4. *Carbon* 0.50 *to* 0.62. Used primarily for springs and tools.
   Medium hardenability AISI 4068, 5150, 5152, 6150, 8650, 9254, 9255, 9260, 9261
   High hardenability — AISI 8653, 8655, 8660, 9262
5. *Carbon* 1.02. Used mainly for ball and roller bearings but also for other machine parts requiring high hardness and wear resistance.
   Low hardenability — AISI 50100
   Medium hardenability — AISI 51100, 52100.

Medium-carbon alloys (4140, 4340) are used at hardnesses from 32 to 50 Rockwell C for the majority of parts. From 50 to 60 Rockwell C or over, tough oil and water-hardening high-carbon steels are utilized (9262 etc.): The rich alloys (4340, 2345, etc.) show little or no marked difference in physicals over the medium-alloy group (3140, 4140, etc.) in this hard-tough area unless over 321 Brinell (34 Rockwell C) in hardness and $3^1/_2$ inches in section.

Owing to the many variations present in this area, it is only possible to suggest practical specifications which can be used effectively, with due consideration to the other factors previously discussed. These can be for parts machined after heat treatment to a hardness of 32 to 38 Rockwell C:

1. *Comparatively chunky or symmetrical design*
   Heat treat
   Machine

2. *Comparatively long or intricate design*
   Heat treat
   Straighten
   Stress relieve
   Machine

For parts to be heat treated to a hardness of 32 to 50 Rockwell C after machining:

1. *Comparatively chunky or symmetrical design*

Anneal	or	Anneal
Machine		Rough machine
Heat treat		Heat treat
Grind		Finish machine
		Grind

2. *Comparatively long or intricate design*

Anneal	or	Anneal
Rough machine		Rough and finish machine
Heat treat		Stress relieve
Finish machine		Heat treat
Grind		Grind

or

Anneal
Rough and finish machine
Stress relieve (if necessary)
Martemper
Grind.

Considering flame or induction hardening in this category, although for hardness only steels in the carbon range of 0.40 to 0.50 are generally used, for applications where hard-tough properties are desirable steels from 0.30 to 0.52 per cent may be employed. Hardnesses from 41 to 63 Rockwell C can be obtained.

General Design. To promote greatest economy in heat treatment, the design details of parts should be carefully considered. Sharp corners are a problem; hardening stresses may result in cracking where sharp corners are used. Fillets should be employed generously, instead, on all internal corners and junctions. Radical changes in section thickness should also be avoided.

Wherever possible, heavy and light sections should be designed as separate parts to help maintain uniformity in individual pieces. Openings or holes should not be clustered together; wherever practicable such features should be set up to give a balanced layout. If threaded sections are used, the portion adjacent to the threaded length should be relieved to eliminate the sharp stress raiser created by the last thread. Again, on shafts, good practice is to provide pads or portions of enlarged diameter for seating gears or wheels. In no case should changes in section be located in a line so as to weaken the section.

Selection of Materials

The process of selection of the right steel for any part is so closely concerned with design considerations for satisfying service conditions that the major portion of this problem has been covered. As a rule, more than one steel can be found to satisfy the necessary design requirements. Final selection, therefore, will involve consideration of price, processability and availability, etc. A generalized grouping of some typical standard steels, listed from top down according to cost, starting with the lowest, is given in TABLE 59-III. Those in any one group may be used interchangeably, in varying degrees, of course, depending upon the specific design requirements.

Specifications. It is seldom good practice to specify both a hardness range and a physical property range. Often, one may be incompatible with the other. The best method is to select one or the other to provide maximum flexibility and economy.

Carburizing Steels. For carburizing processes, steels with 0.25 per cent carbon, maximum, are generally preferred. The best steels are said to be the nickel-molybdenum alloys. In order of desirability, carburizing steels might be listed as follows:

AISI 4820
4617
8620
1117
1020

TABLE 59-III — **Standard AISI Similar Steels** *
(according to cost, least expensive at top)

Group I	Group II	Group III	Group IV	Group V
1015 to 20	5120	1035 to 45	5140	4340
1112	8620	1137 to 41	4145	3250
1117	3115 to 20	1335	8642	2345
12214	4615 to 20		3140	
	4820		4640	
			6145	

*Groups are not complete. Others may be added to suit particular requirements.

Normalizing. Certain steels respond quite well to the normalizing treatment; i.e., chromium-molybdenum, manganese-vanadium, manganese chromium-vanadium, and chromium-molybdenum-vanadium steels. At 0.15 to 0.25-carbon level the steels are suitable for carburizing, at 0.25 to 0.35-carbon for quenching and drawing, and at 0.35 to 0.50-carbon for normalizing and drawing. The use of only a normalizing and drawing treatment is employed for obtaining desired strength in larger sections such as massive turbine rotors, ship shafting, and big forgings far too large for regular quenching to martensite. The triple-alloy chromium-nickel-manganese steels provide normalized properties markedly superior to those of plain-carbon or manganese steels normalized in small sizes. To develop optimum toughness in the molybdenum 43, 86, 87 and 9800 steels, the normalizing treatment must be followed by tempering.

Surface hardening. Selection of a particular steel for surface hardening requires some consideration of the distortion permissible as well as depth desirable. Martin and Van Note have shown[1] that various steels respond with differing depths of hardness and hardening at different temperatures, *Fig. 59.13*. Here it can be seen that specifications may vary widely depending upon conditions.

Maraging Steels. For applications requiring ultrahigh strength, relatively easy fabricating, and simple heat treatment with good dimensional stability there is a group of low-carbon, high-nickel steels that can attain yield strengths up to 270,000 psi with good fracture toughness. This group of maraging steels develop

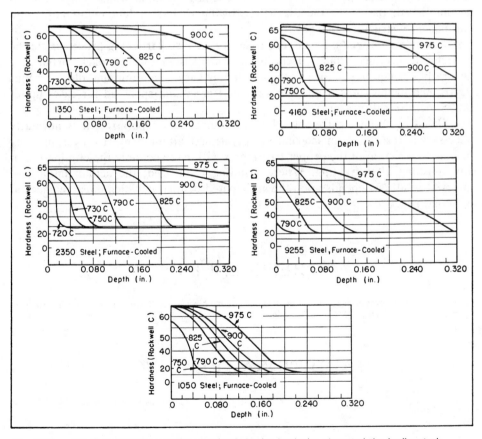

Fig. 59.13 — Groups of penetration curves showing induction hardening characteristics for five steels.

their physical properties without liquid quenching — that is heating and cooling in air which is particularly suitable for parts too large for usual available heat treating equipment. Since these steels are rather high in initial cost and not as readily available as other more common grades it is well to consult with suppliers when considering their use.

Specification of Tolerances

Accuracy of heat-treated parts has been covered to a considerable extent in the preceding discussion. The major factor to be dealt with is distortion. This factor can be great; herringbone gears of 24 1/8-inch diameter and 4-inch face width, for instance, showed a physical distortion of 0.030 to 0.050-inch when pack hardened conventionally, whereas only 0.010-inch was experienced when flame hardening and martempering were utilized. Cost was reduced from $4.80 per gear to $0.90 per gear during the process of selecting a treatment to overcome excessive scrap and finish machining. Rings for ball races, 80 1/2 inches in OD and 68 1/2 inches in ID, have been flame hardened with a maximum out-of-roundness of only 0.030-inch and out-of-flatness of 0.025-inch, even though complicated with holes, slots, gear teeth, etc. Rings 30 inches in diameter by 1/4-inch in section show less than 0.010-inch distortion, out-of-roundness or out-of-flatness. Generally, the lower the temperature and/or the less drastic the quenching media at which temperature is accomplished, the lower the distortion, owing to less violent temperature and consequently structure changes.

Case Hardening. As a rule, most case or surface hardening processes can be expected to produce a fairly accurate, uniform case. However, pack carburizing generally is found to produce a case which may vary, at a minimum, plus or minus 0.010-inch in thickness. Hence, this method is seldom specified for cases under 0.025-inch for this reason.

Grinding. Where extremely close tolerances are imperative, it is normally necessary to resort to a finish grinding operation to attain them. However, in many instances, selection of the proper treatments will make possible the elimination of any finishing after treating. Where grinding is utilized, the actual amount of material to allow depends upon the anticipated distortion which must be checked in actual practice.

REFERENCES

[1] D. L. Martin and W. G. Van Note — "Induction Hardening and Austenitizing Characteristics of Several Medium-Carbon Steels", *Transactions ASM*, Vol. 36, 1946.

ADDITIONAL READING

[1] J. T. Sponzilli, C. H. Sperry and J. L. Lytell, Jr., "A System for Classifying Heat-Treatable Steels," *Metal Progress*, February 1976.

60

Heat Treating Nonferrous Metals

WITHOUT question, the combination of metal selection and the accompanying heat treatment that may be desirable is equally as complex in the nonferrous area as it is for steels. In this chapter two major categories are covered: aluminum alloys and copper and copper alloys. Each category presents the important considerations necessary for practical understanding and specification.

Heat Treating Aluminum Alloys

David S. Thompson, Ogle R. Singleton,
Robert D. McGowan, and Grant E. Spangler

There are two prerequisites for a heat-treatable alloy to be considered age or precipitation hardenable: 1. Solid solubility of major alloying elements shall decrease with decreasing temperature. 2. Guinier-Preston (GP) zone solvus shall be sufficiently high to be able to form GP zones in a reasonable time.

Three major systems produce practical heat-treatable aluminum alloys — Aluminum Assn. designations for wrought alloys are used:

Al-Cu	2XXX series (2219, 2021, 2014, 2024)
Al-Mg-Si	6XXX series (6063, 6061)
Al-Mg-Zn	7XXX series (7005, 7039, 7075, 7178, 7079)

Other systems include Al-Ag, Al-Zn, Al-Mg, Al-Mg-Ag, or Al-Si, but none is practical for reasons such as economics, limited hardening effects, or inability to nucleate homogeneous precipitation.

General Characteristics. The degree of strengthening due to precipitation hardening is much greater than that due to solid solution hardening. The 7000 series (Al-Zn-Mg alloys) have the highest mechanical properties because of the high volume fraction of $MgZn_2$ available for precipitation. The hardening effect depends on the volume fraction and also the size of the precipitates. GP zones are

Mr. Thompson is Director, Alloy Development; Mr. Singleton is Research Engineer, Fabrication Technology; and Mr. Spangler is Director, Metallurgy. Metallurgical Research Div., Reynolds Metals Co., Richmond, Va. Mr. McGowan is Metallurgical Engineer. Aluminum Mill Products Div., American Metal Climax Inc., Morris, Ill.

the most effective hardeners because of their small size and large numbers. Generally, it is desired to age to maximum hardness. Tempers are designated T5, T6, or T8. In certain circumstances, overaging may enhance the stress-corrosion resistance; and this temper is designated T7.

Many factors can influence precipitation and resultant properties such as: quench conditions, deformation after quenching, minor addition elements, and aging practice.

It is interesting to note that a compositional connection exists between the three major systems and their subgroups. This connection is magnesium and to a lesser extent copper, as shown in *Fig.* 60.1. The relative positions of some principal alloys are shown on this diagram, although distances are not proportional to composition.

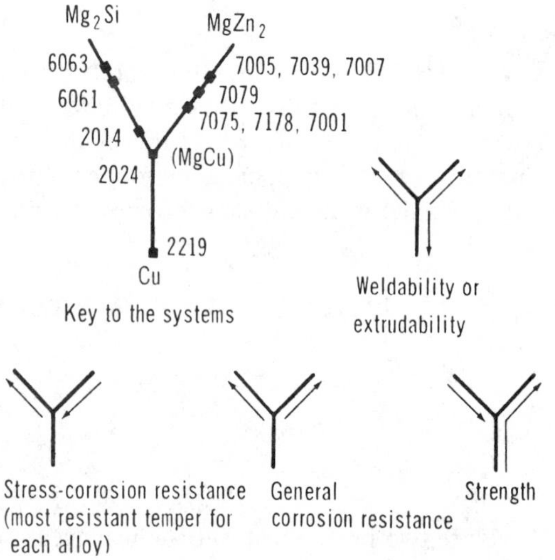

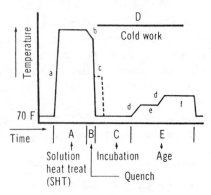

Mg_2Si $MgZn_2$

6063   7005, 7039, 7007
6061   7079
2014   7075, 7178, 7001
2024   (MgCu)

            2219
        Cu
    Key to the systems

            Weldability or
            extrudability

Stress-corrosion resistance   General   Strength
(most resistant temper for   corrosion resistance
each alloy)

Fig. 60.1 — Diagrams illustrate general properties of heat-treatable aluminum alloys. Arrows indicate direction of improving properties, but not magnitude. No arrow indicates no change.

The Process of Heat Treatment. Heat treating is a time-temperature process with or without cold working of the alloy. A typical cycle is shown in *Fig.* 60.2. Vital features are the choice of solution heat treatment temperature, quench rate, and aging practice. Aging can consist of holding at ambient temperature (natural aging) until substantially stable properties are achieved, or holding at an elevated

Fig. 60-2 — The principal and secondary features of heat treatment processes for aluminum include (a) heat-up to SHT temperature (850 to 1,000 F); (b) quench delay (cooling between leaving SHT oven and quench); (c) interrupted or step quench (unusual); (d) heat-up during age; (e) first step age (200 to 400 F); (f) second step age (200 to 400 F).

Temperature

D
b   Cold work

a   c

70 F   d   d   f
                    d   e

Time | A |B| C | E |

        Solution   Incubation   Age
        heat treat
        (SHT)   Quench

How Aluminum Association Designates Tempers

-W Solution Heat Treated: An unstable temper applicable only to alloys which spontaneously age at room temperature after solution heat treatment. Specific only when period of natural aging is indicated (-W $^1/_2$ hr).

-T Thermally Treated to Produce Stable Tempers Other than -F, -O, or -H: Applies to products which are thermally treated, with or without supplementary strain hardening, to produce stable tempers. The -T is always followed by one or more digits. Numerals 1 through 10 indicate one specific sequence of basic treatments as follows:

-T1 Partially Solution Heat-Treated and Naturally Aged to a Substantially Stable Condition: Applies to products partially solution heat treated by an elevated-temperature rapid-cool fabrication process (casting or extrusion).

-T2 Annealed (cast products only): Indicates a type of annealing treatment to improve ductility and increase dimensional stability.

-T3 Solution Heat Treated and Cold Worked: Applies to products cold worked to improve strength, or in which effect of cold work in flattening or straightening is recognized in applicable specifications.

-T4 Solution Heat Treated and Naturally Aged to a Substantially Stable Condition: Applies to products not cold worked after solution heat treatment, but in which effect of cold work in flattening or straightening may be recognized in applicable specifications.

-T5 Partially Solution Heat Treated and Artificially Aged: Applies to products artificially aged after an elevated-temperature, rapid-cool fabrication process, to improve mechanical properties and/or dimensional stability.

-T6 Solution Heat Treated and Artificially Aged: Applies to products not cold worked after solution heat treatment, but in which effect of cold work in flattening or straightening may be recognized in applicable specifications.

-T7 Solution Heat Treated and Stabilized: Applies to products stabilized to carry them beyond point of maximum hardness, providing control of growth and/or residual stress.

-T8 Solution Heat Treated, Cold Worked, and Artificially Aged: Applies to products cold worked to improve strength, or in which effect of cold work in flattening or straightening is recognized in applicable specifications.

-T9 Solution Heat Treated, Artificially Aged, and Cold Worked: Applies to products cold worked to improve strength.

-T10 Partially Solution Heat Treated, Artificially Aged, and Cold Worked: Applies to products artificially aged after an elevated-temperature, rapid-cool fabrication process and then cold worked to improve strength.

A period of natural aging at room temperature may occur between or after the operations listed for tempers -T3 through -T10. Control of this period is exercised when it is metallurgically important.

Additional digits may be added to designations -T2 through -T10 to indicate a variation in treatment which significantly alters the characteristics of the product:

-TX51 Stress Relieved by Stretching: Applies to products stress relieved by stretching the following amounts after solution heat treatment: Sheets and Plates: $1^1/_2$ to 3% permanent set. Rods, Bars, and Shapes: 1 to 3% permanent set. Applies to sheets and plates and rolled or cold finished rods and bars. These products receive no further straightening after stretching. Applies to extruded rods, bars, and shapes only when designation is subdivided as follows:

-TX510: Applies to extruded rods, bars, and shapes which receive no further straightening after stretching.

-TX511: Applies to extruded rods, bars, and shapes which receive minor straightening after stretching to comply with standard tolerances.

-TX52 Stress Relieved by Compressing: Applies to products stress relieved by compressing after solution heat treatment to produce a nominal permanent set of $2^1/_2$%.

-TX53 Stress Relieved by Thermal Treatment: The following two-digit -T temper designations have been assigned for some wrought products heat treated by the user:

-T42: Applies to some alloys solution heat treated by the user.

-T62: Applies to some alloys solution heat treated and artificially aged by the user.

temperature (artificial aging) to achieve stable properties in a shorter time. Other details of the cycle may include cold working which can have important effects (discussed later).

The specific details of most heat treatment practices are contained in the military specifications for aluminum alloys (MIL-H-6088D. Amendment 2, Dec. 23, 1968).

Solution Heat Treatment. The primary objective of solution treatment is to obtain as complete a solid solution of the alloying elements as possible without producing melting or undesirable recrystallization. This can be achieved by either mill (or press) quenching or by a separate (formal) solution heat treatment cycle.

Mill quenching comes directly after hot forming (rolling or extrusion) and is carried out at a temperature above the solvus of the alloy. This process can be less expensive than a separate heating cycle.

However, there are two drawbacks: control of temperature during hot working and hot working characteristics above the solvus are poor. Consequently, this process is only suitable for dilute alloys with a large temperature difference between solvus and solidus.

A separate, formal solution heat treatment is generally required to achieve maximum property levels. Holding time is controlled by the metallurgical structure of the material and can vary from as low as 5 min. to as long as 12 to 16 hr. Commercial practices usually do not produce complete solid solution, and recently it has been shown that extended solution treatment times can lead to increased properties, where recrystallization or grain growth presents no problems.

Quench Delay. Military specification MIL-H-6088D details the maximum quench delays permitted for high-strength alloys. This specification is really aimed at controlling the minimum temperature of the metal immediately prior to quenching to avoid premature precipitation.

Quenching. After solution heat treating, quenching is required to retain both the solid solution and a sufficient supersaturation of vacancies for effective aging. Too slow a quench can lead to: 1. Precipitation of large equilibrium precipitates which contribute little to strength. 2. Increased precipitation in grain boundaries which may reduce ductility or corrosion resistance. 3. Annealing out of vacancies which would subsequently alter the precipitation kinetics.

The maximum quench rate at which these symptoms appear will vary from alloy to alloy. If this critical quench rate is high (100 to 10,000 F per sec.), the alloy is quench sensitive. If the rate is low (0.1 to 1 F per sec.), the alloy is quench insensitive.

The influence of quench rate on mechanical properties of various alloys in the T6 temper is shown in *Fig.* 60.3. The more highly alloyed materials are the most sensitive. It is well known that chromium (or the unrecrystallized structure which chromium promotes) greatly increases quench sensitivity of the 7000 series alloys, particularly those with high zinc, magnesium, and copper contents.

One factor almost universally overlooked is that of optimizing the aging practice for slowly quenched material. In *Fig.* 60.4, aging curves show that protracted aging of very slowly quenched Al-Zn-Mg alloys can yield high properties. Also, the beneficial effect of incubation at room temperature prior to aging is shown. These data suggest that poor quenching primarily lowers the vacancy concentration which in turn lowers the maximum temperature for homogeneous nucleation and also slows the diffusion of solute atoms to GP zones.

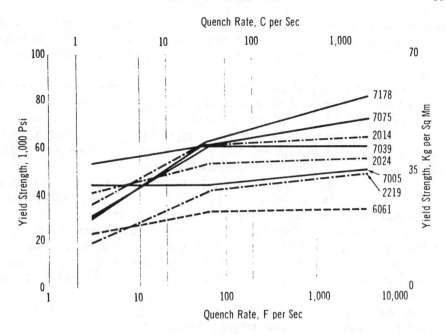

Fig. 60.3 — These data show the effect of quench rate on yield strength of various aluminum alloys after aging to T6 temper.

Note that if only a typical T6 aging practice had been used (say 16 hr. at the aging temperature used in *Fig.* 60.4), then virtually no age hardening would take place for the slow-cooled specimen with no incubation. Thus, a false idea of the quench sensitivity of this alloy would have been formed. A similar, less pronounced effect is observed in some higher-strength alloys.

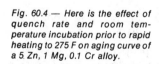

Fig. 60.4 — Here is the effect of quench rate and room temperature incubation prior to rapid heating to 275 F on aging curve of a 5 Zn, 1 Mg, 0.1 Cr alloy.

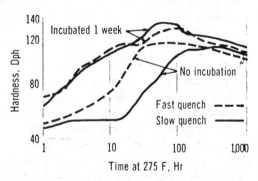

Mechanical properties are generally improved by specifying the fastest possible quench rate. For other properties, the reverse can be true. With copper-free 7000 series alloys, for example, a slow quench rate is particularly beneficial to the stress-corrosion resistance. Also, residual stresses and distortion are generally minimized with a slow quench.

Thus, the selection of a suitable quench medium may depend upon several factors. Cold water, the fastest quench, is also the most convenient and economical. Hot water on heavy sections minimizes residual stresses and distortion of parts.

New quenchants may further reduce quenching stresses. Such stresses are not necessarily caused simply by speed of quench but are related more directly to the change in heat transfer coefficient with surface temperature and from point to point on the surface due to the formation of steam films. One method of breaking down such steam films is the technique called "superquenching," utilizing high pressure and high volumes of coolant.

Another approach is to deposit a thin insulating film on the surface of the part to even out the heat transfer to the quenchant and so increase the over-all speed of quenching in a boiling fluid. The vapor-film mode of cooling is also effective. With liquid nitrogen as the quenchant, a stable vapor-film exists to below room temperature. Distortion in such quenches is low; however, the cost and the low cooling power generally limit use to complex sheet parts.

It is also possible to reduce residual stresses through the use of hot nonaqueous coolants such as oil, liquid metals, or fused salts. Generally, these quenchants have such a high boiling point that vapor phase cooling is not obtained.

These high-temperature quench media are used in the process known as isothermal quench aging (IQA). Although IQA has shown definite advantages, industry has been reluctant to take on the problems such as fire hazards, postquench cleanup, and bath maintenance generally associated with nonaqueous quenchants.

Stress Relief. Stress relief is generally done by stretching; however, some reduction can be achieved by roller leveling. Surface residual stress can be reduced by shot peening or planishing.

Thermal treatments above about 425 F will dimensionally stabilize and stress relieve parts, although these treatments result in lower mechanical properties. Wrought products are stress relieved only partially during artificial aging treatments for maximum hardening. For example: residual quenching stresses are reduced about 10% for 7075 alloy aged at 240 F and by about 30% for 2014 aged at 320 F.

A method especially suitable for relief of residual stresses in complex castings or forgings is that of "up-quenching." The method involves strain reversal which is accomplished by reversing the quenching process — that is, the part is cooled to cryogenic temperatures, then up-quenched with steam. The method can result in tensile stresses on the surface, however, which may be undesirable.

Incubation. Incubation is merely a portion of the aging cycle and if no elevated-temperature aging is to be used, is synonymous with natural aging.

In the case of Al-Zn-Mg-Cu alloys, T6 mechanical properties can vary with incubation time prior to aging. No satisfactory explanation has been offered for this effect, though it is almost certain that reversion is involved. The use of a slow heating rate between room temperature and the aging temperature or the use of a two-step aging practice can eliminate these variations.

In the case of the 6000 series alloys, increased incubation leads to a continual decrease in final properties for highly alloyed material or to an increase for dilute alloys. In these alloys, interruption of the quench at a temperature above the GP zone solvus or brief high-temperature treatments above the GP zone solvus can retard the influence of incubation by annealing out vacancies at the high temperature.

Postquench Working. As was pointed out above, working is useful for the relief of quenching stresses. Postquench working can also serve other purposes including simple straightening, flattening, increasing strength, improving corrosion resistance, reducing incubation effects, and providing nucleation sites for the later stages of precipitation.

In alloys 2021, 2024, and 2219, the T8 temper (1 to 8% cold work, followed by artificially aging) results in markedly increased strength and stress corrosion resistance. Cold work also accelerates the aging process. These effects may be obtained by any method of cold working, including explosive shocking which can be applied so that no dimensional change takes place. In most cases, cold working results in the nucleation of fine precipitates on the dislocations introduced during plastic deformation. Alloy 2014 is one exception.

In the Al-Mg$_2$Si system, cold work can be added before aging, between steps of a two-step age, or after aging with increasing improvement in mechanical strength.

In the instance of Al-Zn-Mg alloys, even 50% cold work prior to aging produces little change in final mechanical properties. The aging process is merely accelerated.

The combination of low-temperature aging, cold working, and final aging at a higher temperature substantially increases strength without loss of elongation for 7000 alloys. Cold working after complete aging leads to similar increases in strength but elongation is greatly reduced. Thus, an age-cold work-age cycle appears to offer the best improvement in strength and ductility.

Aging. Isothermal aging can be used up to the GP zone solvus, provided the heating rate is not too rapid. To age above the GP zone solvus, a two-step practice or a slow heating rate must be specified. Approximate ranges for the GP zone solvus of the main systems are:

2000 series	350 to 410 F
6000 series	350 to 425
7000 series	230 to 325

In general, the highest mechanical properties are achieved at the lowest practical aging temperature. However, the recently developed stress-corrosion-resistant temper (T73 for 7075) and exfoliation-resistant tempers (T76 for 7075 and 7178), require some overaging. To do this in reasonable times, it is necessary to go to aging temperatures above the GP zone solvus. Hence, either a two-step aging practice (with the first step below the GP zone solvus) or a slow heatup cycle must be adopted.

The beneficial effects of these practices on mechanical properties are shown in *Fig.* 60.5. Optimum properties are obtained with either a two-step aging practice or a heat-up rate of less than approximately 1 F per min. Peak properties using these aging practices are in fact close to those obtained using a normal T6 aging practice. To attain the stress-corrosion resistance of the T73 tempers, however, it is necessary to age beyond the peak. Typical maximum heating rates for fully loaded commercial aging ovens are in the range 0.3 to 1.5 F per min. Therefore, in the plant, two-step aging may be unnecessary. As a precaution, though, the hold at a low temperature is specified for these tempers.

Commercial Heat Treating. Two basic furnace types for solution heat treating are the salt bath and the gaseous atmosphere furnace. Salt baths offer high heat capacity, fast-heat-up rates, and uniform temperature distribution. The major user of the salt bath furnace is the aircraft industry because of the complex shapes of castings, extrusions, and formed wrought products treated. Disadvantages are related to cost, dragout, and the corrosive nature of the medium. In addition, it is necessary that the piece to be heat treated be clean to avoid contamination of the salt and be free of moisture to avoid explosion.

Handling of parts, especially heavy sections, can be a problem. The nature of

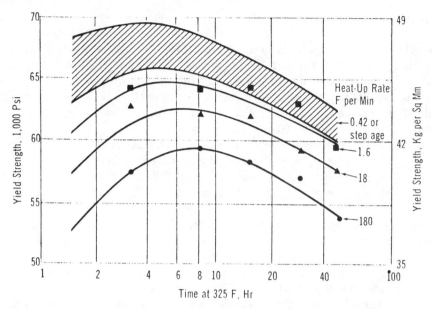

Fig. 60.5 — The influence of rate of heating to aging temperature on aging curves for 7075 plate (2 in. thick) is depicted by these curves.

the salt bath furnace requires that parts be lifted upward to clear the top during transfer to a quench tank. A safe distance must be maintained between the bath and the quench tank due to the incompatibility of molten salt and water. Naturally, this distance cannot be too far because the temperature drop could negate the heat treatment. Transferring is usually done manually or with the aid of an overhead hoist. Only through the use of sophisticated handling equipment does the salt bath lend itself to high production rates. Dragout from the furnace combined with the corrosive nature of salt dictates that an effective rinse be employed after quenching.

Most aluminum sheet and plates are heat treated in either air or controlled-atmosphere furnaces. Controlled atmospheres may be achieved by purging with bottled gas such as dry nitrogen or argon. However, for large-scale operations, the controlled atmosphere is obtained through the combustion of air and natural gas. It is desirable to eliminate oxygen and water vapor from the atmosphere because both are highly reactive with aluminum and can cause stains, blisters, and high-temperature oxidation (HTO). Inasmuch as water vapor is one of the products of combustion, it is necessary to separate it from the other gases, and this can be done by cooling. A proper mix of air and natural gas will yield, after drying, an atmosphere composed mainly of nitrogen and carbon dioxide with less than 0.3% O_2.

Both air and controlled-atmosphere furnaces may utilize electricity, gas, or oil as the heat source. However, direct firing is rarely employed with gas and oil due to the detrimental effects of the products of combustion, especially sulfur dioxide. These fuels are fired through radiant tubes, and the gas is circulated past these tubes. With all atmosphere furnaces, rapid circulation of air is essential for uniform temperature distribution and this is provided by fans located within the furnace. Proper design and baffling is also required to eliminate hot and cold spots.

There are two basic types of atmosphere furnaces. In one, the metal is held stationary within the furnace while in the other the metal is propelled through a long, multizoned unit — coils are handled this way. Cut-to-length sheets and plates can be processed by both types of furnaces.

In a vertical heat treatment furnace, metal is supported vertically during the thermal operation. This type is normally erected directly over a quench tank, and access is through a sealable opening in the bottom. Metal is lifted upward by a hoist. After sufficient time at temperature, the furnace bottom is opened, and the load is plunged into the quench tanks.

Vertical furnaces are relatively small in comparison with continuous types. Typical over-all dimensions are 10 ft. by 10 ft. by 30 ft. Production rates are also relatively low due to the sequence of operations (load, heat, soak, quench, and unload). The furnace has the advantages of simple design, very few zones, and the ability to monitor temperature readily.

In a continuous heat treatment furnace, metal is propelled through a series of zones at a predetermined speed which is a function of the alloy itself, furnace heat input, gas velocity, and furnace length.

In a continuous furnace for cut-to-length sheets and plates, the metal is heat treated in the horizontal position. It is supported on and moved along by a series of drive rolls or cables. The load enters at one end of the furnace, is heated and soaked, and then, at the opposite end, quenched and unloaded. The drive system in the last few zones, separate from the balance of the furnace, provides for the rapid exit of metal into the quench system. A series of water sprays floods the load on both top and bottom surfaces.

The length of the furnace depends to a large extent upon the type of product to be heat treated — the lighter the gage, the shorter the furnace. A furnace 60 ft. long is sufficient for the heat treatment of 0.010 to 0.125 in. sheet gages. For plates as thick as 6 in., furnaces 200 ft. long may be required. Typical line speeds are in the range of 10 to 15 ft. per min. for sheets. Line speeds for plates range from 3 in. per min. to 6 ft. per min.

The same type of furnace is used in the continuous heat treatment of coiled sheet, but it requires more sophisticated auxiliary equipment than one treating cut-to-length material. Faster line speeds are necessary because a part of the coil is always entering the quench. The unwind and take-up reels must be synchronized with the line speed of the furnace, and provisions must be made for the rapid changing and attachment of succeeding coils. This is achieved with accumulator towers at both ends of the furnace.

A recent development is "air flotation." The force needed to propel the coil is provided by "pinch" rolls on the entry and exit sides of the furnace. The sheet in the furnace is supported by high velocity air streams impinging on the sheet surfaces. The streams are recirculated past burner tubes, also serving as a heat source. The high velocity required to support the sheet also provides rapid heating rates. Because relatively fast line speeds can be attained (in excess of 100 ft. per min.), a furnace of this type is capable of high production rates. It also has the advantage of eliminating contact marks.

Flattening Operations. One of the possible consequences of quenching is severe distortion of the metal. Usually, the lighter the gage, the more distortion encountered. Therefore, a flattening operation is normally employed on heat-treated metal.

Cut-to-length material is flattened by stretching, rolling on a planishing mill, or by reverse flex leveling. Material in coils is flattened by tension or roll flex leveling or both and by light reductions on a cold rolling mill ($^1/_2$ to 6%, depending on final temper).

Applications. While the 6000 series of heat-treatable alloys are applied in a wide range of products, largely as extruded shapes, the principal applications of the 2000 and 7000 series alloys are in the aircraft and aerospace industries. Modest inroads are being made by titanium alloys, beryllium, and composite materials, but aluminum is still the principal structural material in both aircraft and missiles. Even in the new giant transport and passenger planes, as well as the supersonic *Concorde,* designers working with materials engineers have chosen aluminum as the major structural material. The alloys that are popular in these aircraft are 2014, 2024, 7075, and 7079 — 2618 is used in the *Concorde* for its better elevated-temperature properties.

The aluminum industry and government research agencies are actively working to improve the general and stress-corrosion resistance and mechanical properties of present alloys and tempers. One important demand in the new large aircraft is for heavier sections. Aircraft design has moved from riveted structures made from thin sheets and extrusions to integrally stiffened skins machined from heavy plate and large extrusions. Forgings up to 8 in. thick are being made and heat treated.

In our laboratory and others, development work is aimed at reducing quench sensitivity in the highest-strength 7000 alloys to maintain high properties in thick sections. One new alloy meeting some of the above requirements is 7080, which is currently being evaluated.

The more recently developed medium-strength, weldable 7000 series alloys, such as 7004 and 7005, have promising futures. They are particularly suited to the ground transportation field, where high strength is not as critical as it is in aircraft, but where a medium-strength, readily weldable material can be effectively employed. The major limitation lies in poor stress-corrosion resistance in the short transverse direction. However, this problem can be avoided by recognizing it during design and construction. These alloys have been successfully applied in trucks, buses, trains, LP gas tanks, and portable bridges.

Heat Treating Copper & Copper Alloys

Mary W. Covington

Fabricators of copper and copper alloy products may require three heat treatments in their metalworking operations:

1. Annealing — heating cold-worked metal to a temperature that causes recrystallization and grain growth.

Miss Covington is Supervisor, Information Systems. Copper Development Association Inc., New York.

TABLE 60-I — **Annealing Temperatures for Widely Used Coppers and Copper Alloys**

Copper and Copper Alloy No.	Name	Temperature, F
101-103	Oxygen-free copper	700-1,200
104-107	Oxygen-free silver-bearing	900-1,400
108	Oxygen-free low phosphorus	700-1,200
110	Electrolytic tough pitch	700-1,200
111	Electrolytic tough pitch, anneal resistant	900-1,400
113-116	Silver-bearing tough pitch	900-1,400
120-122	Phosphorus deoxidized	700-1,200
125, 127-130	Fire-refined tough pitch with silver	750-1,200
142-147		800-1,200
155		900-1,000
162	Cadmium copper	800-1,400
192		1,300-1,500
194		700-1,200
195		750-1,100
210, 220	Gilding, commercial bronze	800-1,450
226	Jewelry bronze	800-1,400
230	Red brass	800-1,350
240	Low brass	800-1,300
260	Cartridge brass	800-1,400
268, 270	Yellow brass	800-1,300
280	Muntz metal	800-1,100
314, 316, 330, 332, 340, 349	Leaded commercial bronzes and brasses	800-1,200
335, 342, 353	Low, medium, and high leaded brasses	800-1,300
350, 356, 360, 365-368, 370, 377, 385	Leaded brasses, free-cutting brass, leaded Muntz, forging brass, and architectural bronze	800-1,100
411		800-1,400
413		900-1,400
425		800-1,300
443-445, 464-467, 482, 485	Inhibited admiralty and naval brasses	800-1,100
505	Phosphor bronze, 1.25% E	900-1,200
510-511, 521, 524, 544	Phosphor bronzes	900-1,250
608	Aluminum bronze, 5%	1,000-1,200
610		1,100-1,250
613		1,125-1,600
614	Aluminum bronze, D	1,125-1,650
618, 623-625		1,100-1,200
619		1,000-1,450
630, 642		1,100-1,300
632		1,150-1,300
638		750-1,100
651	Low-silicon bronze, B	900-1,250
655	High-silicon bronze, A	900-1,300
667	Manganese brass	930-1,300
674-675, 687		800-1,100
688		750-1,100
694	Silicon red brass	800-1,200
706	Copper nickel, 10%	1,100-1,500
710, 715	Copper nickels, 20% and 30%	1,200-1,500
725		1,200-1,475
745, 752	Nickel silvers, 65-10 and 65-18	1,100-1,400
754, 757, 770	Nickel silvers, 65-15, 65-12, and 55-18	1,100-1,500
782	Leaded nickel silver	930-1,150
953-955	Aluminum bronze castings	1,150-1,225

Source: Copper Development Assn. Inc.

2. Stress relieving — heating to reduce residual stresses remaining in some copper alloys as a result of nonuniform plastic deformation below the recrystallization temperature.

3. Solution treatment and precipitation heat treatment — techniques applicable to some specialized copper alloys that substantially increase hardness and strength. A major advantage of these processes is that they may be done after the manufacturer has utilized copper's ductility to form complex shapes.

Annealing. The fabrication of components from copper alloy sheet and strip may require annealing. Temperatures commonly used for coldworked copper and copper alloys are given in TABLE 60-I.

Annealing is primarily a function of metal temperature and time at temperature. The object is to soften the metal for subsequent cold working, obtaining the optimum combination of ductility, strength, and surface texture.

For copper and copper alloys, grain size is the standard test for annealed material which has been previously cold worked. Grain size requirements are spelled out in the applicable purchasing specifications. For convenience, Rockwell-type hardness testers are often used to approximate grain size.

Annealing schedules are based on the annealing characteristics of the copper or copper alloy, and the effects of variables such as furnace design, furnace atmosphere, heat source, part shape, and the ability to accurately control treatment conditions.

TABLE 60-II — Typical Stress Relieving Temperatures for Wrought Coppers and Copper Alloys

Copper and Copper Alloy No.	Name	Temperature, F						
		Sheet and Strip		Rod and Wire			Tube	
		Flat Products*	Parts	Rod†	Wire‡	Parts	Tube§	Parts
110	Electrolytic tough pitch	355	355	355	355	355	—	—
120	Phosphorus deoxidized DLP	—	—	—	—	—	430	390
122	Phosphorus deoxidized DHP	—	—	—	—	—	465	430
142	Phosphorus deoxidized DPA	—	—	—	—	—	500	465
210	Gilding 95%	525	525	—	—	—	—	—
220, 226	Commercial bronze, jewelry bronze	525	525	570	500	525	—	—
230, 240	Red brass, low brass	525	525	570	500	525	625	525
260	Cartridge brass	500	500	555	480	500	610	500
270	Yellow brass 65%	500	500	555	480	500	555	500
314	Leaded commercial bronze	—	—	570	500	525	—	—
330, 332	Low and high leaded brasses	—	—	—	—	—	610	500
335	Low leaded brass	—	—	555	480	500	—	—
340, 350	Medium leaded brasses	500	500	—	—	—	—	—
353, 356, 360, 377	Leaded, free cutting, and forging brasses	—	—	555	480	500	—	—
430		525	525	570	500	525	—	—
434		525	525	—	—	—	—	—
443, 445	Admiralty	—	—	—	—	—	610	500
462, 464, 485	Naval brasses	—	—	555	480	500	—	—
510	Phosphor bronze, A	525	525	570	500	525	—	—
521	Phosphor bronze, C	—	—	570	500	525	—	—
544	Phosphor bronze, B-2	—	—	570	—	525	—	—
651, 658	Silicon bronzes	—	—	570	525	525	—	—
681		—	—	555	480	500	—	—
687	Aluminum brass, arsenical	—	—	—	—	—	625	555
697		—	—	680	680	680	—	—
706	Copper nickel, 10%	790	790	—	—	—	895	790
715	Copper nickel, 30%	860	860	—	—	—	970	860
735		715	715	750	660	715	—	—
745	Nickel silver, 65-10	—	—	645	555	610	—	—
752	Nickel silver, 65-18	715	715	—	—	—	—	—
754	Nickel silver, 65-15	—	—	750	660	715	—	—
757	Nickel silver, 65-12	—	—	660	570	645	—	—
770	Nickel silver, 55-18	645	645	—	—	—	—	—

*Extra hard; †Half hard; ‡Spring; §Hard drawn. **Note:** Time is 1 hr at temperature with the exception of tube which is 20 min.
Source: Copper Development Assn. Inc.

TABLE 60-III — **Typical Stress Relieving Temperatures for Cast Copper Alloys**

Copper Alloy No.	Temperature, F
813-822	500
824-828	390
833-948	500
952-958	600
966-978	500
933	950

Note: Time is 1 hr per in. of section thickness except for Copper Alloy 993 which is 4 hr per in.
Source: Copper Development Assn. Inc.

Factors. Optimization of the annealing operation involves consideration of several additional factors. Among them are the effects of prior cold work, time, hydrogen embrittlement, oxidation, and staining.

Increasing the amount of cold work prior to annealing lowers the recrystallization temperature. The smaller the degree of prior deformation, the larger the grain size after annealing.

In commercial brass mill practice, copper alloys are usually annealed at successively lower temperatures as the material approaches the final anneal, with intermediate cold reductions of 35% and more wherever possible. This gradual reduction of grain size gives better control over a uniform final grain size.

The normal practice is to give a full anneal and then obtain desired properties by controlled cold working.

It is not always practical to anneal for a specific tensile strength or hardness between the normal cold worked and fully softened conditions because of the difficulties of accurately controlling heating rate and temperature. A small change in metal temperature, for example, results in an extremely rapid change in properties.

Time. Furnace variations and charge size affect the required time at temperature. Usually, however, $1/2$ to 1 hr. is sufficient. Because of rapid grain growth, brasses are generally more sensitive to time than coppers.

To prevent hydrogen embrittlement in electrolytic tough pitch coppers, avoid heating in hydrogen above 750 F (400 C). Instead, choose a suitable reducing or slightly oxidizing atmosphere to prevent the internal formation of steam by the combination of cuprous oxide and hydrogen.

Oxygenfree, deoxidized coppers, and all copper alloys are not subject to "gassing" at any temperature.

Oxidation can be minimized by specifying bright annealing-type furnaces or special protective atmospheres.

Staining is avoided by the thorough removal of sulfur-containing lubricants or by using low-sulfur furnace fuels.

Stress Relieving. Residual internal stresses in cold-worked copper alloys are relieved by low-temperature heat treatments below the recrystallization temperature. Typical stress relieving temperatures for wrought and cast alloys are given in TABLES 60-II and III.

Thermal stress relief is employed to safeguard against stress-corrosion cracking (SCC) in copper alloys, particularly brasses containing 20% Zn or more.

TABLE 60-IV — **Precipitation Heat Treatments for Heat Treatable Wrought and Cast Copper Alloys**

Copper and Copper Alloy No.	Name	Solution Treating Temperature, F	Time, Min	Precipitation Treating Temperature, F	Time, Hr
Wrought					
150	Zirconium copper				
	Solution heat treated and aged	1,650-1,700	5-30	930-1,020	1-4
	Solution heat treated, cold worked and aged	1,650-1,700	5-30	710-880	1-4
170, 172	Beryllium copper				
	Solution annealed (AT)	1,425-1,475	10-30	600	3
	¼ hard (¼ HT)			600	2
	½ hard (½ HT)			600	2
	Hard (HT)			600	2
175	Beryllium copper				
	Solution annealed (AT)	1,675-1,725	10-30	900	3
	½ hard (½ HT)			900	2
	Hard (HT)			900	2
182, 184-185	Chromium copper				
	Solution heat treated and aged	1,800-1,850	10-30	800-930	2-4
	Solution heat treated, cold worked and aged	1,800-1,850	10-30	800-930	2-4
190-191	Copper-nickel-phosphorus				
	All solution annealing and age hardening annealing tempers	1,300-1,450	10-30	800-900	2-4
—	PD-135	1,700-1,760	5-120	850-1,050	¼-2
619	Aluminum bronze				
	Hard, spring, extra spring	Low-temperature heat treat at 450 F for 1 hr			
647	Copper-nickel-silicon				
	Solution annealed and drawn	1,375-1,475	10-30	850-900	1½
717	———	1,800-1,850	10-30	950	2-3½
Cast			Hr/In.		Hr/In.
813	———	1,800-1,850	1	900	2
814	———	1,830-1,850	3	900	2
815	———	1,830-1,850	1	900	3
817	———	1,650-1,700	1	850	3
818	———	1,650-1,700	1	900	3
820	———	1,650-1,700	3	900	3
821	———	1,650-1,700	1	850	3
822	———	1,650-1,700	1	835-850	2
824	———	1,450-1,500	1	650	3
825	———	1,450-1,475	1	650	3
826	———	1,450-1,475	1	650	3
827	———	1,450-1,475	3	650	3
828	———	1,450-1,500	1	650	3
947	———	1,425-1,475	2	600	5
948	———	—	—	600	6-10
953	———	1,585-1,635	1	—	—
954	———	1,600-1,675	1	—	—
955	———	1,585-1,635	1	—	—
966	———	1,800-1,850	1	950	3

Source: Copper Development Assn. Inc.

Although different copper alloys vary in their susceptibility to stress-corrosion cracking, the presence of even mild corrodents can contribute to failure. Thermal stress relieving avoids this possibility. Test procedures to determine freedom from high residual stresses are covered by ASTM B154 and other techniques under development.

Copper itself (with rare exceptions) or fully annealed products are not subject to SCC failures.

Practically, the ideal stress relief cycle to retain mechanical properties is either the highest possible temperature for the shortest time, or the lowest temperature for the longest time. Interestingly, hardness and strength of severely cold-worked parts show slight increases under a low-temperature heat treatment.

Hardening. Mechanical properties of most coppers and copper alloys are achieved by cold working. But for those heat-treatable copper alloys containing small amounts of beryllium, chromium, zirconium, nickel, silicon, or phosphorus, unusually high strength and hardness can be obtained by precipitation hardening.

All precipitation-hardening copper alloys have similar metallurgical characteristics: they can be obtained in their soft condition by quenching from a high temperature (solution treating), and then subsequently hardened by heating to a moderate temperature (precipitation treating).

Chief advantages are:

1. Customer fabrication is easily performed in the soft, mill-supplied, solution-annealed condition.

2. The hardening heat treatment performed by the fabricator is relatively simple. It is carried out at moderate temperatures in an uncontrolled atmosphere, controlled cooling is not needed, and the treatment is not critical with respect to time.

3. Different combinations of properties — including strength, hardness, ductility, conductivity, impact resistance, and anelasticity — can be obtained by varying hardening times and temperatures. This particular application's requirements determine the type of hardening treatment.

Mill. Further hardening of the alloy after fabrication is not necessary if the mill performs the precipitation heat treatment operation. A stress relief may, however, be desirable to remove stresses induced in the part during fabrication.

TABLE 60-IV gives time and temperature procedures for precipitation hardening wrought and cast heat-treatable copper alloys.

Another type of hardening mechanism is associated with aluminum bronzes containing more than about 10% Al. These alloys are hardened by cooling rapidly from a high temperature to produce a martensitic structure, and then tempering at a lower temperature.

ACKNOWLEDGEMENT

This chapter is included by permission of *Metal Progress*, from the September 1970 and the May 1974 issues, copyright American Society for Metals.

61

Shot Peening

U SE of surface compression to improve strength is undoubtedly ages old and was practiced as cold hammering by old Toledo swordsmiths, for example, to achieve phenomenal results. Benefits derived from such cold working of metals, often incidental with many manufacturing processes such as rolling, heading, bending, stretching, etc., consist mainly of increased fatigue strength imparted to machine parts when their metal structure is stretched, twisted or compressed beyond the elastic limit.

Shot Blasting. Selective surface compression imparted by a rain of metal shot impelled either from the rotating blades of a wheel or by air blast achieves like beneficial results on a great variety of machine parts such as crankshafts, springs,

Fig. 61.1 — Arrangement for production shot peening Allison engine connecting rods at Cadillac Motor Car Co. plant. Photo, courtesy Whellabrator-Frye Corp.

shafts, gears, torsion bars, connecting rods, etc., *Fig.* 61.1. Each metallic shot has the effect of a tiny peen hammer to distort the surface structure of the metal and set up compressive stresses. This compressively stressed layer usually extends 0.005 to 0.010-inch below the surface and is balanced by residual tensile stresses within the interior.

Cold Working. Inasmuch as most fatigue cracks have a tendency to start at the tension point of a stressed section, a compressively stressed outer layer at such a point will act to reduce the net applied working stress. Cold work thus increases the strength of the part but sacrifices a portion of the ductility. Intensity of peening therefore must be limited because an excess of working acts to reduce the resistance of a part under repeated stress by exhausting the ductility and producing incipient cracking. As little as 3 or 4 per cent ductility in a machine part after assembly, however, is considered sufficient for all probable readjustments of stress distribution.

Equipment and Processing. Although the equipment utilized for shot peening is essentially the same as that used for blast cleaning, it necessarily requires more accurate control and auxiliary equipment for consistent results. In rare cases blast cleaning of certain parts results in an increase in fatigue strength but more often the opposite is the case, *Fig.* 61.2. Increased fatigue life has been achieved with certain aluminum wing spars sand-blasted for finish, but most all investigations of ordinary sand blasting show no appreciable increase.

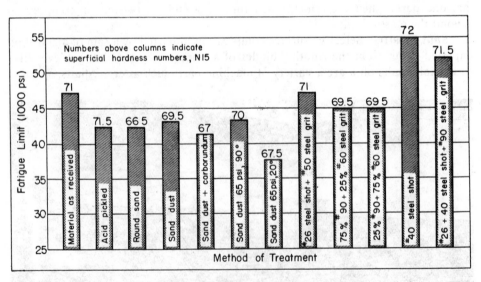

Fig. 61.2 — Graph of fatigue limits resulting from various methods of surface treatment using a specimen of 0.054 per cent carbon steel heat treated to 115,000 psi tensile. From "Prevention of Fatigue of Metals" — Battelle Memorial Institute, John Wiley & Sons, Inc., New York, 1941, Page 59.

The important difference therefore lies in the type of blasting material and its condition. For successful peening, whole unbroken shot is necessary. Broken shot, having a much lighter mass, gives a lower or varying peening intensity and inconsistent results. In addition, the sharp edges of broken shot pit and abrade the surface creating scars which act as stress raisers with a resultant decrease in fatigue life. Production peening equipment thus requires continuous removal of broken shot and a rationing system to introduce new shot at the same rate at which the fines are removed. Most recent and successful type of shot from the standpoint of

UNCONVENTIONAL METHODS

62

Chemical Machining

 M ANUFACTURED parts that are produced by means of machining operations generally are shaped through the removal of unwanted metal areas with ship cutting tools. These are of direct cutting sharp edged form tools or abrasives. However, demand for the precise manufacture of complex parts in materials which cannot be cut successfully in production and parts of extremely fine detail has led to the development of processes that do not rely on mechanical cutting but rather employ electrical and/or chemical action to remove the necessary metal in forming parts. Chemical machining processes fall into this category of unconventional but highly successful methods.

Scope of Process. In general, the chemical machining processes are of sophisticated character and suited primarily to the production of parts in limited quantities or the production of parts that can be most easily made by these methods. Size range and complexity of designs that fall within the range of the chemical machining methods is growing as new and larger equipment is made available.

Falling into the group of processes utilizing chemical etching action on metals to create the needed shape are variations such as: photoforming, chemical milling, electrochemical milling, electrochemical machining, and electrochemical grinding.

All of these processes provide answers to manufacturing problems that simply are unobtainable by other methods. In many instances, the unit cost compared to that resulting from conventional methods is highly favorable, so their use should not be overlooked in the design of unique or critical parts in which the advantage of stress-free forming is desirable.

Photoforming. Basically an extension of the technology of photoengraving, this process utilizes a photo-resist mask made by photographic processes to permit selective removal of metal. Metal is removed by etching away the areas of metal not protected by the photo-resist coating, *Fig.* 62.1. The process lends itself to production of parts from very thin, very brittle or very soft materials.

Generally, photoforming is competitive with stamping or blanking operations but is best suited to small or medium-size runs on parts too small or too intricate for successful stamping, *Fig.* 62.2. Dimensional tolerances are exact and fine detail is easily reproduced in production. Quantities practical may run from a few pieces to thousands, *Fig.* 62.3. Two thousand stainless steel retainer rings have been produced from one single 6-inch by 12-inch sheet 0.015-inch thick. Metal gage may

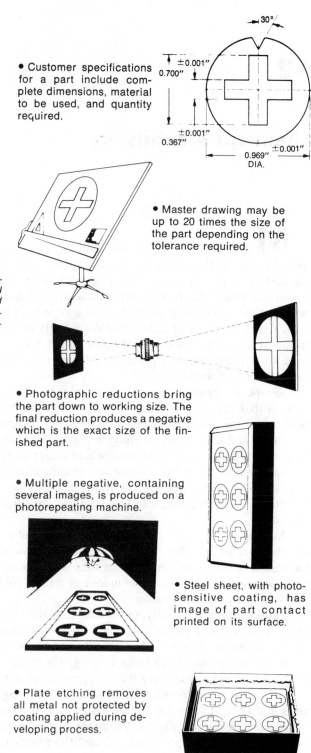

• Customer specifications for a part include complete dimensions, material to be used, and quantity required.

• Master drawing may be up to 20 times the size of the part depending on the tolerance required.

Fig. 62.1 — Steps required in photoforming parts in production. Small and intricate parts can be produced competitively by this process. Photos, courtesy Eastman Kodak Co.

• Photographic reductions bring the part down to working size. The final reduction produces a negative which is the exact size of the finished part.

• Multiple negative, containing several images, is produced on a photorepeating machine.

• Steel sheet, with photosensitive coating, has image of part contact printed on its surface.

• Plate etching removes all metal not protected by coating applied during developing process.

Fig. 62.2 — Typical group of photoformed parts showing complexity and range of sizes. The inset photo shows how 130 parts are produced on a single sheet. Photos, courtesy Eastman Kodak Company.

run from 0.00008-inch to a maximum of 1/8-inch.

Chemical Milling. This variation of the process differs from photoforming in that it deals with much larger amounts of metal removal. The patterns used in chemical milling are less precise than those for photoforming inasmuch as control

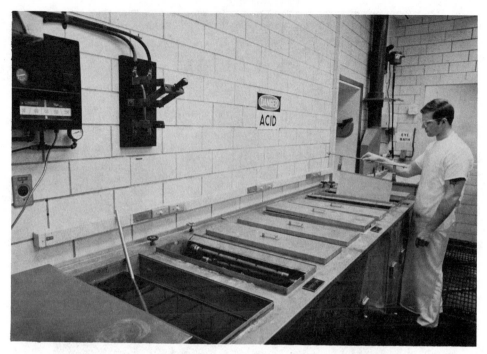

Fig. 62.3 — Photoforming production in process. Even substantial volumes can be readily made. Courtesy, Eastman Kodak Co.

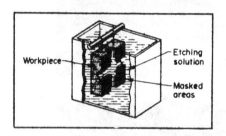

Workpiece — Etching solution

Masked areas

Fig. 62.4 — Arrangement for chemical milling operation. Almost any size bath can be set accommodate components as large as aircraft wing spars and skins.

of accuracy is largely a function of etching bath control. Large, heavy parts with pockets, tapers, ribs, grid patterns, etc. are typical of production. A common application is in the aircraft industry as a means of weight reduction on forgings, extrusions and wing skin sheets.

As with photoforming, the work is immersed in a tank containing an acid or alkali etching solution which uniformly eats away the metal from all exposed areas. The depth of etch is controlled by the time of immersion in the solution, *Fig.* 62.4. Metal removal of about 1/2-inch thickness is considered the practical maximum.

Equipment for both process variations may be rather simple, but production machines are already in use and automated equipment is available. Present ones are straight-line continuous for production from a coil of metal and of circular operating type.

Electrochemical Milling. Wherever the speed of metal removal in chemical milling is to be increased for greater efficiency, a dc power source can be added to the etching system. This variation includes the prefix "electro" to designate this feature but results otherwise are similar.

Electrochemical Machining. This process is often confused with chemical milling and electrochemical milling. It is a distinctly different method and is used for much different component production.

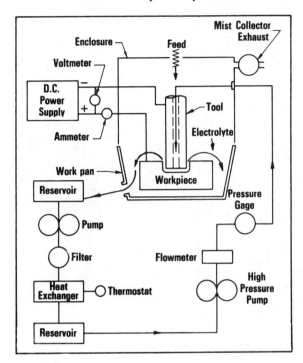

Fig. 62.5 — Typical electrochemical machining system showing basic elements of the machines. A standard work enclosure measures 41 by 24 by 19 inches. Diagram, courtesy Chemform, Div. KMS Ind. Inc.

Unlike the former two processes, this one uses fairly expensive equipment with costly tooling. No photo-resist masks are used. Instead, a shaped cathode tool is used which is connected to the negative terminal of a high-amperage dc power source. The workpiece is connected to the positive terminal. In operation, the tool is advanced into the workpiece while immersed in an electrolyte (usually a sodium chloride solution) that completes an electrical circuit between the two. Metal is removed through a reverse plating action or electrolysis. As the current passes from the work to the tool, metal particles or ions are caused to go into solution due to electrochemical reaction, *Fig.* 62.5.

With this system there is no tool wear and nearly all metals can be machined with ease regardless of hardness. Speed of metal removal is roughly 1 cubic inch per minute for every 10,000 amperes of electric current. Machines using up to 40,000 amperes are not uncommon.

Electrochemical Grinding. This process is much the same as electrochemical machining. Only the application and the workpiece requirement are different. A schematic diagram of a typical ECG system is shown in *Fig.* 62.6 and the machine is similar to a regular table grinder with an electrical power supply and electrolyte pumping system.

Whereas in electrochemical machining the tool never touches the workpiece, in grinding, the wheel touches the work surface lightly. With the wheel and work connected to a direct current source, the electrolyte wets the contact area in such a manner that the added electrochemical action depletes or dissolves the surface.

Only about 10 per cent of conventional grinding wheel pressure is needed in this process and, hence, an average of 90 per cent of the metal removal is created by

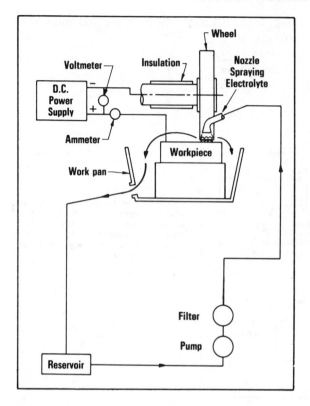

Fig. 62.6 — Electrochemical grinding is similar to electrochemical machining except it is applied to standard grinders with modifications to accept the process. This diagram shows a typical system. Diagram, courtesy Chemform, Div. KMS Ind. Inc.

electrochemical action. The metal bonded conducting abrasive wheel removes but 10 per cent of the stock and minimizes wheel dressing.

This method of grinding is much faster than conventional grinding and also permits fast metal removal regardless of material hardness. In addition, there is the relative absence of heat. This along with the light wheel contact makes the process ideal for finishing fragile parts that must be stress free or are subject to thermal damage.

Face wheel grinding, peripheral or surface grinding, cone wheel grinding and, form wheel or square grinding can be done. Because standard machines can be easily converted to this method, it is the easiest of the nonconventional machining processes to apply in production.

Design Considerations

Precision parts may be designed for production from sheet as thin as 0.0005-inch and as thick as 1/8-inch, maximum, with *photoforming*. In fact, fairly heavy parts are being presently produced competitively, *Fig.* 62.7.

One of the major advantages of this process is in the manufacture of preproduction or limited volume components when design change is imminent. Modifications are easily made when and as needed, *Fig.* 62.8.

Hole diameter that can be produced is a function of the etching process and the metal thickness. The smallest diameter hole that can be etched is equal to two times the thickness of the stock. This is because of the character of the etching action which has an undercutting or "etch factor." For this reason, also,

From a design standpoint *electrochemical machining* offers a highly flexible method for creating certain otherwise "impossible" features. Small and odd-shaped holes and cavities in materials too hard, too soft or too fragile to machine by conventional methods are ideal for the process. In fact, it is more competitive on any irregular shaped holes, *Fig.* 62.10.

This process also can produce complex, three-dimensional shapes with single-axis movement of the tool feed. Operations can be classed in the areas of hole sinking, both through and blind, cavities, deburring, planing and surfacing, trepanning, turning, and external shaping, *Fig.* 62.11.

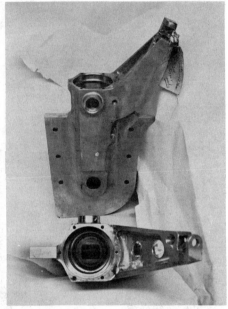

Fig. 62.11 — Typical electrochemical machine setup and the component showing special hole produced. With modular heads and indexing fixtures, suitable production levels are easily achieved. Photos, courtesy Chemform, Div. KMS Ind. Inc.

Holes may range from as small as 0.020-inch diameter on up to the capacity of the equipment. Shapes such as those required on turbine and compressor blades are readily produced with typical thickness dimensions of 0.025 to 0.005-inch, chord dimensions of 1.5 inches and lengths of several inches and over.

Hole depths may run to as much as 200:1 in length-to-diameter ratio for critical design requirements, *Fig.* 62.12. Holes are drilled burr-free with entry and exit corners smoothly radiused through control of the machine cycle. Blind holes or pockets may have taper or straight sides but bottom corners must have a radius. These features and other feasible external shaping operations are shown in *Fig.* 62.13.

Easiest of the processes, *electrochemical grinding* offers similar design possibilities and restraints as regular grinding. However, it extends these into the range of materials not economical to grind ordinarily. In the more everyday materials electrochemical grinding may be as much as 80 per cent faster without the heat effect problem.

Fig. 62.12 — Electrochemical drilling machine with 32-inch stroke. Single and multiple-head and N/C machines are available for almost any requirement. Sectioned turbine blade shows 9-inch deep longitudinal holes of 0.050-inch diameter drilled for cooling. Holes can be produced up to a 200:1 depth to diameter ratio, plus or minus 0.002-inch diameter tolerance and within 0.001-inch straightness. Surface finish is 16 rms. Photos, courtesy General Electric Co.

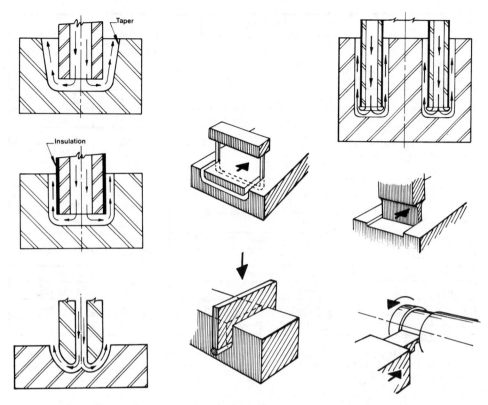

Fig. 62.13 — Sketches showing character of blind hole shaping via electrochemical machining operations. Variation in tooling provides control of internal accuracy and shape. External shaping, sawing, wire cutting, surfacing and turning can be effected. Drawings, courtesy Chemform, Div. KMS Ind. Inc.

Selection of Materials

Any material that can be etched can be processed by chemical milling. Thus besides the high-strength, high hardness materials, fragile materials such as honeycomb cores can be processed.

A list of typical operations and typical materials is shown in *Fig.* 62.14 for electrochemically milled components.

Virtually any commercially available metal or alloy of any hardness can be worked.

Specification of Tolerances

In all probability the matter of tolerances in production range from chemical milling on the high side to photoforming on the low.

Generally, in chemical milling, metal is etched away uniformly and surface marks, dents and scratches are reproduced in the chemically milled surfaces. In magnesium irregularities, however, tend to vanish. Thus, tolerances are usually from + or − 0.002-inch to 0.006-inch for such work.

In electrochemical work the accuracy is somewhat better, with ideal

TYPICAL RESULTS FROM A VARIETY OF ECM APPLICATIONS			
Operation Performed	Material	Tolerance Held	Surface Finish Obtained RMS
Cavity Machining	High Temperature Alloys	± 0.003	20
Cavity Machining	3046	± 0.002	200
Cavity Machining	Aluminum, Steel Superalloys	± 0.005	12 to 60
Cavity Machining	Nickel Alloys 52100, 4340	± 0.002	65
Cavity Machining	4340	± 0.0015	20 to 80
Hogging Cut	High Temperature Alloys	± 0.010	20
Hogging Cut	1020	± 0.010	125
Forging Dies	Tool Steel	± 0.001	100
Non-circular Holes	304L	± 0.002	200
Non-circular Holes	Aluminum, Steel Superalloy	± 0.005	12 to 60
Forging Dies	Die Steel	± 0.002	10
Billet Samples	Tough Alloy	± 0.005	—
Drilling	Varied	± 0.002	10 to 125
Drilling	Aluminum, Steel Superalloy	± 0.005	12 to 60
Drilling Rotating Electrode	52100, 4340 Brass, Aluminum	± 0.001	5 to 65

Fig. 62.14 — List of typical operations and materials shaped by electrochemical milling. Production tolerances and surface finish achieved are given for each application to show the general ranges. Chart, courtesy Chemform, Div. KMS Ind. Inc.

THICKNESS-TOLERANCE COMPARISONS

Metal	Thickness			
	.0005 in.	.001-.003 in.	.005-.007 in.	.010-.020 in.*
Aluminum Alloys	± .0005	± .0008	± .0015	± .0030
Brass	± .0002	± .0005	± .0010	± .0020
Copper	± .0002	± .0005	± .0010	± .0020
Carbon Steel	± .0004	± .0005	± .0010	± .0050
Invar	± .0004	± .0005	± .0010	± .0050
Stainless Steels (300 and 400 series)	± .0004	± .0005	± .0010	± .0050
Nickel Silver	± .0002	± .0003	± .0007	± .0030
Tool Steel	± .0004	± .0005	± .0010	± .0050

*Note: Metal thicknesses greater than .020 inch can be used. However, a proportional increase in tolerance (usually one-quarter to one-half the over-all metal thickness) must be considered.

Fig. 62.15 — Chart of typical photoforming metal thickness and tolerances held in general production lots. Chart, courtesy Eastman Kodak Co.

conditions + or − 0.001-inch or less can be held. Generally, the expectation is for high repeatability tolerances since tool wear is virtually nonexistant. Average work will range from + or − 0.001 to + or −0.005-inch. Fig. 62.14 shows some typical jobs and tolerances. Depth of slots or cavities can be held to + or − 0.020-inch.

Photoforming similarly provides high accuracy and repeatability. Tolerances normally are within 10 per cent of material thickness and, thus, tolerances from + or − 0.000 to + − 0.005-inch are attainable in the range of thickness up to

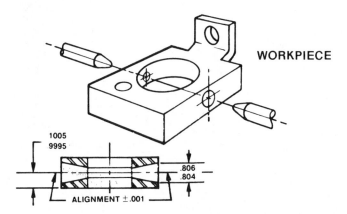

WORKPIECE

1005
9995

.806
.804

ALIGNMENT ±.001

Fig. 62.16 — Electrochemically machined component showing locational and size tolerances held in production. Drawing, courtesy Chemform, Div. KMS Ind. Inc.

about 0.020-inch, *Fig.* 62.15. Control of depth of etch can be held to + or − 0.002 to + or − 0.010-inch.

Tolerances with electrochemical grinding is generally + or − 0.001-inch for all types of surfaces and shapes at maximum metal removal rates.

Locational tolerances with these processes is excellent. *Fig.* 62.12 shows how precisely holes can be located. *Fig.* 62.16 indicates typical accuracy of alignment.

Surface Finish. Capabilities of these processes in producing acceptable surface finish is excellent. Under good conditions finish can be as fine as 5 microinches. Bottoms of ECM cavities will be under 32 microinches while generated surfaces will be from 50 to 125. *Fig.* 62.14 shows a range of surface finish results with typical parts.

Finishes on holes will be as fine as 4 microinches on parts such as those shown in *Fig.* 62.12. For electrochemical grinding, finishes are easily held in the range from 8 to 10 microinches with little or no pattern.

63

Electrical Discharge Machining

ELECTRIC metal disintegrators were developed many years ago for the purpose of removing broken drills, taps, studs, punches, and like items so as to salvage expensive components. Today, however, what was once a prosaic maintenance tool has been evolved into a highly sophisticated and truly unconventional manufacturing production process of growing value, *Fig.* 63.1. As such, it is now known as Electric Discharge Machining or EDM.

In the process of machining via EDM, *Fig.* 63.2, the workpiece is placed in a bath of dielectric fluid which is nonconductive. In the bath of fluid, usually oil, a tool or electrode, shaped to suit, is advanced toward the work. When the tool comes close to the workpiece, a multiplicity of tiny spark discharges leap across the gap to dislodge particles of the metal, *Fig.* 63.3. Variation in the number and size of the discharges varies the rate of metal removal.

So called disintegrators hold a vibrating tool in the automatic head and as the process of spark discharge proceeds, a coolant pumped through the electrode or tool washes away the metal particles. However, the majority of EDM machines employ a bath in which the workpiece is immersed so as to better control the concentration of electrical pulse energy and maintain constant temperature and flow around and from the work area.

Speed of metal removal is variable with this process by changing the current draw and the spark discharges per second. Fast roughing work is possible using high current densities and lower spark discharges but for fine finish and tolerances on size, modest amperage and high volume of spark density is desirable. Fine finishes are possible and nominal tool overcut per side is on the order of 0.0007-inch.

Scope of the Process. As with any unconventional production method, EDM has evolved to solve problems in the manufacture of otherwise "impossible" design requirements. The size of parts worked now ranges from those requiring holes as small as 0.002-inch diameter to water passages for pump impellers 6 feet in diameter. The usual design parameters that make EDM worthy of careful consideration are: materials that are hard to cut, fragile or extremely difficult to hold during working; designs that are extremely complex in contours; and designs that call for holes or shapes that defy conventional tooling, *Fig.* 63.4.

Although EDM until recently has been relegated to the category of low

Fig. 63.1 — EDM machine set up for slotting erase heads utilizing a rotating spindle, table servo and a fine finishing attachment. Photo, courtesy Ex-Cell-O Corp.

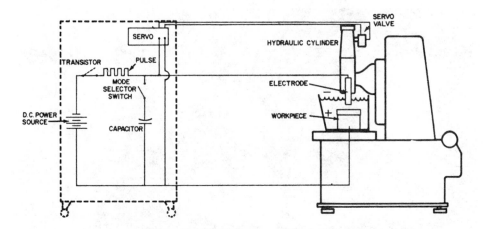

Fig. 63.2 — Diagram of an EDM setup showing the machine tool, power supply and dielectric oil reservoir. Diagram, courtesy South Bend EDM, Amsted Industries.

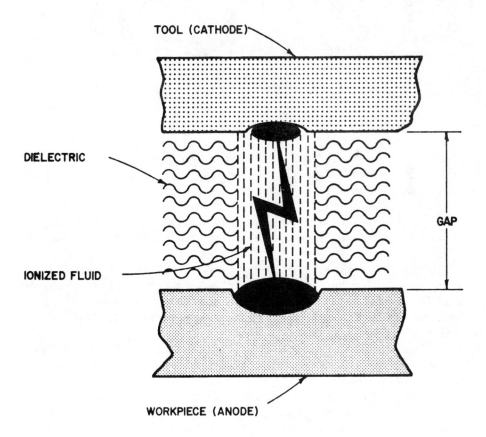

Fig. 63.3 — Enlarged view showing the EDM process as it occurs when attraction between the tool or electrode and the workpiece ionizes the dielectric. Diagram, courtesy South Bend EDM, Amsted Industries.

Fig. 63.4 — Aluminum impeller with 14 blades produced at a rate equaling that of a 4-spindle profiler costing four times as much. Graphite electrode machines 4 parts simultaneously at 28 minutes per cut to 330 to 350 rms finish. Photo, courtesy Cincinnati Milacron.

production — tool making and die sinking in hard materials — because of its limited metal removal ability, it is now entering the higher production area. The old limitation of EDM was a maximum metal removal rate of about one cubic inch per hour. However, the new multihead machines permit much higher energy levels without impairment of accuracy or surface finish and offer removal rates from 25 to 50 cubic inches per hour. Some rates to as high as 100 cubic inches per hour have been achieved.

In considering EDM for production lots, it should be noted that automotive firms presently drill carburetor pump shooter holes and nozzle tips with automatic machines. Fast cycle times and burr-free holes of high accuracy make EDM highly attractive even though special machines are required or at least special adaptations, *Fig.* 63.5.

As a rule production lots are more modest, quantities in the 100's are typical. This may run on up to lots of 10,000 depending on the economics of the design/production setup.

Electrodes. Originally, metal electrodes were used in EDM work. These consisted mainly of copper, copper-tungsten, silver-tungsten, or alloys of zinc or tin. However, graphite is becoming the most popular electrode material due to low cost, easy machining or molding to the needed shape, and adaptability to the "no-wear" mode of operation in which the electrode is positive.

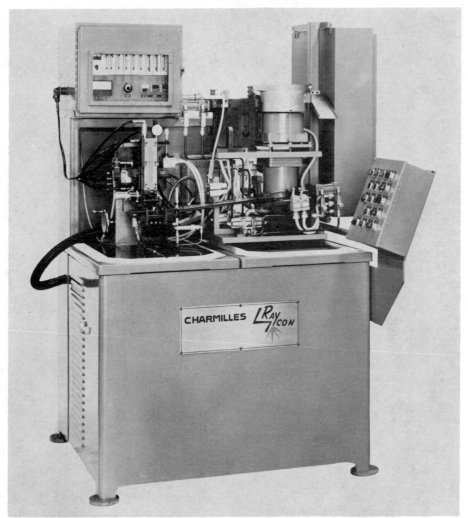

Fig. 63.5 — Special automatic EDM machine for drilling 0.020-inch orifice holes in automotive brass metering rod at 2000 per hour. The machine drills ten holes simultaneously. Wire electrodes are cartridge fed into position automatically. Photo, courtesy Raycon Corp.

Fig. 63.6 — Line of four standard EDM machines for handling parts from 12 by 21 inches to 41 by 83 inches in size. Photo, courtesy Ex-Cell-O Corp.

Fig. 63.7 — Giant four-poster machine offering workpiece capacity to 100,000 pounds in tanks up to 86 by 146 by 42 inches. Photo, courtesy Cincinnati Milacron.

Range of Machines. A wide variety of machines of standardized design is available for EDM work today. One line of four models is shown in *Fig.* 63.6. These machines reach considerable size, *Fig.* 63.7, and are used in lieu of specially designed machines unless production runs justify otherwise. Power packs and electrode heads are available to arrange complex specials when such is the case.

Design Considerations

For design purposes, one of the greatest advantages of EDM is its disregard for material hardness. If the material is a conductor of electricity, it can be machined.

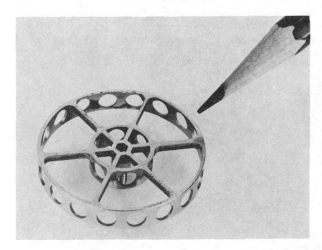

Fig. 63.8 — Complex injector support cage fabricated from Inconel 600. All but the OD and thickness are finished by EDM. Photo, courtesy Rocket Research Corp.

Fig. 63.9 — Amazingly fragile and complex designs can be done by EDM. This small copper part is machined in detail over its 0.007-inch of cross sectional area. Double walled section has 0.008-inch walls. Production rate was increased 10 times via EDM. Photo, courtesy South Bend EDM, Amstead Industries.

Fig. 63.10 — Molydenum rocket motor part with 187 holes EDM produced simultaneously. Photo, courtesy Rocket Research Corp.

Fig. 63.11 — Corner radius of 0.020-inch, maximum, is held on this incrementer unit. Tools shown produce rectangular holes to 0.3325-inch, plus 0.0004-inch, minus 0.0000-inch by 0.619-inch, plus 0.002-inch. Thickness is 0.125-inch. Photo, courtesy International Business Machines Corp.

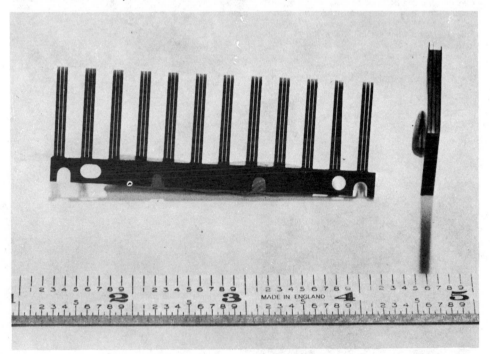

Fig. 63.12 — Punched card sensing member of spring steel with a carbide coating on the pin tips. Part is produced by blanking the twelve prongs over-width and EDM finishing to size. This dimensionally locates the 12 prongs and slots each one into 3 pins. The graphite electrode prongs are milled to 0.0034-inch, plus or minus 0.0003-inch. Photo, courtesy International Business Machines Corp.

Parts requiring extremely fine, minute details are excellent subjects for this process, *Fig.* 63.8. Insofar as the electrode tool can be made to create the shape

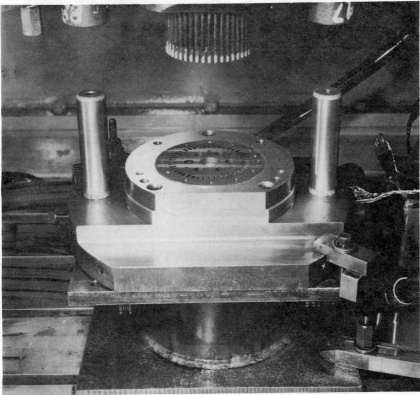

Fig. 63.13 — Close-up of a lamination die (above). Produced on the EDM setup. Sharp reproduction of detail and fine finish is evident. Actual punches are tipped with special electrode material that is removed after the operation is completed. Photos, courtesy General Electric Co.

needed and the surfaces to be machined are accessible to the electrode as required, a part can be machined, *Fig.* 63.9.

Parts may be designed with a large variety of holes, unusual holes as to cross-sectional shape or many holes perforating the piece, *Fig.* 63.10. Since the process is essentially stress-free, parts such as these can be produced accurately and without need for removal of burrs.

Design considerations are somewhat different in EDM than in conventional processes regarding the character of the opening left by the tool. For example, the opening cut is always larger than the cutting tool and a slight taper is inherent in the process. The overcut or metal removal beyond the electrode is on the order of 0.0007 to 0.002-inch where the gap is well flushed with clean coolant. In general, taper is a problem primarily with "blind" holes, but is not severe, amounting to about 0.0001-inch per side per 1/8-inch of depth.

TABLE 63-I — **Wear Ratio Chart for EDM†**

Electrode Material	Workpiece Material	Polarity	Wear Ratio*
Aluminum	Steel	Reverse	7
Brass	Brass	Standard	.5
Brass	Carbide	Standard	7
Brass	Steel	Standard	2
Brass	Tungsten	Standard	7
Carbide	Carbide	Standard	1.5
Copper	Steel	Standard	2
Copper Tungsten	Carbide	Standard	1.5
Copper Tungsten	Copper Tungsten	Standard	.5
Copper Tungsten	Steel	Reverse	.3
Copper Tungsten	Tungsten	Standard	.75
Zinc-Tin	Steel	Standard	5
Graphite	Steel	Standard	.5
Graphite	Carbide	— —	1.5 w/caution
Silver Tungsten	Steel	Standard	.2
Steel	Steel	Reverse	1.5

*Electrode length required to cut 1" of work piece (90° corner and .010" radius

†Chart, courtesy South Bend EDM, Amsted Industries.

The most significant point is electrode wear which creates rounded corners and bottoms in blind holes. The caution is to avoid pockets where metal debris removed can become trapped. With stepped or irregular holes, more than one electrode may be needed. However, with average work very sharp corners can be produced as compared with other processes, *Fig.* 63.11.

Selection of Materials

As mentioned, one of the leading attributes of EDM lies in its ability to machine almost any metal, hard or soft, exotic, heat treated, brittle, dense, porous, and even dissimilar metals at one pass. Parts are burr-free, stress-free and with a multidirectional random-lay surface.

Care must be used with some metals, as the surface is usually a recast white layer and in heat treatable metals this will be very hard with a sublayer slightly annealed. If extra high amperage is used, minute cracks may occur. Very sensitive tool steels may be damaged in extra thin sections or sharp corners unless a post-tempering operation is used to relieve stresses.

The data in TABLE 63-I shows a typical range of workpiece materials and the electrode material considered appropriate for machining.

Specification of Tolerances

This process, not unlike that of ECM, is noted for its accuracy. It is generally possible to produce shapes uniformly to within plus or minus 0.0002-inch to 0.0003-inch when conditions are constant, *Fig.* 63.12. In fact, with mating parts, a near perfect match is possible, *Fig.* 63.13. Where less demanding tolerances are possible production speed can be maximized to achieve an ideal relationship between cost and quality.

One important aspect of EDM, as previously indicated, is the surface finish achieved. Surfaces machined are non-directional and, depending on the economics of production, may range from as low as 5 rms to 150 to 200 rms. Good grades of graphite electrodes will produce 30 to 50 rms regularly in production. It is necessary, however, to use a rotating electrode or switch to more expensive metal electrodes for the finer finishes, so cost may be higher.

SECTION

PRODUCTION

11

PROCESSES

FINISHING
METHODS

64

Drawing Finish Notes

J. B. Mohler

BEFORE final release of drawings for parts and assemblies, it must be decided whether the surfaces of the items, as detailed, meet design requirements. Do the processes which give parts their shapes and dimensions leave functional surfaces on the parts? If so, finish might be "as cast," "as rolled," or "as extruded." Ordinarily, however, the mill, foundry, or plant finish is not the final finish of a part. Usually, the part is painted, plated, or otherwise coated. In any case, finish is specified by drawing notes — a shorthand technique for expression of surface finish requirements.

At times, a finish may necessitate redesign of a complex part because of the problems of applying a coating — for instance, an electroplate. At other times, the finish will be specified after the part is made. The customer may have a choice of finishes for an item carried in a manufacturer's stock without a finish. The finish should still be called out on the drawing — in this case "as required" or "as specified" (by the customer at the time of sale).

The drawing may specify "none." This call-out indicates that the designer considered the finish and decided that none was required. The call-out could also be "as finished" or "as machined," indicating that the finish resulting from the normal machining operation is acceptable.

When a finish is necessary, the call-out will encompass all requirements essential to the integrity and life of the part. When the finish is conventional, it can be indicated by reference to an acceptable (or required) military, federal, trade, industry, or company specification. If no suitable specification is available, then it may be advisable to include on the drawing a long note giving detailed requirements for surface preparation, pretreatments, coating, and post-treatment.

The drawing will call out:

- The material of which the part is made.
- The configuration and dimensions of the part.
- The finish or finishes applied to the part.

The finish can be that of the material "as received," that of the configuration "as manufactured," or a specified mechanical finish, conversion coating, or applied finish.

J. B. Mohler is a Consultant in Seattle, Wash.

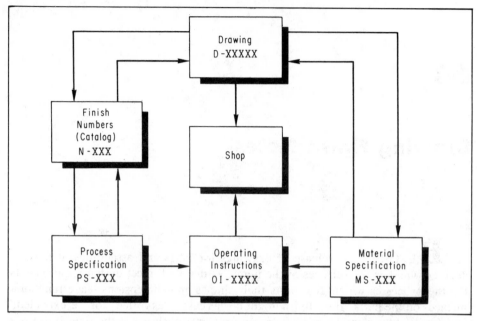

Fig. 64.1 — Types and movement of finishing documents.

Mechanical finishes consist of the surface texture or pattern produced by machining, grinding, or polishing as defined by an RMS or RHR designation.

Conversion coatings are chemical or electrochemical coatings formed by reaction with the substrate. Examples are phosphating, chromating, oxidizing, and anodizing.

Applied coatings include paint and metallic coatings such as those formed by electroplating, hot dipping, or vacuum depositing.

TABLE 64-I — **Types of Finishing Documents**

Designation	Document	Function
D-XXXXX	Drawing	Describe the part
None	Design Guide	Recommend drawing requirements
N-XXX	Finish Number Catalog	Standardize finishes
PS-XXX	Process Specification	Control processing
MS-XXX	Material Specification	Control material
OI-XXXX	Operating Instructions	Control manufacturing

Finish Number Documentation

Finish call-outs intended to be used repeatedly can be standardized by assigning numbers to finishes and cataloging the call-outs. The identification system itself — whether it is digital, sequential, letters and numbers, or types and classes — is of minor importance. Simply "Finish No. XXX" or possibly "N-XXX" relates the drawing to a standard finish.

A definition of each number must be made and recorded in a specification, design guide, or catalog. After this is done, the number is established as a standard. It can be modified, revised, or cancelled.

All the drawings which use it can be changed just by retaining the number but changing its definition. Or all drawings after a specified date can be changed by adding a note to the cataloged finish number.

Finish numbers can be useful in organizing a design guide. The guide will describe the properties, advantages, limitations, and application of finishes. It will give types, combinations, and thicknesses of coatings recommended for outdoor exposure, resistance to wear, decoration, solderability, and other properties.

Types and functions of finishing documents are listed in *Fig.* 64.1 and TABLE 64-I. The documents are related as indicated.

Finish Number Sources

Many finish number systems and the description of processes they identify originate in specifications and standards prepared by government, societies, and individual companies.

MIL-STD-171: Military Standard 171, "Finishing Of Metal And Wood Surfaces" is a good example of the use of finish numbers. The standard covers paint, chemical conversion coatings, electrodeposits, and other metal coatings. It establishes requirements for finishing and treating metal and wood surfaces and serves as a guide for selection of finishing materials and procedures. For example, to finish a part with chromated zinc plate 0.0010 in. thick, the call-out is

Finish 1.9.2.1 of MIL-STD-171

Mil Std 171 gives hundreds of finish numbers which describe metallic coating, black oxide finishing, cleaning methods, phosphating, passivating of stainless steel, anodizing, chromating, pickling, etching, abrasive blasting, and other processes.

Decimal numbering provides useful categories:

Inorganic Finishes, Metallic Coatings
1.1 Cadmium
1.2 Chromium
1.3 Lead

through 1.14 Rhodium, including the common metals applied by electroplating, vacuum depositing, electroless plating, and hot dipping.

Under cadmium, 24 numbers are listed to apply coatings of various thicknesses by plating or vacuum deposition and to apply supplementary chromate or phosphate when desired. These numbers, in part, are listed as

Finish No.	Requirements
1.1	Cadmium coatings
1.1.1	Plating Specification QQ-P-416, Type I, without supplementary treatment

1.1.1.1	Class 1, 0.0005 in. thick
1.1.1.2	Class 2, 0.0003 in. thick
1.1.1.3	Class 3, 0.0002 in. thick

These finish numbers are related to Federal Specification QQ-P-416 "Cadmium Plating (Electrodeposited)" which gives requirements for workmanship, thickness, adhesion, baking of high strength steels after plating, salt spray resistance of chromated coatings, and paint adhesion of phosphated coatings.

In Mil Std 171, Table IV, "Cleaning Methods," provides 12 numbers for special cleaning where desired or where adequate cleaning is not provided by the finish call-out:

Finish No.	Requirements
4.1	Abrasive blasting
4.2	Hot alkaline cleaning
4.3	Solvent clean
etc.	

MIL-F-14072: Military Specification MIL-F-14072-A "Finishes For Ground Signal Equipment" is another good example of the use of finish numbers. It gives requirements for plated and other finishes and details for application of organic finishes to metallic and wood surfaces. The finish numbers have three digits preceded by a letter. An example is M212 which applies 0.6 mil of nickel per QQ-N-290, Type II over 0.65 mil of copper over a steel substrate.

Fed. Std. 141: Federal Standard 141 "Paint, Varnish and Related Materials, Methods of Inspection, Sampling and Testing" is a comprehensive manual of detailed test procedures to control paints and painted surfaces. Many of the tests correspond to ASTM testing but many others do not.

Fed. Std. 595: Federal Standard 595 "Colors" provides 362 color chips each of which is identified with a 5-digit number that can be referenced on a drawing. Before a particular number is called out, paint suppliers should be contacted to see if they can match the color chip.

Architectural Finishes: The Aluminum Association (AA) has provided designations for mechanical finishing, chemical finishing, anodizing, resinous coating, vitreous coating, electroplating, and laminating designated as M, C, A, R, V, E, and L respectively. The letter is followed by a number. For example, polished finishes are

M20 — unspecified
M21 — highly specular
M22 — medium specular
M2x — other (to be specified)

An example of the use of the designation system produces a matte anodized finish. The code specifies a matte finish, then chemical cleaning, followed by architectural class II natural anodizing, thus:

AA — Aluminum Association
M32 — Mechanical finish, directional textured, medium satin
C12 — Chemical treatment, inhibited alkaline cleaning
A31 — Anodic coating, architectural class II (0.4 to 0.7 mil thick) clear (natural)

The designation on the drawing would be

AA — M32C12A31

Finish Call-Outs

Plating: Examples of complete finishing call-outs, incorporating finish numbers, can be given for cadmium plating designated by Federal Specification QQ-P-416, "Cadmium Plating (Electrodeposited)." There are advantages to using this specification. It is well known and commonly cited in reference works. Three classes and three types are available:

Class	Thickness (in.)
1	0.00050
2	0.00030
3	0.00020

Type	Supplementary Treatment
I	none
II	chromate
III	phosphate

These types provide nine possible finishes. In addition, the plating can be bright or dull. Furthermore, the product can be tested to the quality requirements of the specification.

There are disadvantages to the use of this specification. The finishes are limited to the classes and types specified. A call-out implies that all requirements of the specification will be met, including responsibility for specific sampling, inspection, and test procedures. In addition, the specification is revised periodically. Thus, QQ-P-416b will become QQ-P-416c. This poses the problem of working to an obsolete version or accepting a new version and changing all drawings, specifications, and practices to conform to it.

A call-out for plating small items of hardware for military purposes, defined by a finish number, might be

N-123 Cadmium plate per QQ-P-416 Class I, Type II.

If it is desired to reveal the meaning of the class and type, the call-out can state

N-124 Cadmium plate 0.0005 in. thick and chromate per QQ-P-416.

The same finish expressed as a commercial callout, avoiding the Federal Specification, would read

N-125 Cadmium plate 0.0005 in. thick and chromate treat.

Even though the finish is the same, a different finish number is used to simplify the quality requirements.

A combined requirement is written as

N-126 Cadmium plate per QQ-P-416 Class 1, Type I or equivalent. Note: The requirements of QQ-P-416 apply to all military contracts.

The "equivalent" statement and note allows interpretation of the requirements for nonmilitary contracts. If the "Note" is dropped, the call-out implies that the thickness and chromate requirements of QQ-P-416 will be met and other requirements of that specification may be waived.

Painting. A call-out for painting of aluminum, based on military specifications, might read

N-234 Vapor degrease, deoxidize and then anodize per MIL-A-8625 Type I, Class 1, except do not seal; apply one coat of wash primer MIL-C-8514 and two coats of lacquer MIL-L-7178 Color No. XXXXX per Fed. Std. 595.

If a process specification were written covering the same procedure the call-out might read

N-235 Clean and anodize per PS-123, wash prime per PS-124 and apply two coats of lacquer per PS-125, Color No. XXXXX.

This N-235 call-out would provide much better control than N-234 because all necessary details would be included in the process specifications.

One process specification can be written that covers all procedures or it can reference other specifications that provide the details. A single specification that covers all the details wholly or by reference will simplify the N-235 call-out to

N-236 Apply two coats of lacquer per PS-126 Color No. XXXXX.

Another variation is a listing of modifications of the N-number to include pretreatments, primers, and possibly colors in the process specification or the design guide, by an extended number, so that the call-out now becomes

Finish: N-236-123

Special Considerations

Commercial Coatings and Trade Names: Most of the coatings that are applied commercially are either proprietary or use proprietary chemicals in the cleaning, etching, and coating solutions. Sometimes a complete proprietary process carries a trade name. These processes usually come in a series of modified names and numbers and should be identified specifically. Finish numbers will aid in detailed identification of trade name processes:

Finish No.	Supplier	Trade Name
N-238-001	A Company	XXX
N-238-002	A Company	XXXXX
N-238-003	B Company	XX

Obsolete Specifications and References: If a call-out is made for the first time, the most recent official specification index should be checked to be sure that the finish is not obsolete. Government and other specifications are occasionally up-dated and even cancelled. When a revision is made, the final letter is changed. Thus, QQ-P-416a became QQ-P-416b on April 5, 1967.

Some specifications, like MIL-STD-171, reference specifications without the letter revision, meaning that the latest revision is acceptable. This is also a common practice in other specifications and references (to avoid automatic obsolescence). Of course, when a letter revision is made, the requirements are changed. If the "b" revision is called out, it becomes obsolete when the "c" revision is issued. On the other hand, it will be clear that the requirements of the "b" revision were originally intended.

Observance of Tolerances: Conventional thickness and roughness requirements apply to a fit or a threaded tolerance that will be affected by plating. MIL-STD-8C points out that when a fit will be affected, one of the following notes should be placed on the drawing:

Dimensional limits apply before plating.
Dimensional limits apply after plating.
Dimensional limits and surface roughness designations apply before plating.
Dimensional limits and surface roughness designations apply after plating.

Equivalents: It is always desirable and in fact often mandatory to have at least two sources of supply and preferably three for finishes that are vital to production. This can be done by providing finishes that can be evaluated to requirements and listed under a single number:

Finish No.	Supplier	Trade Name
N-238-001	A Company	XXXX
N-238-001	C Company	XXXX
N-238-001	D Company	XXXX

Subcontracting: Finish numbers can be a burden to a subcontractor if the meaning of the number is not clear. The finish number will either have to be described in sufficient detail on the drawing or a list of descriptions will have to be provided separately.

Even if finishing is detailed on a drawing, a finish number is useful to identify the description, standardize the finish, and simplify discussions and references to the finish.

Acknowledgement

This chapter is included by permission of *Machine Design*, from the August 10, 1972 issue, copyright, Fenton, Inc.

65

Cleaning Parts Automatically

J. H. McRainey

PARTS CLEANING is one of those art-sciences in which only the experts really appreciate the range of complexities involved. For example, what is the definition of "clean?" How is "clean" measured? What degree of "cleanliness" is required prior to painting? Assembly? Electroplating?

Those who think there are simple answers to these types of questions might be interested in this aside: For over 15 years, the American Electroplaters' Society has sponsored a research program in which a major objective has been to study the degree of cleanliness required for plating and to develop appropriate standards of cleanliness[1].

An analogy can be used to highlight the rather unusual conditions found in the cleaning field. Consider what would follow if "heating" were defined as "raising the heat content to a desirable level." The desirable level, "hot," would be that which made the work acceptable for some further processing. Surveying the various industrial areas in which heat is used, one would find that "hot" applied to molten metal in the foundry industry. In another industry, "hot" wouldn't be sufficient to boil water. There might be a considerable body of knowledge about the mechanics of heat transfer and related subjects. Despite this, so long as key words such as "heated" or "hot" were defined in a relative manner, "heating" would never give the appearance of a unified technology. So it is with "cleaning."

Relative Nature of Soil

As the term is used in industry, "cleaning" is the removal of objectionable soils from the surfaces of parts. Obviously, "objectionable" is a relative term. Something might be objectionable because its presence will interfere with subsequent processing operations. Or its presence might be detrimental to part end-use, or to continual functioning of a part. Or something might be undesirable for esthetic reasons.

However, the point is that the something on the surface which is objectionable in one situation, or for one purpose, can be acceptable in another. Cleanliness, then, is a relative term that only can be used to describe the surface condition of a

J. H. McRainey is Executive Editor, Automation Magazine, Cleveland, Ohio.

part in a particular situation — in other words, within the framework of a particular set of variables. It follows, then, that to appreciate cleaning as a technology, one must essentially deal with the variables.

A listing of the general variables contains those types of things one would expect intuitively: type of soil, cleanliness required, nature of material being cleaned, geometry of part, surface condition, production load, safety and health considerations, waste disposal requirements, energy requirements, maintenance costs, etc.

These general variables can be expanded in some detail. For example, a listing of soils might include: grease, oils, rust, scale, ink, paint, lacquers, drawing compounds, polishing compounds, cutting fluids, loose metal, loose sand, and perspiration. Most often, mixtures of these soils are encountered. And then there is the condition of the soil: Is it aged and hardened? Was it deposited in the presence of heat, as in buffing?

Cleaning in Perspective

As has been implied, there is no standard quantifiable way used in industry to prescribe what is wanted when considering the variable of required cleanliness. Experience is the yardstick used. And experience teaches that there is a difference in the cleaning effort which typically must be applied to obtain satisfactory results for a given purpose.

In this sense, it is of some interest to consider a list prepared by a noted author of articles and books dealing with cleaning[2]. The list shows a number of reasons for cleaning in the order of increasing standards of cleanliness required:

1. Further processing in order to remove gross soils so as to facilitate handling during manufacture and to prevent rusting; considerable oily residue can be tolerated.

2. Application of a rust preventive; many soil residues could be compatible with the rust preventive.

3. Flame welding; minor quantities of soil would not interfere.

4. Painting with or without a phosphate or conversion coating; traces of oily soils and even small amounts of oxide might be tolerated, but acceptable soils could be very selective.

5. Hot tinning, soldering, brazing, galvanizing; only traces of organic soil could remain without interference while oxides, scale, and rust must be almost completely removed.

6. Spot or contact welding; for good welds organic soils should be absent and the surface almost completely free of more than a superficial oxide layer.

7. Porcelain enameling; similar to spot or contact welding but oxide may not be quite as critical.

8. Electroplating; high levels of cleanliness are required with respect to organic soils or oxides.

9. Storing liquid oxygen; all traces of organic soils, particles of inorganic soil, sand, or anything that could react must be absent.

10. Assembling delicate electronic components; surfaces must be completely clean and even traces of atmospheric dust must be avoided.

This listing provides a general perspective of the complex field of metals cleaning.

Another observation can help place parts cleaning in proper perspective. It is useful to think of cleanliness as an unstable condition of a part. This means that not only must effort be applied to overcome a soiled condition but also it must be applied to maintain a part in a cleaned condition.

The relationship between effort and improvement follows an exponential curve in which greater and greater effort is required to obtain smaller and smaller increments of improvement. Here, effort should be understood as embracing such things as time, care, money, labor, etc. As to the upper limit of improvements, it should be realized that from a practical standpoint it is impossible to obtain a metal surface free of all contamination (a freshly exposed metal surface free of contamination can be obtained and maintained only in a perfect vacuum).

This exponential relationship between effort and cleaning improvement has several interesting implications. For one thing, the relationship suggests that in searching for improvements, greater returns might be expected to be found in changing the peripheral variables associated with cleaning rather than the direct variables within the cleaning process. By way of example, instead of expending greater effort to obtain more thorough removal of a certain problem soil, it would be of more benefit to examine and control the source of the soil in the first place. Or perhaps some pretreatment of soil will make cleaning easier.

One of the major problems in industrial parts cleaning is to obtain reproducibility of cleanliness in a production environment. Each part might not be equally soiled but, following the cleaning process, it is hoped that all equally will meet a minimum test of cleanliness. One solution to this problem is to establish a capability to deal with a typical bad condition of soiling. And then to be certain that all such cases will be handled, still more capability might be built into the cleaning process.

The prevalency of this brute-force approach is apparent from the number of "humorous" incidents that can be told about malfunctioning cleaning equipment (clogged spray nozzles, or closed flow valves, and even reversed ultrasonic transducers in equipment) which was not reflected in the usable quality of parts coming out of the cleaning equipment at a user's plant.

If a cleaning operation is conducted as if all parts are equally soiled then it is apparent that the potential for waste can be great. The implication is that there are definite economies in the typical cleaning process which can be uncovered by astute engineering and management.

The room for maneuver in obtaining engineered improvements appears at first glance to be circumscribed. For example, compared to many areas, cleaning as generally practiced in industry appears to have been relatively static over the years. With the exception of ultrasonic equipment, *Fig.* 65.1, and other forms using vibratory forces, a description of metal cleaning equipment used in industry thirty years ago differs in only minor respects from that in use today[3].

In the past, various industries — such as automotive — were faced with a relatively pressing need for quantity of output. The various acid and alkaline cleaners used then were no less dangerous than today. And thirty years ago labor represented 95 percent of the cost of cleaning. The answer to many cleaning problems then was to use labor-saving equipment. Thus, one interpretation of the seeming lack of progress is that the cleaning function was simply forced quite early

Fig. 65.1 — With continual improvements in equipment and knowledge, use of ultrasonics has moved progressively from the laboratory into the light cleaning field, and, in the last few years, into heavy industrial cleaning applications. All indications are that its use will continue to grow.

Advantages of ultrasonics as an aid to cleaning include: a) Improved general cleanness (the latter term is apparently preferred in the ultrasonic industry in lieu of "cleanliness"). b) Reduced cleaning costs. c) Faster cleaning cycles. d) Feasibility of using less pernicious cleaning materials. e) Ability to clean recesses, cavities, even assemblies of parts which are not effectively cleaned by other techniques. f) Improvements can be obtained by adding ultrasonic devices to existing equipment. g) Improvements do not require excessive floor space or maintenance effort.

Equipment required for ultrasonic cleaning includes a generator to convert service power into radio frequency power at a desired frequency; a transducer that converts this electrical energy into mechanical energy at the same frequency; and a tank containing a liquid through which sound energy generated by the transducer can be transmitted to parts to be cleaned. The transducer can be mounted on the outside wall of the tank or immersed in the tank.

General range of ultrasonic frequencies extends from around 16,000 cps to several million cps. Frequencies used in cleaning are typically in the lower range, say up to 50,000 cps.

Mechanical vibrations of sufficient density imparted to a liquid in these lower ranges cause alternate compressions and rarefactions in the liquid. At the interface between the liquid and an object in the liquid, the intense low pressure phase of the vibration cycle causes the solution to be detached locally and form many cavities or bubbles. When the low pressure is suddenly released as the high pressure phase occurs, the cavities collapse. This implosion, or cavitation, of each bubble, creates a powerful shock wave. At the surface of the part, collapse of these millions of small bubbles at thousands of times a second results in a mechanical scrubbing action which effectively cleans the surface. The effect occurs throughout the liquid and as such reaches into holes and crevices where the liquid has penetrated.

From this, it is apparent that without effective cavitation then ultrasonic equipment adds nothing to a cleaning process. Many factors influence cavitation. As to the liquid, variables include surface tension, pressure, density, viscosity, amount of dissolved gases, and temperature. The density of applied ultrasonic energy and its frequency are factors (at very high frequencies there simply isn't enough time for cavities to form). The orientation of the part and its location relative to the transducer are also factors.

Interestingly, this latter aspect favors mechanical handling in ultrasonic cleaning setups. Since precise positioning cannot be obtained because of practical considerations, it has been found beneficial to move parts through a solution. Movement tends to promote cavitation as part travels through areas of both high and low cavitational energy. Relative motion of parts and solution also aids in removing soils.

With development of unit immersible transducers that can be added to existing tanks, industry has available an attractive means of obtaining improvements. The units can be applied in whatever numbers are necessary to meet power requirements. They can be added to solvent systems, aqueous systems, rinsing, and other process baths.

to use both automatic and semiautomatic equipment. And these equipment forms have generally withstood the pressures of change.

It is in the details of equipment design, and operation that improvements are to be found. In other words, this perspective of the cleaning field suggests that manufacturers should concentrate on using existing techniques and technology more efficiently rather than looking for some novel scientific breakthrough that will solve all cleaning problems. There are areas where considerable improvements can be made within the state of existing technology.

It is also apparent that improvements should be sought within the context of the total production system. Only by so doing can a manufacturer achieve his cleaning goal: optimum cleanliness at a minimum total production cost.

Solvent Conservation

The potential for improvement in existing equipment forms is exemplified by a conservation device designed for solvent degreasing equipment. Until relatively recent times it has been assumed that the most important source of solvent loss in well-designed and operating equipment was in dragout. Control of vapor loss was presumed to be taken care of by designing tanks with a generous free-board height to contain the heavier-than-air solvent vapors that form over the liquid solvent bath and to protect the vapor layer from unusual disturbances. Water cooled coils around the tank helped maintain the level of vapors.

It was recognized that some losses occurred through diffusion of some lighter vapors into room air. These fumes were considered to be relatively unimportant except for their irritant effect on workers. Because of this effect, every effort was made to exhaust the vapor-contaminated air.

From present knowledge it is now apparent that in carefully operated equipment, vapor diffusion is the dominant source of solvent loss. And whatever else they might be designed to do, various fume exhaust systems also move air — a highly effective vapor sponge — into and out of contact with an expensive degreasing material.

Major savings can be obtained by preventing the loss of the lighter fumes from the degreasing equipment. That the savings can be "major" is demonstrated from experiences of numerous companies which have installed a "cold trap," a low temperature refrigerated air convector developed by Fume Control Div., AutoSonics Inc. Reductions in solvent loss have been documented in which savings range from 50 per cent to over 90 per cent.

As shown in *Fig.* 65.2, the device includes a peripheral duct that is secured against the degreaser's inside wall. Refrigerating coils are located in the duct. A small refrigeration unit located near the degreaser circulates a refrigerant at -20F

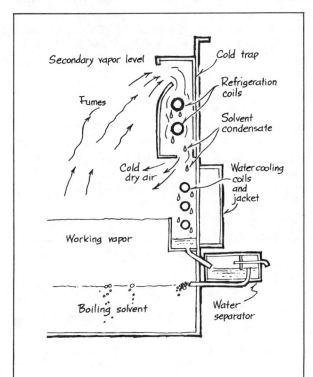

Fig. 65.2 — Peripheral condensing device installed within freeboard of tank containing solvent can prevent significant loss of the solvent. Convection currents are established in fume-saturated air as cold dry air flows back into tank. The simple peripheral device will control fumes rising from center area of even large square tanks.

through the coils. Warm solvent-laden air enters the duct at the top and settles out through the bottom slat, condensing out the solvent enroute. Subsequently, the condensed solvent is returned to the tank through a water separator. The effect of this addition of cold dry air to a degreaser is to establish a secondary vapor level in the tank in the same manner that water cooling coils have been used to control the level of the hot working vapors above the boiling solvent.

It should also be noted that the addition of this efficient means of controlling solvent loss to already accepted techniques could have far-reaching effects. For example, it makes it feasible and economical to use more volatile solvents with lower boiling points. And if heat is not needed to vaporize solvents, then savings would be obtained in utility costs and in minimizing solvent breakdown.

Agitation Always Helps

Both designers and users of cleaning equipment stress the importance of agitation for thorough cleaning[4]. Here agitation should be understood as any technique which results in the movement of parts to be cleaned and the cleaning solution relative to each other. In other words, agitation can be obtained by moving the work relative to the solution or by moving the solution relative to the work.

Agitation provides several benefits. For one, it continually brings fresh solution into contact with a soil. It displaces air pockets and enables the solution to penetrate blind holes in the work. Agitation also can be a considerable aid in dislodging soils from a part.

A variety of techniques are used to obtain agitation. They are not all equally effective and, in some instances, are even misused in the sense that the side effects are more harmful than the good obtained from agitation. For example, compressed air can be used for agitating solutions. It is useful in the mixing of solutions. It is particularly effective as a source of agitation in rinse tanks. But when used in some heated cleaning solutions, air agitation causes undesirable foaming, cools the solution, and can result in chemical reactions that reduce the effectiveness of the cleaner.

Turbulence of convection flow can be used as a source of agitation. The solution can also be agitated by injectors, pumps, propellers, or paddle wheels. Intermittent agitation of work suspended from a hoist can be obtained by simply moving it up and down. Similarly, continuous mechanical agitation of suspended work can be provided by cams, eccentrics, or rockers. Mechanical agitation also can be incidental to work movement, as when parts are transported through a solution by a conveyor.

Agitation might be viewed as the applications of energy to a part surface to aid in overcoming the bond of the soil to the part. In this light, one would expect that the most effective agitation techniques would be those that concentrate or focus the energy application onto the surface. And so it is.

The cleaning industry uses the term "washing machines" for equipment which forcibly discharges cleaning solutions against the work. The efficient cleaning power of washing machines is attributable to the "agitation" obtained as the solution under pressure or other force impinges against the work. For example, with properly spaced and directed spray nozzles, all areas on the part can be subjected to liquid impingement which not only covers the soil but lifts it off with a shearing action.

Electrolytic cleaning is another technique in which agitation is effectively concentrated at the part surface. As the name implies, in electrolytic cleaning the workpiece is connected to a terminal and immersed in a conductive cleaning solution. With the passage of direct current through the solution, soil particles "plate off." This plating off is enhanced by the extensive agitation created by the gases that are evolved on the work surface.

In ultrasonic energy, the chemical cleaning field has discovered what is perhaps the ultimate in agitation techniques. Ultrasonic transducers are used in solvent or detergent solutions. However, the effective cleaning power of ultrasonics is attributed to cavitation — an intense impact on the surface of a part caused by the generation and collapse of countless small bubbles. The effect of cavitation is to "explode" the soil off the surface, after which the soil can be handled efficiently by the cleaning solution, see *Fig.* 65.1.

Ultrasonics is not the answer to every cleaning problem. However, the quality of cleanliness obtained with this powerful agitation technique highlights the importance of agitation in cleaning, especially where particulant soil is the problem. The method of agitation is a critical variable that should not be overlooked in searching for improved results whenever parts and solutions are brought together — whether at the cleaning or rinsing stage.

Handling Considerations

Basic goal of mechanical handling in parts cleaning is to move parts automatically through various processing steps, obtaining for some desired time a desired degree of contact between the parts and processing media.

There are two general forms of the problem. In one the parts must be moved into, through, and out of the processing media. Typically some type of dipping motion is required as parts are transferred in and out of several tanks in which a liquid or vapor is maintained at a working level, *Fig.* 65.3. This form of handling is prevalent in electrolytic, immersion, and ultrasonic cleaning installations.

The second general form of the handling problem involves presentation of the part to thorough impingement by a processing medium, *Fig.* 65.4. Thoroughness is obtained by locating a multiple number of impingement sources such that all sides of a part are treated and/or by manipulating a part to expose all sides to impingement. This form is used in spray washing machines and blast cleaning equipment, for example.

Ideally, the particular equipment form, along with handling techniques, would evolve in any particular situation after considering all variables. The conditions and nature of the part and the cleanliness desired would suggest a certain cleaning technique. This plus desired production rates would then lead to the use of specific equipment items and handling techniques. In this context, then, there would be no serious handling problem since the particular technique would be engineered for a particular application.

Typically, however, the ideal situation does not exist. Practical considerations dictate that existing plant equipment be used in cleaning a variety of parts for a variety of purposes. The result is that attempts are made to clean some parts by one technique when another would be better. And, in making do with existing automatic equipment, effective cleaning of different parts will be obtained only by attention to details, a number of which are concerned with the movement of work

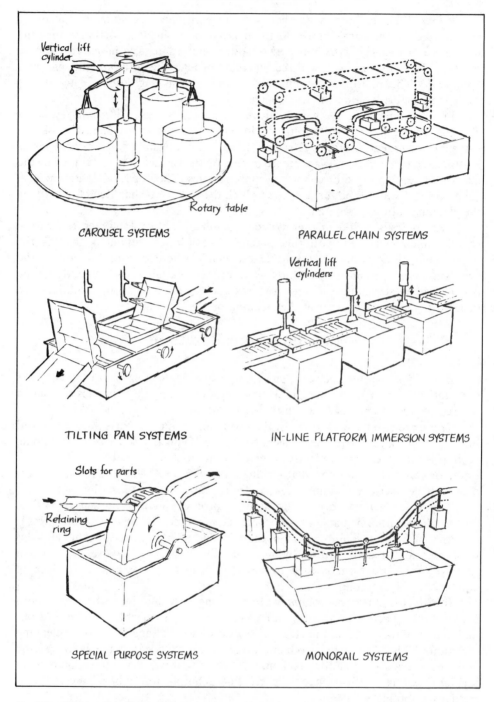

Fig. 65.3 — Variety of handling techniques can be used in moving parts to be cleaned into and out of cleaning solutions and rinse baths. In practice, to obtain benefit of impingement force in cleaning, spray devices also might be incorporated in equipment using illustrated handling techniques. Ultrasonic transducers can be placed in the tanks to obtain agitation.

through the equipment.

For example, a conveyor rate of travel for one set of conditions is not the best

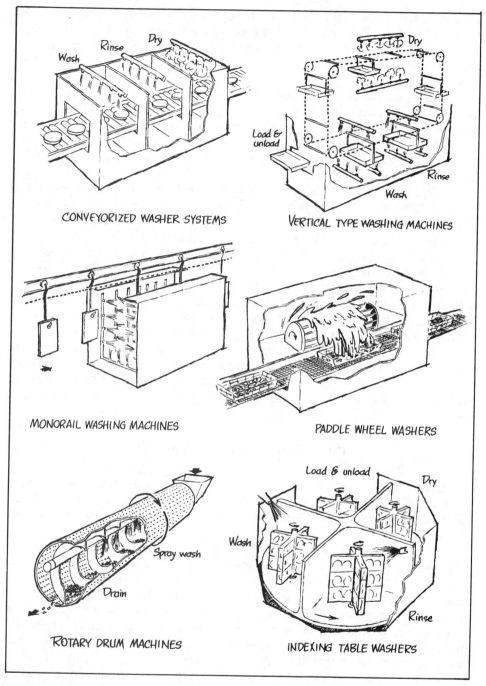

Fig. 65.4 — Techniques in which cleaning and rinsing solutions are force fed are particulary suited to mechanical handling of work. Various stages in the process (as between clean and rinse) must be adequately separated to minimize intermixing of solutions. Maintenance of nozzles is critical to successful operation of spray cleaning machines.

for a different set. Some parts might require several passes through fixed-cycle equipment. The contours of the part determine how effective spray cleaning and rinsing will be. The spray pattern that cleans one type of part will not effectively cover a different part. Different spray pressures are indicated to compensate for different weight of parts processed (changes in the cleaner can also be required). Temperatures that promote drying for one type of part will not be satisfactory for another. Frequency of ultrasonic transducers should be adjusted to type of part surface being cleaned.

Automatic cleaning can be improved in processing batches of parts in baskets by taking care in loading the baskets. If nesting of parts is not avoided, the cleaning solution can be prevented from contacting the parts. Even when ultrasonic agitation is provided, nested and closely packed parts will not be thoroughly cleaned because of scattering and shadowing effects on the irradiation, and because of possible entrapped air pockets.

In addition to the problem of handling parts within cleaning equipment, there is the problem of integrating the equipment into a production line. This requires that a common handling system be used throughout various successive processes or that parts be transferred at the interface between processes.

The latter problem is serious enough to lead manufacturers of some processing equipment (furnaces, for example) to design their own cleaning equipment around some preferred handling means rather than attempting to tie in their equipment with another manufacturer's cleaning equipment. It should be appreciated that integrating cleaning into other processes can involve other than handling problems, e.g., compatibility of conveying equipment or containers with chemicals and contamination of cleaning and rinsing solutions. In many instances, the provisions for or techniques used in integrating cleaning equipment into production lines are what really distinguish one manufacturer's design from another. The happy result for users is that a variety of equipment forms are available to solve particular problems.

Parts Designers Can Help

Many cleaning problems are created on the design engineer's drawing board. Some can best be solved by returning them to that location. It could hardly be otherwise, considering the number of cleaning variables under control of the design engineer.

Part design establishes material specifications and part geometry. These not only influence cleaning directly, they also determine production processes which in turn influence the condition of the part as received by the cleaning operation. Part design also specifies surface finishes which follow cleaning.

It is neither practical or feasible to run to the design engineer every time some difficulty is encountered in cleaning. But certainly the design engineer should be aware that by specifying some materials he might be setting in motion some expensive chain of events throughout the production department, including cleaning and subsequent surface finishing. Cleaning solutions must be tailored to the metal. It is perhaps obvious that different treatments are required for, say, aluminum and steel. But even among the alloys of a given metal there can be significant differences in their sensitivity to a given cleaning solution. A change in material specifications might be made so that existing cleaning processes could be used rather than requiring that production establish special ones.

A design engineer should also appreciate that blind holes, crevices, and similar configurations present cleaning difficulties worthy of some consideration in

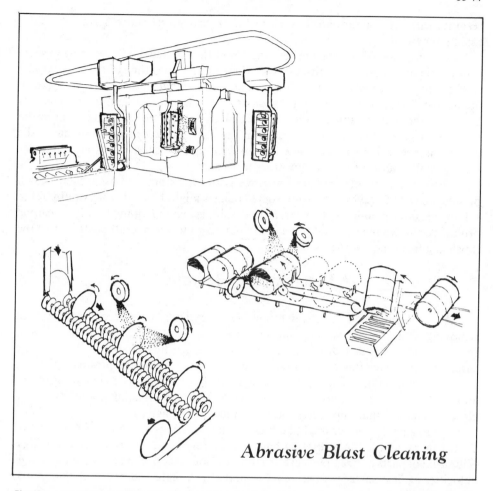

Abrasive Blast Cleaning

Fig. 65.5 — Inherent capability of abrasive blast cleaning is that it can remove tough soils — sand, scale, oxides, etc. — from a given surface area both thoroughly and quickly. And, of course, various other soils — oil, paint, soot, etc. — are removed at the same time. The emphasis on chemical cleaning in this article should not hide the fact that abrasive blasting is highly reliable and among the most widely applied mechanical surface treatments.

Standard equipment designs exist that automatically can handle successive batch loads of castings, with the total load measured in hundreds of pounds. The castings are tumbled within a confined space to present all surfaces to the impingement of the cleaning medium. Cleaning cycle is a matter of minutes.

Continuous strip steel is being cleaned by abrasive blasting. Efficiency of modern equipment and availability of cast steel shot enables abrasive blasting to compete with acid pickling ahead of cold rolling operations. Operating speeds are compatible with other equipment items in a continuous rolling process.

Because of the type of soil for which abrasive blasting is ideally suited, the process is applied in cleaning parts of unusual sizes or shapes — room-sized castings, railroad cars, billets, and structural steel, for example.

The nature of abrasive cleaning and the parts to be processed also put a premium on well-engineered special handling techniques for individual workpieces. This is illustrated in the sketches. In the drum cleaning operation (by Whellabrator Corp.), drums are conveyed to a point below three blast wheels. The drums are rotated to expose all surfaces to the blast effect. Subsequently, each is conveyed to a point where it is upended to dump out any shot collected within the drum, and then is kicked off the line. In the illustrated special lid-cleaning machine, lids are conveyed and turned in the upright position by a feed-screw type of conveyor. The lids are cleaned by two blast wheels.

The engine block processing equipment (by Pangborn Corp.) illustrates a newly developed robot-type system. The feature of this equipment is that the work-handling unit is a self-propelled "robot" which automatically responds to preset controls while traversing the monorail track system. As such, it is independent of other robot units traveling through the system and can selectively present areas of work to the blast effect. Selectivity is obtained by varying the rate of rotation as a part is turned in the presence of the abrasive hurling wheel.

part design — particularly when final finish is critical. Special effort is required in both cleaning and rinsing of such parts to ensure that small and difficult-to-detect residues will not remain on the work to bleed through or mar a surface finish at some later time.

One sure way to increase the cost and lower the efficiency of a cleaning process is to continually dump quantities of cleaning solution into the rinse water. At one stroke, this wastes cleaning solutions and contaminates the rinse, making it ineffective. Frequently, the configuration of parts guarantees that this problem will be encountered in cleaning. Only by special effort can the cleaning solution be removed from pockets and depressions in the part before rinsing. Here again the design engineer can be of assistance — by providing for a small hole that will allow the cleaning solution to drain out prior to rinsing.

The summation of these observations is that it is not violating good design practice to consider whether or not some changes might be made that would aid in solving cleaning problems. The paramount point is that in helping to solve cleaning problems, the engineer is making a contribution to lower overall costs and to the finished appearance of the work.

Mechanical Cleaning

As has been noted, mechanical energy is required, or extremely helpful, in removing some soils from the surface of parts. Scale, rust, mold, sand, weld splatter, and imbedded drawing lubricants are examples of such soils. In some cases, chemical cleaning augmented by agitation can handle the problem (there is very little mechanical energy available for soil removal in chemical solutions alone). In other cases, circumstances dictate that the cleaning cycle include a cleaning method that depends entirely on mechanical action.

For example, brushes might be used to clean castings and splatter from welds. Parts can be tumbled in a rotating barrel to remove scale and sand from castings. Vibration of parts can remove certain soils. Various media can be projected at high velocities against surfaces to remove, in particular, scale and brittle surface soils, see *Fig.* 65.5.

In the various mechanical cleaning methods — because soil is brushed, abraded, or blasted off the surface — the opportunity exists to do something to the surface in addition to cleaning it. For example, wire brushing both cleans and burnishes a surface. When parts and some medium are placed together and subjected to rotation or vibration various surface finishes can be obtained, depending on the medium used. Strength of parts can be increased because of the peening action. Similarly, various effects can be obtained in blasting operations, depending on the medium used.

Mechanical cleaning is sometimes the only solution to a production problem. For instance, ahead of some surface finishing operations, cleaning chemicals can become a soil unless thoroughly removed in rinsing. If the part design makes it impossible to obtain thorough, reproducible rinsing then chemical cleaning cannot be used and mechanical cleaning must be adopted.

Air pollution and waste disposal are currently important industrial problems. The outlook is that they will become more important, even critical. This tends to favor the selection of mechanical cleaning techniques since such factors as preventing escape of fumes or disposing of exhausted chemical baths are avoided. However, it should be appreciated that mechanical cleaning is not always "dry" nor does away in all cases with the need for chemical cleaning. Good practice dictates that in some cases, parts be degreased before mechanical cleaning. In some blasting equipment, the abrasive medium is carried by a water spray — the water being chemically treated in some respect. Then there can be the problem of removing by some cleaning operation the dust and metal particles from the surface of work that has been blasted. Further, the media used in shot blasting, for example, tends to pick up soil and requires some cleaning step.

Soil Prevention

In attempting to assess the future of cleaning technology, an interesting thought emerges. The cleaning field could be at a point where a useful development would be one that articulates and emphasizes the logical concern of "cleaning" with all aspects of soil formation as well as its traditional concern with soil removal. If "cleaning" can be broadened to "soil prevention" or "soil control," such a development might be as significant in some industries as the outgrowth of "preventive maintenance" from "maintenance" or the outgrowth of "quality control" from "inspection."

The foundations exist for formalizing such a development. For example, the sources of most "unnecessary" soils are generally known. The term, "unnecessary" applies to such soils as greases, oils, and other service lubricants that can contaminate parts when improperly applied to machines and conveyors. If fingerprints or perspiration are soils in a particular application then it should be possible to exclude them from parts. Parts which become excessively dusty, rusted, or otherwise contaminated from just lying around can certainly be considered as unnecessarily soiled. A formalized soil control program could be expected to prevent the formation of these types of soils.

Various soils have their origin in materials which must be applied to the surface of a part in some production process. Examples are cutting fluids, drawing lubricants, welding fluxes, buffing compounds, etc. The ease with which these materials can be cleaned off parts is obviously one factor that should be considered before using them. As gross contaminants, most can be easily removed immediately following their use. If allowed to dry and age, their removal is considerably more difficult. A formalized soil control program might conclude that all parts should receive some type of treatment at the workstation. Or perhaps that only heavily soiled parts immediately be spot cleaned in the interest of overall production economy.

The value of various precleaning treatments in the cleaning cycle has been well proved. For example, heavy soils might be removed by steam cleaning prior to the primary cleaning process. Precleaning with vapor solvents can remove oils that would contaminate a primary acid cleaning process. Hot water is used to flush off welding fluxes ahead of an alkaline cleaning process. Heavily rusted areas can be spot treated with an acid cleaner. With emphasis on precleaning immediately following a production process even more effective results might be expected with the proper materials, particularly since the treatment might be made compatible with the time used for parts movement or storage awaiting further processing.

That soil prevention can be valuable in particular situations is shown by the elaborate techniques which have sprung up in some industries. For example, some firms have obtained facilities for heat processing of metal parts in a vacuum or special atmosphere in order to eliminate formation of scale and oxides. In addition to heat treatment to change physical properties, such facilities are also used in brazing processes. An important advantage of vacuum processing is that the work is produced with chemically clean surfaces. Thus, various finish machining operations can be accomplished prior to heat treatment while the metal is soft. The parts retain a clean surface during the heat treatment.

Soil prevention and control programs will not eliminate the need for cleaning. However, they can highlight the fact that soils have a controllable origin in many instances. And one way to eliminate soil from the surface of parts is to take steps to insure that it doesn't occur in the first place.

ACKNOWLEDGEMENT

The assistance of the following companies in preparation is gratefully acknowledged:

Acoustica Associates Inc.
AutoSonics Inc.
Baron Blakeslee Inc.
Cincinnati Cleaning &
   Finishing Machinery Co.
Conrac Corp.
   Powertron Div.
DeVilbiss Co.
E. I. du Pont de Nemours
   & Co.
Electric Furnace Co.
Huffman-Wolfe Co.,
   Cleaning Machine Div.

Industrial Washing
   Machine Corp.
R. C. Mahon Co.
   Industrial Div.
Pangborn Corp.
Phillips Mfg. Co.
Ransohoff Co.
Tronic Corp.
Vapor Blast Mfg. Co.
Westinghouse Electric Corp.,
   Industrial Electronics Dept.
Wheelabrator Corp.

REFERENCES

[1] Henry B. Linford, "Preparation of Metals for Electroplating," *Plating,* December 1965, page 1262.
[2] Samuel Spring, *Metal Cleaning,* Reinhold Publishing Corp., New York, 1963.
[3] R. W. Mitchell, *The Metal Cleaning Handbook,* Magnus Chemical Co., Garwood, N. J., 1935.
[4] A. Pollack and P. Westphal, as translated from German by R. Weiner, *An Introduction to Metal Degreasing and Cleaning,* Robert Draper Ltd., Teddington, England 1963. Original book, *Metall-Reinigung und-Entfettung,* published by Eugen G. Lenze Verlag, Salgau/Wttbg., 1961.

66

Power Brushing, Polishing and Buffing

A. F. Schieder

POWER brushes are tools for conditioning surfaces to achieve specific degrees of smoothness or roughness, removal of burrs and lightly attached inclusions or particles, scale and oxides, finishing of edges by rounding or blending, removing surface film or encrustations, and generally affecting the surface itself or causing the removing of materials from or along a surface. Brushing tools do not find efficient use to remove metal or other materials from a surface to achieve a dimensional size or tolerance, although recent advancements with abrasive filament brushes do remove metal uniformly to a small degree and within specified tolerances.

Brushes in general do a wide range of work which may be categorized as follows:

Removing burrs.
Finishing surfaces for stress relief.
Removing scale and oxides.
Wet scrubbing.
Polishing and buffing.
Stripping insulation from wire.

END BRUSHES

WHEEL BRUSHES

SIDE ACTION BRUSHES

CUP BRUSHES

WIDE FACE BRUSHES

Fig. 66.1 — Power brush types.

A. F. Scheider is Assistant Chief Engineer, Brush Div., Osborn Manufacturing Corp.

For Jobs Requiring Finish Less Than 30 Microinches rms

◄─────────── Brush Flexibility ───────────►

Medium	High

◄─────────── Brush Finishing Ability ───────────►

Very Fine (low rms)	Medium

◄─────────── Brushing Action ───────────►

Medium	Very Fast

Cord & Fabric Brushes	Tampico & Treated Tampico Brushes	Fine Wire Sections Used With Burring Compound

Major Brush Characteristics

Cord & Fabric Brushes	Tampico & Treated Tampico Brushes	Fine Wire Sections Used With Burring Compound
Fast cutting medium flexible wheels. Used with cut and color buffing compounds for producing low microinch finishes	Flexible, fast cutting wheels. Used with burring compounds of tacky composition for producing surface finishes of approx. 8 to 10 microinches rms	For efficient use of wire brushes with compound, select brushes having medium or high density and fine wire size (.008 diam. or finer)

Major Applications in Order of General Usage

Cord & Fabric Brushes	Tampico & Treated Tampico Brushes	Fine Wire Sections Used With Burring Compound
1. Cut and color buffing 2. Surface blending 3. Light scale removal 4. Producing radii 5. Burr removal	1. Burr (small) removal 2. Producing radii 3. Light scale removal 4. Surface blending 5. Cleaning 6. Satin finishing	1. Burr (medium) removal 2. Light scale removal 3. Satin finishing 4. Producing radii

Adjustments for Desired Results

Brush works too slowly	Brush works too fast	Action of brush peens burr to adjacent surface
1. Increase surface speed by increasing OD or rpm 2. Decrease trim length and increase fill density 3. Increase filament diam. 4. Increase grit size	1. Reduce surface speed by reducing rpm or OD 2. Reduce filament diam. 3. Reduce fill density 4. Increase trim length 5. Decrease grit size	1. Decrease trim length and increase fill density 2. If wire brush tests indicate metal too ductile (burr is peened rather than removed), change to non-metallic brush such as abrasive brushes or treated tampico used with burring compound

Fig. 66.2 — Brush selection chart based on surface finish requirements.

For Jobs Requiring Surface Finish More Than 30 Microinches rms

←————————————— Brush Flexibility —————————————→

Low   High

←————————————— Brush Finishing Ability —————————————→

Fine   Coarse

←————————————— Brushing Action —————————————→

Medium Fast   Medium   Fast

Wire Wheels	Wire Sections	Coiled Knot Sections	Abrasive Brushes

Major Brush Characteristics

Wire Wheels	Wire Sections	Coiled Knot Sections	Abrasive Brushes
Wide face, dense fill. Used where fast cutting and fine finishes are required. Best general purpose wire brush	Narrow face, medium density. Used where medium brush flexibility is essential to follow contoured surfaces	Narrow face, low density. Used where a high degree of brush flexibility is needed and where surface impact action is necessary, especially to remove surface encrustations	Narrow to wide face. Medium to heavy density with grit sizes ranging from very fine to coarse.

Major Applications in Order of General Usage

Wire Wheels	Wire Sections	Coiled Knot Sections	Abrasive Brushes
1. Burr (heavy) removal 2. Producing radii 3. Rust and oxide removal 4. Scale removal 5. Cleaning 6. Surface blending	1. Cleaning 2. Medium scale removal 3. Rust and oxide removal 4. Satin finishing 5. Surface blending 6. Burr removal	1. Heavy scale removal 2. Rust and oxide removal 3. Satin finishing 4. Producing radii 5. Burr removal (on very hard metals)	1. Surface finishing 2. Removing surface soils for chemical cleanliness 3. Light deburring 4. Producing radii 5. Metal removal 6. Light polishing and honing

Adjustments for Desired Results

Finer or smoother finish required	Finish too smooth and lustrous	Brushing action not sufficiently uniform	
1. Decrease trim length and increase fill density 2. Decrease wire diam. 3. Try treated Tampico or cord brushes with suitable compounds at recommended speeds 4. Use abrasive brushes	1. Increase trim length 2. Reduce brush fill density 3. Reduce surface speed 4. Increase filament diam. 5. Use coarser grit size	1. Increase trim length and decrease fill density 2. Devise hand held or mechanical fixture or machine which will avoid irregular off-hand manipulation 3. With abrasive increase brush pressure	

Fig. 66.3 — General types of brushes employed.

Cleaning to remove oil, rust, stain, paint and other surface soils and encrustations.

Finishing for appearance.

Finishing to achieve a specific surface of a desired roughness or smoothness.

Surface and juncture blending.

Rubber and plastic removal.

Each application, whether it be removal of sharp edges, tool marks or scratches, or the improvement of surface finish and detection of cracks, must be properly engineered and tooled for brushing. There are few specific rules that can be set up to cover the many variables surrounding the use of brushing wheels as production tools. Each job has its own characteristics, and each must be studied from the standpoint of brushing-wheel technique, construction of wheels, and effects of brush speed, wheel size, and manner of presenting the work to the wheel, which may be used with or without abrasive, and the design and construction of machine tool to be used.

Brush Selection. Selection of a brushing wheel for a given application is done best after compiling information along the following lines:

    1.   What is to be accomplished?

        a. Clean or remove — what material?

        b. Rough or smooth — what kind of surface?

        c. Surface blending — to what degree?

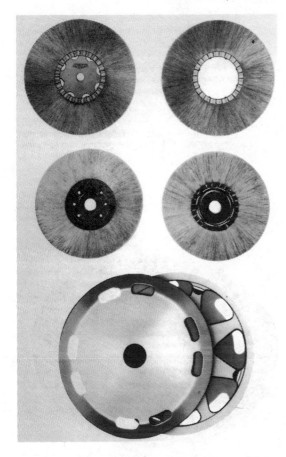

Fig. 66.4 — The strength of the brush is dependent upon the steel retaining member or ring which is shrunk into position. The steel base is accurately formed to permit sections to be used side by side. These sections are readily adaptable by means of pressed-board or steel centers and metal flanges to any suitable standard arbor.

      d. Blending surface junctures or blend surfaces — how much?
2.   What kind of equipment is available?
      a. Brushing lathe — what horsepower and speed?
      b. Semi-automatic machine — how to present work to brush.
      c. Automatic brushing machine — what type?
      d. Portable or flexible shaft equipment?
3.   What are the properties of the material involved, especially hardness?
4.   At what rate and under what conditions must the work be done?

Answers to these basic questions will usually furnish the facts necessary to determine brush type, *Fig.* 66.1 and material and the brushing technique required.

One of the advantages of a brush and brushing method is the possibility of dual usage, ranging from strictly off-hand operations to fully automatic equipment, with practically every degree between these two extremes, depending upon the volume and rate of production required.

Brush selection is best done empirically. A single trial with a brush usually determines the correction action which is necessary to determine the proper brush specifications. The initial brush selection is based on the surface finish requirements. A brush selection chart based on experience can be used to make the first selection. This chart, *Fig.* 66.2, requires that a determination be made regarding the surface requirements. If the job requires a surface finish less than 30 microinches., rms, a certain group of brushes is indicated. On the other hand, if the

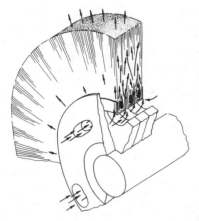

Fig. 66.5 — Ventilation must be considered, particularly when nonmetallic fill material is used, when the brushing face is 3 in. or more, and when the work cycle is continuous. Air vents are located in the steel bases to facilitate cooling, thus ensuring a flow of cool air inward through the central portion of the brush, with the brush material acting as fan blades to throw out the heated air and suck in new, cool air.

Fig. 66.6 — Flexibility of brushes enables them to conform to the many slight irregularities and contours which cannot be accomplished with setup wheels or abrasive belts.

surface roughness is permitted or required to exceed 30 microinches, a second group of brushes is indicated.

There are three major determinations to help further in selecting the correct brush. These factors are brush flexibility, brush finish ability, and brushing action strength. Following the direction of the arrows on the chart, it is possible to select a brush of high or low brush flexibility with faster or slower brushing action.

It is common practice to use one brush for several different kinds of work. This may be considered practical where job lots are manufactured; but where parts are mass-produced, no time should be allowed for unnecessary manipulation of the work or brush. For minimum operational costs, the correct brush for each job must be selected. The essential brush specifications to be considered are as follows:

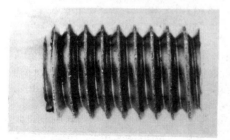
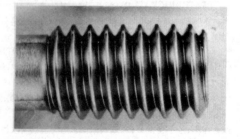

Fig. 66.7 — Before and after views of brushed threads.

Fig. 66.8 — Brushing setup to remove excess rubber from tank-track pins prevented the formation of small microscopic cracks which would have caused fracture through fatigue.

1. Type of fill material (wire, fiber, synthetic, abrasive, etc.).
2. Size of fill material (diameter).
3. Trim length (length of fill material extending beyond face plates or brush back).
4. Density (packing of fill material).
5. Quality of fill material.
6. Outside diameter of brush.
7. Surface speed of brush.

Fig. 66.9 — Gear before and after brushing to remove burrs and sharp edges.

BEFORE BRUSHING

AFTER BRUSHING

Fig. 66.10 — The treated tampico brush finds its place in surface blending prior to plating.

Fig. 66.11 — Photomicrographs showing the reduction of surface imperfections which are obtainable through brushing a cold-rolled steel surface, top right, before and bottom right, after brushing.

8.   Other treatment or special quality of fill material, such as the crimp in wire or synthetic

Brush Classification. Of equal importance with the brush specifications are the requirements of the job to be done. The most important are these:

1.   Nature of work to be done, i.e., removing burrs or scale, or reducing the finishing of a surface to a definite specification in microinches, rms.

2.   Surface finish requirements. The quality of finish required is a primary consideration. A better-than-necessary finish often will cost more, while a finish

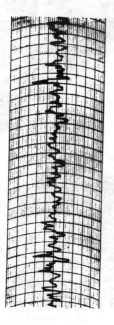

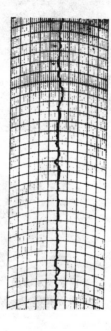

Fig. 66.12 — Analyzer charts of a sample before and after being brushed with a treated tampico wheel. A roughness of 24 to 35 μin., rms, for the original sample, whereas the same surface brushed results in a roughness of 4-1/2 to 7 μin., rms.

Fig. 66.13 — Stainless-steel parts mass-finished with brushes in one pass using three brushing heads. Attention is called to the sharpness of the letters.

Fig. 66.14 — Centerless grinder being utilized for centerless brushing.

not up to specification may result either in rejection at inspection or in failure in actual use.

3. Metal or other material from which the part is made and its condition at the time of brushing.

4. Limitations related to part clearance, size, and restricted areas.

5. Special specifications requiring definite results, such as the degree and character of rounding an edge to 0.015 in.

Brushing wheels can produce a wide variety of finishes, depending on the wheel material and how the wheel is applied. *Fig.* 66.3 shows some of the wide face and wheel-type brushes. For practical selection, brushes may be classified as follows:

Cord or Fabric Brushes: Fast cutting, medium flexible wheels used with cut and color compounds for producing low microinch finishes.

Tampico and Treated Tampico Brushes: Flexible, fast cutting wheels used with burring compounds of tacky composition for producing surface finishes of approximately 8 to 10 μ in., rms.

BRUSH ROLL RECIPROCATION

Axial reciprocation or oscillation may be necessary in limited cases, especially when brushing parts in multiples, as shown on the sketch. When brushing parts in multiples, the amount of reciprocation must exceed the width or diameter of one part plus one half of the space in between parts to keep an even and uniform brush face. Best results have been found when the brush roll is oscillating three to five complete cycles per minute. Oscillation by itself will not prevent streaking of individual flat brush sections that are assembled to make a wide face unit. Offsetting the individual section or going to a helically wound brush is recommended.

BACKUP ROLL

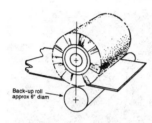

Backup rolls must be designed to support the part being brushed to avoid interference with the brush roll. The brush roll cannot successfully serve as a backup roll for another brush because individual pressure control is lost. For most parts and continuous sheet or strip, a single backup roll immediately opposite the brush, as shown in the sketch, is recommended. For extremely thin parts where brush contact with the ends of the backup roll is to be avoided, it is desirable to offset the single backup roll slightly from the center line of the brush.

NUMBER OF BRUSHING HEADS

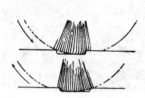

Continuous operations for brushing strip, rod and similar materials require the use of brushing heads in multiples of two. This requirement exists because of surface discontinuities and defects which frequently have a shape that will tend to mask the brushing action out of certain areas as indicated in this sketch. By having two brushing heads operating in reverse directions, it can be easily visualized how a more uniform surface finish can be obtained. For brushing to remove surface encrustations such as scale, it is most important that this two-direction approach with brushes be followed if pits, fissures and other surface imperfections are to be thoroughly cleaned.

WIRE BRUSH FACE CONDITIONING

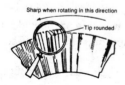

Brushes like other perishable tools must have their working face reconditioned and resharpened. It can be visualized easily how the tip of a brush fill-wire could eventually become dull and lose its working clearance. This clearance must be restored at intervals which are determined by the requirements of the job. Reconditioning and resharpening a brush are accomplished simply and easily by alternately reversing the direction of rotation during production or by periodic regrinding or touch-grinding the face.

BRUSHING SPEEDS

Brushing speeds vary considerably from job to job, depending upon the speed of the work or brushing intervals and such other factors as the amount of material to be removed. While it is difficult to give specific speeds, ranges for various general classification of work are given below in fpm:

Speed
fpm

Removing burrs	5500 to	7500
Blending for mechanical strength	4700 to	7500
Cleaning dry (wire brushes)	4000 to	5500
Scrubbing wet	1900 to	4000
Finishing for appearance:		
Polishing	6400 to	8000
Buffing	8000 to	10,000
Removing scale	7500 to	10,000
Cleaning welds	7200 to	9400

BRUSH FACE CONTACT

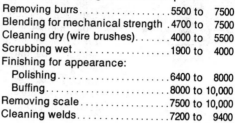

Recommended   Not recommended

In setting up a brushing operation, it is important to have the brushes contact the part being brushed in a way that the full face of the brush is in contact with the part. Except when a reciprocating brush is employed, full brushing face contact is necessary to avoid grooving the brush. Operations which are set up so that the brush face is not in full contact with the work required some provision for dressing the brush face.

CENTERLESS BRUSHING

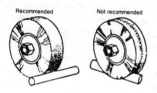

Regulating wheel   Work   Brushing wheel

Work rest
blade

This sketch shows schematically a centerless brushing set-up. Brushes, like grinding wheels, can efficiently do work on the centerless principle. Brushes will not reduce parts to a specific dimension but only change the surface character-istics. Brushes are made interchangeable with grinding wheels so that conversion from centerless grinding to center-less brushing merely requires a change of wheel. In most cases the regulating wheel used is the same as for grinding operations.

WHEEL OR ROLL ADJUSTMENTS

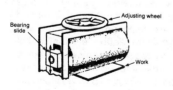

Bearing
slide

Adjusting wheel

Work

Probably the most important feature of any machine concerns the adjustment of the wheel or roll to obtain the required brushing pressure and to compensate for brush wear. The best results have been obtained when the adjustment of a brush or roll is made through one adjusting screw with mi-crometer control. The adjusting may be done with a hand wheel or through a gearhead motor operated by pushbutton control. Experience proves it best to avoid independent adjustment.

Fig. 66.15 — Basic requirements pertaining to effective brushing in production.

BRUSHING PRESSURE RELIEF

Good production brushing requires that provision be made to allow pressure relief under conditions of extreme brushing pressure caused by misalignment, oversize or improper seating of the part. This feature is especially necessary for brushes having a short trim or a dense fill where the brushing face could be damaged because of the inability of the individual filaments to escape being stressed beyond the yield point. There are two ways of accomplishing pressure relief. For small parts it is usually best to allow the part to retract, especially where this can be done so as to allow a straight pass line to be maintained. The second method, which is standard practice for polishing and buffing work, is to have the brushing head floating so as to allow the brush to fall and rise as the contour may require during the operation.

BRUSHING PRESSURE METER

Brushing pressure affects the quality of the work and the life of the brush to a very considerable extent. It is essential to have brushing pressure meters on all brushing machinery since it is not possible to visually estimate brushing pressure. An ammeter connected to the driving motor circuit for each brushing head is recommended for this purpose as a permanent part of the machine. Ammeters have been successfully used for this work and can be quickly calibrated to suit each individual job so as to eliminate guesswork. Quantitative knowledge of brushing pressure during the entire period of operation of the machine is most essential.

BRUSH BALANCE

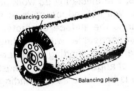

Satisfactory machine operation usually requires the use of balanced brushes. This is especially true for long brushing rolls and for wheels or rolls which operate at speeds in excess of 100 rpm. Best practice requires that the shaft, mounting and the brush be dynamically balanced to within one-half ounce-inch.

FLOW-THRU MOUNTINGS

Internally fed water or solution provides maximum cleaning and lower end of service costs. Internal feeding of rinse water or solution continuously flushes the dirt or accumulation of smut out of brushes which are designed for operation with flow-thru mountings. The flow-thru action eliminates hot spots in the brush core and assures uniform brush operating temperatures. In addition, internal solution feeding provides added stiffness to the brush bristle and also provides lubrication to the bristles during operation. The illustration shows a typical flow-thru mounting with a drilled hole pattern.

Fig. 66.15 — Basic requirements pertaining to effective brushing in production.

KEYWAYS

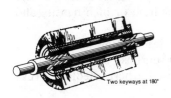

Two keyways at 180°

Keyways should be specified and used for long face brushes, heavy duty work and where the brush reversal method of face conditioning is used. American Standard keyways are recommended. On long face brushes which are adapted to a shaft with a mounting, it is recommended that two keyways, 180° apart, be used the entire length of the shaft. Short keyways on both ends of the brush mounting will produce alignment difficulties.

BRUSH VENTILATION

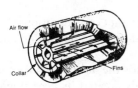

Air flow

Collar

Fins

Brush ventilation is extremely important where brushing rolls or wheels are used under conditions of continuous and/or heavy duty and where the brushing roll face exceeds the length of four to six inches. Radial fin type mountings with ends open to allow free passage of air as shown are good for obtaining the required ventilation. When using such a mounting, it is very important to provide adequate room at each end of the brush so that sheave and bearing housings do not block the openings in the mounting ends. Good practice requires that ventilation be provided also for wheel and sectional brushes of nonmetallic fill materials. Machine flanges must be built with the necessary ventilation openings.

SHAFT SIZE

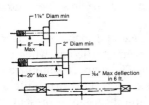

1¼" Diam min
8" Max
2" Diam min
20" Max
⅟₆₄" Max deflection in 6 ft.

While shaft size standards for brushing machines have been developed as a result of many years of usage, it is still essential to calculate the shaft diameter for special machines and operating conditions. In general, for semiautomatic machines having a shaft size length requirement of 8 to 12 inches overhang, a 1¼-inch diameter has been found to be very satisfactory. For this same type of service where the shaft length goes up to a maximum of 20 inches, a 2-inch diameter shaft has been found satisfactory. For a special machine the shaft size is determined on the basis of a uniformly applied load with the shaft fixed at both ends. For brushing operations, a maximum deflection of 1/64" in six feet is a standard recommended allowance.

BRUSH REPLACEMENT

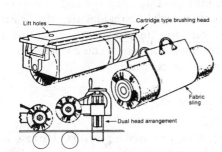

Lift holes
Cartridge type brushing head
Fabric sling
Dual head arrangement

Replacement of worn-out brushes is a very important detail for which provision must be made when the brushing machine is designed. The important thing is to make replacement easy and quickly so as to keep machine downtime at the lowest level possible. In fact it is desirable in many cases to design the equipment so as to permit replacing the brushes while the operation continues. For this purpose a spare head is sometimes used so that as a brush becomes worn the spare head can be placed in service while the regular head is being serviced. This must all be done without damaging the face of the brushing roll. An example of what can be done to facilitate brush replacement is illustrated in these sketches.

Fine Wire Sections Used with Burring Compound: For efficient use of wire brushes with compound, select wire brushes having medium or high density and fine wire size (0.008-inch diameter wire or finer).

Wire Wheels: Wide face, dense fill, used where fast cutting and equal or somewhat improved finishes are required. Best general purpose wire brush.

Wire Sections: Narrow face, medium density, used where medium brush flexibility is essential to follow contoured surfaces.

Coiled Knot Sections: Narrow face, low density wire brush used where high brush flexibility is needed and where surface impact action is necessary, especially to remove surface encrustations.

Abrasive Grit Brushes: Narrow to wide face, medium to heavy density, with grit sizes ranging from very fine to coarse — used for surface finishing, light deburring, metal removal, producing radii, and light polishing.

In evaluating brush construction, there are three important considerations: (1) proper ventilation or liquid cooling, (2) sufficient strength for safe operation, and (3) necessary cushion or flexibility.

Strength of a brush is important because on it depends the ability to hold the brush fibers, bristles, or wires, and to withstand the loads created by centrifugal force and by friction against the work. In certain brushing operations, centrifugal force is the major load while in others it is the load created by the friction or drag of the fibers, bristles, abrasive or wire against the work, or a combination of the two. With heavy loading, it is important that the material be held in tightly in such a way that it will not pull out, break off or slip around so as to cause out of balance, *Fig.* 66.4. In addition, the wheels and rolls must be made so that they can be adapted to various shafts and arbors under conditions of wetness and dryness, heat, and intermittent and continuous loads.

Ventilation is critical when the brushing roll has a face of 3 inches or longer, *Fig.* 66.5. This is especially so when the fill material is nonmetallic and when the work is of a continuous nature, such as the brushing of sheet or strip. It is important that the temperature of the brush be kept at 130° F or cooler, if short brush life or outright burning or scorching is to be avoided. Where a liquid coolant is present, it is highly desirable to utilize flow-thru shafts and pump the liquid coolant through the brush shaft and through the brushes themselves. This will provide a uniformly low temperature, clean and flush out dirt and soil particles, and lubricate the brush bristles.

Flexibility of brushes, *Fig.* 66.6, as compared with setup wheels or abrasive belts is a characteristic which enables them to conform to the many slight irregularities and contours of the part. They act uniformly over the surface, including penetration of slight recesses. In contrast to this, the more rigid wheel or belt works on the high spots alone. By penetrating surface irregularities, the brush is able to blend their edges and surfaces. This blending, in addition to promoting a pleasing appearance, disperses stress concentration points which cause fracture and set up differences in electrical potential to promote corrosion.

Design Considerations

Brushing to Reduce Fatigue. The subject of repeated or reversed stress is important because of the higher speeds which have become common in modern machines. Experience and experiments both have shown clearly that when structural and machine members are subjected to repeated or reversing stresses, the safe operating loads for most of them may be very much less than the safe static loads for the same members.

When a metal part is subjected to repeated stresses of sufficient magnetude, the action of the stresses is such so as to enlarge and even start microscopic cracks which, under continued application of stress, spread until failure occurs. The term "fatigue" is commonly used to describe this behavior, although a better description of the action might be "progressive or gradual fracture."

The failure of any part caused by repeated loads or reversed loads is not characterized by yielding but is accompanied by other faults which are characteristic and for which brushing is finding increasing application as a practical preventive measure. Failure, or progressive fracture, starts at some point in the material where the actual stress is much larger than calculated stress. This high localization of stress, which seems to be caused in many cases or at least is accompanied by sharp edges and burrs, enlarges or starts small microscopic cracks until, as the stress is repeated, the whole member ruptures.

Power-driven brushes have been instrumental in helping solve one phase of the problem of stress concentration and progressive fracture by providing in fast, economical, and high quality production methods a means for reducing and, in many cases, eliminating the real causes. A number of striking examples have been encountered. For instance, with bolts and studs most failures start at the root of the thread where highly localized stresses occur. Blending of sharp edges and the improvement of the surface finish, without appreciably changing the shape of the threads, help spread the concentrated stresses and assist in overcoming this problem. Removal of the burrs at the same time minimizes stress concentration through more equal load distribution, *Fig.* 66.7.

Another example is that of removing excess rubber from pins associated with tank track treads. This is a case where special cutters could be used to remove the rubber in a method almost as economical as the brush method. However, it was actually found that by brushing the pins the surfaces resisted the formation or spreading of small microscopic cracks which would eventually have caused rupture through fatigue, whereas the use of cutting tools promoted breakage of the pins in use. The cutting tools, regardless of the speeds and feeds used, produced tool marks and fragmented the surface metal to a sufficient degree to allow small microscopic cracks to start or spread and eventually caused the pins to rupture, *Fig.* 66.8.

Failures of various machine members due to repeated stress have occurred frequently in service when preventative measures have been omitted. Gears, crankshafts, piston rods, valve rods, threaded parts, springs, and shafts are but a few of the very large number of parts so affected.

It has been estimated that the primary cause of as high as 80 per cent of machine part failures is stress concentration and progressive fracture. The most common occasion of fatigue failure is the presence of an abrupt change of cross section, which may be a square shoulder, a V notch, a screw thread, burrs, toolmarks, scratches, a keyway, or external or internal flaws in the metal. Proper brushing has been shown to be a helpful preventive measure for all of these, with principal use in reducing the effects of burrs, tool marks, sharp edges, and scratches.

Blending Surface Junctures. It is practical to consider a brush for removing burrs from an edge and to produce a uniformly blended surface juncture. Where a

Fig. 66.16 — Brushes and brushing equipment constitute important production tools. The production yield of a brushing operation is influenced by the design of the brushing machine employed to operate the brush.

brush can act on an edge, it is possible to remove the amount of metal required to produce this effect, whereas on the surface adjacent to the edge, it is practical only to produce surface finish requirements which do not involve the removal of more than insignificant amounts of metal. Then, too, there is another factor related to dimensional tolerances — the shape of the part — which has a great deal of influence on the type of brushing tool to be used. For example, a fixed grit wheel, cutter, or file will remove metal from the surface and change the dimensional tolerance of the part, whereas a flexible brushing tool will follow the shape and contour of the surface to be finished without removing an appreciable amount of metal. This all important feature of brushing offers certain advantages for production applications, *Fig. 66.9.*

Cord brushes have shown outstanding results in this type of work, as they can produce commercially surfaces having roughness values of 8 microinches, rms, or less where necessary while removing burrs from all types of metal parts, including hardened, high alloy steels.

Surface Blending and Polishing. Treated tampico brushes find use in surface blending prior to plating. Here the capacity for fast cutting inherent in this type of brush enables the surface to be uniformly brushed, thereby blending imperfections so that a uniform satisfactory plating job may be accomplished at a minimum cost, *Fig. 66.10.*

Hard cord brushes used in finishing stainless steel have much less tendency to develop drag lines in the finish than have conventional methods. It is the fast cutting action and firm cushion of the brush which prevents them from producing these drag lines. The fast cutting action of these cord brushes used with suitable compound permits the production of a satisfactory finish with fewer wheels or passes.

Fiber, cord and abrasive filament brushes can modify and change the condition of metallic surfaces very appreciably. It is this ability which gives these brushes value for polishing and buffing. The action of a brush is that of blending imperfections and surface irregularities, rather than the actual removal of metal as is obtained with abrasive grinding wheels or belts. It has also been demonstrated that the blending action of a brush reduces surface imperfections to a state where a uniformly good plating job can be obtained. This can be done without the need for cutting an entire surface down to the depth of the deepest surface imperfections, and therefore fewer wheels or passes are required.

The results shown in the photomicrographs of *Fig.* 66.11 characterize the reduction in surface imperfections which are obtainable through the use of brushing. Comparison of the two surfaces shown emphasizes the improvements in surface appearance produced by brushing. However, it is of equal importance that these surfaces can be evaluated by examining the charts in *Fig.* 66.12 which show the surface roughness in microinches and the magnitude of the sharpness of the surface marks as measured by a brush surface analyzer.

Evaluation of a surface brushed preparatory to plating should be made in the light of the results obtained after plating. When comparing a brushed surface with a surface obtained by older methods of polishing and buffing, it will be found that brushes tend to produce a somewhat duller surface finish which is not confined to the high spots but actually refine the entire surface. The result is that the brushed surface plates better by virtue of the greater refinement. It might be noted that an inspection of a brushed surface is less deceptive. After brushing has produced a satisfactory finish, fabric brushes or cloth buffs frequently have a place in producing a higher color.

In *Fig.* 66.13 is shown a group of stainless steel parts efficiently mass-finished with brushes in one pass using three brushing heads. Particular attention is called to the sharpness of the letters which is the result of fast cutting action and firm cushion of the brush.

High Production Power Brushing. Brushes and brushing equipment constitute important production tools because they fit into those manufacturing operations where the ratio of labor costs to other production costs is very high, *Fig.* 66.14. The production yield of a brushing operation is influenced by the design of the brushing machine employed to operate the brush. In *Fig.* 66.15 a few suggestions relating to design are outlined. Completely automatic equipment, *Fig.* 66.16, designed to feed and move parts through one or more brushing stations is capable of finishing at the rate of 1,000 to 2,000 parts per hour. On the other hand, a simple work fixture, such as a hand-held tool for holding the part in front of a rotating brush, would result in a much lower production yield per hour.

In any event, brushes are capable of finishing parts economically at a high production rate provided that suitable brushing equipment with automatic machine features are considered. With well designed machines capable of feeding and guiding the work, parts produced on a high production basis can be finished by brushes at a rate comparable to related manufacturing operations. With thorough consideration of the proper application, power brushing methods can be effective in reducing labor costs by minimizing skilled labor requirements through machine design.

67

Automatic Painting

Julian E. Wilburn

$\mathbf{A}$VAILABLE to potential users are many painting techniques that can be used with handling and control equipment to provide automated finishing systems. These techniques include compressed air spraying, airless spraying, hot spraying, electrostatic spraying, dip coating, electrocoating, flow coating, curtain coating, roller coating, and several other coating methods. Products as large as a freight car or as small as a machine screw can be painted automatically with one or more of these techniques.

Selecting the best painting technique for a particular product is not a simple task. Each technique offers significant advantages for the types of products it is capable of painting. At the same time, each technique has disadvantages or limitations that are sometimes overlooked when only one technique is under consideration. To arrive at the most efficient and economical solution to a painting problem, all applicable techniques should be thoroughly investigated and compared with each other.

The purpose of this article is to examine all of the major painting techniques and to impartially point out major advantages and disadvantages. It is realized that such an analysis can only be general in nature and that specific equipment offered by equipment manufacturers may have superior advantages to the ones discussed or may have overcome any disadvantages mentioned. Therefore, while this article may be used as a guide to available painting techniques, it should not be used to rule out any painting technique for a particular product without detailed discussion with a potential equipment supplier.

The most dependable method of properly selecting the best painting technique is to allow the paint material manufacturer and the painting equipment supplier to run actual test trials on the product at their laboratories. In this manner, painting techniques that might be widely dissimilar in application can be tested under simulated production conditions on random lots. Differences in the techniques can be compared and actual results can be used to arrive at the best solution for the specific product. The principal danger or over-all problem in the painting equipment field is the attempt to apply one technique to all problems or to be influenced by a new and exotic technique which may have a best specific application, but is unsuitable when applied to general applications.

Julian E. Wilburn was Associate Editor, Automation Magazine, Cleveland, Ohio.

Material Considerations

The purpose of any painting technique is to apply a coating material to the surfaces of a product. Generally, this material must be applied in a liquid form by means of spraying, flowing, or rolling the material on the product or by dipping the product into the material. Paint manufacturers usually produce and package coating materials at the proper viscosity for the painting technique being used. Most finishing materials contain three ingredients: Pigment which provides color and sheen, vehicle which forms a dry film binding the pigment to the product surface, and thinner which adjusts the viscosity of the material to a point where it can be easily applied.

The important factor in any application of paint is the amount and quality (appearance, gloss, uniformity, etc.) of solid deposit that remains after the liquid material has been applied and allowed to cure or dry. Thinners such as solvents and diluents increase the fluidity and volume of the painting material, but do not add to the solids content of the paint. Once they have been used and evaporated, thinners usually cannot be recovered for re-use.

In the various painting techniques, thinners assist in getting the solid deposits onto the surface of the product being painted. In addition, they serve to modify the paint film flow on the product surface so as to obtain a quality finish. However, once the painting material has been evenly applied on the surface, it is desirable that the material become viscous as soon as possible and remain only sufficiently fluid as to flow out evenly and attain the desired gloss.

Just about every type of painting material can be applied by an automatic painting technique. Some materials can only be applied by certain techniques. For example, two-component epoxy materials are usually restricted to spraying methods. However, most materials have been thoroughly tested by paint manufacturers before being marketed and are sold with proper recommendations as to methods of application.

Many companies use a painting material that previous experience has shown works best on the product. If a new material is being considered, the user and the paint manufacturer generally work in conjunction with each other in its development for a particular use. By the time a painting technique is to be selected, the material factor has usually been eliminated as a basic consideration. However, the painting equipment supplier needs to know the type of material, the manufacturer, the solvent required and the recommended reduction rate, the viscosity of the material, the number and type of coats to be applied, the wet and dry film thickness required, the quality of finish desired, the per cent solids of the material, the flash time of the material, and the recommended baking time and temperature.

One factor stressed in any comparison of painting techniques is material utilization, or how much painting material used by a technique is actually applied to the product. For example, cold air spraying is said to lose some material due to rebound and overspray. On the other hand, electrostatic spraying is said to be more efficient since material is electrically attracted to the product surface. Flow coating is said to have as high as 97 per cent material utilization.

If the selection of a painting technique is based on material utilization alone, that is, how many products can be painted per gallon of material, then cold air spraying might appear to be the most wasteful and expensive painting method.

However, some products that are painted with electrostatic methods may require extensive hand reinforcing or touchup since the technique does not penetrate into deep recesses or sharp corners. Extra cost is involved in this touchup. Flow coating may require as much as 2¹/₂ gallons of solvent and thinner for each gallon of paint applied due to excessive evaporation of the solvent in the process. This cost should be added to the cost of the painting technique.

Therefore, while material utilization is a factor to be considered in evaluating painting techniques, its importance should not be over-valued. Each painting technique should be evaluated in terms of the total cost of applying paint to a product. This cost should include the cost of equipment, painting materials, solvents and thinners, direct labor, maintenance labor, hand reinforcing labor and materials, and other factors. The separate sections devoted to the various painting techniques will discuss important factors that should be considered in evaluating painting as a production process.

Painting Methods

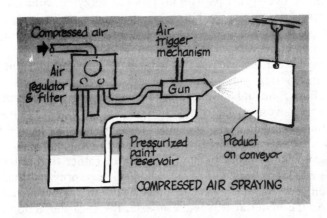

Compressed Air Cold Spray Painting. This technique is known as conventional spraying and has been used to apply almost every sprayable material. The basic elements of cold air spraying are an air compressor, spray guns, paint supply by pressure tank or pump, and hoses. Dozens of gun designs and hundreds of cap/tip combinations offer a great range of flexibility and adaptability to automatic production requirements.

Compressed air is used to feed paint from a tank or a pump to the gun nozzle. A separate line feeds air to the nozzle. When the gun is actuated the material and air streams join together either inside or outside the gun nozzle where the stream of air atomizes the material into tiny particles of paint. The velocity of the air stream carries the material to the surface to be painted.

Painting material usually contains a balanced combination of high and low boiling thinners. The low boiling thinners aid in atomization at the gun nozzle. The high boiling thinners aid in leveling the material after it reaches the surface. The goal in conventional spraying is to get maximum atomization of paint particles with the most efficient vaporization of the thinners. This goal is achieved by using the proper ratio of air pressure to fluid pressure. Today's low pressure air caps and easily atomized materials make high atomization pressures unnecessary.

Conventional spray is often criticized for producing overspray. Technically, overspray is that portion of the spray pattern that passes over the edge of the product being painted, and a certain amount of it is present in all types of spray painting systems. If atomizing air pressure is too high, misting can occur. When both fluid and air pressures are too high, conventional spraying can cause rebound or the bouncing of paint particles off the surface being painted. These conditions can be corrected by proper adjustment of the spray equipment.

Application Considerations. Conventional spraying is universally known and understood. It is simple in operation, versatile in use, and adaptable to a wide range of standard and optional accessories. It provides more selectivity of spray pattern size, a greater range of atomization, and greater flexibility of use than any other spraying technique.

Cold spray produces a fine paint finish, although it requires a careful balance of paint material, thinner, and solvent. It can be used in both air cure and heat cure operations, depending upon the thinners and solvents used. Speed of application is very fast, but film builds must be kept small to prevent material runs and sags. Depending upon the product configuration, material utilization may be only 30 to 60 per cent due to overspray. As with all types of spraying, cold spraying usually requires air scrubbers and spray booths due to overspray, misting, and volatile vapor conditions. Maintenance costs can be high if frequent spray booth and scrubber cleaning are necessary.

Cold spraying can be used with all types of painting machines to provide automation of painting processes. Its greatest advantage is its flexibility of adjustment of both fluid and air pressures to provide spraying patterns for all types of products. It is lowest in cost of all the spraying systems. Commonly, it is not efficiently adjusted for proper atomization at lowest pressures, and wastes paint material and solvent — yet it will still yield acceptable results.

Airless Cold Spray Painting. Hydraulic force rather than compressed air is used to produce atomization of paint material in airless spray painting. The coating material is pumped at high pressure through a hose to a spray nozzle or tip in an airless gun. As the material leaves the nozzle, the shearing action of the liquid passing at high velocity over the sharp edge of the nozzle orifice causes the material to atomize into tiny particles instantaneously. The momentum of each particle carries it to the surface being painted.

Pumping pressure of the material is usually between 500 and 3000 psi with most systems operating at about 2000 psi. The degree of atomization and the spray pattern are determined by two variables: Size and shape of the orifice in the spray nozzle and pressure of material at the nozzle. Spray nozzles are available in equivalent orifice diameters of from 0.007 to 0.109-inch and in spray angles ranging from 5 to 95 degrees. Some nozzles are available with pre-orifices to obtain a better feather edge of the spray pattern and to help reduce the tendency of the material to spit when the gun is triggered.

An airless gun is either full on or full off. Any adjustment to flow rate of material, fan pattern, or degree of atomization can only be made by changing the hydraulic pressure or by changing the spray nozzle. Spray nozzles are usually made of tungsten carbide and are very accurately machined to produce the desired spray pattern. If worn to any degree by abrasive pigments in the paint, spraying results will begin to deteriorate and cannot be corrected except by changing the spray nozzle. In contrast, compressed air spraying has infinite adjustment of fluid pressure and atomizing pressure to adjust for varying conditions in a manufacturing plant.

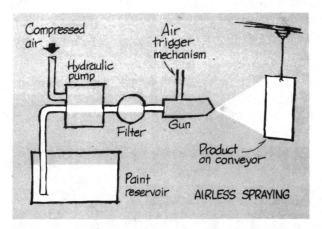

Application Considerations. Airless cold spray painting eliminates possible contamination of paint material from compressed air impurities and water. The average size of paint particles formed by airless atomization is slightly larger than those formed by conventional atomization. However, there are fewer tiny particles formed by airless atomization, which cuts down misting and overspray. Since the airless particles are not carried to the work by an air stream, there is less rebound of the particles from the product surface.

In many applications, airless spraying attains higher material utilization than conventional spraying, particularly in spraying large, broad surfaces or rough and irregular surfaces. In spraying smooth surfaces it can produce as fine a finish as conventional spraying.

With less overspray and rebound of paint particles, there is less need for spray booths and air scrubbers with airless spraying. Maintenance is also reduced since booths and air scrubbers do not have to be cleaned as frequently when used.

Airless spraying equipment is more expensive than compressed air equipment since it requires the use of high-pressure hydraulic equipment — pumps, fittings, hoses, and accessories. Due to the small size of orifices in airless spray nozzles, they are more susceptible to clogging than compressed air nozzles. High pressure and abrasive materials in paints can cause more and faster wear in airless nozzles.

Airless spraying can be used with all types of painting machines to automate coating processes. While it is simpler to operate than conventional spraying, its lack of adjustment can limit its application. For example, when conditions such as conveyor speed, temperature, product size, and paint viscosity have been established for a certain nozzle size, any change in conditions may require a change in nozzle size to obtain proper atomization and coverage. Such condition changes can be accommodated in conventional spraying by changing fluid or air pressures.

Hot Spray Painting. One of the major variables in painting is the change in viscosity of painting material due to temperature variation. For example, if room temperature changes 15 degrees between morning and mid-day, material viscosity may change as much as one-third. Therefore, cold material that is at the right viscosity for spraying in the morning may be too thin by noon. By heating the material, the paint can be delivered to spray guns at a uniform viscosity throughout an operating shift.

Hot spray can be applied to both conventional and airless spraying systems. For viscosity control only, the material is heated to about 100F. For viscosity and hot spray application, the paint is heated from 120 to 170F, with most applications being in the 150-160F range.

Several methods are used to heat the paint material. In air heating, an electrically heated hot air interchanger heats material circulating in coils surrounding the heater unit. The material is then fed to spray guns through a dual hose assembly with paint in the inner hose and hot air in the outer hose. Another arrangement immerses the coils in an electrically heated hot water tank, and heated

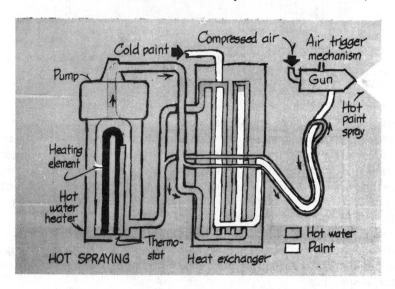

material is pumped to spray guns through a single hose assembly. In more elaborate systems, a separate hot water heater circulates water to a heat exchanger where cold paint is heated to proper temperature. The material is then pumped through a dual hose assembly to the spray guns. Hot water is constantly recirculated through the outer hose to maintain material temperature.

Application Considerations. Any material with a viscosity reducible by heat will give better results when sprayed hot. Most standard lacquers, enamels, and oil paints benefit from hot spray. Materials such as two-component epoxys and water-emulsions receive little benefit from hot spray.

In cold spraying, viscosity is reduced by thinners which do not increase the solids content of the material and must be evaporated. When material is heated to 150-160F, its viscosity is reduced without thinners or with less thinner. Solids content is higher, and less volatile material must be evaporated to produce final film. In addition, with heated material the low boiling thinners are released more rapidly during atomization.

With heavier bodied material, a much greater film build can be sprayed. One coat of hot sprayed material will usually give greater coverage than two coats of cold sprayed material. The faster setting hot sprayed film may be sprayed faster and handled sooner. With lower workable viscosities than are practical using only thinner, less pressure is needed to atomize the heated paint, thereby reducing paint losses due to over atomization, misting, and paint rebound. Since there is less thinner to be evaporated from the product surface, hot spraying provides less film shrinkage during drying, particularly on edges and corners.

Both conventional and airless spraying systems can benefit by the addition of hot spray heaters. Of course, the cost of either system is increased by the cost of the hot spray equipment. However, this cost can be recovered by reduction in the cost of thinners, higher material utilization, and reduction in the number of coats required to attain high film build. Waste of paint in overspray and misting can be reduced generally. Lower atomizing pressures of hot spray material can reduce air consumption substantially. Although spray booths and air scrubbers will still be needed, less maintenance will be required to clean them.

Electrostatic Spray Painting. Electrostatic spray painting is based on the principle that unlike electrical charges attract each other. At the point of atomization, the paint material is given a high negative charge. By electrically grounding the product, the charged material is attracted to its bare or primed surfaces.

There are many different techniques used in commercial equipment to produce an electrostatic spray. Basically, atomization of material in an electrostatic gun is achieved by compressed air (conventional spray), by hydraulic force (airless spray), or by rotating disks or bells. In compressed air and airless guns, the atomized material is usually charged by being exposed to a high voltage needle extending through the nozzle orifice. Application techniques are similar to those discussed under conventional and airless spraying.

In guns that use a rotating disk or bell to atomize the material and apply the electrical charge, paint is fed to the disk or bell and atomizes from the periphery of the device under the influence of an electrostatic field. If a flat disk is used, conveyorized products must circle around the disk to receive a full coating. If a rotating bell is used, the products can be conveyed past it in a straight line.

Charging voltages for electrostatic systems vary from about 60,000 to 150,000 volts with currents limited to milliampere levels. Some electrostatic systems are adaptable to spraying heated materials which permit heavier film builds per coat.

Application Considerations. Electrostatic spraying has a great advantage over other spraying systems because it recognizes area rather than shape and size of product. Material utilization can be as high as 80 to 90 per cent when the proper product is electrostatically coated. The attraction between the charged particles of paint and the grounded product is responsible for the high efficiency of the technique. With other systems, atomized paint may bounce off and bypass the product and be lost in the exhaust air. With electrostatic systems, the atomized particles are held onto the product. Some of the paint that would normally miss the product is drawn to it, producing what is called electrostatic attraction and "wrap-around." On tubular products, the entire product can be generally coated while spraying from one side. In some electrostatic systems, the electrostatic charge provides partial atomization, and air or hydraulic force is not needed for atomization. Many of the limitations discussed under conventional and airless spraying still apply when these methods are used with electrostatic spraying.

Electrostatic spraying is not a cure-all for painting problems. Since the electrically charged particles or paint are attracted to the *nearest* conductive surface, products that have sharp inner corners, deep cavities, blind holes, and close together fins may not be adequately covered in these areas and might require hand touchup or reinforcing by hand spray guns. If hand reinforcing is extensive, it

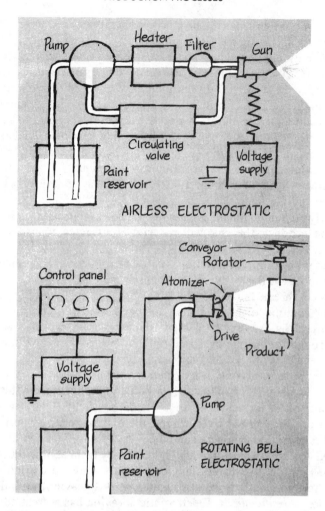

might be more economical to paint the product by another spray technique that provides complete coverage in one application.

Another restriction of electrostatic spraying is that the product should be electrically conductive. This restriction can be overcome by dipping the product in a conductive coating. Again, the over-all cost of using two or more operations as against using one conventional or airless spray operation should be evaluated.

The cost of electrostatic systems is generally higher than other spraying systems. In industrial plants, spray booths are usually required, but the over-all cost of a system can be lower. Maintenance is also lower if booths require less cleaning. Electrostatic spraying can provide a finish as high in quality as any other spraying system, but not necessarily at lower costs.

Dip Coating. Dip coating is used to apply prime or finish coats to products that require thorough coverage. It is capable of producing a smooth, reasonably uniform coating for a wide variety of products ranging in size from small parts of complex shape to automobile frames and bodies.

In most applications of dip coating, an overhead conveyor is used to carry the products into and out of a tank containing the coating material. Dip coating can also be accomplished by raising and lowering the dip tank to and from the products, by raising and lowering the level of material in a stationary tank, and by lowering and raising groups of pole-like products in and out of a vertical dip tank. After the products have been dipped, they are withdrawn at a predetermined rate and allowed to drain excess material under controlled conditions. The drained material can be reused in the painting system.

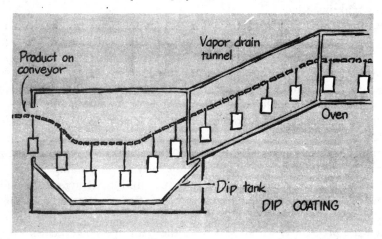

Application Considerations. In dip coating, control of painting material viscosity in the dip tank and in the drain area is necessary to maintain a desired dry film thickness. Proper control usually requires agitation of paint in the tank to prevent settlement, heating of the paint to maintain a constant application temperature, and using a vapor drain tunnel to control the rate of evaporation of thinners and solvents from the paint film. The vapor drain tunnel is a vapor saturation controlled unit connected to the exit end of the dip tank. The unit is air sealed at its exit so that a high concentration of solvent vapors build up in the tunnel. By controlling the concentration of the vapors, the coating material is allowed to flow and drain from the product for a longer period than it would in free air.

Dip coating finishes do not approach the surface quality of spray finishes. Therefore, this technique is used for prime coats or for finish coats on products that do not require a fine finish such as internal brackets and supports for washers and dryers. Greatest advantage of dip coating is that it will penetrate into hidden corners and crevices that may not be reached by spray methods. It provides a complete film coverage and is used in many cases to apply corrosion resistant materials. However, the product must have adequate drain holes so that material is not trapped in interior pockets.

When solvent base material is used in a dip tank, the system must be protected against fire. The material surface is usually protected by built-in CO_2 fire extinguishers. In addition, a dump tank facility should be provided so that the material can be dumped into a sealed tank in a few seconds in case of fire.

Depending upon the design of the system, material utilization can be quite high, usually over 70-80 per cent. Solvent loss due to evaporation is much lower than in flow coating since there is little atomization of material and only the surface of the material in the dip tank is exposed to air. However, a dip coating operation may require several thousand gallons of paint in a tank to do the same job as a flow coating operation that uses 10 to 15 per cent of the same material. When a large volume of material is required in a dip tank, there is danger of material loss due to deterioration and aging.

Color changes can be handled in a dip coating line by having tanks mounted on tracks. When a change is desired, the tanks are covered to prevent evaporation and formation of a film on the surface. Then the old tank is moved out of position and a new tank installed.

Electrocoating. Electrocoating is a relatively new dipping technique for applying organic coatings through the use of electrical energy. The process is known by many other names such as electrophoresis, electropainting, electrodeposition, electric dipping, and anodic hydrocoating.

Basically, the product to be coated is immersed in a specially-formulated, low solids content, water-soluble paint. The product is connected to a direct current power supply and is generally made the anode. The cathode is the reservoir or electrodes in the tank. When voltage is applied (usually 50-500 vdc with a current density from 2-5 amps per sq ft), resin and paint pigment migrate to the product and form a uniform film. Coating time varies depending upon product configuration, film thickness desired, pH of solution, temperature and solids content of bath. It may take 15 seconds to coat a flat panel or up to 3 minutes for complex enclosed products.

When the product is withdrawn from the bath, it is covered with a uniform coating which has been irreversibly deposited. The film is virtually dry and well-adhered to the product. It could be scratched off, but would not leave paint on a finger touching the film. The low solids content of the bath ensures that runs, sags, and heavy edges are almost non-existent. The product is rinsed to remove any undeposited particles of paint. Drain-off of water is rapid, and the product may be baked immediately or may be overcoated wet-on-wet with a conventional paint and then baked.

Application Considerations. Most conductive products can be electrocoated, regardless of size or shape. Coating of internal surfaces is possible with special electrodes. Satisfactory results have been obtained on mild steel, stainless steel, phosphated steel, zinc diecastings, surface-treated magnesium alloys, and certain aluminum alloys. Any standard pretreatment for steel may be used, but adhesion is also outstanding without pretreatment.

In spray, dip, and flow coating techniques, film thickness is controlled by factors such as rate of application, material viscosity, solvents and thinners, and material flow characteristics. Electrocoating eliminates these variables while imposing a self-regulating feature that coats in areas not covered by initial currents. As the paint film builds up, it insulates the metal and causes the film to creep into less exposed areas, such as recessed areas and welded seams. Since the deposited film is generally a poor conductor, maximum film thickness averages about 1.0 mil.

Best results have been obtained by using a 5 to 12 per cent solids dispersion in the water solution. Material utilization is about 95 per cent. Since water-base materials are used, no organic solvents are needed. Fire hazards and toxic vapors are avoided, and fire extinguisher equipment and dump tanks are not required. In addition, vapor drain tunnels are not required.

The water-base solutions used are temperature sensitive so a constant temperature must be maintained to provide a constant deposition rate. Mixers are required to recirculate the materials and keep the paint particles in suspension. At

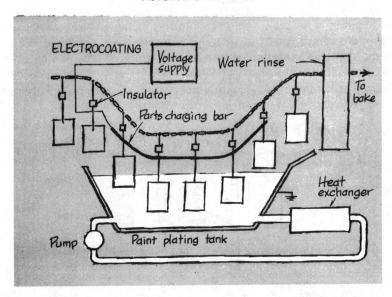

present, there are a limited number of resins and colors available. These are mostly dark colors such as brown, gray, red iron oxide, and black. Pastel colors are becoming available such as yellow, beige, and light gray. Some whites have been developed, but these must be applied on galvanized steel to prevent discoloration.

Several installations are now in operation in the United States and England. Ford Motor Co. is using electrocoating to deposit a red primer on bodies and to coat wheels in a one-coat black job.

Flow Coating. Flow coating is an automatic operation in which products to be painted are conveyed through a chamber equipped with low pressure nozzles that completely flood the products with paint. Depending upon the configuration of the product, the nozzles can flood downward from the top and sides of the chamber, can flood upward from an H-pattern in the bottom, or can flood upward from a bottom manifold tube that oscillates around and reciprocates along its lineal axis. After being flooded, the products are in the same general state as dip coated products. Vapor drain tunnels are usually required to control drain and flow-out times. Excess paint is collected and reused by the system.

Application Considerations. Flow coating provides a finish that is comparable with a dip coating finish. When compared to dip coating, flow coating can be used to apply materials with a higher solids content. Film thickness is slightly higher and is fairly uniform. Due to the flow-out process in the vapor tunnel, the film thickness will be slightly heavier at the bottom of the product than it will be at the top, also a characteristic of dip coating.

When compared to dip coating, flow coating uses a much smaller facility for the same size product. Only about 10 to 15 per cent of the material normally required for automatic dipping is used for flow coating the same volume of work. Although fire extinguisher equipment and dump tanks are still required in flow coating systems, they have to protect a much smaller area, and dump facilities can be considerably smaller.

Flow coating can usually handle all jobs that can be dip coated. In addition, it can handle many products that cannot be dip coated such as tanks that would float in a dip tank and large products where the size of a dip tank or the amount of material required in the tank would be a limiting factor. Solvent loss can be higher than that of dip coating since the material is constantly recirculated and is subject to atomization and evaporation. However, solvent loss can be minimized by using low pressure nozzles and as few as possible to completely flood the product.

Material utilization in flow coating can be up to 97 per cent on all products. It

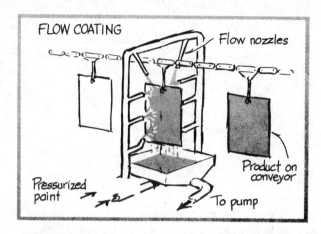

is the same whether an open wire mesh is being coated or a flat steel panel. Since all excess material and solvent can be collected and reused, material utilization does not vary from product to product. Spraying methods will vary widely in material utilization when spraying wire mesh and steel panels.

Main advantage of flow coating is that it can provide complete coverage of product at high production rates. When a two-coat finish is required, the product is usually reversed to minimize the heavier bottom film thickness effect. Flow coating is considered much safer than dipping with respect to material loss, fire, explosion, and danger to personnel. A flow coater can be more completely protected against fire than an open dip tank with a large volume of paint.

Curtain Coating. Curtain coating is a modern, high-speed production technique for applying smooth films of lacquers, paints, varnishes, adhesives, and other materials to flat, curved, concave, fluted, or molded surfaces of wood, metal, and plastic products. It uses the flow coating principle of conveying the product through a continuous falling stream of coating material. However, the stream is accurately controlled in thickness and rate of flow so that there is no excess material runoff from the product.

Two methods are used to accomplish curtain coating. In one method, material is pumped from a reservoir to a coating head where it is uniformly distributed and allowed to flow over a weir or dam. The material then drops from the skirt of the weir by gravity and falls either onto the product or into a gutter for return to the reservoir. The product is conveyed through the curtain by a flat bed conveyor which extends horizontally away from two sides of the gutter.

In a second method, material is pumped from a reservoir to an air-tight coating head which has an adjustable orifice across its V-shaped bottom. The width of the orifice opening, the speed of the pump, and the viscosity of the coating material control the pressure of an air cushion formed in the upper part of the coating head. When the pressure is greater than atmospheric, the curtain of coating material is forced through the orifice at a velocity greater than that caused by gravity alone. When the pressure is less than atmospheric, the curtain of material is restrained. Products are conveyed through the curtain by a flat belt conveyor. Excess coating material falls into a gutter and is returned to the reservoir.

Application Considerations. Most coating materials can be successfully applied by this technique without special formulations. It is almost 100 per cent efficient since material either is deposited on the product or is collected for recirculation. Solvent loss is small as only the curtain itself is exposed to evaporation. Applied coating thickness may be as small as $1/2$ mil or as great as 25 mils. By placing products at an angle on the conveyor, the coating may be applied to an end, two edges, and the top in one pass. Depending upon the equipment, conveying speeds may be adjusted from 50 to 800 feet per minute. Some machines are equipped with conveyors that may be tilted from 0 to 45 degrees to provide better coverage of contoured products.

Hot melt machines are available for applying melted plastics and waxes. When the hot fluid is applied to a room temperature surface, the coating material cools and solidifies fast, eliminating need for drying equipment. Catalyzed coatings can be applied by using two heads to apply resin and catalyst in one pass, using one head to apply premixed resin and catalyst in one pass, or by using one head to apply resin and catalyst separately in two passes.

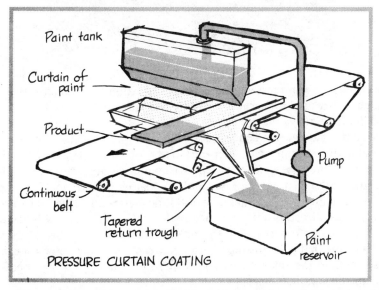

PRESSURE CURTAIN COATING

Curtain coating applies material with a high-solids content and reduces the need for solvents and thinners. By using less solvent than other techniques, a better quality of finish is obtained. Gloss is improved, and drying is accelerated. Machines are available with interchangeable paint reservoirs so that color changeovers can be made in less than 10 minutes.

Roller Coating. Two techniques are employed in roller coating — direct and reverse. In direct roller coating, the product is coated as it moves between a coating roll and a drive roll. The coating roll may be either on the top or bottom of the product for single coat jobs, or the drive roll may be a coating roll for painting top and bottom surfaces simultaneously. In any case, the coating roll rotates in the same direction as the product moves.

In reverse roller coating, the product is driven independently of the coating

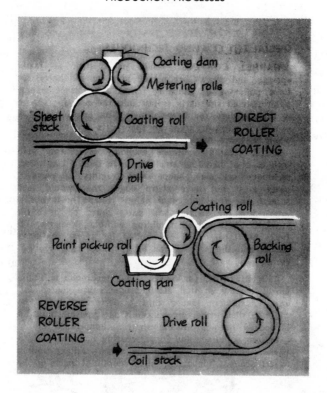

roll which can rotate in the same or opposite direction to product travel. Direct coating is generally used for sheet stock while reverse coating is used for continuous web or strip materials.

Both direct and reverse coating use three or more rolls to transfer material from a manually or pump-fed reservoir to the product. In direct coating, the reservoir is formed by the space between two counter-rotating rolls having dams at each end. In reverse coating, the reservoir is a coating pan from which material is picked up by a single roll. In either case, after the material is adhered to a roll, it is either transferred directly to the product by the roll or transferred to one or more intermediary rolls before reaching the product. In direct coating, material film thickness is controlled by adjusting the gap between the two counter-rotating rolls. In reverse coating, the thickness is controlled by adjusting the gap between the paint pickup roll and its adjacent transfer roll.

Application Considerations. Roller coating can be used on any flat product such as metal, veneer, plywood, fabric, paper, cardboard, rubber, plastic, glass, hardboard, or cork. It can apply any fluid material such as paint, glue, adhesive, drawing compound, latex, wax, oil, sizing, and plastic.

Roller coating applies a uniform thickness film that is precisely metered. Thickness ranges depend upon the material used and can be as small as 1/10 mil or more than 20 mils. Application speeds depend on material and drying equipment and generally vary from 30 to 300 feet per minute. Roller coating is being used in many industries to coat strip material which is then sheeted and formed into finished products. A few products being made from precoated stock are drapery hardware, aluminum siding, venetian blinds, toys, shelving, kitchen cabinets, washers, television sets, and awnings.

Roller coating can also be used to achieve different effects such as wood graining, multi-color patterns, and conventional lettering. It also can be used to emboss patterns. Embossing is usually limited to plastisol and organsol coatings and is done generally on coil stock.

Major advantages of roller coating are fine quality finishes, up to 98 per cent material utilization since unused material can be collected and recirculated, low rate of evaporation of thinner, occupies small area, and requires no special booths or exhausts. When used in coil lines, roller coating is usually combined with cleaning, rinsing, treating, and oven equipment to produce finished stock in one operation. As such, the initial capital investment in equipment can be quite high. For sheet products, the over-all cost is comparable with other painting methods.

Specialized Coating Techniques

Barrel Coating. Small parts such as fasteners, springs, and screws are placed in a tapered barrel similar to a polishing or finishing barrel. An exact amount of paint to coat the parts is put into the barrel. The barrel is then rotated for a set time. The coating material is usually a quick-dry type so that parts are essentially dry when they are dumped out of the barrel.

Barrel coating combines economy of coating material with ease of handling small parts in bulk. Limitations of technique are size, shape, and fragility of parts. The parts must not have sharp edges that would tend to scratch the freshly-applied paint. Shape must not cause interlocking in barrel, and parts must be of sufficient strength to withstand tumbling without breakage. Many types of finishes can be applied by barrel coating techniques and the process can be followed by a heating and curing process.

Centrifugal Coating. This technique is similar to barrel coating in that small products are coated in bulk lots. Parts are loaded into a perforated metal basket which is then loaded into a centrifuge-type machine. Basket is locked onto a driving shaft. Compressed air enters a bottom container and forces coating material up into the basket, completely flooding the parts. Filling time is 15 to 30 seconds, depending upon depth of parts in basket and viscosity of material. When all parts are covered, coating material is allowed to drain by gravity into bottom container, requiring about 15 to 30 seconds. Basket is then spun, and centrifugal force removes excess material from parts, leaving thin uniform film. Spin time is normally 10 to 30 seconds. Machine is then stopped, and parts are dumped on wire mesh for air or oven drying.

Centrifugal Spraying. High-speed rotary sprayhead centrifugally atomizes any sprayable material with a high degree of efficiency and consistency. It can atomize 1/2 to 30 gallons of material per hour with a degree of atomization comparable to compressed air guns. One or more sprayheads are mounted in a collecting chamber through which the conveyed product passes. This chamber confines the spray field and collects overspray for reuse in the system. The centrifugal spraying system eliminates need for spray booths, exhaust systems, and water curtains.

Electrostatic Powder Spraying. This technique is similar to electrostatic spraying except a dry powder is used. The product to be coated must have adequate electrical conductivity to serve as a collecting electrode for charged powder. Products may be preheated or not, depending upon type of powder used and film thickness desired. Powder does not fall off prior to fusing or curing since the electrostatic charge bonds the powder to the product, despite handling vibrations or air currents. Powder coating materials are available in a variety of colors. Organic powders most commonly used for electrostatic powder spraying are epoxy resins, polyethylenes, nylons, polyvinyl chlorides, cellulose acetate, Teflon, and polyesters.

Fluidized Bed Coating. Parts to be coated are preheated in an oven to a temperature above the melting point of the coating to be applied. The parts are then immersed with suitable motion in a fluidized bed or tank. A rising current of air, passing through a porous plate at the bottom of the tank, fluidizes the dry powder material. As the powder particles contact the heated surface, they fuse and adhere. After removal from the fluidized bed, the parts are often briefly postheated to completely fuse the coating or to cure the resin if it is a thermosetting compound. This technique can apply most of the materials mentioned under electrostatic powder spraying.

Trichlorethylene Finishing. This technique is based on using room temperature trichlorethylene as a solvent for black lacquer. Color is obtained by a dye in the resin, so there is no tendency for the pigment to settle out. Finish is a high gloss black. The trichlor solvent evaporates within 60 seconds, leaving a relatively dry film. Products are baked after coating. The material can be sprayed but products usually are dipped.

DuPont has developed another trichlorethylene process offered by several companies. Its process is designed for cleaning, phosphatizing, and painting parts continuously on a production line basis in a single medium of trichlorethylene. In the DuPont process, the vapor degreasing, phosphatizing, and painting is accomplished at the boiling point of the solvent (188 F). Either dipping or spraying can be used in the phosphatizing and painting sections of a system. Parts leave the system completely dry when lacquer is used, or ready for a short baking step when curing paints are used. The trichlorethylene is nonflammable, providing plant safety, lower insurance rates, and reduced cost for fire protection equipment.

Automatic Painting Systems

In the painting techniques discussed, the product handling method is dictated by the nature of the technique in many cases. Dip coating, electrocoating, and flow coating techniques generally use an overhead conveyor to move the products through the processes. Curtain coating is usually available as a complete machine which includes a horizontal belt conveyor to move products to and from the coating head. Various types of horizontal conveyors are used to feed products to and from direct roller coaters. In reverse roller coating, the handling system for the web or strip material feeds the product past the coating head.

However, many types of automatic machines are available for use with spray painting techniques (compressed air, airless, hot spray, and electrostatic). Some machines are designed to automatically spray products as they are being carried on various types of overhead and floor conveyors. Other machines are individual units that paint small and medium sized products after they have been removed from a conveyor. In either case, the machines are designed to provide a motion or a combination of motions to one or more spray guns so as to adequately paint a product. Some of the more commonly used motions are as follows:

Oscillating motion. Gun is oscillated on a vertical or horizontal axis through its center so as to spray in an arc.

Swinging motion. Gun is physically moved in an arc by an oscillating arm or by a cam mechanism.

Rotating motion. Gun is mounted on an arm and rotated in a full circle about the axis of the arm support.

Reciprocating motion. Gun is fixed to a carriage that is reciprocated in a horizontal or vertical direction.

Following motion. Gun is moved in parallel with a product so as to allow more time for spray coverage. After a specified stroke, the gun is cut off and retracted to pick up another product.

Telescoping motion. Gun is moved horizontally or vertically into the interior of a product and sprays as it is retracted from the product.

Various combinations of these motions can be used in automatic machines to paint particular products. For example, a following gun motion may be combined with a telescoping gun motion to move a gun into the interior of a product on a conveyor. Then an oscillating motion can be applied to the gun to attain better coverage of the interior area.

Many spray machines also use fixed spray guns. In this case the product must be moved through the spray area in a horizontal or vertical plane. The product can also be rotated as it is sprayed to attain full coverage.

While automatic spray painting machines can be designed for almost any product configuration, there are many standard machines that are commercially available. These machines fall into the following categories:

Horizontal reciprocating machine. Unit is designed for flat or nearly flat products that can be handled on horizontal conveyors. One to four automatic guns are mounted on a carriage that reciprocates back and forth across the work at a uniform rate of speed. Guns are turned on and off at the start and end of each stroke to reduce loss of paint. Conveyor can handle large flat pieces or many small pieces together or in trays.

Horizontal contour-following reciprocating machine. Unit is designed to follow curved or irregular contoured parts, keeping the guns normal to work surface. Reciprocating carriage holding guns is guided by cam surfaces matching contour of work. Various gun motions can be incorporated to handle extreme variations in contours.

Vertical reciprocating machine. Unit is designed for flat or nearly flat products being carried on overhead conveyors. If both sides of a product are to be painted, machines can be installed in a staggered pattern, spraying from both sides of a conveyor. Some small products may be painted on both sides with only one vertical reciprocator. For example, on the down stroke of the carriage, the products are painted on one side by one spray gun. The product hanger is then rotated 180 degrees to expose the other side of the product to a second spray gun on the up stroke of the carriage.

Vertical contour-following reciprocating machine. Unit is designed to paint curved or contoured surfaces hanging from an overhead conveyor. Cam surfaces are used to match the carriage movement to the shape of the product surface.

Rotary arm spray machine. Unit is designed to spray relatively flat products being carried on a horizontal conveyor. Machine has a horizontal bridge over conveyor on which is mounted a rotating head equipped with 4, 6, or 8 gun arms. As each gun rotates in a circular path above the conveyor, its downward spray is automatically turned on at the edge of the product and turned off as it leaves the product. Its main advantage over horizontal reciprocating machines is its ability to apply coatings at much higher conveyor speeds.

Chain-on-edge machine. Unit is designed to paint small and medium-sized products being carried on fixed or rotating spindles attached to an endless roller chain. Various combinations of spray gun motions can be used to paint both exterior and interior surfaces of products. The machine can be arranged to spray parts, flash dry, and oven dry. The parts can then be conveyed by the chain to another spray unit and oven. Depending upon the weight of the products, chain lengths up to 300 feet or more can be used.

Rotary table machine. Unit is designed for painting products that are fast drying or that can be removed while wet for batch oven drying. Circular table is equipped with rotating spindles mounted on its periphery. As table rotates, products in workholders are spun as they pass spray position. Only loaded workholders and spindles actuate spray guns. In some cases, spray guns are oscillated to follow product and extend spraying time.

In addition to the more standard type machines mentioned, there are a wide variety of special painting machines for handling all types of products. These include revolving gun machines for applying circular spray patterns, rotating part machines in which the guns are stationary and the part revolves on a spindle, horizontal and vertical oscillator machines for coating the interior surfaces of tubular parts. There also is a wide selection of special machines available for masking parts automatically during spraying.

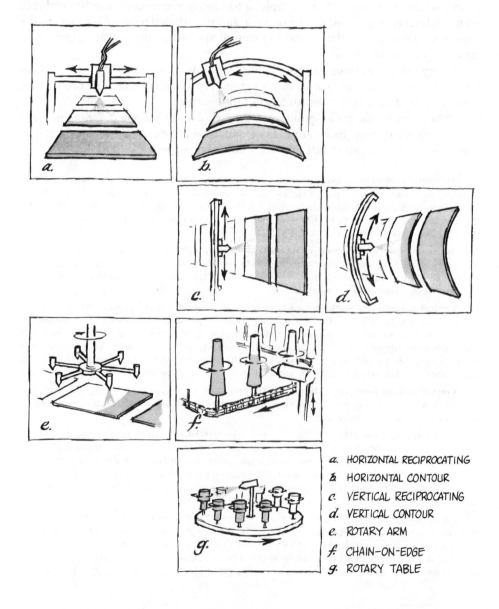

a. HORIZONTAL RECIPROCATING
b. HORIZONTAL CONTOUR
c. VERTICAL RECIPROCATING
d. VERTICAL CONTOUR
e. ROTARY ARM
f. CHAIN-ON-EDGE
g. ROTARY TABLE

Very large products can also be handled with automatic spraying machines. For example, freight cars can be automatically sprayed by a traveling spray booth that travels the length of the car on the rails. As the booth moves under its own power, automatic gun machines spray up and down the sides of the cars while fixed guns spray the roof. The unit is complete with water-wash exhaust chambers, an air compressor, paint pumps, and fire extinguisher system. When not in use, it requires minimum floor space for storage.

Many control devices are available for incorporation in automatic painting systems. Some of the more common control devices are as follows:

1. Automatic skip spray controls to shut off spray guns if no work is in spray area.

2. Interrupted spray stroke controls for shutting off spray guns when passing over openings in products.

3. Automatic controls for changing length of spray stroke when spraying short and long pieces hung at random on the same conveyor line.

4. Variable speed spinner drives for rotating work hanging on an overhead conveyor.

5. Program controllers which sense the outline of the product and activate only the spray guns that would cover the product.

6. Solenoid valve controls which shut off spray guns in case of power failure or conveyor stoppage.

SUPPLIERS OF AUTOMATIC PAINTING EQUIPMENT, MACHINES, AND SYSTEMS

Listed are major sources for automatic painting equipment, machines, and systems. Boldface letters in parentheses denote each supplier's capability according to the following code:

CAS—Compressed Air Spraying

AS—Airless Spraying

HS—Hot Spraying

ES—Electrostatic Spraying

DC—Dip Coating

EC—Electrocoating

FC—Flow Coating

KC—Curtain Coating

RC—Roller Coating

BC—Barrel Coating

CC—Centrifugal Coating

CS—Centrifugal Spraying

FB—Fluidized Bed Coating

EP—Electrostatic Powder Spraying

TF—Trichlorethylene Finishing

CO—Supplies Components Only

PS—Supplies Painting Systems

PM—Supplies Painting Machines

Alemite Div., Stewart-Warner Corp., 1826 Diversey Pkwy., Chicago, Ill. 60614 (CAS, AS, PS)
Aro Corp., Industrial Pumping Div., 400 Enterprise St., Bryan, Ohio 43506 (CAS, AS, FC, PS)
Ashdee Corp., P. O. Box 325, Evansville, Ind. 47704 (ES, EC, KC, EP, PS)

Balcrank Inc., Disney St., Cincinnati, Ohio 45209 (CAS, AS, ES, PS)
Baron-Blakeslee Inc., 1620 S. Laramie Ave., Chicago, Ill. 60650 (TF, PS)
Leon J. Barrett Co., 1660 Grafton Rd., Worcester, Mass. 01601 (CC, PS)
Belco Industries Inc., 9138 Belding Rd., Belding, Mich. (DC, FC, PS)
Binks Mfg. Co., 3114 W. Carroll Ave., Chicago 12, Ill. (CAS, AS, HS, ES, PS, PM)
Black Bros. Co. Inc., P. O. Box 191, Mendota, Ill. 61342 (KC, RC, CO)

Campbell-Hausfeld Co., 801 Moore St., Harrison, Ohio 45030 (CAS, CO)
Capitol Machine & Switch Co., 36 Balmforth Ave., Danbury, Conn. 06813 (PM)
Cincinnati Cleaning & Finishing Machinery Co., 3280 Hageman St., Cincinnati, Ohio 45241 (DC, FC, RC, BC, PS)
Cleanola Co., 1116 William Flinn Hwy., Glenshaw, Pa. 15116 (AS, HS, PS for pipe)
Conforming Matrix Corp., 830 New York Ave., Toledo, Ohio 43611 (CAS, HS, PS, PM)

Deco Tools Inc., 1541 Coining Drive, Toledo, Ohio 43612 (CAS, AS, HS, ES, FC, PS, PM)
Despatch Oven Co., 619-8th St. S. E., Minneapolis 14, Minn. (ES, DC, EC, FC, PS)
DeVilbiss Co., 300 Phillips Ave., Toledo, Ohio 43601 (CAS, AS, HS, ES, DC, FC, PS, PM)
Dualheet Inc., 3350 W. 137th St., Cleveland, Ohio 44111 (CAS, AS, HS, DC, FC, CO)
DuBois Machine Co., Box 186, Jasper, Ind. (KC, RC)
Dumatic Devices Inc., 1731 E. 11 Mile Rd., Madison Heights, Mich. 48071 (CAS, AS, HS, FC, PS, PM)

Eclipse Air Brush Co., 390 Park Ave., Newark, N. J. 07107 (CAS, HS, ES, PS, PM)
Equi-Flow Div., Vibro Mfg. Co. Inc., 61-17 Roosevelt Ave., Woodside, N. Y. 11377 (FC, KC, RC, PS)

Finish Engineering Co. Inc., 921 Greengarden Rd., Erie, Pa. 16501 (FC, PS, PM)
H. G. Fischer & Co., 9451 W. Belmont Ave., Franklin Park, Ill. (ES, PS, PM)

Gasway Corp., 6463 N. Ravenswood Ave., Chicago 26, Ill. (KC, RC, PS)
Granco Equipment Inc., 1055 Fulton E., Grand Rapids, Mich. 49503 (PS spray)
Gray Co. Inc., 60 11th Ave., N. E., Minneapolis, Minn. 55413 (PS spray, flow and barrel coating)
Gyromat Corp., P. O. Box 4157, Newfield Station, Bridgeport, Conn. 06607 (CAS, ES, CS, PS)

Henderson Bros. Co., Waterbury, Conn. 06720 (BC, CO)
Herr Equipment Corp., 1201 Vine St. N. E., Warren, Ohio 44484 (RC, PS)

Industrial Heat Systems, 517 W. Garfield Ave., Glendale 4, Calif. (DC, FC, RC, PS)

George Koch Sons Inc., 10 S. 11th Ave., Evansville, Ind. 47704 (DC, FC, KC, PS)

Lincoln Engineering Co., 4010 Goodfellow Ave., St. Louis 20, Mo. (AS, CO)

R. C. Mahon Co., Industrial Div., 6565 E. Eight Mile Rd., Detroit, Mich. 48234 (CAS, DC, FC, PS)
McKay Machine Co., P. O. Box 180, Youngstown, Ohio 44501 (RC, PS)
Metalwash Machinery Corp., 902 North Ave., Elizabeth 4, N. J. (PS)

Nordson Corp., Amherst, Ohio 44001 (AS, HS, ES, PS)

Paasche Airbrush Co., 1909 W. Diversey Ave., Chicago, Ill. 60614 (CAS, HS, PS, PM)
Partswash Equipment Co. Inc., 150 Willard Ave., Newington, Conn. (FC, PS)
Phillips Mfg. Co., 3475 W. Touhy Ave., Chicago 45, Ill. (DC, EC, FC, PS)
Polymer Corp., Coating Resins Div., 2120 Fairmount Ave., Reading, Pa. 19603 (FB, PS)
Pyles Industries Inc., 20855 Telegraph Rd., Southfield, Mich. 48075 (CAS, AS, CO)

Ransburg Electro-Coating Corp., 3939 W. 56th St., Indianapolis, Ind. 46208 (ES, EP, PS, PM)
Ransohoff Co., N. 5th & Ford Blvd., Hamilton, Ohio (ES, DC, FC, PS)
Ross Engineering Div., Midland-Ross Corp., P. O. Box 751, New Brunswick, N. J. 08903 (FC, RC, PS)

Schweitzer Equipment Co., 3764 Ridge Rd., Cleveland 9, Ohio (CAS, AS, HS, ES, PS, PM)
Sepanski & Associates, 900 Clancy Ave., N. E., Grand Rapids, Mich. 49503 (CAS, PM)
Spee-Flow Mfg. Corp., 6614 Harrisburg Blvd., Houston 11, Texas (CAS, AS, HS, PS)
Spra-Con Co., 3600 N. Elston Ave., Chicago, Ill. 60618 (DC, EC, FC, PS)
Spray Engineering Co., 100 Cambridge St., Burlington, Mass. (CAS, AS, HS, PS, PM)
Spraymation Inc., 25 Amity St., Little Falls, N. J. (CAS, AS, HS, PS, PM)
Static-Spray Systems, 467 Connecticut Ave., South Norwalk, Conn. 06856 (ES, EC, PS)

Udylite Corp., 1651 E. Grand Blvd., Detroit 11, Mich. (EC, PS)
Union Tool Corp., 222 E. Market St., Warsaw, Ind. 46580 (RC, PS)

L. R. Wallace & Co. Inc., 172 N. Vernon Ave., Pasadena, Calif. (RC, PS)

7. Individual timer controlled spray guns that actuate at different times in multiple gun installations.

8. Many types of sensing, identifying, memory, and readout devices for controlling spray gun actuation in all types of situations.

Which System to Choose? There is no one technique or piece of painting equipment that can handle all jobs, and every technique and machine has a best use. The important point about selecting a painting system is to consult with your paint manufacturer and equipment supplier who can explore ways and means of developing the most efficient and economical paint system for your particular product. It is a matter of testing the product, material, technique, and equipment under simulated production conditions and then evaluating the best methods against each other.

If you have existing painting equipment and believe you should make a change, be sure that you evaluate your present method thoroughly before you decide on a change. Many times, a paint manufacturer or an equipment supplier will find that the material or the equipment is not being used as efficiently as possible. Adjustment and training may improve your existing method to the point where new equipment or techniques are not necessary. Get the most out of the existing equipment before you decide to change to a different method or different equipment.

Acknowledgement

This chapter is included by permission of *Automation*, from the May 1965 issue, copyright, Penton, Inc.

Additional Reading

[1] A.A. Conte, Jr., "Choosing a Powder for Coating Applications," paper presented at the ASME Design Engineering Conference, April 9-12, 1973, New York, N.Y.
[2] Bernard Feinberg, "Electrostatic Powder Coating," *Manufacturing Engineering & Management*, June 1975.
[3] Bob Baldwin, "Selecting Coatings for Plant Painting Projects, "*Plant Engineering*, February 6, 1975.

MATERIALS
AND METHODS

68

Materials Design Data

ALL THROUGHOUT this Handbook it has been evident that product design, production method selected and ultimate unit cost are inextricably and closely related. Change material to be used and, immediately, you may have changed the processing method or methods available to be economically utilized. Again, should a radically different method of processing be decided upon the materials eminently suited for the process may be altered.

Hence, this Chapter is devoted solely to materials data as related to methods of processing. Because of the overlapping of the materials and processes in actual application it is desirable to study this aspect carefully. Changing costs as well as change in design can dictate a switch in materials.

Careful and continued study of the design, processing method, quantity and materials equation is imperative if optimum cost results are expected. The possibilities are endless, *Fig. 68.1.*

Fig. 68.1 — Redesign to simplify assembly resulted in this combination mower bearing flange and seal which replaced three parts; a new selection of materials and processes. Courtesy International Packings Corp.

TABLE 68-1 — **Machining Characteristics of Metals**[**]

Base Metal	Material Representative Alloy	Cutting Nature[*]	Machining Characteristics Expected Finish[†]	Approximate Ratings[§]	Precaution[‡]
Aluminum	C113-F	F to G	E to G		2,3,4,7
	220-T4	F	E		2,4,7
	2017-T4	F	E	300	2,4,7
	1100-H12	F to G	E to G	300	2,3,4,7
	1100-O	G, S	G	300	2,3,4,7
	220-F	G	G		2,3,4,7
	2017-O	G, S	G		2,3,4,7
Beryllium	Unalloyed	A, B	F	50	6,7,8
Columbium	Unalloyed, wrought	G	G		3
Copper	Brass, leaded FC	F	E	200	4
	Brass, med leaded	F	E	140	4
	Tin bronze, high lead	F	E	140	4
	Yellow brass	G, S	G	80	3,4
	Navy bronze		G	80	4
	Aluminum bronze	G, S	G	40	3,4
	Aluminum bronze		G	40	3,4
Iron	Carbon Steels				
	B1113	F	E	130	7
	C1213	F	E	130	7
	C1212	F	E	100	7
	B1112	F	E	100	7
	C1119	F	E	100	7
	B1111	F	E	90	7
	C1117	F	E	90	7
	C1144	F	E	80	7
Iron	Stainless Steels				
	416	F, R	G	95	1,2,3,4,5,7
	430 F	F, R	G	90	1,2,3,4,5,7
	203	F, R	G	85	1,2,3,4,5,7
	303	F, R	G	70	1,2,3,4,5,7
	410	S, R	G	60	1,2,3,4,5,7
	430	S, R	G	55	1,2,3,4,5,7
	431	S, R	G	50	1,2,3,4,5,7
Magnesium	ASTM-AZ 61	F	E	500	1,2,4,9
	ASTM-AZ 91	F	E	500	1,2,4,9
Molybdenum	Unalloyed, wrought	A	G	8	2,3,6,8
Nickel	Nickel 200 to 233	H, S	G	55	5
	Monel Alloy 400 to 404	H, S	G	55	5
	Monel Alloy 501	H, S	G	40	5
	Monel Alloy K500	H, S	G	35	5
	Inconel Alloy 600	H, S	G	35	5
	Hastelloy B	H, S	G	35	5
	Inconel X750	H, S	G	25	5
	Superalloys	H, R, S	G	6 to 10	3,5
Tantalum	Unalloyed, wrought	G	G		3
Titanium	Commercially pure	R	E	40	1,2,3,4,5,9
	Ti-6Al-4V	R	E	20 to 30	1,2,3,4,5,9
	Ti-13V-11Cr-3Al	R	E	15	1,2,3,4,5,9
Tungsten	Unalloyed, wrought	R	G	8	3,6
	Unalloyed, sintered	A	F	8	
Zirconium	Commercially pure	G, R	G	35	1,2,3,4,5,9
Zinc	ASTM-AG40A (XXIII)	F	E	200	

[*]F = free cutting; G = gummy; H = strain hardens excessively; R = reactive to tools; S = stringy; A = abrasive; B = brittle.
[†]E—Excellent, G—Good, F—Fair.
[§]Based on B1112 steel as 100.
[‡]Precautions: 1. Avoid dull tools to prevent distortion. 2. Avoid overheating for best dimensional accuracy. 3. Minimize galling, smearing, and tool build-up by using proper tool geometry, preparation, set up, and fluids. 4. Minimize deflection by proper supports. 5. Use positive feeds and avoid dwelling in the cut to minimize work hardening. 6. Minimize spalling by proper set ups and avoiding exit cuts. 7. Minimize build-up of edge formation by proper speeds and fluids. 8. Hood all machine tools and exhaust to dust collectors to avoid toxic dust. 9. Chips are pyrophoric.
[**]Olofson, Metals Reference Issue, Machine Design, Feb. 12, 1970, p. 114.

TABLE 68-11 — **AISI - SAE Steel Specifications**

S. A. E. Steel Specifications

The following numerical system for identifying carbon and alloy steels of various specifications has been adopted by the Society of Automotive Engineers.

COMPARISON
A.I.S.I.—S.A.E. STEEL SPECIFICATIONS

The evergrowing variety of chemical compositions and quality requirements of steel specifications have resulted in several thousand different combinations of chemical elements being specified to meet individual demands of purchasers of steel products.

The S.A.E. developed an excellent system of nomenclature for identification of various chemical compositions which have symbolized certain standards as to machining, heat treating and carburizing performance. The American Iron and Steel Institute has now gone further in this regard with a new standardization setup with similar nomenclature but with restricted carbon ranges and combinations of other elements, which have been accepted as standard by all manufacturers of bar steel in the steel industry, because it has become apparent that steel producers must concentrate their efforts upon a smaller number of standardized grades. The society of Automotive Engineers have as a result revised most of their specifications to coincide with those set up by the American Iron & Steel Institute.

PREFIX LETTERS—
> No prefix for basic open-hearth alloy steel.
> **(B)** Indicates acid Bessemer carbon steel.
> **(C)** Indicates basic open-hearth carbon steel.
> **(E)** Indicates electric furnace steel.

NUMBER DESIGNATIONS—
> **(10XX series)** Basic open-hearth and acid Bessemer Carbon Steel grades, non sulphurized and non-phosphorized.
> **(11XX series)** Basic open-hearth and acid Bessemer Carbon Steel grades, sulphurized but not phosphorized.
> **(1300 series)** Manganese 1.60 to 1.90%
> **(23XX series)** Nickel 3.50%
> **(25XX series)** Nickel 5.0%
> **(31XX series)** Nickel 1.25% — Chromium .60%
> **(33XX series)** (Nickel 3.50% — Chromium 1.60%
> **(40XX series)** Molybdenum
> **(41XX series)** Chromium Molybdenum
> **(43XX series)** Nickel-Chromium-Molybdenum
> **(46XX series)** (Nickel 1.65% — Molybdenum .25%
> **(48XX series)** Nickel 3.25% — Molybdenum .25%
> **(51XX series)** Chromium
> **(52XX series)** Chromium and High Carbon
> **(61XX series)** Chromium Vanadium
> **(86XX series)** Chrome Nickel Molybdenum
> **(87XX series)** Chrome Nickel Molybdenum
> **(92XX series** Silicon 2.0% — Chromium
> **(93XX series)** Nickel 3.0% — Chromium — Molybdenum
> **(94XX series)** Nickel — Chromium — Molybdenum
> **(97XX series)** Nickel — Chromium — Molybdenum
> **(98XX series)** Nickel — Chromium — Molybdenum

Carbon Steels

SAE No.	C	Mn	P Max	S Max	AISI Number
....	0.06 max	0.35 max	0.040	0.050	C1005
1006	0.08 max	0.25-0.40	0.040	0.050	C1006
1008	0.10 max	0.25-0.50	0.040	0.050	C1008
1010	0.08-0.13	0.30-0.60	0.040	0.050	C1010
....	0.10-0.15	0.30-0.60	0.040	0.050	C1012
....	0.11-0.16	0.50-0.80	0.040	0.050	C1013
1015	0.13-0.18	0.30-0.60	0.040	0.050	C1015
1016	0.13-0.18	0.60-0.90	0.040	0.050	C1016
1017	0.15-0.20	0.30-0.60	0.040	0.050	C1017
1018	0.15-0.20	0.60-0.90	0.040	0.050	C1018
1019	0.15-0.20	0.70-1.00	0.040	0.050	C1019
1020	0.18-0.23	0.30-0.60	0.040	0.050	C1020
....	0.18-0.23	0.60-0.90	0.040	0.050	C1021
1022	0.18-0.23	0.70-1.00	0.040	0.050	C1022
....	0.20-0.25	0.30-0.60	0.040	0.050	C1023
1024	0.19-0.25	1.35-1.65	0.040	0.050	C1024
1025	0.22-0.28	0.30-0.60	0.040	0.050	C1025
....	0.22-0.28	0.60-0.90	0.040	0.050	C1026
1027	0.22-0.29	1.20-1.50	0.040	0.050	C1027
....	0.25-0.31	0.60-0.90	0.040	0.050	C1029
1030	0.28-0.34	0.60-0.90	0.040	0.050	C1030
1033	0.30-0.36	0.70-1.00	0.040	0.050	C1033
1034	0.32-0.38	0.50-0.80	0.040	0.050	C1034
1035	0.32-0.38	0.60-0.90	0.040	0.050	C1035
1036	0.30-0.37	1.20-1.50	0.040	0.050	C1036
1038	0.35-0.42	0.60-0.90	0.040	0.050	C1038
....	0.37-0.44	0.70-1.00	0.040	0.050	C1039
1040	0.37-0.44	0.60-0.90	0.040	0.050	C1040
1041	0.36-0.44	1.35-1.65	0.040	0.050	C1041
1042	0.40-0.47	0.60-0.90	0.040	0.050	C1042
1043	0.40-0.47	0.70-1.00	0.040	0.050	C1043
1045	0.43-0.50	0.60-0.90	0.040	0.050	C1045
1046	0.43-0.50	0.70-1.00	0.040	0.050	C1046
1050	0.48-0.55	0.60-0.90	0.040	0.050	C1050
....	0.45-0.56	0.85-1.15	0.040	0.050	C1051
1052	0.47-0 55	1.20-1.50	0.040	0.050	C1052
....	0.50-0.60	0.50-0.80	0.040	0.050	C1054
1055	0.50-0.60	0.60-0.90	0.040	0.050	C1055
....	0.50-0.61	0.85-1.15	0.040	0.050	C1057
....	0.55-0.65	0.50-0.80	0.040	0.050	C1059
1060	0.55-0.65	0.60-0.90	0.040	0.050	C1060
....	0.54-0.65	0.75-1.05	0.040	0.050	C1061
1062	0.54-0.65	0.85-1.15	0.040	0.050	C1062
1064	0.60-0.70	0.50-0.80	0.040	0.050	C1064
1065	0.60-0.70	0.60-0.90	0.040	0.050	C1065
1066	0.60-0.71	0.85-1.15	0.040	0.050	C1066
....	0.65-0.75	0.40-0.70	0.040	0.050	C1069
1070	0.65-0.75	0.60-0.90	0.040	0.050	C1070
....	0.65-0.76	0.75-1.05	0.040	0.050	C1071
1074	0.70-0.80	0.50-0.80	0.040	0.050	C1074

Alloy Steel

AISI Number	C	Mn	P Max	S Max	Si	Ni	Cr	Other	SAE Number
1320	0.18-0.23	1.60-1.90	0.040	0.040	0.20-0.35	.....	.....	.....	1320
1321	0.17-0.22	1.80-2.10	0.050	0.050	0.20-0.35	.....	.....	.....	.....
1330	0.28-0.33	1.60-1.90	0.040	0.040	0.20-0.35	.....	.....	.....	1330
1335	0.33-0.38	1.60-1.90	0.040	0.040	0.20-0.35	.....	.....	.....	1335
1340	0.38-0.43	1.60-1.90	0.040	0.040	0.20-0.35	.....	.....	.....	1340
2317	0.15-0.20	0.40-0.60	0.040	0.040	0.20-0.35	3.25-3.75	.....	.....	2317
2330	0.28-0.33	0.60-0.80	0.040	0.040	0.20-0.35	3.25-3.75	.....	.....	2330
2335	0.33-0.38	0.60-0.80	0.040	0.040	0.20-0.35	3.25-3.75	.....	.....	.....
2340	0.38-0.43	0.70-0.90	0.040	0.040	0.20-0.35	3.25-3.75	.....	.....	2340
2345	0.43-0.48	0.70-0.90	0.040	0.040	0.20-0.35	3.25-3.75	.....	.....	2345
E 2512	0.09-0.14	0.45-0.60	0.025	0.025	0.20-0.35	4.75-5.25	.....	.....	2512
2515	0.12-0.17	0.40-0.60	0.040	0.040	0.20-0.35	4.75-5.25	.....	.....	2515
E 2517	0.15-0.20	0.45-0.60	0.025	0.025	0.20-0.35	4.75-5.25	.....	.....	2517
3115	0.13-0.18	0.40-0.60	0.040	0.040	0.20-0.35	1.10-1.40	0.55-0.75	.....	3115
3120	0.17-0.22	0.60-0.80	0.040	0.040	0.20-0.35	1.10-1.40	0.55-0.75	.....	3120
3130	0.28-0.33	0.60-0.80	0.040	0.040	0.20-0.35	1.10-1.40	0.55-0.75	.....	3130
3135	0.33-0.38	0.60-0.80	0.040	0.040	0.20-0.35	1.10-1.40	0.55-0.75	.....	3135
3140	0.38-0.43	0.70-0.90	0.040	0.040	0.20-0.35	1.10-1.40	0.55-0.75	.....	3140
3141	0.38-0.43	0.70-0.90	0.040	0.040	0.20-0.35	1.10-1.40	0.70-0.90	.....	3141
3145	0.43-0.48	0.70-0.90	0.040	0.040	0.20-0.35	1.10-1.40	0.70-0.90	.....	3145
3150	0.48-0.53	0.70-0.90	0.040	0.040	0.20-0.35	1.10-1.40	0.70-0.90	.....	3150
E 3310	0.08-0.13	0.45-0.60	0.025	0.025	0.20-0.35	3.25-3.75	1.40-1.75	.....	3310
E 3316	0.14-0.19	0.45-0.60	0.025	0.025	0.20-0.35	3.25-3.75	1.40-1.75	.....	3316
								Mo	
4017	0.15-0.20	0.70-0.90	0.040	0.040	0.20-0.35	.....	.....	0.20-0.30	4017
4023	0.20-0.25	0.70-0.90	0.040	0.040	0.20-0.35	.....	.....	0.20-0.30	4023
4024	0.20-0.25	0.70-0.90	0.040	0.035-0.050	0.20-0.35	.....	.....	0.20-0.30	4024
4027	0.25-0.30	0.70-0.90	0.040	0.040	0.20-0.35	.....	.....	0.20-0.30	4027
4028	0.25-0.30	0.70-0.90	0.040	0.035-0.050	0.20-0.35	.....	.....	0.20-0.30	4028
4032	0.30-0.35	0.70-0.90	0.040	0.040	0.20-0.35	.....	.....	0.20-0.30	4032
4037	0.35-0.40	0.70-0.90	0.040	0.040	0.20-0.35	.....	.....	0.20-0.30	4037
4042	0.40-0.45	0.70-0.90	0.040	0.040	0.20-0.35	.....	.....	0.20-0.30	4042
4047	0.45-0.50	0.70-0.90	0.040	0.040	0.20-0.35	.....	.....	0.20-0.30	4047
4053	0.50-0.56	0.75-1.00	0.040	0.040	0.20-0.35	.....	.....	0.20-0.30	4053
4063	0.60-0.67	0.75-1.00	0.040	0.040	0.20-0.35	.....	.....	0.20-0.30	4063
4068	0.63-0.70	0.75-1.00	0.040	0.040	0.20-0.35	.....	.....	0.20-0.30	4068
.....	0.17-0.22	0.70-0.90	0.040	0.040	0.20-0.35	.....	0.40-0.60	0.20-0.30	4119
.....	0.23-0.28	0.70-0.90	0.040	0.040	0.20-0.35	.....	0.40-0.60	0.20-0.30	4125
4130	0.28-0.33	0.40-0.60	0.040	0.040	0.20-0.35	.....	0.80-1.10	0.15-0.25	4130
E 4132	0.30-0.35	0.40-0.60	0.025	0.025	0.20-0.35	.....	0.80-1.10	0.18-0.25	.....
E 4135	0.33-0.38	0.70-0.90	0.025	0.025	0.20-0.35	.....	0.80-1.10	0.15-0.25	.....
4137	0.35-0.40	0.70-0.90	0.040	0.040	0.20-0.35	.....	0.80-1.10	0.15-0.25	4137
E 4137	0.35-0.40	0.70-0.90	0.025	0.025	0.20-0.35	.....	0.80-1.10	0.18-0.25	.....
4140	0.38-0.43	0.75-1.00	0.040	0.040	0.20-0.35	.....	0.80-1.10	0.15-0.25	4140
4142	0.40-0.45	0.75-1.00	0.040	0.040	0.20-0.35	.....	0.80-1.10	0.15-0.25	.....
4145	0.43-0.48	0.75-1.00	0.040	0.040	0.20-0.35	.....	0.80-1.10	0.15-0.25	4145
4147	0.45-0.50	0.75-1.00	0.040	0.040	0.20-0.35	.....	0.80-1.10	0.15-0.25	.....
4150	0.48-0.53	0.75-1.00	0.040	0.040	0.20-0.35	.....	0.80-1.10	0.15-0.25	4150
4317	0.15-0.20	0.45-0.65	0.040	0.040	0.20-0.35	1.65-2.00	0.40-0.60	0.20-0.30	4317
4320	0.17-0.22	0.45-0.65	0.040	0.040	0.20-0.35	1.65-2.00	0.40-0.60	0.20-0.30	4320
4337	0.35-0.40	0.60-0.80	0.040	0.040	0.20-0.35	1.65-2.00	0.70-0.90	0.20-0.30	.....
4340	0 38-0.43	0.60-0.80	0.040	0.040	0.20-0.35	1.65-2.00	0.70-0.90	0.20-0.30	4340
4608	0.06-0.11	0.25-0.45	0.040	0.040	0.25 Max	1.40-1.75	.....	0.15-0.25	4608
4615	0.13-0.18	0.45-0.65	0.040	0.040	0.20-0.35	1.65-2.00	.....	0.20-0.30	4615
.....	0.15-0.20	0.45-0.65	0.040	0.040	0.20-0.35	1.65-2.00	.....	0.20-0.30	4617
E 4617	0.15-0 20	0.45-0.65	0.025	0.025	0.20-0.35	1.65-2.00	.....	0.20-0.27	.....
4620	0.17-0.22	0.45-0.65	0.040	0.040	0.20-0.35	1.65-2.00	.....	0.20-0.30	4620
X 4620	0.18-0.23	0.50-0.70	0.040	0.040	0.20-0.35	1.65-2.00	.....	0.20-0.30	X 4620
E 4620	0.17-0.22	0.45-0.65	0.025	0.025	0.20-0.35	1.65-2.00	.....	0.20-0.27	.....
4621	0.18-0.23	0.70-0.90	0.040	0.040	0.20-0.35	1.65-2.00	.....	0.20-0.30	4621
4640	0.38-0.43	0.60-0.80	0.040	0.040	0.20-0.35	1.65-2.00	.....	0.20-0.30	4640
E 4640	0.38-0.43	0.60-0.80	0.025	0.025	0.20-0 35	1.65-2.00	.....	0.20-0.27	.....
4812	0.10-0.15	0.40-0.60	0.040	0.040	0.20-0.35	3.25-3.75	.....	0.20-0.30	4812
4815	0.13-0.18	0.40-0.60	0.040	0.040	0.20-0.35	3.25-3.75	.....	0.20-0.30	4815
4817	0.15-0.20	0.40-0.60	0.040	0.040	0.20-0.35	3.25-3.75	.....	0.20-0.30	4817
4820	0.18-0.23	0.50-0.70	0.040	0.040	0.20-0.35	3.25-3.75	.....	0.20-0.30	4820

Alloy Steel (Continued)

AISI Number	C	Mn	P Max	S Max	Si	Ni	Cr	Other	SAE Number
5045	0.43-0.48	0.70-0.90	0.040	0.040	0.20-0.35	. .	0.55-0.75	....	5045
5046	0.43-0.50	0.75-1.00	0.040	0.040	0.20-0.35	. .	0.20-0.35	... .	5046
.....	0.13-0.18	0.70-0.90	0.040	0.040	0.20-0.35	....	0.70-0.90	...	5115
5120	0.17-0.22	0.70-0.90	0.040	0.040	0.20-0.35	..	0.70-0.90	..	5120
5130	0.28-0.33	0.70-0.90	0.040	0.040	0.20-0.35	..	0.80-1.10	...	5130
5132	0.30-0.35	0.60-0.80	0.040	0.040	0.20-0.35		0.80-1.05	...	5132
5135	0.33-0.38	0.60-0.80	0.040	0.040	0.20-0.35		0.80-1.05	..	5135
5140	0.38-0.43	0.70-0.90	0.040	0.040	0.20-0.35	.	0.70-0.90	. ..	5140
5145	0.43-0.48	0.70-0.90	0.040	0.040	0.20-0.35		0.70-0.90	....	5145
5147	0.45-0.52	0.75-1.00	0.040	0.040	0.20-0.35	..	0.90-1.20	....	5147
5150	0.48-0.53	0.70-0.90	0.040	0.040	0.20-0.35	. .	0.70-0.90	...	5150
5152	0.48-0.55	0.70-0.90	0.040	0.040	0.20-0.35	...	0.90-1.20	...	5152
E 50100	0.95-1.10	0.25-0.45	0.025	0.025	0.20-0.35	....	0.40-0.60	.....	50100
E 51100	0.95-1.10	0.25-0.45	0.025	0.025	0.20-0.35	. .	0.90-1.15	....'	51100
E 52100	0.95-1.10	0.25-0.45	0.025	0.025	0.20-0.35	.	1.30-1.60	....	52100
								V	
6120	0.17-0.22	0.70-0.90	0.040	0.040	0.20-0.35	.....	0.70-0.90	0.10 Min	.....
6145	0.43-0.48	0.70-0.90	0.040	0.040	0.20-0.35	...	0.80-1.10	0.15 Min	.. ...
6150	0.48-0.53	0.70-0.90	0.040	0.040	0.20-0.35	.....	0.80-1.10	0.15 Min	6150
6152	0.48-0.55	0.70-0.90	0.040	0.040	0.20-0.35	.....	0.80-1.10	0.10 Min	....
								Mo	
8615	0.13-0.18	0.70-0.90	0.040	0.040	0.20-0.35	0.40-0.70	0.50-0.60	0.15-0.25	8615
8617	0.15-0.20	0.70-0.90	0.040	0.040	0.20-0.35	0.40-0.70	0.40-0.60	0.15-0.25	8617
8620	0.18-0.23	0.70-0.90	0.040	0.040	0.20-0.35	0.40-0.70	0.40-0.60	0.15-0.25	8620
8622	0.20-0.25	0.70-0.90	0.040	0.040	0.20-0.35	0.40-0.70	0.40-0.60	0.15-0.25	8622
8625	0.23-0.28	0.70-0.90	0.040	0.040	0.20-0.35	0.40-0.70	0.40-0.60	0.15-0.25	8625
8627	0.25-0.30	0.70-0.90	0.040	0.040	0.20-0.35	0.40-0.70	0.40-0.60	0.15-0.25	8627
8630	0.28-0.33	0.70-0.90	0.040	0.040	0.20-0.35	0.40-0.70	0.40-0.60	0.15-0.25	8630
8532	0.30-0.35	0.70-0.90	0.040	0.040	0.20-0.35	0.40-0.70	0.40-0.60	0.15-0.25	8632
8635	0.33-0.38	0.75-1.00	0.040	0.040	0.20-0.35	0.40-0.70	0.40-0.60	0.15-0.25	8635
8637	0.35-0.40	0.75-1.00	0.040	0.040	0.20-0.35	0.40-0.70	0.40-0.60	0.15-0.25	8637
8640	0.38-0.43	0.75-1.00	0.040	0.040	0.20-0.35	0.40-0.70	0.40-0.60	0.15-0.25	8640
8641	0.38-0.43	0.75-1.00	0.040	0.040-0.060	0.20-0.35	0.40-0.70	0.40-0.60	0.15-0.25	8641
8642	0.40-0.45	0.75-1.00	0.040	0.040	0.20-0.35	0.40-0.70	0.40-0.60	0.15-0.25	8642
8645	0.43-0.48	0.75-1.00	0.040	0.040	0.20-0.35	0.40-0.70	0.40-0.60	0.15-0.25	8645
8647	0.45-0.50	0.75-1.00	0.040	0.040	0.20-0.35	0.40-0.70	0.40-0.60	0.15-0.25	8647
8650	0.48-0.53	0.75-1.00	0.040	0.040	0.20-0.35	0.40-0.70	0.50-0.80	0.15-0.25	8650
8653	0.50-0.56	0.75-1.00	0.040	0.040	0.20-0.35	0.40-0.70	0.40-0.60	0.15-0.25	8653
8655	0.50-0.60	0.75-1.00	0.040	0.040	0.20-0.35	0.40-0.70	0.40-0.60	0.15-0.25	8655
8660	0.50-0.65	0.75-1.00	0.040	0.040	0.20-0.35	0.40-0.70	0.40-0.60	0.15-0.25	8660
8720	0.18-0.23	0.70-0.90	0.040	0.040	0.20-0.35	0.40-0.70	0.40-0.60	0.20-0.30	8720
8735	0.33-0.38	0.75-1.00	0.040	0.040	0.20-0.35	0.40-0.70	0.40-0.60	0.20-0.30	8735
8740	0.38-0.43	0.75-1.00	0.040	0.040	0.20-0.35	0.40-0.70	0.40-0.60	0.20-0.30	8740
8742	0.40-0.45	0.75-1.00	0.040	0.040	0.20-0.35	0.40-0.70	0.40-0.60	0.20-0.30	.....
8745	0.43-0.48	0.75-1.00	0.040	0.040	0.20-0.35	0.40-0.70	0.40-0.60	0.20-0.30	8745
8747	0.45-0.50	0.75-1.00	0.040	0.040	0.20-0.35	0.40-0.70	0.40-0.60	0.20-0.30	.....
8750	0.48-0.53	0.75-1.00	0.040	0.040	0.20-0.35	0.40-0.70	0.40-0.60	0.20-0.30	8750
.....	0.50-0.60	0.50-0.60	0.040	0.040	1.20-1.60	.....	0.50-0.80	.....	9254
9255	0.50-0.60	0.70-0.95	0.040	0.040	1.80-2.20	.....	.....	.....	9255
9260	0.55-0.65	0.70-1.00	0.040	0.040	1.80-2.20	.....	.....	.....	9260
9261	0.55-0.65	0.75-1.00	0.040	0.040	1.80-2.20	.....	0.10-0.25	.....	9261
9262	0.55-0.65	0.75-1.00	0.040	0.040	1.80-2.20	.....	0.25-0.40	.....	9262
E 9310	0.08-0.13	0.45-0.65	0.025	0.025	0.20-0.35	3.00-3.50	1.00-1.40	0.08-0.15	9310
E 9315	0.13-0.18	0.45-0.65	0.025	0.025	0.20-0.35	3.00-3.50	1.00-1.40	0.08-0.15	9315
E 9317	0.15-0.20	0.45-0.65	0.025	0.025	0.20-0.35	3.00-3.50	1.00-1.40	0.08-0.15	9317
9437	0.35-0.40	0.90-1.20	0.040	0.040	0.20-0.35	0.30-0.60	0.30-0.50	0.08-0.15	9437
9440	0.38-0.43	0.90-1.20	0.040	0.040	0.20-0.35	0.30-0.60	0.30-0.50	0.08-0.15	9440
9442	0.40-0.45	1.00-1.30	0.040	0.040	0.20-0.35	0.30-0.60	0.30-0.50	0.08-0.15	9442
9445	0.43-0.48	1.00-1.30	0.040	0.040	0.20-0.35	0.30-0.60	0.30-0.50	0.08-0.15	9445
9747	0.45-0.50	0.50-0.80	0.040	0.040	0.20-0.35	0.40-0.70	0.10-0.25	0.15-0.25	9747
9763	0.60-0.67	0.50-0.80	0.040	0.040	0.20-0.35	0.40-0.70	0.10-0.25	0.15-0.25	9763
9840	0.38-0.43	0.70-0.90	0.040	0.040	0.20-0.35	0.65-1.15	0.70-0.90	0.20-0.30	9840
9845	0.43-0.48	0.70-0.90	0.040	0.040	0.20-0.35	0.85-1.15	0.70-0.90	0.20-0.30	9845
9850	0.48-0.53	0.70-0.90	0.040	0.040	0.20-0.35	0.85-1.15	0.70-0.90	0.20-0.30	9850

TABLE 68-III — **Typical Characteristics of Aluminum Rod, Bar and Wire Alloys**[1]

Alloy and temper	Resistance to corrosion	Workability (cold)	Machinability	Brazeability	Weldability			Forgeability
					Gas	Arc	Resistance (spot and steam)	
1100-0	A	A	D	A	A	A	B	A
-H14	A	A	C	A	A	A	A	A
-H18	A	C	C	A	A	A	A	A
2011-T3	C	C	A	D	D	D	B	n.a.
-T8	C	D	A	D	D	D	B	n.a.
2014-T6	C	D	B	D	D	B	B	C
2017-T4	C	C	B	D	D	B	B	C
2024-T4	C	C	B	D	D	B	B	D
-T36	C	D	B	D	D	B	B	D
2117-T4	C	B	C	D	D	B	B	B
3003-0	A	A	D	A	A	A	B	A
-H14	A	B	C	A	A	A	B	A
-H18	A	C	C	A	A	A	A	A
5005-0	A	A	D	B	A	A	B	A
-H18	A	C	C	B	A	A	A	A
-H32	A	A	D	B	A	A	A	A
6061-0	A	A	D	A	A	A	B	A
-T6	A	C	C	A	A	A	A	A
6063-0	A	A	D	A	A	A	B	A
-T5	A	B	C	A	A	A	B	A
-T6	A	C	C	A	A	A	A	A
6262-T9	A	C	B	A	A	A	A	n.a.
7075-T6	C	D	B	D	D	D	B	D

(1) Resistance to Corrosion. Workability (Cold). Machinability and Forgeability (Hot) ratings A. B. C and D are relative ratings in decreasing order of merit. Weldability and Brazeability ratings A. B. C and D are relative ratings defined as follows.

A. Generally weldable by all commercial procedures and methods.

B. Weldable with special technique or on specific applications which justify preliminary trials or testing to develop welding procedure and weld performance.

C. Limited weldability because of crack sensitivity or loss in resistance to corrosion and lowering of all mechanical properties.

D. No commonly used welding methods have so far been developed.

n.a. Not applicable.

Courtesy. Aluminum Association

TABLE 68-IV — Basic Design Properties and Applications of Cast Steels

STRUCTURAL GRADES—CARBON STEELS

Tensile Strength, psi	60,000	65,000	70,000	80,000	85,000	100,000
Indicated Application	Low electric resistivity. Desirable magnetic properties carburizing and case hardening grades weldability	Excellent Weldability Medium strength with good ductility	High strength carbon steels with good machinability and excellent fatigue resistance			Wear resistance hardness
Current Specifications	ASTM: A27-46T U60-30 40-30; ASTM: A216-44T WCA; AAR: M201-46 Grade AU, Grade AA; Federal: QQ-S-681b Class 1; Navy: 46S1 Class 1; A.B.S. Class 1	ASTM: A27-46T 65-30 65-35; SAE: Auto. 0030; Federal: QQ-S-681b Class 2; ABS Class 2; Lloyd's Class A	ASTM: A27-46T 70-36; ASTM: A95-44; ASTM: A216-44T; AAR: M201-46 Grade B; AREA	SAE Automotive 0050; Federal: QQ-S-681b Class 3; Navy: 46S1 Class 3	SAE Automotive Class 0050	SAE Automotive Class 0050
A Typical Specification for the Tensile Grade with Requirements listed below	ASTM A27-46T Class 60-30	ASTM A27-46T Class 65-35	AAR M201-46 Grade B	Federal QQ-S-681b Class 3	SAE Automotive Class 0050	SAE Automotive Class 0050

All values listed below are specification minimum values.

	60,000	65,000	70,000	80,000	85,000	100,000
Tensile Strength, psi	60,000	65,000	70,000	80,000	85,000	100,000
Yield Point, psi	30,000	35,000	38,000	40,000	45,000	70,000
Elongation in 2", %	24	24	24	24	16	10
Reduction of Area, %	35	35	36	25	24	15
Brinell Hardness No.	—	—	—	—	170	207

Values listed directly below are those normally expected in the production of Steel Castings for the tensile strength values given in the upper portion of the chart.†† The values listed below are only to be used as design or specification limit values.

	60,000	65,000	70,000	80,000	85,000	100,000
Tensile Strength, psi	60,000	65,000	70,000	80,000	85,000	100,000
Yield Point, psi	30,000	35,000	38,000	45,000	50,000	70,000
Elongation in 2", %	30	30	28	26	24	20
Reduction of Area, %	50	53	50	43	40	43
Brinell Hardness No.	120	130	140	160	175	215
Charpy Impact** at 70° F. ft. lbs.	35	35	30	35	30	25
Charpy Impact** at −50° F. ft. lbs.	8	12	10	12	12	15
Endurance Limit psi	25,000	28,000	31,000	35,000	38,000	47,000
Modulus of Elasticity	30 million psi	30 million psi	30 million psi	30 million psi	30 million psi	30 million psi
Machinability Rating‡	55	60	65	70	70	65
Type of Heat Tr.	Annealed	Normalized	Normalized	Normalized and Tempered	Normalized and Tempered	Quenched and Tempered

ENGINEERING GRADES—LOW ALLOY STEELS*

Tensile Strength, psi	70,000	80,000	90,000	100,000	110,000	120,000	150,000	175,000	200,000
Indicated Application	Excellent Weldability Medium strength with high toughness and good machinability		Certain steels of these classes have excellent high temp properties and deep hardening properties Toughness		High resistance to impact Excellent low temp properties for certain steels Deep hardening Excellent combination strength and toughness		Deep hardening High strength Wear resistance Fatigue resistance	High strength Wear resistance High hardness High fatigue resistance	None specified
Current Specifications	ASTM: A157-44 C1; ASTM: A217-46T WC 2 WC 3; Navy: 46S33 [int] A	ASTM: A148-46T 80-40 80-50; ASTM: A217-46T WC4; SAE: Automotive 080; Federal: QQ-S-681b 4A1	ASTM: A148-46T 90-60; ASTM: A157-44 C3; Federal: QQ-S-681b 4A2, 4B1, 4B2, 4C1; Navy: 49S1 [int] Grade F; AAR: M201-46 Grade C	ASTM: A157-44 C11; Federal: QQ-S-681b 4B3	ASTM: A148-46T 105-85; SAE: Automotive 0105; Federal: QQ-S-681b 4C2	ASTM: A148-46T 120-100; SAE: Automotive 0120; Federal: QQ-S-681b 4C3	ASTM: A148-46T 150-125; SAE: Automotive 0150; Federal: QQ-S-681b 4C4	ASTM: A148-46T 175-145; SAE: Automotive 0175	None specified
A Typical Specification for the Tensile Grade with Requirements listed below	ASTM A157-44 Class C1	ASTM A148-46T Class 80-50	ASTM A148-46T Class 90-60	Federal QQ-S-681b Class 483	ASTM A148-46T Class 105-85	ASTM A148-46T Class 120-100	ASTM A148-46T Class 150-125	ASTM A148-46T Class 175-145	None specified

All values listed below are specification minimum values.

	70,000	80,000	90,000	100,000	110,000	120,000	150,000	175,000	200,000
Tensile Strength, psi	70,000	80,000	90,000	100,000	105,000	120,000	150,000	175,000	—
Yield Point, psi	45,000	50,000	60,000	65,000	85,000	100,000	125,000	145,000	—
Elongation in 2", %	22	22	20	17	14	9	6	—	—
Reduction of Area, %	35	35	40	30	35	22	—	—	—
Brinell Hardness No.	—	—	—	—	217†	248†	311‡	363‡	—

Values listed directly below are those normally expected in the production of Steel Castings for the tensile strength values given in the upper portion of the chart.†† The values listed below are only to be used as design or specification limit values.

	70,000	80,000	90,000	100,000	110,000	120,000	150,000	175,000	200,000
Tensile Strength, psi	70,000	80,000	90,000	100,000	110,000	120,000	150,000	175,000	200,000
Yield Point, psi	42,000	50,000	60,000	65,000	85,000	97,000	130,000	148,000	175,000
Elongation in 2", %	28	27	24	21	18	16	12	8	5
Reduction of Area, %	55	50	50	46	42	38	25	15	11
Brinell Hardness No.	140	160	190	215	235	260	325	380	420
Charpy Impact** at 70° F. ft. lbs.	35	30	26	22	28	24	18	10	—
Charpy Impact** at −50° F. ft. lbs.	15	15	17	15	22	18	15	6	—
Endurance Limit psi	33,000	38,000	41,000	45,000	49,000	55,000	65,000	77,000	85,000
Modulus of Elasticity	30 million psi	30 million psi	30 million psi	30 million psi	30 million psi	30 million psi	30 million psi	30 million psi	10 million psi
Machinability Rating‡	65	70	70	65	60	50	30	—	—
Type of Heat Tr.	Normalized and Tempered	Normalized and Tempered	Normalized and Tempered	Quenched and Tempered	Quenched and Tempered	Quenched and Tempered	Quenched and Tempered	Quenched and Tempered	Quenched and Tempered

* Value in percent total alloy content.
** [impact value]
‡ Machinability Rating by Research Committee on Cutting Fluids—Metal Progress Oct. 1945, p. 627-634. Cold rolled screw steel equals 100.
†† BHN measurement.

(Courtesy, Steel Founders' Society of America)

TABLE 68-V — Standards For Aluminum Sand and Permanent Mold Castings

Selection of an alloy for a particular application requires consideration not only of mechanical properties, but also of numerous other characteristics, such as behavior in the casting process or subsequent treatments in the course of manufacture, and response to the environmental conditions of service.

The following table includes several significant characteristics which deserve consideration in the selection of an alloy. The characteristics are comparatively rated from 1 to 5 in decreasing order of perforance.

Alloy	Product	Fluidity	Resistance to Hot Cracking	Pressure Tightness	Normally Heat Treated	Strength at Elevated Temperatures	Corrosion Resistance	Machinability	Polishing	Anodizing Appearance	Weldability
208.0	S	2	2	2	Optional	3	4	3	3	3	2
213.0	P	2	3	3	No	3	4	2	2	3	3
222.0	S&P	3	3	3	Yes	1	4	1	2	3	3
242.0	S&P	3	4	4	Yes	1	4	2	2	3	4
295.0	S	3	4	4	Yes	3	4	2	2	2	3
B295.0	P	3	4	3	Yes	2	4	3	2	3	3
308.0	P	2	2	2	No	3	3	3	3	4	2
319.0	S&P	2	2	2	Optional	3	3	3	4	4	2
328.0	S	1	1	2	Optional	2	3	3	3	4	1
A332.0	P	1	2	2	Yes	1	3	4	4	4	3
F332.0	P	1	2	2	Yes	1	3	4	4	4	2
333.0	P	1	2	2	Yes	2	3	3	3	4	3
354.0	P	1	1	1	Yes	2	3	4	4	4	3
355.0	S&P	1	1	1	Yes	2	3	3	3	4	1
C355.0	S&P	1	1	1	Yes	2	3	3	3	4	1
356.0	S&P	1	1	1	Yes	3	2	3	4	4	1
A356.0	S&P	1	1	1	Yes	3	2	3	4	4	1
357.0	S&P	1	1	1	Yes	3	2	3	4	4	1
A357.0	S&P	1	1	1	Yes	2	2	3	4	4	1
359.0	S&P	1	2	2	Yes	2	2	4	4	4	1
B443.0	S&P	1	1	1	No	4	2	5	4	4	1
514.0	S	4	4	5	No	3	1	1	1	1	3
A514.0	P	4	4	4	No	3	1	1	1	1	3
B514.0	S	3	3	4	No	3	1	2	2	2	3
520.0	S	4	4	5	Yes	5	1	1	1	1	4
535.0	S	5	4	5	Optional	3	1	1	1	1	4
705.0	S&P	4	4	4	No	4	2	1	2	2	4
707.0	S&P	4	4	4	No	4	2	1	2	2	4
A712.0	S	4	5	4	No	4	4	1	2	2	4
D712.0	S	3	5	4	No	4	3	1	2	2	4
713.0	S&P	3	4	4	No	4	3	1	1	1	4
771.0	S	3	4	4	Yes	4	3	1	1	1	4
850.0	S&P	4	5	5	Yes	5	4	1	3	*	5
A850.0	S&P	4	5	5	Yes	5	4	1	3	*	5
B850.0	S&P	4	5	5ʹ	Yes	5	4	1	3	*	5

*Information not available.   Note: Aluminum Association designation system.

TABLE 68-VI — Compositions, Characteristics, Properties, and Typical Applications of Aluminum Sand and Permanent Mold Casting Alloys

Alloy Desig. AA No.	Former Desig.	Si	Fe	Cu	Mn	Mg	Cr	Ni	Zn	Ti	Others Each	Others Total	Pressure Tightness	Heat Treatable	Elevated-Temperature Strength	Corrosion Resistance	Machinability	Weldability	Product (c)	Temper	Tensile Strength 1,000 Psi (d)	Elongation % x 2 in.	Typical Applications
208.0	108	2.5-3.5	1.2	3.5-4.5	0.50	—	—	—	1.0	0.25	—	0.50	2	Optional	3	4	3	2	S	F	19.0 / 21.0	1.5	General purpose sand castings, manifold and valve bodies.
213.0	C113	1.0-3.0	1.2	6.0-8.0	0.6	0.10	—	0.35	2.5	0.25	—	0.35	3	No	3	4	2	3	P	F	19.0	—	Washing machine agitators, automotive cylinder heads, and timing gears.
222.0	122	2.0	1.5	9.2-10.7	0.50	0.15-0.35	—	0.50	0.8	0.25	—	0.35	3	Yes	1	4	1	3	P	T2 / T61	23.0 / 30.0	—	Primarily a piston alloy, also used for air cooled cylinder heads and valve tappet guides; bushings, bearing caps, and meter parts; engine parts.
242.0	142	0.7	1.0	3.5-4.5	0.35	1.2-1.8	0.25	1.7-2.3	0.35	0.25	0.05	0.15	4	Yes	1	4	2	3	S/P	T571 / T61 / T77	30.0 / 40.0 / 23.0 / 29.0 / 34.0 / 40.0	—	Air cooled cylinder heads, pistons in high-performance gasoline engines; aircraft generator housings.
295.0	195	0.7-1.5	1.0	4.0-5.0	0.35	0.03	—	—	0.35	0.25	0.05	0.15	4	Yes	3	4	2	3	S	T4 / T6 / T62 / T7	29.0 / 32.0 (20.0) / 36.0	6.0 / 3.0	General structural castings requiring high strength and shock resistance, machinery and aircraft structural members; crankcases, flywheel housings, aircraft wheels, bus wheels.
B295.0	B195	2.0-3.0	1.2	4.0-5.0	0.35	0.05	—	0.35	0.50	0.25	—	0.35	4	Yes	2	4	2	3	P	T4 / T6 / T7	33.0 / 35.0 / 36.0	4.5 / 2.0 / 3.0	Permanent mold version of 295.0; aircraft fittings, gear housings, gun control parts, fuel pump bodies, wheels, railway car seat frames.
308.0	A108	5.0-6.0	1.0	4.0-5.0	0.50	0.10	—	—	1.0	0.25	—	0.50	2	No	3	3	3	2	P	F	24.0	—	General purpose permanent mold castings and ornamental grilles.
319.0	319, Allcast	5.5-6.5	1.0	3.0-4.0	0.50	0.10	—	0.35	1.0	0.25	—	0.50	2	Optional	3	3	3	2	S/P	F / T5 / T6	23.0 / 31.0 / 28.0 / 34.0	1.5 / 1.5 / 2.0	General purpose alloy used for engine parts, automotive cylinder heads, piano plates.
328.0	Red X-8	7.5-8.5	1.0	1.0-2.0	0.20-0.6	0.20-0.6	0.35	—	1.5	0.25	—	0.50	2	Optional	2	3	3	1	S	F / T6	25.0 / 34.0	1.0	General use where high strength and pressure tightness are required, such as pump bodies and liquid-cooled cylinder heads; intricate castings requiring good strength and ductility.
A332.0	A132	11.0-13.0	1.2	0.50-1.5	0.35	0.7-1.3	—	2.0-3.0	0.35	0.25	0.05	—	3	Yes	1	4	4	3	P	T551 / T65	31.0 / 40.0	—	Automotive pistons; diesel engine pistons; pulley sheaves and engine parts operating at elevated temperatures.
F332.0	F132	8.5-10.5	1.2	2.0-4.0	0.50	0.50-1.5	—	0.50	1.0	0.25	—	0.50	2	Yes	2	3	3	2	P	T5	31.0	—	Automotive pistons, parts requiring elevated temperature strength and good abrasion resistance.
333.0	333	8.0-10.0	1.0	3.0-4.0	0.50	0.05-0.50	—	0.50	1.0	0.25	—	0.50	2	Yes	2	3	3	2	P	F / T5 / T6 / T7	28.0 / 30.0 / 35.0 / 31.0	—	General purpose alloy used for engine parts, meter housings, flywheel housings, and regulator parts.
354.0	354	8.6-9.4	0.20	1.6-2.0	0.10	0.40-0.6	—	—	0.10	0.20	0.05	0.15	3	Yes	2	3	4	3	P	T6	25.0	1.0	Aircraft, missile, and other applications requiring high-strength castings.
355.0	355	4.5-5.5	(0.6/1.0)	1.0-1.5	0.50	0.40-0.6	0.25	—	0.30	0.25	0.05	0.15	3	Yes	1	3	3	3	S/P	T51 / T6 / T61 / T7 / T71	25.0 / 32.0 (20.0) / 35.0 / 30.0 / 27.0 / 37.0 / 42.0 / 34.0	2.0 / 1.5	Similar to 328.0; general use where high strength and pressure tightness are required, such as pump bodies and liquid-cooled cylinder heads; crankcases, accessory housings and aircraft fittings; stressed castings, such as blower housings, snow removal equipment, and scaffold pedestals.
C355.0	C355	4.5-5.5	0.20	1.0-1.5	0.10	0.40-0.6	—	—	0.10	0.20	0.05	0.15	3	Yes	2	3	3	3	S/P	(e) / T61	40.0	3.0	Similar to 355.0, but stronger and more ductile; aircraft, missile, and other structural uses requiring high strength; parts requiring high strength-to-weight ratios, such as crankcases and wheels in aerospace applications.
356.0	356	6.5-7.5	0.6	0.25	0.35	0.20-0.40	—	—	0.35	0.25	0.05	0.15	2	Yes	3	2	3	1	S/P	T51 / T6 / T7 / T71	23.0 / 30.0 (20.0) / 31.0 (29.0) / 25.0 / 33.0 / 29.0	3.0 / 3.0 / 4.0	Similar to 328.0; intricate castings requiring good strength and ductility, transmission cases, truck axle housings, machine tool and aircraft parts, cylinder blocks, railway tank car fittings, marine hardware, valve bodies, and bridge railing parts; outboard motor parts, fan blades, pneumatic tools, storage tank fittings, gray anodized architectural components.
A356.0	A356	6.5-7.5	0.20	0.20	0.10	0.20-0.40	—	—	0.10	0.20	0.05	0.15	2	Yes	3	2	3	1	S/P	T6 / T61	37.0 / 45.0	5.0 / 3.0	Similar to 356.0, but stronger and more ductile; aircraft and missile components requiring strength, ductility, and corrosion resistance.
357.0	357	6.5-7.5	0.15	0.05	0.03	0.45-0.6	—	—	0.05	0.20	0.05	0.15	2	Yes	3	2	3	1	S/P	T6	45.0	3.0	Highly stressed castings requiring a high strength-to-weight ratio and excellent corrosion resistance; aircraft and missile components, machine parts, high-velocity fan blades.
A357.0	A357	6.5-7.5	0.20	0.20	0.10	0.40-0.7	—	—	0.10	0.10-0.20	0.05(d)	0.15	2	Yes	4	2	5	4	S&P	(e)	—	—	Aircraft and other structural applications.
359.0	359	8.5-9.5	0.20	0.20	0.10	0.50-0.7	—	—	0.10	0.20	0.05	0.15	2	Yes	4	2	5	4	S&P	(e)	—	—	High-strength aircraft, missile, and other structural applications.
B443.0	43	4.5-6.0	0.8	0.15	0.35	0.05	—	—	0.35	0.25	0.05	0.15	2	No	3	3	5	1	S/P	F	17.0 / 21.0	3.0 / 6.0	General-purpose casting alloy, cooking utensils, architectural and marine applications, pipe fittings.
514.0	214	0.35	0.50	0.15	0.35	3.5-4.5	—	—	0.15	0.25	0.05	0.15	5	No	3	1	1	4	S	F	22.0	6.0	Moderate strength, high corrosion resistance, food handling and marine equipment; marine hardware, and architectural uses; fittings, cooking utensils.
A514.0	A214	0.30	0.40	0.10	0.30	3.5-4.5	—	—	1.4-2.2	0.20	0.05	0.15	5	No	3	1	2	5	P	F	22.0	2.5	Cooking utensils and ornamental hardware, castings requiring anodic finishes, parts requiring corrosion resistance such as marine fittings, pipe couplings, and automotive trim.
B514.0	B214	1.4-2.2	0.6	0.35	0.30	3.5-4.5	—	—	0.35	0.25	0.05	0.15	4	No	3	1	2	4	S	F	17.0 (10.0)	3.0	Cooking utensils and pipe fittings.
520.0	220	0.25	0.30	0.25	0.15	9.5-10.6	—	—	0.15	0.25	0.05	0.15	5	Yes	3	1	2	5	S	T4	42.0 (22.0)	12.0	Strong and ductile structural uses, requires special foundry practices, aircraft structural members requiring strength and shock resistance.
535.0	Almag 35	0.15	0.15	0.05	0.10-0.25	6.2-7.5	—	—	—	0.10-0.25	0.05(m)	0.15	4	No	3	1	2	4	S	F	35.0 (18.0)	9.0	Ornamental castings, machine tool parts, electrical hardware.
705.0	603, Tenzaloy 5	0.20	0.8	0.20	0.40-0.6	1.4-1.8	0.20-0.40	—	2.7-3.3	0.25	0.05	0.15	4	No	4	2	1	2	S/P	F / For T5	30.0 (17.0) / 37.0	5.0 / 10.0	High strength, general-purpose alloy, excellent machinability and dimensional stability, good corrosion resistance, anodizable.
707.0	607, Tenzaloy 7	0.20	0.8	0.20	0.40-0.6	1.8-2.4	0.20-0.40	—	4.0-4.5	0.25	0.05	0.15	4	No	4	2	1	2	S/P	F / For T5	33.0 (22.0) / 42.0	2.0 / 3.0	Similar to 705.0.
A712.0	A612	0.15	0.50	0.35-0.65	0.05	0.6-0.8	—	—	6.0-7.0	0.25	0.05	0.20	4	No	4	3	1	4	S	F	32.0 (20.0)	4.0	Similar to 705.0, easily polished, used for cast castings that are brazed, often used as a substitute for medium-strength alloys, such as 356.0 to eliminate heat treating.
D712.0	D612, 40E	0.30	0.50	0.25	0.50-0.65	0.50-0.65	0.40-0.6	—	5.0-6.5	0.15-0.25	0.05	0.20	4	No	4	3	1	5	S	F	34.0 (25.0)	4.0	General purpose structural castings that develop strength equivalent to 295.0 without requiring heat treatment.
713.0	613, Tenzaloy	0.25	1.1	0.40-1.0	0.6	0.20-0.50	0.35	0.15	7.0-8.0	0.25	0.10	0.25	4	No	4	3	1	4	S/P	For T5	32.0 (22.0) / 32.0	3.0 / 4.0	Similar to A712.0.
771.0	Precedent 71A	0.15	0.15	0.10	0.10	0.8-1.0	0.06-0.20	—	6.5-7.5	0.10-0.20	0.05	0.15	4	Yes	5	3	4	3	S	T6 / T71	42.0 (35.0)	4.0	Aircraft and missile guidance system components, computing device parts.
850.0	750	0.7	0.7	0.7-1.3	0.10	0.10	—	0.7-1.3	—	0.20	—	0.30	5	Yes	5	4	1	5	S/P	T5	16.0 / 18.0	5.0 / 8.0	Torque converter blades and brazed parts.
A850.0	A750	2.0-3.0	0.7	0.7-1.3	0.10	0.10	0.30-0.7	—	—	0.20	(i)	0.30	5	No	5	4	1	5	S/P	T5	17.0	3.0	Bearings; rolling mill bearings.
B850.0	B750	0.40	0.7	1.7-2.3	0.10	0.6-0.9	—	0.9-1.5	—	0.20	(i)	0.30	5	No	5	4	1	5	S/P	T5	24.0 (18.0) / 27.0	5.0 / 3.0	Bearings and bushings which will be used under heavy loads; truck roller bearings.

(a) Composition in per cent maximum, unless otherwise shown as a range. (b) Alloys comparatively rated from 1 to 5 in decreasing order of performance. Pressure tightness is the degree of soundness attainable in a casting as indicated by the absence of leakage, through its walls or sections, of air or other media under pressure. Elevated temperature strength is based on tests made after prolonged exposure at temperatures up to 500 F. (c) S, sand cast; P, permanent mold cast. (d) Yield strength, if determined, shown in parentheses.

(e) Consult individual foundries for tempers and properties of these alloys. (f) If iron exceeds 0.45%, manganese content shall not be less than ½ the iron content. (g) Also contains 0.04 to 0.07% Be. (h) Also contains 0.003 to 0.007 Be, 0.002 B maximum. (i) Also contains 5.5 to 7.0% Sn.

Source: Jobbing Foundry Div., Aluminum Assn, New York. Typical applications summarized from Society of Automotive Engineers, Aluminum Co. of America, and Alcan Aluminum Corp. publications.

Courtesy, Metal Progress, July 1974

TABLE 68-VII — Composition, Properties, and Applications of Magnesium Alloys

Form	Alloy	Al	Mn Min	Zn	Zr	Other (a)	Melting Point, F	Temper (b)	Tensile Strength, Psi (c)	Yield Strength, Psi (c)	Elongation, % (c)	Shear Strength, Psi (c)	Brinell	Rockwell E	Applications
Casting Alloys															
Sand and permanent mold castings	AM100A	10.0	0.10	—	—	—	1,101	F	20,000	10,000	—	18,000	53	64	Pressure-tight sand and permanent mold castings. Good combination of room-temperature properties.
								T4	34,000	10,000	6	20,000°	52	62	
								T6	34,000	15,000	2	—	69	80	
								T61	34,000	17,000	—	—	50	59	
	AZ63A	6.0	0.15	3.0	—	—	1,130	F	26,000	11,000	4	16,000	55	66	Sand castings requiring good room-temperature strength with ductility.
								T4	34,000	11,000	7	17,000	55	66	
								T5	34,000	16,000	3	17,000	73	83	
								T6	34,000	16,000	7	19,000	55	66	
	AZ81A	7.6	0.13	0.7	—	—	1,132	T4	34,000	11,000	7	17,000	60	66	Tough, leak-proof sand castings.
	AZ91C	8.7	0.13	0.7	—	—	1,105	F	23,000	11,000	2	—	55	62	Sand and permanent mold castings requiring good room temperature strength with ductility.
								T4	34,000	11,000	7	17,000	55	66	
								T6 (d)	34,000	12,000	2	—	70	77	
	AZ92A	9.0	0.10	2.0	—	—	1,100	F	23,000	11,000	3	19,000	65	76	Pressure-tight sand and permanent mold castings. Good combination of room-temperature properties.
								T4	34,000	11,000	6	18,000	63	75	
								T5	34,000	12,000	1	17,000	69	80	
								T6 (d)	34,000	18,000	2	20,000°	81	88	
	EZ33A	—	—	2.6	0.7	3.2 RE	1,189	F	20,000	14,000	4	20,000	50	59	Pressure-tight sand and permanent mold castings. For applications at 350-500 F.
								T5	20,000	14,000	4	19,000°	55	66	
	HK31A	—	—	—	0.7	3.2 Th	1,200	T6	27,000	13,000	14	21,000°	55	66	Sand castings used at 400 F and higher.
	HZ32A	—	—	2.1	0.7	3.2 Th	1,198	T5	27,000	13,000	4	—	—	—	Sand castings used at 400 F and higher.
	K1A	—	—	—	0.6	—	1,205	F	24,000	6,000	14	8,100°	—	—	Sand castings. Excellent damping capacity.
	QE22A	—	—	—	0.7	2.2 RE, 2.5 Ag	1,190	T6 (d)	35,000	25,000	4	—	78	86	Sand castings. Good room and elevated-temperature properties.
	ZE41A	—	—	4.2	0.7	1.2 RE	1,185	T5 (d)	29,000	19,000	2	22,000°	62	72	Sand castings. Good strength at room temperature. Improved castability over ZK alloys.
	ZH62A	—	—	5.7	0.7	1.8 Th	1,169	T5 (d)	35,000	22,000	5	—	70	77	Sand castings. Good strength at room temperature. Improved castability over ZK alloys.
	ZK51A	—	—	4.6	0.7	—	1,185	T5 (d)	34,000	22,000	5	22,000°	65	77	Sand castings. Good strength at room temperature.
	ZK61A	—	—	6.0	0.8	—	1,175	T6 (d)	40,000	26,000	5	—	—	—	Sand castings. Good strength and ductility at room temperature.
Die Castings (e)	AM60A	6.0	0.13	—	—	—	1,140	F	38,000*	20,000°	12°	—	—	—	Automobile wheels.
	AS41A	4.2	0.35°	—	—	1.0 Si	1,150	F	35,000°	20,000°	10°	—	—	—	Automobile engines and housings.
	AZ91A and AZ91B	9.0	0.13	0.7	—	—	1,105	F	39,000°	24,000°	6°	—	—	—	Die castings. Parts for cars, lawnmowers, luggage, business machines, tools.
Wrought Alloys															
Sheets and plates	AZ31B	3.0	0.20	1.0	—	—	1,160	O	32,000	15,000-18,000	9-12	17,000	56	67	General purpose sheet and plate alloy.
								H24	34,000-39,000	18,000-29,000	6-8	18,000	73	83	
	HK31A	6.5	0.15	1.0	—	3.2 Th	1,145	F.O	33,000-34,000	24,000-26,000	4	23,000	57	68	Reserve type batteries
	HM21A	—	0.80	—	0.7	2.0 Th	1,200 1,202	H24	30,000-33,000	18,000-21,000	6	20,000	—	—	Missile and aircraft skins; for other uses at 400 to 700 F.
								T8	34,000	25,000	4	—	—	—	Missile and aircraft uses up to 800 F.
								T81	—	—	—	—	—	—	
	LA141A	1.2	0.15	—	—	14.0 Li	1,075	T7	18,000-19,000	13,000-15,000	10	13,000	54	65	Formed parts requiring the lightest possible alloy.
Extruded bars, rods, shapes, tube, and wire	AZ21A	2.0	0.15 max	1.0	—	0.2 Ca	1,160	F	32,000-37,000	16,000-25,000	4.8	17,000	47	54	Anode material for primary batteries.
	AZ31B	3.0	0.20	1.0	—	—	1,160	F	36,000-40,000	16,000-24,000	7.9	19,000	55	66	General purpose extrusions.
	AZ61A	6.5	0.15	1.0	—	—	1,145	F	42,000-43,000	27,000-28,000	4.9	19,000	66	77	Extrusions requiring higher properties than AZ31B
	AZ80A	8.5	0.12	0.5	—	—	1,130	F	45,000-47,000	30,000-33,000	2.4	20,000	80	88	Highly stressed extrusions.
	HM31A	—	1.20	—	—	3.0 Th	1,202	T5	37,000	26,000	4	22,000°	75	84	Missile and aircraft use as high as 800 F.
	ZK60A	—	—	5.5	0.45	—	1,175	T5	40,000-43,000	28,000-31,000	5.6	22,000°	82	88	For highly stressed extrusions requiring better ductility than AZ80A
								T5	43,000-46,000	31,000-38,000	4.6	22,000	—	—	
Forgings	AZ31B	3.0	0.20	1.0	—	—	1,160	F	34,000	19,000	6	19,000°	50	59	For low stressed forgings.
	AZ61A	6.5	0.15	1.0	—	—	1,145	F	38,000	22,000	6	20,000	55	66	For forgings with higher strength than AZ31B
	AZ80A	8.5	0.12	0.5	—	—	1,130	T5	42,000	26,000	5	20,000	69	80	Heat-treatable forgings.
	HM21A	—	0.80	—	—	2.0 Th	1,202	T5	34,000-42,000	22,000-30,000	2	20,000°	72	82	Aircraft and missile forgings up to 800 F.
	ZK60A	—	—	5.5	0.45	—	1,175	T5	33,000	25,000	3	18,000°	66	66	For forgings of maximum strength in aircraft and military applications.
									38,000-42,000	20,000-26,000	7	24,000°	65	77	

(a) RE = rare earths (added as didymium in QE22A); (b) Temper: F = as fabricated; H24, H26 = strain-hardened then partially annealed; O = annealed; T4 = solution heat treated; T5 = artificially aged; T6, T61 = solution heat treated and artificially aged; T7 = solution heat treated and stabilized; T8, T81 = solution heat treated, cold worked, and artificially aged. (c) Properties of separately cast bars. All are minimums or ranges except those typical properties marked by asterisk (*) which spells out premium strength in designated areas of castings; (e) Tentative typical properties determined by four laboratories with separately cast ASTM ¼ in. diameter bars cast by NL Industries Inc. Properties may vary as more data are amassed from other producers. (d) These alloys are included in MIL-M-46062 which spells out premium strength in designated areas of castings;

Source: Dow Chemical Co.; Kaiser Magnesium, Div of Kaiser Aluminum & Chemical Corp.; Magnesium Research Center, Battelle Columbus; Magnesium Div., NL Industries Inc.

Courtesy, Metal Progress, June 1974.

TABLE 68-VIII — Compositions and Properties of Die Casting Alloys

Composition, % (b)

Aluminum Alloys (a)

SAE	ASTM B85	AA	Cu	Fe	Si	Mn	Mg	Zn	Ni	Sn	Other, Total	Al
305	S12A	A413.0	1.0	1.3	11.0-13.0	0.35	0.10	0.50	0.50	0.15	0.25	Bal
—	S12B	413.0	1.0	2.0	11.0-13.0	0.35	0.10	0.50	0.50	0.15	0.25	Bal
306	SC84A	A380.0	3.0-4.0	1.3	7.5-9.5	0.50	0.10	3.0	0.50	0.35	0.50	Bal
308	SC84B	380.0	3.0-4.0	2.0	7.5-9.5	0.50	0.10	3.0	0.50	0.35	0.50	Bal
309	SG100A	A360.0	0.6	1.3	9.0-10.0	0.35	0.40-0.60	0.50	0.50	0.15	0.25	Bal
—	SG100B	360.0	0.6	2.0	9.0-10.0	0.35	0.40-0.60	0.50	0.50	0.15	0.25	Bal
303	SC114A	384.0	3.0-4.5	1.3	10.5-12.0	0.50	0.10	3.0	0.50	0.35	0.50	Bal
—	G8A	518.0	0.25	1.8	0.35	0.35	7.5-8.5	0.15	0.15	0.15	0.25	Bal
—	SC102A	383.0	2.0-3.0	1.3	9.5-11.5	0.50	0.10	3.0	0.30	0.15	0.50	Bal
304	S5C	443.0	0.6	2.0	4.5-6.0	0.35	0.10	0.50	0.50	0.15	0.25	Bal

Magnesium Alloys

SAE	ASTM B94		Al	Mn	Zn	Si	Cu	Ni	Other, Total	Mg
501	AZ91A	—	8.3-9.7	(0.13)	0.35-1.0	0.50	0.10	0.03	0.3	Bal
501A	AZ91B	—	8.3-9.7	(0.13)	0.35-1.0	0.50	0.35	0.03	0.3	Bal
—	AM60A	—	5.5-6.5	(0.13)	0.22	0.50	0.35	0.03	0.3	Bal
—	AS41A	—	3.5-5.0	0.20-0.50	0.12	0.50-1.50	0.06	0.03	0.3	Bal

Zinc Alloys

SAE	ASTM B86		Cu	Al	Mg	Fe	Pb	Cd	Sn	Zn
903	AG40A	—	0.25	3.5-4.3	0.02-0.05	0.100	0.005	0.004	0.003	Bal
925	AC41A	—	0.75-1.25	3.5-4.3	0.03-0.08	0.100	0.005	0.004	0.003	Bal

Copper Alloys

SAE	ASTM B176	CDA	Si	Pb	Sn	Mn	Al	Fe	Mg	Other, Total(g)	Zn	Cu
—	879	879	0.75-1.25	0.25	0.25	0.15	0.15	0.15	—	0.50	Bal	63.0-67.0
—	878	878	3.75-4.25	0.15	0.25	0.15	0.15	0.15	0.01	0.25	Bal	80.0-83.0
—	858	858	0.25	1.50	1.50	0.25	0.25	0.50	—	0.50	(30.0)	(57.0)

Typical Mechanical Properties (c)

Aluminum Alloys (a)

SAE	ASTM B85	AA	Tensile Strength, Psi	Yield Strength, Psi	Elongation, %	Shear Strength, Psi	Charpy Impact Strength, Ft-Lb	Fatigue Strength, Psi	Hardness Brinell	Hardness Rockwell	Modulus of Elasticity, 10⁶ Psi
305	S12A	A413.0	42,000	19,000	3.5	25,000	—	19,000	80	—	10.3
—	S12B	413.0	43,000	21,000	2.5	25,000	—	19,000	80	—	—
306	SC84A	A380.0	47,000	23,000	3.5	27,000	3.5	20,000	80	—	10.3
308	SC84B	380.0	46,000	23,000	2.5	28,000	—	20,000	80	—	—
309	SG100A	A360.0	46,000	24,000	3.5	26,000	4.2	18,000	75	—	10.3
—	SG100B	360.0	44,000	25,000	2.5	28,000	—	20,000	75	—	—
303	SC114A	384.0	48,000	24,000	2.5	29,900	2.0	20,000	80	—	10.3
—	G8A	518.0	45,000	28,000	5.0	29,000	8.3	20,000	65	—	10.3
—	SC102A	383.0	45,000	22,000	3.5	—	—	—	—	—	—
304	S5C	443.0	33,000	14,000	9.0	19,000	4.5	17,000	50	—	10.3

Magnesium Alloys

SAE	ASTM B94		Tensile Strength, Psi	Yield Strength, Psi	Elongation, %	Shear Strength, Psi	Charpy Impact Strength, Ft-Lb	Fatigue Strength, Psi	Hardness Brinell	Hardness Rockwell	Modulus of Elasticity, 10⁶ Psi
501	AZ91A	—	34,000	23,000 / 22,000 (d)	3.0	20,000	2.0	14,000	63	E 75	6.5
501A	AZ91B	—	34,000	23,000 / 22,000 (d)	3.0	20,000	2.0	14,000	63	E 75	6.5
—	AM60A	—	30,000	17,000 / 17,000 (d)	6	—	—	—	—	—	—
—	AS41A	—	32,000	22,000 / 17,000 (d)	4	—	—	—	—	—	—

Zinc Alloys

SAE	ASTM B86		Tensile Strength, Psi	Yield Strength, Psi	Elongation, %	Shear Strength, Psi	Charpy Impact Strength, Ft-Lb	Fatigue Strength, Psi	Hardness Brinell	Hardness Rockwell	Modulus of Elasticity, 10⁶ Psi
903	AG40A	—	41,000	60,000 (d)	10.0	31,000	43.0	—	82	—	95,000 (e)
925	AC41A	—	47,000	87,000 (d)	7.0	38,000	48.0	—	91	—	105,000 (e)

Copper Alloys

SAE	ASTM B176	CDA	Tensile Strength, Psi	Yield Strength, Psi	Elongation, %	Shear Strength, Psi	Charpy Impact Strength, Ft-Lb	Fatigue Strength, Psi	Hardness Brinell	Hardness Rockwell	Modulus of Elasticity, 10⁶ Psi
—	879	879	70,000	35,000	25.0	—	50.0	—	120	B 68-72	15.0
—	878	878	85,000	50,000	25.0	—	70.0	—	160	B 85-90	20.0
—	858	858	55,000	30,000	15.0	—	40.0	—	—	B 55-60	15.0

Nominal Physical Properties

Aluminum Alloys (a)

SAE	ASTM B85	AA	Poisson's Ratio	Specific Gravity	Density, Lb/Cu In.	Melting Range, F	Electrical Conductivity, % IACS	Thermal Conductivity at 25 C, Cal/Sq Cm/Cm C Sec	Thermal Expansion, Micro-In./In./F 68-212 F	68-392 F	68-572 F	Solidification Shrinkage, In./Ft
305	S12A	A413.0	0.33	2.65	0.096	1,065-1,080	31	0.30	11.5	12.0	12.6	—
—	S12B	413.0	—	2.65	0.095	1,065-1,080	31	0.29	—	—	—	—
306	SC84A	A380.0	0.33	2.74	0.099	1,000-1,100	27	0.24	11.7	12.2	12.6	—
308	SC84B	380.0	—	2.74	0.099	1,000-1,100	23	0.23	—	—	—	—
309	SG100A	A360.0	0.33	2.64	0.095	1,035-1,105	37	0.29	11.8	12.4	12.8	—
—	SG100B	360.0	—	2.63	—	1,035-1,105	—	—	—	—	—	—
303	SC114A	384.0	0.33	2.74	0.099	960-1,080	23	0.23	11.3	11.8	12.3	—
—	G8A	518.0	0.33	2.57	0.093	995-1,150	24	0.24	13.3	14.0	14.3	—
—	SC102A	383.0	—	—	—	960-1,080	—	—	—	—	—	—
304	S5C	443.0	0.33	2.65	0.096	1,065-1,170	37	0.34	12.3	12.9	13.4	—

Magnesium Alloys

SAE	ASTM B94		Poisson's Ratio	Specific Gravity	Density, Lb/Cu In.	Melting Range, F	Electrical Conductivity, % IACS	Thermal Conductivity at 25 C	Thermal Expansion 68-212 F	68-392 F	68-572 F	Solidification Shrinkage, In./Ft
501	AZ91A	—	0.35	1.81	0.066	875-1,105	12	0.12	14.5	15.1	16.0 (65-750 F)	(f)
501A	AZ91B	—	0.35	1.81	0.066	875-1,105	12	0.12	14.5	15.1	16.0	(f)

Zinc Alloys

SAE	ASTM B86		Poisson's Ratio	Specific Gravity	Density, Lb/Cu In.	Melting Range, F	Electrical Conductivity, % IACS	Thermal Conductivity at 25 C	Thermal Expansion 68-212 F	68-392 F	68-572 F	Solidification Shrinkage, In./Ft
903	AG40A	—	—	.6.6	0.240	717-728	27	0.27	15.2	—	—	0.14
925	AC41A	—	—	6.7	0.240	717-727	26	0.26	15.2	—	—	0.14

Copper Alloys

SAE	ASTM B176	CDA	Poisson's Ratio	Specific Gravity	Density, Lb/Cu In.	Melting Range, F	Electrical Conductivity, % IACS	Thermal Conductivity at 25 C	Thermal Expansion 68-212 F	68-392 F	68-572 F	Solidification Shrinkage, In./Ft
—	879	879	—	8.5	0.308	1,650-1,700	15.0	—	—	—	—	0.187
—	878	878	—	8.3	0.300	1,510-1,680	6.7	16.0	—	—	10.9(h)	0.187
—	858	858	—	8.4	0.305	1,600-1,650	20.0	—	—	—	—	0.187

(a) AA = Aluminum Assn.; ASTM = American Society for Testing & Materials; CDA = Copper Development Assn.; SAE = Society for Automotive Engineers; other data supplied by American Die Casting Institute and Dow Chemical Co. (b) Maximum, except as noted. Minimum values are in parentheses. (c) Yield strength at 0.2% offset; elongation in 2 in.; fatigue strength at 500,000,000 cycles; Brinell hardness (500 kg load; 10 mm ball). Specimens for AM60A and AS41A data were from 0.080 in. thick walls of special test box castings; dimensions listed as acceptable to ASTM. (d) Compressive yield strength. (e) Modulus of rupture. (f) Specific heat: 0.25 Btu/lb/F for both magnesium alloys. (g) Of these, the amount of arsenic, antimony, and sulfur shall not exceed 0.5% for each, and phosphorus shall not exceed 0.01%. (h) To 500 F.

Courtesy, *Metal Progress*, June 1974.

TABLE 68-IX — Characteristics of Standard Plaster-Mold Casting Alloys

Alloy	Specifications	Tensile Strength (psi)	Hardness (Rockwell)	Characteristics	Applications
Yellow Brass No. 10	Similar Fed. Spec. QQB-726-Class D	55,000 to 70,000	55 to 75 B	High-strength brass, lowest in cost, good machining, high fatigue resistance.	Used for brackets, brush holders, frames, retainers, gears, levers, hardware.
Aluminum Bronze No. 20		75,000 to 85,000	56 to 80 G	High-strength, lighter than steel or other copper-base alloys, high wear and corrosion resistance.	Used for gears, sprockets, bearings, marine parts, dies, pump parts, universal joints.
Manganese Bronze No. 70	Fed. Spec. QQB-726-Class A Navy Spec. 49-B-3	65,000 to 80,000	60 to 80 B	High-strength bronze, good corrosion resistance, good ductility.	Used for levers, fasteners, valve bodies, gears and marine parts.
Manganese Bronze No. 100	Similar Navy Spec. 46-B-29 Class B	100,000 to 115,000	60 to 85 G	Super strength bronze, no heat treatment necessary, good ductility, good corrosion resistance.	Similar uses to No. 70 bronze where superior physical properties are desired.
Silicon-Aluminum-Bronze		59,000	65 to 75 B	High copper-silicon alloy, excellent corrosion resistance and wearing qualities.	For outdoor applications, electrical parts, connectors, hardware, etc.
Nickel Brass (15 and 22% nickel)		50,000	80 to 90 B	Good ductility and corrosion resistance, white satin finish.	Used for corrosion resistant parts, dairy and food processing machinery parts, lock parts. 22% nickel largely used in dairy & food equipment.
Aluminum Alloy No. 60	Alcoa No. 355 SAE No. 322	16,000 to 21,000 (as cast)	....	Casts in thin sections, no skin hardness, good for plating, can be age hardened to 34,000 psi.	Used for instrument and similar parts where highest strength is not required.
Aluminum Alloy No. 375	(Universal Castings Corp.)	25,000	70 to 80 E	High-strength alloy specially developed for plaster-mold casting, superior castability and physicals, good wear resistance.	Used where high strength and corrosion resistance is desired, for aircraft fittings, instrument parts, pump parts, pistons, cylinder heads, etc.

Table 68-X — Some Commercial Extrusions and their Characteristics

Alloy	Cu	Si	Mn	Mg	Zn	Ni	Cr	Pb	Al	Te	Fe	Sn	Other	Applications	Design	Extrudability
Copper	99.9													Electrical parts. Machinability 20%	Heavy sections, wide variety of shapes, min gage 1/8-in. av. Ext'd & drawn only	C
Copper (Free cutting)	Rem.									0.5				Electrical parts. Machinability 60%	Same as for copper	C
Yellow Brass	62.0				Rem.			1.5						For moderate to severe bending and flaring. Machinability 40%	Simple section. Min. gage 3/32-in. desirable. Ext'd & drawn only	D
Brass (Free cutting)	61.5				Rem.			3.0						Machinability 100% same as for SAE 72 brass	All types sections in gages down to 1/8-in. Ext'd and drawn only	B
Muntz Metal	60.0				Rem.									Moderate bends, good punching and brazing	Same as for yellow brass	C
Brass (Forging)	59.5				Rem.			2.0						Some bending with large radii. Good hot forming. Machinability 80%	Same as for free-cutting brass	B
Bronze (Architectural)	58.0				Rem.			3.0						Machines like free-cutting brass. No bending. Color match Muntz metal	Similar to free-cutting brass. Gages to 0.062-in. can be ext'd and drawn	B
Bronze (Architectural)	57.0				Rem.			2.25						Machinability 80%. No bending. Color match Muntz metal	Same as for arch. Bronze. Plain ext'd only	P
Bronze (Manganese)	58.5		0.1		Rem.						1.50	1.0		Good wear, high strength. Machinability 30%. Used in lieu of Naval brass for strength	Heavy, simple sections only. Min gage 3/16-in. Plain ext'd only	C
Naval Brass	60.0				Rem.							0.75		Little to moderate bending. Poor machinability 30%	Same as for free-cutting brass. Plain ext'd and drawn	C
Naval Brass (Leaded)	60.0				Rem.	13.0		1.75						Good strength. Machinability 70%	Same as for Naval brass	C
Nickel Silver	43.5				Rem.	13.0								Some bending, hard and strong. White color	Similar to free-cutting brass. Gages to 3/32-in. Plain ext'd only	E
Aluminum Alloy 3003			1.2						Rem.					Good welding and corrosion resistance. Not heat treatable. Strength only fair—lowest of aluminum alloys	Minimum section 0.040-in. Good for complicated and hollow parts. Hollow shapes in sizes up to those circumscribed by 8-in. circle. Used in "as ext'd" condition	A — Extrusion speed up to 300 fpm

Alloy	Composition (per cent)	Characteristics	Minimum section*	Extrudability	Extrusion speed
Aluminum Alloy 2014	4.4 / 0.8 / 0.8 / 0.4 / .. / .. / .. / Rem. / .. / .. / ..	High mechanical properties and hardness. Used where strength of 75S is not required. Heat treatable. Used for machine members	Minimum section ⅛-in.	A	
Aluminum Alloy 2017	4.0 / .. / 0.5 / 0.5 / .. / .. / .. / Rem. / .. / .. / ..	Not ordinarily extruded. Used where machinability is a prime consideration, 200 to 1000%. Usually supplanted by 14S	Minimum section ⅛-in.	B	
Aluminum Alloy 2024	4.5 / .. / 0.6 / 1.5 / .. / .. / .. / Rem. / .. / .. / ..	Strength not quite as good as 14S. Used for machine members	Better for thin or heavy sections than 14S. Min section 0.040-in.	C	10 to 14 fpm
Aluminum Alloy 6061	0.25 / 0.6 / .. / 1.0 / .. / 0.25 / .. / Rem. / .. / .. / ..	Excellent forming qualities. High yield strength and corrosion resistance. Strength not quite as good as 24S	For complicated and hollow shapes. Hollow shapes in sizes up to those circumscribed by 8-in. circle. Min section thickness 0.040-in.	B	
Aluminum Alloy 6063	.. / .. / .. / 0.7 / 0.4 / .. / .. / Rem. / .. / .. / 0.3	Excellent working qualities. Strength fair. Lowest priced heat-treatable aluminum alloy. Good welding and corrosion resistance. Poor machinability.	For complicated and hollow shapes. Hollow shapes in sizes up to those circumscribed by 8-in. circle. Min section thickness 0.040-in.	A	
Aluminum 7075	1.6 / .. / .. / 2.5 / 5.6 / .. / 0.3 / Rem. / .. / .. / ..	Highest strength aluminum alloy	Minimum section thickness 0.050-in.	C	2 to 5 fpm
Magnesium Alloy AZ31B-F	0.05 / 0.3 / 0.2 / Rem. / 0.7-1.3 / 2.5-3.5 / 0.005 / .. / 0.005 / .. / 0.3	Best elongation and cold formability. High machinability, 350 to 1500%.‡	All types of complicated and hollow sections. Minimum section thickness 0.035-in.	B	15 to 40 fpm
Magnesium Alloy AZ61A-F	0.05 / 0.3 / 0.15 / Rem. / 0.4-1.5 / 5.8-7.2 / 0.005 / .. / 0.005 / .. / 0.3	General-purpose extrusion alloy. Moderate cost and better strength than ASTM AZ31X.‡	Same as for AZ31	C	7 to 20 fpm
Magnesium Alloy M1A	0.05 / 0.3 / 1.2 / Rem. / .. / .. / 0.01 / .. / .. / .. / 0.3	Moderate-strength alloy. Best weldability, hot formability, corrosion resistance, and lowest cost. Lightest of the magnesium alloys.‡	Same as for AZ31	A	20 to 100 fpm
Magnesium Alloy AZ80A-F	0.05 / 0.3 / 0.15 / Rem. / 0.2-0.8 / 7.8-9.2 / 0.005 / .. / 0.005 / .. / 0.3	Highest strength in extruded condition. Can be aged after extrusion to improve strength.‡	Solid shapes only	C	4 to 7 fpm

* Minimum section thickness will vary with size of extrusion. Minimum section normally can be obtained with shapes which do not exceed 2 inches in longest dimension. In larger shapes minimum section is greater, reaching about ¼-inch with the largest shapes. Only materials so noted are adaptable for extruding hollow shapes. Machinability ratings are based on free-cutting brass. Extrudability is designated according to the relative ease with which the particular metal can be extruded. ‡ Machinability of all magnesium alloys varies from 350 to 1500% depending upon the operation, design characteristics, tooling, etc.

TABLE 68-X1 — Extruded Magnesium Alloys and Forms Available

Extruded Form	AZ31B, C	AZ61A	AZ80A	HM31A	ZK60A
Rod, Bar, Solid Shapes	X	X	X	X	X
Structural Shapes	X	X	X	X	X
Wire	X	X	X	—	—
Tube	X	X	—	—	X
Hollow, Semi-hollow Shapes	X	X	—	—	X
Pipe	X	X	—	—	X
Precision Extrusions	X	X	X	X	X

Nominal Composition and Physical Properties of Magnesium Extrusion Alloys

Alloy and Temper	Nominal Composition—%						Density lb./cu. in. −70°F	Melting Point —°F	Thermal Conductivity cgs units −70°F	Electrical Resistivity microhm - cm. −70°F
	Al	Mn, min.	Th	Zn	Zr	Mg				
AZ31B, or C-F	3.0	—	—	1.0	—	Bal.	0.0642	1160	0.18	9.2
AZ61A-F	6.5	—	—	1.0	—	Bal.	0.0649	1140	0.14	12.5
AZ80A-F	8.5	—	—	0.5	—	Bal.	0.0651	1130	0.12	14.5
-T5	8.5	—	—	0.5	—	Bal.	0.0651	1130	0.12	14.5
HM31A-T5	—	1.2	3.0	—	—	Bal.	0.0652	1202	0.25	6.6
ZK60A-F	—	—	—	5.7	0.5	Bal.	0.0660	1175	0.28	6.0
-T5	—	—	—	5.7	0.5	Bal.	0.0660	1175	0.29	5.7

Courtesy, Dow Chemical Co.

TABLE 68-XII — Steels used for Hot Extrusion

| | Commonly used grades | | | | | | | Typical mechanical properties | | | | | | | |
| | CHEMICAL COMPOSITION | | | | | | | AS EXTRUDED—CONDITION° | | | | | HEAT TREATED CONDITION | | |
GRADE**	CR	NI	C	MN	SI	MO	OTHER	TENSILE PSI	YIELD PSI	ELONGATION %	REDUCTION IN AREA %	HARDNESS	TENSILE PSI	YIELD PSI	HARDNESS
C-1018†			.15-.20	.60-.90				55,000	35,000	30	50	126 BHN			
C-1020†			.18-.23	.30-.60				55,000	35,000	30	50	126 BHN			
C-1040			.37-.44	.60-.90				75,000	50,000	30	48	170 BHN	100,000	70,000	217 BHN
C°-1050			.48-.55	.60-.90				85,000	52,000	28	45	179 BHN	120,000	80,000	241 BHN
C-1095			.90-1.03	.30-.50				95,000	55,000	22	38	192 BHN	150,000	95,000	311 BHN
C°1117†			.14-.20	1.0-1.3			S .08-.13	60,000	40,000	32	54	124 BHN			
C-1213†			.13 Max	.70-1.0	.10 Max		P .07-.12 S .24-.33	52,000	35,000	28	48	118 BHN			
4023†			.20-.25	.70-.90	.20-.35	.20-.30		65,000	35,000	28	48	156 BHN			
4130	.80-1.10		.28-.33	.40-.60	.20-.35	.15-.25		70,000	40,000	28	48	159 BHN	100,000/160,000	70,000/140,000	207/331 BHN
4140	.80-1.10		.38-.43	.75-1.0	.20-.35	.15-.25		85,000	55,000	26	54	179 BHN	100,000/170,000	70,000/150,000	207/341 BHN
4320†	.40-.60	1.65-2.0	.17-.22	.45-.65	.20-.35	.20-.30		80,000	60,000	30	54	174 BHN			
4620†	.40-.60	1.65-2.0	.17-.22	.45-.65	.20-.35	.20-.30		78,000	56,000	30	48	170 BHN			
5120†	.70-.90		.17-.22	.70-.90	.20-.35			70,000	40,000	28	48	159 BHN	90,000/150,000	65,000/130,000	200/311 BHN
5140	.70-.90		.38-.43	.70-.90	.20-.35			75,000	50,000	30	48	170 BHN			
8620†	.40-.60	.40-.70	.18-.23	.70-.90	.20-.35	.15-.25		78,000	55,000	32	60	170 BHN			
8640	.40-.60	.40-.70	.38-.43	.75-1.0	.20-.35	.15-.25		90,000	60,000	25	48	192 BHN	100,000/180,000	70,000/160,000	207/375 BHN
8750	.40-.60	.40-.70	.48-.53	.75-1.0	.20-.35	.20-.30		100,000	65,000	23	44	212 BHN	120,000/180,000	90,000/160,000	235/375 BHN
52100	1.30-1.60		.95-1.10	.25-.45	.20-.35			89,000	51,000	31	60	180 BHN	126,000/205,000	112,000/187,000	262/415 BHN
302†	17-19	8-10	.15 Max	2. Max	1. Max			80,000	35,000	50	70	146 BHN			
303 Se‡	17-19	8-10	.15 Max	2. Max	1. Max		Se 0.15 Min	80,000	35,000	50	70	146 BHN			
304†	18-20	8-12	.08 Max	2. Max	1. Max			80,000	35,000	50	70	146 BHN			
304 L	18-20	8-12	.03 Max	2. Max	1. Max			80,000	35,000	50	70	146 BHN			
309	22-24	12-15	.20 Max	2. Max	1. Max			80,000	35,000	50	70	146 BHN			
309 S	22-24	12-15	.08 Max	2. Max	1. Max			80,000	35,000	50	70	146 BHN			
310	24-26	19-22	.25 Max	1.5 Max	1.5 Max			80,000	35,000	50	70	146 BHN			
316	16-18	10-14	.08 Max	2. Max	1. Max	2-3		80,000	35,000	50	70	146 BHN			
316 L	16-18	10-14	.03 Max	2. Max	1. Max	2-3		80,000	35,000	50	70	146 BHN			
321	17-19	9-12	.08 Max	2. Max	1. Max		Ti 5 x C Min	80,000	35,000	50	70	146 BHN			
347	17-19	9-13	.08 Max	2. Max	1. Max		Cb-Ta 10 x C Min	80,000	35,000	50	70	146 BHN			
403	11.5-13		.15 Max	1. Max	.50 Max			95,000	55,000	30	70	192 BHN	100,000/150,000	70,000/135,000	207/311 BHN
405	11.5-14.5		.08 Max	1. Max	1. Max		Al 0.10-0.30	80,000	50,000	28	65	170 BHN	Non-Hardenable Grades		
410	11.5-13.5		.15 Max	1. Max	1. Max			95,000	55,000	30	70	192 BHN	100,000/150,000	70,000/135,000	207/311 BHN
416 Se	12-14		.15 Max	1.25 Max	1. Max		Se 0.15 Min	95,000	55,000	30	70	192 BHN	100,000/150,000	70,000/135,000	207/311 BHN
430	14-18		.12 Max	1. Max	1. Max			75,000	40,000	28	48	146 BHN			

†Carburizing Grades—Would be heat treated by fabricator. ‡Austenitic Stainless—Hardenable only by work hardening.
°These values represent typical properties in the annealed condition for the hardenable grades (the condition in which they are shipped unless otherwise agreed upon).
**Other grades available upon request.

Courtesy, Babcock & Wilcox Company.

Table 68-XIII — Cold Heading Materials

C1008-C1010 Steel — Purchased as killed steel with closely controlled chemical properties. Used for parts requiring heavy amounts of extrusion or upsets and/or parts to be staked or riveted.

C1015-C1018 Steel — Used for machine screws and wood screws, and specials where material is specified as low carbon and is not heat treated. The guaranteed minimum tensile strength is 55,000 psi. Low carbon steel is also used for sheet metal and thread-cutting screws where no mechanical properties are specified by the customer, but a case hardened surface is required.

C1019-C1022 Steel — Purchased as killed steel with closely controlled chemical limits and grain size for good combination of formability and reaction to heat treatment. Used primarily for sems and thread cutting or forming screws requiring closely controlled mechanical properties.

C1038 Steel — This grade is considered a medium carbon steel and is used on Grade 5 bolts and special parts requiring a minimum tensile strength of 120,000 psi. Close control of surface finish, chemistry and hardenability.

C1062 Steel — A high carbon grade with limited cold workability, it is used in applications where extremely high strength and high hardness (50 Rc and above) are required. Close control of surface finish, chemistry and hardenability.

4037-H Alloy Steel — This grade is in the medium carbon, low alloy group. Parts made of 4037-H have a high yield to tensile strength ratio and very consistent properties. Used for Grade 7 and Grade 8 bolts and special applications requiring 133,000 psi minimum tensile strength. Close control of surface finish, chemistry and hardenability.

CDA 260 Brass — A 70/30 alloy of copper and zinc, CDA 260 brass is a tough, rustless, non-magnetic material possessing mechanical properties similar to low carbon steel. It can be easily formed into intricate shapes and is relatively inexpensive compared to other copper base alloys.

Copper — Copper in various grades such as Electrolytic Tough Pitch, Oxygen Free and Tellurium Bearing is used extensively in architectural, automotive, electrical, and hardware applications. It can be easily formed by many methods and has relatively high corrosion resistance.

Silicon Bronze — High tensile strength superior to mild steel. It has good corrosion resistance to salt and fresh water, atmospheric conditions and a variety of industrial gases and chemicals. It can be readily cold worked and is non-magnetic and non-sparking.

T305 Stainless Steel (18-8) — Has excellent corrosion resistance, heat resistance and high mechanical properties, but has a fairly high rate of work hardening. This enables parts only with rather modest upsets to be made. This type has slightly better corrosion resistance than 302HQ but is higher in cost.

302HQ Stainless Steel (18-8) — A modified austenitic stainless with a 3/4% copper addition to lower the cold working tendencies of the conventional 305. Its uses include parts that resist corrosion from a wide variety of organic and inorganic chemicals and foodstuffs. Corrosion resistance compares favorably with 305.

T410 Stainless Steel — The most widely used of all hardenable types of corrosion resisting steels. It is capable of attaining high mechanical properties when heat treated. For maximum corrosion resistance, T410 should be used in the hardened condition. Since it is also ductile and workable, it is particularly suited to cold-heading. Not recommended for applications in which severe corrosion is encountered.

T430 Stainless Steel — A ductile, non-hardening stainless steel exhibiting mechanical properties higher than mild steel with good resistance to general corrosion and oxidation at temperatures up to 1400° F. All forms are magnetic.

Aluminum — While there are many grades of aluminum used, the most common for us is 2024. It actually meets the tensile strength of mild steel, at only 1/3 the weight. 2024 is fabricated in the H-13 temper but heat treated to T-4 temper in finished stage.

Nickel #400 (Monel) — Monel is approximately 2/3 Nickel and 1/3 copper with a small amount of iron. Monel makes an excellent fastener material because of its high strength, resistance to heat and corrosion and good cold workability. Monel has wide use in chemical, electrical, marine and construction applications.

ALUMINUM ALLOYS

GRADE	1100	2024 - T4	6061 - T6
Composition	Al 99.0 Min	Cu 3.8 - 4.9 Mn 0.3 - 0.9 Mg 1.2 - 1.8 Al Balance	Mg 0.8 - 1.2 Si 0.4 - 0.8 Cu 0.15 - 0.40 Cr 0.15 - 0.35 Al Balance
PHYSICAL PROPERTIES			
Density lb/cu in	0.098	0.100	0.098
Melting Temperature Range, °F	1190 - 1215	935 - 1180	1080 - 1200
Ther Cond (77F, ann) Btu/hr/sq ft/°F/ft	128	111	99
Coef of Therm Exp (68-212F) per °F	13.1 x 10⁻⁶	12.6 x 10⁻⁶	13.0 x 10⁻⁶
Spec Heat (212F) Btu/lb/°F	0.22	.22	0.23
Elec Res (68F) microhm — cm annealed	2.92	3.45 (0), 5.75 (T4)	3.8 (0)
MECHANICAL PROPERTIES			
Tensile Strength	13,000	55,000 (T4)	45,000 (T6)
Yield Strength (.2% Elong)	5,000	50,000 (T4)	40,000 (T6)
Hardness (Rockwell)	F20-25	B53-64	B60

NICKEL ALLOYS

GRADE	NICKEL 400 MONEL	INCONEL 600
Composition	Ni 66 Fe 1.35 Cu 31.5	Ni 77.0 Cr 15.0 Fe 7.0
PHYSICAL PROPERTIES		
Density lb/cu in	.319	0.304
Melting Temp Range, °F	2370 - 2460	2540 - 2600
Ther Cond (80-212) BTU/hr/sq ft/°F/ft	15	8.7
Coef of Ther Exp (77-212), per °F	7.8 x 10⁻⁶	6.4 x 10⁻⁶
Spec Ht (70-750F) Btu/lb/°F	0.127	0.109
Elec Res (68F) microhm-cm	482	98.1
Magnetic	Slightly Magnetic	Non-Magnetic
MECHANICAL PROPERTIES		
Tensile Strength	80,000 Min	80,000 Min
Yield Strength (0.2% offset)	60,000 Min	45,000 Min
Hardness	B90	B72 Min

STEEL

GRADE	0	1	2	3	5	8
Material ▶	1008 1010	1015 1018	1019 1022	1035 1038	1035 1038 1040 1045 1062	8635 8640 4140 4037
PHYSICAL PROPERTIES						
Density lb/cu in	0.283	0.283	0.283	0.283	0.283	0.283
Melting Temp Range, °F	2750 - 2775	2750 - 2775	2750 - 2775	2700 - 2750	2700 - 2750	—
Coef of Ther Exp (70 - 1200 F), per °F	8.4×10^{-6}	8.4×10^{-6}	8.4×10^{-6}	8.3×10^{-6}	8.3×10^{-6}	8.3×10^{-6}
Spec Ht Btu/lb/°F	0.10 - 0.11	0.10 - 0.11	0.10 - 0.11	0.10 - 0.11	0.10 - 0.11	0.10 - 0.11
Elec Res (68 F) microhm-cm	14.3	14.3	14.3	19	19	19
Magnetic	Yes	Yes	Yes	Yes	Yes	Yes
MECHANICAL PROPERTIES						
Tensile Strength	No Req.	55,000	69,000	110,000	120,000	150,000
Yield Strength	—	—	55,000	85,000	85,000	120,000
Brinell Hardness	—	207 Max	241 Max	207 - 269	241 - 302	302 - 352
Rockwell Hardness	—	B95 Max	B100 Max	B95 - 104	C23 - 32	C32 - 38
Other Requirements	—	—	—	Stress Relieved	Min. Tempering Temp. 800°F	Min. Tempering Temp. 800°F

COPPER ALLOYS

GRADE ▶	CDA110 COPPER (ELECTROLYTIC TOUGH PITCH)	CDA270 YELLOW BRASS	CDA260 CARTRIDGE BRASS	CDA220 COMMERCIAL BRONZE	CDA230 RED BRASS
PHYSICAL PROPERTIES Composition ▶	Cu 99.9 min. O₂ .04	Cu 63-68.5 Zn Bal	Cu 68.5-71.5 Zn Bal	Cu 89 -91 Zn Bal	Cu 84-86 Zn Bal
Density lb/cu in	.321 - .323	.306	.308	.318	.316
Melting Temp Range °F	1949 - 1981	1660 - 1710	1680 - 1750	1870 - 1910	1810 - 1880
Ther Cond (77°F, ann) Btu/hr/sq ft/°F/ft	226	67	70	109	92
Coef of Ther Exp (68-572F) per °F	9.8×10^{-6}	11.3×10^{-6}	11.1×10^{-6}	10.2×10^{-6}	10.4×10^{-6}
Spec Heat (68°F) Btu/lb/°F	0.092	0.09	0.09	0.09	0.09
Elec Res (68°F) microhm-cm	1.71	6.4	6.2	3.9	2.8
MECHANICAL PROPERTIES					
Tensile Strength	35,000 - 45,000	60,000 Min	60,000 Min	45,000 Min	50,000 Min
Yield Strength (.5% Elong)	10,000 - 35,000	40,000 Min	40,000 Min	40,000 Min	35,000 Min
Rockwell Hardness	F40 - 80	B55 - 65	B55 - 65	B40 - 50	B40 - 55

COPPER ALLOYS

GRADE ▶	CDA651 LOW SILICON BRONZE B	CDA220 PHOSPHOR BRONZE A
PHYSICAL PROPERTIES Composition ▶	Cu 96 Min Si .8 - 2.0 Zn 1.5 Max	Cu 95 Sn 3.5 - 5.8 P .03 - .35
Density lb/cu in	.316	.320
Melting Temp Range °F	1890 - 1940	1750 - 1920
Ther Cond (77°F, ann) Btu/hr/sq ft/°F/ft	33	47
Coef of Ther Exp (68-572F) per °F	9.9×10^{-6}	9.9×10^{-6}
Spec Heat (68°F) Btu/lb/°F	.09	.09
Elec Res (68°F) microhm-cm	14.4	9.6
MECHANICAL PROPERTIES		
Tensile Strength	70,000 Min	70,000 Min
Yield Strength (.5% Elong)	35,000 Min	40,000 Min
Rockwell Hardness	B65- 75	B70 - 80

STAINLESS STEEL

GRADE	305	302HQ	316	410	430
Approximate Chemical Composition ▶	C 0.12 max Mn 2.00 max Si 1.00 max Cr 17.00 - 19.00 Ni 10.00 - 13.00	C 0.08 max Mn 2.00 max Si 1.00 max Cr 17.00 - 19.00 Ni 8.00 - 10.00 Cu 3.00 - 4.00	C 0.08 max Mn 2.00 max Si 1.00 max Cr 16.00 - 18.00 Ni 10.00 - 14.00 Mo 2.00 - 3.00	C 0.15 max Mn 1.00 max Si 1.00 max Cr 11.50 - 13.50	C 0.12 max Mn 1.00 max Si 1.00 max Cr 14.00 - 18.00
PHYSICAL PROPERTIES					
Density lb/cu in	0.29	0.29	0.29	0.28	0.28
Mean coefficient of thermal expansion per °F x 10⁻⁶ 32-212°F	9.6	9.6	8.9	5.5	5.8
32-600°F	9.9	9.9	9.0	5.6	6.1
32-1000°F	10.2	—	9.7	6.4	6.3
32-1200°F	10.4	10.4	10.3	6.5	6.6
Melting point range	2550 - 2650°F			2700 - 2790°F	2600 - 2750°F
Magnetism (*In fully annealed condition)	*Non-magnetic	*Non-magnetic	*Non-magnetic	Magnetic	Magnetic
MECHANICAL PROPERTIES					
Brinell, Hardness Annealed	130 - 150	120 - 140	130 - 150	135 - 165	140 - 160
Ultimate, Tensile Strength, PSI					
Annealed	80,000 - 95,000	60,000 - 90,000	80,000 - 100,000	65,000 - 85,000	70,000 - 90,000
Cold Worked	100,000 - 180,000	80,000 - 160,000	100,000 - 180,000	100,000	90,000 - 130,000
Yield, point, PSI					
Annealed	35,000 - 45,000	25,000 - 45,000	35,000 - 50,000	35,000 - 45,000	40,000 - 55,000
Cold Worked	50,000 - 150,000	45,000 - 120,000	50,000 - 150,000	85,000	65,000 - 130,000
Elongation in 2 inches					
Annealed	60 - 55%	60 - 70%	40 - 50%	30 - 20%	30 - 20%
Heat Resistance					
Scaling temperature					
Continuous Service	1600°F	1600°F	1650°F	1250°F	1500°F
Intermittent Service	1450°F	1450°F	1500°F	1400°F	1600°F
COLD-HEADING PROPERTIES	Fair to Good	Good	Fair	Good	Good
Corrosion Resistance	Very good resistance to atmosphere. Good resistance to many organic and inorganic chemicals.	Similar to 305	More resistant to sulfuric, acetic and tartaric acids in chloride, bromide and iodide solutions. Generally better than 302-305	In both the hardened and unhardened condition it has very good resistance to atmosphere, water, carbonic acid, crude oil, gasoline, blood, perspiration, alcohol, ammonia, mercury, soap, sugar solutions and many other reagents.	Excellent resistance to weather, water, good resistance to most chemicals and food products.

TABLE 68-XIV — Typical Mechanical Properties of Standard P/M Materials**

IRON BASE

APM Material Designation	Density gm-cc	Nominal Composition	Ultimate Tensile Strength (PSI)	Elongation %	Transverse Rupture Strength (PSI)	Apparent Hardness Rockwell (Typical)	ASTM No.	ASTM Type	ASTM Grade	ASTM Class	SAE	MIL-B 5087-C Type	MIL-B 5087-C Comp	MPIF No. 35-68
F-1000	5.6-6.0 / 6.0-6.4 / 6.4-6.8 / 7.2-7.6	Fe-100 / C-3 Max.	16,000 / 25,000 / 30,000 / 40,000	2 / 3 / 6 / 11	32,000 / 50,000 / 60,000 / 80,000	F30 / F40 / F50 / F60 / F80	310	I / II / III / IV	—	A / A / A / A	850 / 853	II	A1	F-0000-N / F-0000-P / F-0000-R / F-0000-S / F-0000-T
F-1000 (Sulphur)	5.6-6.0 / 6.0-6.4 / 6.8-7.2	Fe-99 / S-2/.5	21,000 / 27,000 / 38,000	—	42,000 / 54,000 / 58,000 / 74,000	F35 / F50 / F65 / F75	—	—	—	—	—	—	—	—
FC-9010	5.8-6.2	Fe-90, Cu-10	30,000	.5	60,000	F70	222 & 439	—	3	—	862	II	B	FC-0000-N
FG-1000	5.6-6.0 / 6.0-6.4 / 6.4-6.8 / 6.8-7.2	Fe-99, C-1	29,000 / 35,000 / 44,000 / 60,000	0-.5 / 1.0 / 2.0	52,000 / 68,000 / 88,000 / 108,000	B35 / B50 / B65 / B80	310 / 310 / 310	I / II / III / IV	—	C / C / C / C	852 / 855	II	A3	FC-0008-N / FC-0008-P / FC-0008-R / FC-0008-S
FG-1000 (Hdn. & Tempered)	5.6-6.0 / 6.0-6.4 / 6.4-6.8 / 6.8-7.2	—	40,000 / 50,000 / 65,000 / 80,000	0-.5 / 0-1.0 / 1.0 / 2.0	70,000 / 80,000 / 100,000 / 130,000	C20 / C35 / C40 / C50	—	—	—	—	—	—	—	—
FGS-9902	5.6-6.0 / 6.0-6.4 / 6.4-6.8	Fe-97, Cu-2 / C-1	33,000 / 45,000 / 60,000	0-.5 / 1.0 / 2.0	66,000 / 90,000 / 120,000	B45 / B60 / B70	426	I / II	—	A / A	864A / 864B	—	—	FC-0208-N / FC-0208-P / FC-0208-R
FGS-9902 (Cu2)	6.0-6.4 / 6.8-7.2	Fe-97, Cu-2 / C-1.5,3	48,000 / 75,000	—	96,000 / 150,000	B55 / B75	426	—	—	—	—	—	—	—
FG-9505 (Hdn. & Tempered)	5.6-6.0 / 6.0-6.4 / 6.4-6.8	Fe-97 / Cu-2 / C-1	40,000 / 60,000 / 90,000	0.5 / 0.5	—	C20 / C34 / C41	—	—	—	—	—	—	—	—
FCS-9505	5.6-6.0 / 6.0-6.4 / 6.4-6.8	Fe-94 / Cu-5 / C-1	45,000 / 60,000 / 70,000	0-.3 / 0-1.0	90,000 / 105,000 / 135,000	B60 / B75 / B85	426 / 426	I / II	—	C / C	866A / 866B	—	—	FC-0508-N / FC-0508-P / FC-0508-R
FCS-9505 (Hdn. & Tempered)	5.6-6.0 / 6.4-6.8	Fe-94 / Cu-5 / C-1	68,000 / 90,000	0-.3 / 0-.5	—	C25 / C35 / C41	—	—	—	—	—	—	—	—
FCS-9007	5.6-6.0 / 6.0-6.4 / 6.4-6.8	Fe-92 / Cu-7 / C-1	36,000 / 55,000 / 65,000	0-.5 / 1.5	90,000 / 135,000 / 150,000	B60 / B75 / B80	426 / 426	I / II	—	C / C	866A / 866B	—	—	FC-0708-N / FC-0708-P / FC-0708-R
FCS-9007 (Hdn. & Tempered)	5.6-6.0	Fe-52, Cu-7 / C-1	45,000 / 55,000	0-.5	—	C30	—	—	—	—	—	—	—	—
FN-9902	6.0-6.4 / 6.4-6.8 / 6.8-7.2 / 7.2-7.6	Fe-98 / N-2	28,000 / 38,000 / 45,000	4.0 / 7.0 / 10.5	56,000 / 76,000	B38 / B42 / B51	484 / 484	I / II / III	—	A / A	—	—	—	FN-0200-P / FN-0200-R / FN-0200-T
FN-9902 + 5G	6.0-6.4 / 6.4-6.8 / 7.2-7.6	Fe-97.5 / N-2 / C-5	37,000 / 50,000 / 61,000	3.0 / 3.5 / 4.5	70,000 / 100,000 / 110,000	B57 / B74 / B85	484 / 484	I / II / III	—	B / B	—	—	—	FN-0205-P / FN-0205-R / FN-0205-S / FN-0205-T
FN-9902 + 5G (Hdn. & Tempered)	6.4-6.8 / 6.8-7.2 / 7.2-7.6	Fe-97.5 / N-2 / C-5	80,000 / 110,000 / 134,000	0-.5 / 1.0 / 2.0	120,000 / 140,000 / 160,000	C32 / C42 / C46	—	—	—	—	—	—	—	—
FNG-9902	6.0-6.4 / 6.4-6.8 / 6.8-7.2 / 7.2-7.6	Fe-97 / N-2 / C-1	33,000 / 48,000 / 65,000 / 79,000	2.0 / 3.5	66,000 / 80,000 / 110,000 / 130,000	B45 / B62 / B79 / B87	484 / 484	I / II / III	—	C / C	—	—	—	FN-0208-P / FN-0208-R / FN-0208-S / FN-0208-T
FNG-9902 (Hdn. & Tempered)	6.0-6.4 / 6.4-6.8 / 6.8-7.2 / 7.2-7.6	Fe-97 / N-2 / C-1	100,000 / 135,000 / 160,000	0.5 / 0.5	130,000 / 144,000 / 180,000	C34 / C45 / C47	484 / 484	I / II	—	C / C	—	—	—	FN-0208-P / FN-0208-R / FN-0208-S / FN-0208-T

APM Material Designation	Density gm-cc	Nominal Composition	Ultimate Tensile Strength (PSI)	Elongation %	Transverse Rupture Strength (PSI)	Apparent Hardness Rockwell (Typical)	ASTM No.	ASTM Type	ASTM Grade	ASTM Class	SAE	MIL-B 5087-C Type	MIL-B 5087-C Comp	MPIF No. 35-68
FNG-9604	6.0-6.4 / 6.4-6.8 / 6.8-7.2 / 7.2-7.6	Fe-95, N-4 / C-1	57,000 / 77,000 / 93,000	1.5 / 3.0 / 4.5	90,000 / 120,000 / 140,000	B72 / B88 / B95	—	—	—	—	—	—	—	FN-0405-P / FN-0405-R / FN-0405-S / FN-0405-T
FNCG-9541	6.4-6.8 / 6.8-7.2 / 7.2-7.6	Fe-94, N-4 / Cu-1 / C-1	80,000 / 95,000	1.5 / 3.0 / 4.0	105,000 / 145,000 / 180,000	B75 / B85 / B95-	—	—	—	—	—	—	—	—
FNCG-9541 (Hdn. & Tempered)	6.4-6.8 / 6.8-7.2 / 7.2-7.6	Fe-94, N-4 / Cu-1 / C-1	95,000 / 150,000	1.0 / 1.0	—	C30 / C45 / C47	—	—	—	—	—	—	—	—
4640 Steel	6.8-7.2 / 7.2-7.6	Fe-Bal., Mo-.5 / C-.45	75,000 / 80,000	3.0 / 3.5	90,000 / 115,000 / 125,000	B60 / B75 / B80	—	—	—	B	—	—	—	—
4640 Steel (Hdn. & Tempered)	6.4-6.8 / 6.8-7.2 / 7.2-7.6	Fe-Bal., Mo-.5 / C-.45	150,000	0 - 3	125,000 / 180,000 / 225,000	C22 / C30 / C35	—	—	—	—	—	—	—	—

IRON BASE INFILTRATED

APM Material Designation	Density gm-cc	Nominal Composition	Ultimate Tensile Strength (PSI)	Elongation %	Transverse Rupture Strength (PSI)	Apparent Hardness Rockwell (Typical)	ASTM No.	ASTM Type	ASTM Grade	ASTM Class	SAE	MIL-B 5087-C Type	MIL-B 5087-C Comp	MPIF No. 35-68
FG-1000 + infilt.	7.2-7.6	Fe-80, Cu-Bal.	65,000	0 - 1.0	105,000	B60	303	—	—	A	870	—	—	FX-2008-T
FG-1000 + infilt.	7.2-7.6	Fe-80, C-1 / Cu-Bal.	85,000	0 - 1.0	170,000	B80	303	—	—	C	872	—	—	FX-2008-T
FG-1000 + infilt. (Hdn. Tempered)	7.2-7.6	Fe-80, C-1 / Cu-Bal.	125,000	0 - .5	180,000	Rc40	—	—	—	—	—	—	—	—

STAINLESS STEELS

APM Material Designation	Density gm-cc	Nominal Composition	Ultimate Tensile Strength (PSI)	Elongation %	Transverse Rupture Strength (PSI)	Apparent Hardness Rockwell (Typical)	ASTM No.	ASTM Type	ASTM Grade	ASTM Class	SAE	MIL-B 5087-C Type	MIL-B 5087-C Comp	MPIF No. 35-68
303L-SS	6.0-6.4 / 6.4-6.8	C-.18, N-9 / S-.15, Mo-.6	35,000 / 52,000	1.0 / 2.0	85,000 / 125,000	B50 / B65	—	—	—	—	—	—	—	SS-303-P / SS-303-R
316L-SS	6.0-6.4 / 6.4-6.8	Cr-17, N-12 / Mo-2.5, Cu-2 max / Fe-Bal.	38,000 / 54,000	2.0 / 4.0	85,000 / 125,000	B50 / B65	—	—	—	—	—	—	—	SS-316-P / SS-316-R
410-SS (Hdn. As-sint.)	5.6-6.0 / 6.0-6.4 / 6.4*-6.8	C-.12, S / Mn-1, Fe-Bal.	40,000 / 55,000 / 80,000	0-1.0 / 0-1.0 / 0-1.0	80,000 / 110,000 / 160,000	B90 / B95 / B98	—	—	—	—	—	—	—	SS-410-H / SS-410-P / SS-410-R

BRONZE

APM Material Designation	Density gm-cc	Nominal Composition	Ultimate Tensile Strength (PSI)	Elongation %	Transverse Rupture Strength (PSI)	Apparent Hardness Rockwell (Typical)	ASTM No.	ASTM Type	ASTM Grade	ASTM Class	SAE	MIL-B 5087-C Type	MIL-B 5087-C Comp	MPIF No. 35-68
CS-9010	5.6-6.0 / 6.0-6.4 / 6.4-6.8 / 6.8-7.2	Cu-90 / Sn-10	8,000 / 10,000 / 14,000 / 18,000	1.0 / 1.0 / 2.0 / 2.5	15,000 / 20,000 / 28,000 / 36,000	H40 / H52 / H35 / H40	438 / 438 / 255	—	I / II / III	—	840 / 841 / 842	—	A	CT-0010-N / CT-0010-P / CT-0010-R / CT-0010-S
CFS-65251	6.0-6.4 / 6.4-6.8	Cu-65 / Zn-25, Zn-6, Sn-1	11,000 / 14,000	—	22,000 / 24,000	H35 / H40	—	—	—	—	—	—	—	—
FCS-6040	5.6-6.0 / 6.0-6.4	Fe-60 / Cu-36, Sn-4, C-1	13,000 / 21,000	—	26,000 / 42,000	H45 / H49	—	—	—	—	—	—	—	—

BRASS

APM Material Designation	Density gm-cc	Nominal Composition	Ultimate Tensile Strength (PSI)	Elongation %	Transverse Rupture Strength (PSI)	Apparent Hardness Rockwell (Typical)	ASTM No.	ASTM Type	ASTM Grade	ASTM Class	SAE	MIL-B 5087-C Type	MIL-B 5087-C Comp	MPIF No. 35-68
C2-8020	7.2-7.6 / 7.6-8.0 / 8.0-8.4	Cu-78.5 / Zn-20 / Pb-1.5	24,000 / 28,000 / 32,000	13 / 19 / 22	40,000 / 50,000 / 70,000	H55 / H68 / H75	282 / 282	—	—	A / B	890	—	—	CZP-0218-T / CZP-0218-U / CZP-0218-W
C2-9010	7.2-7.6 / 7.6-8.0	Zn-10, Cu-88.5 / Pb-1.5	10,000 / 15,000	—	20,000 / 30,000	H40 / H60	—	—	—	—	—	—	—	CZP-0210-T / CZP-0210-U

NICKEL SILVER

APM Material Designation	Density gm-cc	Nominal Composition	Ultimate Tensile Strength (PSI)	Elongation %	Transverse Rupture Strength (PSI)	Apparent Hardness Rockwell (Typical)	ASTM No.	ASTM Type	ASTM Grade	ASTM Class	SAE	MIL-B 5087-C Type	MIL-B 5087-C Comp	MPIF No. 35-68
CZN-18	7.2-7.6 / 7.6-8.0 / 8.0-8.4	Cu-64, N-18 / Zn-Bal.	25,000 / 30,000 / 37,000	9 / 10 / 12	—	H73 / H79 / H85	458	—	—	—	—	—	—	CZN-1818-T / CZN-1818-U / CZN-1818-W

Presinter, repress, hardened as-sintered.

Courtesy American Powdered Metals.

Table 68-XV — Properties of Some Ceramics †

Material	Water Absorption (%)	Spec. Gr.	Density (lb per cu in.)	Standard Colors	Alternative Colors	Safe Temp. (F)	Hardness (Moh)	Tensile Strength (psi)	Comp. Strength (psi)	Flexural Strength (psi)	Properties and Applications
Vitreous Ceramics											
Commercial Porcelain[1]	.5 to 2	..	...	White	Green, Brown, etc.	2300	..	3,000 to 3,500	30,000 to 40,000	.....	Good quality and most economical type of porcelain for use in most applications.
Steatite AlSiMag 35	0 to .02 impervious	2.5	.090	White	206 Gray 207 Brown	1832	7.5	8,500	75,000	18,000	High-strength insulators with good electrical characteristics, made to close tolerances for appliances, etc.
Steatite AlSiMag 196	0 to .02 impervious	2.6	.094	White	....	1832	7.5	10,000	85,000	20,000	Lower dielectric losses than AlSiMag 35. High-strength insulators made to accurate dimensions for r-f circuits.
Steatite AlSiMag 228	0 to .02 impervious	2.7	.098	White	....	1832	7.5	10,000	85,000	20,000	Excellent mechanical properties with lower electrical losses than AlSiMag 196. For use in high frequency equipment.
Steatite AlSiMag 197	.02 to 1 Vitrified to Tech. Vitreous	2.6	094	White	209 Gray 210 Brown	1832	7.5	8,500	75,000	20,000	Mechanically strong with high electrical resistance at elevated temperatures. For heater wire supports in appliances.
Forsterite AlSiMag 243	0 to .02 impervious	2.8	.101	Buff	....	1832	7.5	10,000	85,000	20,000	For service where very low loss insulators are required. For ceramic to metal seals. Close tolerances by grinding.
Titanium Dioxide AlSiMag 192	0 to .02 impervious	4.0	.144	Tan	....	1832	8	7,500	80,000	20,000	Ceramic for polished thread guides. AlSiMag 193 is the same mechanically but made conductor to dissipate static.
Zircon AlSiMag 475	0 to .02 impervious	3.7	.134	White	....	2012	8	12,000	90,000	22,000	Vitrified ceramic with excellent electrical properties, high strength and good resistance to thermal shock.
Alumina AlSiMag 491[2]	0 to .02 impervious	3.5	.125	Blue	....	2552	9	.....	100,000	45,000	Fine texture, extremely hard, resistant to abrasion, chipping and corrosion. Best adapted to small machined or pressed parts.
Alumina AlSiMag 513[2]	0 to .02 impervious	3.4	.122	Gray	....	2462	9	.....	100,000	45,000	High-strength alumina ceramic for somewhat larger size articles than possible in 491, by machining or pressing.

TABLE 68-XV — Properties of Some Ceramics (Continued)

Lavas											
Aluminum Silicate Grade A Lava	2 to 3 Technically Vitreous	2.3	.085	Pink	....	2012	6	2,500	20,000	9,000	Good electrical and heat resistance, excellent for close tolerances. Available in larger sizes than other lavas.
Magnesium Silicate Lava 1136	2 to 3 Technically Vitreous	2.8	.102	Tan	White	2282	6	.....	25,000	9,000	Excellent electrical properties, made to close tolerances. For insulating spacers in small size vacuum tubes.
Refractories											
Cordierite AlSiMag 202	[3]	2.1	.076	[3]	[3]	2282	7	3,500	40,000	8,000	Low coefficient of expansion, high resistance to heat shock. For heating elements, thermocouple insulators, burner tips.
Zirconium Oxide AlSiMag 550[2]	6 to 10 Semi-vitreous	3.8	.136	Orange	....	2912	..	.....	30,000	10,000	Fine grain, high mechanical strength refractory for simple extruded shapes. Has excellent resistance to heat shock.
Zircon AlSiMag 504	8 to 14 Highly Porous	2.9	.105	White	.....	2462	..	.....	16,000	13,000	Very fine grain refractory with good resistance to heat shock. Suitable for firing plates and supports, brazing jigs, etc.
Alumina AlSiMag 393	12 to 18 Highly Porous	2.4	.087	White	.....	2552	..	.....	11,000	10,000	Easily regassed refractory with good electrical resistance at high temperature. Porous body principally used for vacuum tube insulators.
Magnesium Silicate AlSiMag 222[2]	14 to 18 Highly Porous	2.0	.072	Light Brown	....	2372	6	2,500	25,000	8,000	Refractory which is machinable in the fired state. Sold in rods, tubes, plates and blocks. For models or repair parts.
Zirconium Oxide AlSiMag 508[2]	12 to 16 Highly Porous	2.9	.106	Orange	....	2912	..	.....	11,000	4,000	Refractory body excellent for heat shock. Made only in plates, rods and simple shapes by pressing or casting.
Metal-Ceramics											
Metamic[4]	..		0.22	....	....	3100	35[5]	17,500[6]	.....	38,000	Relatively new. Used for flame nozzles, burner tips, pinions for high temperatures, turbine blades, etc.

† Courtesy American Lava Corp.
1 Courtesy Star Porcelain Co.
2 Economic limitations on sizes and shapes.
3 Furnished in various densities: 202, porous, 10 to 15% (tan), 178, porous, 10 to 15% (white), 547, semi-vitreous, 2 to 7% (brown)
4 Courtesy Haynes Stellite Div.
5 Rockwell C.
6 At 1800 F, 3410 psi at 2400 F.

TABLE 68-XVI — Relative Properties of Natural and Synthetic Rubber

	Natural Rubber	GR-S	Butyl	Butadiene or Isoprene	Ethylene Propylene	Nitrile	Neoprene	Urethane	Thiokol	Silicone	Fluoro-Silicone	Fluoro-Elastomer	Poly-Acrylic
ASTM Symbol	NR	SBR	IIR	BR or IR	EP	NBR	CR						
Durometer (Shore A)	30-90	40-90	20-80	30-90	40-80	40-95	20-90	50-90	50-80	10-90	40-80	60-90	50-90
Tensile Strength (PSI)	3000	3000	2000	3000	2500	2500	3000	5000	1000	1400	900	2000	2000
Tear Resistance	E	VG	VG	E	G	G	VG	E	F	NR	NR	G	G
Abrasion Resistance	E	E	G	E	G	G	G	VG	F	NR	NR	G	VG
Fire Resistant Fluids	NR	NR	VG	NR	VG	NR	F	NR	F	G	VG	F to E	NR
Lubricating Oils	NR	NR	NR	NR	NR	VG	G	VG	E	G	E	E	VG
Fuel Oils	NR	NR	NR	NR	NR	VG	F	VG	E	F	E	E	VG
Hydraulic Oils	NR	NR	NR	NR	NR	E	G	VG	E	G	E	E	E
Vegetable Oils	NR	F	E	F	F	E	G	VG	F	G	E	E	E
Animal Fats	NR	F	E	F	F	E	G	VG	F	G	E	E	E
Gasoline (Reg.)	NR	NR	NR	NR	NR	VG	F	VG	E	F	E	E	VG
Gasoline (Hi Octane)	NR	NR	NR	NR	NR	G	NR	VG	E	NR	VG	E	G
Jet Fuel	NR	NR	NR	NR	NR	VG	F	VG	E	F	E	E	VG
Aromatic Hydrocarbons	NR	NR	NR	NR	NR	F	NR	G	E	NR	E	E	F
Aliphatic Hydrocarbons	NR	NR	NR	NR	NR	VG	G	VG	E	F	E	E	VG
Water (Below 180°F.)	G	VG	E	G	E	VG	F	F	F	E	E	E	NR
Water (Above 180°F.)	NR	F	G	NR	E	F	NR	NR	NR	E	VG	G	NR
Alcohols	G	G	E	G	E	VG	E	F	VG	G	NR	NR	NR
Ketones	NR	NR	VG	NR	G	NR	NR	NR	NR	F	G	VG	NR
Acids (Conc.)	NR	F	G	NR	F	F	NR	NR	F	G	VG	G	NR
Acids (Dilute)	F	G	E	F	G	F	G	F	NR	G	F	E	NR
Alkalies	G	NR	E	G	VG	F	G	NR	VG	NR	E	G	NR
Chlorinated Solvents	NR	NR	NR	NR	NR	F	NR	G	VG	NR	F	E	F
Ozone and Sunlight	F	F	F	F	E	F	VG	VG	VG	E	E	E	VG
Heat (Max. Cont. Use)	160°F.	220°F.	250°F.	180°F.	275°F.	220°F.	220°F.	160°F.	150°F.	500°F.	400°F.	450°F.	350°F.
Electrical Properties	E	VG	E	E	E	VG	VG	NR	E	E	VG	F	NR
Compression Set	VG	VG	G	VG	G	VG	VG	NR	NR	E	VG	F	G
Flame Resistance	No	No	No	No	No	No	Yes	No	No	No	No	Yes	No

KEY: E = Excellent, VG = Very Good, G = Good, F = Fair, NR = Not Recommended

Courtesy, Minnesota Rubber Co.

TABLE 68-XVII — Specific Physical Properties of Electroformed Metals*

	Units	Silver	Copper	Iron	Nickel
Density	lb./cu. in.	.379	.32	.284	.322
Specific Gravity, 20°/4°			8.9	7.87	8.908
Melting Point†	°F °C	1761 961	1980 1080	2795 1535	2650 1455
Specific Heat (32–212°F) (0–100°C)	B.t.u./lb./°F Cal./g./°C	.0559 .0559	.092 .092	.108 .108	.11 .11
Linear Coef. of Expansion (32–212°F) (0–100°C)	10^{-6} in./in./°F 10^{-6} cm./cm./°C	10.4 18.8	9.2 16.5	6.7 12.0	7.5 13.5
Thermal Conductivity (32–212°F) (0–100°C)	B.t.u./sq.ft./hr./°F./in. Cal/sq.cm./sec./°C./cm.	2960 1.0	2700 .93	470 .16	470 .16
Specific Resistivity (20°C.)	microhm-cm.	1.59	1.7	9.9	7.0
Temp. Coef. of Sp. Resistivity (0–100°C)	(per °C)	.0041	.004	.005	.006
Modulus of Elasticity in Tension	10^6 lb./sq.in.	.10	16	30	28
Crystal Structure		F.C.C.	F.C.C.	B.C.C.	F.C.C.
Lattice Constant (A_0)	A.U. = 10^{-8} cm.	4.086	3.608	2.861	3.517

Courtesy, Gar Electroforming Div., Mite Corp.

TABLE 68XVIII — **Joinability of Coppers and Copper Alloys***

Former Commonly Accepted Trade Name	Copper or Copper Alloy No.	Soft soldering	Brazing	Oxyacetylene welding	GMAW or GTAW	SMAW
Electrolytic tough pitch copper (ETP)	110	E	G	NR	F	NR
Deoxidized copper, high residual phosphorus (DHP)	122	E	E	G	E	NR
Oxygen-free copper (OF)	102	E	E	F	G	NR
Silver-bearing tough pitch coppers ......113, 114, (STP)	116	E	G	NR	F	NR
Gilding, 95%	210	E	E	G	G	NR
Commercial bronze, 90%	220	E	E	G	G	NR
Jewelry bronze, 87.5%	226	E	E	G	G	NR
Red brass, 85%	230	E	E	G	G	NR
Low brass, 80%	240	E	E	G	G	NR
Cartridge brass, 70%	260	E	E	G	F	NR
Yellow brass 268, 270		E	E	G	F	NR
Muntz metal	280	E	E	G	F	NR
Leaded commercial bronze	314	E	G	NR	NR	NR
Low-leaded brass (tube)	330	E	G	NR	NR	NR
Low-leaded brass	335	E	G	NR	NR	NR
Medium-leaded brass	340	E	G	NR	NR	NR
High-leaded brass (tube)	332	E	G	NR	NR	NR
High-leaded brass 342, 353		E	G	NR	NR	NR
Extra-high-leaded brass	356	E	G	NR	NR	NR
Free-cutting brass	360	E	G	NR	NR	NR
Leaded muntz metal 365-368		E	G	NR	NR	NR
Free-cutting muntz metal	370	E	G	NR	NR	NR
Forging brass	377	E	G	NR	NR	NR
Architectural bronze	385	E	G	NR	NR	NR
Inhibited admiralty 443-445		E	E	G	G	NR
Naval brass	464	E	E	G	F	NR
Leaded naval brass	485	E	G	NR	NR	NR
Manganese bronze, A	675	E	E	G	F	NR
Aluminum brass	687	F	G	NR	F	NR
Phosphor bronze, 5% A	510	E	F	F	G	F
Phosphor bronze, 8% C	521	E	E	F	G	F
Phosphor bronze, 10% D	524	E	E	F	G	F
Phosphor bronze, 1.25% E	505	E	E	F	G	F
Free-cutting phosphor bronze	544	E	G	NR	NR	NR
Copper nickel, 30%	715	E	E	F	F	F
Copper nickel, 10%	706	E	E	F	E	G
Nickel silver, 65-18	752	E	E	G	F	NR
Nickel silver, 55-18	770	E	E	G	F	NR
Nickel silver, 65-15	754	E	E	G	F	NR
Nickel silver, 65-12	757	E	E	G	F	NR
Nickel silver, 65-10	745	E	E	E	F	NR
High-silicon bronze, A	655	G	E	G	E	F
Low-silicon bronze, B	651	E	E	G	E	F
Aluminum bronze, D 613, 614		NR	F	NR	E	G

*E, excellent; G, good; F, fair; NR, not recommended.

TABLE 68-XIX — Guide for Selecting Minimum Thickness for Quality Coatings

Plated Zinc and Cadmium Coatings

Service Conditions	Zinc on Iron and Steel Parts				Cadmium on Iron and Steel Parts			
	Min Thickness, Mils*	Chromate Finish	Salt Spray Hr to White Corrosion	Typical Applications	Min Thickness, Mils*	Chromate Finish	Salt Spray Hr to White Corrosion	Typical Applications
Mild — Indoor atmospheres involving infrequent condensation and minimum wear or abrasion.	0.2	None Clear Iridescent yellow Olive drab	— 12-24 24-72 72-100	Screws, nuts and bolts, buttons, wire goods, fasteners.	0.2	None Clear Iridescent yellow Olive drab	— 24 24-72 72-100	Springs, lock washers, fasteners, tools, electronic and electrical parts.
Moderate — Mostly dry indoor atmospheres, plus occasional condensation, wear, or abrasion.	0.3	None Clear Iridescent yellow Olive drab	— 12-24 24-72 72-100	Tools, zipper pulls, shelves, machine parts.	0.3	None Clear Iridescent yellow Olive drab	— 24 24-72 72-100	TV and radio chassis, threaded parts, screws, bolts, radio parts, instruments.
Severe — Exposure to condensation, perspiration, infrequent wetting by rain, cleaners.	0.5	None Clear Iridescent yellow Olive drab	— 12-24 24-72 72-100	Tubular furniture, insect screens, window fittings, builders' hardware, military hardware, washing machine parts, bicycle parts.	0.5	None Clear Iridescent yellow Olive drab	— 24 24-72 72-100	Washing machine parts, military hardware, electronic parts for tropical service.
Very severe† — Exposure to bold atmospheric conditions; frequent exposure to moisture, cleaners, and saline solutions; likely damage by denting, scratching or abrasive wear.	1.0	None	—	Plumbing fixtures, pole line hardware.	1.0	None Clear Iridescent yellow Olive drab	— 24 24-72 72-100	

Decorative Bright Nickel-Chromium Plated Finishes

Service Conditions	Steel and Iron Parts, Zinc Die Castings‡					Copper and Copper Alloy Parts				
	Classification §		Min Thickness, Mils		Typical Applications	Classification §		Min Thickness, Mils		Typical Applications
	Ni	Cr	Ni	Cr		Ni	Cr	Ni	Cr	
Mild — Normally warm, dry indoor atmospheres; coatings subject to minimum wear or abrasion.	10b 10p 10d 10b 10p 10d 10b 10p 10d	r r r mc mc mc mp mp mp	0.4 0.4 0.4 0.4 0.4 0.4 0.4 0.4 0.4	0.005 0.005 0.005 0.03 0.03 0.03 0.01 0.01 0.01	Toaster bodies, rotisseries, waffle bakers, oven doors and liners; interior auto hardware; trim for major appliances, hair driers, fans; inexpensive utensils; coat and luggage racks; standing ash trays; interior trash receptacles; inexpensive light fixtures.	5b 5p 5d 5b 5p 5d 5b 5p 5d	r r r mc mc mc mp mp mp	0.2 0.2 0.2 0.2 0.2 0.2 0.2 0.2 0.2	0.005 0.005 0.005 0.03 0.03 0.03 0.01 0.01 0.01	Toaster bodies and similar appliances, oven doors and liners; interior auto hardware; trim for major appliances; receptacles; light fixtures.
Moderate — Indoor exposure where condensation of moisture may occur, as in kitchens and bathrooms.	20b 20p 20d 15b 15p 15d 15b 15p 15d	r r r mc mc mc mp mp mp	0.8 0.8 0.8 0.6 0.6 0.6 0.6 0.6 0.6	0.01 0.01 0.01 0.03 0.03 0.03 0.01 0.01 0.01	Steel and iron: stove tops; oven liners; home, office, and school furniture; bar stools, golf club shafts. Zinc alloys: bathroom accessories, cabinet hardware.	15b 15p 15d 10b 10p 10d 10b 10p 10d	r r r mc mc mc mp mp mp	0.6 0.6 0.6 0.4 0.4 0.4 0.4 0.4 0.4	0.01 0.01 0.01 0.03 0.03 0.03 0.01 0.01 0.01	Plumbing fixtures, bathroom accessories; hinges; light fixtures, flashlights, and spotlights.
Severe — Occasional or frequent wetting by rain or dew or possibly strong cleaners and saline solution.	30d 25d 25d 40p 30p 30d	r mc mp r mc mp	1.2 1.0 1.0 1.6 1.2 1.2	0.01 0.03 0.01 0.01 0.03 0.01	Patio, porch, and lawn furniture; bicycles; scooters; wagons; hospital furniture; fixtures; and cabinets.	25d 20d 20d 25p 20p 20p 30b 25b 25b	r mc mp r mc mp r mc mp	1.0 0.8 0.8 1.0 0.8 0.8 1.2 1.0 1.0	0.01 0.03 0.01 0.01 0.03 0.01 0.01 0.03 0.01	Patio, porch, and lawn furniture; light fixtures; bicycle parts; hospital and laboratory fixtures.
Very severe — Service includes likely damage from denting, scratching, or abrasive wear in addition to corrosive media.	40d 30d 30d	r mc mp	1.6 1.2 1.2	0.01 0.01 0.01	Auto bumpers, grills, hub caps, and lower body trim; light housings.	30d 25d 25d	r mc mp	1.2 1.0 1.0	0.01 0.01 0.01	Boat fittings; auto trim, hub caps, lower body trim.

* Thickness specified is after chromate treatment, if employed.

† There are some applications for heavy electroplated coatings, but for very severe service they are most usually satisfied by hot dipped or sprayed coatings.

‡ 0.2 mil copper or yellow brass required on zinc and zinc alloys prior to nickel-chromium.

§ The classification numbers shown in the table indicate the thickness in microns and the type of deposit to be provided in the nickel and chromium layers of the coating. The number indicating the thickness of total nickel in microns is approximately the same as the minimum thickness in mils shown in column under "Min Thickness".
The type of nickel is designated as follows:
b — nickel deposited in the fully bright condition.

p — dull or semibright nickel relatively free from sulfur requiring buffing to give full brightness.
d — a double-layer or triple-layer nickel coating in which the bottom layer contains less than 0.005% S and the top layer more than 0.04% S. If there are three layers, the intermediate layer shall contain more sulfur than the top layer.
The type of chromium is designated as follows:
r — regular (conventional) chromium.
mc — microcracked chromium having more than 750 cracks per inch in any direction over significant surfaces.
mp — microporous chromium containing a minimum of 64,500 pores per square inch which are invisible to the unaided eye.
Copper may be used as an undercoat for the nickel, but it cannot be substituted for any part of

the nickel thickness specified. On zinc or zinc alloy parts, copper is customarily applied to a thickness of 0.2 to 0.3 mil minimum prior to nickel plating.
By combining semibright and bright leveling nickel and chromium, it is possible to produce a coating that provides a superior finish plus wear and corrosion resistance. The system, called double-layer nickel-chromium plating, consists of a combination of a layer of bright nickel over a heavier layer of semibright nickel, as an undercoating for a final flash of chromium. The result is a highly lustrous finish. Recommended applications include auto trim, porch and lawn furniture, bicycles, and other items that require a bright durable finish in severe outdoor exposure.

Source: "MFSA Quality Metal Finishing Guide — Zinc and Cadmium Coatings" and "MFSA Quality Metal Finishing Guide — Copper-Nickel-Chromium and Nickel-Chromium Coatings," Metal Finishing Suppliers' Assn. Inc.

TABLE 68-XX — Coating Methods for Metallic Substrates

Coating Methods for Metallic Substrates

Usual Coating Method	Ag	Al	Be	Cb	Cr	Cu	Fe	Mg	Mo	Ni	Pb	Sn	Ta	Th	Ti	U	W	Zn	Zr
									Metal or Alloy Being Coated										
Anodizing	No	Yes	Yes	No	No	No	No	Yes	No	No	No	No	No	No	Yes	No	No	Yes	No
Cementation	No	Yes	Yes	Yes	Yes	Yes	Yes	No	Yes	Yes	No	No	Yes	Yes	Yes	Yes	Yes	No	Yes
Chemical conversion	No	Yes	Yes	No	No	No	Yes	Yes	No	No	No	Yes	No	Yes	Yes	Yes	No	Yes	No
Chemical deposition	Yes	Yes	Yes	No	Yes	Yes	Yes	Yes	No	Yes	No	No	No	No	Yes	Yes	No	Yes	Yes
Chemical displacement	No	Yes	Yes	No	No	Yes	Yes	Yes	No	No	No	No	No	Yes	No	No	No	Yes	Yes
Diffusion coating	No	Yes	Yes	Yes	No	No	Yes	No	Yes	Yes	No	No	Yes	Yes	No	Yes	Yes	No	No
Electrodeposition	Yes	Yes	Yes	Yes	Yes	Yes	Yes	Yes	Yes	Yes	Yes	Yes	Yes	Yes	Yes	Yes	Yes	Yes	Yes
Hot dipping	No	No	No	Yes	No	Yes	Yes	No	No	Yes	Yes	No	Yes	No	No	Yes	No	No	No
Mechanical plating	No	No	No	No	No	No	Yes	No	No	No	No	No	No	No	No	No	No	No	No
Metallurgical cladding	No	No	Yes	Yes	Yes	Yes	Yes	No	Yes	Yes	No	No	Yes	Yes	Yes	Yes	No	No	No
Spray metallizing	No	No	Yes	Yes	No	No	Yes	No	Yes	Yes	No	No	Yes	No	No	No	Yes	No	No
Vacuum metallizing	No	No	No	No	No	Yes	Yes	No	No	Yes	No	No	No	Yes	No	Yes	No	No	No
Vapor deposition	No	No	No	Yes	No	No	Yes	No	Yes	Yes	No	No	Yes	Yes	No	Yes	Yes	Yes	No
Vitreous slurry	Yes	Yes	No	Yes	Yes	Yes	Yes	No	Yes	Yes	No	No	Yes	Yes	No	Yes	Yes	No	No

Courtesy, Machine Design, February 14, 1974.

69

Tables

MEASURES and WEIGHTS

UNIT	EQUIVALENTS IN OTHER UNITS OF SAME SYSTEM	METRIC EQUIVALENT
● LENGTH ●		
mile	5280 feet, 320 rods, 1760 yards	1.609 kilometers
rod	5.50 yards, 16.5 feet	5.029 meters
yard	3 feet, 36 inches	0.914 meters
foot	12 inches, 0.333 yards	30.480 centimeters
inch	0.083 feet, 0.027 yards	2.540 centimeters
● AREA ●		
square mile	640 acres, 102,400 square rods	2.590 square kilometers
acre	4840 square yards, 43,560 square feet	0.405 hectares, 4047 square meters
square rod	30.25 square yards, 0.006 acres	25.293 square meters
square yard	1296 square inches, 9 square feet	0.836 square meters
square foot	144 square inches, 0.111 square yards	0.093 square meters
square inch	0.007 square feet, 0.00077 square yards	6.451 square centimeters
● VOLUME ●		
cubic yard	27 cubic feet, 46,656 cubic inches	0.765 cubic meters
cubic foot	1728 cubic inches, 0.0370 cubic yards	0.028 cubic meters
cubic inch	0.00058 cubic feet, 0.000021 cubic yards	16.387 cubic centimeters
● WEIGHT ●		
ton – short ton	20 short hundredweight, 2000 pounds	0.907 metric tons
long ton	20 long hundredweight, 2240 pounds	1.016 metric tons
hundredweight		
short hundredweight	100 pounds, 0.05 short tons	45.359 kilograms
long hundredweight	112 pounds, 0.05 long tons	50.802 kilograms
pound	16 ounces, 7000 grains	0.453 kilograms
ounce	16 drams, 437.5 grains	28.349 grams
dram	27.343 grains, 0.0625 ounces	1.771 grams
grain	0.036 drams, 0.002285 ounces	0.0648 grams
● CAPACITY ●		
U. S. liquid measure		
gallon	4 quarts (231 cubic inches)	3.785 liters
quart	2 pints (57.75 cubic inches)	0.946 liters
pint	4 gills (28.875 cubic inches)	0.473 liters
gill	4 fluidounces (7.218 cubic inches)	118.291 milliliters
fluidounce	8 fluidrams (1.804 cubic inches)	29.573 milliliters
fluidram	60 minims (0.225 cubic inches)	3.696 milliliters
minim	1/60 fluidram (0.003759 cubic inches)	0.061610 milliliters
cup	1/2 pint (8 oz.)	
tablespoon	1/2 ounce	
teaspoon	1/6 ounce	
U. S. dry measure		
bushel	4 pecks (2150.42 cubic inches)	35.238 liters
peck	8 quarts (537.605 cubic inches)	8.809 liters
quart	2 pints (67.200 cubic inches)	1.101 liters
pint	1/2 quart (33.600 cubic inches)	0.550 liters

Decimal and Metric Equivalents

1″ = 25.4 millimeters   **1 millimeter = .03937″**

FRACTIONAL INCH	DECIMAL INCH	MILLIMETERS	FRACTIONAL INCH	DECIMAL INCH	MILLIMETERS
1/64	.0156	0.3968	33/64	.5156	13.0966
1/32	.0312	0.7937	17/32	.5312	13.4934
3/64	.0468	1.1906	35/64	.5468	13.8903
1/16	.0625	1.5874	9/16	.5625	14.2872
5/64	.0781	1.9843	37/64	.5781	14.6841
3/32	.0937	2.3812	19/32	.5937	15.0809
7/64	.1093	2.7780	39/64	.6093	15.4778
1/8	.125	3.1749	5/8	.625	15.8747
9/64	.1406	3.5718	41/64	.6406	16.2715
5/32	.1562	3.9686	21/32	.6562	16.6684
11/64	.1718	4.3655	43/64	.6718	17.0653
3/16	.1875	4.7624	11/16	.6875	17.4621
13/64	.2031	5.1592	45/64	.7031	17.8590
7/32	.2187	5.5561	23/32	.7187	18.2559
15/64	.2343	5.9530	47/64	.7343	18.6527
1/4	.250	6.3498	3/4	.750	19.0496
17/64	.2656	6.7467	49/64	.7656	19.4465
9/32	.2812	7.1436	25/32	.7812	19.8433
19/64	.2968	7.5404	51/64	.7968	20.2402
5/16	.3125	7.9373	13/16	.8125	20.6371
21/64	.3281	8.3342	53/64	.8281	21.0339
11/32	.3437	8.7310	27/32	.8437	21.4308
23/64	.3593	9.1279	55/64	.8593	21.8277
3/8	.375	9.5248	7/8	.875	22.2245
25/64	.3906	9.9216	57/64	.8906	22.6214
13/32	.4062	10.3185	29/32	.9062	23.0183
27/64	.4218	10.7154	59/64	.9218	23.4151
7/16	.4375	11.1122	15/16	.9375	23.8120
29/64	.4531	11.5091	61/64	.9531	24.2089
15/32	.4687	11.9060	31/32	.9687	24.6057
31/64	.4843	12.3029	63/64	.9843	25.0026
1/2	.500	12.6997	1	1.000	25.3995

Standard Drawing Symbols

ITEM	AMERICAN	SYMBOLS
ROUNDNESS	ROUND WITHIN .OOI ON R OR ROUND WITHIN .OO2 ON DIA.	⊂ O .OOI ⊃
TRUE POSITION DIMENSIONING	BASIC	⊕
TRUE POSITION	LOCATED AT TRUE POSITION WITHIN .OIO DIA. OR LOCATED WITHIN .OO5 R OF TRUE POSITION	⊕ .OIO
TRUE POSITION WITH DATUM	LOCATED AT TRUE POSITION WITHIN .OIO DIA. REGARDLESS OF FEATURE SIZE IN RELATION TO DATUM A (REGARDLESS OF DATUM SIZE)	⊕ A Ⓢ .OIO Ⓢ
TRUE POSITION AT MMC	LOCATED AT TRUE POSITION WITHIN .OIO DIA. AT MMC OR LOCATED WITHIN .OO5 R OF TRUE POSITION AT MMC	⊕ .OIO
TRUE POSITION WITH DATUM AT MMC	LOCATED AT TRUE POSITION WITHIN .OIO DIA. AT MMC IN RELATION TO DATUM A AT MMC	⊕ A .OIO
TRUE POSITION WITH TWO DATUMS	LOCATED AT TRUE POSITION WITHIN .OO5 DIA. AT MMC IN RELATION TO DATUM A AT MMC AND DATUM B REGARDLESS OF SIZE OF DATUM B	⊕ A B Ⓢ .OO5
STRAIGHTNESS	STRAIGHT WITHIN .OO3 TOTAL	⎯ .OO3 or ⌒ .OO3
FLATNESS	FLAT WITHIN .OO3 TOTAL	⎯ .OO3 ⌒ .OO3
PARALLEL	PARALLEL TO A WITHIN .OO5 TOTAL	II A .OO5 or // .OO5 A *
SQUARENESS OR PERPENDICULARITY	SQUARE WITH A WITHIN .OO2 TOTAL	⊥ A .OO2
ANGULARITY	ANGULAR TOL. .OO6 TOTAL DATUM A	∠ A .OO6
CONCENTRICITY	CONCENTRIC TO A WITHIN .OO3 FIR	◎ A .OO3
SYMMETRY	SYMMETRICAL WITH A WITHIN .OO4 TOTAL	≡ A .OO4

METRIC CONVERSION TABLES

INCHES TO MILLIMETERS

The values in the following Tables are based on 25.4 mm. = 1 Inch

Inches	mm.	Inches	mm.
	.000010	.000254	13/32	.40625	10.3187
	.000020	.000508	27/64	.421875	10.7156
	.000030	.000762	7/16	.4375	11.1125
	.000040	.001016	29/64	.453125	11.5094
	.000050	.001270	15/32	.46875	11.9062
	.000055	.001397	31/64	.484375	12.3031
	.000060	.001524	1/2	.5	12.7000
	.000065	.001651	33/64	.515625	13.0969
	.000070	.001778	17/32	.53125	13.4937
	.000075	.001905	35/64	.546875	13.8906
	.000080	.002032	9/16	.5625	14.2875
	.000085	.002159	37/64	.578125	14.6844
	.000090	.002286	19/32	.59375	15.0812
	.000095	.002413	39/64	.609375	15.4781
1/64	.015625	.3969	5/8	.625	15.8750
1/32	.03125	.7937	41/64	.640625	16.2719
3/64	.046875	1.1906	21/32	.65625	16.6687
1/16	.0625	1.5875	43/64	.671875	17.0656
5/64	.078125	1.9844	11/16	.6875	17.4625
3/32	.09375	2.3812	45/64	.703125	17.8594
7/64	.109375	2.7781	23/32	.71875	18.2562
1/8	.125	3.1750	47/64	.734375	18.6531
9/64	.140625	3.5719	3/4	.75	19.0500
5/32	.15625	3.9687	49/64	.765625	19.4469
11/64	.171875	4.3656	25/32	.78125	19.8437
3/16	.1875	4.7625	51/64	.796875	20.2406
13/64	.203125	5.1594	13/16	.8125	20.6375
7/32	.21875	5.5562	53/64	.828125	21.0344
15/64	.234375	5.9531	27/32	.84375	21.4312
1/4	.25	6.3500	55/64	.859375	21.8281
17/64	.265625	6.7469	7/8	.875	22.2250
9/32	.28125	7.1437	57/64	.890625	22.6219
19/64	.296875	7.5406	29/32	.90625	23.0187
5/16	.3125	7.9375	59/64	.921875	23.4156
21/64	.328125	8.3344	15/16	.9375	23.8125
11/32	.34375	8.7312	61/64	.953125	24.2094
23/64	.359375	9.1281	31/32	.96875	24.6062
3/8	.375	9.5250	63/64	.984375	25.0031
25/64	.390625	9.9219			

DECIMAL EQUIVALENTS OF FRACTIONS

1/64 —	.015625
1/32 —	.03125
3/64 —	.046875
1/16 —	**.0625**
5/64 —	.078125
3/32 —	.09375
7/64 —	.109375
1/8 —	**.125**
9/64 —	.140625
5/32 —	.15625
11/64 —	.171875
3/16 —	**.1875**
13/64 —	.203125
7/32 —	.21875
15/64 —	.234375
1/4 —	**.25**
17/64 —	.265625
9/32 —	.28125
19/64 —	.296875
5/16 —	**.3125**
21/64 —	.328125
11/32 —	.34375
23/64 —	.359375
3/8 —	**.375**
25/64 —	.390625
13/32 —	.40625
27/64 —	.421875
7/16 —	**.4375**
29/64 —	.453125
15/32 —	.46875
31/64 —	.484375
1/2 —	**.5**

33/64 —	.515625
17/32 —	.53125
35/64 —	.546875
9/16 —	**.5625**
37/64 —	.578125
19/32 —	.59375
39/64 —	.609375
5/8 —	**.625**
41/64 —	.640625
21/32 —	.65625
43/64 —	.671875
11/16 —	**.6875**
45/64 —	.703125
23/32 —	.71875
47/64 —	.734375
3/4 —	**.75**
49/64 —	.765625
25/32 —	.78125
51/64 —	.796875
13/16 —	**.8125**
53/64 —	.828125
27/32 —	.84375
55/64 —	.859375
7/8 —	**.875**
57/64 —	.890625
29/32 —	.90625
59/64 —	.921875
15/16 —	**.9375**
61/64 —	.953125
31/32 —	.96875
63/64 —	.984375
1 —	**1.**

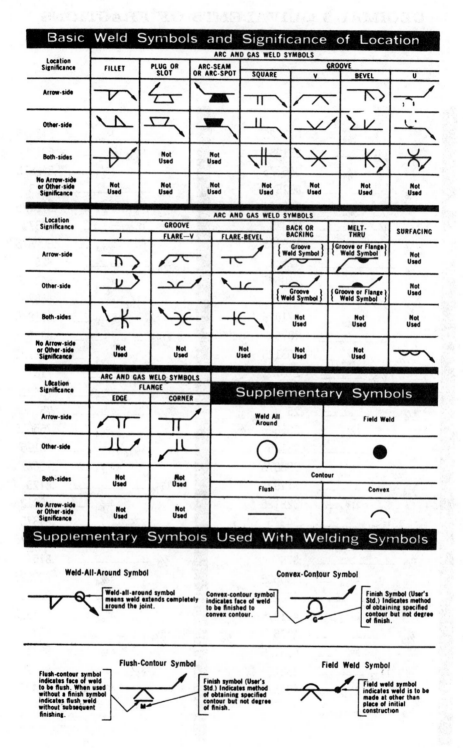

Basic Weld Symbols and Significance of Location

Weld-All-Around Symbol

Weld-all-around symbol means weld extends completely around the joint.

Convex-Contour Symbol

Convex-contour symbol indicates face of weld to be finished to convex contour.

Finish Symbol (User's Std.) Indicates method of obtaining specified contour but not degree of finish.

Flush-Contour Symbol

Flush-contour symbol indicates face of weld to be flush. When used without a finish symbol indicates flush weld without subsequent finishing.

Finish symbol (User's Std.) Indicates method of obtaining specified contour but not degree of finish.

Field Weld Symbol

Field weld symbol indicates weld is to be made at other than place of initial construction

INDEX